Pharmaceutical Chemistry

VOLUME 2

Pharmaceutical Chemistry

EDITED BY

LESLIE G. CHATTEN

PROFESSOR OF PHARMACEUTICAL CHEMISTRY
FACULTY OF PHARMACY
UNIVERSITY OF ALBERTA
EDMONTON, ALBERTA, CANADA

VOLUME 2
Instrumental Techniques

CBS

CBS Publishers & Distributors Pvt. Ltd.

New Delhi • Bengaluru • Chennai • Kochi • Kolkata • Mumbai
Hyderabad • Nagpur • Patna • Pune • Vijayawada

ISBN: 978-81-239-1584-5 (PB)
ISBN: 978-81-239-1601-9 (HB)

First Indian Reprint: 2008

Published by **Satish Kumar Jain** and produced by **Varun Jain** for
CBS Publishers & Distributors Pvt. Ltd.,
4819/XI Prahlad Street, 24 Ansari Road, Daryaganj, New Delhi - 110002
delhi@cbspd.com, cbspubs@airtelmail.in • www.cbspd.com
Ph.: 23289259, 23266861, 23266867 • Fax: 011-23243014

Corporate Office: 204 FIE, Industrial Area, Patparganj, Delhi - 110 092
Ph: 49344934 • Fax: 011-49344935
E-mail: publishing@cbspd.com • publicity@cbspd.com

Branches:
• *Bengaluru:* 2975, 17th Cross, K.R. Road, Bansankari 2nd Stage,
  Bengaluru - 70 • Ph: +91-80-26771678/79 • Fax: +91-80-26771680
  E-mail: cbsbng@gmail.com, bangalore@cbspd.com
• *Chennai:* No. 7, Subbaraya Street, Shenoy Nagar, Chennai - 600030
  Ph: +91-44-26681266, 26680620 • Fax: +91-44-42032115
  E-mail: chennai@cbspd.com
• *Kochi:* Ashana House, 39/1904, A.M. Thomas Road, Valanjambalam,
  Ernakulum, Kochi • Ph: +91-484-4059061-65
  Fax: +91-484-4059065 • E-mail: cochin@cbspd.com
• *Kolkata:* 6-B, Ground Floor, Rameshwar Shaw Road, Kolkata - 700014
  Ph: +91-33-22891126/7/8 • E-mail: kolkata@cbspd.com
• *Mumbai:* 83-C, Dr. E. Moses Road, Worli, Mumbai - 400018
  Ph: +91-9833017933, 022-24902340/41 • E-mail: mumbai@cbspd.com

Representatives:
• Hyderabad: 0-9885175004   • Nagpur: 0-9021734563
• Patna: 0-9334159340   • Pune: 0-9623451994
• Jharkhand: 0-9811541605   • Uttarakhand: 0-9716462459

Printed at:
India Binding House, Noida, UP (India)

Preface

The increasing complexities of pharmaceutical preparations and the marked emphases on quality control by the ethical manufacturer have placed a greater load on the ingenuity of the control chemist. Such an individual must be so well trained that he is capable of appreciating the merits of the known techniques before selecting the one most suited to the task at hand. If established procedures are not suitable, the control chemist must be capable of developing new methods. Therefore, this new and comprehensive two-volume textbook on pharmaceutical chemistry has been written for use in the practical training of pharmacy students to enable them to take their rightful place in the pharmaceutical industry and in government laboratories concerned with the quality of medicinal preparations.

Texts dealing with analytical chemistry that have been written for and by analytical chemists do not, in general, adequately treat the subject of analytical pharmaceutical chemistry. When one applies quantitative techniques to the analyses of pharmaceutical dosage forms which, even in their simplest state, are complex entities, one must employ special considerations. Books on analytical chemistry lack the proper emphasis and suitable applications. Existing textbooks in pharmaceutical chemistry, while considering the pharmaceutical aspects, often fail to treat the subject in the depth which it deserves. This text is meant to fulfill these needs.

The editor, based upon his experience with the Food and Drug Directorate of the Government of Canada as well as his teaching of pharmaceutical chemistry at the University of Alberta, and the contributors, who, for the most part, are all experienced teachers of pharmaceutical chemistry or a related discipline of pharmacy, have consciously attempted to arrive at a satisfactory blend of procedures in order to provide a broad basis on which the senior undergraduate and graduate student can build. The focal point of each chapter is the presentation of the theory. Practical experiments have been carefully selected to demonstrate an application of the theoretical considerations. Questions and problems, together with a list of references for supplementary reading, have been included at the end of most chapters.

v

Since the aim of this book is to provide a depth of understanding not evident in other books, it was decided to separate what is generally thought of as classical analytical techniques from those of instrumentation. The first volume, therefore, deals with theoretical and practical considerations of gravimetric analysis, acid-base titrimetry and pH, precipitation and complex formation, acidimetry and alkalimetry, nonaqueous titrimetry, complexo- metric analysis, alkaloidal assay, miscellaneous methods, ion exchange, chromatography, and the analysis of fixed and volatile oils. The second volume presents the theory and application of the following instrumental techniques: visible and ultraviolet spectrophotometry, fluorescence spectro- photometry, turbidimetry and nephelometry, infrared spectrophotometry and Raman spectroscopy, flame photometry and atomic absorption analysis, x-ray diffraction and optical crystallography, mass spectrometry, refractom- etry and interferometry, polarimetry and optical rotatory dispersion, gas chromatography, radioactivity, nuclear magnetic resonance, potentiometric titrations and instrumental determination of pH, coulometric methods and chronopotentiometry polarography, amperometry, and conductance and high frequency.

The editor is indebted to the various authors for their contributions. Their efforts are responsible for the quality of this text.

L. G. C.

Contributors to Volume 2

JOHN A. BILES, Ph.D., *Professor of Pharmacy, School of Pharmacy, University of Southern California, Los Angeles, California*

CARMAN A. BLISS, Ph.D., *Associate Professor of Pharmacy, College of Pharmacy, University of Saskatchewan, Saskatoon, Saskatchewan*

A. CHISHOLM,* D.Sc., *Analytical Research, Wm. S. Merrell Co., Division of Richardson-Merrell, Inc., Cincinnati, Ohio*

RONALD T. COUTTS, Ph.D., *Professor of Pharmaceutical Chemistry, University of Alberta, Edmonton, Alberta*

STUART ERIKSEN, Ph.D., *Director, Medical Research, Allergan Pharmaceuticals, Santa Ana, California*

DAVID E. GUTTMAN, Ph.D., *Professor of Pharmaceutics, School of Pharmacy, State University of New York at Buffalo, Buffalo, New York*

J. GEORGE JEFFREY, Ph.D., *Professor of Pharmacy, College of Pharmacy, University of Saskatchewan, Saskatoon, Saskatchewan*

PETER KABASAKALIAN, Ph.D., *Physical & Analytical Chemical Research, Schering Corporation, Bloomfield, New Jersey*

D. S. LAVERY, Ph.D., *Lash Miller Chemical Laboratories, University of Toronto, Toronto, Canada*

ROBERT A. LOCOCK, Ph.D., *Assistant Professor of Pharmaceutical Chemistry, Faculty of Pharmacy, University of Alberta, Edmonton, Alberta, Canada*

MATHIAS P. MERTES, Ph.D., *Associate Professor of Pharmacy, Department of Medicinal Chemistry, School of Pharmacy, University of Kansas, Lawrence, Kansas*

ANTOINE A. NOUJAIM, Ph.D., *Associate Professor of Pharmacy, Faculty of Pharmacy, University of Alberta, Edmonton, Alberta, Canada*

* Present address: Food and Drug Directorate, Toronto, Canada.

vii

M. PERNAROWSKI, Ph.D., *Professor of Pharmacy, Faculty of Pharmacy, University of British Columbia, Vancouver, British Columbia*

E. A. ROBINSON, Ph.D., *Associate Dean, Erindale College, University of Toronto, Toronto, Canada*

J. W. ROBINSON, Ph.D., *Professor of Chemistry, Louisiana State University, Baton Rouge, Louisiana*

FRED T. SEMENIUK, Ph.D., *Professor of Pharmaceutical Chemistry, School of Pharmacy, University of North Carolina, Chapel Hill, North Carolina*

JOHN W. SHELL,† Ph.D., *Director of Research, Allergan Pharmaceuticals, Santa Ana, California*

FRED W. TEARE, Ph.D., *Associate Professor of Pharmaceutical Chemistry, Faculty of Pharmacy, University of Toronto, Toronto, Ontario, Canada*

† Present address: University of Kansas, Lawrence, Kansas.

Contents of Volume 2

Pharmaceutical Chemistry

VOLUME 2

CHAPTER **1**

Absorption Spectrophotometry

M. Pernarowski

FACULTY OF PHARMACY
UNIVERSITY OF BRITISH COLUMBIA
VANCOUVER, BRITISH COLUMBIA

Absorption spectrophotometry is of fundamental importance to the pharmaceutical analyst. Many drugs absorb electromagnetic radiation. By using the proper instruments and techniques, the analyst can, in many instances, determine the amount and nature of a drug in a dosage form, in a reaction vessel, or in a biological system. These determinations are of importance in the assessment of the drugs produced by the industry, in research, and in the clinical evaluation of the effectiveness of many pharmaceuticals. Without spectrophotometry, quality control would be impossible and many of the research programs of the drug industry and the universities would not have yielded the drugs that are now available to the public.

Radiant energy is energy transmitted as electromagnetic radiation. *Absorption spectrophotometry* is the measurement of the absorption of radiant energy by various substances. It includes the measurement of the absorptive capacity for radiant energy in the ultraviolet, visible, and infrared regions of the spectrum. However, the techniques and instrumentation associated with infrared spectrophotometry differ, in some respects, from those used to study the absorption of light and ultraviolet energy. The subject is, therefore, discussed in Chapter 2.

The sun is our most important *source* of radiant energy. Many naturally occurring substances—droplets of water, the leaves on trees, chemicals imbedded in rocks—may act as *dispersing, absorbing,* or *reflecting* devices for radiant energy. Man's eyes are the most important *detectors* of radiant energy. The italicized words are, in essence, the component parts of a spectrophotometer or colorimeter. Man, therefore, has utilized the principles of spectrophotometry from the day that he became aware of his surroundings. Without light, life would be impossible. Without the subtleties of light absorption, reflection, and transmission, life would be a continuous psyche-delic experience. Without the differentiating capabilities of the eye, the world would be a study in gray.

The first "analyst" to utilize the principles of spectrophotometry was probably the trader who visually compared the color quality of the materials that he bought and sold. The first historical reference to product control may be found in the writings of Pliny, the Roman encyclopedist.[1] He reported that iron in vinegar could be detected by dipping a piece of papyrus soaked in an aqueous extract of gallnut tannins into the solution being tested. The papyrus turned blue or black if iron was present. However, absorption spectrophotom-etry was not used extensively by the pharmaceutical analyst until after World War II. Inexpensive but sophisticated instruments became available at that time and contributed to our understanding of the applications of spectrophotometry to the point that many textbooks are devoted solely to the subject matter covered in this chapter. Reviews in scientific journals are published periodically[2,3] and are an important source of information for the novice and the expert in spectrophotometry. The treatment in this chapter may whet the readers' appetite, but can hardly satisfy the student who wishes to learn about the subtleties of the subject. References 4–6 are, therefore, reserved for those who may wish to pursue any of the material covered herein in more detail.

1.1 ELECTROMAGNETIC RADIATION

Electromagnetic radiation may be described in terms of its wavelike prop-erties. The *wavelength* (λ) of a beam of electromagnetic radiation is the linear distance, measured along the line of propagation, between two points which are in phase on adjacent waves. The unit of wavelength is the *angstrom* (Å). It is equal to 1/6438.4696 of the wavelength of the red line of Cd. This is almost, but not exactly, equal to 10^{-8} cm. Therefore, 10 Å are equal to 1 mμ (millimicron). One centimeter is equal to 10,000 μ (microns) or to 10^7 mμ.

The *frequency* (ν) of the beam is the number of cycles occurring per second (H_z). The relationship between wavelength (in centimeters) and frequency is stated mathematically in Eq. (1.1).

$$\nu = \frac{c}{\lambda} \tag{1.1}$$

The velocity (c) of electromagnetic radiation in a vacuum is equal to 3.0 × 10^{10} cm/sec. This value is used in most calculations, but will vary if the electromagnetic radiation passes through something other than a vacuum.

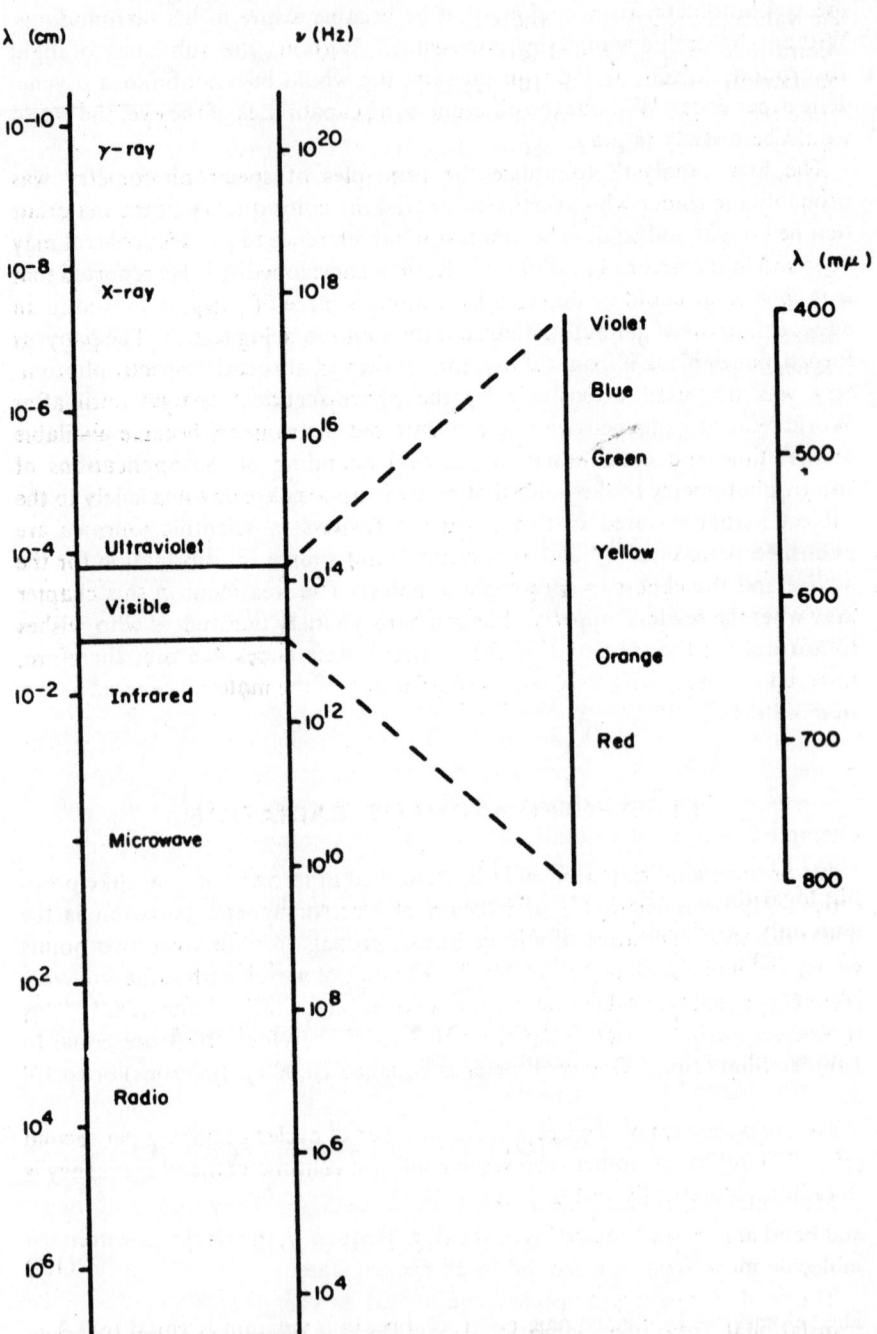

FIGURE 1.1: The electromagnetic spectrum.

The pharmaceutical analyst calculates, but rarely uses, frequency values. The values are large and, therefore, cumbersome to use when presenting spectral data in a graphical form. For example, a wavelength of 300 mμ is equal to a frequency of 1 × 10^{15} H_z. For this reason, most analysts prefer to define the spectrum in terms of wavelength or *wave number* (σ). Wave number is the number of waves per unit length or the reciprocal of the wavelength (in centimeters). The unit is the reciprocal centimeter.

$$\sigma = \frac{1}{\lambda} \tag{1.2}$$

In some instances, radiant energy behaves as if it were made up of discrete packets of energy or *photons*. The energy (E) of a photon depends on the frequency of the radiation.

$$E = h\upsilon \tag{1.3}$$

or

$$E = \frac{hc}{\lambda} \tag{1.4}$$

Planck's constant (h) is equal to 6.62 × 10^{-27} erg sec. Several simple calculations will show that the energy of a photon increases as the wavelength decreases. X-rays, for example, are much more energetic than ultraviolet energy.

Sir Isaac Newton, in 1666, described in some detail the visible region of the electromagnetic spectrum. When a beam of light is passed through a prism, it is dispersed and, by projecting this dispersed radiation onto a screen, the six major colors of the electromagnetic spectrum can be observed. However, this region is but a small part of the entire spectrum. For example, if we attempted to draw it to scale, it would be 100,000,000 ft long if we let 1 μ equal 1 ft and drew the spectrum to 100 m. (The scale in Fig. 1.1 is not linear but logarithmic.) However, the subject matter in this chapter is concerned with only two regions of this spectrum. The ultraviolet region extends from 10 to 380 mμ. Special instruments are required for studies in the far or vacuum ultraviolet region (10 to 200 mμ). The analyst, therefore, depends upon visible (380 to 780 mμ) and ultraviolet energy (200 to 380 mμ) for the routine analysis of pharmaceuticals.

1.2 ABSORPTION OF RADIANT ENERGY

Molecules are as energetic as the modern teenager. They rock, roll, twist, and bend and, if the "music" is of the right frequency, the electrons within the molecule move from the ground to an excited state.

The total energy in a molecule is the sum of the energies associated with the translational, rotational, vibrational, and electronic motions of the molecule or the electrons and nuclei in the molecule. This classical explanation of particle motion and its relationship to energy is limited by the Heisenberg uncertainty principle, which states that the "exact" orbit for the motion of a particle cannot be determined because position and velocity cannot be measured at the same time. With this limitation in mind, the energies that contribute to total molecular energy can be defined.

Translational energy is associated with the motion (velocity) of the molecule as a whole.

Rotational energy is associated with the overall rotation of the molecule.

Vibrational energy is associated with the movement of atoms within the molecule.

In general, these atomic and molecular motions are related to the absorption of infrared energy and are not pertinent to the subject matter in this chapter.

Electronic energy is associated with the motion of electrons around the nuclei.

When a molecule absorbs visible or ultraviolet energy, an electron or electrons will be raised to a higher energy level if the energy requirement for that transition is equal to the energy of the incoming photon. This statement is valid only for certain types of electrons. The electrons in the inner shells of atoms and those that are shared by two adjacent atoms are not affected to the same degree by incoming radiation as those that cannot be localized within the molecule. Electrons of the latter type give rise to spectra in the ultraviolet and visible regions of the electromagnetic spectrum. Such electrons are found in conjugated double bonds. Saturated molecules, therefore, will not absorb ultraviolet energy.

An excited electron returns to the ground state in about 10^{-9} to 10^{-8} sec. Energy must now be released to compensate for the energy absorbed by the system. If the electron returns directly to the ground state, heat is evolved. If it returns to the ground state by way of a second excited state, energy is released in the form of heat and light.

If a large amount of energy is absorbed by certain substances, bonds may be ruptured and new compounds formed. For example, ergosterol is converted to calciferol by ultraviolet radiation. However, dramatic changes are the exception rather than the rule. Changes are usually minimal and, for this reason, ultraviolet spectrophotometry is considered to be a nondestructive method of analysis.

A. ABSORBING GROUPS

A *chromophore* is a group which, when attached to a saturated hydrocarbon, produces a molecule that absorbs a maximum of visible or ultraviolet energy at some specific wavelength (or wavelengths). A number of simple chromophores are listed in Table 1.1. Although the type of information in Table 1.1 is useful to the research chemist, its value to the pharmaceutical analyst has decreased since the advent of inexpensive and sophisticated infrared spectrophotometers. Functional groups can be readily identified by subjecting the unknown to infrared energy.

Many molecules contain two or more chromophores. The interaction of radiant energy with the molecule then depends upon the relative positions of the two chromophores in the molecule.

TABLE 1.1: Absorption Bands for Simple Chromophoric Groups[a]

Chromophore		Example	λ_{max}, mμ	e[b]
Acetylene	—C≡C—	Acetylene	173	6,000
Amide	—CONH$_2$	Acetamide	< 208	—
Azo	—N=N—	Azomethane	347	5
Azomethine	> C=N—	Acetoxime	190	5,000
Carbonyl	RHC=O	Acetaldehyde	293	12
Carbonyl	RR'C=O	Acetone	271	16
Carboxyl	—COOH	Acetic acid	204	60
Ethylene	RCH=CHR	Ethylene	234	0.05
			193	10,000
Nitrate	—ONO$_2$	Ethyl nitrate	270	12
Nitrile	—C≡N	Acetonitrile	< 160	—
Nitrite	—ON=O	Octyl nitrite	370	55
			230	2,200
Nitro	—NO$_2$	Nitromethane	271	19
Sulfone	> SO$_2$	Dimethyl sulfone	< 180	—
Sulfoxide	> S=O	Cyclohexyl methyl sulfoxide	210	1,500
Thiocarbonyl	> C=S	Thiobenzophenone	620	70

[a] This table is reproduced, with permission, from an application data sheet (No. UV-77-MI) published by Beckman Instruments, Inc., Fullerton, California.
[b] Molar absorptivity.

a. When two chromophores are separated by more than one carbon atom, total absorption is the sum of the absorption of each of the two chromophores.

b. When two chromophores are adjacent to each other, the absorption maximum shifts to longer wavelengths (*bathochromic shift*) and the intensity of absorption is increased (*hyperchromic effect*). The opposite effects can be produced by changing the structure of the organic substance. A shift to a shorter wavelength is called a *hypsochromic shift*; a reduction of intensity is called a *hypochromic effect*.

c. When two chromophores are attached to the same carbon atom, there is a summation of absorption and a shift toward longer wavelengths, but the degree of change is less than that shown by conjugated chromophores.

There are two kinds of chromophores. The *simple* chromophores (see Table 1.1) give rise to *R bands*. The molar absorptivity value (see Section 1.3 for a definition of ϵ) for this type of band is usually less than 100. There are two types of *complex* chromophores. The first type is that found in aromatic compounds whose structures contain a benzene ring. These chromophores give rise to *B bands*. The molar absorptivity values for these bands range from 250 to 3000. The second type has the following formula:

$$A-(CH=CH)_n-CH=B$$

A is equal to H, R, OR, SR, NR_2, O—, S—, or —NR. B is equal to CH_2CHR, CR_2, NR, O, S, $^+NR_2$, ^+OR, or ^+SR. These chromophores give rise to K bands. The molar absorptivity values for these bands are more than 10,000.

Auxochromes are groups which, when introduced into an absorbing system, cause a bathochromic shift. Auxochromes are either coordinatively saturated (e.g., $—^+NH_3$) or coordinatively unsaturated (e.g., $—NH_2$). The hydroxyl group, amino groups and their substituted derivatives, alkyl groups, and halogens will produce changes in absorption spectra.

B. SOLVENT EFFECTS

The absorption spectrum of a drug depends, in part, on the solvent used to solubilize the substance. A drug may absorb a maximum of radiant energy at one wavelength in one solvent, but will absorb little at the same wavelength in another solvent. These changes in spectrum are due to:

    a. The nature of the solvent
    b. The nature of the absorption band
    c. The nature of the solute

These effects can be correlated with the polarity of the solvent. As the polarity of the solvent increases, R bands are displaced to shorter wavelengths. K bands are displaced to longer wavelengths. Polar solutes are more

TABLE 1.2: Wavelength Ranges for Commonly Used Solvents

Solvent	Wavelength range, $m\mu$
Acetonitrile	200–800
Carbon tetrachloride	265–800
Chloroform	245–800
Cyclohexane	210–800
Dimethylformamide	270–800
Ethanol	210–800
Ether	220–800
n-Heptane	210–800
n-Hexane	210–800
Isooctane	210–800
Isopropanol	210–800
n-Pentane	210–800
Water	200–800

sensitive to solvent change than are nonpolar solutes. Molar absorptivity values will vary with changes in solvent. The absorption spectrum of a drug is, therefore, meaningful to the pharmaceutical analyst only if the solvent that has been used to solubilize the solute is specified.

The most widely used solvents are listed in Table 1.2. Solvents will begin to absorb ultraviolet radiant energy at some specific wavelength. For example, chloroform absorbs ultraviolet energy so strongly below 245 mμ that

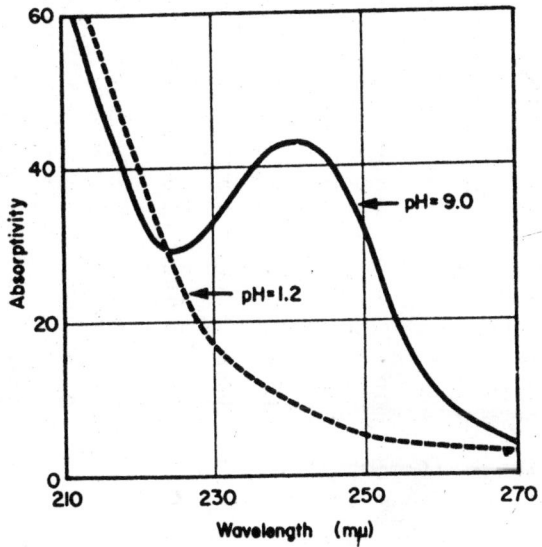

FIGURE 1.2: Absorption spectrum for phenobarbital at pH 1.2 and 9.

it cannot be used for a solvent for spectral studies at that or at a lower wavelength setting. Effective solvent ranges are given in Table 1.2. Moreover, solvents must not contain trace impurities. Many impurities (e.g., the benzene in absolute alcohol prepared by azeotropic distillation) absorb radiant energy and complicate the analysis. Spectral grade solvents are sold by chemical supply houses, but, in many instances, the analyst can readily purify those that he normally uses in the laboratory.

The effect of pH on the absorption spectrum for phenobarbital is shown in Fig. 1.2. These pH effects are of vital importance to the pharmaceutical analyst because they enable him to carry out the analysis of one substance in the presence of another (see Sections 1.3 and 1.7) and to determine the structural characteristics of certain drugs. Like the solvents listed in Table 1.2, buffers must transmit ultraviolet energy if they are to be used for the determination of the spectral characteristics of drugs. Many buffers do not meet this requirement, but those listed in Table 1.3 transmit a maximum of ultraviolet energy at most wavelengths.

TABLE I.3: Absorbance Values for Buffers[a]

pH	Ml of 0.2 M solutions of buffer constituents in 100 ml of buffer	Wavelength, mμ									
		220	230	240	250	260	270	280	290	300	310
1.1	HCl, 47.4 ml KCl, 2.72 ml	0.011	0.007	0.004	0.003	0.003	0.002	0.001	0.001	0.002	0.001
2.0	HCl, 5.95 ml KCl, 44.1 ml	0.006	-0.003	-0.005	-0.002	-0.002	-0.003	-0.003	-0.003	-0.004	-0.003
2.8	KH2PO4, 25.0 ml HCl, 3.50 ml	0.004	0.003	0.003	0.003	0.005	0.001	0.002	0.001	-0.001	-0.001
4.0	KH2PO4, 25.0 ml HCl, 0.25 ml	0.004	0.004	0.001	0.001	0.002	0.002	0.001	0.001	0.000	0.000
5.1	KH2PO4, 25.0 ml NaOH, 0.25 ml	0.027	0.024	0.023	0.023	0.022	0.015	0.015	0.014	0.011	0.010
5.8	KH2PO4, 25.0 ml NaOH, 2.82 ml	0.039	0.029	0.021	0.016	0.015	0.016	0.015	0.011	0.010	0.006
7.0	KH2PO4, 25.0 ml NaOH, 14.8 ml	0.042	0.030	0.021	0.018	0.017	0.015	0.014	0.011	0.007	0.004
7.9[b]	H3BO3-KCl, 25.0 ml NaOH, 2.00 ml	0.076	0.050	0.032	0.026	0.023	0.023	0.023	0.018	0.013	0.009
9.0[b]	H3BO3-KCl, 25.0 ml NaOH, 10.7 ml	0.147	0.105	0.077	0.067	0.063	0.058	0.053	0.045	0.036	0.030
10.0[b]	H3BO3-KCl, 25.0 ml NaOH, 22.0 ml	0.128	0.063	0.044	0.036	0.031	0.025	0.023	0.021	0.018	0.022
10.8	Na2HPO4, 25.0 ml NaOH, 4.13 ml	0.209	0.101	0.074	0.065	0.057	0.053	0.050	0.045	0.038	0.032
11.8	Na2HPO4, 25.0 ml NaOH, 21.6 ml	0.545	0.163	0.104	0.089	0.080	0.070	0.061	0.052	0.045	0.038

[a] This table is reproduced through the courtesy of the Division of the State Racing Commission, Dept. of State of New York, 148-07, Hillside Avenue, Jamaica, N.Y., 11435.

[b] H_3BO_3-KCl solutions are 0.2 M with respect to each constituent.

Many researchers have characterized drugs by studying their absorption spectrum in buffered solutions. Vandenbelt and Doub[7] were the first researchers to identify the absorbing groups of the sulfanilamido derivatives by determining their spectral characteristics in acidic and basic solutions. The spectral band associated with the sulfanilamido portion of the molecule is observed at 257–259 mμ; with the thiazole structure in sulfathiazole, at 280–283 mμ and at 258 mμ; with the pyridine in sulfapyridine, at 311 mμ; with the sulfanilyl portion, at 261 mμ; and with the pyrimidine ring, at 241 mμ.

In 1 N hydrochloric acid solution, sulfonamides with the following structure:

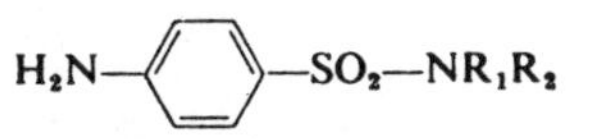

$$H_2N-\underset{}{\bigcirc}-SO_2-NR_1R_2$$

absorb little of the radiant energy above 230 mμ. The auxochrome, under these conditions, is $-^+NH_3$. In 1 N sodium hydroxide solution, a more efficient auxochrome is formed ($-NH_2$), and the substance now exhibits an absorption maximum at 251 mμ.

Under acidic conditions, phenobarbital does not absorb ultraviolet energy, to a significant degree, at 240 mμ. The tautomeric forms of phenobarbital are shown in the following equation:

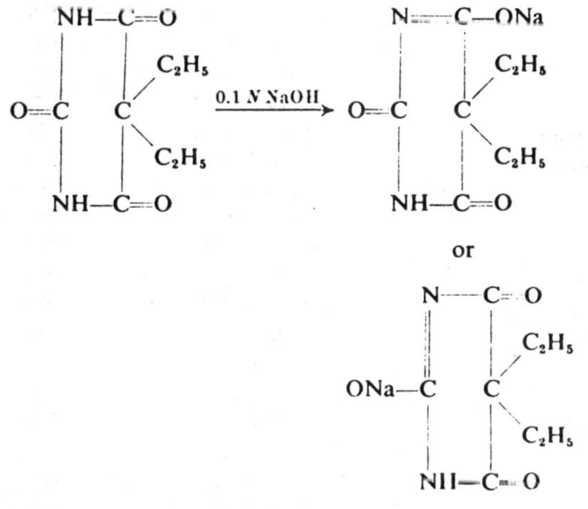

Under alkaline conditions, the chromophoric system is $-C{=}N-C{=}O$. At a pH of 9, a high intensity absorption band may be observed at 240 mμ. However, in 0.1 N sodium hydroxide solution, this band appears at 255 mμ.

Solvents and buffers are, therefore, useful adjuncts in structural determination studies. The literature on the subject is comprehensive. The annual reviews cited earlier and the major textbooks on spectrophotometry provide the levels of information required for advanced study.

Absorption maxima for a selected group of common drugs are given in Table 1.4. Absorption spectra for many drugs may be found in a recently published manual.[8]

TABLE 1.4: Spectrophotometric Data for Common Drugs[a]

Drug	Solvent	λ_{max}	a
Acetazolamide	0.1 N HCl	265	48
Acetophenetidin	0.5 N NaOH	245	70
Amobarbital sodium	0.5 N NaOH	255	35
Antazoline HCl	Water	242	52
Apomorphine HCl	Water	273	67
Ascorbic acid	0.05 N HCl	243	35
Benzocaine	0.5 N NaOH	278	120
Caffeine	Water	273	51
Chloramphenicol	Water	278	30
Chlorcyclizine HCl	Alcohol	230	44
Chloromethapyrilene citrate	Alcohol	242	41
Chlorpheniramine maleate	Water	262	15
Cortisone acetate	Alcohol	240	40
Cyanocobalamin	Water	273	51
Cycloserine	Water	226	40
Dibucaine HCl	0.05 N HCl	246	59
Dimenhydrinate	0.05 N HCl	276	30
Epinephrine	0.5 N NaOH	296	57
Ergotamine	Alcohol	316	18
Hydrocortisone	Alcohol	240	44
Isoniazid	Dilute HCl	266	38
Levarterenol bitartrate	Dilute HCl	279	82
Methapyrilene HCl	Alcohol	240	63
Methyl Salicylate	0.05 N HCl	237	66
Morphine	1 N NaOH	258	31
Nicotinic acid	0.05 N HCl	260	37
Novobiocin	0.1 N NaOH	307	60
Oxytetracycline HCl	Buffer (pH 2.0)	353	28
Phenylbutazone	Water	255	45
Prednisone	Water	245	44
Procaine HCl	Water	290	66
Promethazine HCl	Water	298	93
Pyridoxine HCl	0.1 N HCl	292	42
Salicylamide	0.05 N HCl	235	59
Sulfamethazine	0.05 N HCl	244	62
Tetracaine HCl	Water	311	75
Tetracycline HCl	0.25 N NaOH	380	37
Thiamine HCl	Water	270	28
Thonzylamine HCl	Alcohol	244	74
Tripelennamine HCl	Dilute HCl	314	27

[a] The values in this table are illustrative. Exact values should be determined in the laboratory before carrying out an analysis on an unknown solution.

1.3 QUANTITATIVE SPECTROPHOTOMETRY

Electromagnetic radiation is absorbed by gases, liquids, and solids. If the absorption is total, the laws derived in this section are of no value to the pharmaceutical analyst. If, however, the substance transmits a part of the radiant energy that reaches it, the qualitative and quantitative characteristics of that substance can be easily determined.

A. BEER'S LAW

Beer's law is actually a combination of two laws. The first law relates absorptive capacity to the thickness of the absorbing medium. It was first enunciated by Bouguer in 1729 and restated by Lambert in 1768. According to this law, each layer of equal thickness absorbs an equal fraction of the radiant energy which passes through it. Mathematical statements relating the quantities involved may be presented in two different ways. In the first instance the absorptive capacity varies directly as the logarithm of the thickness. In the second instance, the intensity of the transmitted energy decreases exponentially as the thickness of the absorbing medium increases arithmetically.

The second law relates absorptive capacity to the concentration of the solute in the solvent. Beer, in 1852, enunciated a law which, in rather obscure language, stated that the absorptive capacity of a system is directly proportional to the concentration of solute in that system. Historical accuracy need not concern us here. However, for those who are interested in a critical examination of Beer's paper, the paper by Pfeiffer and Liebhafsky[9] gives the details of the origin of this law.

The intensity of a beam of monochromatic radiation dP decreases as it passes through each increment of absorbing material db. Therefore:

$$-\frac{dP}{db} \alpha P \tag{1.5}$$

where P is the incident radiant power. A proportionality constant may now be introduced into Eq. (1.5).

$$-\frac{dP}{db} = KP \tag{1.6}$$

Rearrange Eq. (1.6).

$$-\frac{dP}{P} = K db \tag{1.7}$$

Integrate Eq. (1.7).

$$-\log P = Kb + C \tag{1.8}$$

If P_0 is equal to the incident radiant power reaching a given area per second, b, the thickness, is then equal to zero and C is equal to $-\log P_0$.

$$-\log P = Kb - \log P_0 \tag{1.9}$$

$$\log \frac{P}{P_0} = -Kb \tag{1.10}$$

Bouguer's law [Eq. (1.10)] may also be presented in its exponential form.

$$P = P_0 e^{-kb} \tag{1.11}$$

Beer's law, which deals with the relationship between absorptive capacity and the concentration c of the solute in the solution, may be derived in the manner illustrated previously.

$$-\frac{dP}{dc} = K'P \tag{1.12}$$

and

$$\log \frac{P}{P_0} = -K'c \tag{1.13}$$

The exponential form is

$$P = P_0 e^{-k'c} \tag{1.14}$$

The constants in Eqs. (1.11) and (1.14) may now be combined and incorporated into an equation which includes the concentration and thickness symbols.

$$P = P_0 e^{-k''bc} \tag{1.15}$$

Rearrange and convert to base-10 logarithms.

$$\log \frac{P_0}{P} = k''bc \tag{1.16}$$

Substitute A for $\log P_0/P$ and a for k''. The Bouguer-Beer law is, therefore, written in the following form:

$$A = abc \tag{1.17}$$

This is the equation of a straight line with an intercept value of zero and a slope value of ab.

B. SPECTROPHOTOMETRIC TERMINOLOGY

The author of this chapter was first exposed to spectrophotometric nomenclature in a course on spectrophotometry given by Dr. M. G. Mellon, Professor of analytical chemistry at Purdue University. The following statement is copied from the notes taken at that time: "The literature in this field is a fine example of writers doing about as they pleased." Attempts have been made to standardize nomenclature, but too many authors still do as they please. The pharmaceutical analyst depends, however, on the *United States Pharmacopeia* (USP) and the *National Formulary* (NF) for guidance in many areas, and it is only natural that he accepts the nomenclature given in those books. Fortunately, this terminology is acceptable to most scientists.

Transmittance T is the quotient of the radiant power P transmitted by a sample divided by the radiant power P_0 incident upon the sample. The per cent transmittance is equal to 100 T.

Absorbance A is the negative logarithm, to the base 10, of the transmittance. Spectrophotometers and colorimeters record either the absorbance or the transmittance of the solution.

Absorptivity a is the quotient of the absorbance divided by the concentration c of the solution (in grams per liter) and the absorption path length b in centimeters. The absorptivity value varies with the wavelength of the incident energy. However, at a specified wavelength, the absorptivity value for a drug is a constant if Beer's law is obeyed.

Molar absorptivity ϵ is the quotient of the absorbance divided by the concentration c of the solution (in moles per liter) and the absorption path length b in centimeters. It is also the product of the absorptivity and the molecular weight M of the substance.

The definitions and symbols given herein are preferable to those listed in the last column of Table 1.5. Such terms are confusing and unacceptable to the editors of most scientific journals.

TABLE 1.5: Spectrophotometric Terminology[a]

Term	Symbol	Definition	Obsolete terms
Absorbance	A	$-\log T$	Absorbancy Extinction Extinctance Optical density
Absorptivity	a	$a = A/bc$ (c = concentration in grams per liter)	Absorbancy index Extinction coefficient Specific absorption coefficient Specific extinction coefficient
Molar absorptivity	ϵ	$\epsilon = aM$ (M, molecular weight)	Molar absorbancy index Molar absorption coefficient Molar extinction coefficient
Path length	b	Length of cell in centimeters	l or d
Transmittance	T	P/P_0	Transmission Transmittancy

[a] USP terminology.

C. GRAPHICAL PRESENTATION OF DATA

An *absorption spectrum* is a graphic representation of A, log A, a, ϵ, log ϵ, or T plotted against wavelength. Visible and ultraviolet spectra are, in general, plotted on graph paper which is divided along the abscissa into millimicrons. Frequency or wave number plots are rare.

1. Absorbance vs. Wavelength

The absorbance data obtained from a spectrophotometer can be plotted directly on a graph. However, the absorbance varies with the concentration of the solute in the solution. Therefore, the concentration of the solute in the solution must always be included in the caption of such spectrum.

2. Log A vs. Wavelength

Beer's law can be further modified by taking the logarithm of both sides of Eq. (1.17).

$$\log A = \log a + \log bc \qquad (1.18)$$

The logarithm of bc is a constant over the wavelengths at which the absorbance is recorded. At each wavelength, log A is decreased by a fixed quantity. This means that the spectra of solutions containing the same substance but in different quantities will be parallel to one another. If they are not, the system does not obey Beer's law.

A recording spectrophotometer plots absorbance vs. wavelength. However, certain instruments can be modified to plot log A vs. wavelength. Such modified instruments can be used to determine reaction rates in kinetic studies. If the reaction is first order, a plot of log A vs. time is a straight line whose slope is equal to the reaction rate constant.

3. Absorptivity vs. Wavelength

This is, in effect, a plot of a constant vs. wavelength. The absorptivity value, at each wavelength, is calculated by dividing A by bc. This type of plot produces one spectrum over the entire concentration range and is illustrated in Fig. 1.2.

4. Molar Absorptivity vs. Wavelength

This is also a plot of a constant versus wavelength. Because the molecular weight of the substance is included in the calculation, substances with similar absorptivity values but different molecular weights will yield different spectra.

5. Log ε vs. Wavelength

If a spectrum exhibits two absorption maxima and if one is very weak and the other is very strong, a plot of ε vs. wavelength may require a change of scale. The log ε plot increases the magnitude of the weaker bands and decreases the magnitude of the stronger bands.

6. Transmittance vs. Wavelength

Transmittance and absorbance values may be obtained from the scales of most spectrophotometers. However, transmittance values are usually converted to absorbance values and plots of T vs. wavelength, are, therefore, rare.

D. RELIABILITY OF THE MEASUREMENT

All spectrophotometric analyses are based on a comparison of the amount of energy transmitted by the solvent with the amount transmitted by the solution. The T value so determined is then used in subsequent calculations. The transmittance scale of a spectrophotometer is usually divided into 100 equal units. It is, therefore, possible to read per cent transmittance to one decimal place. However, this does not mean that the concentration can be determined with the same precision over the entire scale. The transmittance can be determined with a constant absolute error over much of this range, but the effect of this error on the analysis can be substantial if the T value is more than 0.6 or less than 0.2.

Beer's law states that $-\log T$ is equal to abc. If $-\log T$ is converted to a natural logarithm, the derivative of the equation is

$$-\frac{0.434}{T} dT = abdc \tag{1.19}$$

Divide Eq. (1.19) by $-\log T = abc$ and rearrange:

$$\frac{dc}{c} = \frac{0.434}{T \log T} dT \tag{1.20}$$

Using finite increments,

$$\frac{\Delta c}{c} = \frac{0.434}{T \log T} \Delta T \tag{1.21}$$

For all practical purposes, the relationship between concentration error $\Delta c/c$ and absorbance error is nearly constant between 20 and 60% T. If ΔT is equal to ± 0.005 and T is equal to 0.95, the error in the concentration will be of the order of $\pm 10\%$. Similar errors will result if the solution absorbs most of the radiant energy. However, if T is equal to 0.40, the error in concentration will be less than $\pm 1.4\%$.

If the derivative of Eq. (1.21) is made equal to zero, it can be shown that the per cent error in concentration is a minimum when the absorbance value is equal to 0.434. Solutions should, therefore, be prepared in such a way that the absorbance value is about 0.45. Values of not less than 0.3 and not more than 0.6 are generally acceptable.

Many scientists have studied the accuracy and precision of photoelectric spectrophotometry. The papers by Gridgeman[10,11] and Edisbury[12] cover, in detail, the subject matter presented in this section.

E. DETERMINATION OF A DRUG IN A DOSAGE FORM

Procedures for the spectrophotometric determination of drugs in dosage forms are published in the scientific literature or in the pharmacopeias. The pharmaceutical analyst must, therefore, refer to these sources constantly, but it is possible to generalize and outline a procedure that can be used to determine the amount of active ingredient in many pharmaceuticals.

1. The drug to be analyzed must be available in a relatively pure form. If its purity is not known, it must be characterized and, if necessary, recrystallized or purified in some other way. Reference standards are available for many of the drugs described in the USP[23] and NF.[13]

2. An optically transparent solvent is selected (see Tables 1.2 and 1.3) and, if necessary, its purity is checked. A cell is filled with the solvent, transferred to a spectrophotometer, and the instrument is balanced to read $100\% T$.

3. The drug is dissolved in the solvent and a portion of the solution is transferred to a spectrophotometric cell. The absorption spectrum for the drug is determined by using either a manually operated or a recording spectrophotometer. As a general rule, the final concentration should be about 10 mg of drug per liter of solution. However, the scientific literature or the pharmacopeias should be consulted if it is known that absorptivity values are published therein. For example, the absorptivity value for phenylbutazone at 265 mμ in pH 8 buffer is about 66. If the cell length is equal to 1 cm, a solution containing 10 mg of drug per liter of solution will give an A value of 0.66 at 265 mμ, the wavelength at which the drug absorbs a maximum of ultraviolet energy. This concentration will, therefore, yield a suitable spectrum.

The spectrum is now examined and the wavelength at which the drug absorbs a maximum of energy is determined. All subsequent determinations are carried out at this wavelength.

4. It is important, at this point, to consider the effect of the solvent on the analysis. For example, the A value for phenylbutazone changes rapidly (see Fig. 1.3) between a pH of 3 and 6. The drug could be determined in a pH 1 or 2 buffer. However, this acidic substance is more soluble in alkaline media. If the pH of the solution is 7 or more, the absorptivity value is

relatively constant at 265 mμ, the wavelength of maximum absorption. Analyses should not be attempted if the pH of the buffer is more than 3 or less than 6.

Solvent selection must also be based on the possible presence of interfering substances in the dosage form. Phenylephrine HCl can be dissolved in a pH 6 buffer and determined at 272 mμ. However, the absorptivity value at this wavelength is low and, more important, methylparaben, which is added to certain aqueous preparations containing the vasoconstrictor, absorbs some of the radiant energy at this wavelength. The absorbance value is, therefore,

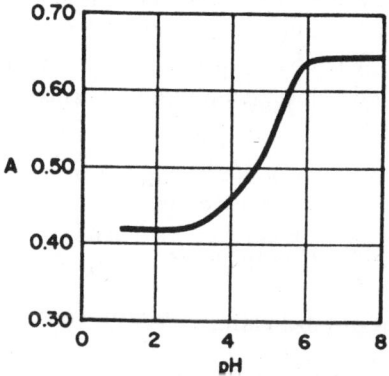

FIGURE 1.3: Effect of pH on the absorbance value for phenylbutazone. (Each liter of buffered solution contains 10 mg of phenylbutazone.)

due to both the preservative and the phenylephrine HCl. Concentrations cannot be calculated under these circumstances. However, by changing the pH to 13, the 272-mμ absorption maximum shifts to higher wavelengths and a new and more intense maximum appears at 237 mμ. Methylparaben absorbs little energy at this wavelength. The concentration of the phenylephrine HCl in the solution can now be determined with considerable accuracy.

5. The absorptivity value for the drug at the wavelength of maximum absorption is determined by measuring the A values of solutions containing different quantities of drug. The results are plotted (A vs. c) and, if Beer's law is obeyed, there will be a linear relationship between absorbance and concentration.

Some analysts will use published absorptivity values. These values, for the most part, have been accurately determined, but, because there are differences between instruments, most researchers prefer to determine their own values. Most of the spectrophotometric tests and assays in the USP and NF call for comparison against the appropriate reference standard.[14] The analyst is, therefore, determining the absorptivity value and the absorbance of the unknown at the same time and under the same circumstances.

6. The drug in the dosage form is now separated from any interfering substances. In the example given in Section 4, the methylparaben was not separated from the phenylephrine HCl, but many dosage forms contain contaminants which absorb energy at all wavelengths and under all circumstances. These must be removed by extraction, chromatography, distillation, or by other means before dissolving the drug in the chosen solvent. The absorbance of this solution is determined and the amount of drug in the dosage form calculated.

Here is an example of a typical calculation. A tablet is purported to contain 100 mg of phenylbutazone. The analyst reduces the tablet to a fine powder, extracts it with ethanol, filters the extracting solvent, and dilutes to 100.0 ml. A 10.0-ml aliquot is then diluted to 1000.0 ml with a pH 8 buffer (see Table 1.3). The absorbance A for the solution at 265 mμ is 0.64. The absorptivity a, under these conditions, is known to be 66. The cell length b is 1 cm.

$$c = \frac{A}{ab} \text{ g/liter}$$

$$\text{Milligrams phenylbutazone in the tablet} = \frac{0.64}{66} \times 10 \times 1000 = 97.0 \text{ mg}$$

$$\text{Per cent of label claim} = \frac{97.0}{100} \times 100 = 97.0\%$$

F. EXPERIMENT 1.1: DETERMINATION OF PROCAINE HCl IN AN INJECTION

It is assumed that the injection contains no interfering substances.

Weigh accurately 100 mg of procaine HCl and dissolve in 250.0 ml of water. Dilute a 25-ml aliquot of this stock solution to 1 liter with water. Mix well and transfer a portion of the solution to a 1-cm sample cell. Fill the reference cell with water. Transfer the cells to the cell compartment of the spectrophotometer (Beckman model DU spectrophotometer, or the equivalent). Record the absorbance of the solution at 220 mμ and at 10-mμ intervals thereafter. The last reading should be taken at 340 mμ. Plot A vs. wavelength. Determine the wavelength at which the drug absorbs a maximum of ultraviolet energy. Check the results by recording the absorbance of the solution at 1-mμ intervals in the vicinity of the wavelength which shows maximum absorption.

Dilute a 25.0-ml aliquot of the stock solution to 1000.0 ml with water. Record the absorbance of this solution at the wavelength at which procaine HCl exhibits maximum absorption. Repeat the procedure using 23.0-, 21.0-, 19.0-, and 17.0-ml aliquots of the stock solution. Calculate absorptivity and molar absorptivity values. Plot A vs. c to check for compliance with Beer's law.

Dilute 10.0 ml of the procaine HCl injection (unknown) to 100.0 ml with water. Dilute a 10.0-ml aliquot of this solution to 1000.0 ml with water. (These dilutions are based on an injection which is purported to contain 1 % w/v procaine HCl.) Record the absorbance of the solution at the wavelength at which procaine HCl exhibits maximum absorption. Calculate the concentration of procaine HCl in the injection.

1.4 DEVIATIONS FROM BEER'S LAW

There are no known exceptions to Bouguer's law. If the concentration of a solution is fixed, there is a linear relationship between absorbance and path length. However, deviations from Beer's law are frequently observed by the analyst in the laboratory. A plot of absorbance vs. concentration is not always a straight line or is a straight line only within well-defined limits. Some of these deviations are due to the nature of the solution being examined; others are due to chemical changes in the solution or to the type of radiant energy used in the measurement process.

A. REAL DEVIATIONS

Beer's law describes the absorption process within the solution accurately only if the concentration of solute is kept below that which leads to molecular (or ionic) interaction. The degree of molecular interaction depends upon the concentration. As the concentration increases, the charge distribution on the molecule changes. Each solute molecule in a concentrated solution does not, therefore, absorb radiant energy in the same manner as does the same molecule in a dilute solution. If interaction occurs at higher concentrations, Beer's law will not be obeyed.

Beer's law should take into consideration not only the concentration but also the refractive index (n) of the solution. It is generally assumed that the absorptivity value remains constant as the concentration increases. However, it is known that this constant does change with concentration (that is, Beer's law is not obeyed) and that the value will remain constant[15] only if it is multiplied by $n/(n + 2)^2$. This limitation of the law is not noticeable in dilute solutions.

B. INSTRUMENTAL DEVIATIONS

Beer's law assumes that monochromatic radiation is available for the determination. All spectrophotometers isolate, in theory, the wavelengths specified on the scale of the monochromator. However, under actual operating conditions, solutions are exposed to several wavelengths of radiant energy.

The *spectral slit width* is the total range of wavelengths emerging from the exit slit of a spectrophotometer. It is this width that determines the "purity" of the radiant energy used for the absorbance measurement. By definition, the spectral slit width is equal to two times the dispersion in millimicrons per millimeter times the slit width in millimeters. The *effective band width*, however, is the span of wavelengths emerging from the exit slit whose end wavelengths have half of the energy of the central wavelength.

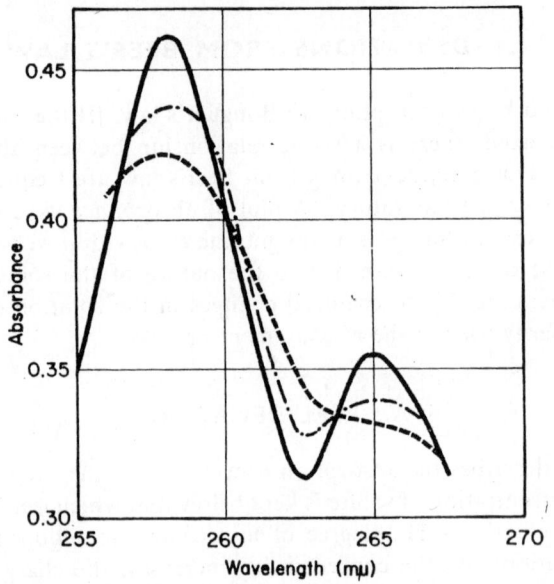

FIGURE 1.4: Absorption spectrum for a 0.08% aqueous solution of procyclidine HCl for slit widths 0.4 mm (———), 1.2 mm (–·–·–), and 2.0 mm (– – – –). Reproduced with permission from *J. Pharm. Pharmacol.* See Ref. 16.

If the slit width is fixed, Beer's law will be obeyed even though the spectrophotometer delivers polychromatic radiation to the solution. The assumption in the latter statement is that the absorptivity value remains relatively constant over the wavelengths isolated by the slit. This situation is true if the absorption maximum is flat, that is, there is little change of absorbance with wavelength around the wavelength of maximum absorption. However, if the absorbance is measured on the slope of an absorption spectrum, absorbance changes rapidly with wavelength. Under these circumstances, deviations from Beer's law may occur.

If the slit width is changed, the energy reaching the solution is not only polychromatic but polychromatic to a different degree. This means that the same solution may absorb more (or less) radiant energy at the second slit setting. This does not usually happen, but certain drugs are sensitive to spectral slit-width changes. There will be no linear relationship between absorbance and concentration for these drugs unless the slit width is kept constant during the entire operation.

Figure 1.4 illustrates the effect of slit width on absorbance values. The narrower the slit, the greater the absorbance. Both Rogers[16] and Gibson[17] discuss the theory associated with this phenomenon.

C. CHEMICAL DEVIATIONS

The solute in the solution may associate, dissociate, or react with the solvent. These processes produce two or more species and, in some instances, these have different absorptivity values at the wavelength of maximum absorption. The analyst is, therefore, being asked to analyze a solution which contains two or more substances. Under these circumstances, there can be no linear relationship between absorbance and concentration.

If a weak acid (e.g., phenylbutazone) is dissolved in a suitable solvent, the following reaction occurs.

$$HX \rightleftharpoons H^+ + X^-$$

Since absorbances are additive,

$$A = a_{HX}bc_{HX} + a_{X^-}bc_{X^-} \tag{1.22}$$

Divide by c:

$$\frac{A}{c} = \frac{a_{HX}bc_{HX}}{c} + \frac{a_{X^-}bc_{X^-}}{c} \tag{1.23}$$

If the pH of the solution is kept constant, the ratio A/c is independent of c. There is, therefore, a linear relationship between absorbance and concentration. If the pH of the solution is changed between determinations, Beer's law will not be obeyed.

A spectrophotometric determination is not normally carried out at those pH's at which absorbance is changing rapidly with pH (see Section 1.3E). However, this change of absorbance with pH can be used analytically for the calculation of pK_a values. A weak acid, at or below a certain pH, will be in the molecular form. Similarly, at some higher pH and above, it will be in the ionic form. A solution whose pH is between these two values contains both forms. The amount of acid transformed to its salt form (α) can be calculated from spectrophotometric data.

$$\alpha = \frac{\epsilon'' - \epsilon'''}{\epsilon' - \epsilon'''} \tag{1.24}$$

The molar absorptivity value for the ionic form is equal to ϵ', for the molecular form, to ϵ''', and for the mixture, to ϵ''. By using the appropriate buffers, the absorption spectrum for the various species can be prepared and the molar absorptivity values at the wavelength at which the acid absorbs a maximum of radiant energy can be calculated. The apparent dissociation constant can be determined graphically (A vs. pH) or can be calculated by using a modified form of the Henderson-Hasselbach equation.

$$pK_a = pH - \log \frac{\alpha}{1 - \alpha} \qquad (1.25)$$

The method is described in detail in a paper by Sager et al.[18]

In the example just cited, two species with different absorptivity values at the same wavelength are formed in the solution. Depending on the pH,

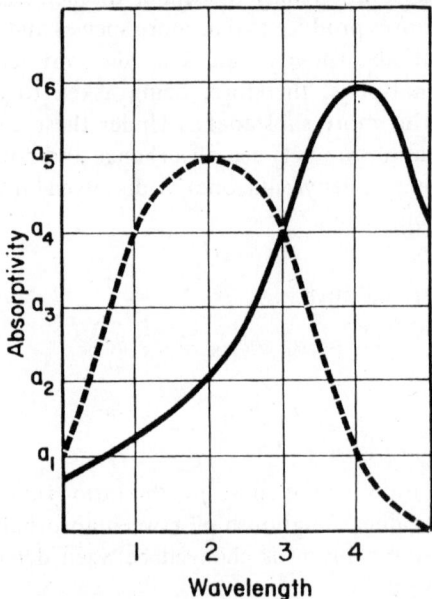

FIGURE 1.5: Hypothetical absorption spectrum for X (———) and Y (– – – –), illustrating spectral characteristics necessary for binary analysis.

aqueous solutions of potassium chromate may contain CrO_4^{-2}, $HCrO_4^{-}$, and $Cr_2O_7^{-2}$.

$$Cr_2O_7^{-2} + H_2O \rightleftharpoons 2HCrO_4^{-} \rightleftharpoons 2H^{+} + 2CrO_4^{-}$$

Since the molar absorptivity values for the dichromate ion and the two chromate species are different at the wavelength of maximum absorption, chromate solutions, when diluted with water, deviate from Beer's law. However, the absorbance values for solutions containing the chromate (or dichromate) ion are directly proportional to their molar concentrations. If the solution is made alkaline (with 0.05 N potassium hydroxide solution), Beer's law is obeyed. Solutions of this type (that is, potassium chromate in 0.05 N sodium hydroxide solution) not only obey the law but have been so accurately characterized that they are used to check photometric scales for accuracy (see Section 1.8).

1.5 ANALYSIS OF MIXTURES

Absorption spectrum for two hypothetical substances, X and Y, are shown in Fig. 1.5. Solutions containing both X and Y cannot be analyzed by measuring an absorbance value at any one wavelength because this value represents the absorbance due to both of the components in the solution. However, if X does not react with Y and each component follows Beer's law, a binary mixture containing both components can be easily analyzed by measuring absorbances at two wavelengths.[1]

All derivations herein are based on the spectral characteristics of X and Y (see Fig. 1.5). The b symbol in Beer's law has been dropped because it is assumed that the length of the cell is equal to 1 cm. Absorbances are usually additive. However, this additivity should be checked experimentally before any of the equations derived in this section are used to determine the concentrations of the components of a binary mixture.

A. SOLUTION OF SIMULTANEOUS EQUATIONS

At λ_2, the total absorbance (A_2) is equal to the sum of the absorbances due to X and Y. Therefore:

$$A_2 = a_5 C_y + a_2 C_x \tag{1.26}$$

Similarly, at λ_4,

$$A_4 = a_1 C_y + a_6 C_x \tag{1.27}$$

Multiply Eq. (1.26) by a_1 and Eq. (1.27) by a_5.

$$a_1 A_2 = a_1 a_5 C_y + a_1 a_2 C_x \tag{1.28}$$

$$a_5 A_4 = a_5 a_1 C_y + a_5 a_6 C_x \tag{1.29}$$

Subtract Eq. (1.28) from Eq. (1.29). Therefore:

$$a_5 A_4 - a_1 A_2 = a_5 a_6 C_x - a_1 a_2 C_x \tag{1.30}$$

$$C_x = \frac{a_5 A_4 - a_1 A_2}{a_5 a_6 - a_1 a_2} \tag{1.31}$$

Similarly:

$$C_y = \frac{a_6 A_2 - a_2 A_4}{a_5 a_6 - a_1 a_2} \tag{1.32}$$

Before a mixture can be analyzed, the analyst must determine the absorptivity values for X and Y at λ_2 and λ_4. These values are substituted into Eqs. (1.31) and (1.32). The absorbance of the mixture at λ_2 and λ_4 is now determined and Eqs. (1.31) and (1.32) solved for C_x and C_y.

The procedure for the determination of salicylic acid in the presence of acetylsalicylic acid[19] is representative of the many methods of analysis in the pharmaceutical literature which are based on Eqs. (1.31) and (1.32).

An accurately weighed sample of acetylsalicylic acid (360 mg) is dissolved in 100.0 ml of chloroform. The absorbance of this solution is determined at 308 mμ, the wavelength at which salicylic acid absorbs a maximum of radiant energy. A 1-ml aliquot of this solution is diluted, with chloroform, to 100.0 ml. The absorbance of this solution is determined at 272 mμ, the absorption maximum for acetylsalicylic acid. Concentrations are calculated by using the following equations:

$$\text{mg acetylsalicylic acid} = 1347A_{278} - 2.3A_{308}$$
$$\text{mg salicylic acid} = 3.58A_{308} - 3.49A_{278}$$

The researchers claim that over a range of 98–100% purity for acetylsalicylic acid the method has an error of less than 0.20%.

Binary mixtures cannot be analyzed unless:

a. Spectral data for the pure components are available.

b. The absorptivity values for the components can be easily and accurately determined.

c. The absorptivity values for the components are sufficiently different at the chosen wavelengths to permit an accurate solution of Eqs. (1.31) and (1.32).

d. The absorbance values for the mixture are accurately determined.

Requirements a and d are more easily met than are b and c. Absorptivity values are difficult to determine if one of the components is a poor absorber of radiant energy at one of the wavelengths chosen for the analysis. Moreover, if the spectral characteristics of the one component approximate those of the other, the absorptivity values at the wavelengths chosen for the analysis will be similar. Under these circumstances, Eqs. (1.31) and (1.32) cannot be accurately solved.

The grouped absorptivity values in Eqs. (1.31) and (1.32) can be determined by measuring the absorbances of solutions containing both X and Y. Nibergall and Mattocks[20] claim that the values so determined are more accurate than those obtained by examining solutions containing only X or Y. Eqs. (1.31) and (1.32) can be modified by combining a values.

$$C_x = K_1A_4 - K_2A_2 \tag{1.33}$$

and

$$C_y = K_3A_2 - K_4A_4 \tag{1.34}$$

If the analyst prepares a series of solutions containing both X and Y and measures the absorbances of these solutions at the chosen wavelengths, the unknowns in Eqs. (1.33) and (1.34) are K_1, K_2, K_3, and K_4. Using the absorbance data obtained for the mixtures, the K values can be calculated by the least squares method. Solutions containing unknown quantities of X and Y can now be quickly analyzed by measuring absorbances and substituting these into Eqs. (1.33) and (1.34).

The multiple regression method is described in detail in the paper by Niebergall and Mattocks.[20] The authors analyzed several solutions containing salicylic acid and salicylamide and concluded that their method was more accurate than that described in this section.

All spectrophotometric methods are based on the assumption that the component (or components) being determined is the only absorber in the solution. However, trace impurities are not always held back by the separatory processes used in the laboratory. Irrelevant absorption is a distinct possibility if these impurities contain chromophoric groups which absorb energy at or near the wavelength chosen for the analysis. Such absorption can produce spurious results and is particularly troublesome when binary mixtures are being analyzed.

Irrelevant absorption in two-component spectrophotometric analysis can be corrected for by using orthogonal functions. The details associated with the use of such functions are presented in a paper by Glenn.[21] Glenn states that harmonic analysis is based on the fact that a given function can be expanded in terms of a set of orthogonal functions of the same variable λ. In other words, the function can be broken down into a set of fundamental shapes, that is, orthogonal functions. An absorption spectrum $f(\lambda)$ can be decomposed into many fundamental shapes (g_0, g_1, g_2, etc.). Each shape is coupled to an appropriate coefficient, which is proportional to concentration. Absorbances are measured at several wavelengths and substituted into the appropriate equations. The author tested the derived equations by analyzing a mixture which contained phenol and epinephrine. The analytical results were not affected by absorbing impurities. The mixture could be analyzed by measuring absorbance values at 270 and 283 mμ, the wavelengths at which the components absorb a maximum of radiant energy, and substituting these into Eqs. (1.31) and (1.32). However, if the solution contains an absorbing impurity which increases the absorbance value at 270 mμ by one-quarter and at 283 mμ by one-third, the per cent recovery of epinephrine and phenol from solutions containing known amounts of the drug and the preservative will be 137.8 and 118.2%, respectively. It is obvious that Glenn's method is superior to that based on Eqs. (1.31) and (1.32). Although the method is complicated, it does deal with a serious spectrophotometric problem. The paper should be read by those who are interested in a more sophisticated approach to the analysis of binary mixtures.

B. CONSTANCY OF THE ABSORBANCE RATIO VALUE

If Beer's law is obeyed at all wavelengths, it may be easily shown that the ratio of two absorbance values determined at two wavelengths is a constant. The spectral characteristics for X (see Fig. 1.5) are used to illustrate the constancy of the absorbance ratio value. At λ_4,)

$$A_4 = a_6 b C_x \qquad (1.35)$$

At λ_3,

$$A_3 = a_4bC_x \tag{1.36}$$

Divide Eq. (1.35) by Eq. (1.36).

$$\frac{A_4}{A_3} = \frac{a_6bC_x}{a_4bC_x} = \frac{a_6}{a_4} = Q:4:3 \tag{1.37}$$

The ratio of two absorbance values is equal to the ratio of two constants (that is, absorptivity values) and is, therefore, equal to a constant. The symbol for the value is Q. For the purposes of this chapter, $Q:290:280$ indicates that this particular constant is calculated by dividing the absorbance value at 290 mμ by that at 280 mμ, the solution and the cell being the same in both cases.

The Q value is independent of concentration and thickness of the solution and can, therefore, be used to assess the purity of pharmaceutically important substances. Q values are given in the NF and the USP for many drugs (aminosalicylic acid, noscapine, promazine HCl, etc.), but the use of such values for the identification of pharmacopeial drugs is a relatively recent development. In actual fact, absorbance ratios have been used for this and other purposes since the turn of the century.[22]

C. EXPERIMENT 1.2: IDENTIFICATION OF AMINOSALICYLIC ACID USP[23]

Dissolve 250 mg of aminosalicylic acid in 3 ml of 4% w/v sodium hydroxide solution. Transfer to a 500-ml volumetric flask, dilute to volume, and mix. Transfer a 5-ml aliquot to a 250-ml volumetric flask containing 12.5 ml of pH 7 phosphate buffer (see USP, p. 913), dilute to volume, and mix.

This solution, when compared in a suitable spectrophotometer against a blank of the same buffer in the same concentration, exhibits absorbance maxima at 265 $\pm$ 2 mμ and 299 $\pm$ 2 mμ, and the $Q:265:299$ value is between 1.50 and 1.56.

D. APPLICATION OF ABSORBANCE RATIOS TO THE ANALYSIS OF BINARY MIXTURES

About 1900, Hüfner[22] showed that a smooth curve resulted when the ratio of the absorbances determined at 540 and 560 mμ were plotted against the per cent reduced hemoglobin in a sample of oxyhemoglobin. The analytical implications of Hüfner's approach to the determination of one component in the presence of another were not immediately apparent. However, the theory associated with this method of analysis is now well understood,[24–26] and several pharmaceutically important mixtures have been analyzed[27,28] by using absorbance ratios rather than absorptivity values.

From Beer's law, the total absorbance at λ_4 (see Fig. 1.5) is equal to the sum of the absorbances due to X and Y. Therefore:

$$A_4 = a_6 b C_x + a_1 b C_y \tag{1.38}$$

Similarly, at λ_3,

$$A_3 = a_4 b C_x + a_4 b C_y \tag{1.39}$$

The two wavelengths chosen for the analysis are the isoabsorptive wavelength and the wavelength at which X absorbs a maximum of radiant energy. (An isoabsorptive point is the wavelength at which two substances have similar absorptivity values.) Divide Eq. (1.38) by Eq. (1.39) and each term in the resulting equation by $C_x + C_y$. Substitute Fx for $C_x/(C_x + C_y)$ and Fy for $C_y/(C_x + C_y)$. Fx and Fy are equal to the fraction of the respective components present in the mixture. Therefore:

$$\frac{A_4}{A_3} = \frac{a_6 Fx + a_1 Fy}{a_4 Fx + a_4 Fy} \tag{1.40}$$

However:

$$Fy = 1 - Fx \tag{1.41}$$

Therefore:

$$\frac{A_4}{A_3} = Fx \left(\frac{a_6}{a_4} - \frac{a_1}{a_4} \right) + \frac{a_1}{a_4} \tag{1.42}$$

A_4/A_3 is equal to Q_o, the absorbance ratio for the binary mixture. Similarly, a_6/a_4 is equal to Qx, the absorbance ratio for pure X, and a_1/a_4 is equal to Qy, the absorbance ratio for Y. Therefore:

$$Q_o = (Qx - Qy)Fx + Qy \tag{1.43}$$

Equation (1.43) is the equation of a straight line having a slope value of $Qx - Qy$ and an intercept value of Qy. The *relative* analysis of a binary mixture can, therefore, be carried out by determining the absorbance ratio values for the pure substances only. More important, Eq. (1.43) is concentration independent. The analysis of a binary mixture does not, therefore, require careful dilution or accurate weighings. However, the analyst can only determine relative concentrations, that is, the per cent of X or Y in the mixture.

If the analyst must determine C_x and C_y, numbers can be substituted into Eq. (1.46). At λ_3,

$$A_3 = a_4 (C_x + C_y) \tag{1.44}$$

and

$$C_x + C_y = A_3/a_4 \tag{1.45}$$

Therefore:

$$C_x = \frac{Q_o - Qy}{Qx - Qy} \times \frac{A_3}{a_4} \tag{1.46}$$

C_y may be determined in a similar manner.

If a wavelength other than an isoabsorptive wavelength is chosen as the second analytical wavelength, a plot of Q_o vs. Fx results in a smooth curve. The analyst must, therefore, prepare a series of solutions containing X and Y, measure absorbances, calculate Q_o values, plot the results, and then use this graph for subsequent analyses. For this reason, an isoabsorptive wavelength should be chosen as the second analytical wavelength.

This method of analysis depends upon the use of an isoabsorptive point. The point must, therefore, be accurately isolated. This can be done by superimposing absorptivity vs. wavelength curves for the components (for example, as in Fig. 1.5) and then determining absorptivity values at and around the wavelength at which the curves cross. The method is tedious. However, the point can be isolated accurately and quickly by preparing two solutions, one containing X and the other Y. The solutions are so prepared that the concentrations of X and Y are the same. A portion of the one solution is placed in the sample cell; the other is placed in the reference cell. The cells are transferred to a spectrophotometer and absorbance values are determined over the range of wavelengths straddling the isoabsorptive point. The wavelength at which the absorbance value is zero represents an isoabsorptive point.

E. EXPERIMENT 1.3: SIMULTANEOUS ANALYSIS OF PROCAINE HCl AND TETRACAINE HCl[25]

Absorption spectra for procaine HCl, tetracaine HCl, and a procaine HCl-tetracaine HCl mixture are shown in Fig. 1.6.

Weigh accurately 160 mg of tetracaine HCl and dissolve in 500.0 ml of water. This is solution A. Weigh accurately 160 mg of procaine HCl and dissolve in 500.0 ml of water. This is solution B. Prepare six solutions in the manner indicated (Table 1.6).

Calculate the per cent tetracaine HCl in each solution. For example, solution 1 contains 100% tetracaine HCl and solution 6 contains 0% tetracaine HCl. Using a recording spectrophotometer, prepare absorption spectra for the six solutions. The curves should intersect at or near 297.5 mμ, the isoabsorptive point. Tetracaine HCl absorbs a maximum of radiant

TABLE 1.6

Solution	Milliliters solution A per liter of water	Milliliters solution B per liter of water
1	25.0	0.0
2	20.0	5.0
3	15.0	10.0
4	10.0	15.0
5	5.0	20.0
6	0.0	25.0

energy at or near 311 mμ. Calculate Q:311:297.5 values. Using the data obtained for each of the six solutions (Q and Fx values for solutions 1–6), calculate the slope and intercept values by the method of least squares.[29] Substitute these values into Eq. (1.43). Calculate the same values from absorbance data obtained for solutions 1 and 6. Do the values agree?

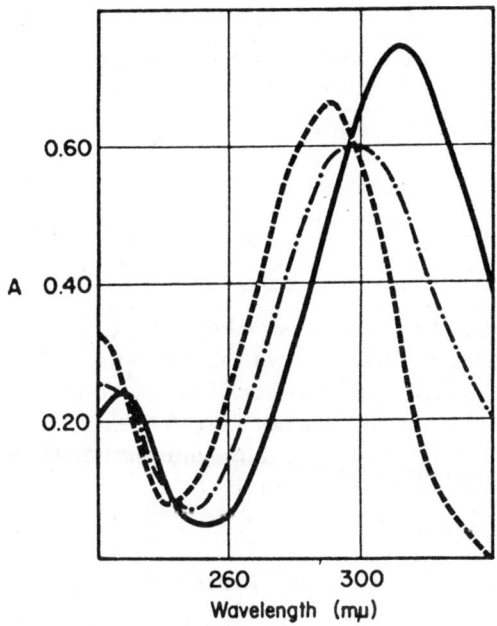

FIGURE 1.6: Absorption spectrum for procaine HCl (– – – –), tetracaine HCl (————), and a procaine HCl-tetracaine HCl mixture (– · – · –). (The solutions contain 10 mg of drug per liter of water. The mixture contains 5 mg of procaine HCl and 5 mg of tetracaine HCl per liter of water.)

Using graduated cylinders, dilute a solution containing unknown quantities of procaine HCl and tetracaine HCl to the point that absorbance values at 311 mμ and 297.5 mμ can be accurately measured. Substitute the Q:311:297.5 value for this solution into the numerical form of Eq. (1.43). Calculate the per cent tetracaine HCl (and the per cent procaine HCl) in the solution.

F. EXPERIMENT 1.4: SIMULTANEOUS ANALYSIS OF THEOBROMINE AND CAFFEINE[30]

Weigh accurately 100 mg of caffeine and dissolve in 100.0 ml of water. This is solution A. Weigh accurately 100 mg of theobromine and dissolve in 100.0 ml of water. This is solution B. Prepare six solutions in the manner indicated (Table 1.7).

Using a recording spectrophotometer, prepare absorption spectra for the six solutions. Calculate a_{240} and a_{273} values for caffeine and theobromine and substitute these into Eqs. (1.31) and (1.32).

The binary mixture is analyzed by measuring absorbances at 240 and 273 mμ. Why did Miles and Englis[30] select these wavelengths?

TABLE I.7

Solution	Milliliters solution A per liter of 0.1 N NaOH	Milliliters solution B per liter of 0.1 N NaOH
1	10.0	—
2	9.0	—
3	8.0	—
4	—	8.0
5	—	9.0
6	—	10.0

Dilute an aliquot of a solution containing unknown quantities of theobromine and caffeine to 1000.0 ml with 0.1 N sodium hydroxide solution. The volume of the aliquot should be such that the absorbance values at 240 and 273 mμ are not less than 0.2 and not more than 0.6. Measure absorbances at 240 and 273 mμ. Substitute these values into the numerical forms of Eqs. (1.31) and (1.32) and solve for C_x and C_y.

1.6 COLORIMETRY

Colorimetry is the determination of the light absorptive capacity of a system. A quantitative determination is, therefore, carried out by subjecting a *colored* solution to those wavelengths of visible energy which are absorbed by that solution. Since such determinations are based on the absorption of energy, the mathematical principles associated with colorimetry are analogous to those described in Sections 1.3 and 1.5.

Determinations can also be carried out by comparing a solution containing an unknown amount of a colored substance with a solution containing a known amount of the same substance. This approach to colorimetry is characteristic of the *limit tests* in the pharmacopeias for a number of organic and inorganic contaminants in certain drugs. The method is comparative and is based on the analyst's ability to distinguish between different intensities of the same color. Ideally, the colored substance should transmit a maximum of energy over the 500- to 600-mμ range, because the eye functions best at these wavelengths. However, the pharmaceutical analyst is primarily concerned with the chemistry of color formation. For this reason, limit tests are based on the formation of substances whose colors range from violet to red.

A drug cannot be analyzed colorimetrically unless it falls into one of three categories:

a. The drug must be self-colored. Some pharmaceutically important substances—for example, pyrvinium pamoate, phenolsulfonphthalein, methylene blue, dithiazanine iodide, etc.—are colored. The substance may be dissolved in a suitable solvent and its absorbance determined at the wavelength at which it absorbs a maximum of visible energy.

b. The drug must react with a reagent to produce a substance which is colored. For example, a reddish substance is formed when norepinephrine is reacted with sodium 1,2-naphthaquinone-4-sulfonate in alkaline media.[31]

c. The drug must be converted to a derivative which reacts with a reagent to produce a substance which is colored. For example, acetophenetidin can be converted to phenetidin, which reacts with chromic acid to produce a colored substance.[32]

A. EXPERIMENT 1.5: LIMIT TEST FOR IRON IN HEAVY MAGNESIUM CARBONATE BP[33]

The test is based on the reaction between iron and thioglycollic acid in a solution buffered with ammonium citrate.

$$2Fe^{3+} + 2CH_2SHCOOH \rightarrow 2Fe^{2+} + COOH—CH_2—S—S—CH_2—COOH + 2H^+$$

Apparatus. A Nessler cylinder is made of clear glass with a nominal capacity of 50 ml. The overall height of the cylinder is about 150 mm; the external height to the 50-ml mark is 109 to 124 mm; the thickness of the wall is 1.0 to 1.5 mm; the thickness of the base is 1.0 to 3.0 mm.

Standard Iron Solution. Dissolve 173 mg of ferric ammonium sulfate in 100 ml of water, add 5 ml of dilute hydrochloric acid solution (10% w/w), and dilute to 1000 ml with water.

Test Solution. Dissolve 0.10 g of heavy magnesium carbonate in 5 ml of water and 0.5 ml of hydrochloric acid, add 2 ml of a 20% w/v solution of citric acid, two drops of thioglycollic acid, mix, and make alkaline with ammonia solution (10% w/w). Dilute to 50 ml with water.

Standard Solution. Dilute 2 ml of the standard iron solution with 40 ml of water. Add 2 ml of a 20% w/v solution of citric acid, two drops of thioglycollic acid, mix, and make alkaline with ammonia solution (10% w/w). Dilute to 50 ml with water.

Test. Compare the color of the two solutions in Nessler cylinders after 5 min. The color of the test solution is not more intense than the color of the standard solution.

B. GENERAL REQUIREMENTS FOR THE COLORED SUBSTANCE

Solutions containing a colored substance are not always amenable to colorimetric measurement. Solutions should, therefore, meet certain general requirements. If one or more of these requirements are not met, the conditions of assay must be carefully spelled out. If they are not, the analyst may not be able to attain the accuracy reported in the literature for that method of analysis.

I. The Solutions Should Be Intensely Colored

Color intensity and sensitivity go hand in hand. If absorbance values change rapidly as the concentration of the constituent in the solution increases, the analyst will be able to determine relatively small quantities of drug. However, if concentrations must be changed drastically (e.g., from 1 to 10 mg) to produce a measurable change in absorbance, the method is not sensitive enough for routine use. In other words, the intensity of the color has changed little, even though there has been a tenfold increase in concentration.

2. The Solutions Must Be Unaffected by pH Changes

Acid-base indicators, potassium chromate, and many other substances (see Section 1.4C) are affected by the pH of the solution. Measurements, in such instances, must be carried out on solutions whose pH is such that the absorbance does not change when the pH of the solution is altered by 1 or 2 units.

3. The Solutions Should Be Stable

When colorimetric methods are developed in the laboratory, it is customary to determine the absorbance value of a solution at various times. If the absorbance value changes with time, the solution must be measured at some specified time or the system must be chemically stabilized.

4. The Constituent Should React Quickly and Quantitatively with the Reagent at Room Temperature

If the colored substance forms slowly, heat may be used to drive the reaction. However, excessive heat may lead to the production of secondary substances. If such substances absorb visible energy at or near the measuring wavelength, Beer's law will not be obeyed.

5. The Solutions Should Obey Beer's Law

Colored solutions will deviate from Beer's law for the reasons given in Section 1.4. One other factor should be considered herein. Filter photometers are often used to determine absorbance values. Depending on the type of filter in the instrument, the band width may be as much as 50 to 60 $m\mu$. Under such circumstances, a plot of absorbance vs. concentration is not always linear. If the deviation from linearity is not too drastic, the graph can be used to determine the concentrations of solutions containing unknown amounts of substance.

To check for compliance with Beer's law, the analyst plots absorbance vs. concentration. However, even if this type of plot is linear, it gives no indication of the accuracy of the method over a certain concentration range. For this reason, some analysts prefer to plot per cent T vs. log c. Such a plot (a *Ringbom* plot) is illustrated in Fig. 1.7. The concentrations at which the analyst can measure absorbance most accurately are defined by the linear portion of the curve. For the example given, best accuracy will be attained for solutions containing from 1 to 6 mg of the substance

C. CHEMISTRY OF COLORIMETRY

Colored substances are formed by reacting the constituent with either an inorganic or an organic reagent. Reaction may, therefore, be classified on the basis of the type of reagent used and on the type of reaction which occurs when the constituent is reacted with the reagent.

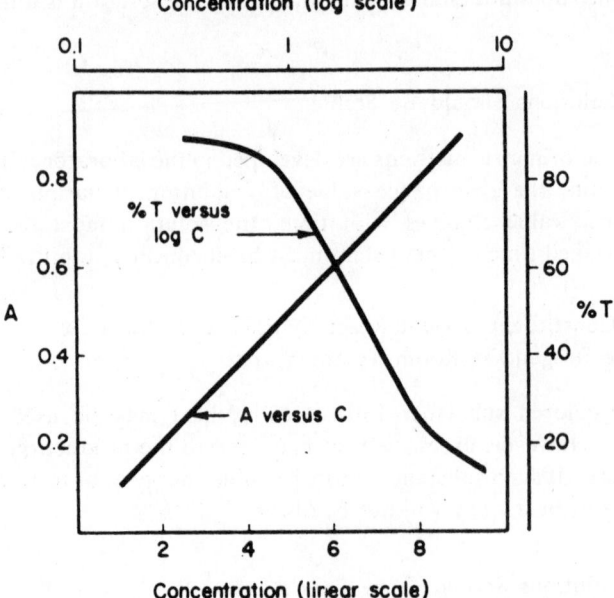

FIGURE 1.7: A comparison of the A vs. c plot with the Ringbom plot.

1. Inorganic Reagents

The reagent may oxidize, reduce, or complex with the constituent. Several of the more common reactions are outlined here:

a. Identification of Copper[34]

$$Cu^{2+} + 4NH_3 \rightarrow Cu(NH_3)_4^{2+}$$
$$\text{(blue)}$$

b. Identification of Iron[35]

$$4Fe^{3+} + 3Fe(CN)_6^{4-} \rightarrow Fe_4(Fe(CN)_6)_3$$
$$\text{ferrocyanide} \qquad\qquad \text{prussian blue}$$

This reaction was first reported in the literature in 1841 by Harting.[36]

c. The Determination of Diphenhydramine HCl[37]

$$HC-O-CH_2-CH_2-N\overset{CH_3}{\underset{CH_3}{|}} \cdot HCl + NH_4(Cr(NH_3)_2(SCN)_4)H_2O \longrightarrow$$
$$\text{ammonium reineckate}$$

$$C_{17}H_{21}NO \cdot H(Cr(NH_3)_2(SCN)_4) + NH_4Cl$$
$$\text{red precipitate}$$

d. Limit Test for Salicylic Acid in APC Tablets[38]

$$\text{salicylic acid} + Fe(NH_4)(SO_4)_2 12H_2O + HCl \longrightarrow \text{Violet iron-salicylic acid complex}$$

ferric ammonium sulfate

2. Organic Reagents

These reagents form salts, couple, or chelate with the constituent.

a. The Determination of Sulfisoxazole

vanillin

yellow Schiff base

b. The Determination of Iron

2,2'-bipyridine   (pink)

c. Limit Test for Lead[39]

dithizone   (violet)

d. The Determination of Sulfonamides.[40,41]

This reagent was first used by Bratton and his co-workers and is now known as the Bratton-Marshall reagent. The analysis is carried out by diazotizing the amine group on the aryl amine, coupling the product to the reagent, and determining the absorbance of the diazo compound at 545 mμ.

$$RHN_2 \quad = \quad H_2N-\!\!\left\langle\!\!\!\bigcirc\!\!\!\right\rangle\!\!-SO_2NH-\!\!\left\langle\!\!\!\bigcirc\!\!\!\right\rangle\!\!\begin{smallmatrix}CH_3\end{smallmatrix}$$

sulfamerazine

$$RNH_2 + HNO_2 + HCl \longrightarrow R-N_2^+Cl^- + 2H_2O$$

$$HNO_2 \text{ (excess)} + H_2NSO_3H \longrightarrow H_2SO_4 + N_2 + H_2O$$
sulfamic acid

$$R-N_2^+Cl^- + \left\langle\!\!\!\bigcirc\!\!\!\right\rangle\!\!-NH-CH_2-CH_2-NH_2 \longrightarrow$$

N-(1-naphthyl)ethylenediamine

$$R-N\!=\!N-\!\!\left\langle\!\!\!\bigcirc\!\!\!\right\rangle\!\!-NH-CH_2-CH_2-NH_2 + HCl$$

No attempt has been made to either spell out the details associated with a particular analysis or to list all the reagents that are described in the USP or the NF. Analytical details and a list of reagents may be found in either the references cited or in *Remington's Pharmaceutical Sciences.*[42]

Like the product, the reagent must meet certain general requirements, listed here, but few of the many organic and inorganic color-forming reagents can meet all of these requirements. The ideal reagent should (a) be colorless, (b) develop the color rapidly, (c) be involved in a reaction whose mechanism is known, (d) react stoichiometrically with the constituent, (e) produce a colored substance with an absorption maximum at a specific wavelength, and (f) react only with the substance being analyzed.

D   EXPERIMENT 1.6: ANALYSIS OF DIPHENHYDRAMINE HCl TABLETS[37]

Ammonium Reineckate Solution. Shake 1 g of ammonium reineckate with 50 ml of water. Filter.

Preparation of a Calibration Curve. Weigh accurately 100 mg of diphenhydramine HCl and dissolve in 50.0 ml of water. Pipet exactly 3.0, 4.0, 5.0, 6.0, and 7.0 ml of this solution into five 30-ml beakers. Adjust the volume in each beaker to 10 ml with 1% H_2SO_4 solution. Place the beakers in a shallow ice bath and add to each, dropwise, 5 ml of ammonium reineckate solution. Allow the precipitate to digest in the ice bath for 1 hr. Collect the precipitate in a sintered-glass crucible of medium porosity. Wash the precipitate with two 5-ml portions of cold water (5°C), dissolve in acetone, transfer the solution to a 25-ml volumetric flask, and dilute to 25.0 ml with acetone. Using a Spectronic 20 spectrophotometer, or the equivalent, measure the absorbance of each of the solutions at 525 mμ. Plot absorbance vs. concentration.

The Analysis of Tablets. Weigh accurately 20 tablets and reduce to a fine powder using a mortar and pestle. Accurately weigh a sample of the powder containing the equivalent of 100 mg of diphenhydramine HCl. Transfer to a beaker, add 50 ml of 1% H_2SO_4 solution, and digest on a steam bath for 15 min. Cool and filter into a 100-ml volumetric flask. Add sufficient 1% H_2SO_4 solution to make 100.0 ml of solution. Subject 10.0 ml of this solution to the procedure just described, that is, add, dropwise, 5 ml of ammonium reineckate solution, etc. Determine, from the calibration curve, the quantity of diphenhydramine HCl in the solution. Calculate the number of milligrams of diphenhydramine HCl in a tablet of average weight.

1.7 SPECIAL METHODS

Spectrophotometric methods of analysis (or, for that matter, any method of analysis) should be simple, specific, and precise. Unfortunately, simplicity, lack of specificity, and poor precision go hand in hand. The pharmaceutical analyst must, therefore, use more elaborate methods if he wishes to analyze for a substance in the presence of an impurity or to increase the precision with which the component is determined. Several of the more popular approaches to standard analytical problems are given herein. However, these should be judged on the basis of what they will do under a particular set of circumstances. If the gain in specificity or precision is relatively small, the pharmaceutical analyst will not use these methods because they are time-consuming and not generally applicable to the analysis of drugs in a quality control laboratory.

A. PRECISION SPECTROPHOTOMETRY

If a solution transmits less than 20% or more than 60% of the incident radiant energy, spectrophotometric errors may be excessive. (Section 1.3D should be reviewed at this point.) A concentrated solution can be diluted or a dilute solution can be concentrated. However, several researchers[13-17] have described techniques that yield a high degree of precision even if the solution contains amounts of drug in excess of or less than those usually subjected to spectrophotometric analysis.

The analyst operates most spectrophotometers by manipulating the slit width, sensitivity, and dark-current knobs. (The wavelength, absorbance-transmittance, and shutter controls must also be moved, but these operations are not directly pertinent to this discussion.) The dark-current knob is used to zero the instrument when the phototube is in darkness. The slit width and sensitivity knobs are used to set 100% T with solvent only in the beam of radiant energy. However, these controls can be manipulated under different circumstances and it is these circumstances that form the basis for precision spectrophotometry.

Reilley and Crawford[43] describe four spectrophotometric methods. The first method is that described in Section 1.3E.

1. Ordinary Method

The instrument is set to read zero with the phototube in darkness, that is, the shutter is closed, and to read 100 when exposed to energy which has passed through pure solvent.

2. Transmittance Ratio Method

This method differs from the ordinary method in that the 100% T setting is made with a reference solution somewhat more dilute than the solution being analyzed.

3. Trace Analysis Method

The instrument is set to read 100 when exposed to energy which has passed through pure solvent and to read zero when exposed to energy which has passed through a reference solution somewhat more concentrated than the solution being analyzed.

4. General Method

The instrument is set to read zero and 100 using reference solutions.

In essence, therefore, the absorbance-transmittance scale on the spectrophotometer can be expanded at will. This results in a gain in precision, but the methods do suffer from a number of disadvantages. Beer's law is not always obeyed and, in some cases, absorption is such that a solution cannot be zeroed or set to 100 by using the dark current or slit-width knobs. On the plus side, Reilley and Crawford show in their paper that a limiting reference transmittance of 78% by method 3 yields a 4.5-fold increase in precision over method 1.

If methods 3 and 4 are to be used, the following procedure should be followed: (a) the range of concentrations over which the measurements are to be made is determined and the extreme points of the range are selected as reference solutions; (b) a series of solutions of known concentration lying between and including the reference solutions is prepared; (c) the instrument is set to its highest operating sensitivity; (d) the reference solution of highest concentration is used to zero the instrument and the reference solution of lowest concentration is used to set the instrument to 100, using dark-current and slit-width knobs alone; (e) the remaining solutions are measured and the data is used to plot a calibration curve; (f) the solution containing an unknown quantity of drug is measured and the concentration is determined graphically.

A colorimetric method of analysis for fluoride is based on one of the methods just described.[48] The researcher analyzed 10 samples, each containing 400 μg of fluoride, and found that the coefficient of variation was only 0.2%.

B. DIFFERENTIAL ANALYSIS

This method of analysis may be used whenever the absorptive properties of a given chromophore can be modified selectively in the presence of a mixture of chromophores. A differential spectrum is obtained by subtracting the spectrum of the starting material from the spectrum of the product. The absorbances of any unmodified chromophores are thereby cancelled out. The method can be used, therefore, to determine the quantity of drug in a mixture of absorbing substances without separating the one from the others. Structurally related substances (e.g., the barbiturates) can also be determined in the presence of each other by using this method of analysis.[49]

A drug can be quickly and easily analyzed by the differential method. For example, Demetrius and Sinsheimer[50] have used this method for the determination of eugenol in pharmaceuticals. All absorbance measurements were carried out at 296 mμ. (Solutions containing known quantities of eugenol were investigated and, on this basis, all subsequent measurements were carried out in solutions having pH values of 3.0 and 12.0. See Section 1.2B. Similarly, the characteristics of the differential spectrum dictate the wavelength at which absorbance is measured.) The concentration of eugenol in pharmaceuticals can be determined by using the following equation.

$$\text{g/liter} = (\Delta A \times 164.2)/\Delta\epsilon$$

$\Delta\epsilon$ is determined by comparing spectrophotometrically two solutions containing identical quantities of eugenol. The pH of the first solution is 3.0. A portion of this solution is transferred to a cell and placed in the reference beam of the spectrophotometer. (In a single beam instrument, this solution acts as the "blank.") The pH of the second solution is 12.0. This solution is transferred to the sample side of the spectrophotometer. The absorbance is measured and then calculated to a molar basis. This constant, according to Demetrius and Sinsheimer, is equal to 3886. ΔA is equal to the observed absorbance at 296 mμ of a given concentration of eugenol in basic solution less the absorbance of the same concentration of eugenol in acid medium. The molecular weight of eugenol is equal to 164.2.

It should be obvious, therefore, that differential analysis is based on the transmittance ratio method (see previous section) and on a change in chromophoric characteristics with a change in solution pH. The method does assume that the chromophoric groups in the contaminating substances are not affected by the change in solution pH.

C. EXPERIMENT 1.7: ANALYSIS OF MORPHINE SULFATE TABLETS[51]

A differential spectrum for morphine, that is, the curve obtained by subtracting the absorbances of the ultraviolet spectrum in acid solution from those of the ultraviolet spectrum in alkaline solution, shows absorption maxima at 256 and 298 mμ. All absorbances are measured at 298 mμ.

Determination of the $\Delta\epsilon$ Value for Morphine. Weigh accurately 80 mg of morphine and dissolve in 100.0 ml of alcohol. Dilute a 10.0-ml aliquot of this solution and 1.0 ml of 1.0 N sodium hydroxide solution to 100.0 ml with water. Dilute a second 10.0 ml-aliquot and 1.0 ml of 0.10 N sulfuric acid solution to 100.0 ml with water. Measure the absorbance of the alkaline solution at 298 mμ relative to the acid solution in the reference cell. Calculate $\Delta\epsilon$.

The $\Delta\epsilon$ value should be equal to approximately 2346.

Assay of Morphine Sulfate Tablets. Weigh accurately 20 $\frac{1}{4}$-gr tablets and reduce to a fine powder using a mortar and pestle. Accurately weigh a sample of the powder containing the equivalent of 90 mg of morphine sulfate. Dissolve the drug in water, filter, and dilute the filtrate to 100.0 ml with water. Dilute two 10.0-ml aliquots as described under the determination of the $\Delta\epsilon$ value for morphine. Measure the relative absorbance of the alkaline solution at 298 mμ. Calculate the amount of morphine sulfate pentahydrate in the aliquot by using the following formula.

$$g/100 \text{ ml} = 379.4\Delta A/(10\Delta\epsilon)$$

This method of analysis can be used to determine morphine in tincture of opium and in camphorated tincture of opium. However, the assay of such preparations requires preliminary separation because narceine, narcotine, papaverine, and some of the nonalkaloidal constituents of opium show absorbance changes with changes in the pH of the solution. The separatory procedure is described in the paper cited.

D. SPECTROPHOTOMETRIC TITRATIONS

A titration, in its simplest form, involves the reaction of the constituent with a titrant to form a product:

$$X + Y \rightarrow Z$$

If an indicator is used to locate the end point, a slight excess of Y, the titrant, will produce a color change. In a spectrophotometric titration, the absorbance of the solution at a specified wavelength is measured after each addition of titrant. The results are plotted (A vs. milliliters of titrant), and the end point is determined graphically. The point at which the two straight lines intersect is the end point of the titration. As in conductometric titrations, the graph is constructed on the basis of data obtained well before and well after the end point. The addition of small increments of titrant at and around the end point is, therefore, unnecessary. However, absorbance values must be corrected because the volume changes as titrant is added to the solution. Instruments for carrying out such titrations automatically are now available.

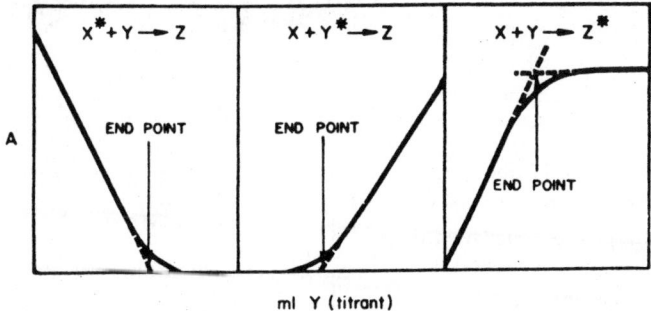

FIGURE 1.8: Spectrophotometric titration curves. (The asterisk indicates the species which absorbs radiant energy.)

If X, the substance being analyzed, absorbs radiant energy at a specified wavelength and its absorptivity value is known, an analysis can be carried out directly. A spectrophotometric titration serves no useful purpose. However, if the constituent is contaminated with other absorbing substances, a spectrophotometric titration can be carried out if a titrant can be found which reacts selectively with X. If X and the contaminants are the only species that absorb radiant energy, the absorbance will decrease as titrant is added. The absorbance will remain constant after the end point, that is, after X has been completely converted to Z, a nonabsorbing substance. Typical plots are shown in Fig. 1.8.

If Y is the only species in solution which absorbs radiant energy, the solution will not absorb radiant energy until the end point is reached. At that point, the absorbance values increase as the concentration of Y increases.

If Z is the only species in solution which absorbs radiant energy, the absorbance will increase as product is formed. The absorbance will remain constant after the end point because the amount of Z in solution is constant.

The plots presented in Fig. 1.8 are for hypothetical substances. The papers

FIGURE 1.9: A Beckman model DU-2 spectrophotometer. Courtesy of Beckman Instruments, Inc., Fullerton, Calif.

by Goddu and Hume[52] and by Higuchi et al.[53] give specific examples of spectrophotometric titrations and discuss, in detail, the advantages and disadvantages of the technique. The method can be used when dilute solutions (10^{-4} M or less) are being analyzed, when binary mixtures of weak acids or bases are being titrated, and when the color change at the end point is poor.

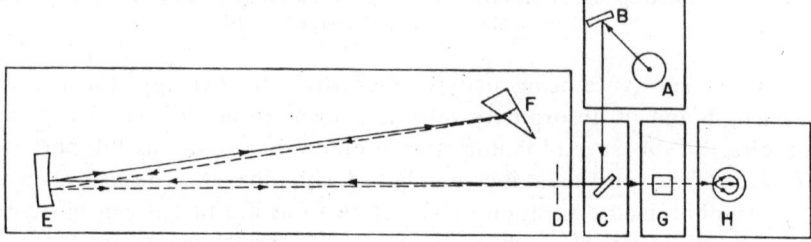

FIGURE 1.10: Schematic diagram of the Beckman model DU spectrophotometer. Light from the source (A) is focused on the condensing mirror (B) and directed in a beam to the 45° slit-entrance mirror (C). The slit-entrance mirror deflects the beam through the slit (D) and into the monochromator to the collimator mirror (E). Light falling on the collimator mirror is rendered parallel and reflected to the prism (F), where it undergoes refraction. The back surface of the prism is aluminized so that light refracted at the first surface is reflected back through the prism, undergoing further refraction as it emerges. The desired wavelength of light is selected by rotating the wavelength selector on top of the monochromator case. This control adjusts the position of the prism. The spectrum from the prism is directed back to the collimating mirror which centers the chosen wavelength of light on the slit and the sample (G). Light passing through the sample strikes the phototube (H), causing a voltage to appear across a load resistor. Voltage is amplified and registered on the null meter. Courtesy of Beckman Instruments, Inc., Fullerton, Calif.

1.8 SPECTROPHOTOMETERS AND COLORIMETERS

A spectrophotometer is an instrument which is capable of isolating "monochromatic" radiation. The desired wavelength is isolated by using a prism or grating and auxiliary mirrors and slits which, collectively, form the monochromator of the instrument. The wavelength dial on a spectrophotometer is set to a specific value, but the radiation leaving the exit slit is rarely monochromatic. However, some of the more sophisticated instruments can isolate, at the exit slit, a band of energy which is 1 mμ or less in width.

A colorimeter or filter photometer isolates several wavelengths of radiant energy by using a filter. Three or four filters are supplied with each instrument and each filter will pass a maximum of light at certain specified wavelengths. These filters isolate "polychromatic" radiant energy. However, both terms—that is, "monochromatic" and "polychromatic"—have real meaning only when the characteristics of the dispersing device are clearly stated. For example, a colorimeter equipped with an interferometric filter will isolate a bandwidth of 10 to 20 mμ. The Spectronic 20, a grating-type instrument with fixed slits, produces a band pass of 20 mμ.

All instruments must be equipped with a radiation source, a device for isolating the desired wavelength, a container or cell for the solution to be examined, and a detector of radiant energy. Diagrams of the optical systems of several of the more common instruments are shown in Figs. 1.10, 1.12, 1.13, and 1.14.

A. RADIATION SOURCES

The radiation source must meet three requirements:

a. The radiation should be continuous, that is, its spectrum should include all of the wavelengths required for the analysis.

b. The power of the beam should be such that the solution will transmit, under normal circumstances, some or all of the radiant energy at all wavelengths.

c. The source should be stable. It should be obvious that the power of the beam must remain constant throughout the measurement. If the solvent is subjected to more radiant energy (P_0 at the detector) than the solution (P at detector), the absorbance value has little meaning. Most instruments are now equipped with regulators to prevent fluctuations in beam intensity, and some instruments are so designed that P_0 and P are measured simultaneously.

Shorter wavelengths of radiant energy are emitted as the temperature of the radiation source is increased. Ultraviolet radiant energy can be obtained from a tungsten lamp by increasing voltage. However, the lamp would burn out quickly and, for this reason, the better spectrophotometers are equipped with dual sources.

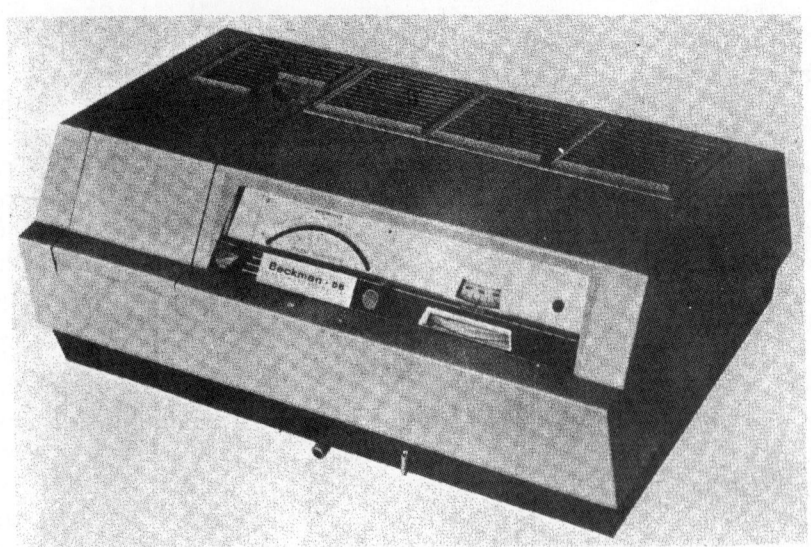

FIGURE 1.11: A Beckman model DB spectrophotometer. Courtesy of Beckman Instruments, Inc., Fullerton, Calif.

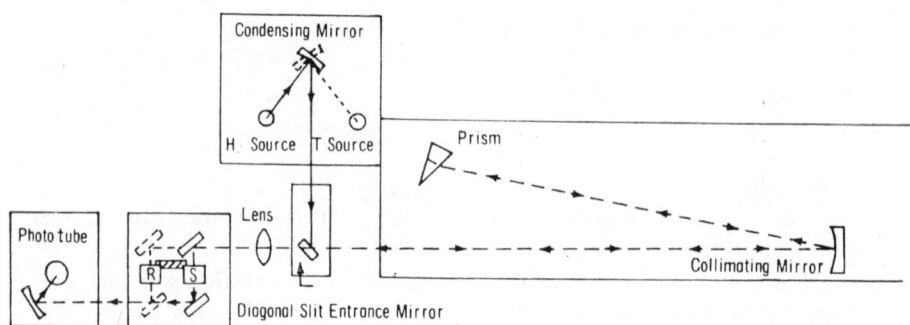

FIGURE 1.12: Schematic diagram of the Beckman model DB spectrophotometer. Light from the source is focused by the condensing mirror and directed to the monochromator diagonal mirror. This mirror directs the light through the entrance slit into the monochromator to the collimating mirror. Light falling on the collimating mirror is collimated and reflected to the prism, where it undergoes refraction. The back surface of the prism is aluminized so that light refracted at the first surface is reflected back through the prism, undergoing further refraction as it emerges from the prism. The desired wavelength is selected by rotating the wavelength control. The spectrum is directed back to the collimator, which centers the chosen wavelength on the exit slit. This light is directed alternately through the sample and reference path by the vibrating mirror. Courtesy of Beckman Instruments, Inc., Fullerton, Calif.

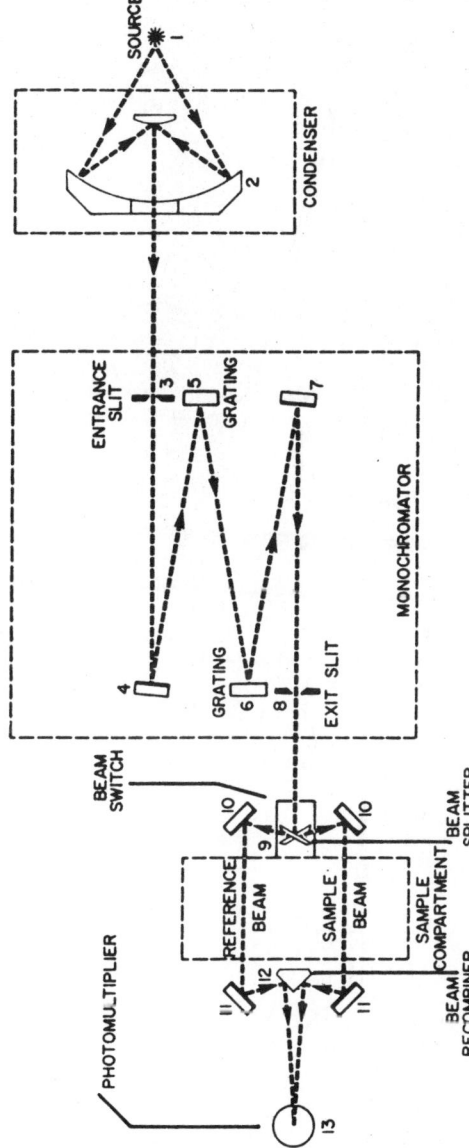

FIGURE 1.13: Schematic Diagram of the Spectronic 505. The beam from the source (1) enters the condenser (2), which focuses a ten-times-enlarged, achromatic image of the lamp filament on the entrance slit (3) of the double grating monochromator. The beam passes through the slit to a collimating mirror (4), is reflected to the first reflectance grating (5) where it is dispersed, then to a second reflectance grating (6) where it is further dispersed. The twice-dispersed beam is reflected from the second collimating mirror (7), passes through the monochromator exit slit (8), and strikes the beam splitter (9) where it is divided into reference beam and sample beam. The two beams are alternately chopped by the rotating beam switch and reflected by mirrors (10) through the sample and reference cells. Leaving the cell compartment, the unabsorbed light from the two beams is reflected by mirrors (11) to the beam recombiner (12) and brought back into coincidence on the photomultiplier (13). The alternate sample and reference beams give electrical pulses proportional to the intensity of the beams.   Courtesy of Bausch & Lomb Inc., Rochester, N.Y.

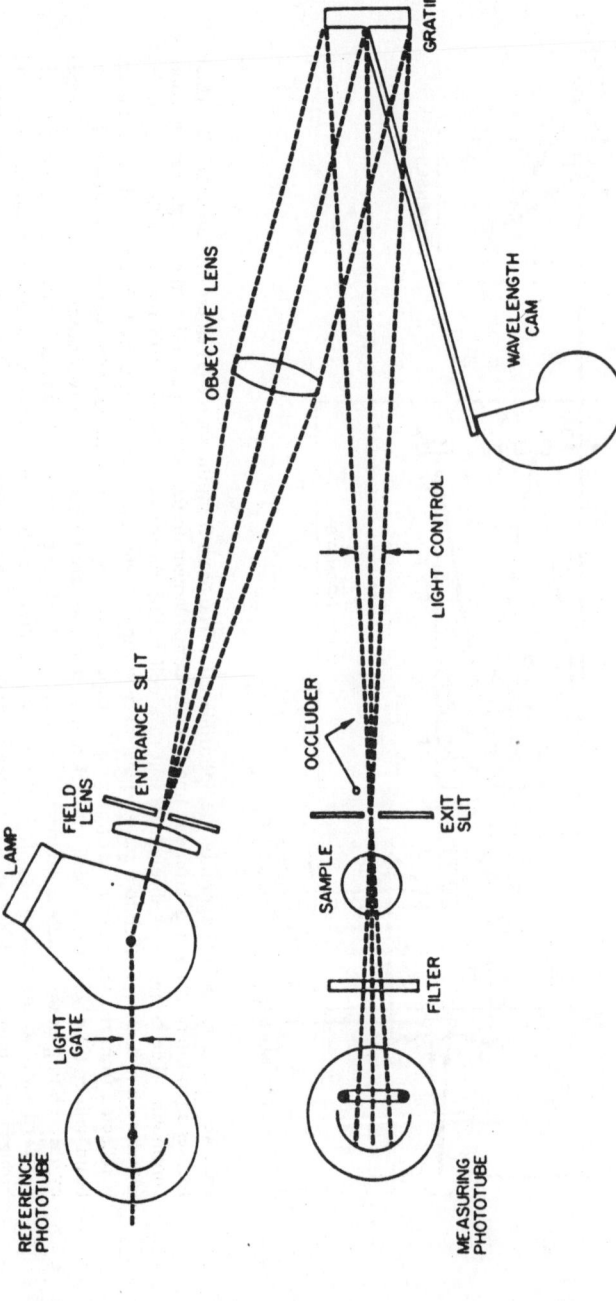

FIGURE 1.14: Schematic Diagram of the Spectronic 20. Courtesy of Bausch & Lomb, Inc., Rochester, N.Y.

Sources of Visible Radiation. In the visible region of the spectrum, the source is an electric light bulb with an incandescent filament. The lamp may be operated from a storage battery or by a 110-V source. If a 110-V source is used, a constant voltage transformer should be used to stabilize lamp output. The tungsten lamp emits radiant energy in the region between 350 and 2500 mμ. Its relative intensity decreases as the wavelength decreases.

Sources of Ultraviolet Radiation. A hydrogen discharge lamp consists of a pair of electrodes in a glass tube with a quartz window. The lamp contains hydrogen gas at a reduced pressure. When a voltage is applied to the electrodes, the hydrogen molecules are excited and produce radiant energy between 185 and 365 mμ.

Both the Beckman DB and DU-2 spectrophotometers are equipped with both hydrogen and tungsten lamps. The Spectronic 505, however, is equipped with a combination light source. This contains a tungsten lamp, a mercury lamp (for wavelength calibration), and a deuterium lamp. The latter lamp contains deuterium gas and is approximately three times brighter than a hydrogen lamp of comparable design and wattage. The Spectronic 505 is a grating type instrument with fixed slits. The bandwidth is equal to 5 Å. In an instrument with variable slits, the slit width is increased if there is not enough energy at the detector to produce an instrument reading. In the Spectronic 505, the output of the source must be increased in order to accomplish the same purpose.

B. FILTERS AND MONOCHROMATORS

The desired wavelength can be isolated by using either a filter or a monochrometer.

Filters. Filters may be made of glass, gelatin, or of two thin semitransparent metallic films separated by a thin film of cryolite or other dielectric material.

Glass Filters. Colored glass filters will transmit certain wavelengths of light, that is, each filter has a characteristic absorption spectrum. The filter is made by incorporating the oxides of certain metals into the glass. For example, cobalt produces a blue filter, manganese, a purple filter, and iron, a green filter.

The filter must be matched to the solution being analyzed. For example, if the colored solution absorbs a maximum of energy at 550 mμ, the filter must transmit a maximum of energy at that wavelength. It is difficult, if not impossible to match transmittance maxima with absorption maxima. The effective bandwidth varies from filter to filter, but some glass filters will transmit a band whose width approaches 150 mμ. Some filters have bandwidths of 25 to 50 mμ, but these transmit no more than 25% of the radiant energy emitted by the source. Little energy reaches the solution (and less, the detector) and, for this reason, some analyses cannot be carried out by using filter photometers.

Gelatin Filters. A gelatin filter is made of two pieces of glass separated by a thin sheet of dye suspended in gelatin. The bandwidth of the filter may be as much as 50 mμ. Gelatin filters are not as stable as glass filters.

Interferometric Filters. Narrow bandwidths (10 to 20 mμ) can be obtained by using an interferometric filter. The filter consists of two thin semi-transparent metallic films separated by a thin film of dielectric material. When light strikes the filter, it is partially reflected by the first metallic layer. The portion that is transmitted passes through the transparent dielectric material and is then partially reflected by the second metallic surface. If this reflected portion is of the right wavelength, it will be partially reflected from the first surface in phase with light of the same wavelength that is entering the surface at this point. That particular wavelength is, therefore, reinforced. All other wavelengths interfere with each other.

The wavelength at which an interference filter passes light can be calculated from the following equation.

$$n\lambda = 2t \tag{1.47}$$

The thickness of the dielectric material is equal to t, and n is an integer. For example, if the thickness of the transparent material is equal to 500 mμ and n is equal to 1, λ is 1000 mμ. If n is equal to 2, λ is 500 mμ.

Monochromators. A monochromator resolves polychromatic radiation into its component wavelengths and focuses several of these wavelengths onto the solution in the cell. Radiation enters the monochromator through an entrance slit, is collimated with a lens or mirror, dispersed by a prism (or grating), returned to a lens or mirror, and focused upon an exit slit. The entrance and exit slits may be one and the same thing or may be separate entities.

The effective bandwidth striking the solution depends upon the nature of the dispersing device, the wavelength, and on the widths of the entrance and exit slits. In a prism spectrophotometer, the dispersion is nonlinear with wavelength. This means that any given slit width will pass a much wider band of wavelengths in the higher than in the lower wavelength regions. In a grating instrument, dispersion is very nearly a linear function of wavelength. Any given slit width will, therefore, pass nearly the same band of wavelengths in all regions.

The Prism. Polychromatic light can be dispersed with a prism. The velocity of light in a medium depends upon the refractive index of that medium.

$$v = c/n \tag{1.48}$$

where v is the velocity of light in the medium, n is the refractive index, and

c is the velocity of light in vacuo. For example, the velocity of light in glass, ($n = 1.5$) is $\frac{2}{3}$ the velocity in vacuo. Because its velocity depends upon the medium, light will be refracted and reflected when it passes from one isotropic medium to another isotropic medium. However, refraction occurs only when the refractive indices of the two media are different.

Snell's law equates the angle of incidence θ to the angle of refraction θ'.

$$n \sin \theta = n' \sin \theta' \qquad (1.49)$$

The angle of refraction depends, therefore, upon the angle of incidence, the refractive indices of the two media (n and n'), and, last, upon the wavelength of the radiant energy. This means that the shorter wavelengths are bent to a greater extent than are the wavelengths in the upper regions of the spectrum. With white light, this dispersion results in the usual visible spectrum.

Many spectrophotometers are equipped with a 30° Littrow prism with a reflecting back. Light enters and emerges from the same face and, in effect, produces the same dispersion as that which occurs with a 60° Cornu prism. Moreover, any birefringence within the prism is cancelled out since light passes in both directions. The lens in a monochromator of this type serves as the collimator and the objective lens. The instrument is so designed that the entering and refracted rays travel the same path except that one lies above the other.

Glass prisms absorb ultraviolet radiant energy. Ultraviolet spectrophotometers are, therefore, equipped with quartz prisms. The useful range of the latter prism is from 180 to 4000 mμ; of the former, from 350 to 2000 mμ.

The Grating. Polychromatic light can also be dispersed with a transmission or a diffraction grating. Grating spectrophotometers are usually equipped with diffraction gratings.

A diffraction grating consists of a large number of parallel, equidistant lines ruled on a polished surface. Each inch of grating may have as few as 5000 and as many as 50,000 parallel grooves. For example, the Spectronic 20 is equipped with a grating which has 600 grooves/mm. The Spectronic 505, however, is equipped with two gratings with 1200 grooves/mm.

A grating may be visualized as a plane or concave surface with a large number of parallel equidistant slits. Each illuminated slit acts as a source. At any angle θ', and at some distance greater than x, the distance between the centers of two adjacent slits, the path difference between rays coming from these slits is equal to $x \sin \theta'$. If the path difference is equal to an integral number of wavelengths, reinforcement of the radiation will occur. At any other angle, destructive interference occurs. The latter statements assume parallel perpendicular incident radiation and an infinite number of slits. The mathematical relationship between the quantities involved is given in Eq. (1.50).

$$x(\sin \theta' + \sin \theta) = \pm m\lambda \qquad (1.50)$$

θ' is the angle of diffraction, θ is the angle of incidence, and m is an integer. Spectra with m equal to ± 1, ± 2, . . . , are called first-, second-, . . . , order spectra. Each value of m gives rise, therefore, to a spectrum. If m is equal to zero, white light will be undispersed. However, with each succeeding value of m, dispersion occurs. The grooves can be shaped in such a way that as much as 80 to 90% of the diffracted radiant energy is concentrated into a specified order.

This brief description of gratings may be supplemented by the information in Meehan's treatise on optical methods.[4] A grating monochromator has certain advantages over those that disperse radiation with prisms. Dispersion is linear, and construction of recording spectrophotometers is simplified because prism instruments require complex cam arrangements. However, prism instruments give slightly better dispersion in the 200- to 250-mμ range.

C. ABSORPTION CELL

The solution which is subjected to radiant energy is held in a transparent cell or cuvette. This cell is an integral part of the instrument's optical system and must be kept scrupulously clean at all times. Cells are usually washed with distilled water or a mild sulfonic detergent solution. If a stronger cleaning agent is required, the cells may be soaked in a 50:50 solution of 3 N HCl and alcohol.

The optical windows of cuvettes are made from silica (for measurements in the ultraviolet and visible regions) or pyrex (for measurements in the visible region). Cuvettes are available in all shapes and sizes. However, the most common size is the 1-cm cell which holds between 5 and 6 ml of solution. Standard cells are usually sold in pairs. Most analysts mark one of the cells of the set and reserve it for solvent only. The other cell is filled with the solution being analyzed. Cells are usually matched to within 2% T. The manufacturers' specifications are acceptable for most purposes, but it may be necessary, at times, to correct for differences in cell length.

D. RADIATION DETECTORS

Phototubes, photomultiplier tubes, and barrier layer (photovoltaic) cells may be used to detect and measure radiant energy. Radiant energy must be converted into electric energy, and the signal produced by the detector must be directly proportional to the amount of radiant energy striking it.

The Phototube. A phototube consists of a concave photoemissive surface (the cathode) and a wire anode. The cathode and the anode are enclosed in an evacuated tube. When radiant energy strikes a cathode coated with potassium, cesium, or an alkaline-earth oxide, electrons are emitted and accelerated toward the anode. However, a potential must be applied (90 V for most phototubes) in order to achieve maximum collection of electrons at the anode. If the applied potential is sufficiently high, the current produced is proportional to the amount of energy reaching the detector. The current produced by the phototube must be amplified before the actual measurement is made. Phototubes emit electrons even if they are in the dark (thermal emission of electrons) and a small current will flow. This is compensated for in most instruments by a "dark current" control.

Most spectrophotometers are equipped with two phototubes. The blue-sensitive phototube responds to ultraviolet and visible energy (to about 600 mμ) and the red-sensitive phototube detects energy of longer wavelengths.

The Photomultiplier Tube. A photomultiplier tube consists of a photosensitive cathode, an anode, and a number of other electrodes called "dynodes." Each dynode functions both as a cathode and an anode—a cathode for the dynode ahead of it and an anode for the dynode behind it. This is brought about by maintaining each dynode at a potential somewhat more positive than that of the preceding electrode.

When the electrons from the cathode strike the first dynode, several additional electrons are liberated. These, in turn, strike the next dynode and again liberate additional electrons. This process produces a cascade of 10^6 or more electrons at the anode.

Standard spectrophotometers are not normally equipped with photomultiplier tubes, but they can be readily installed if the analyst finds it necessary to detect and measure minimal amounts of radiant energy.

The Barrier Layer Cell. A barrier layer cell consists of an iron or copper plate upon which is deposited a layer of selenium or cuprous oxide. This semiconducting material is covered with a transparent film of gold, lead, or silver. The metallic film serves as the collector electrode. The plate, the semiconducting material, and the metallic film are enclosed in a transparent envelope.

The electrons in selenium or cuprous oxide are not mobile under normal circumstances. However, when light strikes the cell, electrons are liberated, penetrate the interface or barrier between the semiconducting material and the metallic film, and are, in this way, transferred to the collector electrode. If the metallic film is connected by way of an external circuit to the plate, a current, which is proportional to the amount of light striking the cell, will flow. This current (10 to 100 mA) can be measured with a galvanometer or microammeter.

The barrier layer cell is used to detect and measure visible radiation. The sensitivity of the cell is similar to that of the human eye, that is, it detects radiation best in the 500- to 600-mμ region. However, its sensitivity is poor at low levels of illumination and, if used for prolonged periods of time, it exhibits fatigue. Under such circumstances, the transmittance reading will decrease with time. This can be rectified by keeping the cell in the dark.

The cell is cheap, rugged, and requires no external source of electric energy. The inexpensive colorimeters are usually equipped with detectors of this type.

E. ACCURACY OF WAVELENGTH AND ABSORBANCE SCALES

The absorbance-transmittance and the wavelength scales of the better spectrophotometers are reasonably accurate. However, the wise analyst calibrates the instrument before carrying out crucial measurements.

The Wavelength Scale. This scale may be checked by inserting a holmium oxide filter into the sample beam, recording absorbance values at various

TABLE 1.8: Absorbance and Transmittance Values for
Standard Potassium Chromate Solution[a]

Wavelength, mμ	T	A	Wavelength, mμ	T	A
210	0.000	—	330	0.715	0.145
215	0.037	1.432	335	0.605	0.218
220	0.350	0.456	340	0.485	0.314
225	0.600	0.222	345	0.380	0.420
230	0.680	0.168	350	0.280	0.553
235	0.620	0.208	355	0.202	0.695
240	0.509	0.293	360	0.148	0.830
245	0.408	0.389	365	0.115	0.939
250	0.319	0.496	370	0.102	0.991
255	0.268	0.572	375	0.103	0.987
260	0.232	0.635	380	0.118	0.928
265	0.201	0.697	385	0.152	0.818
270	0.180	0.745	390	0.207	0.684
275	0.173	0.762	395	0.300	0.523
280	0.189	0.724	400	0.410	0.387
285	0.254	0.595	404.7	0.520	0.284
290	0.372	0.430	410	0.635	0.197
295	0.527	0.278	420	0.748	0.126
300	0.705	0.152	430	0.824	0.084
305	0.830	0.081	435.8	0.861	0.065
310	0.900	0.046	440	0.884	0.054
315	0.905	0.043	450	0.928	0.033
320	0.867	0.062	460	0.961	0.017
325	0.810	0.092	470	0.981	0.008
			480	0.992	0.004
			490	0.998	0.001
			500	1.000	0.000

[a] Dissolve 0.0400 g of potassium chromate in sufficient 0.05 N potassium hydroxide to make 1 liter of solution. The solution is measured in a 1-cm cell. (Exact absorbance values, that is, to four decimal places, may be found in Refs. 17 and 54. This table is reproduced through the courtesy of the National Bureau of Standards, Washington, D.C.)

wavelengths, and comparing the position of absorption bands with standard values. The holmium oxide filter exhibits absorption maxima at 279.3, 287.6, 333.8, 360.8, 418.5, 536.4, and 637.5 mμ.

The best single source of visible and ultraviolet energy for wavelength calibration is the quartz-mercury arc. The lines at 253.7, 302.25, 313.16, 334.15, 365.48, 404.66, and 435.83 mμ may be used to check the wavelength scale. The 486.13- and the 656.28-mμ lines of the hydrogen lamp may be used for the same purpose.

The Absorbance Scale. The spectral characteristics of copper sulfate, cobalt ammonium sulfate, and potassium chromate solutions have been determined by the National Bureau of Standards.[17] These solutions can be used to check the absorbance (or transmittance) scale of the spectrophotometer. Absorbance and transmittance values, at various wavelengths, are given for potassium chromate in Table 1.8.

QUESTIONS

Q1.1. If the absorbance value, at 275 mμ, for the standard potassium chromate solution (see Section 1.8E) is 0.762, what is the transmittance value for a solution containing 0.03 g potassium chromate per liter of solution?

Q1.2. Prove mathematically that

$$\alpha = \frac{\epsilon'' - \epsilon'''}{\epsilon' - \epsilon'''}$$

See Eq. (1.24). *Hint:* Divide an absorption cell into two compartments. The first compartment contains the ionic form and the second the molecular form of the acid. If the incident radiant power is P_1, the intensity of the beam entering the second compartment is P_2 and P_3 is the intensity of the beam leaving the cell. The total transmittance of the cell is equal to P_3/P_1. For the first compartment,

$$P_2 = P_1 e^{-k''bc\alpha}$$

or

$$-\log P_2/P_1 = \epsilon'bc\alpha$$

where ϵ' is the molar absorptivity of the ionic form and α is the fraction of acid that has ionized. Continue.

Q1.3. If the absorbance value for a solution in a 1-cm cell is 0.210, what is the transmittance value for the same solution in a 2-cm cell? What is the transmittance value in a 4-cm cell?

Q1.4. Absorptivity values for drugs X and Y at two different wavelengths are:

	$a_{240 \text{ m}\mu}$	$a_{275 \text{ m}\mu}$
X	60	0
Y	20	40

A mixture of X and Y contains twice as much X as Y. Suggest a method of analysis for the mixture and give the necessary equations.

Q1.5. A substance absorbs a maximum of radiant energy at 350 mμ. What are the frequency and wave number values at this wavelength?

Q1.6. How would you determine barbiturates spectrophotometrically? Can barbiturates be resolved spectrophotometrically? If so, how? (See Section 1.7B.)

Q1.7. An analyst is given a solution which is purported to contain drugs X and Y. He prepares a solution of X and one of Y and measures absorbances at 250 mμ, the absorption maximum for X, at 265 mμ, an isoabsorptive point, and at 290 mμ, the absorption maximum for Y.

Solution	$A_{250 \text{ m}\mu}$	$A_{265 \text{ m}\mu}$	$A_{290 \text{ m}\mu}$
10 mg X/liter	0.720	0.400	0.150
9 mg Y/liter	0.090	0.360	0.468
Unknown	0.612	0.340	0.128

Calculate the amounts of X and Y in the unknown solution.

Q1.8. Criticize the simultaneous equation method for the analysis of binary mixtures.

Q1.9. Criticize the absorbance ratio method for the analysis of binary mixtures.

Q1.10. What type of source would you recommend for a recording spectrophotometer with narrow fixed slits? Why?

Q1.11. Look up Refs. 40 and 41 and describe in detail the procedure for the determination of sulfonamides.

Q1.12. An analyst is asked to analyze phenylbutazone tablets NF. The tablets are purported to contain 100 mg of drug per tablet. The analyst weighs 20 tablets (4.2010 g) and reduces them to a fine powder. A 202.6-mg portion of the powdered tablets is extracted with alcohol, the solution is filtered, and the filtrate is made to 100.0 ml with alcohol. A 10.0-ml aliquot of this solution is diluted to 1 liter with 0.1 N sodium hydroxide solution. The absorbance value for the solution is 0.622. The analyst searches the literature and finds that Pernarowski has reported an absorptivity value of 66 for this drug. All measurements were carried out in a 1-cm cell.

a. Calculate the number of milligrams of drug in a tablet of average weight.

b. Does the brand comply with NF limits for phenylbutazone tablets?

c. Criticize the analyst's procedure.

REFERENCES

1. J. H. Yoe, *Anal. Chem.*, **29**, 1246 (1957).
2. D. F. Boltz and M. G. Mellon, *Anal. Chem.*, **38**, 317R (1966).
3. W. Crummett, *Anal. Chem.*, **38**, 404R (1966).
4. E. J. Meehan, *Optical Methods of Analysis*, Wiley, New York, 1964.
5. A. Weissberger, (ed.), *Physical Methods of Organic Analysis*, Part III, Wiley (Interscience), New York, 1960.
6. M. G. Mellon, *Analytical Absorption Spectroscopy*, Wiley, New York, 1950.
7. J. M. Vandenbelt and L. Doub, *J. Am. Chem. Soc.*, **69**, 2714 (1947).

8. I. Sunshine and S. R. Gerber, *Spectrophotometric Analysis of Drugs*, Thomas, Springfield, Ill., 1963.
9. H. G. Pfeiffer and H. A. Liebhafsky, *J. Chem. Educ.*, **28**, 123 (1951).
10. N. T. Gridgeman, *Photoelec. Spectrometry Group Bull.*, **4**, 67 (1951).
11. N. T. Gridgeman, *Anal. Chem.*, **24**, 445 (1952).
12. J. R. Edisbury, *Photoelec. Spectrometry Group Bull.*, **5**, 109 (1952).
13. *The National Formulary, Twelfth Edition*, Mack Printing, Easton, Pa., 1965, p. 490.
14. *The National Formulary, Twelfth Edition*, Mack Printing, Easton, Pa., 1965, p. 495.
15. G. Kortum and M. Seiler, *Angew. Chem.*, **52**, 687 (1939).
16. A. R. Rogers, *J. Pharm. Pharmacol.*, **11**, 291 (1959).
17. United States Department of Commerce, *Nat. Bur. Std.*, *LC484*, 40 (1949).
18. E. E. Sager, M. R. Schooley, A. S. Carr, and S. F. Acree, *J. Res. Natl. Bur. Std.*, **35**, 521 (1945).
19. R. B. Tinker and A. J. McBay, *J. Am. Pharm. Assoc. Sci. Ed.*, **43**, 315 (1954).
20. P. J. Neibergall and A. M. Mattocks, *Drug Std.*, **28**, 61 (1960).
21. A. L. Glenn, *J. Pharm. Pharmacol.*, **15**, S123 (1963).
22. L. Heilmeyer, *Spectrophotometry in Medicine*, Adam Hilger Limited, London, 1943, p. 65.
23. *United States Pharmacopeia*, Mack Printing, Easton, Pa., 17th rev. ed., 1965.
24. M. Pernarowski, A. M. Knevel, and J. E. Christian, *J. Pharm. Sci.*, **50**, 943 (1961).
25. M. Pernarowski, A. M. Knevel, and J. E. Christian, *J. Pharm. Sci.*, **51**, 688 (1962).
26. A. L. Glenn, *J. Pharm. Pharmacol.*, **12**, 595 (1960).
27. M. Pernarowski, A. M. Knevel, and J. E. Christian, *J. Pharm. Sci.*, **50**, 946 (1961).
28. F. Yokoyama and M. Pernarowski, *J. Pharm. Sci.*, **50**, 953 (1961).
29. W. A. Wallis and H. V. Roberts, *Statistics: A New Approach*, Free Press, New York, 1956, p. 252.
30. J. W. Miles and D. T. Englis, *J. Am. Pharm. Assoc. Sci. Ed.*, **43**, 589 (1954).
31. *British Pharmacopoeia 1963*, Pharmaceutical Press, London, 1968, p. 20.
32. E. F. Degner and L. T. Johnson, *Anal. Chem.*, **19**, 330 (1947).
33. *British Pharmacopoeia 1963*, Pharmaceutical Press, London, 1968, p. 560.
34. *The National Formulary, Twelfth Edition*, Mack Printing, Easton, Pa., 1965, p. 458.
35. *United States Pharmacopeia*, 17th rev. ed., Mack Printing, Easton, Pa., 1965, p. 879.
36. M. G. Mellon, *Anal. Chem.*, **24**, 924 (1952).
37. F. J. Bandelin, E. D. Slifer, and R. E. Pankratz, *J. Am. Pharm. Assoc. Sci. Ed.*, **39**, 277 (1950).
38. *The National Formulary, Twelfth Edition*, Mack Printing, Easton, Pa., 1965. p. 42.
39. *The National Formulary, Twelfth Edition*, Mack Printing, Easton, Pa., 1965, p. 462.
40. *The National Formulary, Twelfth Edition*, Mack Printing, Easton, Pa., 1965, p. 500.
41. K. A. Connors, *Am. J. Pharm. Ed.*, **29**, 29 (1965).
42. E. W. Martin (ed.), *Remington's Pharmaceutical Sciences*, Mack Printing, Easton, Pa., 1965, p. 1578.
43. C. N. Reilley and C. M. Crawford, *Anal. Chem.*, **27**, 716 (1955).
44. R. Bastian, R. Weberling, and F. Palilla, *Anal. Chem.*, **22**, 160 (1950).
45. C. F. Hiskey, *Anal. Chem.*, **21**, 1440 (1949).
46. C. F. Hiskey, J. Rabinowitz, and I. G. Young, *Anal. Chem.*, **22**, 1464 (1950).
47. C. F. Hiskey and I. G. Young, *Anal. Chem.*, **23**, 1196 (1951).
48. J. J. Lothe, *Anal. Chem.*, **28**, 949 (1956).
49. I. A. Williams and B. Zak, *Clin. Chim. Acta*, **4**, 170 (1959).
50. J. C. Demetrius and J. E. Sinsheimer, *J. Am. Pharm. Assoc. Sci. Ed.*, **49**, 523 (1960).
51. J. L. Casinelli and J. E. Sinsheimer, *J. Pharm. Sci.*, **51**, 336 (1962).
52. R. F. Goddu and D. N. Hume, *Anal. Chem.*, **26**, 1679 (1954).
53. T. Higuchi, C. Rehm, and C. Barnstein, *Anal. Chem.*, **28**, 1506 (1956).
54. United States Department of Commerce, *Natl. Bur. Std.*, *LC1017* (1955).

CHAPTER 2

Infrared Spectroscopy

Ronald T. Coutts

UNIVERSITY OF ALBERTA
EDMONTON, ALBERTA

2.1 INTRODUCTION

The literature dealing with the infrared spectroscopy of organic compounds continues to expand at a very rapid rate. Numerous excellent references (e.g., Refs. 1–8) have appeared in recent years, and if an expansive treatment of the subject is desired, the reader is directed to these references. The purpose of this chapter is to give a basic knowledge of the subject and to offer illustrations of a pharmaceutical nature. Although an emphasis has been placed on the determination of the frequencies at which functional groups characteristically absorb, numerous studies have concerned themselves with

TABLE 2.1: Infrared Spectrophotometric Identity Tests of the USP XVII*

Acetazolamide *and* Tablets (KBr)
Bemegride *and* Injection (mineral oil)
γ-Benzene Hexachloride (KBr)
Betazole Hydrochloride *and* Injection (mineral oil)
Bishydroxycoumarin (KBr)
Calcium Disodium Edetate (mineral oil)
Calcium Pantothenate and Racemic Calcium Pantothenate (KBr)
Chlorambucil *and* Tablets (CS₂)
Chlorcyclizine Hydrochloride *and* Tablets (CS₂)ᵇ
Chlorpheniramine Maleate *and* Tablets (CS₂)ᵇ
Chlorpromazine Hydrochloride (KBr)
Chlorpropamide (KBr) *and* Tablets (CHCl₃)
Cholecalciferol (KBr)
Cyclizine Hydrochloride (KBr) *and* Tablets (CS₂)ᵇ
Desoxycorticosterone Acetate (KBr)
Diethyltoluamide *and* Solution (CS₂)
Dihydrotachysterol (KBr)
Dimenhydrinate *and* Syrup *and* Tablets (CS₂)ᵇ
Diphenhydramine Hydrochloride *and* Capsules *and* Elixir and Injection (CS₂)ᵇ
Disodium Edetate *and* Injection (KBr)
Ergocalciferol (KBr)
Estradiol Valerate (KBr)
Ethinyl Estradiol (KBr)
Guanethidine Sulfate (mineral oil)
Hydrochlorothiazide (KBr)
Hydrocortisone *and* Tablets *and* Acetate (KBr)
Sterile Hydrocortisone Acetate Suspension (KBr)
Hydrocortisone Sodium Succinate (KBr)
Hydrocortisone Sodium Succinate for Injection (KBr)
Hydroxystilbamidine Isethionate (KBr)
Sterile Hydroxystilbamidine Isethionate (KBr)
Mecamylamine Hydrochloride *and* Tablets (KBr)
Meclizine Hydrochloride (KBr)
Medroxyprogesterone Acetate *and* Tablets *and* Sterile Suspension (KBr)
Meglumine Iothalamate Injection (KBr)
Methazolamide *and* Tablets (KBr)
Methotrexate (KBr)
Metyrapone (mineral oil)
Phenobarbital *and* Elixir *and* Tablets (KBr)
Phentolamine Mesylate (KBr)
Phentolamine Mesylate for Injection (KBr)
Prednisolone *and* Tablets *and* Acetate *and* Sterile Suspension (KBr)
Prednisone *and* Tablets (KBr)
Primidone (KBr)
Procaine Hydrochloride *and* Injection (KBr)
Sterile Procaine Hydrochloride (KBr)
Prochlorperazine Edisylate *and* Injection (CS₂)ᵇ
Prochlorperazine Maleate *and* Tablets (CS₂)ᵇ
Promethazine Hydrochloride *and* Injection *and* Syrup (CS₂)ᵇ
Propoxyphene Hydrochloride *and* Capsules (CHCl₃)

TABLE 2.1 (continued)

Pyrimethamine (KBr)
Pyrvinium Pamoate (KBr)
Sodium Acetazolamide (KBr)
Sodium Iothalamate Injection (KBr)
Sodium Phenobarbital *and* Injection (KBr)
Sterile Sodium phenobarbital (KBr)
Sodium Warfarin (KBr)
Sodium Warfarin for Injection (KBr)
Spironolactone *and* Tablets (KBr)
Sulfamerazine (KBr)
Sulfamethazine (KBr)
Sulfamethoxypyridazine *and* Tablets (KBr)
Testosterone Cypionate (KBr)
Testosterone Propionate (KBr)
Thiotepa ($CHCl_3$)
Thiotepa for Injection ($CHCl_3$)
Tolbutamide *and* Tablets (mineral oil)
Tripelennamine Citrate (CS_2)[b]
Tripelennamine Hydrochloride *and* Tablets (CS_2)[b]
Tubocurarine Chloride (KBr)

[a] Solvent or dispersing agent is given in parentheses.
[b] The USP "Identification-Organic Nitrogenous Bases" test is employed in these examples. See also practical exercise 2.2.

the application of infrared spectroscopy to pharmaceutical analysis. The review by Price[9] on this subject covers the period up to 1954; Carol's review[10] was published in 1961. Other reviews are also worthy of mention, especially the series on analytical methods used for pharmaceuticals and related drugs which have appeared regularly for many years in *Analytical Chemistry*. Many of the references contained therein are devoted to spectrophotometric methods. The most recent review[11] in this series covers the literature through June 1966. The series of papers (e.g., Refs. 12–14) by Warren, Eisdorfer, Thompson, and Zarembo illustrate the value of infrared spectrophotometry in the identification of organic medicinal compounds. The current editions of the *United States Pharmacopeia* (USP XVII) and the *British Pharmacopoeia* (BP 1963) suggest the extensive use of infrared spectrophotometry as a means of identifying a large number of substances and preparations (Tables 2.1 and 2.2). The USP XVII describes, for the first time, a few assay processes based on infrared absorption spectra (Table 2.3). This number will certainly increase. Reference will be made to typical examples later in this chapter. It is unfortunate that neither the USP nor the BP are consistent in their use of infrared spectroscopy in the identification of medicinal compounds. Some, but not all of the USP official sulfa drugs, for example, are identified by means of their infrared spectra. The BP confines this application of infrared spectroscopy to the identification of steroids, but not all the official BP steroids are identified in this way.

TABLE 2.2: Infrared Spectrophotometric Identity Tests of the BP 1963[a]

Cortisone Acetate *and* Injection *and* Tablets
Deoxycortone Acetate *and* Implants
Deoxycortone Trimethylacetate *and* Injection
Dexamethasone *and* Acetate *and* Tablets
Digoxin
Dimethisterone *and* Tablets
Ethinyloestradiol
Ethisterone *and* Tablets
Fludrocortisone Acetate *and* Tablets
Fluoxymesterone *and* Tablets
Hydrocortisone
Hydrocortisone Acetate *and* Injection
Hydrocortisone Hydrogen Succinate
Hydrocortisone Sodium Succinate *and* Injection
Methylprednisolone *and* Tablets
Methyltestosterone *and* Tablets
Norethandrolone *and* Tablets
Norethisterone
Prednisolone *and* Acetate *and* Tablets
Prednisolone Trimethylacetate *and* Injection
Prednisone *and* Acetate *and* Tablets
Progesterone
Testosterone *and* Implants *and* Propionate

---

[a] Methods of preparation of substance are listed in BP, Appendix IV. Any of these methods can be used.

TABLE 2.3: Infrared Spectrophotometric Assays of the USP XVII[a]

Acetazolamine (pyridine)
Diethyltoluamide *and* Solution (CS_2)
Nitroglycerin Tablets (CS_2)
Propoxyphene Hydrochloride Capsules ($CHCl_3$)
Thiotepa for Injection ($CHCl_3$)

---

[a] Solvent is given in parentheses.

2.2 ABSORPTION SPECTROSCOPY

Absorption spectroscopy, or spectrophotometry, may be defined as the analysis of chemical substances by the measurement of the amount of radiation absorbed by these substances. An infrared absorption spectrum is obtained by passing electromagnetic radiation of the appropriate frequency range through a transparent layer of the substance being examined. The substance may be a solid, a liquid, a gas, or a solution of these. Some of the radiation is absorbed selectively by the substance at certain frequencies. The energy not absorbed, i.e., the transmitted energy, can be analyzed by means of a suitable instrument (the spectrometer, or, if it incorporates a photocell, the spectrophotometer) and it will vary according to the frequency of the incident radiation. In this way, an absorption spectrum over a certain frequency range is obtained. When the molecule under investigation absorbs radiation, some electrons are raised to higher energy levels which are quantized. The high energy or excited states of the molecule are usually short-lived. The absorbed energy is rapidly released in the form of heat. Thus the temperature of the substance or its solution increases while the spectrum is being recorded. If all molecules of a substance were identical, the infrared spectrum would take the form of a series of *lines*. However, a group of molecules exists in a number of different vibrational and rotational states (see later), each state differing from another by a relatively small amount of energy. A group of molecules, therefore, absorbs energy over small wavelength ranges and gives rise to absorption *bands* or *peaks*.

A. UNITS OF MEASUREMENT

Electromagnetic radiation is characterized by either its wavelength (λ) or by its frequency (ν). These quantities are related, $\lambda\nu = C$, where C is the velocity of light, and the energy of radiation is given by the Plank equation, $E = h\nu$. In infrared spectroscopy, the common wavelength unit is the micron (μ; $1 \mu = 10^{-4}$ cm). Wave numbers (σ or $\bar{\nu}$) are also used, where $\bar{\nu} = \lambda^{-1}$, (λ measured in centimeters). Wave numbers are therefore expressed in cm^{-1} units (reciprocal centimeters or Kaysers) and are a measurement of frequency. The position of an absorption peak in an infrared spectrum can be expressed, therefore, in wavelength (measured in microns) or in frequency units (measured in wave numbers). Interconversion is facile:

$$\bar{\nu}(\text{cm}^{-1}) = \frac{10^4}{\lambda(\mu)}$$

B. INFRARED REGION

The range of the electromagnetic spectrum extending from 0.8 to 200 μ is called the "infrared." The infrared region thus extends from just outside the visible region to the microwave region of the electromagnetic spectrum. The infrared region is subdivided. The range 0.8–2.5 μ is the *near infrared* region; the 2.5 to 16 μ is the *infrared* region; the range 16–200 μ is the *far infrared* region. The range 2.5 to 50 μ is sometimes called the "*midinfrared*" region. The range most useful to the organic medicinal chemist stretches from 2.5 to 16 μ, corresponding to 4000 to 625 cm^{-1}. Some infrared spectrophotometers produce tracings which are linear in wavelength units and therefore nonlinear in terms of wave numbers. Most modern instruments, however,

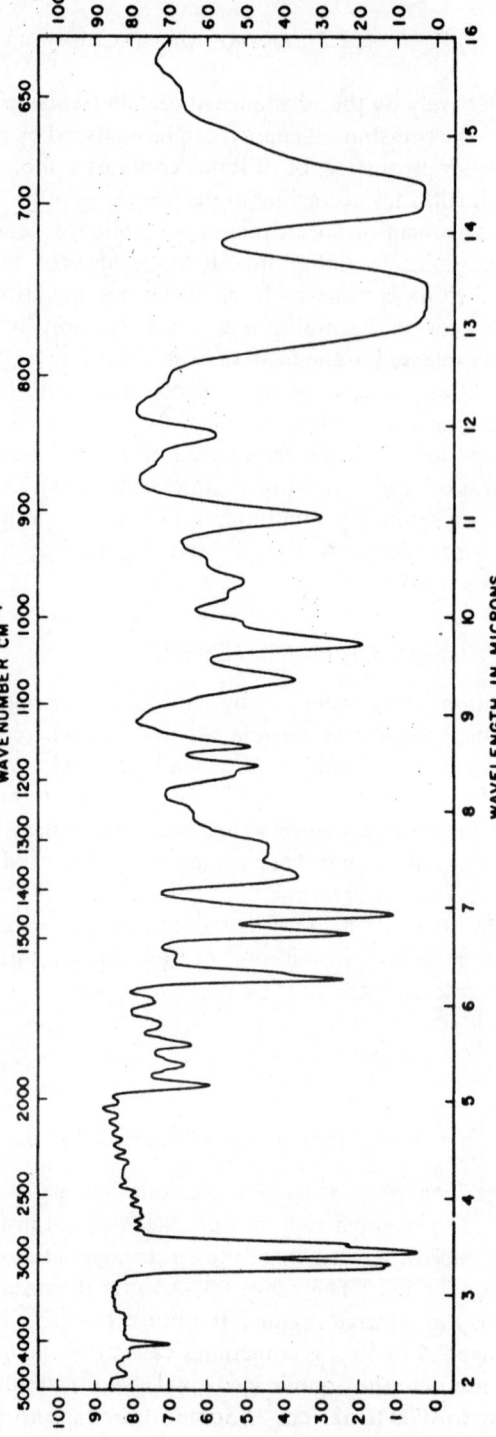

FIGURE 2.1: Infrared spectrum of a polystyrene film recorded on a Beckman IR5A spectrophotometer (linear in wavelength units).

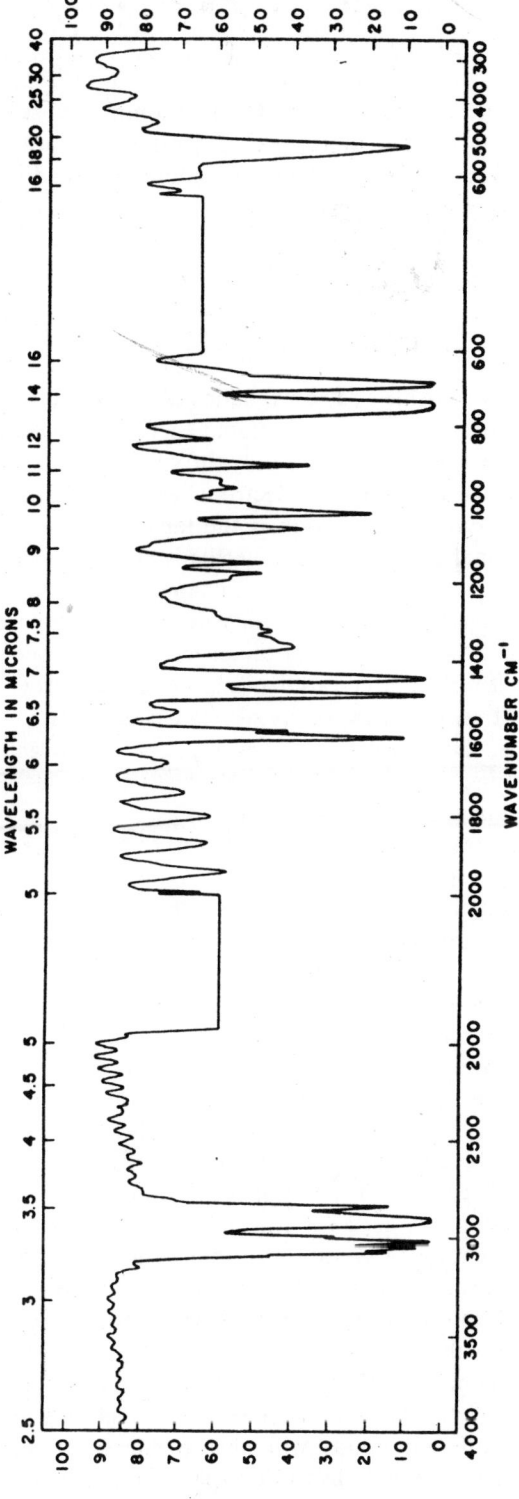

FIGURE 2.2: Infrared spectrum of a polystyrene film recorded on a Beckman IR10 spectrophotometer (linear in frequency units).

65

produce spectra linear in wave numbers, and so it is common to see positions of maximum absorption quoted in reciprocal centimeters rather than in wavelength. The spectrum of a polystyrene film (commonly used to calibrate infrared spectra) has been recorded on the Beckman IR5A spectrophotometer (linear in wavelength units) and on the Beckman IR10 spectrophotometer (linear in frequency units) for comparison purposes (Figs. 2.1 and 2.2, respectively).

C. ENERGY CHANGES

Whereas ultraviolet and visible electromagnetic radiation results in changes in electronic as well as vibrational and rotational energies of a molecule, the weaker energy of infrared radiation affects only the vibrational and rotational energy levels associated with the ground electronic energy state of the irradiated molecule. If the source of radiation is the far infrared or microwave region, only the rotational energy of the molecule is affected, whereas when the energy supplied is on the 2.5–16 μ range, changes in the vibrational as well as the rotational energies of the molecule result. For all intents and purposes however, the 2.5–16 μ region of the electromagnetic spectrum can be considered to affect only the vibrational energy levels of the irradiated molecule. Rotational transitions are rarely seen when solids or solutions are being examined. When they do occur, for example, when molecules in the vapor phase are examined, they appear as fine structure associated with the vibrational absorption bands. For this reason, it is assumed in the discussion which follows that the irradiation of a molecule by means of infrared radiation causes vibrational transitions only

A popular illustration of *stretching and bending vibrations* is to think of the molecule as a system suspended in space in which the component atoms are visualized as balls and the bonds connecting them as springs. The system is imagined to be in constant motion—the springs are stretching and contracting, or bending. These motions are referred to, respectively, as *stretching* and *bending vibrations*, and the frequency with which these vibrations occur depends on the nature of the atoms joined by the bond and on the bond's location in the molecule. In a spring and ball system, if the spring connecting two balls is struck, it vibrates, and this vibration, in turn, influences the rest of the system and is influenced by the rest of the system. In a similar way, if a chemical bond vibrating at a certain frequency is struck by infrared radiation of that same frequency, the vibration of that bond will increase by absorption of the infrared radiation, which causes an electron to be raised to a higher (quantized) vibrational energy level. Because the energy levels are quantized, the electron will be promoted only when the frequency of the infrared radiation corresponds exactly to that of the bond's vibration. If, then, a substance is irradiated by infrared radiation the wavelength of which is constantly changing, different bonds will absorb energy at different frequencies. In this way, an infrared absorption spectrum is obtained in which the percentage intensity of the transmitted infrared light is plotted against the wavelength (or wave number) of that light.

D. VIBRATIONAL MODES

A molecule of N atoms has $3N$ degrees of freedom, which means that the total number of coordinates required to specify the positions of all the atomic nuclei is $3N$. Three of these coordinates are required to indicate the position of the molecule's center of mass, and an additional three (two in linear molecules) describe the rotational motion of the molecule about its center of mass. It follows that to describe the vibrational motions of the atoms relative to one another, $3N$-6 coordinates are required ($3N$-5 if the molecule is linear). A nonlinear molecule containing N atoms, therefore, is said to possess $3N$-6 vibrational degrees of freedom. Another way of expressing this concept is to say that the nonlinear molecule has $3N$-6 possible *normal* or *fundamental vibrational modes* which can be responsible for the absorption of infrared radiation. Thus, the ethanol molecule possesses 21 vibrational modes each of which can, theoretically, give rise to an absorption band. Additional factors, however, must be considered. Molecules displaying partial or total symmetry give rise to simpler spectra than might be anticipated. This is due to the fact that, for a particular vibration to absorb infrared energy, a change in the dipole moment of the molecule must result. The double bonds in symmetrical alkenes ($C{=}C$) and in symmetrical azo compounds ($N{=}N$), for example, do not absorb infrared radiation for this reason. Such bonds are said to be *infrared inactive.*

1. Nonfundamental Vibrations

Absorption bands can occur in addition to those predicted by the number of degrees of freedom possessed by the molecule. These are *nonfundamental absorption bands* and occur for at least two reasons. *Overtones* are sometimes observed at twice the frequency of a strong band. The spectrum of a molecule with a carbonyl group which absorbs strongly near 1700 cm^{-1} may also have a weak intensity band at twice this frequency (near 3400 cm^{-1}). *Combination tones* are also weak bands which appear occasionally at frequencies that are the sum or difference of two or more fundamental bands. Transitions in which the vibrational quantum number changes by unity are permitted to be *infrared active.* Overtones and combination tones require a vibrational quantum change of greater than this value and, therefore, should not be expected. They do occur, however, which indicates that the vibrational motion is not truly harmonic. If an overtone or combination tone occurs near a fundamental frequency, the intensity of the overtone or combination tone is anomalously enhanced. This confusing situation is the result of quantum mechanical resonance (*Fermi resonance*) occurring between the two excited vibrational levels.

A vibrational spectrum is therefore highly characteristic of the particular molecule, but although each molecule has its own particular spectrum, certain *groups of atoms* give rise to absorption bands at or near the same frequencies even though the remaining portions of the molecules are quite different. Most literature references are concerned with studies to establish the ranges within which such *characteristic group frequencies* occur. To predict the presence of a particular group in a molecule is perhaps the most important application of infrared spectrophotometry. Studies in this field, however, can often be complicated by the phenomenon known as "coupling." Sometimes when two strongly absorbing bonds of the same symmetry are closely located within a molecule and absorb in the same region, coupling is said to occur and the absorption bands are shifted outside their characteristic frequency range.

Not only do different bonds or groups of atoms absorb at different frequencies, the *intensity* of the absorption also varies. It has been stated that for a vibrating molecule to absorb infrared radiation, the magnitude of the molecule's dipole moment must change. Generally speaking, the greater the magnitude of this dipole moment change, the greater is the intensity of the absorption band. Other factors, such as Fermi resonance, contribute to the intensity of the absorption band.

2. Fundamental Vibrations

If a group of atoms, $-AB_2$, is represented by

then stretching and bending vibrations can be illustrated as shown (Fig. 2.3).

Stretching vibrations:

asymmetric   symmetric

Bending (or deformation) vibrations:

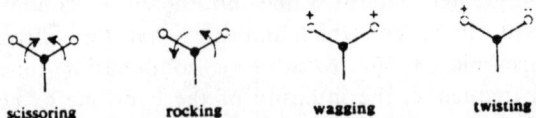

scissoring   rocking   wagging   twisting

FIGURE 2.3: Stretching and bending vibration modes. The arrows indicate the direction of the vibration; the + and − signs represent, respectively, vibrations above and below the plane of the paper.

Typical examples of such a group of atoms include the methylene group

the amino group

the nitro group

and the carboxylate group

Diagrams similar to those shown in Fig. 2.3 can be used to illustrate stretching and bending in other systems, such as the methyl group (Fig. 2.4). The wavelength at which a stretching vibration occurs is dependent on the strength of

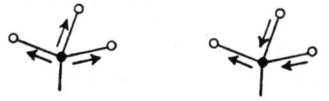

symmetrical stretching   asymmetrical stretching

FIGURE 2.4: Stretching vibrations of the methyl group.

the bond joining the two atoms; the shorter the bond, the shorter is the wavelength (the larger the wave number) of the absorption band (Fig. 2.5). When, however, hydrogen is one of the atoms, stretching vibrations occur at frequencies much higher than those of other single-bond stretching vibrations. Deformation vibrations require less energy and for this reason occur at longer wavelengths (lower wave numbers) than stretching vibrations. Common deformation frequency ranges are shown in Fig. 2.5.

E. EXAMINATION OF AN INFRARED SPECTRUM

A common practice is to examine separately three different areas:

a. The region above 1400 cm⁻¹ (1400–4000 cm⁻¹). The presence or absence of many groups in the molecule, including C=O, NH, OH, C=C,

C=N, etc., is usually readily confirmed. In this region, however, there is a danger of making erroneous assignments. Nonfundamental absorption bands (overtones, combination tones) are often located here and should not be confused with O—H or N—H bands.

b. The region below 900 cm⁻¹. This is generally taken to mean the 900–600 cm⁻¹ region. The infrared spectra illustrated in this chapter cover the range 600–250 cm⁻¹, but it is often not possible to determine the origin of the bands

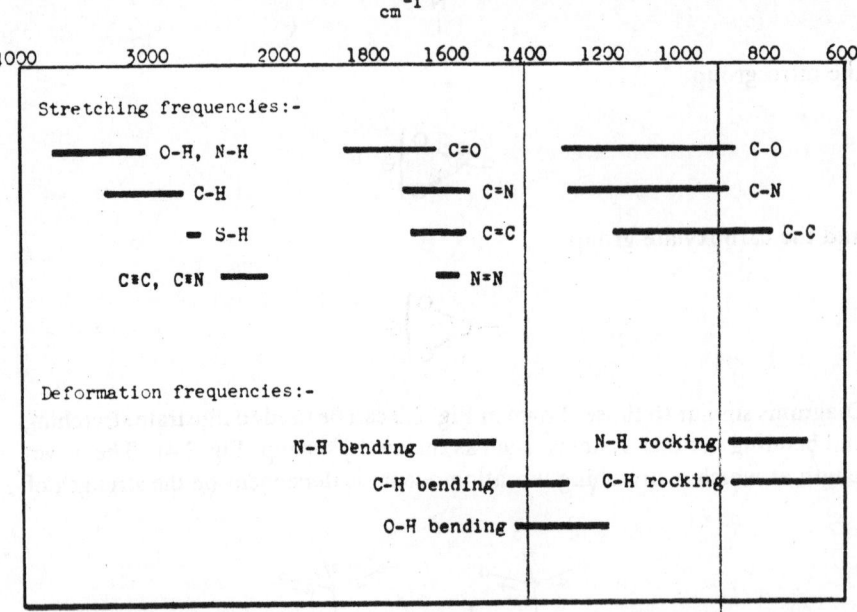

FIGURE 2.5: Common stretching and deformation vibrational frequencies.

in this region because of the lack of information available in the literature. Generally speaking, not many bands are located in the 900–600 cm⁻¹ region and the ones that are can usually be assigned to particular transitions.

c. The region 900–1400 cm⁻¹. This is termed the "finger-print region," where many absorption bands are located, especially those due to bending vibrations as well as C—C, C—O, and C—N stretching vibrations. This area of the spectrum is usually complex, and it is often far from easy to determine the origin of every band. As there are more bending than stretching vibrations in a molecule, the fingerprint region contains many absorption bands of varying intensities which makes this region of particular importance in establishing the identity of a compound by comparison with an authentic sample. Two extremely similar molecules often have virtually identical vibrations in the other two regions (600–900; 1400–4000 cm⁻¹), but almost invariably, there are differences, sometimes slight, in the fingerprint region.

1. Group Frequencies

Numerous authors have constructed detailed tables of characteristic group frequencies. These are readily available, and it would be superfluous to construct yet another series of tables of this type here. Much information can be obtained from correlation charts such as those in the text book by Cross,[3] and by reference to one of the various forms of the Colthup chart, originally published in 1950,[15] modifications of which appear in most textbooks. More comprehensive group frequency correlation charts of this type are published by Beckman. One example is reproduced here* (Table 2.4). Another excellent reference for routine use is the bulletin by Jones,[16] which does not attempt to cover the subject completely, but which includes correlations most commonly used in the identification of naturally occurring organic compounds and their degradation products. With such a wealth of information available, a summary only of the principal absorption bands associated with the more frequently encountered functional groups is presented here (Table 2.5). The frequency of a band and its intensity vary with the nature of the molecule and according to whether a solid or a solution spectrum is recorded.

2. Aromatic Compounds

The presence of an aromatic nucleus in a molecule can usually be readily confirmed by means of its infrared spectrum. Three regions are of particular importance: (a) 1000–625 cm^{-1} (10–16 μ), (b) 1670–1430 cm^{-1} ($\sim$6–7 μ), and (c) 2000–1670 cm^{-1} ($\sim$5–6 μ).

A number of C—H bending vibrations appear in the 10–16 μ region; they vary in intensity. The pattern of absorption depends on the number of adjacent hydrogen atoms possessed by the aromatic nucleus. Of particular interest to the pharmaceutical chemist are the spectra of monosubstituted (**I**), o-disubstituted (**II**), and p-disubstituted (**III**) benzene derivatives. All monosubstituted benzenes exhibit strong absorption near 14.3 μ (700 cm^{-1}). Such is the importance of this strong band that it is true to say that if a molecule does not absorb strongly in this region, it is *not* a monosubstituted benzene. Many, but not all, monosubstituted benzenes also absorb strongly near

(I)   (II)   (III)

13.3 μ (750 cm^{-1}). The spectra of benzoic acid and many benzoates, for example, do not show a strong band near 750 cm^{-1}.

Ortho-disubstituted benzenes absorb strongly near 13.3 μ (750 cm^{-1}) only. A band near 700 cm^{-1} is not observed. Para-disubstituted benzenes, and other phenyl compounds with two adjacent hydrogen atoms on the phenyl ring (i.e., 1,2,4-trisubstituted and 1,2,3,4-tetrasubstituted benzenes), have a strong band in the 860–800 cm^{-1} region, usually located close to

* Reproduced with the kind permission of Beckman Instruments, Inc.

TABLE 2.4^a: Carbonyl Vibrations (all strong bands)

	Wavelength, μ		
	5	6	7
Saturated ketones and acids		⊢⊣	
$\alpha\beta$-Unsaturated ketones		⊢⊣	
Aryl ketones		⊢⊣	
$\alpha\beta$-, $\alpha'\beta'$-Unsaturated and diaryl ketones		⊢⊣	
α-Halogen ketones		⊢⊣	
$\alpha\alpha'$-Halogen ketones		⊢⊣	
Chelated ketones		⊢—⊣	
6-Membered ring ketones		⊢⊣	
5-Membered ring ketones		⊢⊣	
4-Membered ring ketones		⊢⊣	
Saturated aldehydes		⊢⊣	
$\alpha\beta$-Unsaturated aldehydes		⊢⊣	
$\alpha\beta$-, $\alpha'\beta'$-Unsaturated aldehydes		⊢⊣	
Chelated aldehydes		⊢⊣	
$\alpha\beta$-Unsaturated acids		⊢⊣	
α-Halogen acids		⊢⊣	
Aryl acids		⊢⊣	
Intramolecularly bonded acids		⊢⊣	
Ionized acids		⊢—⊣	
Saturated esters 6- and 7-ring lactones		⊢⊣	
$\alpha\beta$-Unsaturated and aryl esters		⊢⊣	
Vinyl esters, α-halogen esters		⊢⊣	
Salicylates and anthranilates		⊢⊣	
Chelated esters		⊢⊣	
5-Ring lactones		⊢⊣	
$\alpha\beta$-Unsaturated 5-ring lactones		⊢⊣	
Thiol esters		⊢⊣	
Acid halides		⊢⊣	
Chlorocarbonates		⊢⊣	
Anhydrides (open-chain)	Separation 60 cm^{-1} ⊢⊣ ⊢⊣		
Anhydrides (cyclic)	Separation 60 cm^{-1} ⊢⊣ ⊢⊣		
Alkyl peroxides	Separation 25 cm^{-1} ⊢⊣⊢⊣		
Aryl peroxides	Separation 25 cm^{-1} ⊢⊣ ⊢⊣		
Primary amides (CO)		Free ⊢⊣ Bonded ⊢⊣	
Secondary amides and δ-lactams (CO)		Free ⊢⊣	
		Bonded ⊢⊣	
Tertiary amides (CO)		⊢—⊣	
γ-Lactams	Fused rings ⊢⊣		
		Unfused ⊢⊣	
β-Lactams	Fused ⊢⊣ Unfused rings ⊢⊣		

^{a} Based on work done by Bellamy.

TABLE 2.5: Principal Absorption Bands of Selected Functional Groups

	C—H stretching
Alkanes	Located in the 2850–3300 cm^{-1}
Alkanes	(3.51–3.03 μ) region. Intensity[a] m to s.
Alkynes	C≡C—H stretching characteristically
Aromatic hydrocarbons	located near 3300 cm^{-1} (3.03 μ) (s)
Aldehyde C—H	Characteristic. Two bands 2800–2900 cm^{-1}
	(3.57–3.45 μ) (w) and 2700–2780 cm^{-1}
	(3.70–3.60 μ) (w)

	Frequency range cm^{-1}	Wavelength range, μ	Intensity
C—H bending			
Alkanes	1340–1480	7.46–6.75	m–s
Alkenes	1300–1420	7.70–7.04	s
	800–1000	12.5–10.0	s
Alkynes	630	15.9	s
Aromatic[b]	700–900	14.3–11.1	v
C—C double- and triple-bond stretching			
Alkenes	1620–1680	6.17–5.95	w–m
Alkynes[c]	2100–2300	4.76–4.35	w–m
Aromatic[d]	1450–1600	6.90–6.25	w–m
Carbonyl C=O stretching			
Aldehydes[e]			
Saturated aliphatic	1720–1740	5.81–5.75	s
α,β-Unsaturated aliphatic	1680–1705	5.95–5.87	s
Aryl	1680–1715	5.95–5.83	s
Ketones			
Saturated aliphatic	1705–1750	5.87–5.71	s
α,β-Unsaturated aliphatic	1660–1685	6.02–5.94	s
Aryl	1680–1700	5.95–5.88	s
Esters			
Saturated	1735–1750	5.76–5.71	s
α,β-Unsaturated	1715–1730	5.83–5.78	s
Aryl	1715–1730	5.83–5.78	s
Saturated γ-lactones	1760–1780	5.68–5.62	s
Esters			
Saturated β-lactones	~ 1820	~ 5.50	s
α,β-Unsaturated γ-lactones	1740–1760	5.75–5.70	s
β,γ-Unsaturated γ-lactones	~ 1800	~ 5.56	s
β-Ketoesters	~ 1650	~ 6.06	s
Carboxylic acids[f]			
Saturated	1700–1725	5.88–5.80	s
α,β-Unsaturated	1690–1715	5.92–5.83	s
Aryl	1680–1700	5.95–5.88	s

TABLE 2.5 (continued)

	Frequency range cm^{-1}	Wavelength range, μ	Intensity

Carboxylate ion

	1550–1610	6.45–6.21	s
	1300–1400	7.69–7.15	s

Amides[g]

Solid or concentrated solution	1630–1680	6.14–5.95	s
Dilute solution	1670–1700	5.99–5.88	s
Lactams	1680–1780	5.95–5.62	s
Ureas	1660–1720	6.02–5.81	s
Imides (two bands)	1700–1730	5.88–5.78	s
	1670–1700	5.99–5.88	s

Alcohols and phenols

O—H stretching

Free OH	3590–3650	2.79–2.74	Sharp bands (v)
Hydrogen-bonded OH	3450–3570	2.90–2.80	Sharp bands (v)

Intramolecular—no change on diluting solution;
Intermolecular—change on diluting solution.

Chelate compounds	2500–3200	4.00–3.10	Broad peak (v)

O—H bending and C—O stretching

1° and 2° alcohols	1050–1100	9.50–9.10	s
	1310–1410	7.60–7.10	s

Amines

N–H stretching

1° (two bands)	∼ 3400	∼ 2.94	Sharp (m)
	∼ 3500	∼ 2.86	Sharp (m)
2°	3300–3500	3.03–2.86	m
3°	—	—	
Amine salts	3000–3150	3.33–3.18	m

N–H bending

1° and 2°	1550–1650	6.45–6.06	m–s
Salts (two bands)	1570–1600	6.37–6.25	s
	∼ 1500	∼ 6.67	s

C—N stretching

Aliphatic (two bands)	1000–1200	10.0–8.30	w
	∼ 1400	∼ 7.14	w
Aromatic	1250–1350	8.00–7.41	s

Nitriles and Isocyanides

C≡N stretching

Nitriles	2220–2280	4.50–4.38	m
Isocyanides	2050–2220	4.87–4.50	m

TABLE 2.5 (continued)

	Frequency range cm^{-1}	Wavelength range, μ	Intensity
	Nitro compounds		
N—O stretching			
Aromatic, asymmetric	1500–1570	6.67–6.37	s
Aromatic, symmetric	1300–1370	7.70–7.30	s
Aliphatic, asymmetric	1500–1570	6.45–6.37	s
Aliphatic, symmetric	1370–1380	7.30–7.25	s
C—N stretching			
Aromatic	~ 860	~ 11.6	m
Aliphatic	usually ~ 915	~ 10.9	m
	Halogen compounds^		
C—F stretching	1000–1400	10.0–7.10	s
C—Cl stretching	600–800	16.6–12.5	s
C—Br stretching	500–650	20.0–15.4	s
C—I stretching	500–600	20.0–16.6	s
	Sulfur compounds		
Thiols; S—H stretching	2550–2600	3.92–3.85	w
Thiocarbonyl compounds;			
C=S stretching	1050–1200	9.52–8.33	s
Sulfones, including sulfonamides;			
S=O stretching (two bands)	1140–1180	8.77–8.48	s
	1300–1350	7.69–7.41	s

ᵃ Intensity: s, strong; m, medium; w, weak; v, variable.

ᵇ Position varies with number of adjacent hydrogen atoms. This subject is considered in more detail on p. 71.

ᶜ If —C≡C—H, characteristic C—H stretching also.

ᵈ Usually four bands. This subject is considered in more detail below.

ᵉ Also display characteristic C—H stretching.

ᶠ O—H stretching also—several bands 2500–2700 cm^{-1} (4.00–3.70 μ) (w).

ᵍ Position of band varies appreciably depending on whether the spectrum is of the solid, of a concentrated solution, or of a dilute solution. Also seen is NH$_2$ and NH stretching, similar to that of amines.

ʰ These bands are strong, but very variable in position.

810 cm^{-1} (12.35 μ). Once again, the presence of a band in these regions does not confirm that the molecule is an ortho- or a para-disubstituted benzene. The absence of strong absorption in these regions is more informative. Substituted benzenes other than those just mentioned are more difficult to characterize by means of infrared spectrophotometry. In these instances, bands due to C—H bending vibrations are weaker and more variable in position.

Aromatic compounds have four characteristic absorption bands in the 6–7 μ (1667–1429 cm^{-1}) region due to C=C stretching vibrations. These bands are often partially masked by other strong absorbing groups. Absence of strong absorption by a compound in the 6–7 μ region is indicative that the substance is not aromatic.

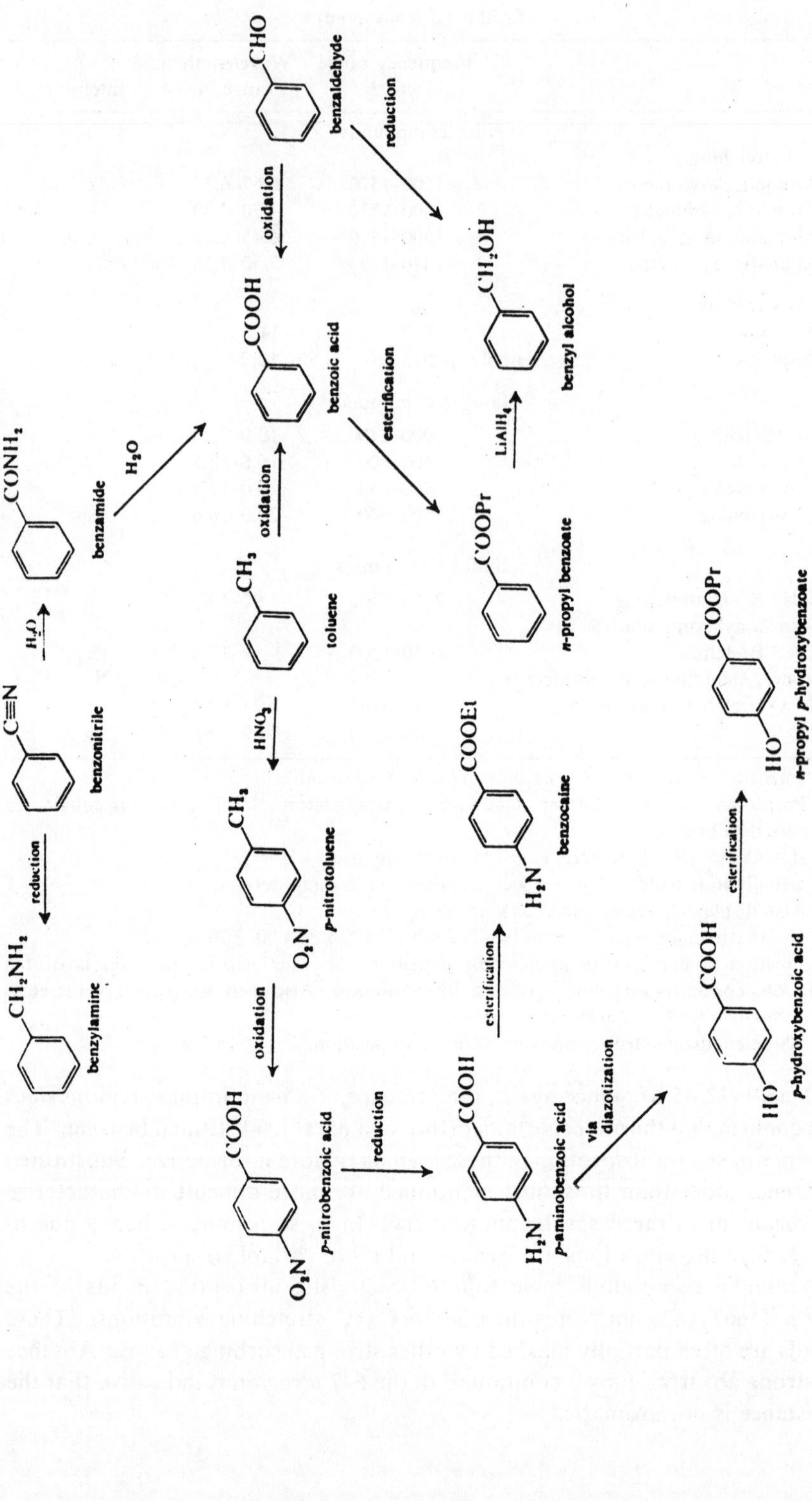

SCHEME 2.1

Low intensity absorption, due to overtone or combination bands, is observed in the 5–6 μ (2000–1667 cm^{-1}) region if the molecule being examined is aromatic. With benzene derivatives, it has been shown that both the number and the positions of these bands depend on the substitution pattern of

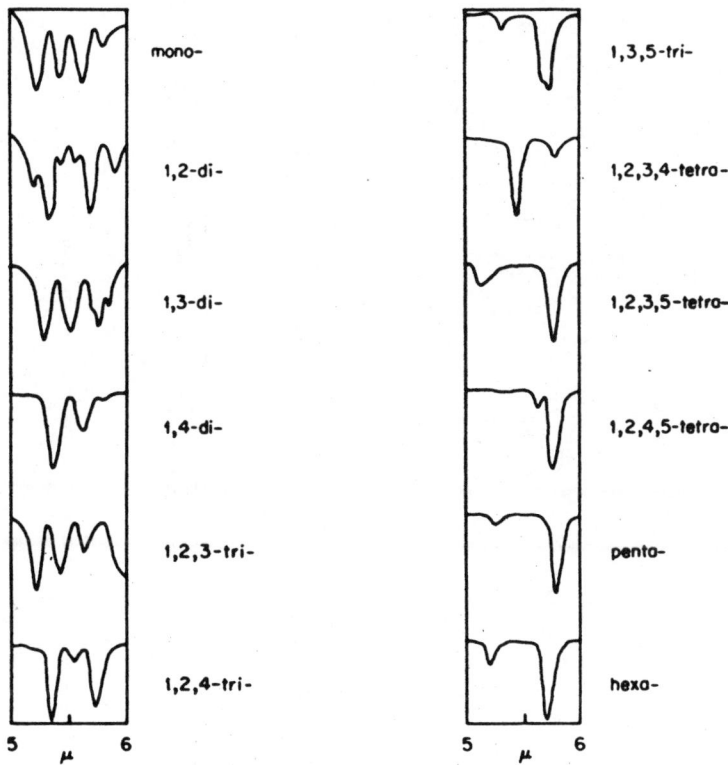

FIGURE 2.6: Diagrammatic representation of absorption bands in the 5–6 μ region for substituted benzenoid compounds.

the benzene ring (Fig. 2.6). A fairly concentrated solution of the substance (e.g., a 5% solution in a suitable solvent; 0.1-mm cells) must be examined if these bands are to be observed.

3. Infrared Spectra of Selected Compounds

An excellent way of gaining some knowledge of the frequencies at which functional groups characteristically absorb is to follow spectrophotographically the course of a reaction sequence in which one functional group is successively replaced by another. The reaction scheme chosen (Scheme 2.1)

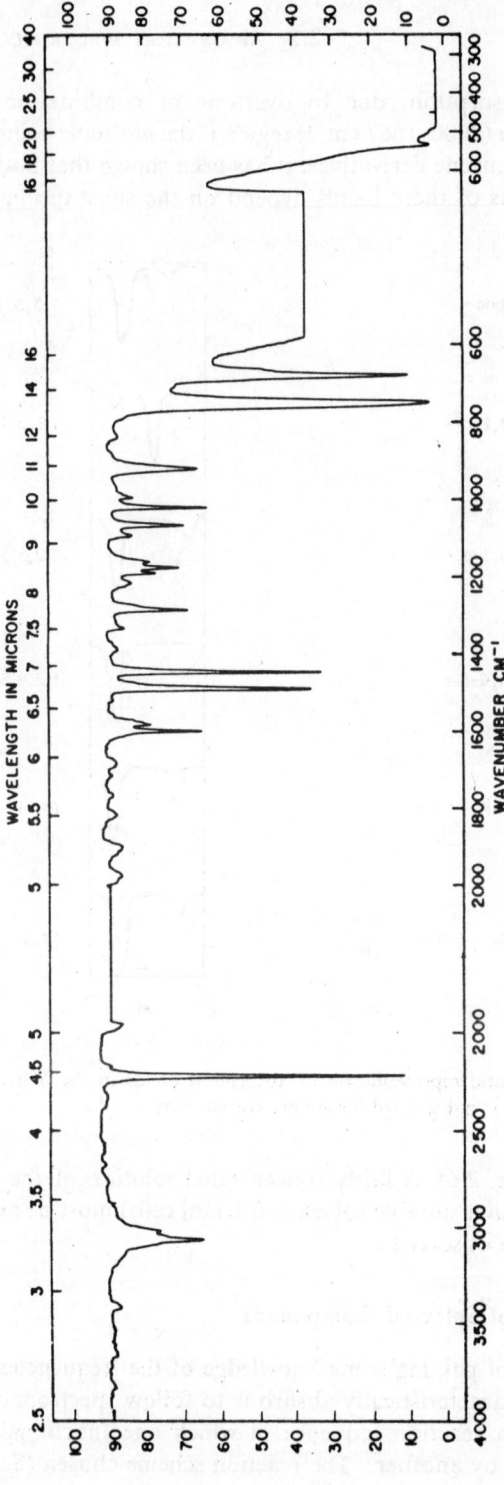

FIGURE 2.7: Infrared spectrum of benzonitrile (thin film).

has been selected for its simplicity and because it includes some examples of the many compounds of medicinal importance which contain substituted benzene rings. In addition, the scheme also illustrates the frequencies at which the absorption bands of carbonyl groups in different chemical environments occur. Brief interpretive comments are made on these spectra, but no attempt is made to identify each peak. The spectra of solids were recorded in the form of KBr disks, and liquids as thin films.

The spectrum of *benzonitrile* (Fig. 2.7) possesses features which suggest readily a monosubstituted benzene nucleus. Characteristic C—H bending occurs at 682 (s) and 752 (s) cm^{-1}, and the absorption pattern in the range 1670–2000 (w) cm^{-1} is expected for such a system. Also characteristic of an aromatic compound are the C—C stretching bands in the 1430–1670 (s or m) cm^{-1} region and the C—H stretching at 3065 (w) cm^{-1}. The strong band at 2220 cm^{-1} is very characteristic of C≡N stretching. The absorption band is noticeably lost on reduction of benzonitrile (to benzylamine) or its hydrolysis (to benzamide or benzoic acid). For the spectrum of *benzylamine* (Fig. 2.8) assignments similar to those described for benzonitrile can be made. Absorption bands due to C—H bending and stretching and aromatic C—C stretching occur at similar frequencies [692 (s), 735 (s); 1670–2000 (w); 1430–1670 (s) cm^{-1}]. The strong bands in the 2900–3100 cm^{-1} region are due to ethylenic and aromatic C—H stretching. A noticeable feature of this spectrum is the occurrence of two sharp bands of medium intensity at 3300 and 3380 cm^{-1}. Such absorption is characteristic of N—H stretching of a primary amino group. The spectrum of *benzamide* (Fig. 2.9) can be interpreted in a similar way. The major feature of this spectrum differing from the previous two is the presence of the strong band at 1655 cm^{-1}. Carbonyl stretching vibrations of primary amides occur in this region in solid-phase spectra. (In a dilute solution spectrum, this band would be located near 1690 cm^{-1}). The band at 1655 cm^{-1} is referred to as the amide I band. The strong band at 1620 cm^{-1} (amide II band) is due to N—H deformation and C—N stretching of a primary amide. The spectrum of *benzaldehyde* (Fig. 2.10) also shows the bands expected of a monosubstituted benzene nucleus though the characteristic substitution pattern in the 1670–2000 cm^{-1} region is partly masked by the strong band at 1700 cm^{-1} (aldehyde C=O stretching). Another noticeable feature of this spectrum is the occurrence of two bands of medium intensity at 2725 and 2810 cm^{-1}. These are very characteristic of aldehydes and arise from symmetric and asymmetric C—H stretching of the aldehyde group.

A single strong peak at 702 cm^{-1} and an ill-defined substitution pattern between 1750 and 2000 cm^{-1} in the spectrum of *benzoic acid* (Fig. 2.11) suggest, once again, a monosubstituted benzene nucleus. Aromatic C—C stretching (1400–1650 cm^{-1}) and C—H stretching ($\sim$ 3000 cm^{-1}) are present. The strong peak at 1698 cm^{-1} in the result of C=O stretching of a conjugated carbonyl group, and the absorption pattern which occurs in the 2500–2900

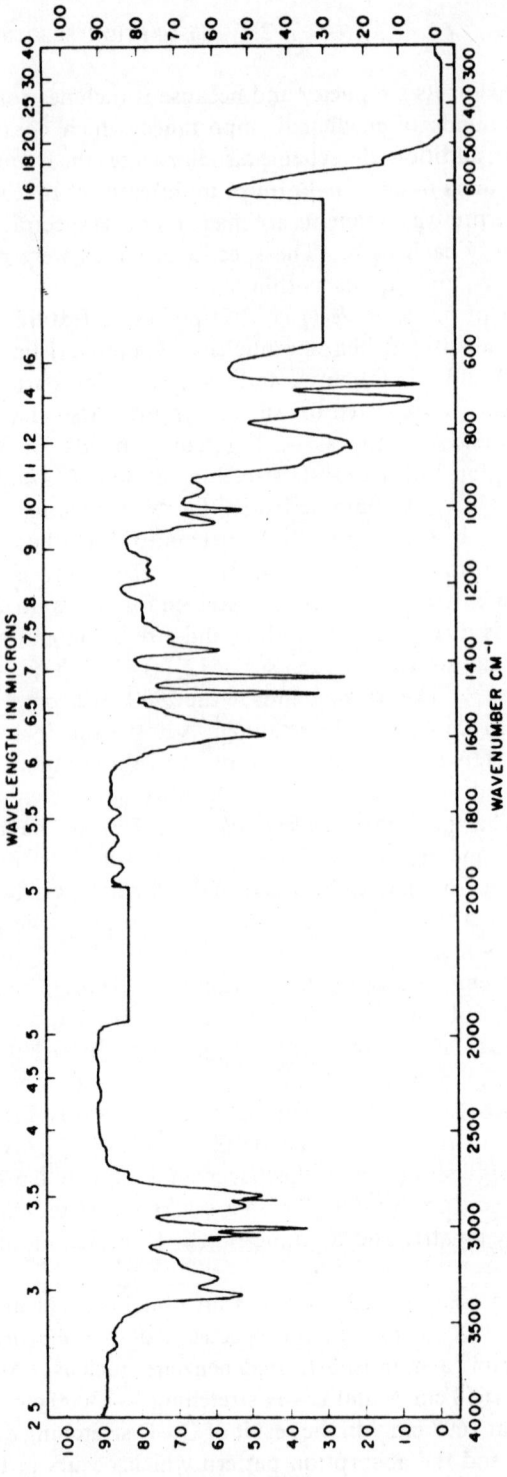

FIGURE 2.8: Infrared spectrum of benzylamine (thin film).

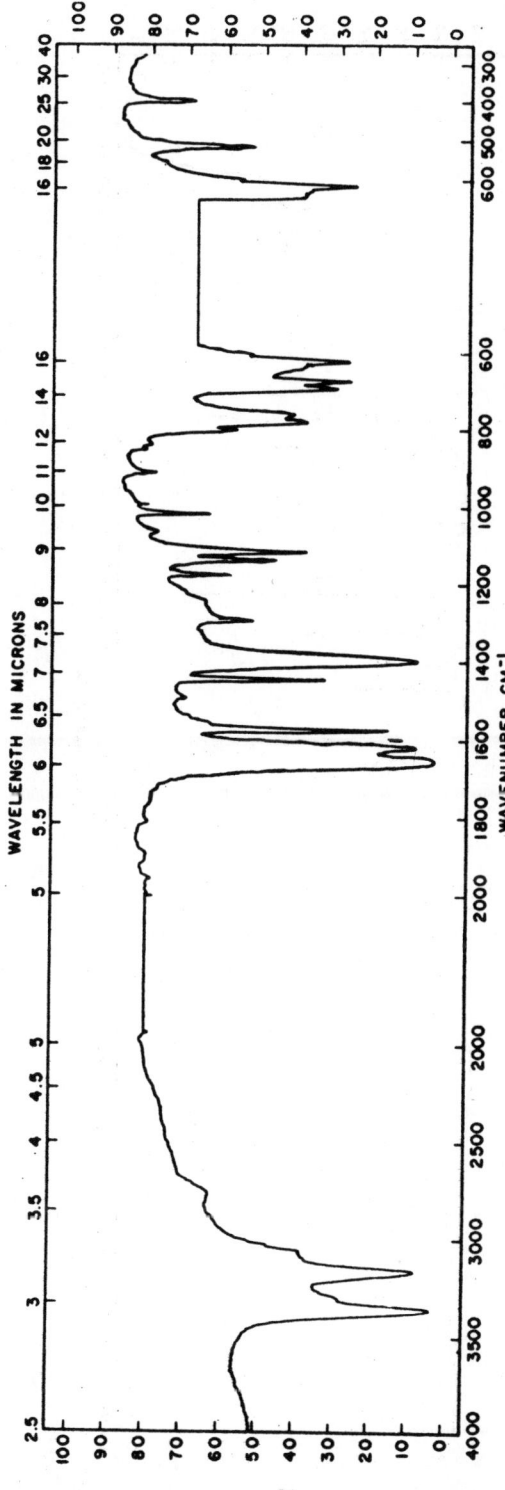

FIGURE 2.9: Infrared spectrum of benzamide (KBr disk).

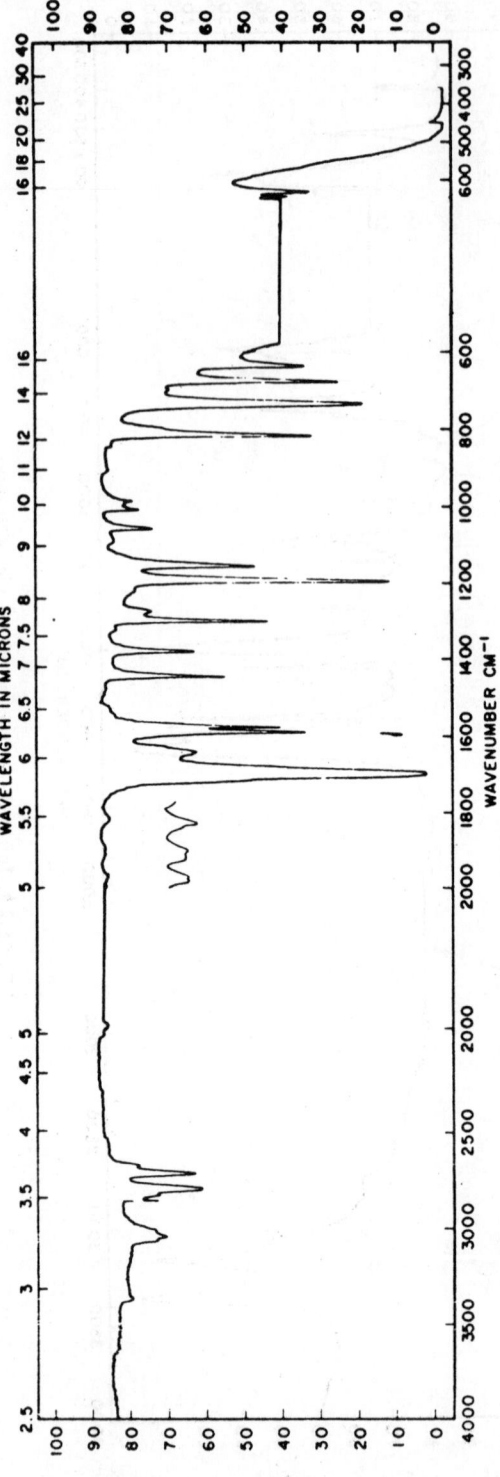

FIGURE 2.10: Infrared spectrum of benzaldehyde (thin film).

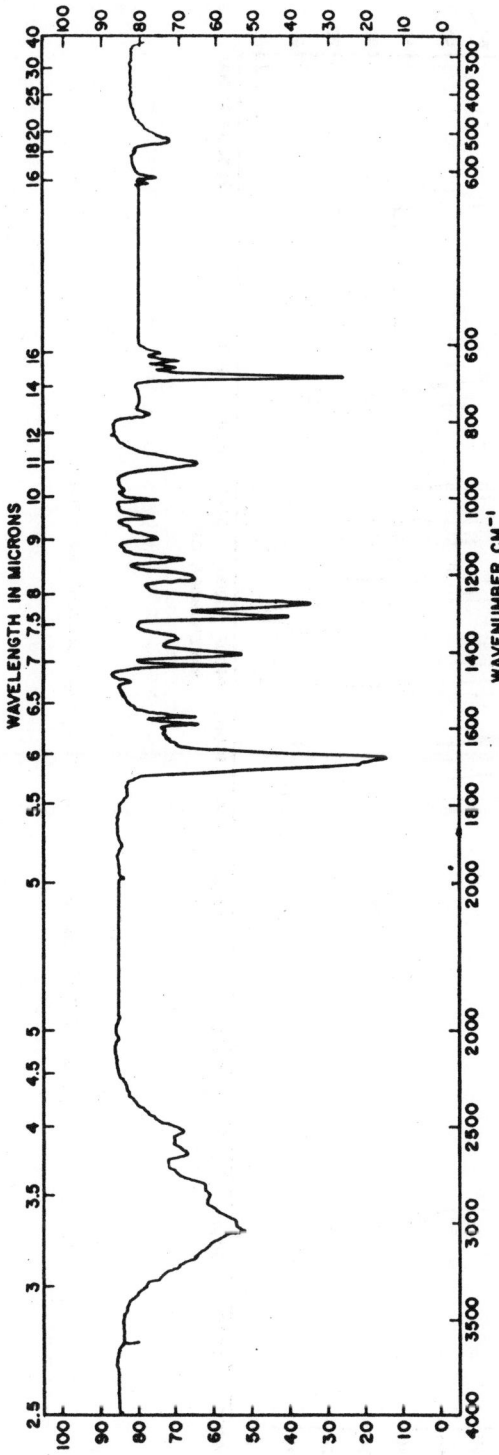

FIGURE 2.11: Infrared spectrum of benzoic acid (KBr disk).

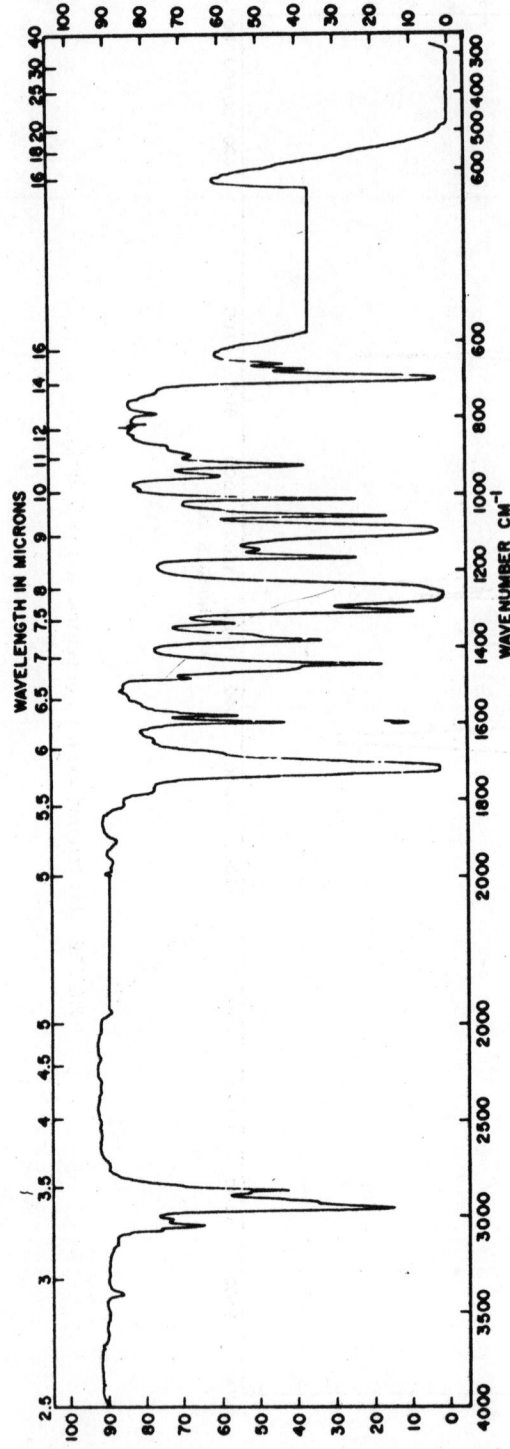

FIGURE 2.12: Infrared spectrum of *n*-propyl benzoate (thin film).

cm^{-1} region is very characteristic of a carboxylic acid dimer. Except in dilute solution, carboxylic acids exist as dimeric species (**IV**). The weak absorption in this region is attributable to bonded O—H stretching, a characteristically broad absorption. (In very dilute solution, O—H stretching is located near 3550 cm^{-1}; the molecule is then in monomeric form.)

(IV)

It is of interest to compare the spectrum of *n-propyl benzoate* (Fig. 2.12) with that of the corresponding acid. Both spectra are very similar in most regions, but are noticably different in the carbonyl and in the 2500–3200 cm^{-1} region. The carbonyl stretching vibration of the ester (1720 cm^{-1}) occurs at a slightly shorter wavelength (higher wave number) than that of the acid. The broad absorption of the acid in the region >2400 cm^{-1} has been replaced by relatively narrow bands attributable to aliphatic and aromatic C—H stretching. The substitution patterns in the 1750–2000 cm^{-1} and ~800 cm^{-1} regions of the spectrum of *p-nitrobenzoic acid* (Fig. 2.13) are consistent with a *p*-disubstituted benzene nucleus. This compound is obviously aromatic (1400–1650 cm^{-1}) and possesses a carboxylic acid group (cf. spectrum of benzoic acid). Two strong absorption peaks are present at 1535 and 1340 cm^{-1} due, respectively, to aromatic asymmetric and symmetric N—O stretching. A weaker peak at 865 cm^{-1} is due to aromatic C—N stretching.

It is informative to compare the spectrum of *benzyl alcohol* (Fig. 2.14) with that of benzaldehyde to see the effect of reducing an aldehyde to the corresponding primary alcohol. The obvious additional feature in the spectrum of the former is the strong, broad absorption band located ~3300 cm^{-1} associated with intermolecular hydrogen bonded (polymeric association) O—H stretching. This band has replaced the C=O and C—H absorption bands of the aldehyde.

The reader is left to interpret along similar lines the spectra of toluene, benzocaine, *p*-aminobenzoic acid, *p*-hydroxybenzoic acid, *p*-nitrotoluene, and *n*-propyl *p*-hydroxybenzoate (Figs. 2.15 to 2.20, respectively). In particular, the effect on the spectrum of replacing one group with another should be rationalized.

All the spectra just mentioned are simple compounds and their spectra are relatively easy to interpret. This is not so with more complex substances with which it is often possible to confirm only the presence of functional groups. With more complex substances, group interaction can occur with resultant displacement of absorption bands. A few examples are now considered. The spectrum of *ethyl (o-nitrophenylthio)acetoacetate* (Fig. 2.21) has a strong band at 730 cm^{-1}, consistent with *o*-disubstitution, but little can be

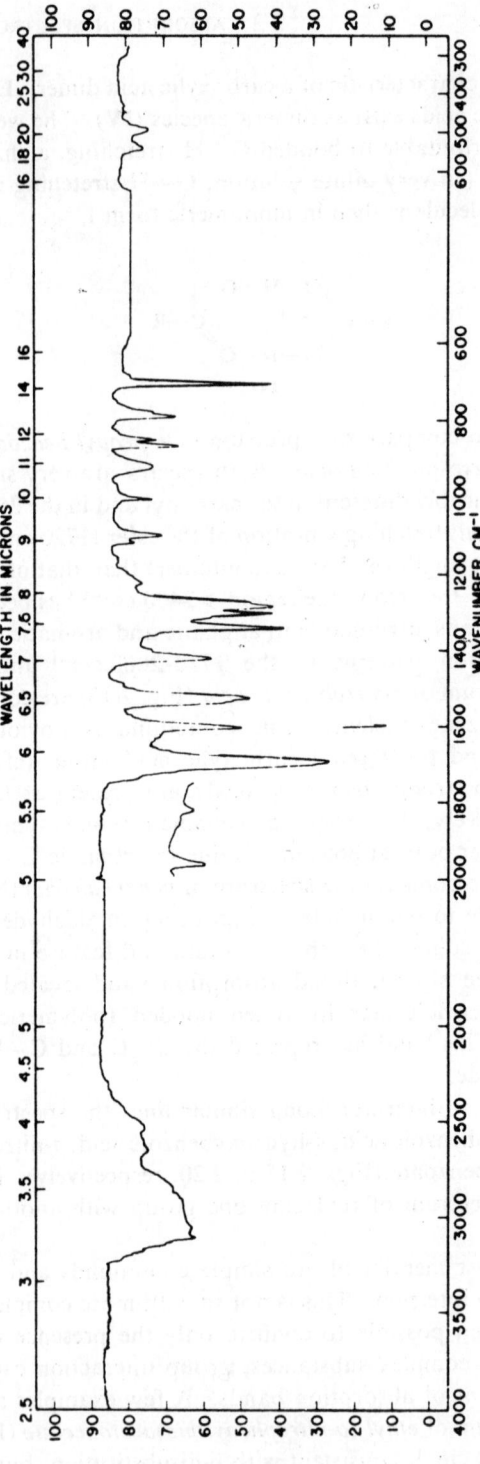

FIGURE 2.13: Infrared spectrum of *p*-nitrobenzoic acid (KBr disk).

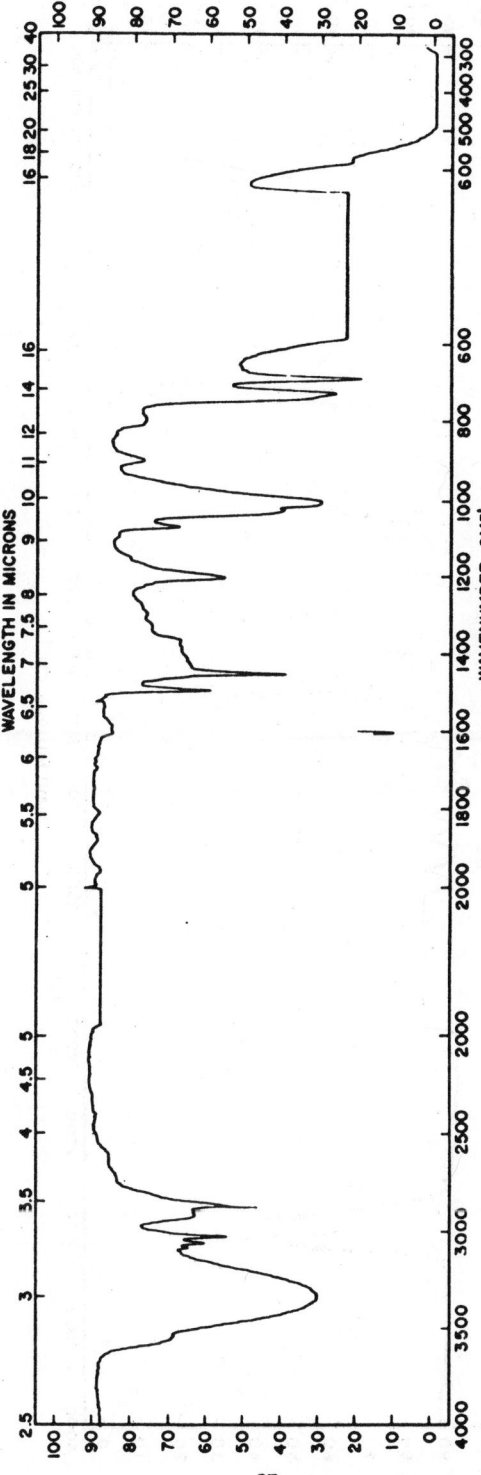

FIGURE 2.14: Infrared spectrum of benzyl alcohol (thin film).

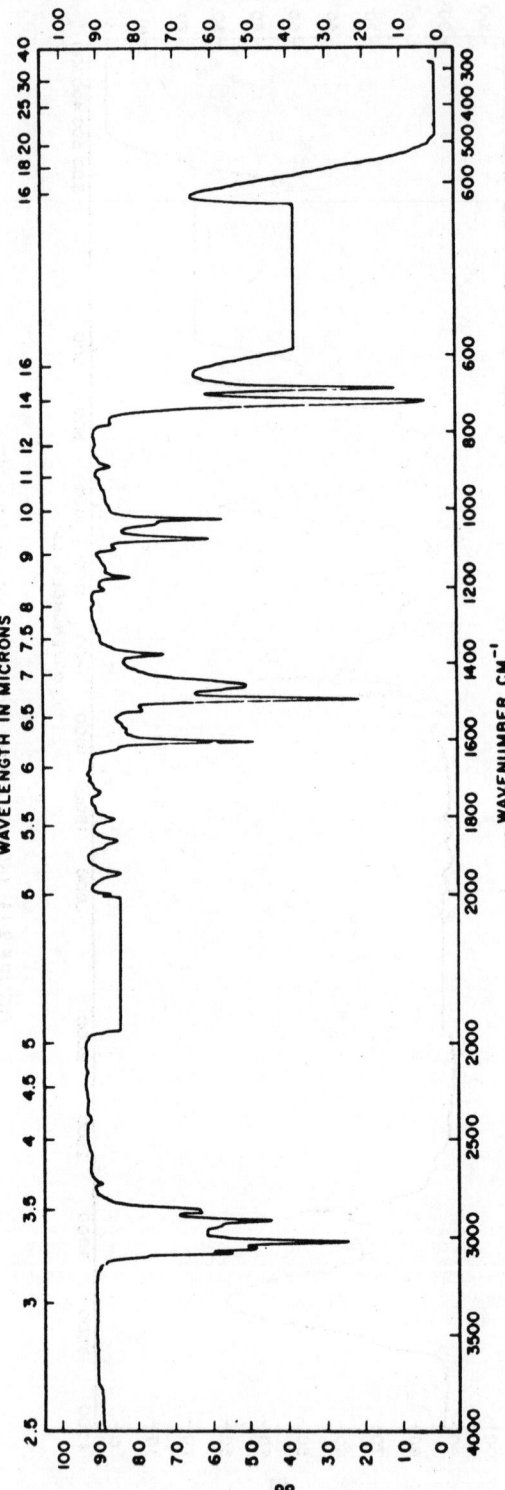

FIGURE 2.15: Infrared spectrum of toluene (0.02-mm thin film).

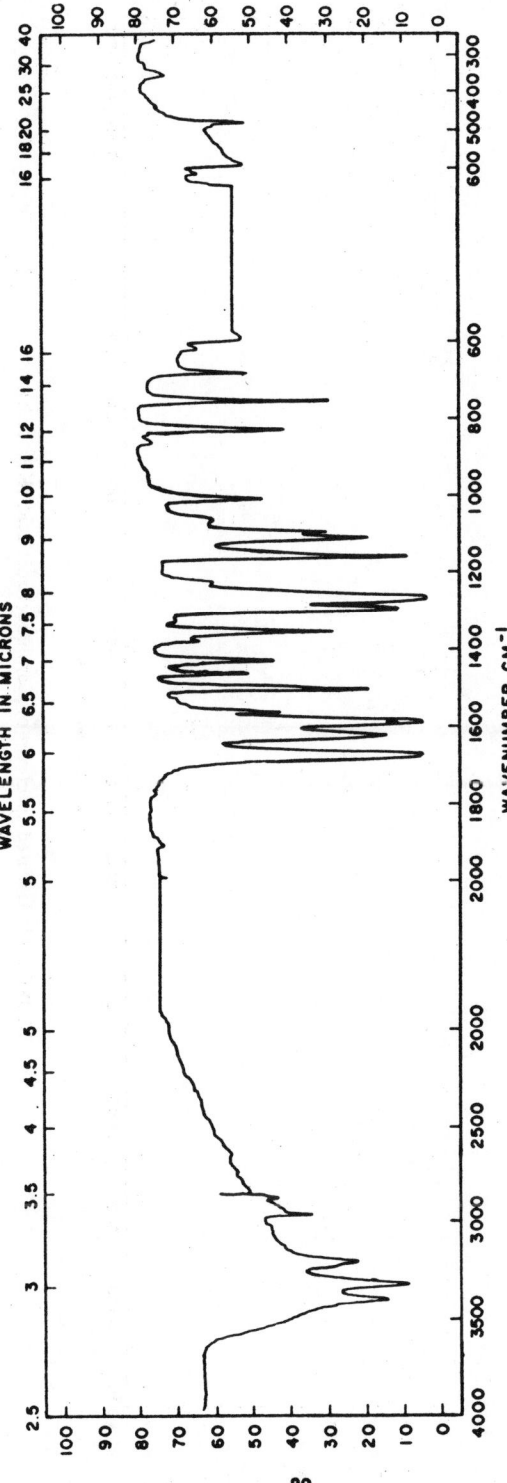

FIGURE 2.16: Infrared spectrum of benzocaine (KBr disk).

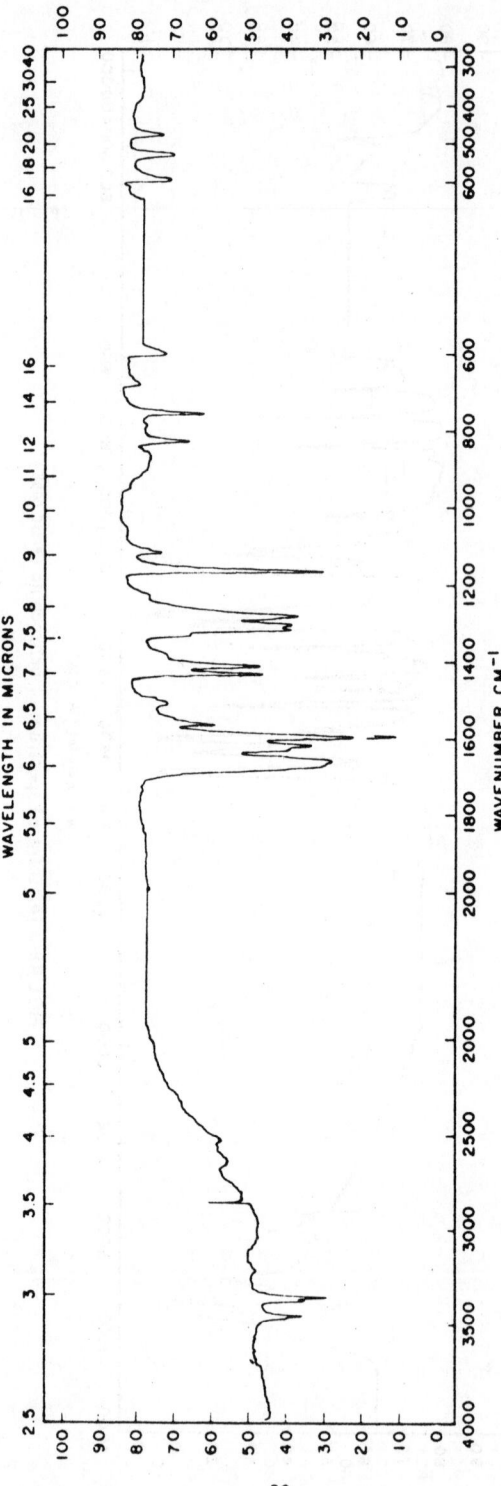

FIGURE 2.17: Infrared spectrum of *p*-aminobenzoic acid (KBr disk).

90

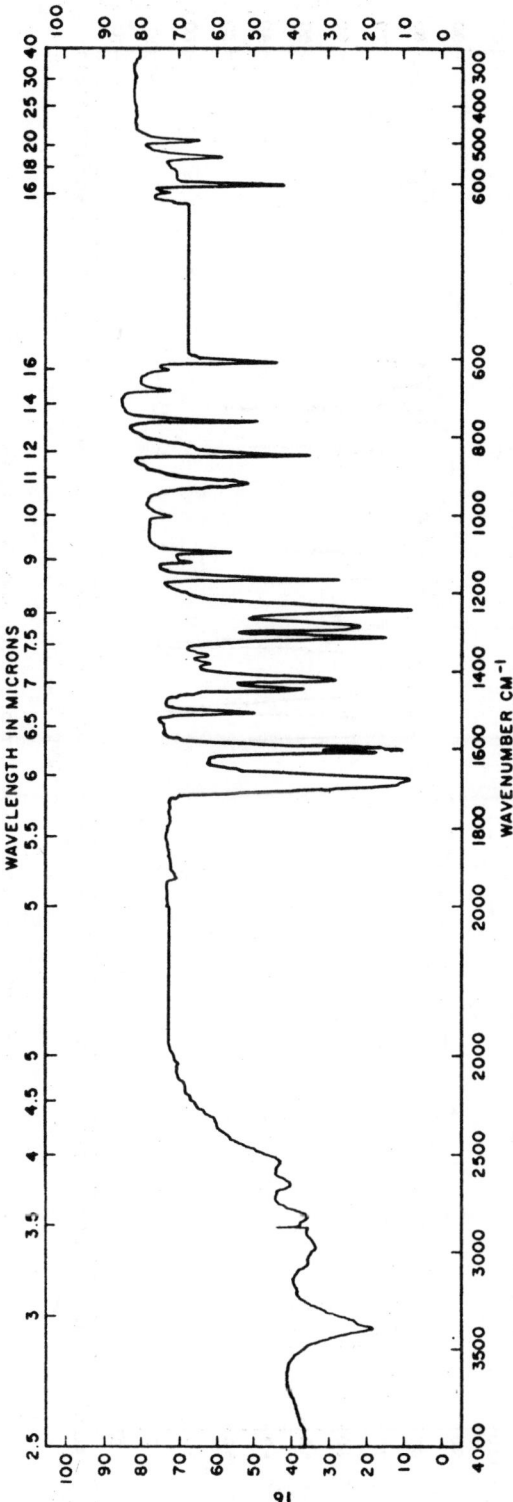

FIGURE 2.18: Infrared spectrum of *p*-hydroxybenzoic acid (KBr disk).

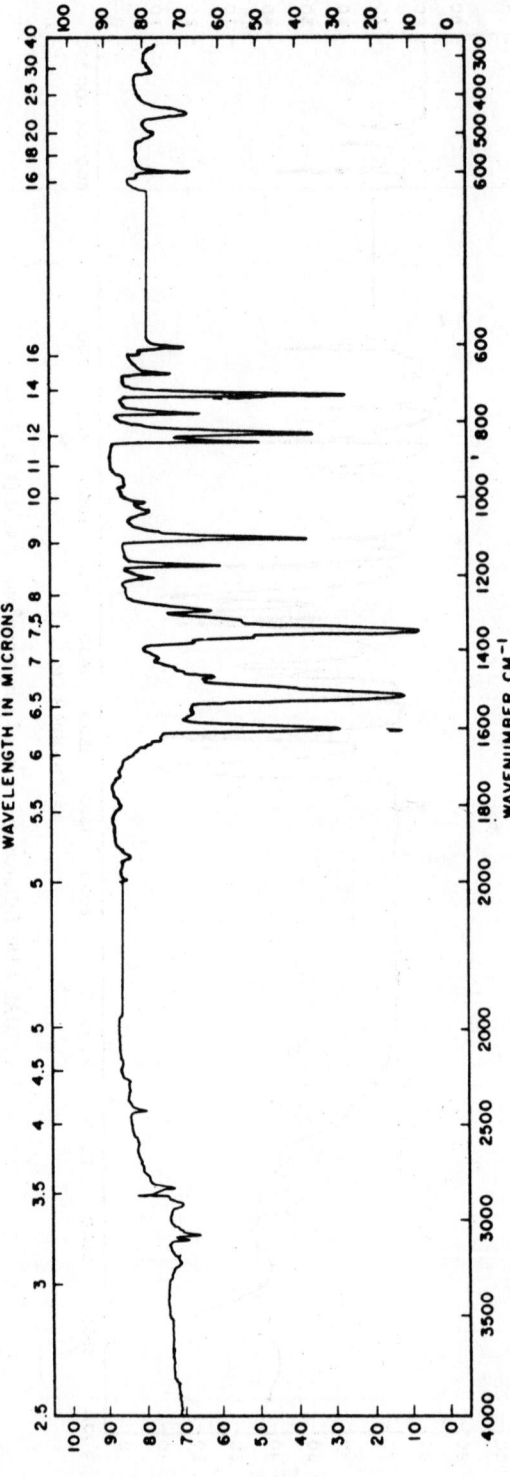

FIGURE 2.19: Infrared spectrum of *p*-nitrotoluene (KBr disk).

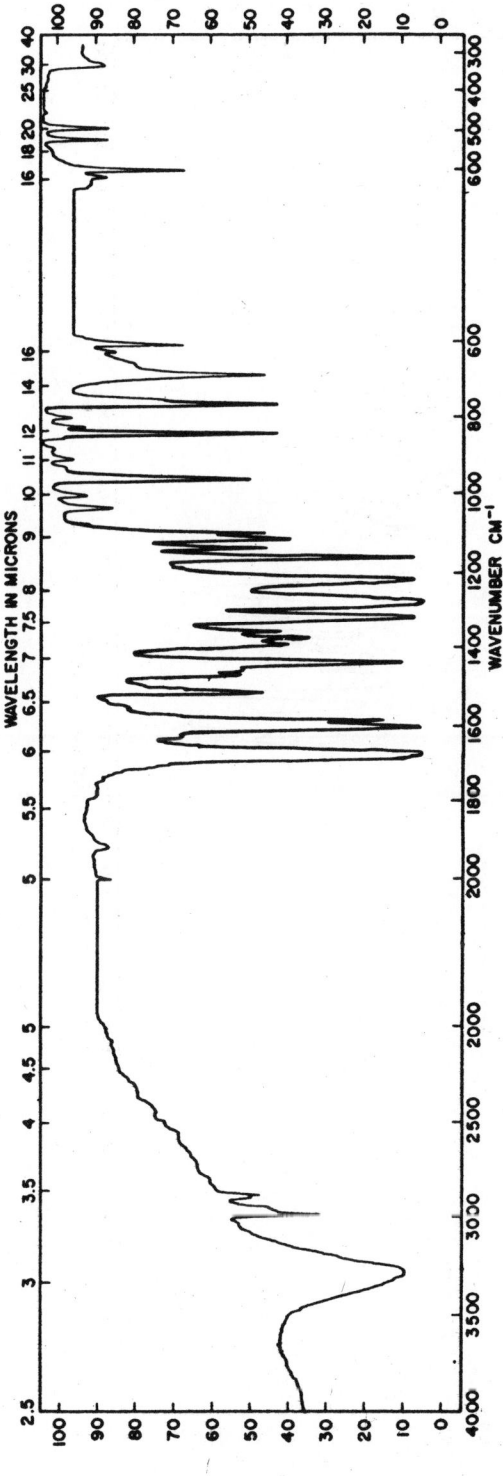

FIGURE 2.20: Infrared spectrum of *n*-propyl *p*-hydroxybenzoate (KBr disk).

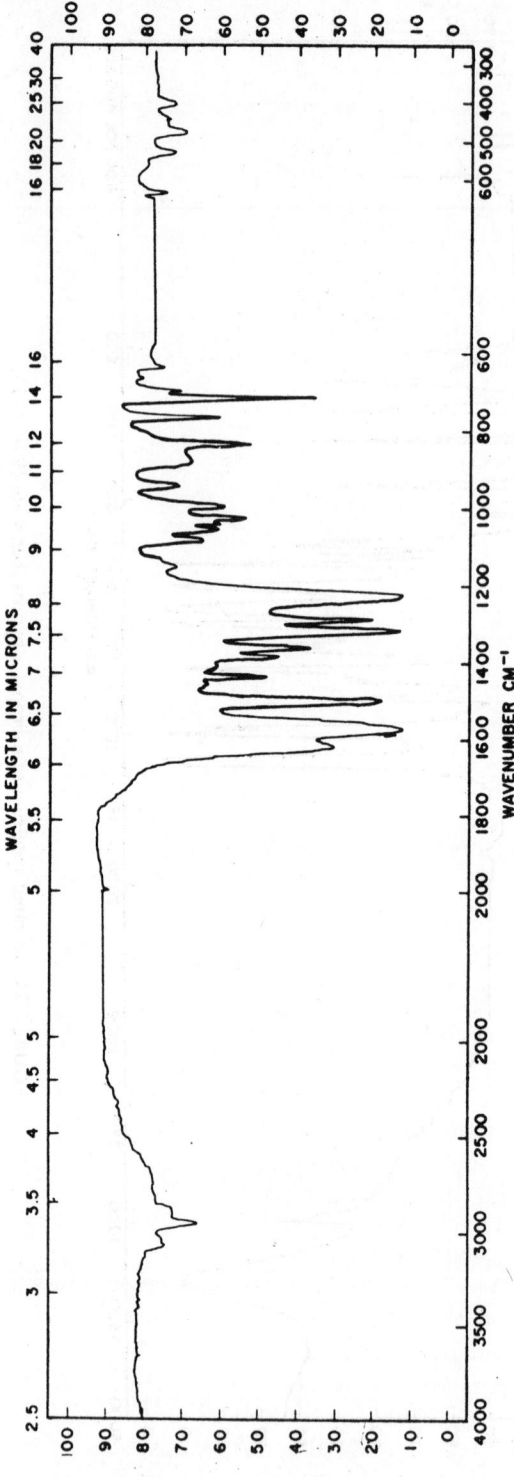

FIGURE 2.21: Infrared spectrum of ethyl (*o*-nitrophenylthio)acetoacetate (KBr disk).

deduced from the usually informative regions, 1670–2000 and 1430–1670 cm⁻¹. In the former region, absorption bands are very weak, and in the latter region, strong bands due to other than C—C stretching occur. The presence of a nitro group is apparent—1512 (s), 1335 (s), 850 (w)—but the absence of absorption bands normally associated with ketone and ester carbonyl groups is noticeable. The reason for this is that this compound is a typical β-ketoester and as such exists mainly in hydrogen-bonded enolic form (VI) rather than as (V). Carbonyl absorption therefore occurs at a longer wavelength than might be expected. Broad weak absorption in the 2600–3000 cm⁻¹ region suggests hydrogen bonded O—H stretching.

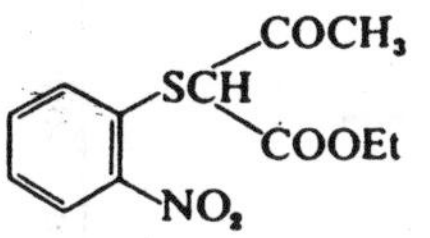

(V)

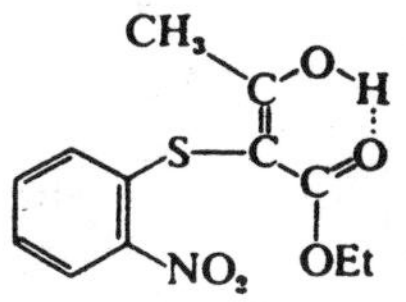

(VI)

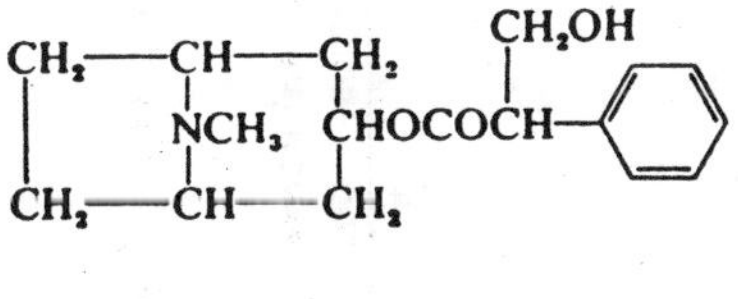

(VII)

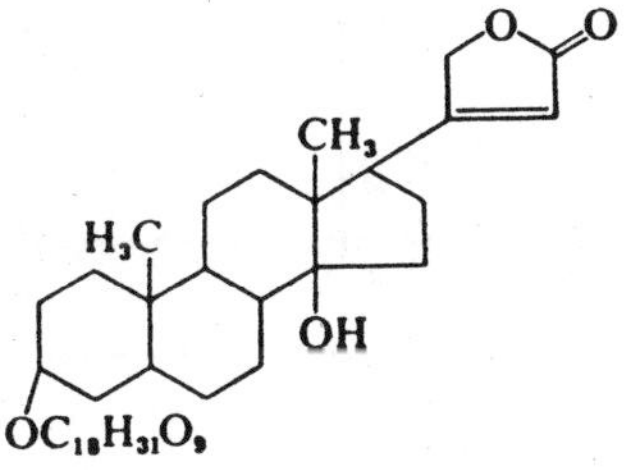

(VIII)

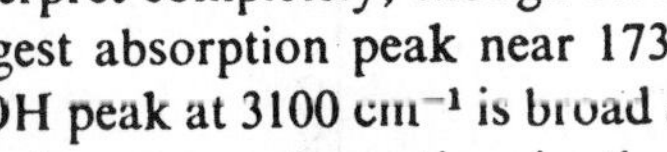

(IX)

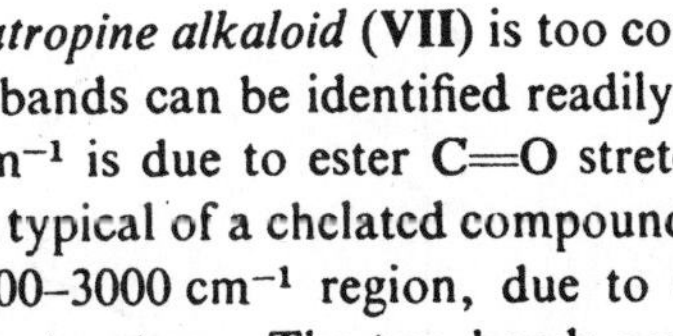

(X)

The infrared spectrum (Fig. 2.22) of *atropine alkaloid* (VII) is too complex to interpret completely, though certain bands can be identified readily. The strongest absorption peak near 1730 cm⁻¹ is due to ester C=O stretching. The OH peak at 3100 cm⁻¹ is broad and typical of a chelated compound, and relatively strong absorption in the 2800–3000 cm⁻¹ region, due to C—H stretching, would be expected of such a structure. The two bands near 685 and 760 cm⁻¹ can be assigned to monosubstituted phenyl C—H deformations; and the substitution pattern in the 1430–1600 cm⁻¹ region, though weak, suggests an aromatic compound. The spectrum (Fig. 2.23) of *digoxin* (VIII) is also complex, but once again it is possible to identify some of the prominent bands in the spectrum. A noticeable feature is the strong O—H stretching band near 3440 cm⁻¹. Digoxin is a triglycoside and therefore a large number

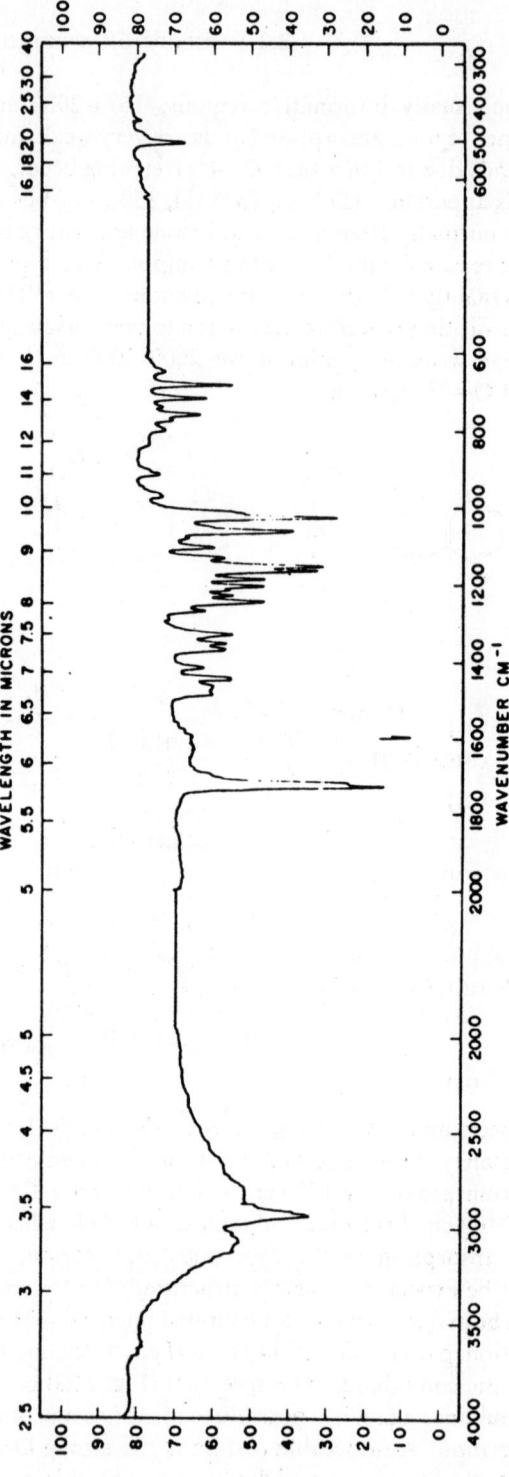

FIGURE 2.22: Infrared spectrum of atropine alkaloid (KBr disk).

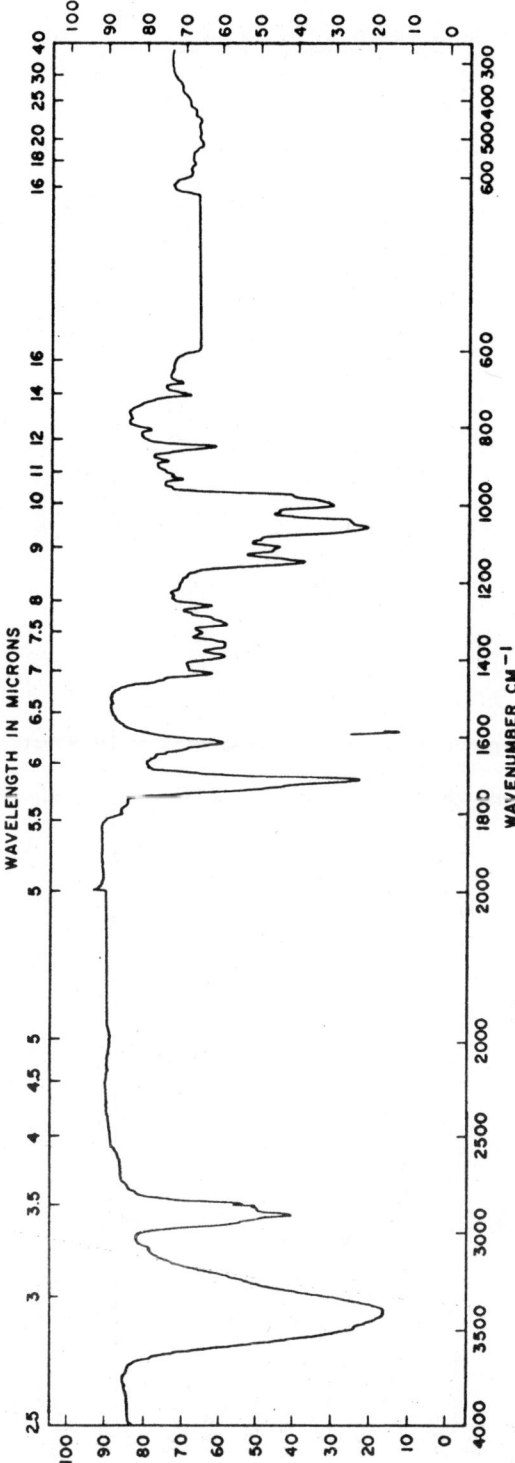

FIGURE 2.23: Infrared spectrum of digoxin (KBr disk).

97

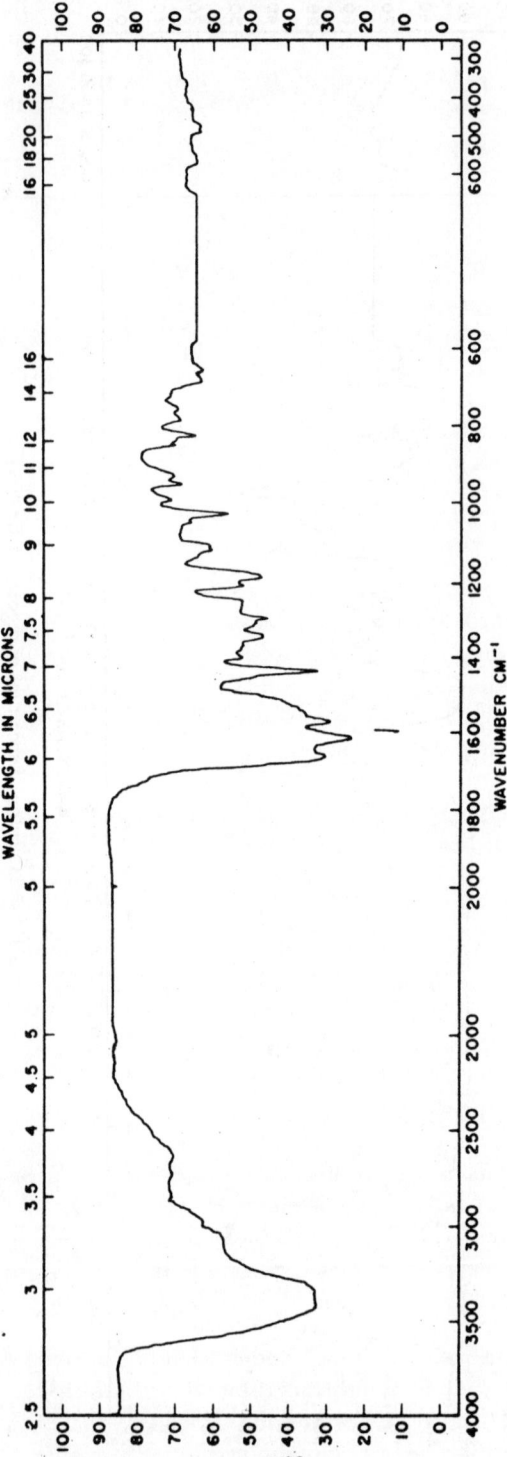

FIGURE 2.24: Infrared spectrum of aureomycin (KBr disk).

98

of hydroxyl groups are located in the C_3-side chain, in addition to the tertiary OH group at the junction of rings C and D. C—H stretching is present (2850–3000 cm⁻¹ region), and the strong band at 1725 cm⁻¹ is due to C=O stretching of an α,β-unsaturated lactone. It is difficult to assign a band to C—O stretching. The strong peak near 1065 cm⁻¹ is probably due to either alcoholic O—H deformations or ether C—O stretching. The absence of absorption bands in the 1470–1610 cm⁻¹ region and the lack of strong bands in the 600–800 cm⁻¹ region indicate a nonaromatic compound.

No strong C=O stretching bands are present in the spectrum (Fig. 2.24) of *aureomycin* (X) at wave numbers greater than 1650 cm⁻¹. Carbonyl stretching bands of typical ketones and amides are usually located near 1700 cm⁻¹, but in this instance, all the carbonyl groups are strongly hydrogen-bonded, which results in a shift of the C=O absorption to lower wave numbers. The very broad and strong O—H stretching, centered at 3420 cm⁻¹ confirms the presence of hydrogen-bonding. The broad band masks the N—H stretching, and the C—H stretching bands are only barely visible Without the aid of reference compounds, it is difficult to interpret the rest of the spectrum except to conclude that it is suggestive of a conjugated system.

The spectrum (Fig. 2.25) of *ephedrine alkaloid hemihydrate* (IX) was recorded as a 5% solution in chloroform. Four weak bands located between 1700 and 2000 cm⁻¹ are readily apparent and are typical of the absorption pattern of a monosubstituted benzene nucleus. The expected strong band near 700 cm⁻¹ for such a system is masked by the solvent, but is apparent in the spectrum of a KBr disk of the alkaloid. Absorption bands due to C—H, N—H, and O—H stretching are present in the 2800–3500 cm⁻¹ region of the spectrum.

F. INSTRUMENTATION

Single-beam and double-beam infrared spectrophotometers are available. The former type has a limited use in routine qualitative and quantitative studies. Most modern infrared spectrophotometers are therefore of the double-beam type, though some of these can be operated as single-beam instruments. The Beckman IR5A spectrophotometer (Fig. 2.26) is a typical example. The basic components of the double-beam instrument are: radiation source, photometer, monochromator, detector system, and recorder. The optical system of a double-beam infrared spectrophotometer is shown (Fig. 2.27).

I. Radiation Source

Infrared radiation sources are black bodies which, when heated electrically to 1200–1800°C, continuously emit radiation of the desired intensity. Common sources are the Nernst glower and the Globar. The Nernst glower is a small rod composed of oxides of zirconium, thorium, and yttrium, which,

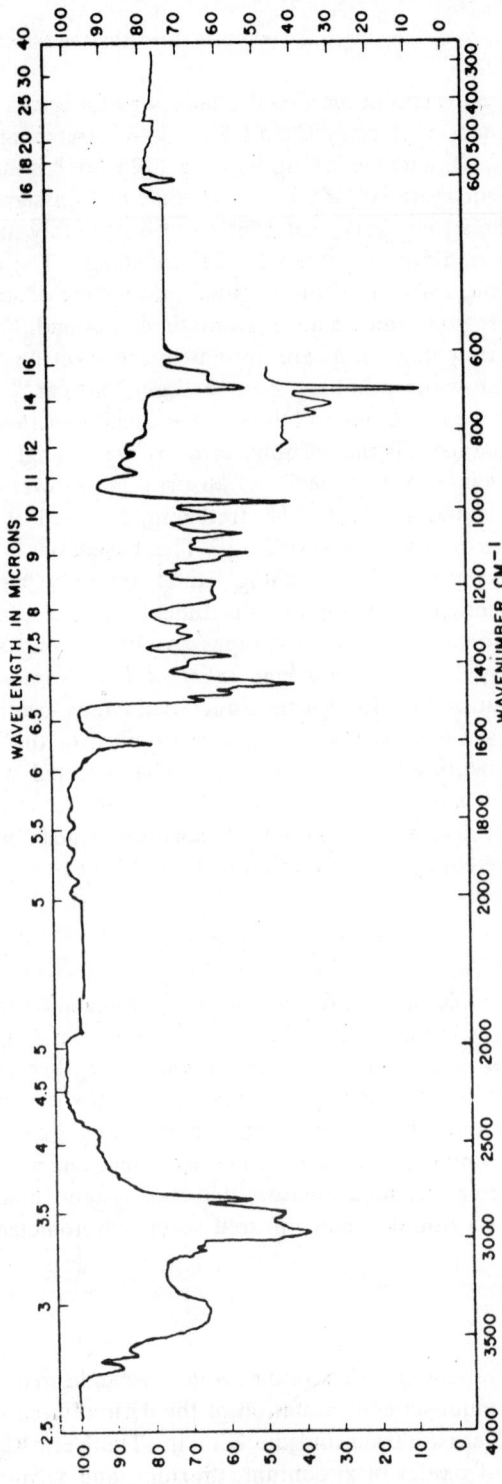

FIGURE 2.25: Infrared spectrum of ephedrine alkaloid hemihydrate (5% w/v in chloroform) (Inset: KBr disk).

FIGURE 2.26: Beckman model IR5A infrared spectrophotometer. Reproduced with the permission of Beckman Instruments, Inc.

when heated to 1800°C, emits infrared radiation. Secondary electrical heaters are required to start the glower because it is nonconducting when cold. The Globar is a small rod of silicon carbide which is an effective infrared source when heated to 1200°C. Although infrared radiation is emitted continuously by both sources, the radiation energy is not constant, but varies with varying wavelength.

The radiation from the source is split by means of mirrors into two equal beams, one of which passes through the reference cell, while the other passes through the sample cell.

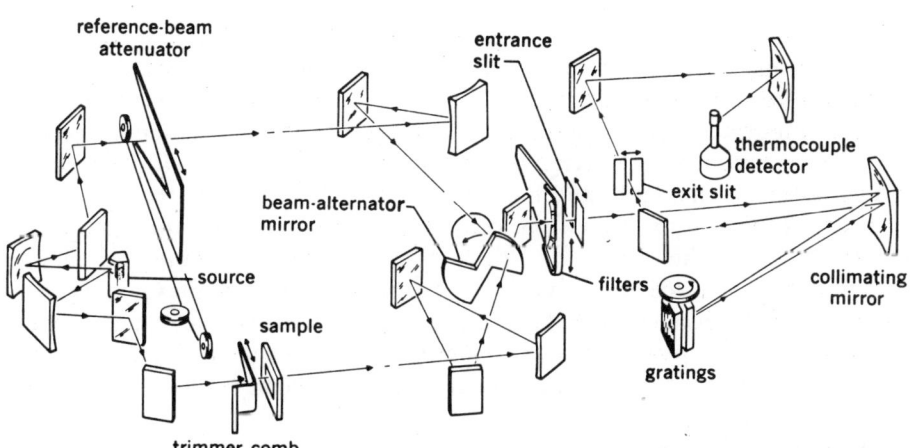

FIGURE 2.27: Optical system of double-beam infrared spectrophotometer.

2. Photometer

The reference and sample beams then enter the photometer area, in which the rotating chopper, i.e., the beam-alternator mirror, alternates the reference and sample beams such that the single beam produced, which passes through the entrance slit into the monochromator, is composed of alternating segments.

3. Monochromator

The beam from the photometer is focused by means of a collimating mirror onto gratings (see Fig. 2.27) or a prism (Fig. 2.28), which disperse the beam

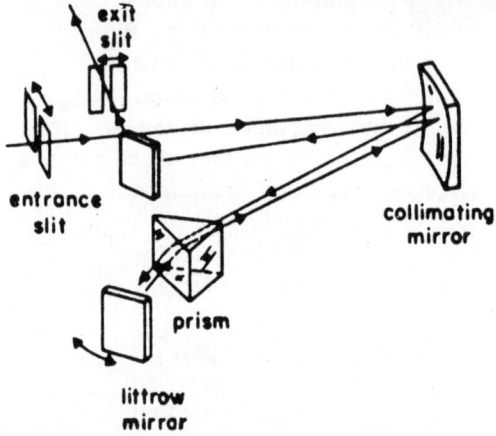

FIGURE 2.28: Prism monochromator.

over a range of frequencies, a small portion of which is reflected back to the collimating mirror. This results in further dispersion. Radiation of a very narrow frequency range emerges into the detector compartment. If the monochromator incorporates a prism, then the frequency of the radiation emerging from the exit slit will be altered by rotating the Littrow mirror. If this mirror is rotated automatically and continuously, the frequency of the radiation reaching the detector will also be continuously changing. Rotating the gratings serves the same purpose. Prisms of calcium fluoride, cesium bromide, and sodium chloride are all employed, but each suffers from the fact that the resolution produced is lower than desirable in certain frequency ranges. The use of gratings permits greater resolution. For this reason, gratings are replacing prisms in all but the cheaper modern instruments.

4. Detector System

Infrared radiation detectors are divided into two categories, thermal detectors and photon detectors. The former category includes thermocouples, bolometers, and pneumatic (or Golay) detectors. Photon detectors employ semiconductors. Thermocouples are basically two dissimilar strips of metal joined together at each end. One junction is heated by the radiant energy of the alternating signal emerging from the monochromator. An emf is produced which is proportional to the degree of heating. Bolometers are similar. They are resistors with a high temperature coefficient of resistance. Both are operated in a vacuum, which decreases noise and increases sensitivity. Golay detectors contain a nonabsorbing gas, which expands and moves a flexible mirror when heated by the radiant energy. The degree of expansion and hence the motion of the mirror is proportional to the degree of heating. The emf produced, and the motion of the mirror, respectively, are very small and to be meaningful have to be amplified. In photon detectors, radiation excites electrons of a suitable semiconductor (e.g., PbSe) from a nonconducting low-energy state to a higher energy level, which can conduct and produce a signal proportional to the amount of radiation. The signal is actually a measure of a decrease in the resistance of the semiconductor. The major disadvantage of such detectors is that they are effective only over short wavelength ranges. The advantages and limitations of various types of detectors have been discussed in detail by Moss.[17]

When the sample under investigation has absorbed some energy, the sample and reference beams will differ in their radiant energies. Then, the detector system generates a signal which is amplified and fed to a servomotor which moves the attenuator comb, blocking part of the reference beam until the energies of the reference and sample beams are again equal. Beam balance (optical null) is restored. The attenuator comb is connected mechanically to the pen of the recorder so that the transmittance of the sample, as a function of wavelength (i.e., an infrared spectrum) can be recorded (see Fig. 2.29, in which the attenuator comb is diagrammatically represented in the form of a solid wedge).

G. PREPARATION OF SAMPLE

Infrared spectra of gases, liquids, and solids can be obtained. The sample must be dry because water absorbs infrared radiation near 3700 and 1630 cm^{-1} and absorption bands in these regions may be erroneously assigned to the substance being analyzed.

1. Solids

Three different methods of preparing a solid for infrared study are used. The solid may be examined neat, in suspension, or in solution.

The least common of these methods is the first mentioned, which is used occasionally if the solid melts without decomposition. A small quantity is melted and placed between two alkali halide plates (which transmit infrared radiation). They are clamped in a suitable holder. The capillary film which forms is allowed to solidify, then the whole, in a suitable holder, is placed in the sample beam path. No reference is necessary, i.e., air is the reference. The method most commonly used is the examination of the solid suspended in an inert oil or inert solid. In both instances, the substance must be reduced to very small particles.* The orientation of a crystalline material in the infrared beam affects the intensity of the absorption bands, but this effect is minimized if numerous very small particles are examined. This is an important consideration in quantitative analysis. In the *mull method*, the finely

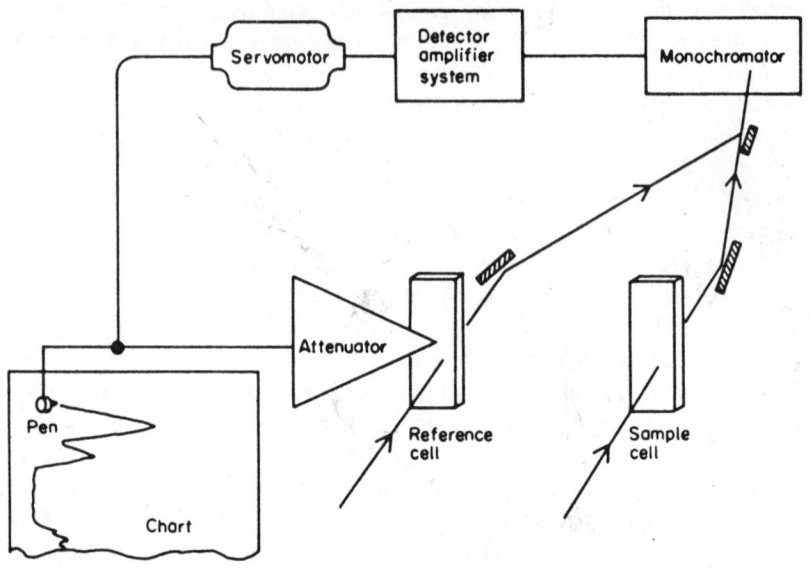

FIGURE 2.29: Beam balance.

powdered sample (ca. 5 mg) is dispersed in a drop of a suitable mulling agent. The use of an agate mortar and pestle or a glass plate and a small, inert spatula are recommended for this purpose. The most popular mulling agent is the mineral oil nujol, a mixture of high molecular weight alkanes. Fluorinated oil, e.g., fluorolube, is also used for this purpose. The disadvantage of using such agents is that they themselves absorb infrared radiation in certain regions and thus obscure some of the absorption bands of the substance being analyzed. Nujol, for example, absorbs strongly near 3000 cm^{-1} (C—H stretching) and less strongly near 1460 and 1375 cm^{-1} (C—H bending). It is transparent in thin layers over the $1375–650 \text{ cm}^{-1}$ region. Fluorolube, however, is transparent in thin layers from $1330–5000 \text{ cm}^{-1}$. Thus, two spectra, one in each medium, would yield maximum information. Hexachlorobutadiene is also used as a mulling agent. It has no hydrogen atoms and therefore will be transparent in the regions where nujol absorbs.

* The Wig-L-Bug amalgamator, manufactured by Crescent Dental Manufacturing Company, Chicago, Ill. is useful for this purpose.

The mull is transferred to the surface of a flat alkali halide plate. A second plate is carefully placed on top and the two are manually pressed together, using a rotatory motion to ensure an even spread of the mull. By varying the amount of pressure applied, films of varying thickness can be obtained. The plates are held together in a suitable holder (Fig. 2.30) which is placed in the sample beam path. No reference cell is necessary.

FIGURE 2.30: Mull holder. Reproduced with the permission of Beckman Instruments, Inc.

In the *pressed disk method*, pure, dry, and finely powdered potassium bromide or chloride is intimately mixed with the sample, preferably in a Wig-L-Bug or its equivalent. Generally a concentration of about 1 % w/w is suitable, i.e., 2 mg of the substance is ground with 200–300 mg of KBr. With high molecular weight substances, a somewhat larger concentration is often required. The mixture is compressed in a die under vacuum at room temperature and at high pressure (40,000–50,000 lb/in.^{2}).* This treatment produces a solid transparent disk, which is mounted in a holder and placed in the sample beam path. A reference disk of pure KBr can be used, but as spectral grade KBr is virtually transparent over the 5000–650 cm^{-1} range, it is the common practice to use no reference cell. One criticism of this method

---

* If the area of the disk is, say, 0.25 in.^{2}, then a pressure at the disk surface of 40,000 lb/in.^{2} is obtained by applying an external pressure of 10,000 lb.

is that during the grinding process, physical and chemical changes can occur
and give rise to anomalous spectra (e.g., see Ref. 18).

Solids can also be examined in solution in a suitable solvent. The three most
commonly used solvents are carbon tetrachloride, chloroform, and carbon
disulfide, each of which has certain disadvantages. They absorb strongly in

FIGURE 2.31: Solution cell. Reproduced with the permission of Beckman Instruments,
Inc.

certain regions (Fig. 2.32) and thus obscure what may be important absorption bands. In addition, many polar substances are not sufficiently soluble in
these solvents. The concentration required depends on the cell path length
and on the molecular weight of the substance being examined. Generally
speaking, a 5% w/v solution of a substance of molecular weight 100 to 500 will
usually give a good spectrum using a cell of 0.1 mm-path length; with 1 mm
cells, a 1.5% w/v solution is often satisfactory. The prepared solution is
transferred to a solution cell (Fig. 2.31) by means of a hypodermic syringe.
Solvent evaporation is avoided, especially in quantitative analysis.

Solution cells are of three types. *Demountable cells*, as the name suggests,
can be dismantled after use, thoroughly cleaned and polished, and reassembled prior to further use. Spacers of varying thicknesses are available for

such cells with the result that cells of varying path lengths can be constructed. *Sealed cells* of fixed cell path length have the advantage of a cell path length which is accurately and permanently known. They are more difficult to clean than the demountable type. This is achieved by repeated flushings with solvent. The *variable-thickness cell* has a micrometer attachment with which path length can be varied. Cost prohibits the routine use of cells of this type.

When a solution is examined, a compensating reference cell of path length

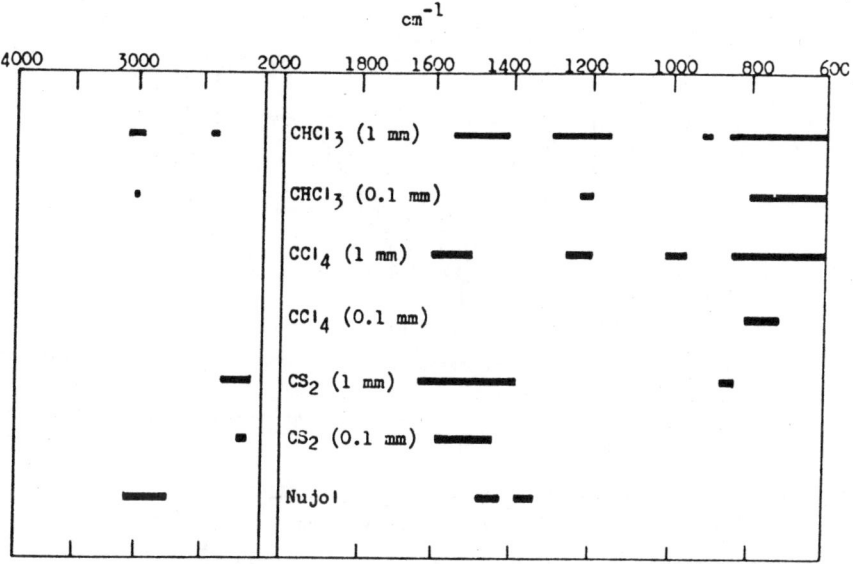

FIGURE 2.32: Strong absorption bands of chloroform, carbon tetrachloride, carbon disulfide, and nujol. (The solvents and the oil absorb strongly in the shaded areas.)

equal to that of the solution cell is necessary. The reference cell contains pure solvent only and is placed in the reference beam. As the path lengths of the sample and reference cells are the same, the recorded spectrum is that of the solute, except in the regions where the solvent absorbs strongly.

Sealed cells of fixed path length are most often employed. A matching set is required. Usually matching pairs of path lengths 0.05, 0.1, 0.5, and 1.0 mm are kept available. Alternatively, one sealed cell of each path length may be available. If this is so, the reference solution is placed in a variable-thickness cell, the thickness of which is adjusted to match that of the sealed cell.

2. Liquids

Liquids are generally examined as thin films to avoid interference of the solvent. They are examined between two alkali halide plates (cf. mulls), and once again control of sample thickness is difficult. Cells with very small path lengths (0.005 mm and upward) are available for quantitative work when the thickness of the liquid film must be accurately known. Volatile liquids must be examined in a sealed cell of this type. Liquids can also be examined in solution. The same limitations as those described under solutions of solids apply.

3. Gases

Molecules of gases, in contrast with liquids, solids, or solutions, are free to rotate. The result is that with simple gas molecules especially, rotational

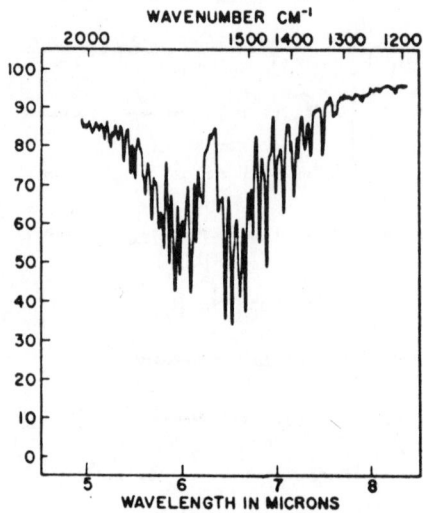

FIGURE 2.33: Atmospheric water measured on a Beckman IR5A spectrophotometer used as a single-beam apparatus.

energy level transitions are observed in their infrared spectra. An abundance of fine structure is obtained corresponding to the rotational energy transitions (Fig. 2.33). The locations of some of these bands are accurately known, and for this reason, various gases are used in the wavelength calibration of a spectrophotometer.

Special techniques are used to weigh very small quantities of gases, and special gas cells with path lengths measured in centimeters (e.g., 10 cm) are employed. Gas cells up to 50 cm in length are used. They are necessary because of the small number of molecules of the gas being examined.

2.3 PHARMACEUTICAL APPLICATIONS OF INFRARED SPECTROSCOPY

Many pharmaceutical substances have been qualitatively identified or quantitatively assayed by means of infrared spectrophotometry. Previous mention has been made to informative reviews in this field of study.[9-11]

A. QUALITATIVE ANALYSIS

In almost every instance, qualitative analysis is performed by comparing the spectrum of the substance with that of an authentic sample. If the spectra of both are identical, then the substance and the authentic sample are the same. There are very few exceptions to this generalization. One of these concerns the spectra of long-chained fatty acids or esters (XI) in which n is large. Such homologs give spectra which are extremely similar.

$$CH_3(CH_2)_nCOOR$$
(XI)

It is well known that when the halide-disk technique is employed, anomalous absorption bands are sometimes observed. These have been summarized by various workers[19-22] who list the reasons as

   a. Variations in crystal size and orientation
   b. Variations in grinding technique
   c. Formation of polymorphic crystalline forms
   d. Formation of hydrates
   e. Formation of amorphic materials
   f. Chemical transformations or combinations.

To identify a compound, it is preferable, therefore, to compare a dilute solution of it with one of an authentic sample. Nevertheless, it is the common practice to perform both qualitative and some quantitative analyses using halide disks. The BP and USP both recognize that spectra vary with the method used to prepare the sample and with the instrument employed, and they require a direct comparison of the spectrum of the substance with that of a similar preparation of an authentic sample. This should not detract from studies such as the one by Warren and co-workers[23] in which the infrared spectra of 23 phenothiazines were recorded and analyzed. Reference to such published spectra enables one to eliminate many possibilities and may even result in a tentative identification which can be confirmed by the USP method. Another example of this approach is the reported study of the infrared spectra of a series of phenols in carbon tetrachloride[24] in which spectra are recorded in a bar-type diagram which, the authors suggest, would be useful in the characterization of unknown phenols.

The presence and identification of impurities in medicinal substances can be confirmed by means of infrared spectrophotometry. The work by Urbanyi, Sloniewsky, and Tishler,[25] to which reference is made in greater detail later, is a good example of this approach.

B. QUANTITATIVE ANALYSIS

Whereas most ultraviolet and visible spectrophotometers provide a chart of absorbance v. wavelength, many infrared spectrophotometers measure, in percentage, the amount of energy transmitted by the substance, i.e., the transmittance T, as a function of wavelength or frequency of the energy supplied. The transmittance is the ratio of the intensity of the radiation transmitted by the sample I to the intensity of the radiation incident on the sample I_0, i.e.,

$$T = I/I_0$$

All quantitative spectrophotometric measurements are governed by the Beer-Lambert law

$$I = I_0 e^{-kcd}$$

where k is an absorption coefficient characteristic of the substance, c is the concentration of the substance (moles/liter), and d is the thickness of the cell containing a solution of the substance. This expression can be rewritten as

$$\log_{10} \frac{I_0}{I} = \epsilon cd$$

where ϵ is the molecular extinction coefficient (or molar absorptivity) of the substance in units of liters centimeter^{-1} mole^{-1}. The expression

$$\log_{10} \frac{I_0}{I}$$

is termed the absorbance (or extinction coefficient, or optical density) A of the substance. As ϵ is characteristic of the sample, and as d can be controlled, it follows that the absorbance A is directly proportional to the concentration of the substance, i.e.,

$$\log_{10} \frac{I_0}{I} \propto C$$

Transmittance T was defined previously as I/I_0. The following relationships are therefore true:

$$A = \log_{10} \frac{I_0}{I} \qquad A = \log_{10} \frac{1}{T} \qquad A = \log_{10} \frac{100}{\% \, T}$$

In quantitative intensity measurements, it must always be first confirmed that absorbance is directly proportional to concentration of solute in the concentration range to be used and at the wavelength selected for the assay. This is done conveniently by preparing a series of dilutions of the substance under investigation and recording

$$\log \frac{I_0}{I}$$

or its equivalent, for each at the chosen wavelength. A plot of this v. concentration should give a straight line.

There are two other sources of error in quantitative infrared analysis. Intensity measurements should be in the 20–60% transmission range. A suitable solution concentration or cell path length should be selected to insure this. Signals can be distorted nearer 0 or nearer 100% transmission.

The second source of error is the uncertainty of the base line I_0. Most organic substances give rise to complex infrared spectra. A suitable band is chosen for the purpose of quantitative analysis. If possible, it should be an isolated symmetrical band because in such an instance (Fig. 2.34a), the value of I_0 can be readily determined. Frequently, bands overlap (Fig. 2.34b), making accurate measurement of I_0 difficult, especially if the neighboring bands arise from impurities or from components of a mixture other than the

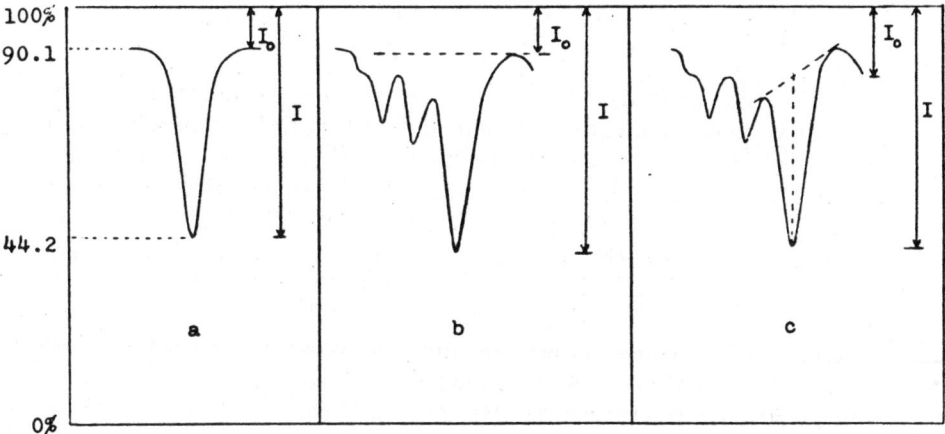

FIGURE 2.34: Selection of base line in quantitative analysis.

substance being determined. In such a situation, the "base line technique" is used—a suitable tangent is drawn to provide the base line.[26] If the neighboring band also arises from the substance being investigated, however, the spectrum will be reproducible, and alternative base lines can be drawn (Fig. 2.34c).

Calculation of the absorbance of a substance (A_s) is illustrated using the ideal situation (Fig. 2.34a). The relationship deduced,

$$A_s = \log_{10} \frac{\% T_0}{\% T}$$

(where T_0 and T are the transmittances of the solvent and the solution of sample, respectively, and correspond to the chosen values of I_0 and I), also holds when the base line technique is employed. From Fig. 2.34a,

$$A_{solvent} = \log \frac{100}{90.1}$$

$$A_{solvent+sample} = \log \frac{100}{44.2}$$

Therefore

$$A_s = \log \frac{100}{44.2} - \log \frac{100}{90.1}$$

$$= \log \left(\frac{100}{44.2} \times \frac{90.1}{100} \right)$$

$$= \log \frac{90.1}{44.2} \doteq 0.3093$$

that is,

$$A_s = \log \frac{\% T_0}{\% T}$$

As dispersion of solids in halide disks can give rise to anomalous absorption bands, most quantitative determinations are carried out on solutions of substances. Because of its superior solubilizing properties, chloroform is commonly employed for this purpose; carbon tetrachloride, carbon disulfide, pyridine, and other solvents are also used. All solvents suffer from the disadvantage that they blank out certain areas of the spectrum (see Fig. 2.32). Some substances, however, are not sufficiently soluble in any of the common solvents. In such instances, provided certain precautions are observed, it is possible to perform quantitative analyses using mulls or potassium bromide disks. Hayden and Sammul[27] have shown that reproducible quantitative results can be obtained using the halide-disk technique, provided that the sample and standard disks are prepared under carefully controlled conditions and a modification of the Beer-Lambert law equation is employed for the calculations. These authors used this method for the quantitative analysis of a large number of pharmaceutical substances. The mull technique has also been adapted to quantitative infrared analysis.[20]

I. Reference Standard

The substance used in the preparation of the standard solution or disk, in both quantitative and qualitative analysis, is almost invariably a pure sample of the substance under investigation. Comer and Ribley,[28] however, used a solution of benzoic acid in chloroform to establish absorbance ratios for several drug substances and found that the accuracy of this technique was equal to the usual method of using the drug substance as a reference standard. Similarly, Hayden[29] has developed a method for the determination of glyceryl trinitrate by infrared analysis using benzoic acid as a secondary standard in place of the normal reference standard (an adsorbate of glyceryl trinitrate on lactose). From these results and others, it would seem that, in future, secondary standards will be employed to a greater extent. Their use would eliminate the need for storing large numbers of different primary reference standard compounds.

C. PRACTICAL INFRARED ANALYSIS

Exercises

E2.1. Identification of Selected Medicinal Compounds. Four single substances chosen from Tables 2.1 and 2.2 will be supplied, as well as authentic pure samples (reference standards) of these substances. Prepare each sample and reference standard in the manner suggested in the USP or BP, obtain the infrared spectrum of each, and compare each pair.

The four substances will be so chosen to require the student to record the spectrum of a thin film, the spectrum of a solution, the spectrum of a dispersion in KBr, and the spectrum of a suspension in mineral oil. If the spectra of the specimens, obtained when KBr or mineral oil is used, differ from those of the reference standards, this may be due to polymorphism (the existence of a substance in more than one crystalline form) or due to other reasons (see p. 109). The procedure suggested by the USP XVII, p. 809, should be adopted, i.e., prepare solutions of both the reference standard and the specimen under test in a suitable solvent, remove the solvent by evaporation, and prepare a new dispersion or suspension of each residue for a second examination. If infrared spectral differences are still observed, it can be concluded that the specimen and reference standard are not identical.

E2.2. Identification of the Medicinal Component of a Tablet. The tablets supplied will contain the salt of an organic base (see examples marked with a superscript *b* in Table 2.1). The appropriate reference standard will also be supplied. Confirm the identity of the medicinal component using the USP test entitled "Identification—Organic Nitrogenous Bases" (USP XVII, p. 908).

E2.3. Infrared Spectra of Primary, Secondary, and Tertiary Amine Salts. Three unnamed pharmaceutically active amine salts, such as those mentioned here, will be supplied. Record their infrared spectra as suspensions in mineral oil, and by reference to these spectra and to the literature[13] decide which one is

(+)-amphetamine sulfate (a primary amine salt)
ephedrine sulfate (a secondary amine salt)
quinine hydrochloride (a tertiary amine salt)

Explain the reasons for your conclusions.
(*Note:* Amine salts other than the ones named may be supplied).

E2.4. Identification of Pharmaceutically Useful Phenothiazines. A medicinally active phenothiazine base or salt will be supplied. (If the latter, dissolve it in a small volume of water, basify the solution with dilute sodium hydroxide solution, and extract the liberated base with ether. Wash the ether solution with water, dry it over sodium sulfate, filter, and remove the ether from the filtrate). Record the spectrum of the base as a thin film. Compare your spectrum with those published in the literature[23] and make a tentative identification.

E2.5. A Study of Complex Formation by Means of Infrared Spectroscopy. Quinine and phenobarbital are known to react to form a 1:1 complex,[18] the formation of which can be confirmed by means of infrared spectroscopy. Prepare the complex by dissolving 1 mmole (324 mg) of quinine alkaloid and 1 mmole (232 mg) of phenobarbital in hot absolute ethanol (5 ml) and cooling the solution to 0°. The crystalline complex will separate within 2 hr. Collect the precipitate and wash it with small quantities of cold ethanol. It should melt at 185°.

Prepare a pure sample of phenobarbital by crystallizing a commercial sample from aqueous ethanol. The mp should be 174°.

Record the mp of a good commercial sample of quinine base. A sample, mp 177°, is suitable for this study without purification.

Prepare a physical mixture of phenobarbital (23 mg) and quinine (32 mg) on a powder paper using a spatula to mix the powders.

Prepare potassium bromide disks of all four specimens, by shaking a mixture of 1 mg of each with 300 mg quantities of potassium bromide in a Wig-L-Bug amalgamator for 5 sec, then pressing the disks at a pressure of 40,000 psi.* Record the infrared spectrum of each over the range 2000–1400 cm^{-1} (5–7 μ). The spectrum of the physical mixture should be an additive spectrum of the individual components. The spectrum of the complex should differ from that of either component.

E2.6. Infrared Absorption Vibrations of Functional Groups. Three, four, or five chemically related compounds will be supplied in unmarked sample tubes. The chemical identity of one will be revealed. Record the infrared spectra of all the compounds, using as many different ways of preparing the samples as possible.

Identify the characteristic functional-group absorption bands in each spectrum, and suggest structures for each compound. Confirm your identification using other simple physical or chemical tests (e.g., bp; mp; refractive index; BP or USP identity tests) or by comparison with spectra of authentic samples which will be supplied on request.

(Typical example: p-nitrophenol; p-aminophenol; p-phenetidin; phenacetin.)

E2.7. Infrared Spectra of Steroids. A selection of the steroids listed in Tables I and II will be available. Each student (or group of students) will record the spectra of two of them as dispersions in KBr, retain a copy of each

---

* These conditions for the preparation of the disk differ from those suggested in the original publication. A pressure of 40,000 psi (i.e., 8000 lb pressure was used to prepare a 0.2-in.² disk) was best when the equipment available to the author was used.

for his own use (see later), and deposit a copy of each with the instructor. In this way, a library of spectra will be compiled. When the library is complete, the student will record the spectrum of another steroid and identify it from the compiled library of spectra, all of which will have been recorded on the same instrument.

On the spectra retained by the student (previous instructions), identify as many as possible of the stretching and deformation vibrations.

E2.8. Qualitative Reaction Sequence Study. The interaction of o-chloronitrobenzene and α-mercaptoacetic acid in the presence of sodium bicarbonate, followed by acidification, gave compound (I), $C_8H_7NO_4S$, which was converted to (II), $C_9H_9NO_4S$, when heated with methanol containing 10% concentrated sulfuric acid. Compound (II) was oxidized with potassium permanganate and glacial acetic acid, which gave (III), $C_9H_9NO_6S$. Hydrolysis of (III) produced compound (IV) which had a molecular formula $C_7H_7NO_4S$. Reduction of (IV) by means of zinc and ammonium chloride gave (V), $C_7H_9NO_3S$, whereas reduction with stannous chloride and hydrochloric acid gave compound (VI), $C_7H_9NO_2S$.[30,31]

The infrared spectra of compounds (I–VI) are reproduced here (Figs. 2.35–2.40, respectively). All were recorded using the KBr disk technique. By means of these spectra, identify the six compounds and suggest the origin of each of the major vibration bands.

E2.9. (a) Confirmation of the Validity of the Beer-Lambert Law Using Solutions of Caffeine in Chloroform, and (b) Determination of the Caffeine Concentration in a Chloroform Solution of the Drug. The band near 10.26 μ is suitable for the quantitative determination of caffeine in chloroform.[32]

Prepare a solution of caffeine in chloroform by dissolving approximately 1 g, accurately weighed, in chloroform in a 25-ml volumetric flask. Make up to volume and mix. This stock solution contains approximately 40 mg/ml. From it, using a 5-ml narrow bore burette and volumetric flasks, make the following dilutions with chloroform:

Stock solution, ml	Final volume, ml	Approximate concentration, mg/ml	Solution identity
5	10	20	A
4	10	16	B
2.5	10	10	C
1	10	4	D

Record the spectra of solutions A–D in 0.5 mm cells, with chloroform in the reference cell. Calculate the absorbance of the band at 10.26 μ for each, using the base line technique and record a plot of absorbance vs. concentration. If the Beer-Lambert law is obeyed over this concentration range, a straight line will be obtained.

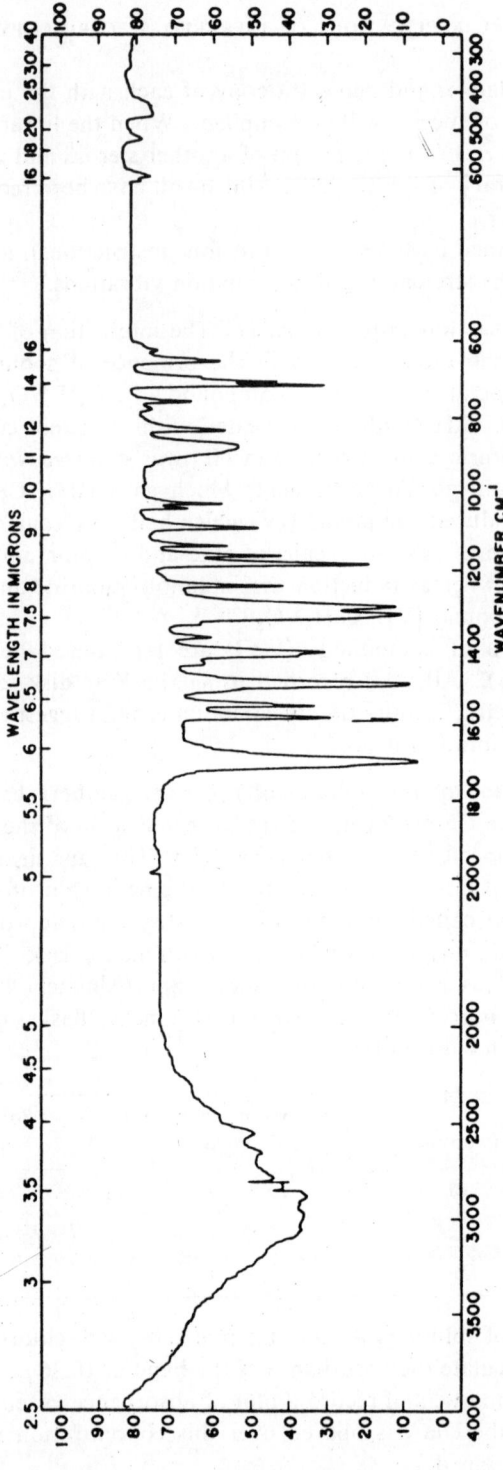

FIGURE 2.35: Infrared spectrum of compound (I) (Exercise 2.8) (KBr disk).

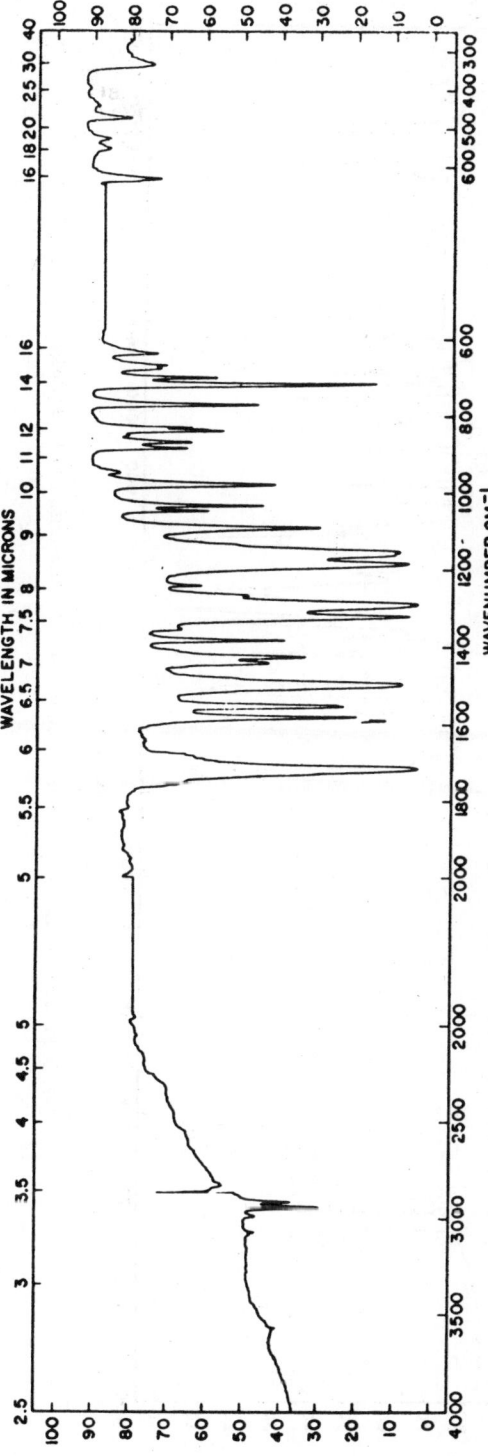

FIGURE 2.36: Infrared spectrum of compound (II) (Exercise 2.8) (KBr disk).

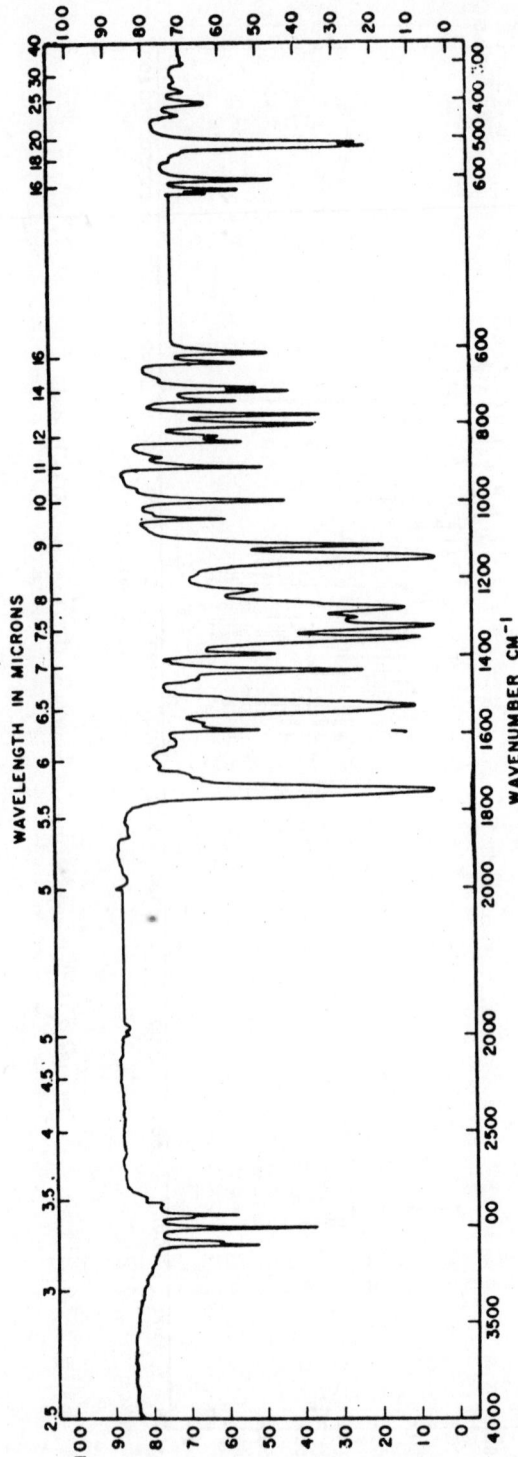

FIGURE 2.37: Infrared spectrum of compound (III) (Exercise 2.8) (KBr disk).

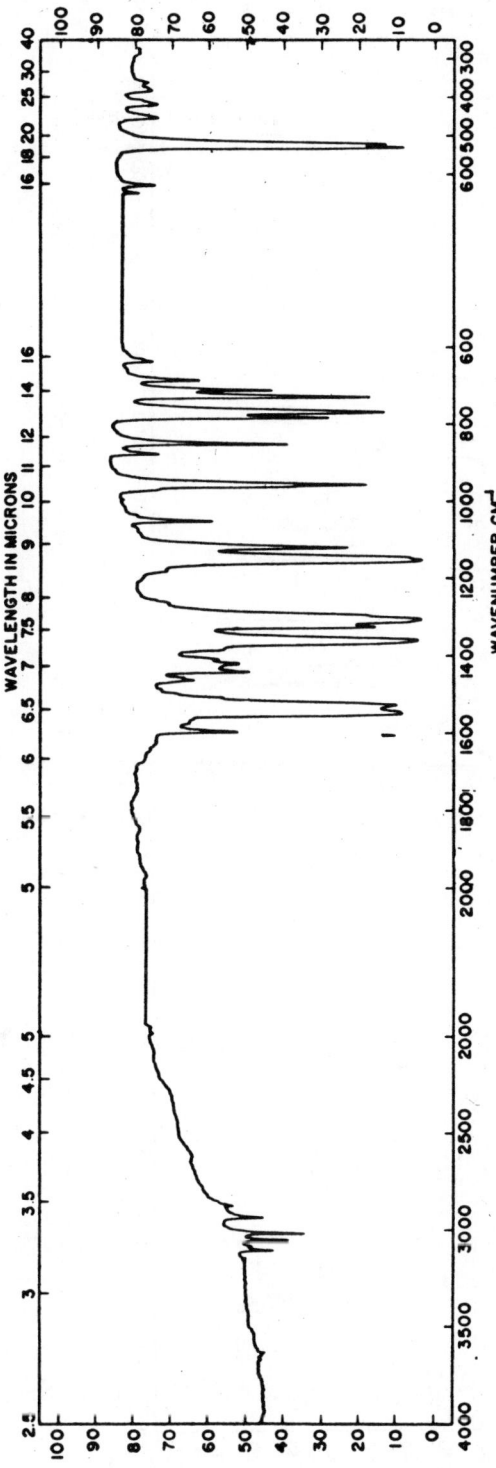

FIGURE 2.38: Infrared spectrum of compound (IV) (Exercise 2.8) (KBr disk).

119

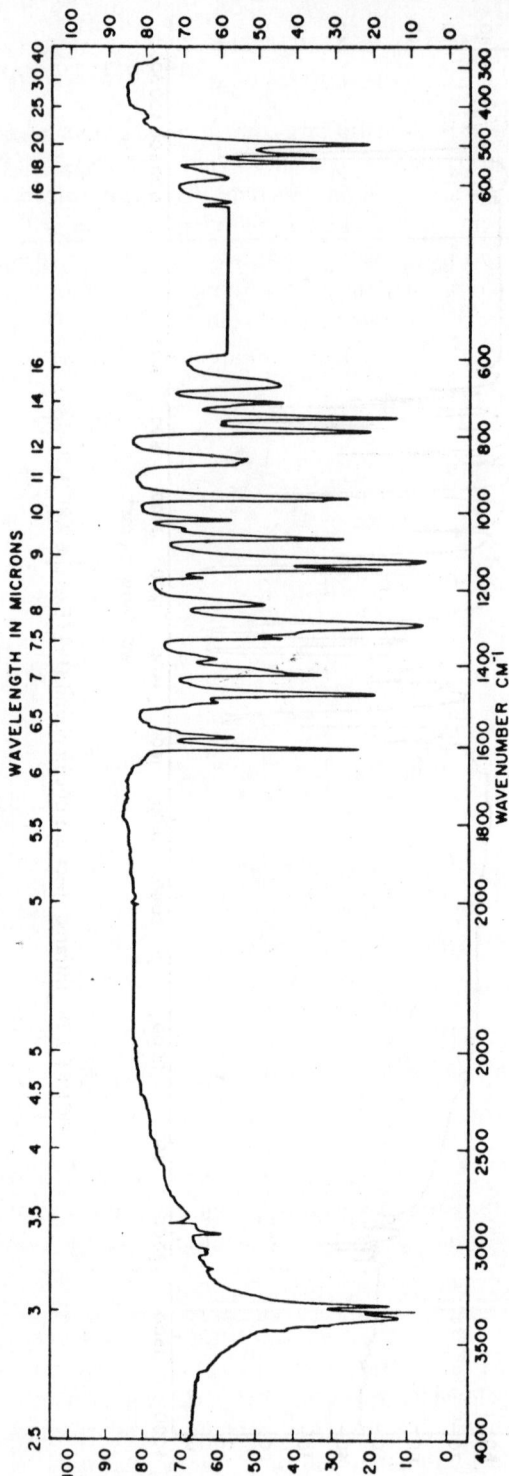

FIGURE 2.39: Infrared spectrum of compound (V) (Exercise 2.8) (KBr disk).

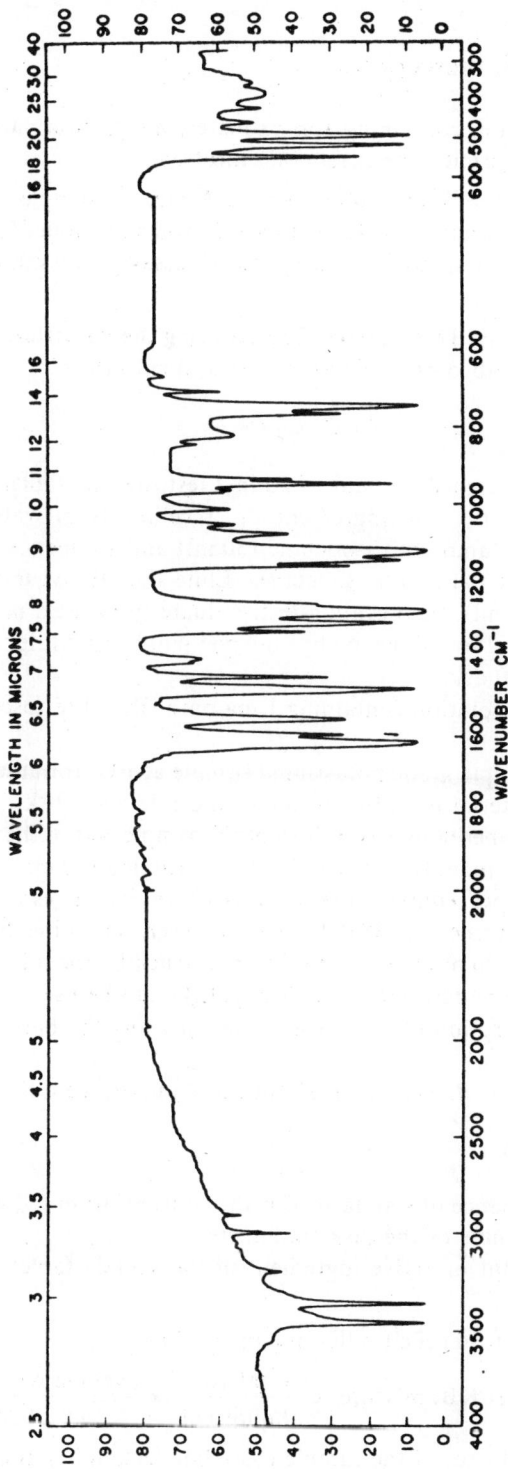

FIGURE 2.40: Infrared spectrum of compound (VI) (Exercise 2.8) (KBr disk).

A chloroform solution containing approximately 8 mg/ml of caffeine will then be supplied. Determine the exact concentration.

E2.10. **Official Infrared Spectrophotometric Assay.** A sample of one of the compounds or preparations listed in Table 2.3 will be supplied, as well as the appropriate reference standard. Carry out the assay process described in the USP.

E2.11. **Assay of Ethinyl Testosterone Tablets using the Potassium Bromide Disk Technique.** The absorption band at 6.03 μ, due to the

$$-\overset{|}{C}=\overset{|}{C}-\overset{|}{C}=O$$

chromophore, was used.[22]

Chromatograph a weighed amount of ethinyl testosterone tablet mixture, equivalent to 25 mg of active ingredient on a celite (10 g)–water (5 ml) column. Wash the column with isooctane (50 ml) and by means of gentle air pressure, blow out the excess isooctane. Elute the ethinyl testosterone with chloroform (150 ml), then evaporate the eluate to dryness in vacuo at less than 50°. Dissolve the residue in absolute methanol and dilute to volume in a 25-ml volumetric flask.

Prepare a reference solution containing 1 mg pure ethinyl testosterone per milliliter.

Dry some spectroscopic grade potassium bromide at 105° for at least 16 hr and use the dried material to prepare sample and reference disks. Mix aliquots (0.10 ml) of the methanolic solutions of the sample and of the reference standard with 200 mg quantities of KBr by hand-grinding (10 min) or vibrator-grinding (6 min) procedures. Press the mixtures at an evacuation of < 1 mm Hg, and at a force of 20,000 lb for 1 min, then determine the average thickness of each disk, in millimeters, in the area struck by the infrared light. Clear, uniform disks are desirable, but cloudy disks can be used.

Determine the absorption of each disk at 6.03 μ using the base line technique.

Calculate the *absorptivity coefficient* K_s for the standard from the equation

$$K_s = \frac{A_s}{CL}$$

where A_s is the absorbance of standard, C is the concentration (% w/w), and L is the average thickness of the disk (millimeters).

Calculate the amount of active ingredient in the sample tablet using the following equation:

Amount of ethinyl testosterone (milligrams) per tablet

$$= \frac{A_u}{K_s \times L_u} \times \text{wt. (mg) KBr mixture} \times \frac{\text{total vol.}}{\text{aliquot vol.}} \times \frac{\text{average wt. per tablet}}{\text{wt. of sample}}$$

where A_u is the absorbance of the sample (base line technique) and L_u is the average thickness of sample disk (millimeters).

E2.12. Quantitative Determination of Phenacetin by Infrared Spectro-photometry Using Benzoic Acid as Reference Standard. In an assay of this type, the absorptivity ratio (pure phenacetin vs. pure benzoic acid) must first be determined. This ratio is constant for a particular infrared spectrophotom-eter over a narrow concentration range, and once it has been determined, the assay of phenacetin no longer requires a phenacetin reference standard. Henceforth, pure benzoic acid can be used as a secondary reference standard.[28]

Determination of Absorptivity Ratio. Prepare a solution of pure benzoic acid by dissolving approximately 1 g, accurately weighed, in chloroform, in a 100-ml volumetric flask. Make up to volume with chloroform and mix. Record the infrared spectrum, using solution cells of path length 0.1 mm, with chloroform in the reference cell. Determine the absorbance of the band near 5.91 μ, and calculate the absorptivity a_b from the expression

$$a_b = \frac{A_b}{C_b \times L}$$

in which a is the absorptivity, A is the absorbance, C is the concentration (grams per liter), L is the absorption path length (centimeters), and subscript b indicates benzoic acid.

Prepare a similar solution of pure phenacetin, using approximately 1 g, accurately weighed, and chloroform to 100 ml, and determine its absorbance at the maximum near 6.61 μ. Calculate the absorptivity from the expression

$$a_p = \frac{A_p}{C_p \times L}$$

in which the symbols are the same as before and the subscript p indicates phenacetin.

Determine the absorptivity ratio, a_p/a_b, which is reduced to

$$a_p/a_b = \frac{A_p \times C_b}{C_p \times A_b}$$

if the benzoic acid and the phenacetin solutions are both examined in the same cell.

Assay of a Sample of Phenacetin, Using Benzoic Acid as Secondary Standard. Prepare a new chloroform solution of pure benzoic acid containing an ac-curately known quantity of benzoic acid (0.8–1.0% w/v). Prepare a chloro-form solution of the phenacetin of unknown purity, containing an accurately known quantity of the drug (0.8–1.0% w/v).

Determine the absorbance of the former solution at the maximum near 5.91 μ, and the absorbance of the latter solution at the maximum near 6.61 μ, and calculate the concentration of $C_{10}H_{13}NO_2$ (pure phenacetin) in the latter solution using the following expression:

$$\text{conc. (g/liter)}C_{10}H_{13}NO_2 = \frac{A_p \times C_b}{A_b} \times \frac{1}{a_p/a_b}$$

From this result, determine the per cent $C_{10}H_{13}NO_2$ in the sample of phenacetin.

E2.13. Quantitative Determination of 5-Chloro-7-iodo-8-hydroxyquinoline, and Qualitative Determination of Common Impurities (5,7-diiodo-8-hxdroxyquinoline; and 5,7-dichloro-8-hydroxyquinoline) in a Commercial Sample of Iodochlorhydroxyquin USP by Infrared Analysis.

C₉H₅ClINO   5-chloro-7-iodo-8-hydroxyquinoline

The USP method for the determination of iodochlorhydroxyquin is based on the compound's halogen content. Its disadvantage is that the method does not distinguish between the pure compound and its intermediates, which may be present as impurities in the synthesized product. The result is that samples which contain less than 60% of 5-chloro-7-iodo-8-hydroxyquinoline can pass the USP tests and could be considered as being of USP quality.[25] A more specific assay procedure is obviously necessary. The reference describes a suitable one, based on the fact that the intensity of the absorption band at 14.4 μ was proportional to the amount of 5-chloro-7-iodo-8-hydroxyquinoline in the sample. A plot of the ratio $\log_{10}(T_{14.9\mu}/T_{14.4\mu})$ vs. concentration was linear over the range investigated (2–10 mg iodochlorhydroxyquin/ml CS_2).

The spectrum of 5-chloro-7-iodo-8-hydroxyquinoline does not have absorption bands at 10.95 μ, 12.15 μ, or 13.45 μ. These bands, if present, are due to 5,7-diiodo-8-hydroxyquinoline (10.95 μ), 5-chloro-8-hydroxyquinoline (12.15 μ), and 5,7-dichloro-8-hydroxyquinoline (13.45 μ) and can be used for quantitative determinations of these impurities.

Method. Dissolve about 50 mg accurately weighed, of iodochlorhydroxyquin USP in carbon disulfide in a 10-ml volumetric flask, add solvent to volume, and mix. Record the complete spectrum vs. a carbon disulfide blank using 3-mm cells. Determine the ratio

$$\log \frac{\%T_{14.9\mu}}{\%T_{14.4\mu}}$$

This is the absorbance A_u of the sample.

In the same way, prepare a solution of pure 5-chloro-7-iodo-8-hydroxy-quinoline (approximately 50 mg, accurately weighed) in carbon disulfide (10 ml) and record its spectrum, using 3-mm cells. Determine the ratio

$$\log \frac{\%T_{14.9\mu}}{\%T_{14.4\mu}}$$

This is the absorbance A_s of the reference standard.

Calculate the quantity in mg of C_9H_5ClINO in the sample of iodochlor-hydroxyquin by means of the formula $10C\,(A_u/A_s)$, where C is the concentration in milligrams per milliliter of 5-chloro-7-iodo-8-hydroxyquinoline in the standard solution. From this, determine the per cent w/w C_9H_5ClINO in the sample of iodochlorhydroxyquin USP.

Examine the spectrum of iodochlorhydroxyquin and determine qualitatively which, if any, of the common impurities are present in the sample.

REFERENCES

1. L. J. Bellamy, *Infrared Spectra of Complex Molecules*, Methuen, London, 2nd ed., 1958.
2. R. N. Jones and C. Sandorfy in *Chemical Applications of Spectroscopy* (W. West, ed.), Wiley (Interscience), New York, 1956, Chap. IV.
3. A. D. Cross, *Introduction to Practical Infrared Spectrophotometry*, Butterworth, London, 1960.
4. K. Nakanishi, *Infrared Absorption Spectroscopy—Practical*, Holden-Day, San Francisco, 1962.
5. J. R. Dyer, *Applications of Absorption Spectroscopy of Organic Compounds*, Prentice-Hall, Englewood Cliffs, N.J., 1965.
6. N. B. Colthup, L. H. Daly, and S. E. Wiberley, *Introduction to Infrared and Raman Spectroscopy*, Academic Press, New York, 1964.
7. D. W. Mathieson, *Interpretation of Organic Spectra*, Academic Press, New York, 1965.
8. R. M. Silverstein and G. C. Bassler, *Spectrometric Identification of Organic Compounds*, Wiley, New York, 1964.
9. W. C. Price, *J. Pharm. Pharmacol.*, **7**, 153 (1955).
10. J. Carol, *J. Pharm. Sci.*, **50**, 451 (1961).
11. J. G. Theivagt, V. E. Papendick, and D. C. Wimer, *Anal. Chem.*, **39**, 191R (1967).
12. R. J. Warren, I. B. Eisdorfer, W. E. Thompson, and J. E. Zarembo, *J. Pharm. Sci.*, **54**, 1806 (1965).
13. W. E. Thompson, R. J. Warren, I. B. Eisdorfer, and J. E. Zarembo, *J. Pharm. Sci.*, **54**, 1819 (1965).
14. I. B. Eisdorfer, R. J. Warren, W. E. Thompson, and J. E. Zarembo, *J. Pharm. Sci.*, **55**, 734 (1966).
15. N. B. Colthup, *J. Opt. Soc. Am.*, **40**, 397 (1950).
16. R. N. Jones, *Infrared Spectra of Organic Compounds: Summary Charts of Principal Group Frequencies*, N.R.C. Bull. 6, Ottawa, 1959.
17. T. S. Moss, *Advan. Spectry.*, **1**, 175 (1959).
18. W. N. French and J. C. Morrison, *J. Pharm. Sci.*, **54**, 1133 (1965).
19. A. W. Baker, *J. Phys. Chem.*, **61**, 450 (1967).
20. R. B. Barnes, R. C. Gore, E. F. Williams, S. G. Lensley, and E. M. Petersen, *Anal. Chem.*, **19**, 619 (1947).

21. J. B. Jensen, *Dansk Tidsskr. Farm.*, **32**, 205 (1958).
22. A. L. Hayden and O. R. Sammul, *J. Am. Pharm. Assoc. Sci. Ed.*, **49**, 497 (1960).
23. R. J. Warren, I. B. Eisdorfer, W. E. Thompson, and J. E. Zarembo, *J. Pharm. Sci.*, **55**, 144 (1966).
24. W. Beckering, C. M. Frost, and W. W. Fowkes, *Anal. Chem.*, **36**, 2412 (1964).
25. T. Urbanyi, D. Sloniewsky, and F. Tishler, *J. Pharm. Sci.*, **56**, 730 (1966).
26. N. Wright, *Ind. Eng. Chem.*, **13**, 1 (1941).
27. A. L. Hayden and O. R. Sammul, *J. Pharm. Sci.*, **49**, 489 (1960).
28. J. P. Comer and A. M. Ribley, *J. Pharm. Sci.*, **52**, 358 (1963).
29. A. L. Hayden, *J. Pharm. Sci.*, **54**, 151 (1965).
30. R. T. Coutts, H. W. Peel, and E. M. Smith, *Can. J. Chem.*, **43**, 3221 (1965).
31. K. B. Shaw, unpublished data, 1968.
32. W. H. Washburn and E. O. Krueger, *J. Am. Pharm. Assoc. Sci. Ed.*, **38**, 623 (1949).

CHAPTER **3**

Raman Spectroscopy

E. A. Robinson and D. S. Lavery

LASH MILLER CHEMICAL LABORATORIES
AND ERINDALE COLLEGE, UNIVERSITY
OF TORONTO, TORONTO, CANADA

3.1 INTRODUCTION

The Raman effect was discovered in 1928 by Sir C. V. Raman[1-3] and complements the method of infrared spectroscopy in the study of the vibrational spectra of molecules. However, since it has a quite different physical basis, being a light-scattering effect rather than one involving the absorption of radiation, it often gives information additional to that obtained from infrared spectroscopy. For example, vibrational frequencies that are inactive in the infrared may often be observed in the Raman spectrum and vice versa.

When a beam of light of suitable frequency, chosen so that it is not absorbed, passes through a translucent medium, free from dust, most of the light is transmitted without change. However at the same time a small fraction is scattered by the molecules of the medium. The total intensity of scattered light is about 10^{-5} of the intensity of the incident light for liquids.

This classical or Rayleigh scattering effect has been known for many years,[4] and its inverse proportionality to the fourth power of the wavelength is the basis for the explanation of the blue color of the sky.[5] In Rayleigh scattering the frequency of the light remains unchanged, i.e., for a monochromatic incident beam the frequency of the scattered light is exactly equal to that of the incident beam.

Raman's discovery was that the scattered light also contains new frequencies (Raman frequencies) to both higher and lower frequencies of the frequency of the Rayleigh scattered light. These Raman lines are very weak (of the order of one-thousandth of the intensity of the Rayleigh scattering) and are therefore relatively difficult to detect. The differences of their frequencies from the frequency of the incident light are characteristic of the molecules of the scattering medium and are found to be equal to vibrational or rotational frequencies of the molecules. Vibrational shifts are readily resolved from the Rayleigh by means of a relatively simple prism spectrograph, while the pure rotational transitions lie so close to the Rayleigh line that their resolution is impracticable in most cases unless a grating spectrograph of very high resolution is used. Thus in normal practice only the vibrational Raman lines are commonly observed, and most of the work of practical importance in Raman spectroscopy is concerned with pure vibrational spectra.

Raman spectra can be observed when light is scattered by solids, liquids, or gases, and it is interesting to note that the method was more widely used in early investigations of the vibrational spectra of molecules than was infrared spectroscopy because it was at that time experimentally more accessible than infrared. In recent years, however, with the introduction of commercial infrared spectrometers, infrared spectroscopy has become widely practiced and Raman spectroscopy has come to be regarded by many as a rather specialized technique. Now, with the development of very high intensity light sources, including the laser, and the use of the photomultiplier tube for the detection of weak intensities, the way is open for rapid advances in technique, and Raman spectroscopy should soon be available as a routine tool.

3.2 THEORY

A. SCATTERING OF LIGHT BY MOLECULES

Only a simple presentation of the basic features of the theory of Raman spectroscopy is given here. More detailed and mathematical accounts are to be found elsewhere (see, for example, Refs. 6–9).

The simplest model that enables us to understand the scattering of light by molecules is to visualize the collision of a photon of incident light of frequency ν_0 (energy $h\nu_0$) with a molecule in the medium. Such a collision can occur *elastically* or *inelastically*. In an elastic collision the photon is scattered without change and is emitted with energy $h\nu_0$ (Rayleigh scattering), while in an inelastic collision (Raman scattering) the internal energy of the molecule changes by transfer of energy with the photon. Thus, the emitted photon has an energy $h\nu'$ different from that of the incident photon. In most instances the change in the internal energy results from a change in the vibrational energy of the molecule. If the transition is from the ground vibrational state ($v = 0$) to the first excited vibrational state ($v = 1$), the change in energy is $h\nu_1$, where ν_1 is a vibrational frequency. Thus,

$$h\nu' = h\nu_0 - h\nu_1$$

and the scattered light shows a frequency shift, relative to the Rayleigh scattering, toward the red end of the spectrum equal in magnitude to a vibrational frequency. If, however, a molecule that is already in its first vibrational state ($v = 1$) collides with a photon, it may transfer energy to the incident photon and fall back into the ground state ($v = 0$). In this event the emitted photon has energy $h\nu''$ where

$$h\nu'' = h\nu_0 + h\nu_1$$

and its frequency is shifted relative to that of the Rayleigh line toward the blue end of the spectrum with the same magnitude of frequency as before.

Thus, Raman frequencies are observed to both higher and lower frequencies, $\pm\nu_1$, of the incident frequency. The latter frequencies are called "Stokes lines" and the former, "anti-Stokes lines." Since at normal temperatures the number of molecules in the ground vibrational state far outnumbers those in the first vibrational state, the intensity of Raman frequencies emitted with frequencies $\nu_0 - \nu_1$ is far greater than those with frequencies $\nu_0 + \nu_1$, and Stokes lines are relatively much more intense than anti-Stokes lines. In fact, the anti-Stokes lines are observed only in favorable cases. A molecular substance that gives a Raman spectrum where both Stokes and anti-Stokes lines are observed using a simple spectrometer is liquid carbon tetrachloride (Fig. 3.1).

For a polyatomic molecule several Raman lines may be observed, the shift of each from the exciting frequency being equal to the frequency of a fundamental mode of vibration or, more rarely, a combination of fundamental vibrational modes of the molecular species. For a polyatomic molecule containing n atoms, the maximum number of fundamental frequencies is $3n - 6$ ($3n - 5$ for a linear molecule). The number of vibrational frequencies observed in the Raman spectrum depends on the symmetry of the molecule.

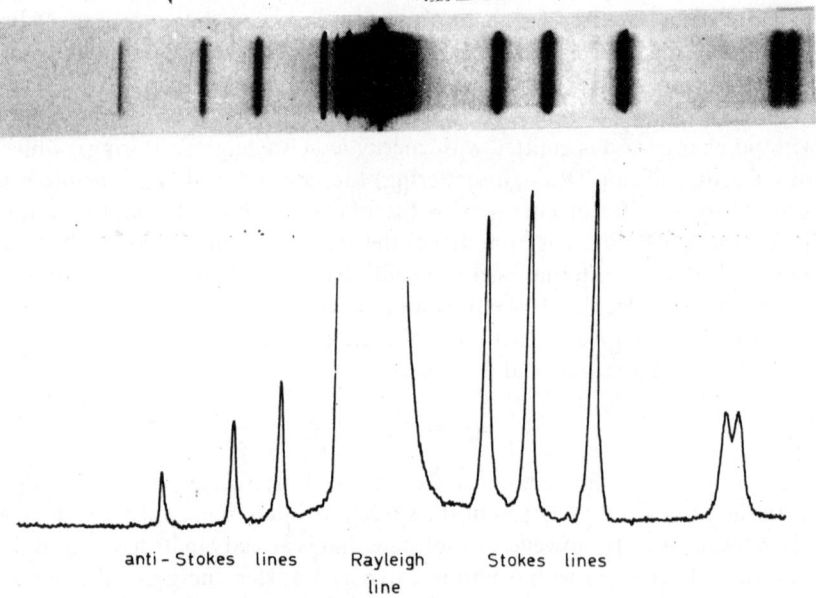

anti-Stokes lines   Rayleigh   Stokes lines
                             line

FIGURE 3.1: Photographically recorded spectrum of carbon tetrachloride showing Stokes
and anti-Stokes lines.

B. VIBRATIONAL RAMAN SPECTRA

In infrared spectroscopy a vibrational frequency is active only if the
molecular vibration involves a change in the permanent dipole moment of the
molecule. Thus, for a homonuclear molecule such as N_2 which has a zero
dipole moment its single vibrational mode, stretching of the $N{\equiv}N$ bond, is
inactive in the infrared. However, the $N{\equiv}N$ stretching frequency is observed
in the Raman spectrum at 2331 cm^{-1}. This comes about because a vibration
is Raman active if it involves a *change in the polarizability* of the electrons in
the molecule, i.e., a change in the induced dipole moment rather than a
change in the permanent dipole moment.

C. POLARIZABILITY

A molecule may be regarded as a collection of positively charged nuclei
embedded in a cloud of negative electrons. In the equilibrium configuration
the electrical centers of the positive and negative charges may coincide, in
which case the molecule has no dipole moment, or may *not* coincide, in which
case the molecule possesses a permanent dipole moment. In both instances
interaction with an electric field, for example with incident light, gives rise to
an *induced dipole moment*, and if the field oscillates with a frequency that is
large compared with nuclear vibrational frequencies, only the electron cloud
(of relatively small inertia) will be distorted by the field and the nuclei will
remain relatively unaffected.

The relation between the applied field strength E and the induced dipole moment M is given by

$$M = \alpha E$$

where α is the *polarizability* of the molecule. In this expression both E and M are vectors and α is a tensor. The nature of α may be visualized by resolving the induced dipole moment into three components M_x, M_y, and M_z in the x, y, and z directions of a coordinate system fixed in space and in the molecule, letting the corresponding components of the field be E_x, E_y, and E_z, i.e.,

$$M_x = \alpha_{xx}E_x + \alpha_{xy}E_y + \alpha_{xz}E_z$$
$$M_y = \alpha_{xy}E_x + \alpha_{yy}E_y + \alpha_{yz}E_z$$
$$M_z = \alpha_{xz}E_x + \alpha_{yz}E_y + \alpha_{zz}E_z$$

The tensor α is then defined by the set of six coefficients α_{xx}, α_{xy}, α_{xz}, α_{yy}, α_{yz}, and α_{zz} which may be used to represent α in a more pictorial way by forming the equation of an ellipsoid:

$$\alpha_{xx}x^2 + \alpha_{yy}y^2 + \alpha_{zz}z^2 + 2\alpha_{xy}xy + 2\alpha_{yz}yz + 2\alpha_{xz}xz = 1$$

This is the so-called polarizability ellipsoid, which can be visualized by drawing arrows from a common origin with lengths proportional to the polarizability in that particular direction. The heads of the arrows define the ellipsoid. If the ellipsoid is oriented for convenience with its principal axes along the x, y, and z axes of the coordinate system, the equation just presented reduces to the form:

$$\alpha_{xx}x^2 + \alpha_{yy}y^2 + \alpha_{zz}z^2 = 1$$

Figure 3.2 illustrates the polarizability ellipsoids of some simple molecules in their equilibrium configurations.[10]

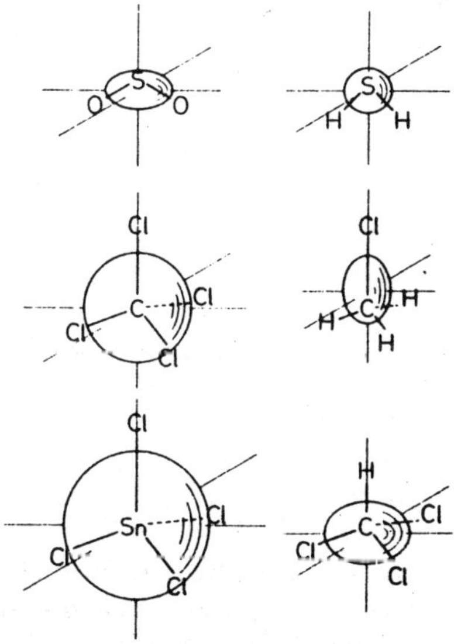

FIGURE 3.2: Polarizability ellipsoids of some simple molecules.

The molecular polarizability α_0 is a measure of the ease with which the electrons in the molecule (the cloud of electrons) are distorted. Thus a vibration that results in a distortion of the electron cloud, e.g., a symmetrical stretching frequency (Fig. 3.3), is Raman active, while, although it may be,

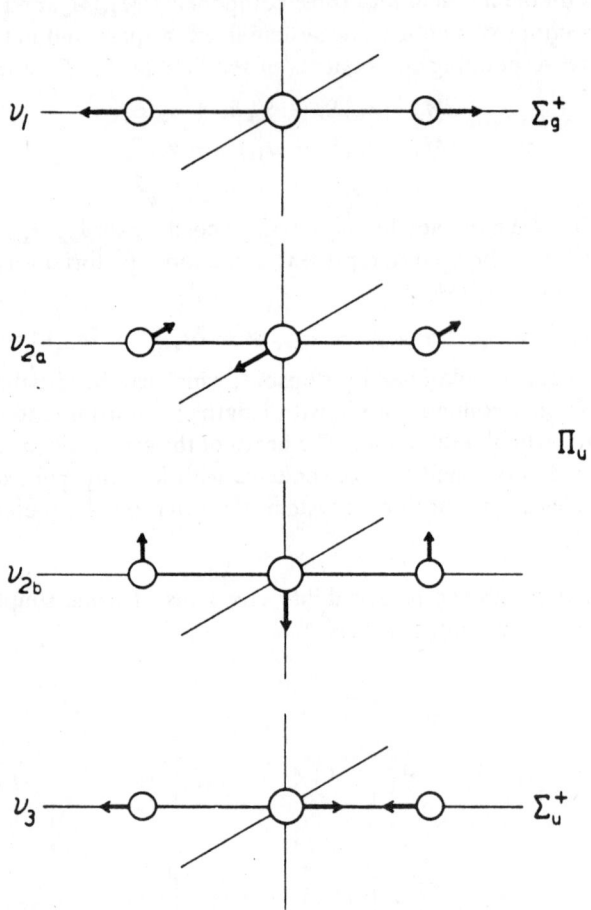

FIGURE 3.3: Normal modes of vibration for CO_2.

it is not necessarily infrared active. Indeed it will not be active in the infrared unless the permanent dipole moment of the molecule changes during the molecular vibration.

By way of illustration we consider the fundamental vibrational modes of two simple triatomic molecules, carbon dioxide and sulfur dioxide, and those of a tetrahedral molecule, carbon tetrachloride.

D.   RAMAN SPECTRA OF SOME SIMPLE MOLECULES

The linear CO_2 molecule has $3n - 5 = 4$ fundamental vibrations (Fig. 3.3), which are conveniently described as a symmetrical stretch ν_1, an antisymmetrical (or antisymmetric) stretch ν_2, and two symmetrical bends ν_3. The two bends have identical frequencies and differ only in their dispositions in space, i.e., experimentally a single bending mode is observed which is doubly degenerate.* Figure 3.4 shows the way in which the polarizability ellipsoid changes during the normal modes of vibration of the CO_2 molecule.

During the symmetric stretching vibration ν_1 the polarizability changes and hence ν_1 is Raman active. However it is inactive in the infrared since the initially zero dipole moment remains zero during the vibration of the molecule. For the asymmetric stretch ν_2 the motion results in a loss of symmetry in the molecule, the dipole moment becomes nonzero, but the polarizability ellipsoid remains unchanged. Hence ν_2 is active in the infrared, but inactive in the Raman. Similarly, for the bending mode ν_3 the vibration is seen to be infrared active, but forbidden in the Raman. These results are perhaps not immediately obvious, but the following more detailed treatment should make the conditions under which the polarizability changes clearer.

In general for a small displacement of the nuclei in the molecule during vibration the distortion can be described by a *normal vibrational coordinate q* constructed so as to express all of the individual displacements of the nuclei involved. For small amplitudes of vibration the polarizability may be described by

$$\alpha = \alpha_0 + (\partial\alpha/\partial q)_0 q$$

where the zero subscript refers to values at the equilibrium configuration.

Thus for the electric moment induced by an applied field

$$M = \alpha E = \alpha_0 E + (\partial\alpha/\partial q)_0 q E$$

For a change in polarizability to occur during a particular mode of vibration the second term in this equation $(\partial\alpha/\partial q) \times q$ must be nonzero, i.e., since q is finite $(\partial\alpha/\partial q)_0$ must be nonzero. Like α_0, $(\partial\alpha/\partial q)_0$ is a tensor and is usually referred to as the "derived polarizability tensor" α_0' of the normal vibration defined by q.

* When two or more molecular vibrations have the same frequency, they are said to be "degenerate." This may arise because of the symmetry of the molecule, in which case the degeneracy will be either twofold (doubly degenerate) or threefold (triply degenerate). Any molecule possessing a threefold, or higher, axis has some doubly degenerate vibrations, and a molecule with more than one threefold, or higher, axis will also have some triply degenerate vibrations.

Degeneracy may also occur accidentally if the frequencies of two or more different modes of vibration happen to coincide. In practice accidental degeneracies are quite rare. They often lead to Fermi resonance.

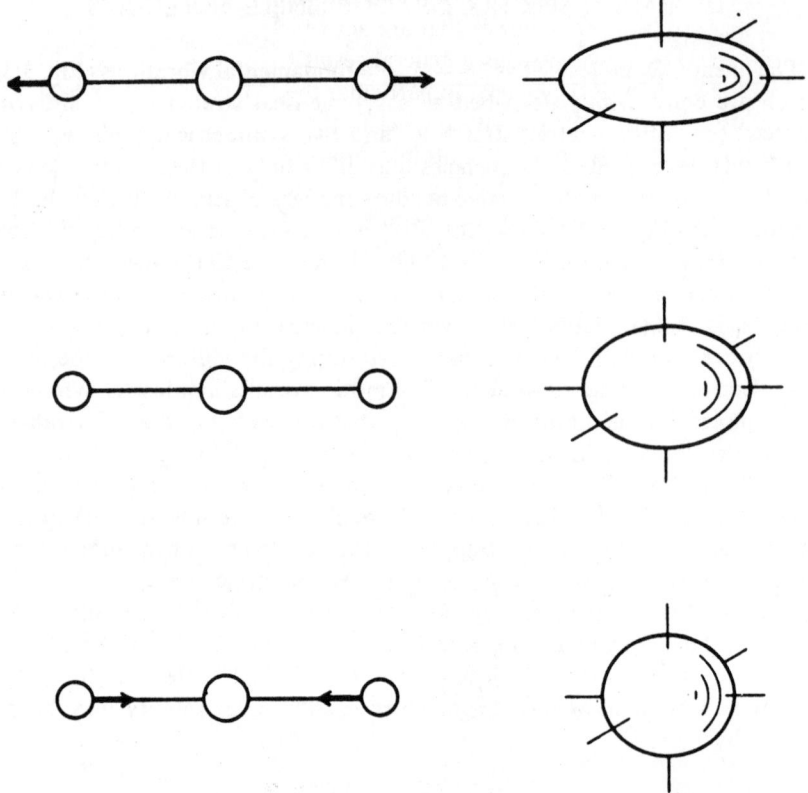

FIGURE 3.4: Change in the polarizability ellipsoid during the symmetrical stretching frequency (v_1) of CO_2: since the other modes v_2 and v_3 (Fig. 3.3) cause no change in polarizability they are not shown.

Deformation of a molecule during a normal vibration may affect the polarizability ellipsoid in several different ways. It may cause an alteration in the size and/or shape of the ellipsoid and/or it may lead to a change in the orientation of the polarizability ellipsoid in space; i.e., either distortion of the ellipsoid or a change in its orientation in space leads to a Raman active vibrational mode.

In the case of the asymmetric stretching mode v_2 of carbon dioxide, and the bending mode v_3, the vibrations do not change the axes of the polarizability ellipsoid away from the fixed x, y, and z axes. Moreover since the polarizability will have, by symmetry, identical values α' at the two extremes of vibration

$$\alpha' = \alpha_0 + (\partial\alpha/\partial q)_0(+q) = \alpha_0 + (\partial\alpha/\partial q)_0(-q)$$

and it follows that $(\partial\alpha/\partial q)_0$ must be zero; hence $\alpha' = \alpha_0$, i.e., the polarizability is unchanged during the vibration.

Carbon dioxide is an example of a molecule with a center of symmetry.

The result that the infrared spectrum and the Raman spectrum are mutually exclusive, i.e., those frequencies that are active in the Raman are inactive in the infrared, and vice versa, is a general result that applies to all molecules possessing a *center of symmetry*. In particular the frequencies that are Raman active are totally symmetric with respect to the center of symmetry.

For sulfur dioxide, a bent triatomic molecule, the $3n - 6 = 3$ normal modes of vibration are shown in Fig. 3.5. They are conveniently described as a symmetric stretch v_1, a symmetric bend v_2, and an asymmetric stretch v_3. In the case of the symmetric stretch, the polarizability ellipsoid breathes with the

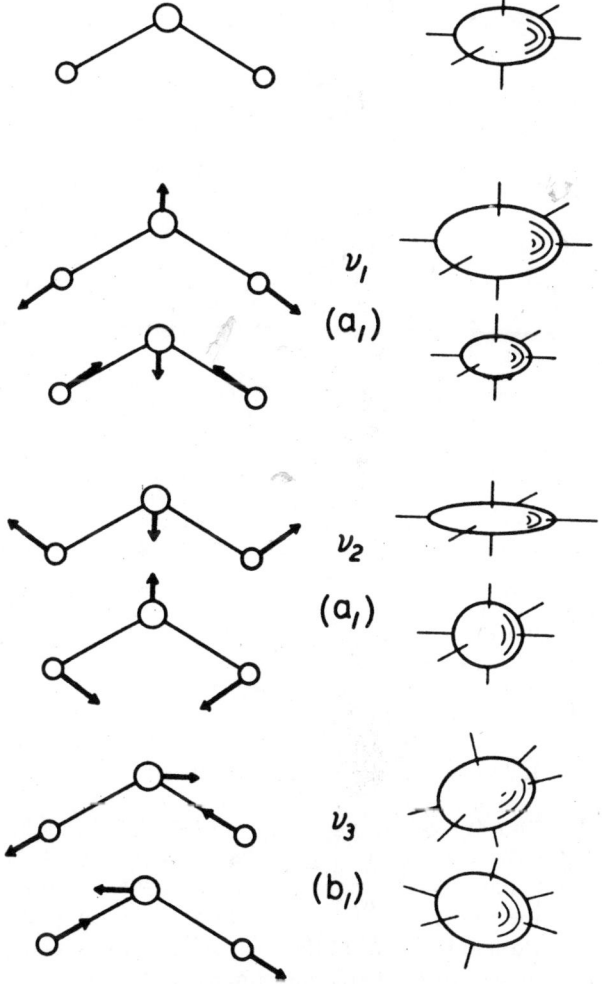

FIGURE 3.5: Normal modes of vibration and polarizability ellipsoids for SO_2; the two extreme displacements are shown for each vibration with the polarizability changes exaggerated.

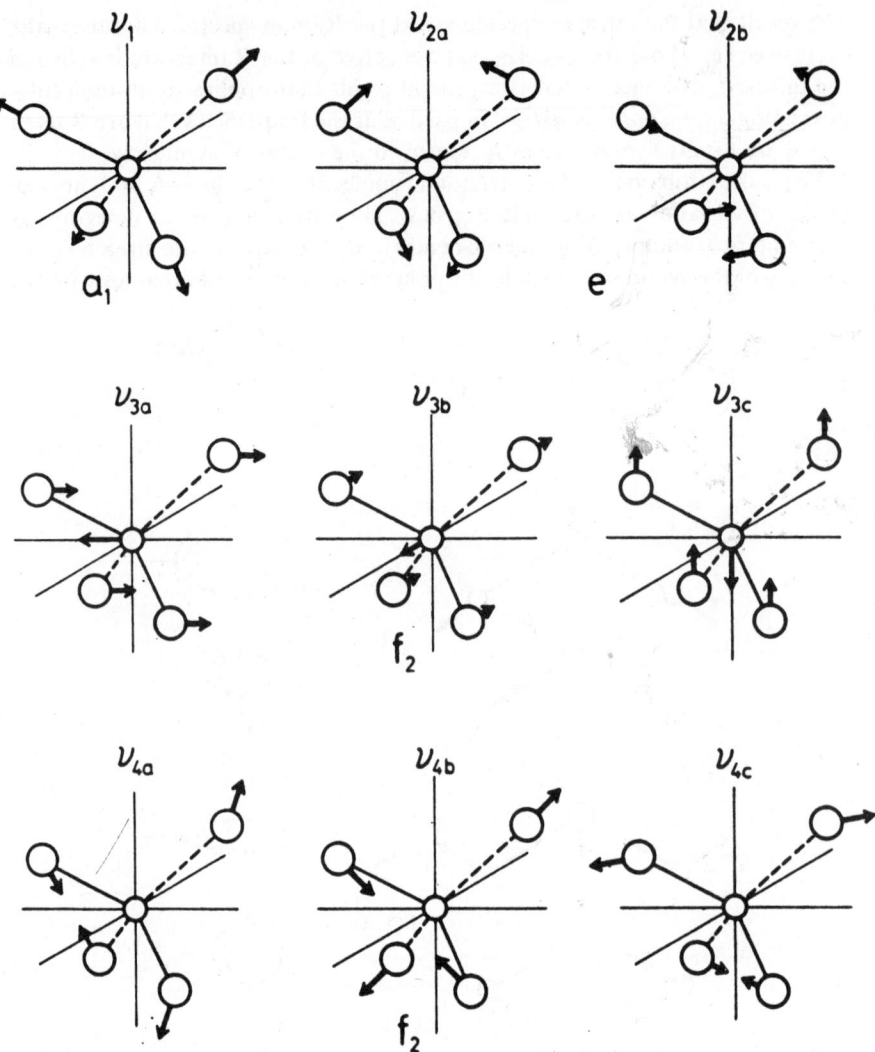

FIGURE 3.6: Normal modes of vibration for CCl_4.

vibrational frequency as does the bend. Therefore both ν_1 and ν_2 are Raman active. For the asymmetric stretching mode ν_3 the size of the polarizability ellipsoid does not change with time, but its axes oscillate with respect to the fixed axes during the vibration. This fundamental vibration mode is therefore also active in the Raman by virtue of the oscillation of the axes of the polarizability ellipsoid relative to the fixed coordinate axes of the molecule. Thus all three normal vibrations are active in the Raman. They are also active in the infrared because in each instance the dipole moment of the molecule changes during the vibration.

A more complicated but common example is that of a tetrahedral molecule such as carbon tetrachloride. For this molecule the $3n - 6 = 9$ fundamental modes give rise to four Raman lines of which two are the only lines observed in the infrared spectrum. The normal vibrational modes are shown in Fig. 3.6.

In carbon tetrachloride, because of the tetrahedral symmetry, one of the Raman-only active vibrations is doubly degenerate (E), and the two frequencies that are active in both the infrared and Raman are each triply degenerate (F_2). In the symmetric stretch ν_1 the polarizability ellipsoid, which is spherical in the equilibrium configuration of the molecule, breathes with the frequency of the vibration so that it remains spherical but changes in size. Hence ν_1 is Raman active, and since the initially zero dipole moment remains zero during the vibration, it is forbidden in the infrared. For the ν_2 bending vibration (doubly degenerate mode), the polarizability ellipsoid is distorted from spherical, but the dipole moment remains zero because the changes in the two halves of the molecule (above and below the xy plane) produce equal and opposite effects (this is most clearly seen in ν_{2a}, Fig. 3.6). Hence ν_2 is Raman active but infrared inactive. In the bending mode ν_4 (triply degenerate), for which there is a threefold axis of symmetry (conveniently chosen as either the x or the y or the z axis) the polarizability ellipsoid becomes distorted and the dipole moment becomes nonzero during the vibration. Thus the vibration is active in both the Raman and infrared. Similar considerations apply to ν_3. The Raman and infrared spectra of CCl_4 are shown in Fig. 3.7.

Thus we have seen that in the vibrational spectrum of a molecule the fundamental vibrational frequencies may be Raman active and infrared inactive if a particular vibration involves a change in the polarizability but no change in the dipole moment; Raman active and infrared active if the vibrational mode causes no change in the polarizability but a change in the permanent dipole moment; active in both Raman and infrared if the vibration causes a change in both the polarizability and dipole moment, and forbidden in both Raman and infrared if neither the polarizability nor the dipole moment changes during the course of the vibration. The total number of observable vibrational frequencies and their activity in the Raman and infrared is determined by the symmetry properties of the particular molecule.

Thus observation of both the Raman and infrared spectra of a molecule provides important information regarding its geometric shape. Indeed in order to obtain a complete vibrational analysis of a molecule, both Raman and infrared spectra are important and usually necessary.

In the analysis of spectra it is convenient to classify the normal vibrations of a molecule into symmetry types by the methods of group theory (see e.g., Ref. 11).

The symmetry type or species of a vibration is indicated by a symbol; those proposed by Mulliken[12] are used almost universally for nonlinear molecules.

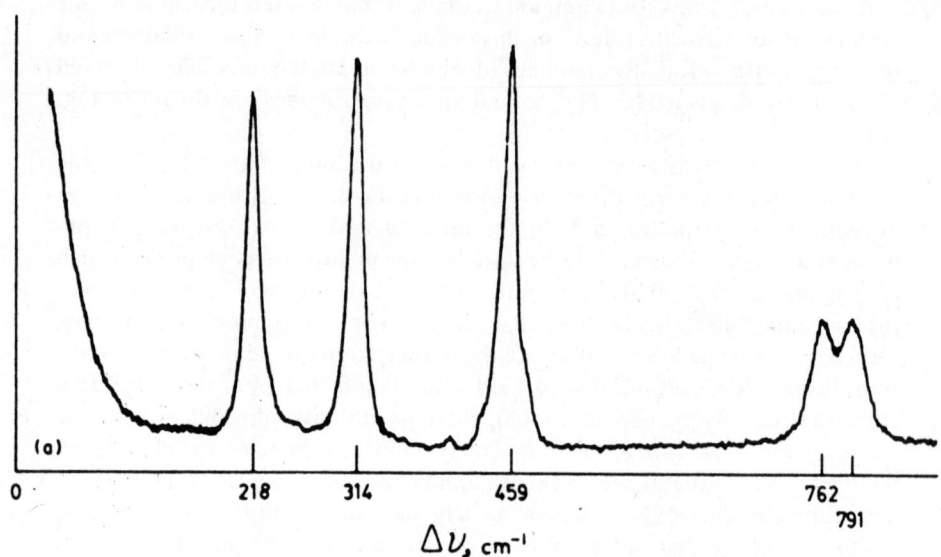

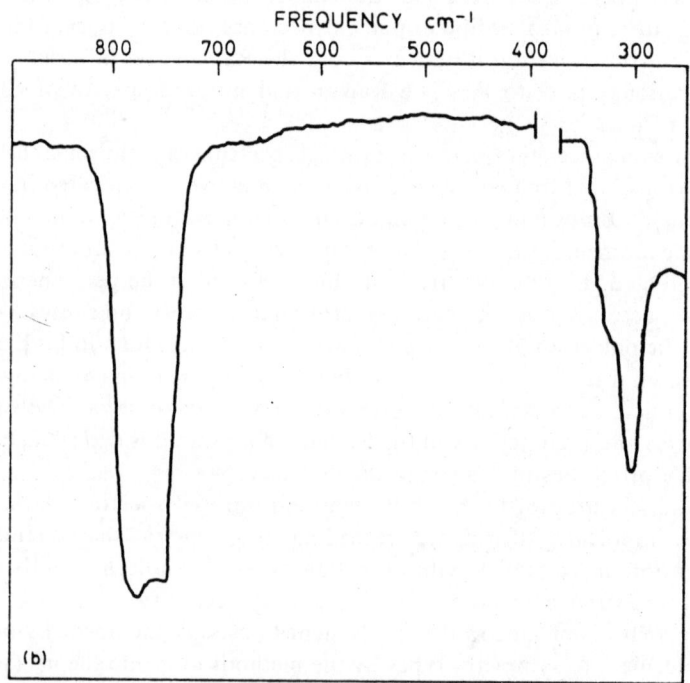

FIGURE 3.7: The Raman and infrared spectra of CCl₄.

For linear molecules a convention established for electronic states is usually followed. Only Mulliken symbols will be discussed since the qualitative organic spectroscopist is rarely concerned with linear molecules. Furthermore no attempt will be made to discuss here the group theoretical arguments used to classify vibrations into species. These are fully covered in several textbooks (e.g., Ref. 11). We attempt only to give the reader a reasonable understanding of such a classification once it has been made.

A Mulliken symbol is one of the letters A, B, E, or F (T is used instead of F by some authors) to which subscripts or superscripts may be added. The symbols A and B are used exclusively for nondegenerate vibrations; E indicates a doubly degenerate vibration and F a triply degenerate vibration. Further distinctions for nondegenerate vibrations are easily understood by using a set of rules after one additional point has been discussed.

As we have discussed previously, molecules in their equilibrium positions may possess certain symmetry elements, but the configurations of the nuclei when displaced in a vibrational mode may not have the same symmetry as in the equilibrium configuration. If a given symmetry element of the equilibrium configuration is also present in the displaced configuration, the element is said to be "preserved," and the vibration is symmetric with respect to that symmetry element. If a symmetry element of the equilibrium configuration is not present in the displaced configuration, the element is not preserved, and the vibration is said to be "asymmetric" with respect to that element. As an example, consider the SO_2 molecule shown in Fig. 3.5. The molecule in its equilibrium configuration has C_{2v} symmetry, i.e., a twofold axis C_2 passing vertically through the sulfur atom, a vertical plane σ_v bisecting the OSO angle and perpendicular to the plane of the molecule and a second vertical plane $\sigma_{v'}$ containing all three atoms. The displaced configuration for each vibration is obtained by moving each atom to the head of the arrow attached to it. (Since the atoms vibrate in phase, any other configuration occurring during the vibration will have the same symmetry.) The modes v_1 and v_2 are clearly symmetric with respect to all three symmetry elements. However the mode v_3 is asymmetric with respect to C_2 and σ_v, but symmetric with respect to $\sigma_{v'}$ since all three atoms remain in this plane throughout the vibration.

The following rules explain the Mulliken symbols for *nondegenerate* vibrations:

1. A designates a vibration that is symmetric with respect to the principal axis (i.e., the axis of highest order). B designates a vibration that is asymmetric with respect to the principal axis.

2. Subscripts 1 and 2 designate vibrations that are, respectively, symmetric and asymmetric with respect to a twofold axis perpendicular to the principal axis, or, if these are absent, with respect to vertical planes.

3. Superscripts ' and " designate vibrations that are, respectively, symmetric and asymmetric with respect to a horizontal plane.

4. Subscripts g and u designate vibrations which are, respectively, symmetric and asymmetric with respect to a center of symmetry.

No more than two of the qualifying symbols in rules 2, 3, and 4 are needed in any one case, i.e., we find Mulliken symbols A, B_2, A_2', B_{1u}, for example, but not symbols of the type A_{1g}''.

The symbols E and F may also carry subscripts and superscripts which follow definite rules. However in these cases the rules can only be stated mathematically. Most experimental spectroscopists do not concern themselves with the origin of the subscripts and superscripts that may accompany E and F, and we shall not do so here.

For linear molecules the symbols Σ^+, Σ^-, and Π are equivalent to A', A'', and E, respectively. The subscripts g and u have the same meaning as in the case of nonlinear molecules.

One of the advantages of symmetry classification is that selection rules need not be worked out for each molecule separately. Once a molecule is classified into the appropriate symmetry class (point group), standard tables for the particular point group can be used.[11] In addition only totally symmetric vibrations (those of species A, A_1, A_1', A_g, A_{1g}, and A_g') will be polarized in the Raman spectrum (Section 3.2E).

Table 3.1 gives the selection rules for some common point groups. The reader should verify that use of this table leads to the same conclusions about Raman and infrared activity of the vibrations shown in Figs. 3.3, 3.5, and 3.6 for CO_2, SO_2, and CCl_4 as do considerations of change in dipole moment and polarizability. Note that a given molecule will not necessarily have vibrations belonging to all the species possible for that point group.

E.   POLARIZATION MEASUREMENTS

An additional feature of Raman spectra is the ability to detect experimentally which of the observed fundamental modes are totally symmetric.

This kind of information is obtained experimentally most easily by irradiating the sample with plane-polarized light and examining the Raman scattering at right angles to the direction of incidence in two cases: where light incident in, say, the z direction is polarized, respectively, parallel to and perpendicular to the xy plane (Fig. 3.8). If the intensity of scattered light from light polarized parallel to the xy plane is $I_{||}$ and that from light polarized perpendicular to the xy plane is $I_\perp$, then the *depolarization factor* ρ is given by

$$\rho = I_\perp / I_{||}$$

Born[13] has shown theoretically that for Rayleigh scattering ρ is given for a random arrangement of molecules in a fluid (liquid or gas) by

$$\rho = 6\gamma^2/(45a^2 + 7\gamma^2)$$

where a, the mean value invariant, is given by

$$a = \tfrac{1}{3}(\alpha_{xx} + \alpha_{yy} + \alpha_{zz})$$

and γ, the anisotropy invariant, is given by

$$\gamma = \tfrac{1}{2}[(\alpha_{xx} - \alpha_{yy})^2 + (\alpha_{yy} - \alpha_{zz})^2 + (\alpha_{zz} - \alpha_{xx})^2]$$

Thus a is, roughly speaking, a measure of the size of the polarizability ellipsoid and γ a measure of the extent of distortion of the shape of the polarizability ellipsoid from spherical.

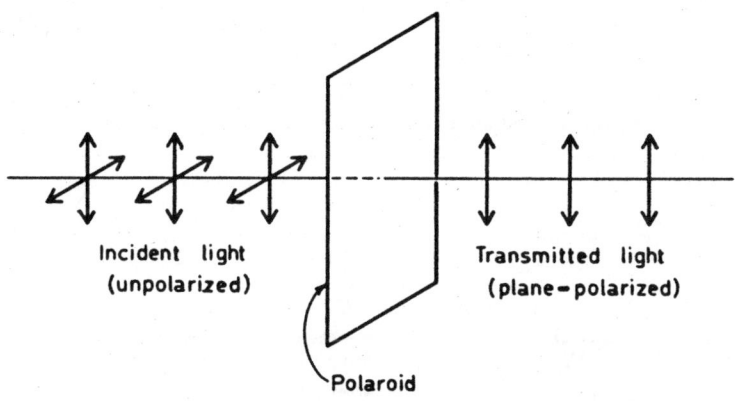

Incident light
(unpolarized)

Transmitted light
(plane-polarized)

Polaroid

FIGURE 3.8: The production of polarized light.

In Raman scattering we are concerned with the derived tensor α_0' rather than with the polarizability α_0, and although it cannot be represented by an ellipsoid, nevertheless it possesses by analogy with α_0 a mean value invariant a' and an anisotropy invariant γ', and the degree of polarization of the Raman line is given by

$$\rho = 6(\gamma')^2/[45(a')^2 + 7(\gamma')^2]$$

Thus the following possibilities are evident:

1. For $a' = \gamma' = 0$ the Raman line is said to be "forbidden."
2. For $a' \neq 0$, and $\gamma' = 0$, $\rho = 0$ and the Raman line is "completely polarized."
3. For $a' = 0$ and $\gamma' \neq 0$, $\rho = \tfrac{6}{7}$ and the Raman line is said to be "depolarized."
4. In all other cases $a' \neq 0$ and $\gamma' \neq 0$ and the value of ρ is between zero and $\tfrac{6}{7}$. The line is said to be "polarized."

Each coefficient of α_0' is related to the corresponding coefficient of α_0 by an equation of the kind $\alpha_{xx}' = (\partial\alpha_{xx}/\partial q)_0$ (see Section 3.2D), i.e., $a' = (\partial a/\partial q)_0$, however $\gamma' \neq (\partial\gamma/\partial q)_0$.

TABLE 3.1

a. Selection rules for some common point groups

Point group	Infrared active species	Raman active species	Example
C_2	A, B	A, B	H_2O_2
C_s	A', A''	A', A''	$CHDCl_2$
C_{2v}	A_1, B_1, B_2	A_1, A_2, B_1, B_2	SO_4
C_{3v}	A_1, E	A_1, E	NH_3
D_{3d}	A_{2u}, E_u	A_{1g}, E_g	C_6H_{12}
D_{6h}	A_{2u}, E_{1u}	A_{1g}, E_{1g}, E_{2g}	C_6H_6
T_d	T_2	A_1, E, T_2	CH_4, CCl_4
$D_{\infty h}$	Σ_u^+, Π_u	$\Sigma_g^+, \Pi_g, \Delta_g$	CO_2

b. Raman and infrared spectra of some simple molecules

Vibrational mode	Frequency, cm^{-1}	
	Infrared	Raman

1. Carbon dioxide, CO_2, $D_{\infty h}$ [a] (gas)

Vibrational mode	Infrared	Raman
$\nu_2 (\Pi_u)$	667	Inactive
$2\nu_2 (\Sigma_g^+)$	Inactive	1286 pol.[b]
$\nu_1 (\Sigma_g^-)$	Inactive	1388 pol.
$\nu_3 (\Sigma_u^-)$	2349	Inactive

2. Sulfur dioxide, SO_2, C_{2v} (gas)

Vibrational mode	Infrared	Raman
$\nu_2 (a_1)$	519	525 (l) pol.
$\nu_1 - \nu_2 (A_1)$	606	—
$\nu_1 (a_1)$	1151	1151 (g) pol.
$\nu_3 (b_1)$	1361	1336 (l) depol.[b]
$\nu_2 + \nu_3 (B_1)$	1871	—
$2\nu_1 (A_1)$	2305	—
$\nu_1 + \nu_3 (B_1)$	2499	—

3. Boron trifluororide, BF_3, D_{3h} (gas)

Vibrational mode	Infrared	Raman
$\nu_4 (e')$	480[c]	480[c] depol.
	482	482
$\nu_2 (a_2'')$	691[c]	Inactive
	720	Inactive
$\nu_1 (a_1')$	Inactive	888 pol.
$\nu_3 (e')$	1446[c]	—
	1497	—

4. Carbon tetrachloride, CCl_4, T_d (liquid)

Vibrational mode	Infrared	Raman
$\nu_2 (e)$	Inactive	218 depol.
$\nu_4 (f_2)$	305	314 depol.
$\nu_1 (a_1)$	Inactive	458 pol.
$\nu_3 (f_2)$	768	762 depol.[d]
$\nu_1 + \nu_4 (F_2)$	797	791 depol.[d]

TABLE 3.1 (continued)

Vibrational mode	Frequency, cm^{-1}	
	Infrared	Raman
5. Chloroform, CHCl$_3$, C_{3v} (liquid)		
$\nu_6\ (e)$	260	262 depol.
$\nu_3\ (a_1)$	364	366 pol.
$\nu_2\ (a_1)$	667	668 pol.
$\nu_5\ (e)$	760	761 depol.
$\nu_4\ (e)$	1205	1216 depol.
$\nu_1\ (a_1)$	3033	3019 pol.
6. Ethylene, C$_2$H$_4$, D_{2h} (gas)		
$\nu_4\ (a_u)^c$	Inactive	Inactive
$\nu_8\ (b_{2g})$	Inactive	943 depol.
$\nu_7\ (b_{1u})$	949	Inactive
$\nu_{10}\ (b_{2u})$	995	Inactive
$\nu_6\ (b_{1u})$	Inactive	(1050)^f
$\nu_3\ (a_g)$	Inactive	1342 pol.
$\nu_{12}\ (b_{3u})$	1443	Inactive
$\nu_2\ (a_g)$	Inactive	1623 pol.
$\nu_{11}\ (b_{3u})$	2990	Inactive
$\nu_1\ (a_g)$	Inactive	3019 pol.
$\nu_9\ (b_{2u})$	3106	Inactive
$\nu_5\ (b_{1g})$	Inactive	3272 depol.

^{a} The species symbols follow a common convention that lower-case symbols are used for fundamentals and upper-case symbols for overtones and combination bands.

^{b} State of polarization of Raman lines: pol., polarized; depol., depolarized.

^{c} All bands except ν_1, in which the boron atom does not move, show isotope splitting due to ^{10}B and ^{11}B.

^{d} Fermi resonance.

^{e} Estimated at 825 cm^{-1} from $2\nu_2\ (A_g) = 1656$ observed in the Raman.

^{f} Should be Raman active, but is not observed. The frequency was estimated from $\nu_6 + \nu_{10}\ (B_{3u}) = 2047$ cm^{-1} and $\nu_{10} = 995$ cm^{-1}.

F. RULES OF SELECTION AND POLARIZATION

We have seen previously that the vibrations of a molecule can affect the polarizability ellipsoid in three ways: by changing its size and/or shape and/or its orientation with respect to the fixed coordinates of the molecule.

1. Asymmetric Vibrations

We saw previously that for the asymmetric stretching vibration of the SO$_2$ molecule ν_3 the size of the polarizability ellipsoid remained unchanged during the vibration, i.e., $a' = 0$, but that its orientation with respect to the fixed axes changed, i.e., $\gamma' \neq 0$. Thus the vibration obeys rule 3 and will have $\rho = \frac{6}{7}$.

This result is generally true for all asymmetric vibrations.

Rule: All asymmetric vibrations give depolarized Raman lines.

If in addition to the size of the polarizability ellipsoid remaining unchanged its orientation is also unaltered, because of some other symmetry property of the molecule, for example, as in the asymmetric stretching vibration of the linear CO_2 molecule, then γ' is also zero and the Raman line is forbidden, i.e., that frequency is inactive in the Raman (rule 1).

2. Degenerate Vibrations

In the degenerate ν_3 bending mode of CCl_4 the size of the polarizability ellipsoid remains unchanged, but the shape changes. Thus $a' = 0$ and $\gamma' \neq 0$ and $\rho = \frac{6}{7}$. i.e., the Raman line is depolarized (rule 3). This is again a general result.

Rule: All degenerate vibrational modes give depolarized Raman lines.

3. Symmetric Vibrations

In symmetric vibrations the size of the polarizability ellipsoid changes during the vibration, $a' \neq 0$ and $\gamma' \neq 0$, and $\rho < \frac{6}{7}$. i.e., the Raman line is polarized (rule 4).

Rule: All totally symmetric vibrations are active in the Raman and polarized.

In the special case where the polarizability ellipsoid is spherical $a' \neq 0$, but $\gamma' = 0$, and the Raman line is completely polarized (rule 2). An example is the totally symmetric breathing frequency ν_1 of CCl_4.

The state of polarization expected for the Raman frequencies of molecules in certain point groups are shown in Table 3.1.

3.3 EXPERIMENTAL

A. LIGHT SOURCES

Since Raman scattering is of relatively low intensity, the first requirement experimentally is for a very intense monochromatic light source of an appropriate wavelength that is not absorbed by the material under investigation. The most widely used source is the mercury emission lamp, which gives strong lines at 2537, 3650, 4057, 4358, 5461, 5770, and 5790 Å. The very intense blue 4358 Å line is usually used for obtaining spectra from colorless liquids and solutions, and the green line at 5461 Å for the spectra of colored (yellow) substances. The 4358 Å line is particularly suitable since it is relatively far away from the line at 5461 Å. To provide a relatively monochromatic source wavelengths other than those required are eliminated by using appropriate filters such as those shown in Table 3.2.

In the early days of Raman spectroscopy simple high pressure lamps were employed which were capable of giving only relatively low intensities so that Raman spectra were obtained only with difficulty by using very sensitive photographic plates for detection, and even then very long exposure times were generally required. In addition this kind of lamp gives an undesirable

Table 3.2: Filters for Mercury Lamps

Exciting line, Å	Filter
2537	Mercury vapor
3650	Corning filter 7-51, or Eastman Kodak glass filter 18A
4047	Sodium nitrite solution, or Corning filter 5-58, or Eastman Kodak Wratten 2B filter, to remove 3650-Å line
	Solution of iodine in carbon tetrachloride, or 0.003 M potassium ferricyanide solution, to remove 4358-Å line
4358	Rhodamine 5GDN dye (DuPont), 0.009% solution slightly acidified with HCl, to remove higher wavelengths
5461	Basic sodium chromate (pH 8.7) for lines below 5461 Å; cupric nitrate solution; saturated neodymium chloride solution

continuous background. However in recent years H. L. Welsh and his co-workers at the University of Toronto[14] have developed low pressure helical mercury lamps which give very high intensities free of continuous background. These enable spectra of, for example, liquid samples to be obtained within a few minutes exposure time.

The so-called Toronto arc requires high currents (of the order of 10 to 30 A), and thus the electrode pools have to be cooled and in some versions of the lamp a cooling coil passes through the center of the helix. The original lamp consisted of a four-turn spiral of Pyrex tubing; however, we have constructed similar lamps in our laboratory in recent years with as many as six turns (Fig. 3.9) and have obtained excellent results using currents up to 15 A without the necessity of cooling the helix. To start the lamp the mercury pools around the electrodes are preheated electrically for a few minutes and the emission then induced by means of a tesla coil held against a piece of aluminum foil that is wound around the helix. As soon as emission starts the cooling water through the electrodes is turned on.

Sources containing other metals and helium arcs have also been developed[15–18] specifically for the study of colored compounds. In particular Ham and Walsh[16] have described a microwave-powered source for use with many types of lamp.

Very recently an important advance in technique has been the introduction of the laser as a Raman source.[19] The He–Ne laser is the most commonly used and gives a strong red monochromatic line a few hundredths of an Angstrom wide free from other emission lines or continuous background. This source has now been introduced in some of the commercial spectrometers such as the Cary model 81 and the Perkin-Elmer models LR-1 and LR-2. It seems quite possible that the introduction of the laser will make it possible to use Raman

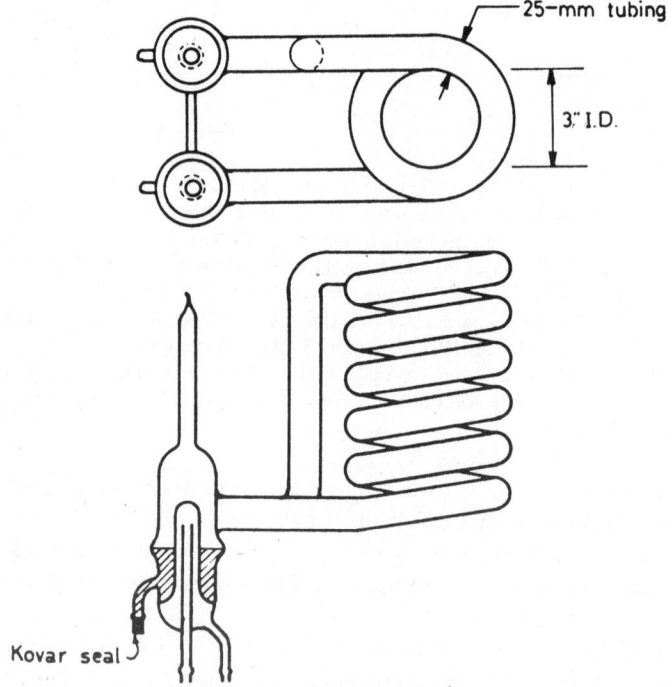

FIGURE 3.9: The Toronto arc.

spectroscopy as a routine tool for the examination of quite small quantities of both colorless and highly colored materials. The most suitable laser presently available appears to be the Spectral-Physics model 125 He–Ne 50-μW source which emits at 6328 Å. However, developments in this field are so rapid that it is to be expected that other sources will be available in the near future.

B. SPECTROMETERS

In this section we shall describe only the most commonly used instruments; namely, the Hilger E612, the Cary model 81, and the Perkin-Elmer LR-1 (and LR-2) spectrometers.

The Hilger E612 instrument (Hilger and Watts, England), is a spectrograph employing two large glass prisms for dispersion and can be used either as a photographic or as a photoelectric recording instrument. The instrument with photographic detection is of moderate expense and therefore perhaps the most widely encountered. It is, however, confined, without adaptation, to the use of a mercury source. A simple schematic layout is shown in Fig. 3.10 and the instrument itself in Fig. 3.11. The Raman source in the commercial instrument is composed of four straight-tube low pressure mercury lamps which surround the sample cell, and relatively high intensity is obtained by enclosing the lamps in a magnesium oxide-coated reflector (Fig. 3.12).

The instrument consists of a solid cast-iron base to which is screwed the collimator and the box casting containing the prisms. Two cameras may be used for recording the spectra and in the purely photographic instrument are interchangeable in position, each being fixed onto an arm that can be rotated to bring either camera into action as desired. Plate holders for each camera are set in frames that can be raised or lowered to allow several spectra to be recorded on the same photographic plate. The spectrometer is calibrated by means of a standard iron or copper arc and functions best in a temperature-controlled room. The adjustable slit, with stainless-steel jaws, is mounted on the front of the collimator and is fitted with a shutter, reducing wedge, and Hartmann diaphragm. The collimator has a lens aperture of 86 mm and a focal length of approximately 60 cm. The camera that gives the greatest dispersion has a relative aperture of $f/5.7$ and gives a spectrum 86 mm in length, covering the range 3900 to 6300 Å. The inverse dispersion on the plate is 16 Å/mm at 4358 Å. Photographic plates, in size $3\frac{1}{4} \times 4\frac{1}{4}$ in., that are most suitable for use with the Hilger E612 are Eastman Kodak type 103a-O or 2a-O for Raman lines near to the 4358-Å mercury exciting line, or 103a-J or 103a-G for higher Raman frequencies in the region of 3000 cm^{-1}.

In the recording instrument the short focus camera is replaced by a photomultiplier scanning unit. The light is dispersed in the first passage through the prisms and reflected at a tilted mirror to obtain double dispersion. The mirror rotates automatically at one of four selected speeds and may be set to a calibrated scale. The dispersion at the photomultiplier is 6.8 cm^{-1}/mm at 4358 Å. The scattered light is monitored relative to the intensity of the source and automatically recorded. The basic photographic instrument may be purchased without the photoelectric recording unit, which can be readily added later.

In our laboratory we have replaced the Hilger-arc assembly by a Toronto arc to give a more flexible unit for some purposes. For example, polarization measurements are easily carried out (Section 3.3D), and a wide variety of Raman sample cells may be used (Section 3.3C).

The Cary model 81 Raman spectrophotometer (Applied Physics Corporation, USA), uses a 3-kw Toronto arc and/or a Spectral-Physics model 125

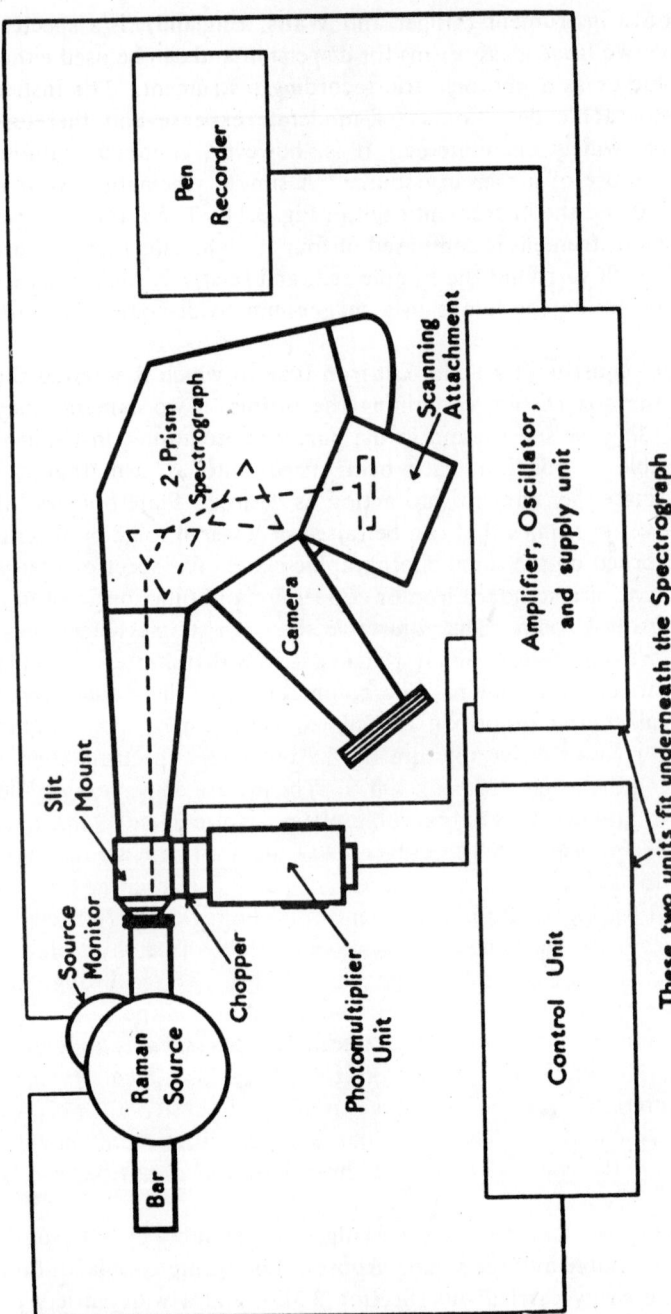

FIGURE 3.10: Schematic layout of the Hilger E612 spectrograph.

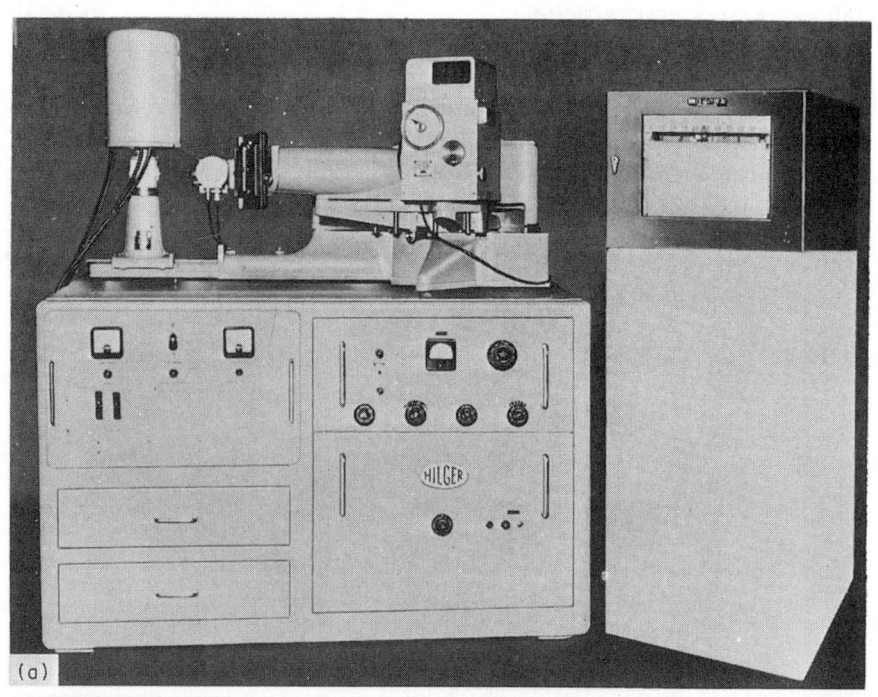

(a)

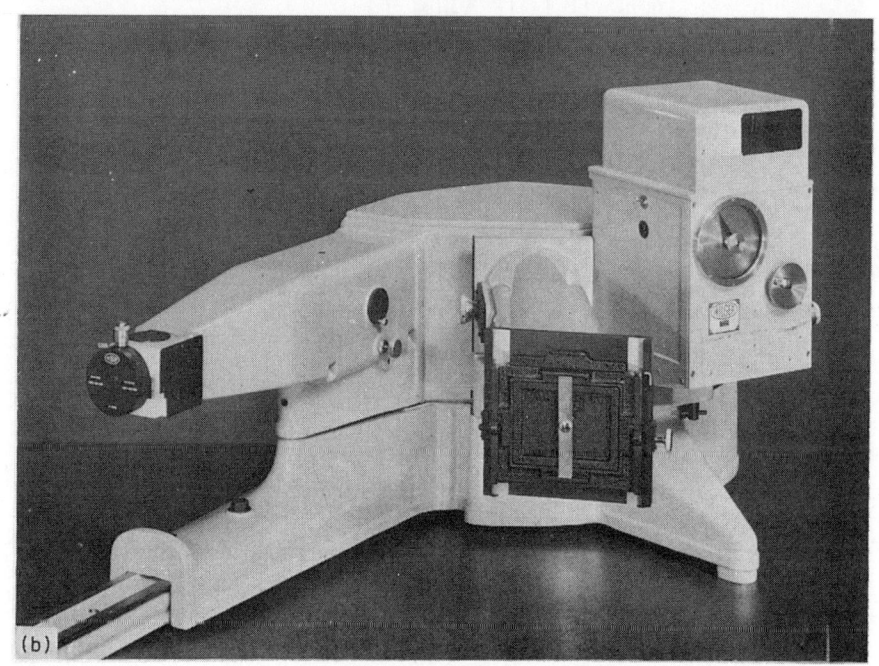

(b)

FIGURE 3.11: The Hilger E612 spectrograph.

He–Ne laser as source. Three versions are available; Toronto arc only (with provision for the later installation of the laser source), laser source only, and a version with both Toronto arc and laser.

The Toronto arc is of Corning 1720 glass tubing in the form of a helix that surrounds the sample cell. The helix is cooled with air from two blowers

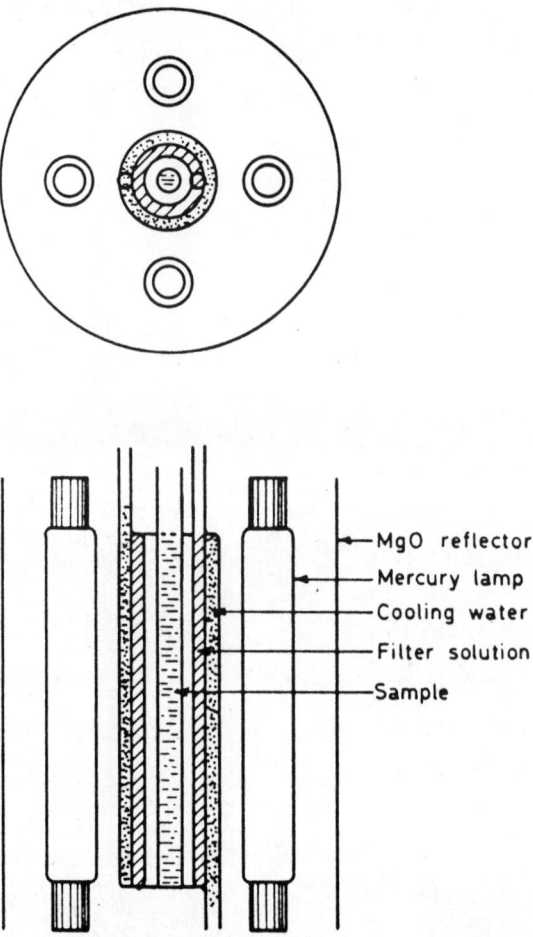

FIGURE 3.12: The Hilger source.

and the electrodes are cooled with thermostated circulating water that keeps the mercury vapor pressure low and thus reduces the lamp continuum. A cylindrical glass-filter jacket is located between the lamp and the sample, and a filter solution is circulated through the jacket. The filter solution is cooled and thermostated to keep the sample cool. Sample cells are mounted horizontally on the axis of the lamp and filter jacket. The optical system is shown schematically in Fig. 3.13 and the instrument in Fig. 3.14.

A unique feature of the Cary model 81 spectrophotometer is its use of image slicers (Fig. 3.15, items G and H) between the sample cell and entrance slit. These make it possible to use the rays from the entire end of the sample cell rather than just a narrow strip down the middle as in conventional spectrometers. The image from the end of the cell is divided into 20 strips which

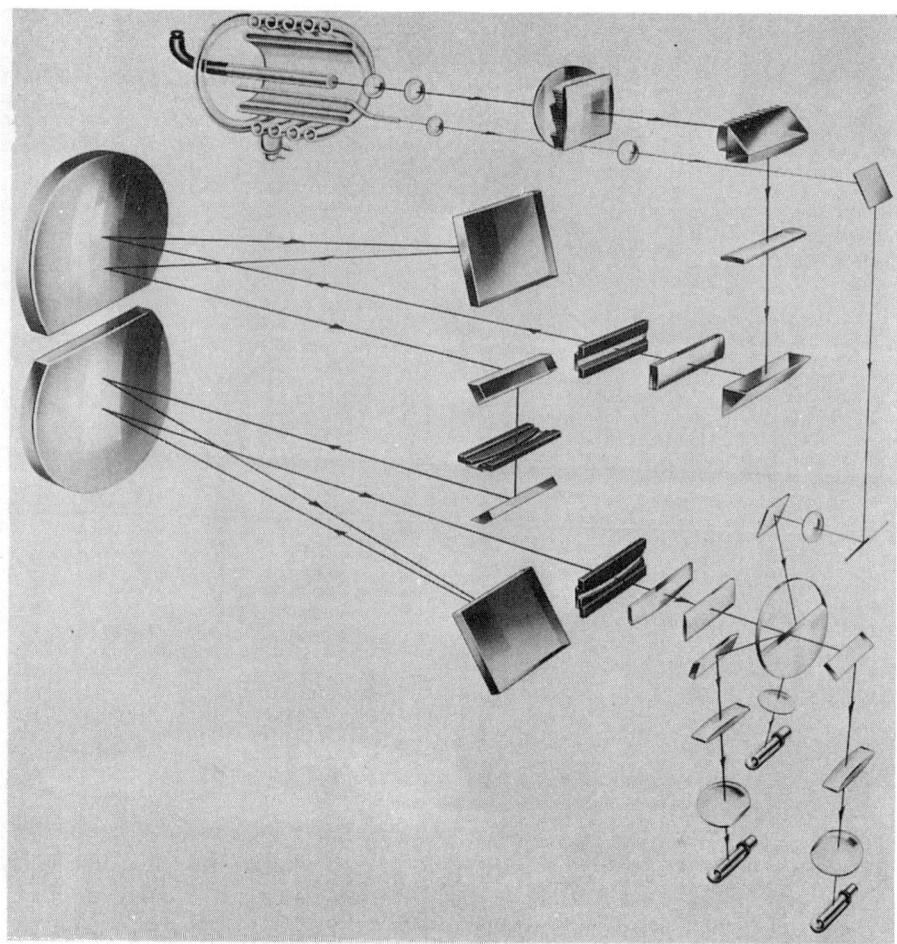

FIGURE 3.13: The Cary model 81 spectrophotometer—schematic layout of the optical system used in conjunction with the Toronto arc.

are superimposed in two sets of 10 strips each and then magnified to the exact size required to fill the double entrance slit. The monochromator is a dual grating, twin slit, double monochromator and has a focal length of 1000 mm. It is designed to give increased light-gathering power and is free from the effects of Tyndall and Rayleigh scattering in the sample. This greatly reduces the difficulty in sample preparation since the double monochromator rejects scattered light, enabling the cell walls and other sources of scattered light to be viewed. Thus advantage may be taken of total reflection at the walls of the sample tubes to effectively increase the length of small diameter tubes many times. The gratings are ruled with 1200 lines/mm. High stability is achieved in the Cary model 81 by means of a chopped-radiation photometric system designed to retain the stability of such systems, but avoiding the 50% loss of light that normally results from chopping. A rotating semicircular

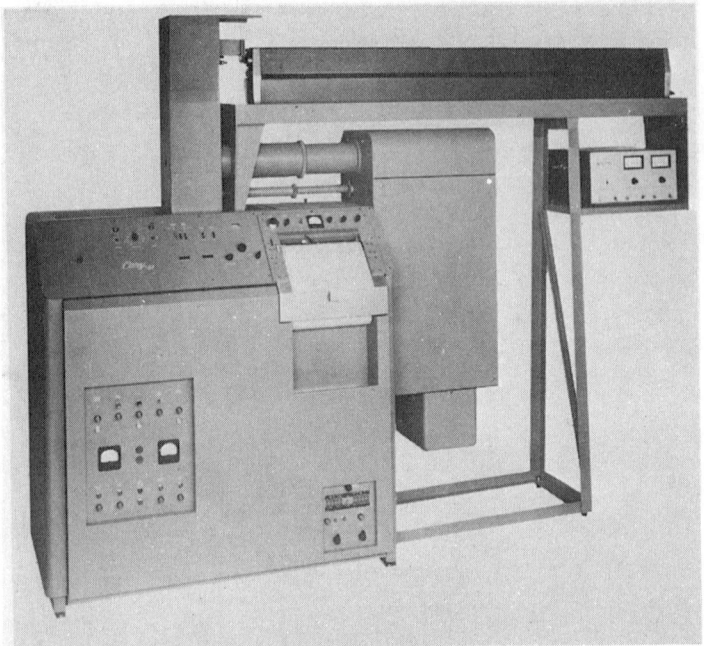

FIGURE 3.14: The Cary model 81 spectrophotometer.

mirror directs the radiation alternately to two phototubes each of which develops an independent Raman signal. The signals are later combined with a consequent improvement in the signal-to-noise ratio. At the same time non-dispersed light from the lamp is chopped by the same mirror and directed to a reference phototube.

The performance and areas of application of the Cary model 81 spectro-photometer have been increased by the introduction of the He–Ne laser source since the narrow intense line at 6328 Å that this emits can give Raman spectra from many fluorescing and most strongly colored compounds. A further advantage is the use of axial excitation in conjunction with a capillary liquid sample cell which permits an order of sample size comparable to that used in microinfrared work. The optical system of this spectrometer is shown in Fig. 3.15.

Two spectrophotometers with He–Ne laser sources have recently been developed by the Perkin-Elmer Corporation. The Model LR-1 is shown in Fig. 3.16 and diagrammatically in Fig. 3.17. It consists of a compact optical

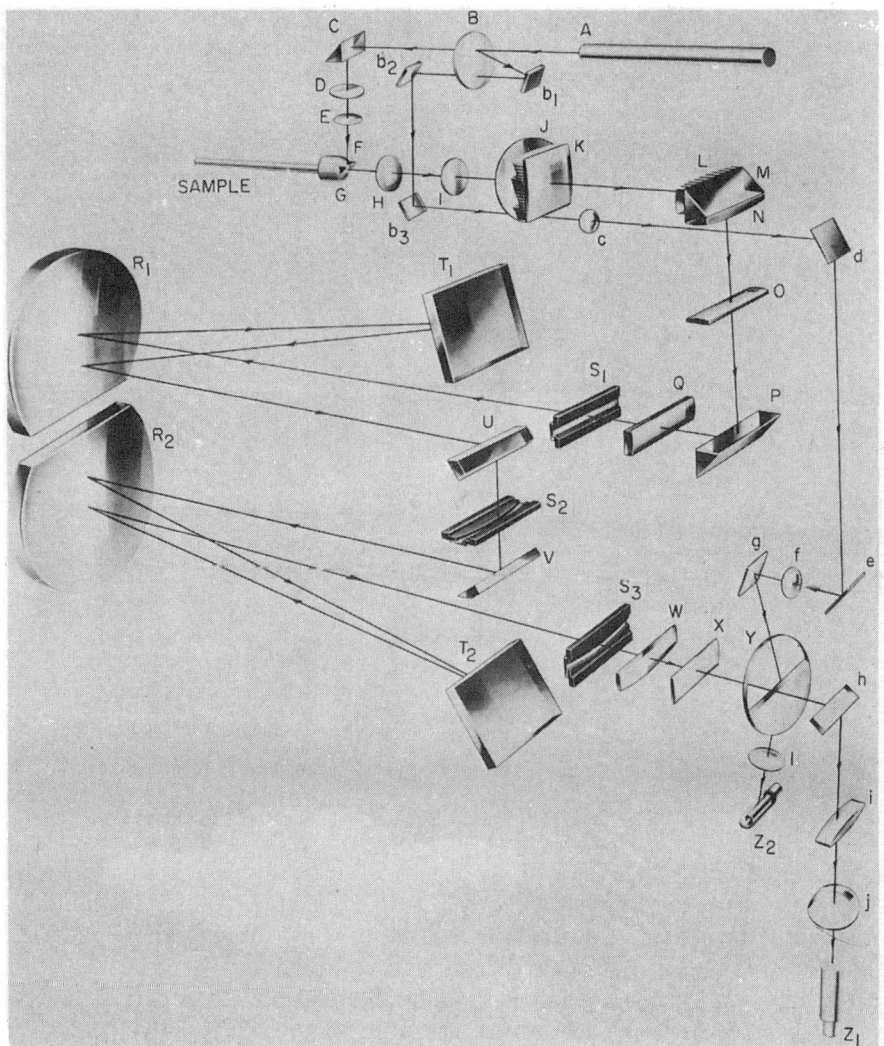

FIGURE 3.15: The Cary model 81 spectrophotometer—schematic layout of the optical system used in conjunction with the laser source.

unit and an associated electronics rack which contains the recorder and other electrical components, which are cable connected to the optical unit. A Perkin-Elmer model 5320 He–Ne gas laser mounted horizontally is used for excitation, and its beam falls on the sample cell, which is slightly wedge-shaped in order to permit multiple traversals of the incident light and hence insure maximum excitation. Approximately 150 passes are thus made in a standard 5 ml cell. Scattered light is viewed at 90° with respect to the incoming beam and is focused by a simple lens onto the entrance slits of a double-pass monochromator equipped with a 1440 lines/mm replica diffraction grating blazed at 6200 Å in the first order. An analyzer prism is mounted between the sample and the monochromator to measure the degree of polarization (Section 3.2E). It is removed when not in use. The monochromator may be

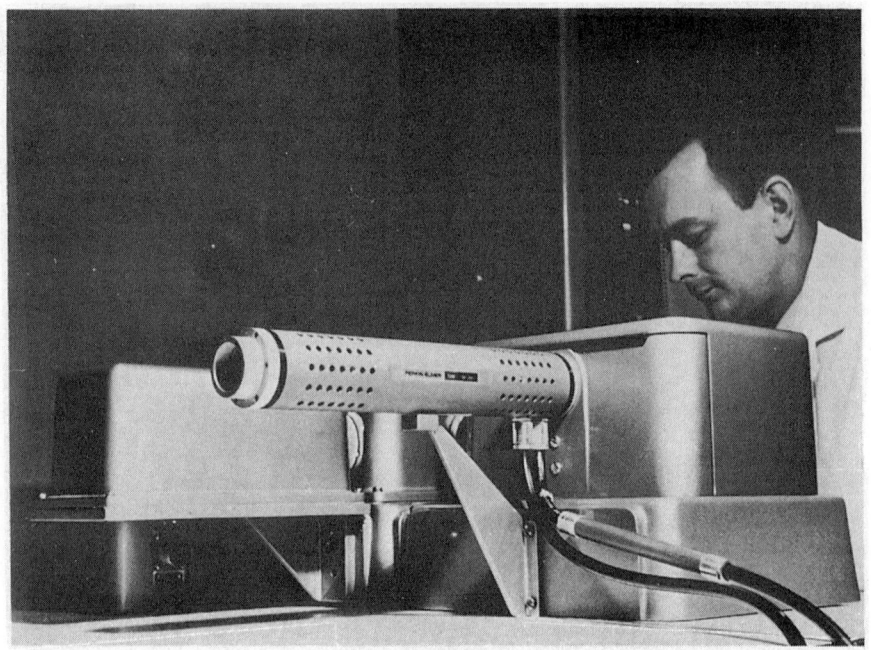

FIGURE 3.16: The Perkin-Elmer LR-1 spectrophotometer.

scanned either manually or by means of a multispeed automatic wavelength drive for Raman shifts from 0 to 3800 cm^{-1}. A 14 stage multiplier phototube is used as a detector. Its output is fed through an amplifier and displayed on a strip-chart recorder.

Recently the Spectro-Physics laser has been introduced as part of the LR-1 and an LR-2 is being developed. The latter features the Spectro-Physics laser and an improved monochromator.

C. SAMPLING TECHNIQUES

Gases, liquids, and solids can be studied by Raman spectroscopy, but different types of cell are normally required to obtain the spectra in each phase. Liquids and solutions present the least difficulties.

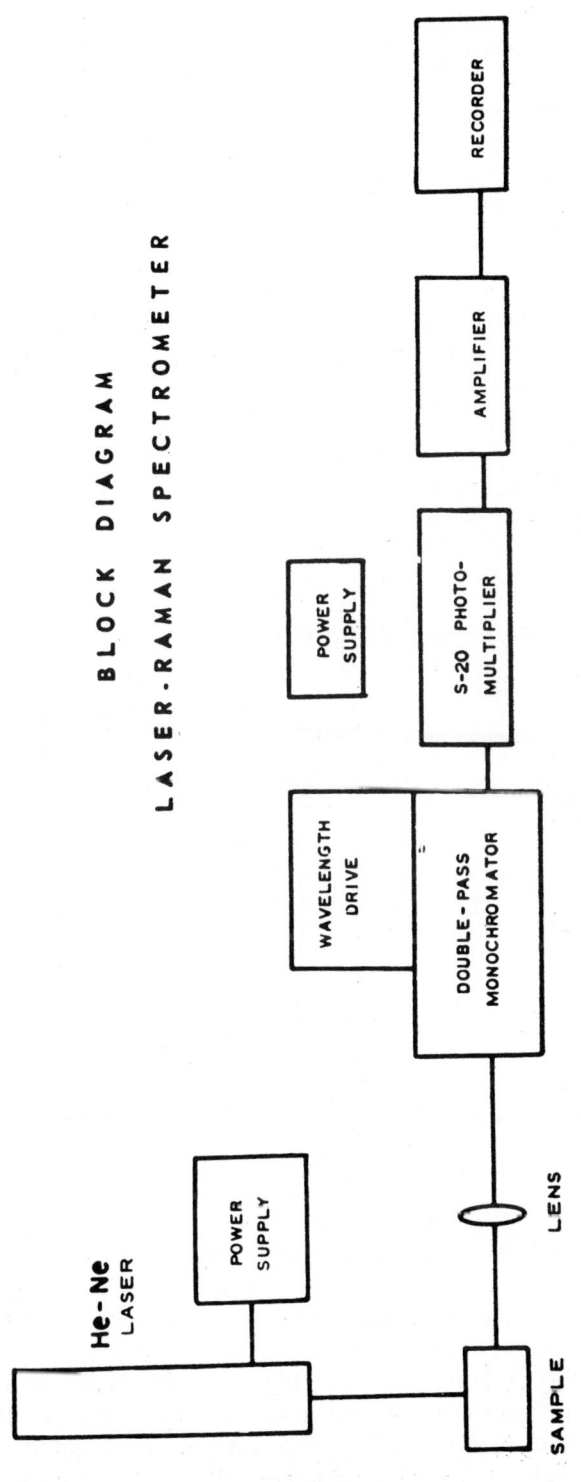

FIGURE 3.17: Schematic diagram of the Perkin-Elmer LR-1 spectrophotometer.

155

The Hilger E612 spectrometer with the Hilger lamp assembly holds a simple glass cell for liquids of 7-ml capacity which fits inside a coaxial jacket. The jacket consists of three isolated concentric tubes and allows both cooling water and a filter solution to be placed between the lamp and sample. The

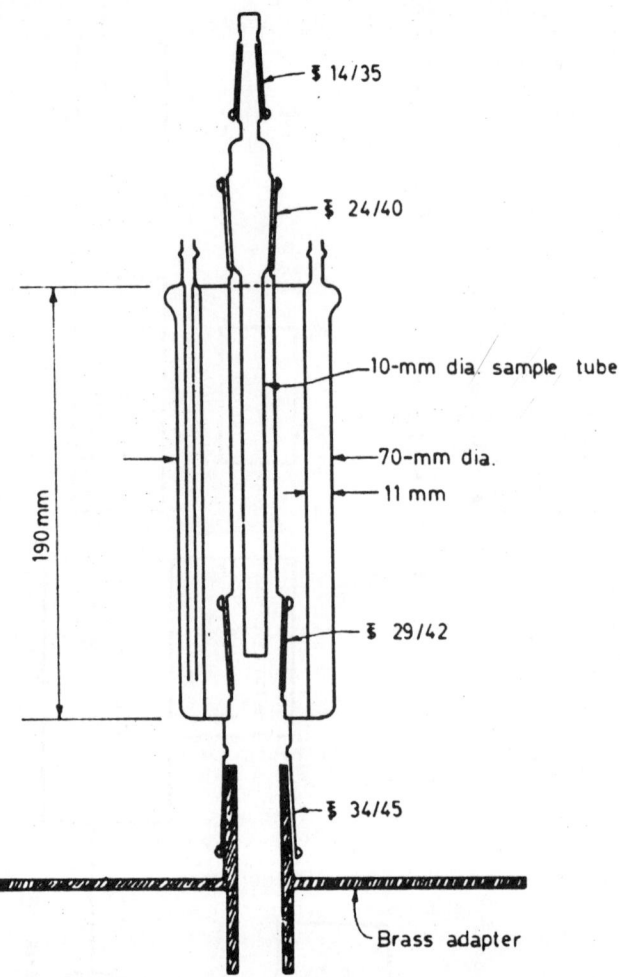

FIGURE 3.18: Raman cell used with a Hilger E612 spectrometer and Toronto-arc source.

Raman cells are straight cylindrical glass tubes with optically flat glass windows fixed to their lower ends.

In our laboratory replacement of the Hilger source by a large Toronto arc has given a more convenient arrangement for some purposes. It can be adapted for the study of liquids at high or low temperatures and for the study of solids. A variety of designs of Raman cells have been used with this apparatus, which is shown schematically in Fig. 3.18. Provision of a standard tapered brass joint at the base of the filter jacket, and a standard glass joint inside it, allows for the rapid interchange of cells and easy access to the sample, and enables the cells to be surrounded by polaroid cylinders for polarization measurements (Section 3.3D).

In the Cary model 81 spectrophotometer with Toronto-arc source a wide range of cells is available. Some of these are shown in Fig. 3.19. The 7-mm

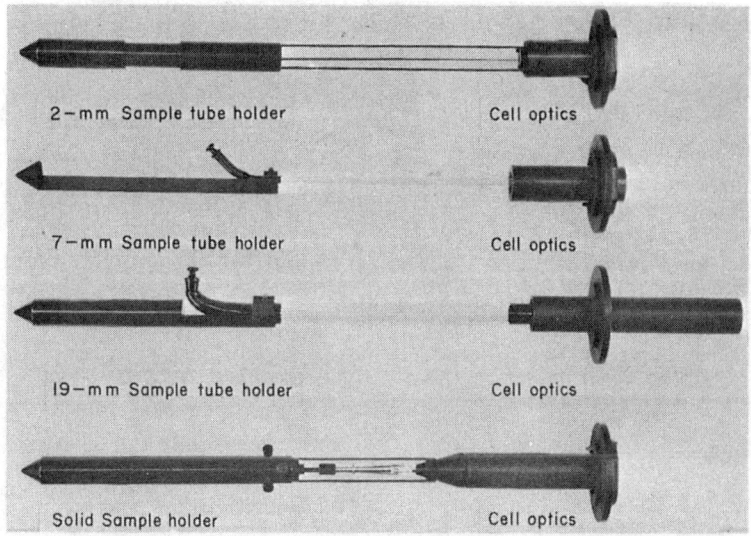

FIGURE 3.19: The Cary model 81 spectrophotometer—Raman cells.

outside-diameter cell for liquids has a capacity of 5 ml and the 19-mm outside-diameter cell has a capacity of 65 ml. Two 2-mm cells are available, one with a volume of 0.2 ml and the other with a volume of 0.6 ml. Selection of the most appropriate cell is important. In the 19-mm cell the monochromator does not *see* the cell walls, whereas the 7-mm and 2-mm cells depend on internal reflection from the wall surface. Therefore the monochromator does *see* the cell walls and any fluorescence that is developed in them. Thus cell selection becomes important when there is need to minimize cell fluorescence. Modification of the 5-ml cell system to permit more effective collection of the Raman radiation by a multiple reflection optical system has been described by Tunnicliff and Jones.[20] R. Norman Jones and co-workers have recently described capillary microcells of two sizes (Table 3.3) for use with the Cary model 81.[21] One end of the capillary is sealed to obtain a symmetrical meniscus seal free from entrapped air, and the commercial cell holder was used to keep the closed end of the tube in contact with a wide-angled hemispherical collecting lens.

TABLE 3.3: Dimensions of Capillary Raman Tubes Used by Jones and Co-workers[21]

	Effective length, mm	Diameter, mm		Volume, ml	Solute required in different solvents, mg[a]		
		Inside	Outside		CS_2	$CHCl_3$	H_2O
Small	225	0.9 ± 0.1	2.0 ± 0.2	0.15	100	200	300
Large	225	1.7 ± 0.1	2.4 ± 0.2	0.50	200	300	300

[a] Under favorable circumstances acceptable spectra were obtained with half these amounts.

In the spectrophotometer with laser source a microcell consisting of a capillary tube 5 cm long with a 0.5-mm inside diameter is used. This has a volume of 10 μliters and can be used with samples as small as 1 μliter.

For the Perkin-Elmer LR-1 spectrometer several cells are available commercially. The standard cell for liquids has a 2.5-ml capacity (Fig. 3.20). An 0.2-ml liquid cell and a special assembly for the sampling of solid powders is also available.

FIGURE 3.20: Cell and cell assembly for use with the Perkin-Elmer LR-1 spectrophotometer.

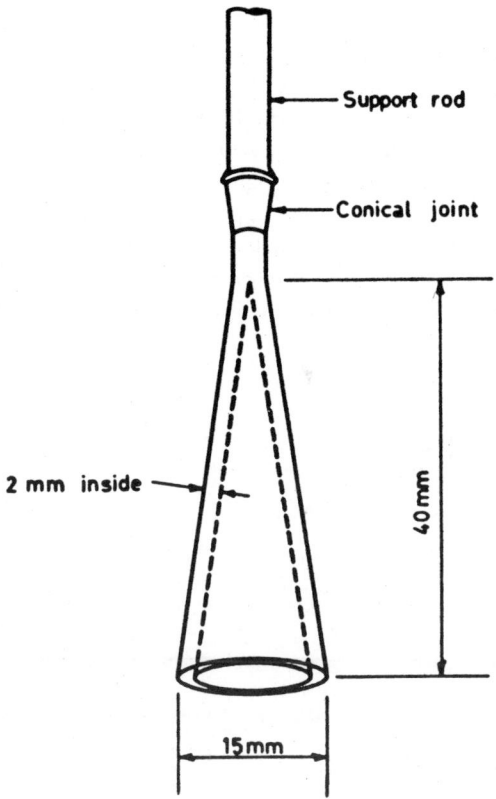

Support rod

Conical joint

2 mm inside

40 mm

15 mm

FIGURE 3.21: A Raman cell for solid powders.

The solid sample holder shown in Fig. 3.19 is used with the Cary model 81 spectrometer with Toronto-arc source, while for the instrument with laser source powders and fine crystals can be packed inside the capillary microcell, and large crystals can be mounted at the surface of the sample lens using a specially designed sample holder. Recently a novel cell for solids that can be used with the Hilger E612 spectrometer has been described.[22] In this cell a solid powder is held between two concentric glass cones (Fig. 3.21).

The gas cell shown in Fig. 3.22 is used with the Cary model 81 equipped with a special source consisting of two Toronto arcs. It is of the multiple-reflection type and allows the light to pass as many as 44 times through the 20-in. path length of the cell. It has a capacity of $3\frac{1}{2}$ liters and is designed to withstand pressures as high as 10 atm.

D. MEASUREMENT OF DEPOLARIZATION RATIOS

We have seen previously that the determination of the depolarization ratio ρ is important in assigning Raman lines to particular normal vibrations. The absolute determination of depolarization ratios is difficult experimentally; however, it is sufficient in many studies to determine ρ only approximately.

With the Hilger E612 spectrometer the experimental method normally used is to compare the intensities of the Raman lines obtained in two separate experiments, one using light from the source polarized parallel to the axis of the Raman tube and the other using light polarized perpendicular to the axis of the Raman tube. To do this two polaroid cylinders constructed to fit around the Raman cell are used. One transmits the parallel, and the other the perpendicular, component. The intensities of the spectral lines in the two experiments may be compared using a suitable microdensitometer.

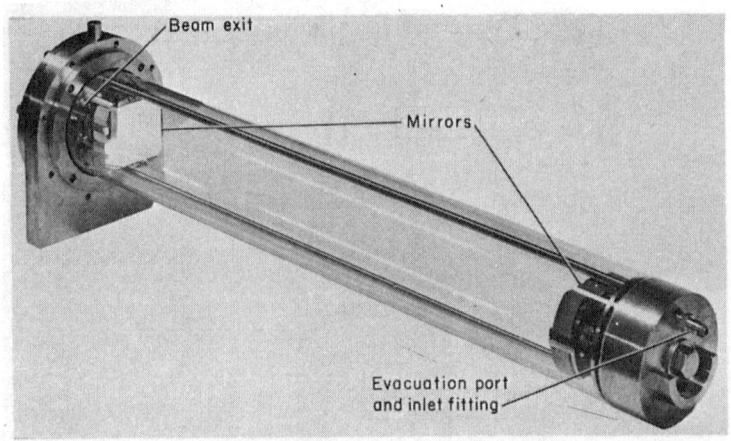

FIGURE 3.22: Gas cell for use with the Cary model 81 spectrophotometer.

The results of such experiments on a sample of carbon tetrachloride are shown in Fig. 3.23. The line at 459 cm^{-1} (v_1) is polarized and therefore corresponds to a totally symmetric vibration, while the lines at 218, 314, 762, and 791 are depolarized and may be assigned to unsymmetrical vibrations.

Various authors have described other methods of measuring depolarization ratios which are more appropriate to a photoelectric instrument (e.g., Refs. 21, 23, and 24).

3.4 APPLICATIONS

Much of the work carried out to date in the field of Raman spectroscopy has been concerned with complete analyses of the vibrational spectra of simple molecules. Work of this kind requires measurement of both Raman and infrared spectra and is usually directed toward the determination of molecular geometry and the force constants of bonds in the molecules.

In contrast, infrared spectroscopy has been widely used in organic chemistry in recent years as a routine tool to give analytical information, largely from group frequencies, concerning the presence of functional groups in molecules. That is not to say that Raman spectroscopy could not be equally useful for this purpose; however, while commercial infrared spectrometers are now required equipment in any organic laboratory, Raman spectrometers are found rather rarely. The situation has to do not only with the rather high cost of Raman spectrometers but also with the difficulty of the technique.

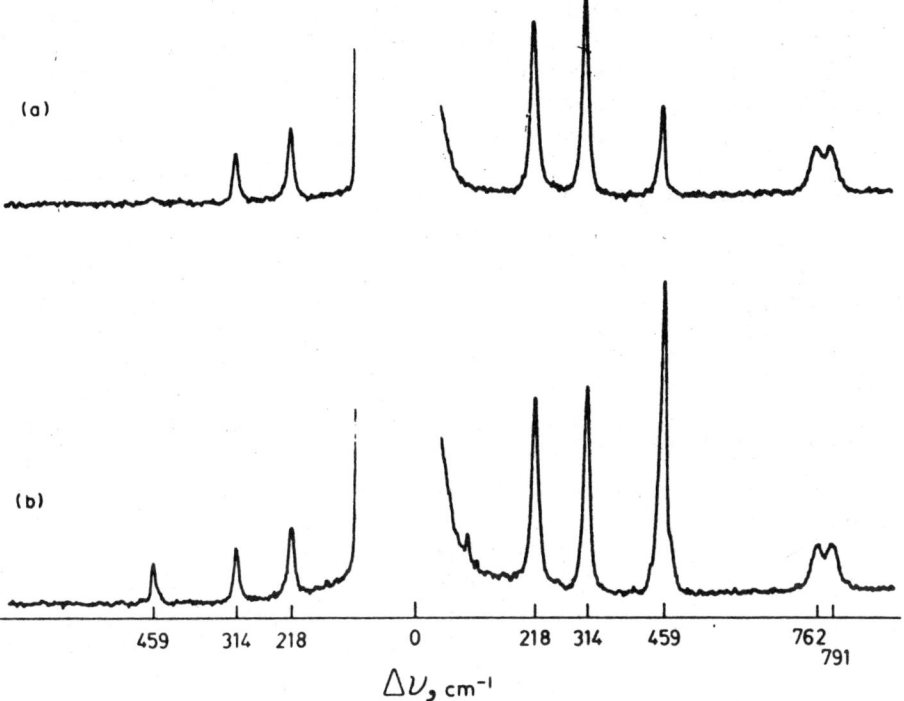

FIGURE 3.23: Polarization measurements on carbon tetrachloride.

While infrared measurements can be performed on very small samples, it has not been until very recently that the Raman spectra of microsamples could be obtained. However there is no doubt that since Raman and infrared spectroscopy provide complementary information about molecular vibrations, the routine accessibility of both kinds of spectra would greatly enhance the value of the group-frequency method of structural analysis.

Much of the work that has been reported on the Raman spectra of organic molecules has been reviewed by Jones and Sandorfy,[25] who give an excellent account of the use of both infrared and Raman spectroscopy in organic chemistry with emphasis not only on infrared but also on those cases where a knowledge of the Raman spectrum of a molecule proves particularly useful. They have discussed the whole scope of vibrational spectrum analysis as it applies to organic chemistry in a systematic way. The reader is referred to their extensive discussion for a detailed account. Here, we shall restrict our comments to some points of general interest.

The literature of Raman spectroscopy is not as extensive as that of infrared, but the subject is well covered in various review articles and monographs. Those of note include the works of Hibben,[26] Kohlrausch,[27] Glockler,[28] Pajenkamp,[29] Braun and Fenske,[30] Stamm,[31] and an up-to-date monograph by Brandmuller and Moser.[32] Recent review articles include those by Rosenbaum,[33] Jones and Tunnicliff,[34] Jones,[35] and Jones and Jones.[36] A collection of spectra also appears in Landolt-Bornstein.[37] Two other articles that deal mainly with inorganic substances but are nevertheless of interest are those by Woodward[6] and Tobias.[38]

One of the important features of Raman spectroscopy that gives it an advantage over infrared spectroscopy is the ease with which low frequency vibrations (50 to 250 cm^{-1}) can be observed. In addition, because of its use of glass cells rather than infrared materials, aqueous solutions and solutions in other solvents such as strong acids can be studied.

In the case of low frequencies most conventional infrared spectrometers cover the region from 650 to 4000 cm^{-1} and more expensive grating instruments the region from 200 to 650 cm^{-1}. In the region below 200 cm^{-1} infrared detection is difficult because of the low energy of infrared radiation in this region. Two spectrometers which allow this region to be covered are the Perkin-Elmer 301 infrared spectrometer and the Beckman IR11 spectrometer, however both spectrometers are very expensive and special cell materials are required. In contrast, even the simplest Raman spectrometer enables lines in the low frequency region to be observed easily.

Solvents such as water present difficulties in the infrared because of their very strong absorption so that the study of the infrared spectra of aqueous solutions presents great difficulties. In addition special cell materials such as polyethylene or silver chloride must be used because of the solubility in water of the more usual cell materials such as sodium chloride and potassium bromide. In the Raman water bands appear with rather weak intensities enabling the spectra of other species in solution to be readily observed.

As yet applications of Raman spectroscopy to organic chemistry have not been seriously exploited except in a few cases. However sufficient has been done to make it clear that some molecular groups are better characterized by their Raman spectra than by their infrared spectra. For example, the infrared spectra of the esters of compounds containing polymethylene chains, e.g., the methyl laurates,[39] are dominated by bands associated with the carbomethoxy group, whereas this group gives only weak bands in the Raman spectra, which are characteristic of the hydrocarbon skeleton. Similarly in the case of crystalline normal-paraffin hydrocarbons study of both the Raman and spectra infrared has shown a characteristic pattern of frequencies which enables a distinction to be made between molecules of odd and even chain length.[40]

Infrared spectra unduly emphasize group frequencies of relatively polar bonds involving hetero atoms and the C—C, C=C, C≡C, and C—H bonds of the structural skeleton which often give weak infrared bands usually give strong Raman lines. With the development of Raman techniques it is certain that useful new developments in structural analysis will result. The recent literature has been reviewed by Jones and Jones.[36]

QUESTIONS

Q3.1. Why is it not possible to use the blue (or indigo) mercury line at 4358 Å to obtain the Raman spectrum of yellow-colored solutions?

Q3.2. What weight of a compound with a molecular weight of 100 would be required to make enough aqueous solution to fill a cylindrical cell 10 mm in diameter and 15 cm long if the solution must be 20% by weight? (Assume that the specific gravity of the solution is 1.) What volume of an ideal gas of the same molecular weight at 10 atm pressure and 25°C would be required in order that the same number of molecules would be irradiated?

Q3.3. The inverse dispersion of a spectrograph is given by $\Delta\lambda/\Delta d$, where $\Delta\lambda$ is the difference in wavelength of two lines and Δd is the distance between the lines, i.e., when photographic detection is used, Δd is the actual distance between the images of the lines on the photographic plate. The Hilger E612 has an inverse dispersion of 16 Å/mm at 4358 Å, and approximately 41 Å/mm at 5461 Å. Tin tetrachloride has four fundamental vibrations which give rise to lines $\nu_1 = 424$ cm^{-1}, $\nu_2 = 150$ cm^{-1}, $\nu_3 = 608$ cm^{-1}, $\nu_4 = 221$ cm^{-1}. How far apart will the images of the ν_2 and ν_4 lines be in a spectrum excited by (a) the 4358-Å mercury line and (b) the 5461-Å mercury line? From which spectrum would you expect to get more accurate values for the Raman shifts?

Q3.4. In the Raman spectrum of carbon tetrachloride the three lowest frequency vibrations (ν_1, ν_2, ν_4) usually give quite strong anti-Stokes lines (see Fig. 3.1). From the data in Table 3.1b calculate the wavelengths of the anti-Stokes lines in a carbon tetrachloride spectrum excited by the 4047-Å mercury line.

Q3.5. The two possible isomers of 1,2-dichloroethylene have vibrational spectra which show the following fundamentals:

Isomer I		Isomer II	
Infrared, cm^{-1}	Raman, cm^{-1}	Infrared, cm^{-1}	Raman, cm^{-1}
570	173	620	349
694	406	820	758
857	563	1200	844
1303	711	3089	1270
1591	876		1576
3086	1179		1626
	1587		1692
	3077		3142
	3160		

Using simple selection rules, deduce the structures of the isomers.

Q3.6. Which of the following functional groups might you expect to be detected more easily in the Raman than in the infrared?

$$\begin{matrix} R \\ \diagdown \\ C{=}O \\ \diagup \\ R' \end{matrix} \qquad \begin{matrix} O \\ \| \\ R{-}C \\ \diagdown \\ OR' \end{matrix}$$

$$R{-}CBr_3 \qquad (CH_3)_3C{-}R$$

$$\begin{matrix} R R' \\ \diagdown \diagup \\ C{=}C \\ \diagup \diagdown \\ R' R \end{matrix} \qquad \begin{matrix} R O \\ \diagdown \| \\ P \\ \diagup \diagdown \\ R' Cl \end{matrix}$$

$$\begin{matrix} O \\ \diagup \diagdown \\ R R' \end{matrix}$$

REFERENCES

1. C. V. Raman and K. S. Krishnan, *Nature*, **121**, 501 (1928).
2. C. V. Raman, *Indian J. Phys.*, **2**, 387 (1928).
3. C. V. Raman and K. S. Krishnan, *Indian J. Phys.*, 2, 399 (1928).
4. H. A. Stùart, *Molekulstruktur*, Springer, Berlin, 1934, p. 169.
5. M. Minnaert, in *Light and Colour*, Dover, New York, 1954, Chap. XI.
6. L. A. Woodward, *Quart. Rev. (London)*, **10**, 185 (1956).
7. A. B. F. Duncan, in *Chemical Applications of Spectroscopy* (W. West, ed.), Wiley (Interscience), New York, 1956.
8. N. B. Colthup, L. H. Daly, and S. E. Wiberly, *Introduction to Infrared and Raman Spectroscopy*, Academic Press, New York, 1964.
9. G. Herzberg, *Infrared and Raman Spectra of Polyatomic Molecules*, Van Nostrand, Princeton, N.J., 1945.
10. H. A. Stuart, *Molekulstruktur*, Springer, Berlin, 1934, p. 221.
11. F. A. Cotton, *Chemical Applications of Group Theory*, Wiley (Interscience), New York, 1963.
12. R. S. Mulliken, *Phys. Rev.*, **43**, 279 (1933).
13. M. Born, *Optik*, Edwards, Ann Arbor, Mich., 1943.
14. H. L. Welsh, M. F. Crawford, T. R. Thomas, and G. R. Love, *Can. J. Phys.*, **30**, 577 (1952).
15. N. S. Ham and A. Walsh, *Spectrochim. Acta*, **12**, 88 (1958).
16. N. S. Ham and A. Walsh, *J. Chem. Phys.*, **36**, 1096 (1962).
17. H. Stammreich, *Phys. Rev.*, **78**, 79 (1950).
18. H. Stammreich, R. Forneris, and H. Sone, *J. Chem. Phys.*, **23**, 972 (1955).
19. R. C. C. Leite and S. P. S. Porto, *J. Opt. Soc. Am.*, **54**, 981 (1964).
20. D. D. Tunnicliff and A. C. Jones, *Spectrochim. Acta*, **18**, 569 (1962).
21. R. N. Jones, J. B. DiGiorgio, J. J. Elliott, and G. A. A. Nonnenmacher, *J. Org. Chem.*, **30**, 1822 (1965).
22. R. H. Busey and O. L. Keller, *J. Chem. Phys.*, **41**, 215 (1964).
23. M. R. Fenske, W. G. Braun, R. V. Wiegand, D. Quiggle, R. H. McCormick, and D. H. Rank, *Anal. Chem.*, **19**, 700 (1947).
24. R. F. Stamm, C. F. Salzman, and T. Mariner, *J. Opt. Soc. Am.*, **43**, 119 (1953).
25. R. N. Jones and C. Sandorfy, in *Chemical Applications of Spectroscopy* (W. West, ed.), Wiley (Interscience), New York, 1956.
26. J. H. Hibben, *The Raman Effect and Its Chemical Applications*, Reinhold, New York, 1939.

27. K. W. F. Kohlrausch, *Ramanspektren, Hand- und Jahr-Buch der Chemischen Physik*, Vol. 9 (A. Eucken and K. L. Wolf, eds.), Becker and Erler, Leipzig (1943); Edwards, Ann Arbor, Mich., 1945.
28. G. Glockler, *Rev. Mod. Phys.*, **15**, 111 (1943).
29. H. Pajenkamp, *Fortschr. Chem. Forsch.*, **1**, 417 (1950).
30. W. G. Braun and M. R. Fenske, *Anal. Chem.*, **21**, 12 (1949); **22**, 11 (1950).
31. R. F. Stamm, *Anal. Chem.*, **26**, 49 (1954).
32. J. Brandmuller and H. Moser, *Einfuhrung in die Ramanspektroskopie*, Steinkopff, Darmstadt, 1962.
33. E. J. Rosenbaum, *Anal. Chem.*, **28**, 596 (1956).
34. A. C. Jones and D. D. Tunnicliff, *Anal. Chem.*, **34**, 261R (1962).
35. A. C. Jones, *Anal. Chem.*, **36**, 296R (1964).
36. R. N. Jones and M. K. Jones, *Anal. Chem.*, **38**, 393R (1966).
37. Landolt-Bornstein, *Zahlenwerte und Funktionen aus Physik, Chemie, Astronomie, Geophysik und Technik* (A. Eucken and K. H. Hellwege, eds.), Springer, Berlin, 1951, pp. 479–551.
38. R. S. Tobias, *J. Chem. Educ.*, **44**, 2, 70 (1967).
39. R. N. Jones and R. A. Ripley, *Can. J. Chem.*, **42**, 305 (1964).
40. J. K. Brown, N. Shepherd, and D. M. Simpson, *Quart. Rev. (London)*, **7**, 19 (1953).

CHAPTER 4

Fluorometry

David E. Guttman

SCHOOL OF PHARMACY
STATE UNIVERSITY OF NEW YORK AT BUFFALO
BUFFALO, NEW YORK

4.1 INTRODUCTION

The mechanisms by which molecules absorb electromagnetic radiation in the visible and ultraviolet regions of the spectrum and are, as a result, raised to excited electronic states were discussed in Chapter 1. The reverse process, loss of energy and concomitant transition of molecules from excited states to ground states of energy, can occur with reemission of radiation. Such emission is known as *luminescence*. The intensity and composition of the emitted radiation can be measured and such measurements form the basis of a sensitive method of analysis called *fluorometry*. Fluorometric methods of analysis have found application in many situations of pharmaceutical interest such as in the analysis of riboflavin, thiamine, and reserpine in drug dosage forms. More significant and widespread, however, has been the application of fluorometric techniques in the analysis of trace amounts of drugs and metabolites in biological tissues and fluids.

4.2 THEORY

Absorption of ultraviolet and visible light by molecules of an irradiated sample generates a population of molecules in excited electronic states. Each excited electronic state has many different vibrational energy levels, and excited molecules will be distributed in the various vibrational energy levels of an excited state. Most usually this state is a *singlet state*, i.e., one in which all of the electrons are paired and in each pair the two electrons spin about their own axes in opposite directions. It is intuitively apparent that since the amount of radiation absorbed by a sample does not decrease if radiation is continued, efficient rapid processes must be operant which result in the loss of energy by excited molecules and their return to the ground state. It has been calculated, in fact, that the average lifetime of a molecule in the singlet excited state is of the order of 10^{-8} sec. Molecules at each vibrational level of the excited state could, for example, lose energy by emitting photons and as a result fall to the original condition of the ground state. The energy and, therefore, the wavelength of emitted light would then be exactly the same as that absorbed. Such a process is termed *resonance fluorescence*. It is an improbable process and is rarely encountered in solution chemistry. Rather, molecules initially undergo a more rapid process, a radiationless loss of vibrational energy, and so quickly fall to the lowest vibrational energy level of the excited state. The vibrational energy is thought to be lost to solvent molecules. The process is known as *vibrational relaxation*. From the lowest vibrational level of the excited state, a molecule can either return to the ground state by photoemission or by radiationless processes. If indeed the former occurs, the emission is a type of luminescence referred to as *fluorescence*. Fluorescence is defined as the radiation emitted in the transition of a molecule from a singlet excited state to a singlet ground state. Because of vibrational relaxation in the excited state and because a molecule may return to a vibrational level in the ground state which is higher than that initially occupied prior to excitation, the radiation emitted as fluorescence is of lower energy and, therefore, of longer wavelength than that originally absorbed.

Other processes involving the excited state can occur to compete with fluorescence emission, and not all of the absorbed energy will be emitted as fluorescence. The extent to which such other processes occur is characterized by a parameter known as the *quantum efficiency of fluorescence* which is symbolized by ϕ and defined as the ratio of the number of light quanta emitted to the number absorbed. Quantum efficiency approaches 1 for highly fluorescent compounds and 0 for those which fluoresce weakly. It is interesting and pertinent to consider the processes which occur to decrease the efficiency of fluorescence. An excited molecule could, for example, undergo a radiationless loss of energy sufficient to drop to the ground state. This process is termed *internal conversion*. With some compounds a process known as *intersystem crossing* can also occur. Here a molecule in the lowest vibrational level of the excited state converts to a *triplet state*, a state lying at an energy level intermediate between ground and excited states and characterized by an unpairing of two electrons. Thus, in contrast to the singlet state, there is a spin reversal involving one electron of a pair and the two electrons spin about their axes in the same direction. Once intersystem crossing has occurred, a molecule quickly drops to the lowest vibrational level of the triplet state by vibrational relaxation. The triplet state is much longer lived than the corresponding singlet state with lifetimes of 10^{-4} to 10 sec. From the triplet state a molecule can drop to the ground state by emission of radiation. This type of luminescence is termed *phosphorescence* and is formally defined as emission of radiation resulting from the transition of a molecule from a triplet excited state to a singlet ground state. Phosphorescence is often characterized by an afterglow, i.e., because of the long life of the triplet state, luminescence can be observed after the source of exciting radiation has been removed. In contrast, no afterglow is observed in fluorescing systems because of the short life of the singlet excited state. A molecule in the triplet state can also undergo radiationless conversion to the ground state. Such a conversion is enhanced by the relatively long life of the triplet state so that collisions fruitful in dissipating energy can occur and by the fact that the energy difference between triplet state and ground state is not inordinately large. These processes are diagrammatically illustrated by Fig. 4.1.

4.3 FLUORESCENCE AND CHEMICAL STRUCTURE

Quantitative aspects of the processes which can involve the excited electronic state are not sufficiently well understood to permit predictions as to whether or not a particular compound will fluoresce to a degree necessary for analytical purposes. Fluorometric methods are, of course, limited to those compounds which possess a system of conjugated double bonds. A compound must absorb radiation in order to fluoresce, and it is the presence within a molecule of the mobile π electrons which is responsible for absorption characteristics in the visible and ultraviolet regions of the spectrum. The presence of a "chromophore" does not necessarily endow a compound with the ability to fluoresce since radiationless processes and intersystem crossing can occur to decrease the quantum efficiency of fluorescence to zero or to a degree that makes practical measurement impossible. It would be expected, however, that structural features which influence the degree of conjugation of a molecule and the delocalization of π electrons might influence the likelihood

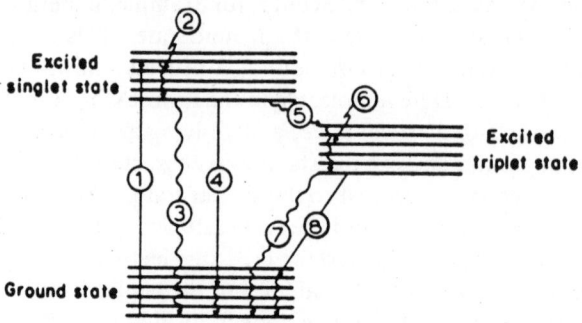

FIGURE 4.1: A diagrammatic representation of the changes in energy levels of a molecule that can occur as a result of the absorption of electromagnetic radiation: (1) absorption of radiant energy boosting molecules to various vibrational energy levels of the excited singlet state; (2) radiationless vibrational relaxation to the lowest vibrational level of the excited singlet state; (3) radiationless internal conversion from excited singlet state to ground state followed by vibrational relaxation; (4) fluorescence followed by vibrational relaxation; (5) intersystem crossing from excited singlet state to excited triplet state; (6) vibrational relaxation to the lowest vibrational level of the excited triplet state; (7) radiationless internal conversion from excited triplet state to ground state followed by vibrational relaxation; (8) phosphorescence followed by vibrational relaxation.

of measurable fluorescence. Thus, for example, saturated compounds such as cyclohexane are nonfluorescent, while benzene is weakly fluorescent, and highly unsaturated polycyclic aromatic compounds such as anthracene are strongly fluorescent. Similarly, the reduced form of riboflavin (I) does not have the degree of conjugation of the parent compound (II) and does not possess the fluorescence characteristics of riboflavin.

Definite correlations between chemical structure and fluorescence cannot be made. However, the work and review of Williams and Bridges[1] does provide some insight to the influence of structural features on the fluorescence of organic compounds. Monosubstitution of benzene with alkyl groupings, for example, was found to have little influence on the intensity of fluorescence of the substituted benzene relative to that of benzene. However, monosubstitution with groups known to increase electron delocalization (ortho-para-directing groups), such as fluoro, amino, alkylaminodialkylamino, hydroxy, and methoxy, yielded compounds that fluoresced more intensely than did the parent compound. Substitution with iodine, chlorine, and bromine resulted in benzene derivatives which either did not fluoresce or fluoresced to a lesser degree than benzene in spite of the fact that these substituents are also ortho-para directing. This influence of halogen substitution was also reported by McClure[2] and is apparently due to an enhanced intersystem crossing process since bromine- and iodine-substituted aromatic compounds exhibit intense phosphorescence but only weak fluorescence. Most meta-directing substituents (which tend to localize π electrons) were found to markedly decrease fluorescent intensity. Thus, benzoic acid, nitrobenzene, benzenesulfonic acid, benzenesulfonamide, and benzaldehyde were found to be nonfluorescent. Benzonitrile, in contrast, fluoresced more intensely than benzene, even though the C≡N group is meta-directing. It was postulated that electrons of the C≡N group interacted with the π electrons of the benzene ring to result in a distribution that favored fluorescence.

The fluorescence characteristics of disubstituted benzenes were also studied, but few generalities could be generated from the results. Observed effects were not predictable and apparently were the result of a combined influence on the mobility of π electrons. For example, it might be expected that substitution of the fluorescent compound, aniline, with a meta-directing group such as $-SO_2NH_2$ would result in a compound which would fluoresce to a lesser degree than aniline. Sulfanilamide, however, was found to be five times as fluorescent as aniline. Similarly, guides to predicting the behavior of heterocyclic compounds could not be made because of the uncertainty of the substituent effect. In general, it was found that a doubly bonded nitrogen (=N—) in a ring tended to decrease the likelihood of fluorescence, while the presence of —NH—, —O—, and —S— appeared to contribute to the likelihood of fluorescence.

Molecular geometry must also be considered in attempting to relate chemical structure and fluorescence. That geometric considerations are important is well illustrated by examples cited by Wehry and Rogers.[3] Fluorescein (III), for example, is highly fluorescent, while phenolphthalein (IV) is nonfluorescent. The oxygen bridge in fluorescein imparts to the molecule a rigidity and

(III)   (IV)

planarity that is not present in phenolphthalein. The absence of a planar, rigid structure permits vibrations and rotations of the aromatic rings to occur which result in radiationless dissipation of excitation energy. Similarly, with a number of compounds that exhibit *cis-trans* isomerism, the *cis* isomer fluoresces much less intensely than the trans. Stilbene (**V**) is such a compound. This can also be ascribed to a planarity effect with the *cis* isomer being non-planar due to the bulkiness of the aromatic rings.

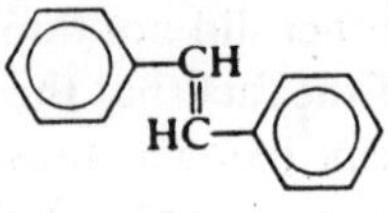

(V)

That a compound does not fluoresce or has a low intensity of fluorescence does not necessarily dismiss fluorometry as a potential tool for the analytical determination of that compound. Many well-accepted and widely used fluorometric procedures are based on chemical conversions of weakly fluorescing compounds to derivatives which fluoresce intensely. For example, tetracycline (**VI**) has a weak native fluorescence, but complexes of the antibiotic with Ca^{2+} and a barbiturate fluoresce quite intensely.[4] Corticosteroids

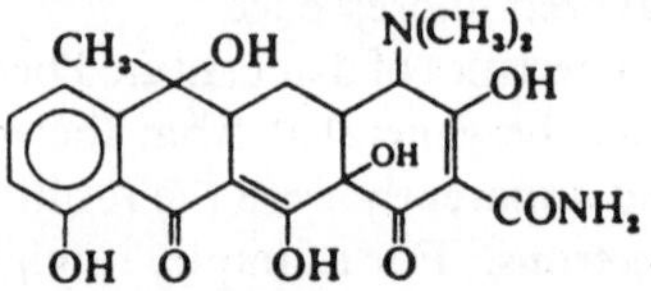

(VI)

such as hydrocortisone (**VII**) do not fluoresce. However, they form, in concentrated sulfuric acid and in the presence of ethanol, strongly fluorescing compounds.[5] *N*-Methylnicotinamide (**VIII**) is determined in biological fluids

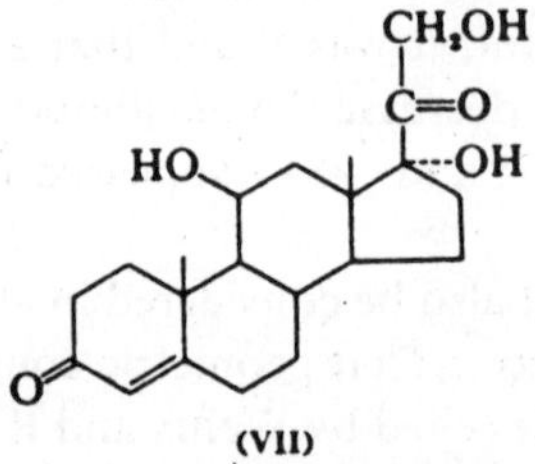

(VII)

by a fluorometric method even though it has little native fluorescence. Here the amide is condensed with acetone and treated with base to yield fluorescent

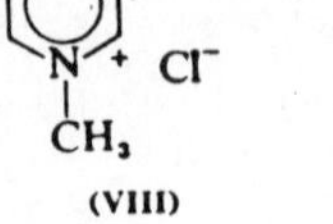

(VIII)

products.[6] Similarly, epinephrine (**IX**) is assayed fluorometrically by measuring the fluorescence of products resulting from oxidation and hydroxylation.[7]

$$HO-\text{}-CH-CH_2-NH-CH_3$$

HO

(IX)

4.4 INSTRUMENTATION FOR FLUOROMETRY

In contrast to spectrophotometry, the intensity of light transmitted by a sample is not of direct concern in fluorometry. Rather, it is the intensity of radiation that is emitted as fluorescence that is measured and related to the

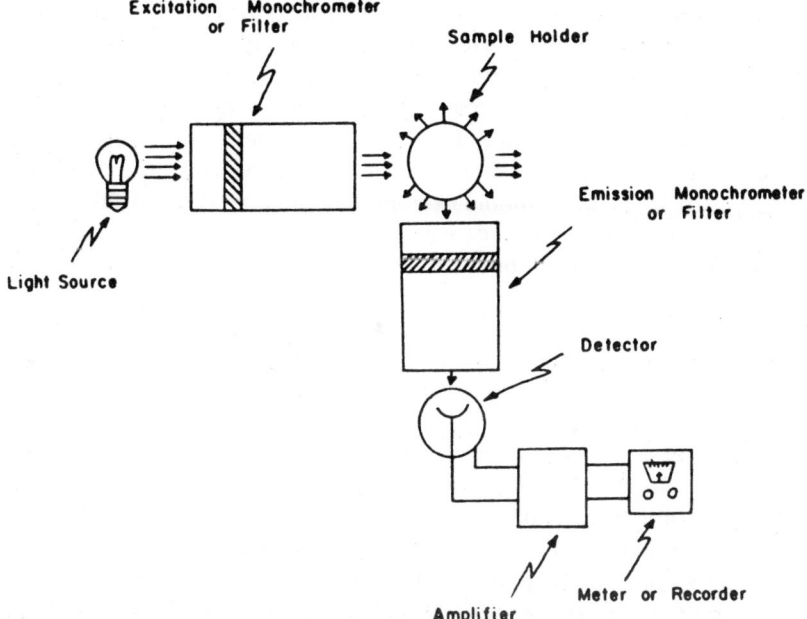

FIGURE 4.2: A diagrammatic representation of an instrument used to measure intensity of fluorescence.

concentration of fluorescing species. The components of instruments which are used in fluorometry are, however, quite similar in design and function to those employed in spectrophotometers and colorimeters. A diagrammatic representation of such a device is shown in Fig. 4.2.

The chief components are: a source of exciting radiation, an excitation filter or monochromator by which a band of exciting light can be isolated to be passed on to the sample, a sample holder, an emission filter or monochromator by which a band of fluorescence can be selected for detection, a detector, and some means for amplifying and indicating the detector response. In most commercially available instruments, the detector is placed at a right angle to the direction of travel of the beam of exciting light. This arrangement has been found to be the most advantageous for measuring the fluorescence of dilute solutions. Other arrangements are possible, however, since fluorescence is emitted in all directions. The light source is usually a mercury or xenon arc. Those instruments which use filters as coarse monochrometers are referred to as *fluorometers*. Those employing more sophisticated and exact grating or prism monochromators are termed *spectrofluorometers*, *fluorescence spectrometers*, or *spectrophotofluorometers*. The first term appears to be the one most commonly used to designate this type of instrument. The other components of an instrument such as sample holders, detectors, amplifying and indicating devices are much the same as those discussed under spectrophotometry.

Spectrofluorometers are used in fluorometric work in a manner analogous to the use of spectrophotometers in absorption spectrophotometry. They enable an investigator to generate two types of spectra which are pertinent to fluorometry. The excitation spectrum is obtained by setting the emission monochromator at a suitable wavelength and measuring the intensity of fluorescence as a function of the wavelength of the exciting radiation. In theory, the maxima and minima exhibited by the excitation spectrum should be at wavelengths which are identical to those found in the absorption spectrum of the compound. In practice, exact coincidence may not be found due to instrumental artifacts. The *emission spectrum* of a compound is obtained by setting the excitation monochrometer at an appropriate wavelength corresponding to strong excitation and measuring the intensity of fluorescence as a function of the wavelength of emitted light. An example of excitation and emission spectra is shown in Fig. 4.3 for grisoefulvin in 1% ethanol. As would be expected from the considerations discussed in the theory section, the emission spectrum is found at longer wavelengths than the excitation spectrum. Some overlap of the two spectra is frequently observed.

Fluorometers are used in a manner somewhat analogous to colorimeters in absorption work. The excitation and emission spectra of a compound dictate the transmittance characteristics of filters that should be employed for a particular analytical determination with a fluorometer. The filters should be as much as possible mutually exclusive. That is, the emission filter should not pass wavelengths which are transmitted by the excitation filter. This precaution is necessary to preclude interferences from light which may be reflected by the sample holder and other parts of the instrument and from light scattered by the solvent used. Fluorometers are somewhat more sensitive than spectrofluorometers since filters pass a more intense radiation than prism or grating monochromators. Fluorometers are recommended for routine quantitative work, while spectrofluorometers are necessary research tools in the development of fluorometric assay methods.

A variety of fluorometers and spectrofluorometers are available from manufacturers of scientific equipment. Figure 4.4 is a schematic diagram of the optical system of a widely used filter fluorometer (model 110, G. K. Turner Associates). Figures 4.5 and 4.6 illustrate the appearance and optical characteristics of a spectrofluorometer (Aminco-Bowman spectrophotofluorometer,

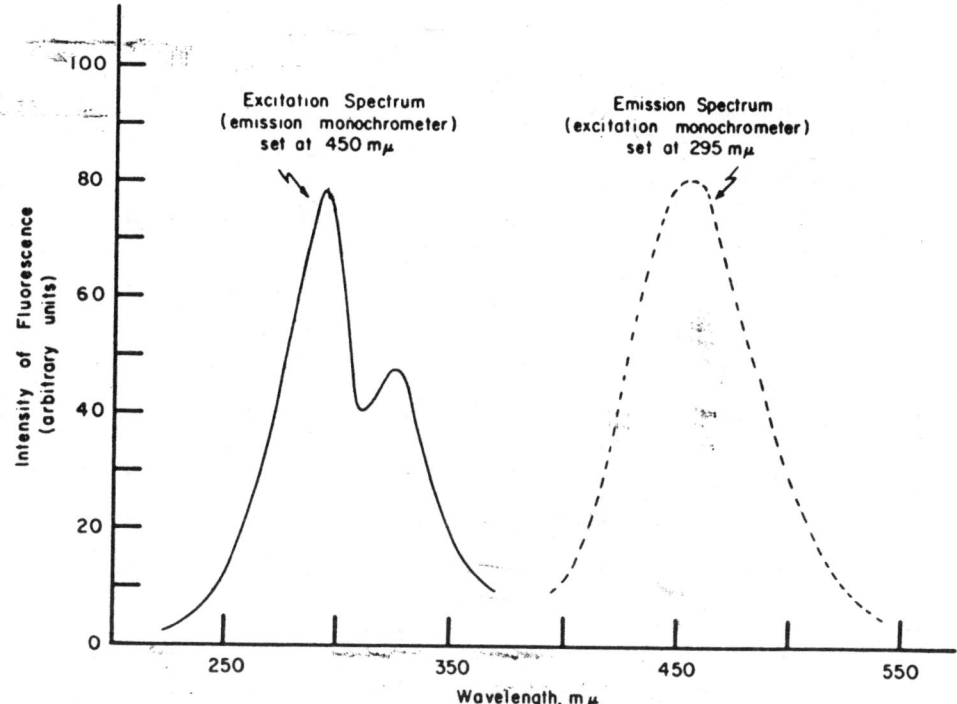

FIGURE 4.3: Excitation and emission spectra of griseofulvin in water containing 1% ethanol. Reprinted from Ref. 41, p. 365, by courtesy of *Nature*.

American Instruments Company, Inc.). The characteristics and features of many available instruments have been the subject of recent excellent reviews.[8,9] It is recommended that the reader consult such reviews and the literature available from manufacturers for specific information on commercially available instruments.

4.5 FACTORS INFLUENCING INTENSITY OF FLUORESCENCE

A. CONCENTRATION OF FLUORESCING SPECIES

The relationship between observed intensity of fluorescence and the concentration of fluorescing species is considerably more complex than that

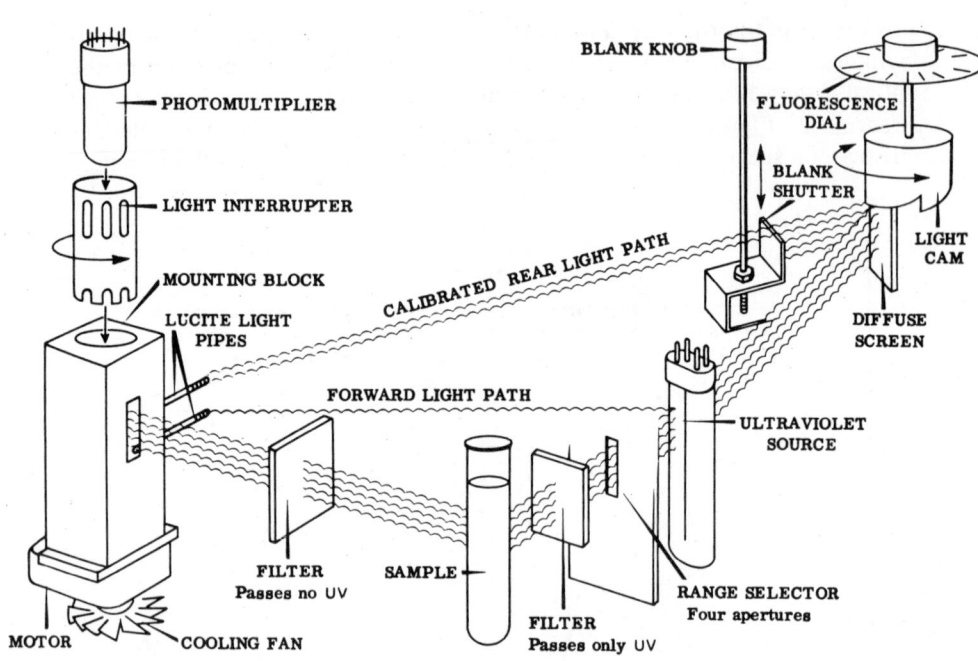

FIGURE 4.4: A schematic diagram of the optics of the Turner, model 110 fluorometer. Courtesy of G. K. Turner Associates.

between absorbance and the concentration of absorbing species which is dictated by Beer's law. Complexities arise from both theoretical considerations and from practical aspects of the instrumentation used to measure fluorescence. It might be intuitively anticipated that a linear relationship might not exist since fluorescence intensity would be expected to be proportional to the concentration of molecules in the excited state and, therefore, proportional to the intensity of radiation responsible for excitation. However,

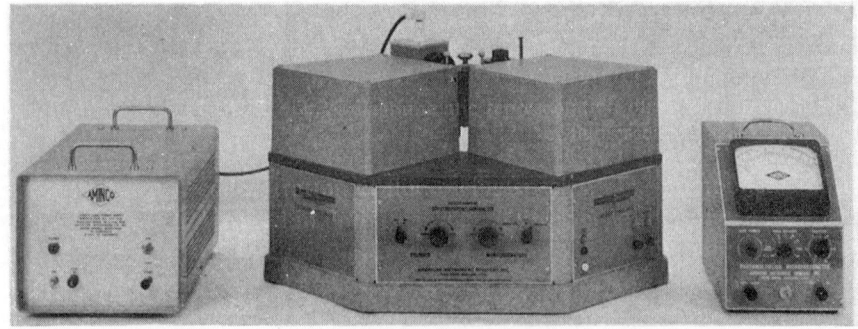

FIGURE 4.5: The Aminco-Bowman spectrophotofluorometer. Courtesy of the American Instrument Company, Inc.

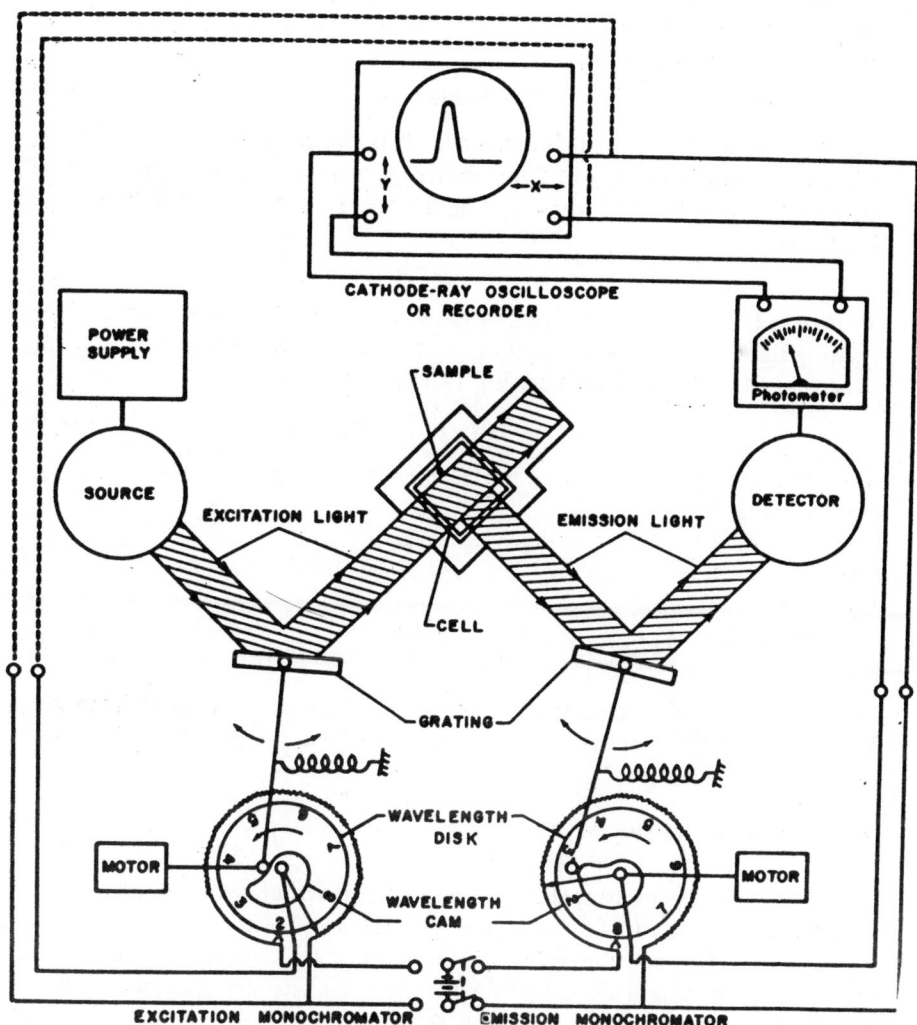

FIGURE 4.6: A schematic diagram of the optics of the Aminco-Bowman spectrophotofluorometer. Courtesy of the American Instrument Company, Inc.

light is absorbed by the sample and the intensity of exciting light does not
remain constant, but diminishes as the light beam traverses a sample. If
the solution is dilute, significant absorption will not occur and the decrease in
intensity of exciting light will not be significant. In a more concentrated
solution, the intensity of exciting light might well be different in different
regions of the sample being irradiated. This consideration was formalized by

Kavanagh[10] and in more detail by Braunsberg and Osborn,[11] who reasoned that fluorescence intensity should be proportional to the amount of light absorbed by the sample, i.e.,

$$F = k\phi(I_0 - I) \tag{4.1}$$

where F is the intensity of fluorescence, k is the proportionality constant, ϕ is the quantum efficiency of fluorescence, I_0 is the intensity of light incident on a sample, and I is the intensity of light transmitted by a sample. Since, by Beer's law,

$$I = I_0 e^{-\epsilon bc} \tag{4.2}$$

where ϵ is the molar absorptivity of the compound at the wavelength of the exciting light, b is the path length along the axis of irradiation, and c is the concentration in moles per liter, Eq. (4.1) can be rewritten:

$$F = k\phi I_0(1 - e^{-\epsilon bc}) \tag{4.3}$$

The term in parentheses can be represented by a series expansion

$$e^x = 1 + x + \frac{x^2}{2!} + \cdots + \frac{x^n}{n!}$$

to yield:

$$F = k\phi I_0 \epsilon bc \left[1 - \frac{\epsilon bc}{2!} + \frac{(\epsilon bc)^2}{3!} - \cdots + \frac{(\epsilon bc)^n}{(n+1)!} \right] \tag{4.4}$$

The detector of a fluorometer does not measure total fluorescence intensity, but rather the intensity from only a segment of the sample. Equation (4.4) must then be modified:

$$S_f = k\phi g\theta I_0 \epsilon bc \left[1 - \frac{\epsilon bc}{2!} + \frac{(\epsilon bc)^2}{3!} - \cdots + \frac{(\epsilon bc)^n}{(n+1)!} \right] \tag{4.5}$$

where S_f is the electric signal generated by the detector, g is the constant reflecting the sensitivity of the detector and the amplification of the detected signal, and θ is the constant reflecting the geometry of the system, particularly the solid angle of light viewed by the detector. As recently discussed by Hercules,[12] there are two concentration regions where Eq. (4.5) can be conveniently simplified. When the concentration c is very small, the equation can be approximated by:

$$S_f = k\phi g\theta I_0 \epsilon bc \tag{4.6}$$

Such an approximation is reasonably valid when $\epsilon bc < 0.05$. Under such conditions, a linear relationship between measured intensity of fluorescence and concentration exists. When the concentration is large, $e^{-\epsilon bc}$ approaches zero and Eq. (4.5) can be approximated by:

$$S_f = k\phi g\theta I_0 \tag{4.7}$$

Under this concentration condition, measured intensity is independent of concentration. At concentrations intermediate to the two extremes, a non-linear relationship would be theoretically expected. It is interesting and relevent to note that the concentration range over which linearity is theoretically expected is dependent on the molar absorptivity of the compound. This is illustrated in Fig. 4.7, which shows the types of intensity-concentration profiles which can be theoretically expected for compounds having different absorptivities. It can be seen that over the concentration range of the graph,

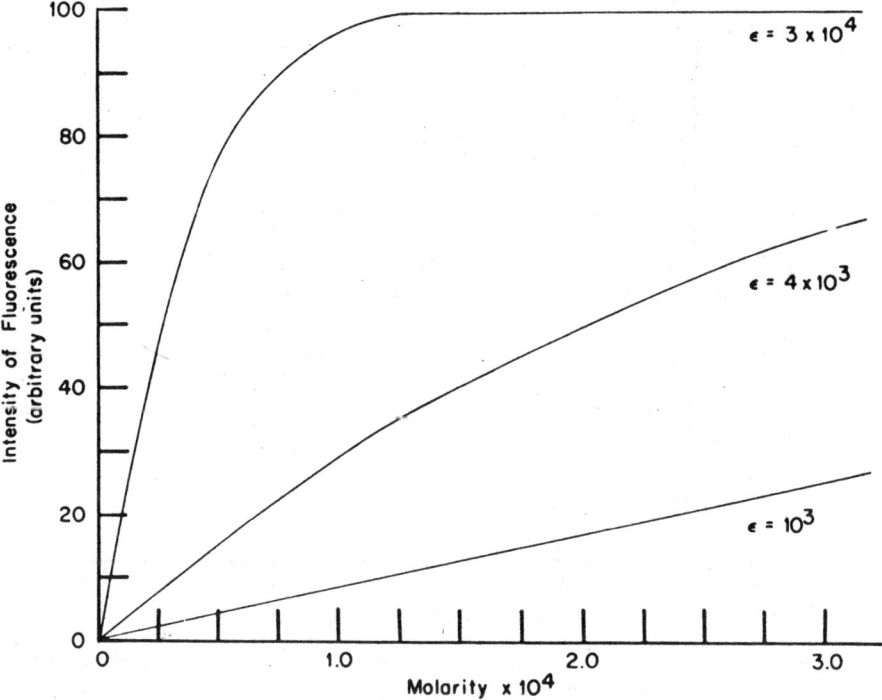

FIGURE 4.7: Intensity-concentration profiles for three different solutes having different molar absorptivities. Reprinted from Ref. 12, p. 31 A, by courtesy of *Analytical Chemistry*.

a linear relationship holds for the compound with the lowest absorptivity, while the compound with a high absorptivity exhibits a linear relationship over a rather small range of concentrations.

A further complication arises if the excitation and emission spectra of the compound overlap. In such a case, photons emitted as fluorescence can be absorbed in exciting other molecules and will not be measured as fluorescence. In dilute solution, such an occurrence will probably not be significant. However, in concentrated solutions, self-absorption of fluorescence radiation can occur and will result in a measured intensity which is less than that predictable on the basis of Eq. (4.5).

A more serious problem which is concentration-related can result due to the geometry of the measuring system. A detector set to view fluorescence at a 90° angle "sees" only a small band in the center of the sample. With dilute solutions, this band emits fluorescence which is representative of the whole cell. However, if the sample is concentrated, sufficient light absorption might occur so that the portion sensed by the detector is only weakly irradiated. This results in the phenomenon of ·concentration reversal, i.e., an

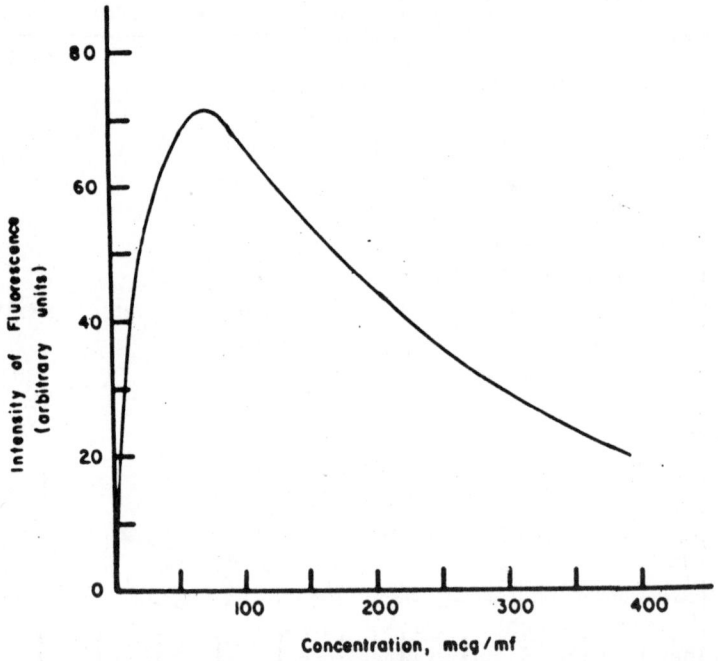

FIGURE 4.8: A plot showing the influence of concentration on the intensity of fluorescence of solutions of phenol.. The excitation wavelength was 295 mμ and the emission wavelength was 330 mμ. Reprinted from Ref. 1, p. 377, by courtesy of the *Journal of Clinical Pathology* by permission of the authors, the editor, and the publishers, B.M.A. House, Tavistock Square, London W.C.1.

increase in concentration results in a decrease in measured fluorescence intensity. Such behavior is illustrated in Fig. 4.8 for phenol.

For these reasons, fluorometric measurements made for assay purposes are restricted to dilute solutions where a linear standard curve can be obtained. In general, a linear response can be expected for solutions which absorb less than 5% of the exciting radiation. It is apparent from Eq. (4.6) that a decrease in measured fluorescence resulting from a reduction in concentration can be compensated for by an increase in the intensity of exciting radiation and/or an increase in the sensitivity of the detector. Because of this, accurate measurements can be made on relatively dilute solutions and a wide range of concentrations can be covered in the linear portion of the intensity-concentration profile. This is in marked contrast to absorption spectrophotometry, where a limited range of concentration is necessary for accurate measurements.

B.   PRESENCE OF OTHER SOLUTES

The presence in a sample of solutes other than the solute being analytically determined can influence the intensity of fluorescence by one or more of a number of different effects.

I. Fluorescent Impurities

An obvious possibility is that another component of the solution might also fluoresce and thus interfere with the determination. Impurities introduced into the sample from solvents, buffers, detergents which are residual on glassware, and from the source of the sample can fluoresce and can introduce error. This possibility should always be anticipated in fluorometric work, particularly if measurements are made on very dilute solutions, and especially if the excitation radiation is in the ultraviolet region of the spectrum. Appropriate precautions must be taken which include the use of pure solvents and chemical reagents and cleanliness in all operations.

2. Inner-Filter Effect

The presence in solution of other solutes which are nonfluorescent can effect fluorescence intensity by the so-called inner-filter effect. The influence here is due to the absorption of light and is similar to the situation discussed previously where excitation and emission spectra overlap. Thus, if the nonfluorescent components absorb either excitation or emission radiation, a reduction in measured intensity of fluorescence will result. Hercules[12] has considered some theoretical aspects of the inner-filter effect and has concluded that if absorption due to other species is constant and if absorption due to the fluorescing species is small, then a linear relationship between measured fluorescence and concentration should still be observed. In such a case, Eq. (4.6) assumes the following form:

$$S_f = k\phi g\theta I_0 \epsilon bc(K) \qquad (4.8)$$

where K is the constant resulting from the presence of other absorbers in the system. It is apparent that if this effect is encountered the concentration of the nonfluorescent absorber must be eliminated or be maintained constant from sample to sample and a standard curve must be used which was determined at that concentration of absorber. An alternative might be to change the wavelength of excitation or emission radiation to minimize this effect.

3. Chemical Quenching

In addition to the effects just discussed, dissolved solutes can result in decreased fluorescence by at least two types of chemical quenching processes.

One is known as *collisional quenching* and results from a diffusion-controlled process in which a molecule of "quencher" interacts with an excited molecule of the potentially fluorescing substance. Interaction results in the dissipation of excitation energy not by fluorescence but by transfer of energy to the quenching molecule. A simplified mechanism can be written to describe this situation:

$$F + h\nu_e \xrightarrow{k_1} F^*$$

$$F^* \xrightarrow{k_2} F + h\nu_f$$

$$F^* + Q \xrightarrow{k_3} F + Q$$

Here, excitation radiation $h\nu_e$ converts the "fluorophor" F to the excited state. The excited molecule F^* can dissipate excitation energy by fluorescence $h\nu_f$ or by interaction with the quencher molecule Q. The rate constants k_1, k_2, and k_3 characterize the rates of the various processes. The intensity of fluorescence will be proportional to the steady-state concentration of molecules in the excited state. In the absence of quenching agent, this concentration is given by:

$$(F^*)_0 = \frac{k_1}{k_2}(F) \tag{4.9}$$

In the presence of quencher, the concentration of excited molecules will be reduced and is given by:

$$(F^*)_q = \frac{k_1(F)}{k_2 + k_3(Q)} \tag{4.10}$$

The ratio of intensities in the absence (f_0) of quencher to that in the presence (f_q) is, therefore,

$$f_0/f_q = 1 + k_3/k_2(Q) \tag{4.11}$$

Equation (4.11) is known as the Stern-Volmer law and predicts that the ratio of the two intensities will be linearly dependent on the concentration of quenching agent.

The molecular basis for dissipation of energy by collisional quenching is poorly understood. A recent hypothesis was discussed by Hercules[13] and involves electron transfer between excited state and quenching agent. It may be represented as follows:

$$F^* + Q \rightarrow F^- : Q^+$$

$$^3F + Q \qquad F + Q$$

The excited molecule F^* interacts with quencher Q, abstracting an electron to form an ion pair $F^- : Q^-$. The ion pair can dissociate to give either a triplet state ^3F and Q or a ground state F and Q. Both processes result in thermal dissipation of energy.

Another type of quenching is called *static quenching*. Here, complex formation occurs between a potentially fluorescing molecule in the ground state and a quencher molecule. If the complexed form of the potentially fluorescing molecule has different spectral characteristics than the free form, it may not undergo excitation or may be excited to a lesser degree that the non-complexed species. Suppose, for example, that complex formation occurred and resulted in the formation of a complex, with a stability constant of K_s, that was not fluorescent, i.e.,

$$F + Q \overset{K_s}{=} F:Q$$

$$F + h\nu_r \xrightarrow{k_1} F^*$$

$$F^* \xrightarrow{k_2} F + h\nu_f$$

The steady-state concentration of F^* in the absence of Q is given by

$$(F^*)_0 = k_1/k_2 (F)_t \tag{4.12}$$

where $(F)_t$ is the total concentration of potentially fluorescing compound. In the presence of quencher,

$$(F^*)_q = \frac{(F)_t}{1 + K_s(Q)} \frac{k_1}{k_2} \tag{4.13}$$

The ratio of fluorescence in the absence to that in the presence of static quencher is then given by:

$$f_0/f_q = 1 + K_s(Q) \tag{4.14}$$

The reduction in fluorescence caused by this type of quenching is thus dictated by the stability constant of the complex and by the concentration of quenching agent.

Many examples of chemical quenching are found in the literature. Halide ions such as iodide and chloride are well known examples of collisional quenchers. Caffeine, related xanthines, and purines[14] have been shown to influence the fluorescence of riboflavin by static mechanisms. In the usual case, quenching is an undesirable effect and the possibility of encountering this type of interference should always be evaluated in developing a fluorometric assay. It is possible to utilize this phenomenon, however, as an analytical means for determining the concentrations of compounds known to quench fluorescence. In addition, it is apparent from Eq. (4.14) that it is possible to employ fluorometric methods for the investigation of complex-forming equilibria.

C. HYDROGEN-ION CONCENTRATION

The intensity of fluorescence emitted by solutions of weak acids and weak bases can exhibit a dependency on the pH of the solution. The effect here may be due simply to a change in the degree of ionization of the weak electrolyte. For example, suppose that the fluorescence of a weak acid HA was investigated in dilute solution using an emission filter or monochromator

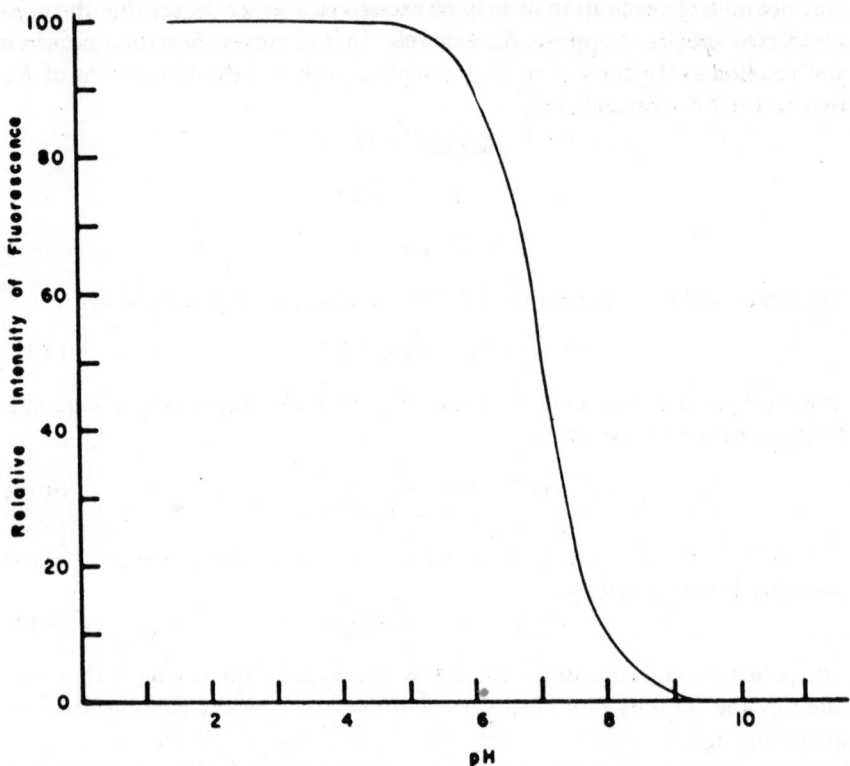

FIGURE 4.9: A plot showing the influence of pH on the fluorescence of a solution of a weak acid with a pK_a of 7. The measured fluorescence is assumed to be due to the conjugate acid. The fluorescence intensity of a strongly acid solution was assigned a value of 100 and the intensities at other values of pH were calculated relative to this.

setting that transmitted fluorescence that was specifically due to the unionized form of the acid. Then

$$S_f = k''(HA) \qquad (4.15)$$

where k'' is a combination of all the constants of Eq. (4.6). However, since

$$(HA) = \frac{C_{HA}(H^+)}{K_a + (H^+)} \qquad (4.16)$$

where C_{HA} is the stoichiometric concentration of weak acid, K_a is the dissociation constant of the weak acid, and (H$^+$) is the concentration of solvated protons, Eq. (4.15) becomes

$$S_f = \frac{k''C_{HA}(H^+)}{K_a + (H^+)} \qquad (4.17)$$

An intensity-pH profile such as that shown in Fig. 4.9 would be expected. Similarly, if fluorescence which is specific for the conjugate base is detected,

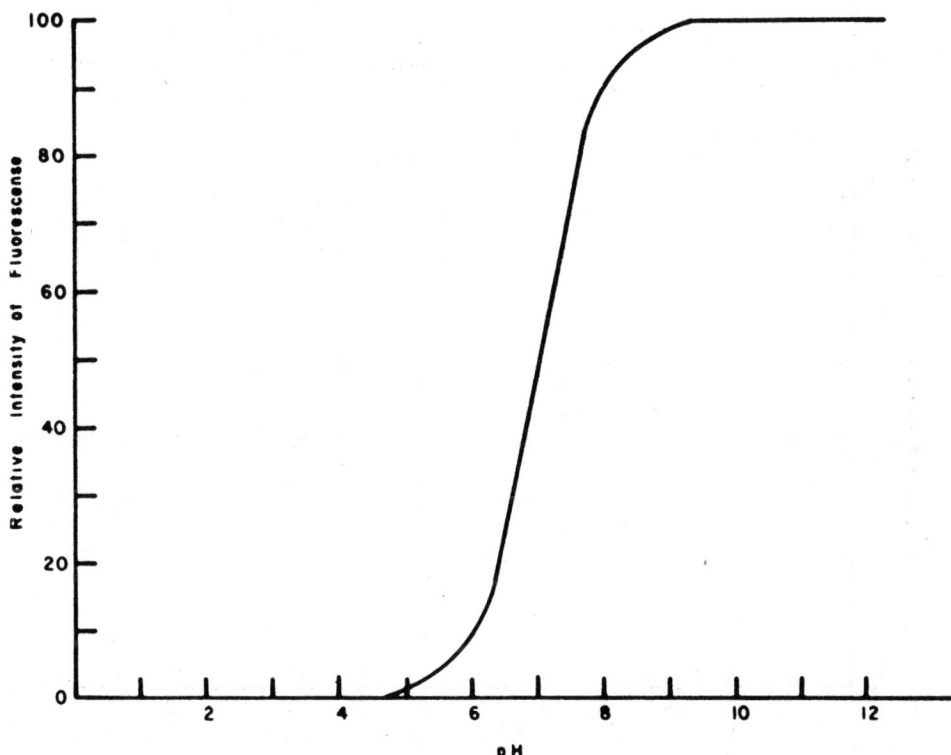

FIGURE 4.10: A plot showing the influence of pH on the fluorescence of a solution of a weak acid with a pK_a of 7. The measured fluorescence is assumed to be due to the conjugate base. The fluorescence intensity of a strongly alkaline solution was assigned a value of 100 and the intensities at other values of pH were calculated relative to this.

then Eq. (4.6) becomes

$$S_f = \frac{k'''C_{HA}K_a}{K_a + (H^+)} \qquad (4.18)$$

and the pH profile illustrated in Fig. 4.10 would be expected.

A more complex pH effect can be observed with some compounds and is due to the acid strength of a molecule in the excited state being different from the acid strength in the ground state. This difference has been shown to be quite marked for a number of compounds. The phenomenon is known as *excited-state dissociation*. It was first studied by Forster[15] and was recently discussed by Ellis.[16] The occurrence is best illustrated by example, and Fig. 4.11 illustrates such a case. Here, the relative intensity of fluorescence for solutions of 2-naphthol was plotted as a function of pH. This phenolic

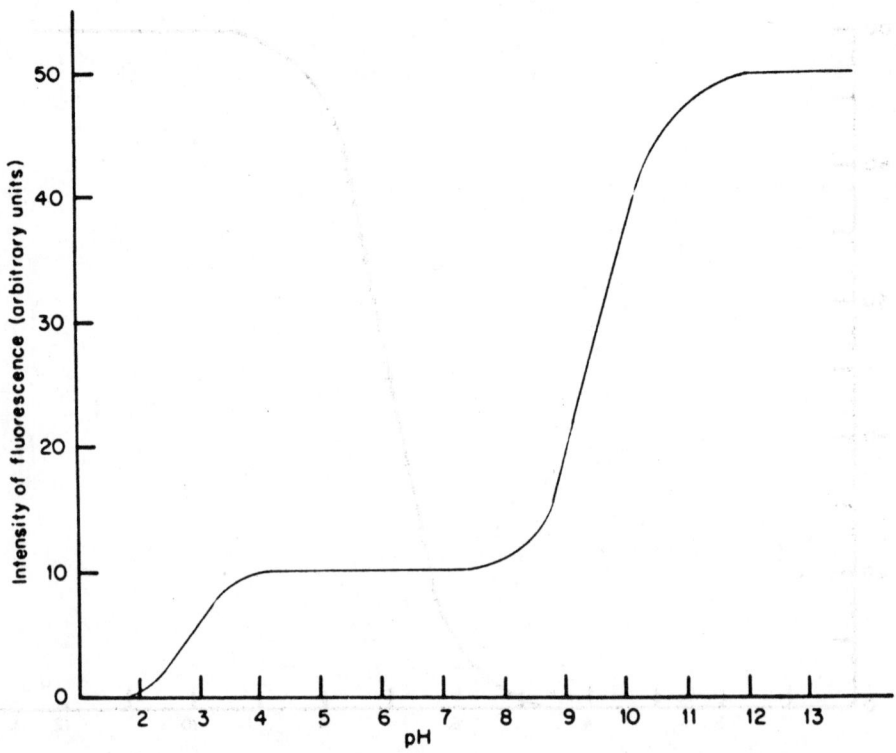

FIGURE 4.11: A plot showing the influence of pH on the intensity of fluorescence of solutions of 2-naphthol. An emission filter which passed wavelengths longer than 415 mμ was employed. The measured fluorescence was, therefore, due to excited-state anions. Reprinted from Ref. 16, p. 261, by courtesy of the *Journal of Chemical Education*.

compound has a pK_a of 9.5. Both ionized and un-ionized species fluoresce, but the fluorescence peak for the ionized form is at 429 mμ, while that for the un-ionized species is at 359 mμ. The fluorescence measurement plotted in Fig. 4.8 was, therefore, due to the 2-naphthoate anion. If the degree of ionization of the acid in the ground state was the only property influenced by a change in the concentration of hydrogen ion, then no observable fluorescence would be expected until a pH was achieved where a detectable degree of ionization occurred, i.e., a pH of 8.5 (pK_a − 1). That fluorescence was observed in the region of 2 to 8.5 is indicative of excited-state dissociation. A possible mechanism, consistent with the data, is given by:

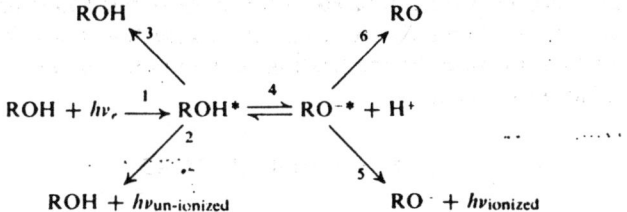

The processes are

1. Absorption of radiation by the un-ionized form
2. Fluorescence of the un-ionized form
3. Radiationless dissipation of energy
4. Excited-state dissociation to produce a proton and an anion in the excited state
5. Fluorescence of the anion
6. Radiationless dissipation of energy

The fluorescence at 429 mμ exhibited by 2-naphthol in the pH range of 2 to 8.5 means that excited-state anions did exist in this range and must have resulted from excited-state dissociation of the molecular species. The first inflection point in the pH profile corresponds to the excited-state dissociation constant, which is approximately 2.8 for 2-naphthol. The second inflection point, of course, corresponds to the ground-state dissociation constant.

Practical consequences of the pH effects on fluorescence are the same as those encountered in spectrophotometry. Thus, with certain compounds, pH must be considered as an experimental variable that is important to control. In some instances, it may be possible to utilize the influence of pH to minimize interferences and to conduct differential fluorometric determinations by appropriate pH adjustment. Additionally, fluorometry might offer a convenient approach to the determination of acidity constants of some compounds.

D. TEMPERATURE

The quantum efficiency of fluorescence is found to decrease with an increase in temperature. Thus, the higher the temperature, the more efficient and effective are radiationless processes in dissipating excitation energy. It is felt that the process of internal conversion is the one whose rate is most significantly influenced by temperature changes. The effect is most probably due to an increase in the thermal motion of molecules at higher temperatures. The increased motion favors the probability of intermolecular collision and subsequent energy loss. In general, a rise in temperature of 1°C results in a decrease in the intensity of fluorescence of about 1%. For some compounds, the sensitivity of fluorescence to temperature is even more pronounced.

Because of the temperature effect, a reasonable degree of temperature control is necessary in fluorometric methods. It is recommended, for example, that analytical samples be equilibrated to the same temperature before measurements are made. Similarly, readings should be taken within a reasonable period of time to preclude the heating of a sample contained in the sample holder of an instrument.

E. OTHER FACTORS

1. Degradation of Sample

The stability of a compound in an analytical sample is of concern in fluorometric methods, as it is in other methods of analysis. Here, in addition to autoxidative and solvolytic degradative routes, the possibility of photolytic degradation should always be anticipated. Many compounds are subject to light-catalyzed degradations and rearrangements. In fluorometric measurements the intensity of light used and the amount of sample irradiated can be sufficiently large to cause measurable loss of material during the time required for measurement. The error introduced by photodecomposition can be reduced by decreasing the intensity of exciting light and by making the measurement in as short a time period as possible.

2. Solvent Effects

The medium in which a potentially fluorescing material is dissolved can influence the intensity and characteristics of fluoresced light. As previously mentioned, impurities in solvents can contribute artifactual fluorescence. In addition, Raman scattering of the exciting light by the solvent might be erroneously measured as being due to the fluorescence of a sample. This effect is usually not of practical significance except when measurements are made on very dilute solutions. Other effects can be encountered such as the quenching of fluorescence by the solvent or by substances such as oxygen dissolved in the solvent. The spectral characteristics of fluoresced light can vary from solvent to solvent due to polarization effects and hydrogen bonding. Since these effects cannot be quantitatively predicted, the solvent cannot be changed at random in fluorometric methods of analysis.

4.6 COMPARISONS OF FLUOROMETRY WITH SPECTROPHOTOMETRY

A. SENSITIVITY

Fluorometry is significantly more sensitive as an analytical tool than is spectrophotometry. In spectrophotometry the intensity of light transmitted by a sample is measured and compared to that transmitted by a blank. The lower limit of detectability is determined by the smallest concentration that will yield a detectable intensity *difference* between sample and blank. As the transmittance of a sample approaches 1, small errors made in measuring the difference between the two intensities result in large errors in calculated concentration. The lowest limit of concentration that can be detected with accuracy is, for all practical purposes, established by the molar absorptivity of the compound under investigation. A fluorometer measures directly the intensity of fluoresced light. Moreover, a decrease in the concentration of fluorescing species can be compensated for by an increase in the intensity of exciting light and/or an increase in the sensitivity of the detector. It is the latter variable that most significantly contributes to the sensitivity of the method. The directly measured intensity can be amplified more readily and accurately than the intensity difference measured in spectrophotometry. The lower limit of concentration here is, therefore, established by characteristics of the instrument and not usually by characteristics of the fluorescing species. In practice, it is the level of inherent "noise" of the instrument relative to the signal caused by sample fluorescence that dictates the lower limit of detectability. It has been calculated[17] that fluorescence measurements can offer sensitivity increases of as high as 10^3–10^4 over absorbance measurements.

B. SPECIFICITY

A fluorometric assay can offer a degree of specificity that might not be attainable with a corresponding spectrophotometric technique. Not all compounds which absorb ultraviolet and visible light fluoresce, and so a potentially interfering compound which absorbs light will not necessarily be a source of interference in a fluorometric method. In addition, the analyst has the ability to vary the wavelengths of both exciting and fluorescing light and to choose a combination of wavelengths which will maximize the measured fluorescence from the compound in question and minimize contributions from interfering substances. It should be noted in this respect that measurements need not be made at wavelengths corresponding to maxima or minima of the spectra. The equations relating fluorescence intensity to concentration hold for any region of the spectrum.

C. EXPERIMENTAL VARIABLES

It is obvious from the previous discussions that there are a larger number of experimental variables that must be controlled in fluorometric methods of analysis than in corresponding spectrophotometric methods. For example, temperature and the intensity of incident light must be maintained reasonably constant in a fluorometric method, but need not be rigidly controlled in a spectrophotometric procedure. Extraneous solutes can markedly affect the intensity of fluorescence by quenching effects, whereas in spectrophotometry it is unusual to encounter a system in which the absorbance of a compound is significantly altered by the presence of other solutes. In addition, the influence of pH on fluorescence can be much more complex than on absorbance and might necessitate closer control of pH in fluorometric procedures than in spectrophotometric assays.

4.7 APPLICATION OF FLUOROMETRY TO PHARMACEUTICAL ANALYSIS

An examination of official compendia might lead to the impression that fluorometry has limited application in the analysis of drugs. There are, for example, only thirteen monographs in USP XVII, BP (1963), and NF XII that specify the utilization of fluorometry as an assay tool and these are concerned with the determination of only two compounds, riboflavin and thiamine. However, a survey of the literature reveals that fluorometry has enjoyed widespread use in the analysis of drugs in systems other than dosage forms. The sensitivity of the method has resulted in its application in a host of pharmacological, biochemical, toxicological, pharmacokinetic, and biopharmaceutical studies for the analysis of small amounts of drugs in biological fluids and tissues. It is not practical nor necessary to review all such applications in this chapter. These have been discussed in some detail by Udenfriend,[9] Phillips and Elevitch,[17] and Williams and Bridges.[1] More recent reviews are found in the annual *Analytical Review* editions of the journal *Analytical Chemistry*. In 1965, Wimer et al.[18] reviewed the literature appearing in the period 1962–1964 which dealt with the analysis of pharmaceuticals. In 1966, White and Weissler[19] reviewed publications pertinent to the subject of fluorometric analysis, including those relating to pharmaceutical analysis and which appeared in the period 1963–1965. It is illuminating and illustrative of the wide applicability of fluorometry to drug analysis to note from these reviews that in the period 1962–1964, publications appeared which described fluorometric procedures for the analytical determination of the following drugs or classes of drugs: adrenaline, aldosterone, androsterone, antihistamines, atropine, barbiturates, chlorpromazine, chlorprothine, codeine, dipyridamole, emetine, ergot alkaloids, estradiol, estriol, esterone, ethinyl estradiol, fursemide, imipramine, isoniazid, mephenesin, mescaline, morphine, narcotine, panthenol, papaverine, quinidine, quinine, reserpine, riboflavin, salicylates, streptomycin, sulfonamides, testosterone, tetracyclines, thebaine, tubocurarine, yohimbine.

A limited number of examples are discussed in the following section to more specifically illustrate the applicability and utility of fluorometry as an analytical tool. These examples will also demonstrate that in most instances the actual determination of the intensity of fluorescence of a sample preparation is the terminal step in a series of operations which demand that the analyst be cognizant of a variety of separatory and chromatographic techniques, and the chemical and physical properties of the constituents of a sample.

Aminocrine. A simple procedure for the determination of aminocrine (**X**) in drug preparations was recently described by Roberts[20] and illustrates a relatively direct fluorometric assay method. Here, the aminocrine was extracted with chloroform from a basic solution; the chloroform was evaporated and the residue was dissolved in acidic ethanol. The fluorescence of the resulting solution was determined using an excitation filter having maximum transmittance at 365 mμ and an emission filter which transmitted light of wavelengths greater than 415 mμ. The concentration of aminocrine was determined by comparing the fluorescence of the sample preparation to that of a standard preparation. The method was applied to a variety of aminocrine-containing dosage forms, including suppositories, creams, ointments, jellies, and tablets. Other constituents of the dosage forms did not interfere

(X)

with the determination. The method appears to be sensitive, as evidenced by the recommended use of a standard preparation having a concentration of 1 μg/ml.

Phenothiazines. Mellinger and Keeler[21] published an interesting study on the fluorescence characteristics of phenothiazine drugs. Thioridazine (**XI**) is representative of the compounds studied. All of the phenothia-

(XI)

zines exhibited similar excitation and emission spectra. The emission spectrum of a solution prepared in 0.2 N sulfuric acid was characterized by a single peak in the range of 450 to 475 mμ. The excitation spectrum was found to have two peaks, one at approximately 250 mμ and the other in the range of 300 to 325 mμ. Addition of potassium permanganate to such solutions resulted in marked changes in spectral characteristics. The wavelength of maximum emission shifted to much lower wavelengths and the intensity of fluorescence at this maximum was much greater (15–20 times) than that exhibited by untreated drug at its maximum. In addition, the excitation spectrum, after permanganate treatment, exhibited four peaks. Evidence was obtained to indicate that oxidation of the phenothiazine to the corresponding sulfoxide was responsible for this behavior. The spectra of phenothiazine-like compounds such as prothipendyl (**XII**), chlorprothixene (**XIII**), and imipramine (**XIV**) were affected by permanganate treatment, but the

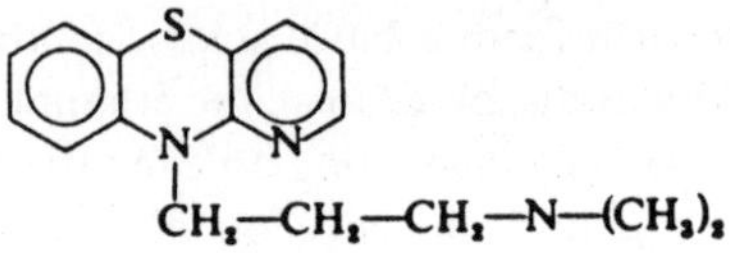

(XII)

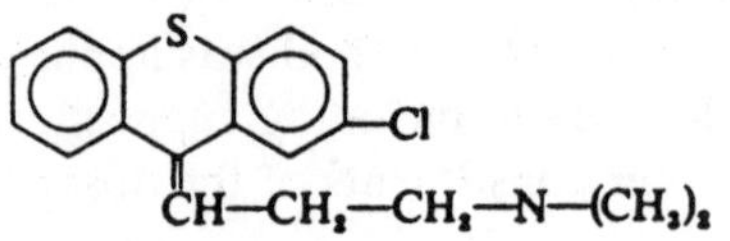

(XIII)

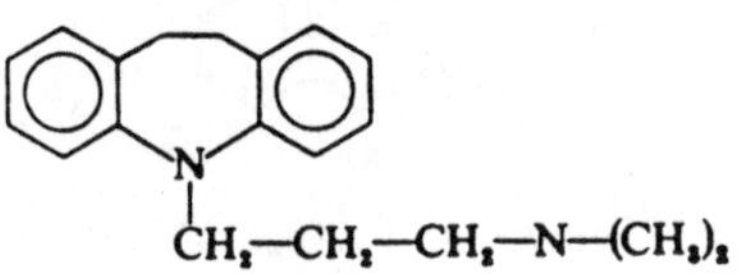

(XIV)

effects were qualitatively much different than those observed with phenothiazines.

Pretreatment of phenothiazines with potassium permanganate permitted quantitative fluorometric analysis of simple solutions in concentrations as low as 0.01 to 0.05 μg/ml. The sensitivity was somewhat less for samples obtained from biological sources but was significantly greater than that afforded by other techniques. For example, thioridazine in urine could be detected fluorometrically at a level as low as 0.8 μg/ml. In contrast, a concentration of 4 μg/ml was required for spectrophotometric detection, while 6 μg/ml was necessary for detection by a colorimetric method which was based on the treatment of a phenothiazine with concentrated sulfuric acid.

Salicylates. In 1948, Saltzman[22] described a fluorometric method for the estimation of salicylates in blood. It was based on the observation that salicylates, in alkaline medium, exhibit a blue fluorescence. The procedure involved precipitation of proteins from a sample with dilute tungstic acid and treatment with strong alkali. Alternatively, a sample of plasma was acidified and extracted with ethylene dichloride. The salicylate was then backextracted into strong alkali. The fluorescence of the resulting solution was measured using a 370-mμ excitation filter and a 460-mμ emission filter. Chirigos and Udenfriend[23] studied the fluorescence characteristics of salicylate in more detail and determined from spectrofluorometric studies that the excitation maximum was at 310 mμ and the emission maximum was at 400 mμ. Salicylate content of biological tissue was determined by extracting a sample with ether, back-extracting into a borate buffer, and fluorometric examination of the borate solution. More recently, Lange and Bell[21] used fluorometry coupled with paper chromatography as the basis for a micro-method for the determination of acetylsalicylic acid and salicylic acid in blood sample. Here, a small sample of blood (100 μl) was extracted with ethylene dichloride. Aliquots of the ethylene dichloride extract were spotted on paper strips and the strips were developed by the ascending technique using 0.75% nitric acid as the solvent system. Segments containing the aspirin and the salicylic acid were cut from the strip and eluted with 5 N sodium hydroxide. The fluorescence of solutions prepared in this manner were determined using as a blank a solution prepared by treating another portion of the chromatograph strip with alkali. Standard curves were prepared for the two compounds by subjecting blood samples containing known amounts of the drugs to the procedure and relating observed fluorescence to concentration. The method was applied to an investigation of aspirin and salicylic acid blood levels after the oral administration of aspirin tablets.

Homovanillic Acid. Homovanillic acid (3-methoxy-4-hydroxylphenylacetic acid, XV) is derived from the metabolism of 3,4-dihydroxyphenylalanine and 3,4-dihydroxyphenylethylamine. The urinary excretion of homovanillic acid (HVA) is elevated in patients with neuroblastoma and pheochromo-cytoma and interest has been expressed in utilizing urinary levels of HVA as a diagnostic tool. Anden et al.[25] and Sharman[26] discovered independently that HVA reacted with oxidizing agents such as potassium ferricyanide and ferric chloride, in alkaline medium, to yield a highly fluorescent solution. Although the product responsible for the fluorescence was not identified, both groups developed fluorometric procedures for the determination of HVA in biological tissues. The procedure of Anden et al. was modified by Sato,[27] who incorporated an ion-exchange treatment and solvent extraction into the procedure to isolate the HVA, which was then treated with ammonia and potassium ferricyanide. The fluorescence of the resulting solution was then determined. Corrodi and Werdinius[28] more recently studied the procedure in more detail and determined that the compound which was responsible for the fluorescence was 2,2'-dihydroxy-3,3'-dimethoxy-biphenyl-5,5'-diacetic acid (XVI). This compound exhibited an excitation maximum

(XV)

(XVI)

at 315 mμ and an emission maximum at 425 mμ.

An interesting application of this oxidative transformation of HVA was suggested by Guilbault et al.[29] They found that the conversion of (XV) to (XVI) could be accomplished enzymatically in a hydrogen peroxide-peroxidase system. They formulated solutions containing HVA, hydrogen peroxide, and peroxidase and measured the rate of change of fluorescence. At a constant concentration of HVA, this rate was found to be directly proportional to the peroxidase concentration and to the concentration of hydrogen peroxide. They recommended this approach for the analysis of oxidative enzymes and of hydrogen peroxide. They reported that as little as 10^{-11} mole/liter of peroxide and 10^{-3} unit/ml of peroxidase are determinable by this method.

Cardiac Glycosides. Jakovljevic[30] reviewed the many methods which have been proposed for the determination of cardiac glycosides. He noted that a number of the reported methods are fluorometric and are based on the generation of fluorophors by the action of dehydrating agents on the steroid moiety of the glycoside. He proposed a new reagent for this purpose, a mixture of acetic anhydride, acetyl chloride, and trifluoroacetic acid. He described the use of this reagent and a procedure for the simultaneous determination of digitoxin (XVII) and digoxin (XVIII) in mixtures by fluorometry. The method is an interesting illustration of how assay specificity can be attained by utilizing a knowledge of both excitation and emission characteristics of fluorescing compounds.

(XVII)   (XVIII)

Digoxin differs from digitoxin by one hydroxyl group at position 12. However, when the two compounds were treated under anhydrous conditions with the dehydrating agent, they yielded products which had significantly different fluorescence characteristics. The fluorophor generated from digitoxin exhibited a single excitation peak at 470 mμ and a single emission peak at 500 mμ. The digoxin fluorophor exhibited two excitation peaks at 345 and 470 mμ. Excitation at 345 mμ resulted in an emission peak at 435 mμ, while excitation at 470 mμ gave an emission peak at 500 mμ. Under the latter conditions, the fluorescence was approximately 30% of that obtained with digitoxin. When the digitoxin preparation was examined at an excitation wavelength of 345 mμ and an emission wavelength of 435 mμ, no fluorescence was observed. The author suggested that the treatment of digitoxin resulted in the formation of a substituted 3,4-benzpyrene, while digoxin yielded a mixture containing compounds related to 3,4-benzpyrene and chrysine. The differences in spectral characteristics permitted the simultaneous determination of digitoxin and digoxin in samples prepared from digitalis leaf and digitalis tincture. Here, the glycosides were isolated by extraction from a sample, treated with the dehydrating agent for 30 min at 45°C, and diluted with dichloromethane. The fluorescence of the resulting solution was determined with a fluorometer using two different filter combinations. One combination was equivalent to an excitation wavelength of 470 mμ and an emission wavelength of 500 mμ. The measured fluorescence under these conditions resulted from both digitoxin and digoxin. The other combination was equivalent to an excitation wavelength of 345 mμ and an emission wavelength of 435 mμ. Fluorescence here was due to digoxin. The later reading could be used to calculate, by comparison to a standard, the concentration of digoxin. A knowledge of this concentration was then used to calculate the fluorescence due to digoxin under the excitation and emission conditions of the first filter combination and to obtain a corrected fluorescence which reflected the digitoxin concentration. The corrected fluorescence was then used to calculate the digitoxin concentration by comparison to that exhibited by a standard. The method was also applied to the determination of digitoxin in tablets and ampoules.

Reserpine. The reaction of nitrous acid with reserpine (**XIX**) to yield a yellow fluorescent pigment was described by Szalkowski and Mader[91] and

(XIX)

forms the basis for a widely used colorimetric method for the determination of reserpine in pharmaceutical preparations. Reserpine has been used as an additive to poultry feeds, where it is found in concentrations as low as 0.0001 %. The low levels encountered in such systems could not be satisfactorily determined by the colorimetric method of analysis. Mader et al.[32,33] utilized the fluorescence characteristics of nitrous acid-treated reserpine to obtain the desired sensitivity. Treatment of reserpine with nitrous acid was found to result in the formation of a product which possessed an excitation maximum at 390 mμ and an emission maximum at 510 mμ. The intensity of fluorescence was linearly related to concentration over a wide range. The reserpine in a feed sample was isolated by a series of extractions and was eventually obtained as an assay preparation in chloroform-methanol. An aliquot of this solution was treated with sodium nitrite and the mixture was acidified with hydrochloric acid. After an appropriate reaction time, sulfamic acid was added to consume the excess nitric acid and the fluorescence of the solution was determined. A blank was prepared by treating an aliquot of the assay preparation in a similar manner, but with the omission of the sodium nitrite. The concentration of reserpine in the assay preparation was calculated by comparison of the measured fluorescence to that found with a standard preparation which was obtained by carrying a known amount of reserpine through the extraction and reaction procedures.

Haycock et al.[34] recently reported the results of a study in which the kinetics and mechanisms of the nitrous acid-induced fluorescence of reserpine were investigated. They presented evidence to show that the fluorescence was due to the formation of 3-dehydroreserpine. The kinetics of the reaction indicated that protonated reserpine initially reacted with nitrous acid to form an intermediate complex, which then underwent an acid-catalyzed reaction to form 3-dehydroreserpine.

Vitamins. Udenfriend[35] reviewed the many fluorometric methods that have been used for the determination of vitamins. He discussed, in some detail, procedures for vitamin A, thiamine, riboflavin and related flavins, nicotinamide, pyridoxine and related compounds, ascorbic acid, vitamin D, folic acid, p-aminobenzoic acid, cyanocobalamin, tocopherols, and vitamin K. The fluorometric determinations of thiamine and riboflavin are of special interest in that they serve as rather classic examples of the application of fluorometry to pharmaceutical analysis.

Thiamine (XX) possesses little native fluorescence, but is readily oxidized to thiochrome (XXI), which is highly fluorescent. Thiochrome has been

(XX)   (XXI)

reported to have an excitation maxima at 365 mμ and an emission maximum of 450 mμ.[36] The procedure described in the seventeenth revision of the USP serves as an example of the fluorometric assay. An aliquot of a sample solution of the vitamin is treated with an oxidizing reagent (an alkaline solution of potassium ferricyanide). The thiochrome which is formed is extracted into isobutanol and the fluorescence of the resulting solution is determined. The fluorescence is corrected by use of a blank and is compared to that of a standard preparation.

Riboflavin has a characteristic pronounced native fluorescence and can be assayed by direct fluorometric examination of a sample solution. As will be discussed in the practical section, fluorometric assays for riboflavin usually employ an internal standard and an internal blank.

4.8 PRACTICAL SECTION

A. GENERAL

The procedures used in fluorometric assays are, in most instances, quite similar to those encountered in colorimetry and spectrophotometry. Treatment of a sample frequently involves dilution, extraction and/or chromatographic separation, chemical reaction, and finally the determination of the intensity of fluorescence of the final assay preparation using a suitable fluorometer fitted with appropriate filters. The intensity is "read out" in arbitrary units and, after correction for the fluorescence contributions for the blank, can be used to estimate the concentration of fluorescing substances. For a nonlinear intensity-concentration profile, a standard curve must be used for this purpose. If the dilution is such that intensity is directly proportional to concentration, the concentration of the assay preparation can be obtained by comparing the intensity reading to that of a single, standard preparation. In such a case, concentration is calculated by a formula, familiar from spectrophotometric assays:

$$C_u = C_s F_u / F_s \tag{4.19}$$

where C_u is the concentration of the assay preparation, C_s is the concentration of the standard preparation, F_u is the fluorometer reading, corrected for blank, obtained with the assay preparation, and F_s is the fluorometer reading, corrected for blank, obtained with the standard preparation. The *International Pharmacopoea*[37] cautions that the ratio F_u/F_s should not be less than 0.04 and not more than 2.50 because of the limited concentration range within which fluorescence is proportional to concentration.

Frequently, internal standards are prescribed in fluorometric procedures. Here, a known quantity of pure material is added to the assay preparation to compensate for quenching effects which might be introduced during the work-up of a sample. The USP XVII assay for riboflavin[38] serves as an example. This assay specifies the treatment of 10 ml of an assay preparation with 1 ml of water and 2 ml of reagents. The fluorescence of the resulting solution is measured and designated I_u. Another 10 ml of assay preparation is treated with 1 ml of a standard preparation containing 0.001 mg of riboflavin per ml and 2 ml of reagents. The fluorescence of this solution is measured and designated I_s. The concentration of vitamin in the assay preparation in milligrams per milliliter is calculated by the formula:

$$C_u = \frac{I_u - I_b}{I_s - I_u} \times 1/13 \times 0.001 \times 13/10 = \frac{I_u - I_b}{I_s - I_u} \times 0.0001 \tag{4.20}$$

where I_b is the fluorescence reading obtained with a blank.

Various procedures are used to obtain blank readings in fluorometry. Conventional blanks are sometimes specified and are prepared by substituting in the final step of an assay a volume of water or buffer for the required volume of assay preparation. When the possibility exists of fluorescent materials being introduced to the assay preparation by the system containing the compound of analytical interest, a more realistic blank is usually recommended. For example, plasma and urine blanks are prepared by carrying a volume of drug-free plasma or urine through the complete procedure. Internal blanks are frequently employed. Here the assay preparation is used as a blank after specifically eliminating, through chemical reaction, the fluorescence due to the drug. In the riboflavin assay, for example, a few crystals of sodium hydrosulfite are added to the cuvette immediately after the fluorescence intensity of a sample is measured. The hydrosulfite rapidly and specifically converts riboflavin to the nonfluorescent, reduced form. The fluorescence of the resulting solution is measured and is used as a blank reading to correct for fluorescence arising from sources other than riboflavin. In other instances, blanks are prepared by omitting a reagent necessary for the generation of a fluorescing species. For example, the official assay for thiamine[39] is based on the oxidation of nonfluorescent thiamine to strongly fluorescent thiochrome. The oxidizing agent employed is alkaline potassium ferricyanide solution. Blanks for both assay and standard preparations are prepared by submitting samples to the full procedure, but with the substitution of a volume of sodium hydroxide solution for the volume of oxidizing reagent which is normally used.

Numerous instrumental variables such as the intensity of exciting light, detector response, signal amplification, etc., influence the measured intensity of fluorescence. Aging of a light source and fatigue of a detector could, for example, result in nonreproducibility of results and assay error. It is important, therefore, to periodically check the sensitivity of a fluorometer and to adjust it to constant sensitivity during the course of assay measurements. A solution of a stable, strongly fluorescing substance is used for this purpose and is known as a comparison standard. The standard chosen should have fluorescence characteristics similar to those of the compound being assayed and, in fact, if that compound is sufficiently stable, no other comparison standard is needed. A solution of quinine sulfate is frequently recommended as a comparison standard. Its use is illustrated by the USP assay for thiamine. A solution of quinine sulfate in 0.1 N sulfuric acid at a concentration of 0.25 μg/ml is recommended since "this solution fluoresces to approximately the same degree as the thiochrome obtained from 1 μg of thiamine hydrochloride and is used to correct the fluorometer at frequent intervals for variations in sensitivity from reading to reading within an assay."

B. LABORATORY PROJECTS IN FLUOROMETRY

The following projects are offered as guides for possible laboratory exercises illustrating some principles and applications of fluorometry. Since different makes of fluorometers differ in sensitivity, ranges of sensitivity, and the manner by which ranges of sensitivity are selected and adjusted, exact experimental details cannot be presented. The student should initially become familiar with the operational characteristics of the fluorometer available for his use by studying the instructional and descriptive literature supplied with the instrument and by appropriate laboratory demonstration.

1. Intensity of Fluorescence of Riboflavin as a Function of Concentration

Prepare a stock solution of riboflavin at a concentration of about 1 $\mu g/ml$ (USP XVII or NF XII may be consulted for directions for preparing this solution). Prepare dilutions of the stock solution to obtain the following concentrations: 0.02, 0.04, 0.06, 0.08, and 0.1 $\mu g/ml$. Determine the relative intensity of fluorescence of each solution with a suitable fluorometer. An appropriate primary (excitation) filter is one that peaks at 360 mμ, while the secondary filter (emission) should pass wavelengths greater than 510 mμ. Present the results in the form of a graph in which fluorometer reading is plotted as a function of concentration. Repeat with solutions ranging in concentration from 0.002 to 0.01 $\mu g/ml$.

2. The Influence of pH on the Fluorescence Intensity of Riboflavin

Prepare buffered solutions of riboflavin ranging in pH from 2 to 11. All solutions should contain the same concentration of the vitamin, which should be such that the solution buffered to approximately pH 7 gives a reading of from 50 to 80% of full scale of the fluorometer. Determine the fluorescence of each solution and plot the fluorometer reading as a function of pH.

3. The Influence of Quenching Agents on the Fluorescence Intensity of Riboflavin

Design and conduct an experiment to demonstrate the influence of potassium iodide concentration on the intensity of fluorescence of riboflavin. Plot the results in a manner suggested by Eq. (4.11). Repeat using caffeine as a quenching agent.

4. The Intensity of Fluorescence of Salicylic Acid as a Function of Concentration

Prepare solutions of salicylic acid in 0.1 N sodium hydroxide to cover a range of concentrations of from 5 to 100 $\mu g/ml$. Determine the relative intensity of fluorescence of each solution and plot fluorometer reading as a function of concentration. The 360-mμ primary filter may also be used in this instance, but a secondary filter transmitting wavelengths greater than 455 mμ should be selected.

5. Assay of Riboflavin Injection

Determine the potency of a sample of riboflavin injection by employing the riboflavin-assay procedure described in USP XVII or NF XII.

6. Assay of Thiamine Hydrochloride Tablets

Employ the thiamine assay procedure described in USP XVII or NF XII to determine whether or not a sample of thiamine hydrochloride tablets meet the label claim.

7. The Determination of 9-Aminoacridine (Aminocrine) in Pharmaceutical Products

Obtain a sample of a pharmaceutical preparation containing aminocrine. Fluorometrically determine the aminocrine content by the method proposed by Roberts.[20]

8. Excited-state Dissociation of 2-Naphthol

Conduct the laboratory experiment described by Ellis[16] to demonstrate the excited-state dissociation of 2-naphthol.

9. Fluorometry in Biopharmaceutical Studies

Determine the physiological availability of riboflavin from a coated tablet using the method of Chapman et al.[40] In this procedure, the amount of riboflavin excreted in the urine following oral ingestion of a coated vitamin tablet is determined fluorometrically and compared with the amount excreted after the ingestion of a rapidly dissolving uncoated tablet.

PROBLEMS

P4.1. The molar absorptivity of riboflavin (molecular weight = 376.36) in aqueous solution at 360 mμ is approximately 7500. Fluorometric examination, using an excitation wavelength of 360 mμ, of a solution of the vitamin containing 0.010 μg/ml yielded a fluorescence intensity of 1.0 unit. Calculate the theoretically expected intensity of fluorescence for a solution containing 1.0 μg/ml. Assume that the equivalent of a 1-cm cell was used.

P4.2. The influence of pH on the intensity of fluorescence of a dilute solution of a weak acid was investigated. Solutions were prepared which varied only in the concentration of hydrogen ion. The fluorescence of each solution was determined with a fluorometer and the following results were obtained:

pH	Fluorescence (arbitrary units)
4.0	80.0
5.0	80.0
6.0	79.5
7.0	74.5
7.5	65.6
8.5	34.4
9.0	25.4
10.0	20.6
11.0	20.0
12.0	20.0

Show that the ratio of the concentration of ionized acid to that of un-ionized acid at any pH is given by $(80 - F)/(F - 20)$, where F is the observed fluorescence at that pH. Plot the logarithm of the ratio as a function of pH and determine the pK_a of the acid from the plot.

P4.3. A standard preparation of riboflavin was prepared by the following procedure. Exactly 48.5 mg of USP riboflavin reference standard was dissolved in sufficient water to make 500 ml. One ml of the resulting solution was diluted to 100 ml with water.

An assay preparation was prepared in the following manner: The riboflavin from 10 riboflavin tablets was dissolved in sufficient water to make 1 liter. One ml of this solution was diluted to 1 liter with water.

A mixture of 10 ml of assay preparation and 1 ml of standard preparation yielded a fluorescence reading of 74.0. Ten ml of the assay preparation and 1 ml of water gave a fluorescence reading of 40.0. The latter mixture was treated with 20 mg of sodium hydrosulfite and the fluorescence was again determined. The reading was found to be 2.0. Calculate the quantity in mg of riboflavin in each tablet.

P4.4. A nonfluorescent complexing agent A is known to form a nonfluorescent complex with a fluorescent compound B. A $1 \times 10^{-6} M$ solution of B was placed in a fluorometer and the meter was adjusted to read 100.0. The fluorescence of a solution which was $1 \times 10^{-6} M$ with respect to B and $1 \times 10^{-2} M$ with respect to A was then determined and was found to be 20.0. Calculate the stability constant of the complex.

P4.5. Exactly 20 mg of USP thiamine hydrochloride reference standard was dissolved in sufficient water to make 1 liter. One ml of the resulting solution was diluted to 100 ml with water. Five ml of this solution was treated with 3 ml of oxidizing agent and the thiochrome which was formed was extracted into 20 ml of isobutanol. The fluorescence of the isobutanol phase was determined to be 67.0. A corresponding blank has a fluorescence of 4.0.

Exactly 1 ml of a thiamine hydrochloride injection was diluted with sufficient water to make 1 liter. Two ml of this solution was diluted to 1 liter with water. Five ml of the final dilution was oxidized and extracted, as previously described, and the fluorescence of the isobutanol extract was 58.0. The fluorescence of a corresponding blank was 8.0. Calculate the quantity, in milligrams, of thiamine hydrochloride in each milliter of injection.

P4.6. A preparation contains riboflavin, thiamine hydrochloride, and a fluorescent coloring agent. Describe how you would approach the problem of developing a fluorometric method for the determination of all three of these components.

REFERENCES

1. R. T. Williams and J. W. Bridges, *J. Clin. Path.*, **17**, 371 (1964).
2. D. S. McClure, *J. Chem. Phys.*, **17**, 905 (1949).
3. E. L. Wehry and L. B. Rogers, in *Fluorescence and Phosphorescence Analysis* (D. M. Hercules, ed.), Wiley (Interscience), New York, 1966, Chap. 3.
4. K. W. Kohn, *Anal. Chem.*, **33**, 862 (1961).
5. D. Mattingly, *J. Clin. Path.*, **15**, 374 (1962).
6. J. W. Huff, *J. Biol. Chem.*, **167**, 151 (1947).
7. H. L. Price and M. L. Price, *J. Lab. Clin. Med.*, **50**, 769 (1957).

8. P. L. Lott, *J. Chem. Educ.*, **41**, A327, A421 (1964).
9. S. Udenfriend, in *Fluorescence Assay in Biology and Medicine*, Academic Press, New York, 1962, Chap. 3.
10. F. Kavanagh, *Ind. Eng. Chem. Anal. Ed.*, **13**, 108 (1941).
11. H. Braunsberg and S. B. Osborn, *Anal. Chim. Acta*, **6**, 84 (1952).
12. D. M. Hercules, *Anal. Chem.*, **38**, 29A (1966).
13. D. M. Hercules, in *Fluorescence and Phosphorescence Analysis* (D. M. Hercules, ed.), Wiley (Interscience), New York, 1966, Chap. 1.
14. H. Harbury and K. A. Foley, *Proc. Natl. Acad. Sci. U.S.*, **44**, 662 (1958).
15. T. Forster, *Naturwiss.*, **36**, 186 (1949).
16. D. W. Ellis, *J. Chem. Educ.*, **43**, 259 (1966).
17. R. E. Phillips and F. R. Elevitch, in *Progress in Clinical Pathology*, Vol. 1 (M. Stefanini, ed.), Grune & Stratton, New York, 1966, Chap. 4.
18. D. C. Wimer, J. G. Theivagt, and V. E. Papendick, *Anal. Chem.*, 185R (1965).
19. C. E. White and A. Weissler, *Anal. Chem.*, **38**, 155R (1966).
20. L. A. Roberts, *J. Assoc. Offic. Agr. Chemists*, 49, 837 (1966).
21. T. J. Mellinger and C. E. Keeler, *Anal. Chem.*, **35**, 554 (1963).
22. A. Saltzman, *J. Biol. Chem.*, **174**, 399 (1948).
23. M. A. Chirigos and S. Udenfriend, *J. Lab. Clin. Med.*, **54**, 769 (1959).
24. W. E. Lange and S. A. Bell, *J. Pharm. Sci.*, **55**, 386 (1966).
25. N. F. Anden, D. E. Roos, and B. Werdinius, *Life Sci.*, **2**, 448 (1963).
26. D. F. Sharman, *Brit. J. Pharmacol.*, **20**, 204 (1963).
27. T. L. Sato, *J. Lab. Clin. Med.*, **66**, 517 (1965).
28. H. Corrodi and B. Werdinius, *Acta Chem. Scand.*, **19**, 1854 (1965).
29. G. C. Guilbault, D. N. Kramer, and E. Hackley, *Anal. Chem.*, **39**, 271 (1967).
30. I. M. Jakovljevic, *Anal. Chem.*, **35**, 1513 (1963).
31. C. R. Szalkowski and W. J. Mader, *J. Am. Pharm. Assoc. Sci. Ed.*, **45**, 613 (1956).
32. W. J. Mader, R. P. Haycock, P. B. Sheth, and R. T. Connolly, *J. Assoc. Offic. Agr. Chemists*, **43**, 291 (1960); **44**, 13 (1961).
33. W. J. Mader, G. J. Papariello, and P. B. Sheth, *J. Assoc. Offic. Agr. Chemists*, **45**, 589 (1962).
34. R. P. Haycock, P. B. Sheth, T. Higuchi, W. J. Mader, and G. J. Papariello, *J. Pharm. Sci.*, **55**, 826 (1966).
35. S. Udenfriend, in *Fluorescence Assay in Biology and Medicine*, Academic Press, New York, 1962, Chap. 7.
36. W. E. Ohnesorge and L. B. Rogers, *Anal. Chem.*, **28**, 1017 (1956).
37. *International Pharmacopoea*, Vol. II, World Health Organization, 1st ed., 1955, Appendix 3, p. 233.
38. *United States Pharmacopeia*, Mack Printing, Easton, Pa., 17th rev. ed., 1965, p. 886.
39. *United States Pharmacopeia*, Mack Printing, Easton, Pa., 17th rev. ed., 1965, p. 888.
40. D. G. Chapman, R. Crisafio, and J. A. Campbell, *J. Am. Pharm. Assoc. Sci. Ed.*, **43**, 297 (1954).
41. C. Bedford, K. J. Child, and E. G. Tomich, *Nature*, **184**, 364 (1959).

CHAPTER **5**

Atomic Absorption Spectroscopy

J. W. Robinson

LOUISIANA STATE UNIVERSITY
BATON ROUGE, LOUISIANA

5.1 INTRODUCTION

Atomic absorption spectroscopy is an analytical technique which has been developed primarily for the determination of metals at low levels of concentration.[1]

It is based on the absorption of radiation by free atoms. Each chemical element in the atomic state absorbs only radiation of well-defined wavelengths characteristic of the particular element involved. It is the reverse physical process to that involved in flame photometry and emission spectroscopy. The relationship between these processes is illustrated in Fig. 5.1.

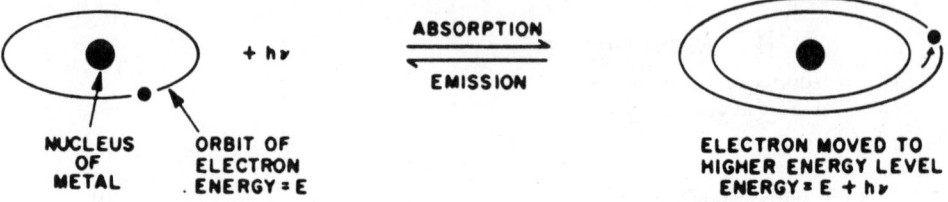

FIGURE 5.1: Relationship between atomic emission and atomic absorption.

Atomic absorption takes place when unexcited atoms absorb energy and become excited atoms. Absorption therefore is carried out by unexcited atoms, whereas emission arises from excited atoms. This relationship leads to the principle that all experimental conditions that affect the total atom population affect the emission and absorption signal in a similar fashion. All variables that increase the percentage of excited atoms (such as an increase in atomizer temperature) increase the intensity of the emission signal, but decrease the absorption signal (very slightly). All variables that affect the number of unexcited atoms affect only the atomic absorption signal. It is important to remember this principle when developing new procedures based on atomic absorption spectroscopy.

5.2 ADVANTAGES OF ATOMIC ABSORPTION SPECTROSCOPY

A. WIDESPREAD APPLICATION

Atomic absorption spectroscopy has been used for the quantitative determination of most of the metals in the periodic table. This illustrates the widespread application of the method to quantitative elemental analysis. In this respect it compares very favorably with other methods of elemental analysis, such as X-ray fluorescence and emission spectrography.

B.   HIGH SENSITIVITY

It has been shown experimentally to be very sensitive compared to other techniques for the determination of trace quantities of metals. Many elements can be determined at the part-per-million concentration level and some can be determined at the part-per-billion level. These facts illustrate the high sensitivity of the method.

C.   FREEDOM FROM INTERFERENCE

A third advantage of the method is its high degree of freedom from interference. No two elements absorb at the same resonance wavelength as each other. As a result, the presence of one element does not directly interfere with the absorption of radiation by another element. Conversely the measurement of absorption is generally a direct measure of the concentration of the absorption element, irrespective of the other elements present.

D.   INDEPENDENT OF FLAME TEMPERATURE AND ABSORPTION WAVELENGTH REGION

Absorption is by unexcited atoms. The ability of the atom to absorb is independent of atomizer temperature and the spectral region of the absorption wavelength. As a consequence, atomic absorption is free from interference from atomizer temperature or the wavelength region of the absorption lines. This is in direct contrast to emission methods, where each of these variables directly affects the emission signal. For example, the sodium resonance line is 5995 Å. This element emits and absorbs strongly at this wavelength. However, the zinc resonance line is at 2138 Å. It absorbs strongly at this wavelength, but the emission is extremely weak.

With these advantages atomic absorption spectroscopy is capable of providing accurate and precise answers when the analysis is performed properly. Further, inasmuch as the equipment is simple to operate, routine analytical laboratories find it an attractive tool.

5.3   DISADVANTAGES OF ATOMIC ABSORPTION SPECTROSCOPY

A.   DOES NOT DETERMINE NONMETALS

In common with all methods of analytical chemistry atomic absorption spectroscopy is subject to some disadvantages. At present, it has not been found useful for the direct determination of nonmetallic elements such as halides, oxygen, and nitrogen. This is because the resonance line of these elements is in the vacuum ultraviolet region of the spectrum and therefore cannot be used with present-day equipment. To operate in this region requires that the optical path be in a vacuum. This itself presents a difficult problem in practice. However, the problem is made much more difficult to solve because the common atomizers use flames and flames absorb strongly in the vacuum ultraviolet. It would be difficult to distinguish between the flame absorption and the absorption by the sample. As a result, no equipment is available to use in this spectral region.

B. DOES NOT ANALYZE SOLID OR GAS SAMPLES DIRECTLY

Another problem in atomic absorption is that it is very useful for analyzing liquid samples, but has not been successfully used for the direct analysis of solids or gases. The difficulty arises in the atomizer stage. All commercial equipment uses atomizers built to handle liquid samples, but not any other phase. Research equipment has been developed to handle solid samples, but these are not commercially available. In routine work, difficulties were encountered in the use of this research equipment and in the interpretation of data obtained using this equipment. Commercial manufacturers have not been encouraged to develop suitable atomizers for handling solid or gas samples.

To carry out a determination, all solid samples must be decomposed or dissolved in a suitable solvent. Similarly, gas samples must be washed or scrubbed so that the metal components are trapped in a suitable liquid. This step may be time-consuming and is always a source of error. However, with proper care these steps can be performed and accurate results obtained.

C. DETERMINES ONLY ONE ELEMENT AT A TIME

The equipment which is currently available is built to determine only one element at a time. This is no problem if a number of samples must be analyzed for a single element. But if one sample must be analyzed for several elements, a difficulty may arise. Each element must be determined separately. This means the use of a different hollow cathode, a change of wavelength, and possibly different flame conditions for each element. These equipment changes are not difficult to perform, but they are time-consuming. It should also be remembered that a separate portion of sample must be used for the determination of each element. This may be expensive in terms of the quantity of sample used. About 1 ml of sample is required for each determination.

D. ANIONIC INTERFERENCE

The major source of analytical interference is from the anions present in the sample. The anions do not absorb the atomic radiation, but they affect the population of free atoms formed in the atomizer.

When a liquid is introduced into a flame atomizer, the liquid is normally in the form of a droplet. The droplet evaporates, leaving a solid residue. The residue is then decomposed by the flame and free atoms are liberated in the process. The last step is directly affected by the predominant anion. The predominant anion determines the chemical form of the metal in the residue. Some chemicals are more easily decomposed than others. For example, a residue of calcium chloride is relatively easily decomposed, but calcium phosphate is much more difficult to break down. More free calcium would be formed in the atomizer from a solution of calcium chloride than from a solution of calcium phosphate of equal calcium concentration. The two solutions would give different absorption signals and therefore different analytical results.

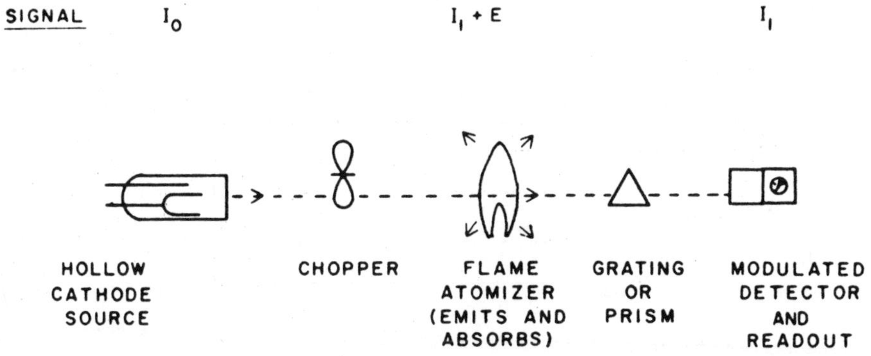

SIGNAL I_0 $I_1 + E$ I_1

FIGURE 5.2: Schematic diagram of equipment.

The problem can be overcome by using a calibration curve prepared from solutions with the same predominant anion as the sample, or by complexing the metal with an organic reagent. In general, the interference is not as severe as in the situation illustrated. However, when setting up a routine procedure, steps should always be taken to eliminate any possible interference from a change in the predominant anion from one sample to the next.

5.4 EQUIPMENT

The equipment used is basically the same as for other spectroscopic absorption methods. It consists of a radiation source, a monochromator, a sampling device, and a detector. A schematic diagram of the equipment is shown in Fig. 5.2.

The individual components are as follows.

A. RADIATION SOURCE

Atoms absorb at a characteristic wavelength and over a very narrow wavelength range. The width of the absorption lines is about 0.01 Å, even under the most adverse conditions.

As a result "continuous" wavelength sources such as the hydrogen lamp or the tungsten lamp are at a severe disadvantage in this equipment. Even under optimistic conditions of resolution and slits, a wave band about 2 Å wide falls onto the detector. The sample atoms can only absorb out a bandwidth of 0.01 Å, consequently, when the atoms absorb all the radiation they can absorb, a large proportion of the radiation still passes unabsorbed to the detector. The method would be insensitive under these conditions and analytical application would be severely limited. The detector would only find a slight change in total radiation intensity, even though the concentration of atoms was high enough to remove all the radiation over its absorption wave band.

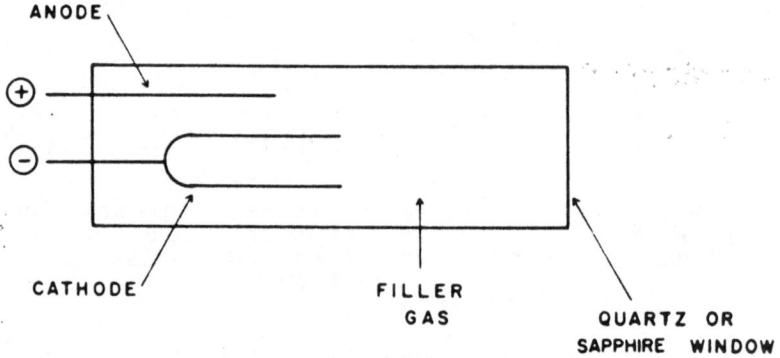

FIGURE 5.3: Diagram of a hollow cathode.

To overcome these problems, the hollow cathode has been developed by Walsh[1] and co-workers as a radiation source. A schematic diagram of this source is shown in Fig. 5.3.

The filler gas used is ionized at the anode. The charged ion is attracted to the cathode. On arrival, it strikes the metal surface and liberates free excited atoms from the surface of the metal. The atoms emit at their own characteristic wavelength. The width of the emission lines emitted by the metal atoms is very narrow and matches the absorption line width of the absorbing atoms from the sample.

If the absorption line is isolated from the rest of the spectrum, it alone will fall on the detector. The absorption line width is similar to the emission line width, hence all the radiation falling on the detector can be absorbed by the atoms. A great increase in sensitivity results.

Recently, high-intensity hollow cathodes have been developed. A diagram is shown in Fig. 5.4. As described in the simple hollow cathode, a cloud of atoms is formed by bombardment of the cathode by filler-gas ions. This creates a cloud of atoms at the mouth of the cathode. Some of these atoms are excited and emit characteristic radiation, but a large proportion are unexcited and do not contribute to the signal. A beam of slow-moving electrons passes through the atom cloud and causes them to become excited. The resonance line is greatly enhanced compared to other spectral lines. The lamp is therefore able to emit at the resonance wavelength with a high intensity.

This lamp is very useful if the resonance line is similar in wavelength to other unabsorbed lines originating from the cathode. For example, the

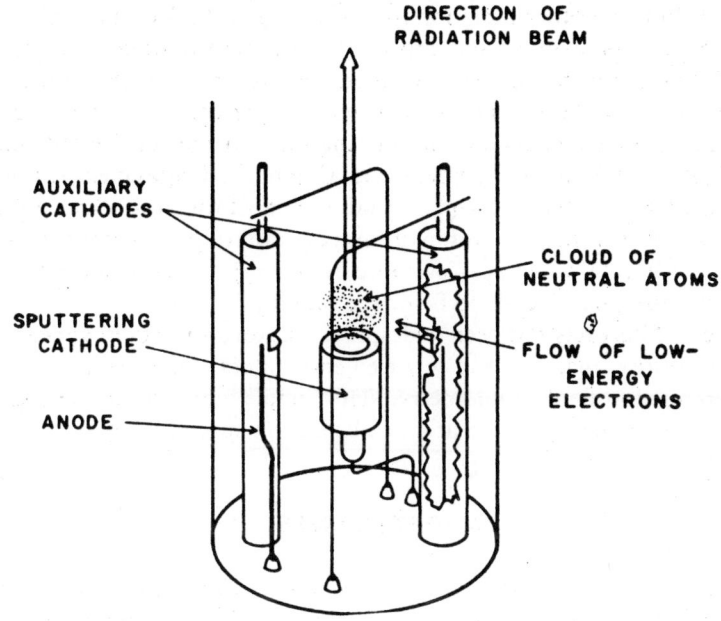

FIGURE 5.4: Schematic diagram of a high intensity lamp.

wavelength of the nickel resonance line is 2320 Å, but the nearby 2321.4 Å is not a resonance line.

Both lines fall on the detector. Even if all the 2320 Å lines were absorbed, the detector would still be exposed to the 2321.4 Å line. Since the 2321.4 Å line is not absorbed, its presence results in a loss of sensitivity of the procedure. Using a high intensity lamp greatly reduces this problem. However, it should be pointed out that high intensity lamps are only useful under these special circumstances. Normally, if the resonance level is strong and isolated from other lines, the advantage of high intensity is limited.

One difficulty involved in using the hollow cathode source is that the cathode must be made of the same element as that being determined. As a consequence, only one element at a time can be determined. As mentioned earlier, multielement hollow cathodes have been developed which will enable three elements to be determined simultaneously. However, in general these sources are not very stable and have a shorter operational life than conventional single-element hollow cathodes.

5.5 MONOCHROMATOR

The optical range of this analytical field is between 2000 and 8000 Å. The function of the monochromator is to select radiation of the correct wavelength and eliminate other radiation from the light path. For many elements such as sodium, potassium, and copper the spectrum is simple and only a low resolution monochromator is required. However, for certain other elements, particularly the transition elements, high resolution is necessary to prevent unabsorbable emission lines originating either from the cathode or the filler gas from falling on the detector. Commercial equipment is usually fitted with a high resolution prism or grating monochromator. Such a prism can be used whether high resolution or low resolution equipment is necessary.

There is little difference in performance between the prism and the grating. Commercial equipment tends to favor the grating.

The use of filters has not been widespread, although it would appear that for very repetitive analysis for a few elements, filters would have been used as they are used in flame photometry. This simple monochromator has not yet been used commercially for atomic absorption spectroscopy.

5.6 DETECTORS

Photomultipliers are used exclusively on commercial equipment. In practice, it is the function of the detector to measure the intensity of radiation before and after absorption by the sample. From this we can calculate how much radiation has been absorbed from the intense beam. If the amount of absorption is small, then the detector must compare an intense beam and a slightly less-intense beam. This measurement cannot be done with film, with any degree of accuracy, but can be done using a photomultiplier.

5.7 OPTICAL SLIT SYSTEM

Two slits are included in the optical system, an entrance slit and an exit slit. The entrance slit serves to obtain a narrow, parallel beam of light from the source. Other radiation, including stray radiation, is physically blocked out by the walls of the slit. The exit slit is used to select radiation of the correct wavelength after it emerges from the monochromator. Other radiation is blocked out and not allowed to continue down the light path. The function of the slits is illustrated in Fig. 5.5.

Because the hollow cathode emits very narrow lines, the actual slit width used is not critical. However, when high resolution is required to eliminate unabsorbed radiation from nearby nonresonance lines, narrow slit widths are necessary. This situation only occurs for a few elements such as iron and nickel. For other elements the slits can be left at a standard width. Most analytical procedures[2,3] specify the slit widths to be used for the particular determination described.

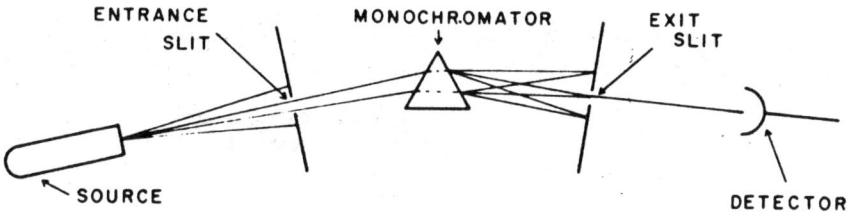

FIGURE 5.5: Function of the slit.

5.8 MODULATION

The wavelength at which atoms are absorbed is usually the "resonance wavelength." Absorption at this wavelength causes a transition from the ground state to the first excited state of the atom. Unfortunately, the atoms of many elements in a flame also emit radiation at this same wavelength. This is particularly so if the resonance wavelength is longer than 3000 Å. Radiation is usually very intense from atoms such as sodium, potassium, lithium, and calcium, which emit at comparatively long wavelengths.

This emission will create a problem in measuring the degree of absorption by the sample. For example, examine the following circumstances:

Intensity of radiation from lamp	Intensity of radiation after absorption by sample	Emission intensity by sample at same wavelength
100 units	50 units	20 units

The "observed" absorption would be: $100 - (50 + 20)$ units $= 30$ units. The true absorption was $100 - 50$ units $= 50$ units.

The emission from the sample in the flame can be, therefore, a serious source of error. It can be overcome by modulating the instrument. This entails using an alternating light source and a detector tuned to the same frequency.

Under these circumstances, the detector "sees" the alternating light from the source and can detect any absorption of that light, however, the dc light from the flame does not give rise to a signal from the tuned detector. The emission signal is thereby eliminated as a source of error and only the absorption signal is measured.

The principal manufacturers of atomic absorption equipment all produce modulated equipment. However, to avoid any unnecessary source of error this point should always be checked when buying a new instrument.

A. ATOMIZER

The function of the atomizer is to convert the combined atoms of the liquid sample into free atoms. The most common atomizer is the flame. In practice

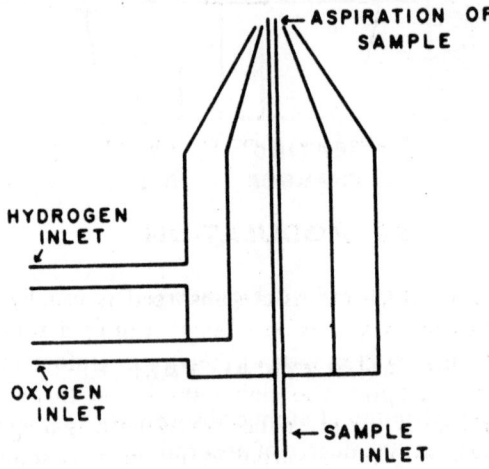

FIGURE 5.6: Total consumption burner.

the liquid sample is introduced into the flame in the form of a droplet. The droplets evaporate and leave a solid residue. The residue, which contains the same atoms, is decomposed by the flame and free atoms are liberated. These free atoms absorb the radiation which is measured in this procedure. The free atoms exist in the flame for varying periods of time, but in most instances quickly become oxidized to the metal oxides, which do not absorb.

The rate of formation and the rate of oxidation of the atoms depends to a large extent on flame conditions such as the fuel used, flame temperature, and fuel-to-oxygen ratio. This series of reactions enters a state of dynamic equilibrium and a steady number of atoms exist at any one time. To get quantitative results it is important that this number remains constant and is reproducible.

Two types of flame atomizers are available, the total consumption burner and the Lundegardh burner. The whole sample is aspirated into the base of the flame and then atomized. The burner is illustrated in Fig. 5.6 and employs a preevaporation chamber which evaporates the sample before it reaches the burner. This burner tends to be more sensitive than the total consumption oration chamber, as illustrated in the Beckman or preheat burner shown in Fig. 5.7.

The total consumption burner is most useful if the sample contains several solvents. It assures that all the sample reaches the flame. The Lundegardh burner is physically quieter and frequently enables more sensitive results to be obtained. Its efficiency is improved because the evaporation step occurs prior to reaching the burner. This helps to conserve the energy of the burner for atomizing the sample.

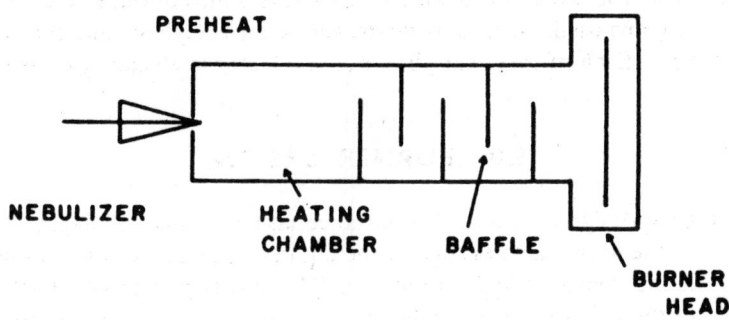

FIGURE 5.7: Preheat burner.

5.9   ANALYTICAL VARIABLES

The *quantitative* application of atomic absorption spectroscopy is based on two relationships. First, the degree of absorption is a function of how many free atoms are present in the light path. Second, the number of free atoms is a function of the concentration of the metal on the original sample.

Consequently, the degree of absorption is proportional to the original concentration of metal. For this relationship to be meaningful all the variables concerned must be carefully controlled. Each step will be considered separately.

The degree of absorption is described mathematically by

$$K_v \, d_v = \frac{\pi e^2}{mc} N f \qquad (5.1)$$

where $K_v \, d_v$ = total amount of radiation absorbed at resonance frequency v

e = charge of the electron (a constant)

m = mass of the electron (a constant)

c = speed of light (a constant)

N = number of atoms of the light path

f = oscillator strength of the absorption line (f varies from element to element, but is constant for a particular element at the resonance frequency)

From this equation it can be seen that for a given line

$$K, d_v = \text{constant} \times N \qquad (5.2)$$

Hence, the degree of absorption is a physical property of the system and depends on the number of atoms in the light path.

However, the relationship between N, the number of free atoms produced in the atomizer, and the concentration of the element in the sample depends on the atomizer efficiency.

The process of atomization depends on several variables. Included are the type of burner used, the rate at which the sample is introduced into the flame, the type of flame used, such as oxyhydrogen or air-acetylene, and the type of solvent used. Each of these variables noted affects the efficiency of atomization.

5.10 BURNER DESIGN

The burner design is of prime importance, as can easily be imagined. The function of the burner is to evaporate the liquid component of the sample and to atomize the element being determined. The efficiency with which a burner carries out these functions depends on its design. It varies greatly from one type of design to another. There is even a significant variation in efficiency between two burners of the same design. It can be seen therefore that for reproducible results to be obtained, the identical burner should be used for analyzing the samples as that used for preparing the calibration curves.

5.11 SAMPLE FLOW RATE

Any given burner atomizes a sample best at the optimum sample flow rate. If the flow rate is too low, too few atoms are formed, if the flow rate is too high, the flame becomes swamped. Between the extremes is the optimum flow rate. In practice, the manufacturers of the burners determine the optimum flow rate of their burner and design it accordingly. No control over this variable is left to the operator. However, should the burner become clogged or distorted, a change in efficiency occurs and erroneous answers are obtained. Care should always be taken to prevent burner clogging during operation.

5.12 TYPE OF FLAME

The type of flame directly affects atomization efficiency. Hot flames decompose the sample more easily than cool flames. Hence, hot flames such as oxyhydrogen flames have a higher efficiency than air-propane flames.

The free atoms formed generally become oxidized shortly after formation.

However, if there is an excess of fuel over oxygen in the flame, the lifetime of the free atoms is extended. By the same reasoning, if there is an excess of oxygen in the flame, the free atoms quickly become oxidized and no longer absorb at their resonance wavelength.

Some elements such as aluminum, titanium, tungsten, and molybdenum form very stable oxides in the flame. Until recently, it was difficult to detect these elements by atomic absorption. However, it has been found by Willis and Amos[1] that by using a nitrous oxide-acetylene flame, the oxide formation is delayed and these elements can be detected at the part-per-million concentration level.

5.13 SOLVENT

The type of solvent used in the sample has a profound effect on atomization efficiency. Aqueous solvents require energy to be evaporated. The residue is usually a hydrated inorganic salt and this requires energy to be decomposed. The whole process of atomization is endothermic and proceeds relatively slowly.

However, if the solvent is organic, it will burn in the flame instead of evaporating. Further, the residue is usually in the form of an organic compound and this burns and decomposes. In short, the reaction is exothermic and atomization is more efficient than with aqueous solvent. The net effect is that when organic solvents are used, a greater absorption signal is observed than when aqueous solvents are used with the same metal concentration.

To obtain reproducible results all these variables must be controlled. In practice, it is not as difficult a task as it appears, but it does require attention in order to avoid errors.

5.14 WAVELENGTH CHOICE

For each element there are several absorption lines originating in the ground state. The most sensitive line is associated with the transition from the ground state to the first excited state. However, a less sensitive absorption line is associated with the transition from ground state to the second excited state. For many elements there are several absorption lines available. If high sensitivity is required, the most sensitive line is used. However, this line could not be used for the determination of a high concentration of the element because most of the radiation will be absorbed by the large concentration of atoms produced from such a sample. Large changes in concentration between samples would result in only small changes in absorption and this would lead to imprecise answers. For high concentrations therefore a less strongly absorbed line is used. The analytical range of the various lines is generally given in the analytical procedure when it is described in the literature.

5.15 QUANTITATIVE METHOD

Equation (5.1) shows the relationship between the degree of absorption and the number of atoms in the light path. However, the variable efficiency of most flames in producing free atoms from the sample makes it impossible to calculate the concentration of the metal in the sample from the determination of N, the number of atoms in the light path.

The conversion of atoms existing in the sample as molecules to free atoms is a very inefficient process. A valid arithmetic calculation of N, and therefore

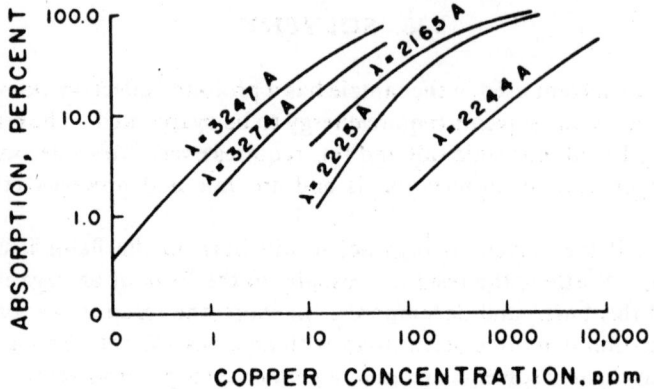

FIGURE 5.8: Calibration curves for copper determination.

the concentration of the element in the sample, cannot be made. Variables such as the efficiency of atomization negate the calculation. The relationship between absorption and concentration is therefore usually determined experimentally by using a calibration curve.

A. PREPARATION OF CALIBRATION CURVES

A series of solutions of the sample element are prepared. The concentration of the element in each solution is predetermined by weighing known amounts of a salt of the metal into a known volume of solution.

The atomic absorption spectrometer is set at the desired resonance wavelength for the metal, and each solution is introduced successively into the flame atomizer. The absorbance is measured for each solution and the relationship is then plotted out. The plot constitutes a "calibration curve." A typical calibration curve is shown in Fig. 5.8.

This curve indicates the relationship between absorbance and concentration of copper in the standard samples.

When a sample of unknown concentration is to be analyzed, it is introduced into the atomizer and the absorbance measured. From the calibration curve the concentration corresponding to the measured absorbance is read off. From this the analysis of the original sample can be determined.

The analytical range of the procedure is determined primarily by the sensitivity limits of the method. Calibration curves can be prepared starting at concentrations about ten times as great as the sensitivity limit. Quantitative determination with lower precision can be obtained at lower concentrations.

The sensitivity of detection of various elements is shown in Table 5.1.

TABLE 5.1: Analytical Sensitivities of the Various Elements[a]

Element	Wavelength, Å	Sensitivity, ppm
Al	3092	2.0
Sb	2175	0.5
As	1937	3.0
Ba	5536	5.0
Be	2349	0.2
Bi	2231	1.0
B	2497	250
Cd	2288	0.04
Ca	4227	0.1
Cs	8521	0.15
Cr	3579	0.15
Co	2407	0.5
Cu	3247	0.1
Ga	2944	4
Au	2428	0.1
Fe	2483	0.3
Pb	2170	0.5
Li	6708	0.07
Mg	2852	0.02
Mn	2798	0.01
Hg	2357	10.0
Mo	3133	0.5
Ni	2320	0.2
Pd	2474	1.0
Pt	2148	2.0
K	7665	0.1
Rb	7800	0.2
Se	1961	0.5
Ag	3281	0.1
Na	5890	0.03
Sr	4607	0.05
Tl	2767	1.0
Sn	2863	5.0
Ti	3653	12.0
V	3184	7
Zn	2138	0.05

[a] The sensitivity is here defined as that concentration of the element which results in the absorption of 1% of the resonant radiation passing through the atomizer.

These sensitivities are reported by different workers in the field and may vary somewhat with equipment and experimental conditions.

The actual experimental conditions to be used for the quantitative determination of each of these elements should be ascertained from the literature.

5.16 CONCLUSIONS

Atomic absorption spectroscopy is an analytical procedure used for the determination of the metals and metalloids. It is sensitive compared to other analytical methods. It enjoys a higher degree of freedom from interference than most methods and is therefore capable of giving accurate and precise analytical results.

The cost of commercial equipment varies from $3500 to $7000. Training of personnel is relatively simple. Also, installation is easily carried out. For highly routine analysis it is a most attractive method.

EXPERIMENTS

E5.1. Prepare a calibration curve for copper salts using copper sulfate in water for the standard solution. Make solutions containing Cu^{2+} in the range 2 to 20 ppm. Measure absorbance of the resonance line at wavelength 3247 Å. Plot absorption curve for the same solution at the resonance line 2024 Å.

E5.2. Prepare a calibration curve for lead salts. Use lead solution (as nitrate) with a concentration range between 4 and 40 ppm. Measure the absorption at 2170 Å. Plot absorbance vs. concentration. Add chloride to the lead solutions; note the depression of the absorption. Add EDTA to the chloride solution; note the absorption depression is removed.

QUESTIONS

Q5.1. Why is it necessary to use a hollow cathode lamp as a light source in atomic absorption spectroscopy?

Q5.2. Why is atomic absorption equipment modulated?

Q5.3. What are chemical interferences?

Q5.4. The data for preparing a calibration curve for gold was as follows:

Gold Conc. ppm	Absorbance
3	0.12
6	0.24
12	0.48
15	0.60
18	0.71
21	0.81

a. Plot the curve relating absorbance and concentration of gold.

b. Does this curve deviate from Beer's law? Several gold solutions of unknown concentration were analyzed. The absorbance was 0.31, 0.51, 0.74, respectively. What were the concentrations of the unknown solutions?

REFERENCES

1. A. Walsh, *Spectrochim. Acta*, 7, 108 (1958).
2. J. W. Robinson, *Atomic Absorption Spectroscopy*, Dekker, New York, 1966.
3. Perkins-Elmer Newsletter, Perkin-Elmer Corp., Norwalk, Conn.
4. J. B. Willis and M. D. Amos, *Spectrochim. Acta*, 22, 1325 (1966).

CHAPTER **6**

Mass Spectrometry

A. Chisholm*

ANALYTICAL RESEARCH
WM. S. MERRELL CO.
DIVISION OF RICHARDSON-MERRELL, INC.
CINCINNATI, OHIO

* Present address: Food and Drug Directorate, Toronto, Canada.

6.1 INTRODUCTION

Mass spectrometry is a technique which complements infrared, ultraviolet, and nuclear magnetic resonance spectroscopy and gas chromatography. It gives information about the structural groups which make up the molecule. The spectrum relates the apparent mass of a fragment of the molecule to its relative intensity based on the strongest peak in the spectrum (Fig. 6.1).

In 1886, Goldstein[9] observed streams of luminous gas in back of a perforated cathode in a discharge tube. The luminosity was due to the passage

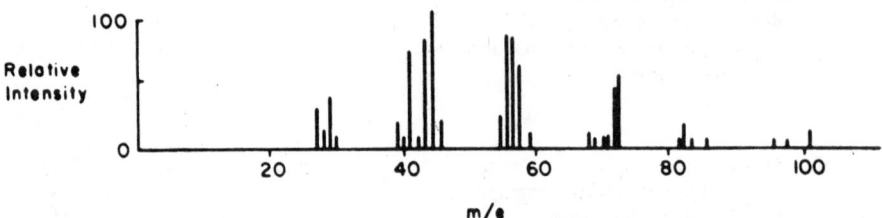

FIGURE 6.1: Idealized mass spectrum.

of ionized particles in the gas. It was found that these particles were positively charged and that they could be deflected by electric or magnetic fields. Since they could be deflected by electric and magnetic fields, it was possible to determine velocity and the charge to mass ratio (e/m) for the particles in the stream.

An early method of positive ray analysis was J. J. Thomson's parabola method.[23] Electric and magnetic fields were aligned parallel to the ray, causing deflections perpendicular to one another (Fig. 6.2). The relationships between the magnetic and electric fields and the ionized particle are basic to mass spectroscopy.

If a uniform electric field R_E is applied to a single positively charged particle moving downward perpendicularly to the plane of the paper toward point 0, (Fig. 6.3), the deflection of the particle can be to the right or left along the x axis, depending on the polarity of the field. The distance the particle is deflected ob is given by:

$$x = k_1 \frac{R_E e}{mV^2} \tag{6.1}$$

where k_1 is a constant depending on the apparatus, e is the charge, m is the mass, and V is the velocity of the particle. Equation (6.1) may be rearranged to give:

$$V^2 = \frac{k_1 R_E e}{mx} \tag{6.2}$$

Now with the electric field off, if the magnetic field (R_H) is applied to deflect

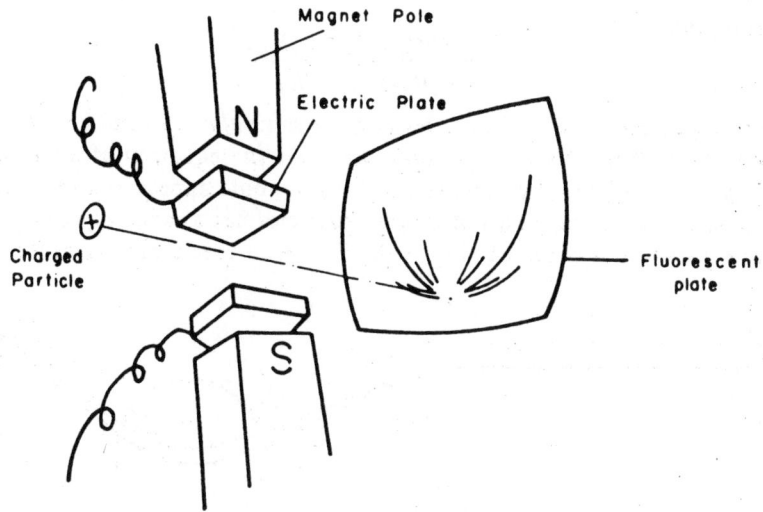

FIGURE 6.2: Positive ray apparatus.

the particle along the y axis to a, the deflection oa is given by:

$$y = \frac{k_2 R_H e}{mV} \tag{6.3}$$

where k_2 also depends on the apparatus. Again rearranging and squaring,

$$V^2 = \left(\frac{k_2 R_H e}{my}\right)^2 \tag{6.4}$$

When both fields are applied simultaneously, the particle is deflected to c or d. Under these conditions, $V_1^2 = V_2^2$, so

$$\frac{k_1 R_E e}{mx} = \frac{k_2^2 R_H^2 e^2}{m^2 y^2}$$

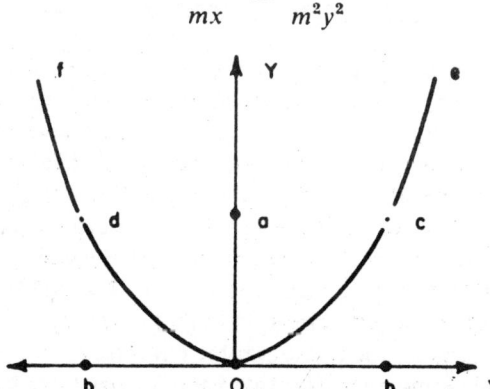

FIGURE 6.3: Path of ionized particle in electric and magnetic fields.

Rearranging,

$$\frac{y^2}{x} = \frac{k_2^2 R_H^2 e^2 m}{k_1 R_E e m^2} = \frac{k_3 R_H^2 e}{R_E m} \qquad (6.5)$$

If now we have a *beam* of particles with the same e/m ratio and hold R_H and R_E constant, then y^2/x is a constant, which is the equation of a parabola. If the particles all had the same velocity, they would appear as a spot on the parabola. Since a real beam of ionized particles has a range of velocities, a definite parabola is obtained (Fig. 6.3). Thomson used a photographic plate

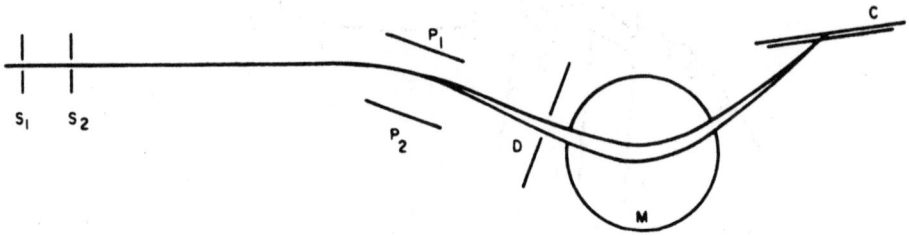

FIGURE 6.4: Aston's mass spectrograph.

to record the positive rays and measure the constants. When the discharge tube contained hydrogen, the e/m value was found to be the same as that found for the hydrogen ion in electrolysis. At the same time, other parabolas were observed at e/m 1/2, 1/12, 1/16, 1/28, and 1/44 which were interpreted as due to hydrogen molecule, carbon atom, oxygen, carbon monoxide, and carbon dioxide. Parabolas were found for any gas introduced into the discharge. When neon ($m = 20$) was studied, a faint line at $m = 22$ was also observed. This was the $m = 22$ isotope of neon. Its discovery marked our first awareness of isotopes.

A. ASTON'S MASS SPECTROGRAPH

A more sensitive type of equipment was developed by Aston.[1] While the magnetic and electric fields in Thomson's positive-ray apparatus produced deflections in planes at right angles, in Aston's arrangement (Fig. 6.4) they were in the same plane, but in opposite directions. The positively charged particles first pass through two very narrow parallel slits S_1 and S_2 and the resulting fine beam is spread out by means of an electric field applied across plates P_1 and P_2. A section of the beam emerges through another small slit at D and then passes through a magnetic field M. This field is so arranged as to bend the rays in the opposite direction to that of the electric field, and bring them to a sharp focus on the photographic collector C.

Aston did not determine the absolute value of e/m from his mass spectrographic data. He took the value of 16.000 for the oxygen isotope $m = 16$

and evaluated all other masses in terms of this. Since there is a significant amount of other oxygen isotopes present in normal oxygen, this led to a small but observable difference between chemical atomic weights based on atmospheric oxygen, a mixture, as 16.000. This difference was resolved in 1961 by an international congress, which took the mass of carbon as 12.000 and compared masses from that basis.

B. DIRECTION FOCUSING

A different principle for separating the fragments was used by Dempster.[5] The positive particles were accelerated by passage through an electric field. When a high potential was used, the particles had uniform kinetic energy. If V is the accelerating potential and e the charge on the particle, the energy is Ve. Kinetic energy of the particle, $1/2\ mv^2$, where m and v are mass and mass and velocity, respectively, is then equal to Ve.

$$Ve = 1/2\ mv^2 \tag{6.6}$$

A thin beam of accelerated ions was directed through a narrow slit into a magnetic field which bends the path of the beam into a semicircular path. By means of a second slit, the beam was directed onto a plate connected to a device for measuring ion currents.

The radius of curvature r of the path of a charged particle moving in a magnetic field H is determined by:

$$e/m = \frac{v}{Hr} \tag{6.7}$$

So, combining Eq. (6.6) and (6.7)

$$e/m = \frac{2V}{H^2r^2} \tag{6.8}$$

Only particles with a definite value of r can pass the second slit and register on the electrometer. The e/m values for the particles having a path of this radius is determined by the accelerating potential V and the strength of the magnetic field H.

Recent developments in the design of mass spectrometers utilize this double focus to extend their usefulness. The ions formed initially are brought to a focus at a slit designed to pass ions with only a particular *mass*. These then are further separated in the magnetic field into beams with a definite mass-to-charge ratio. This leads to extremely high resolution and enables the operator to use a scale linear with respect to mass. The linear mass scale enables chemists to interpret spectra directly in terms of mass (m/e), which is much more readily visualized than the e/m scale which was used to investigate the basic phenomena. Figure 6.5 gives the basic configuration of a modern mass spectrograph.

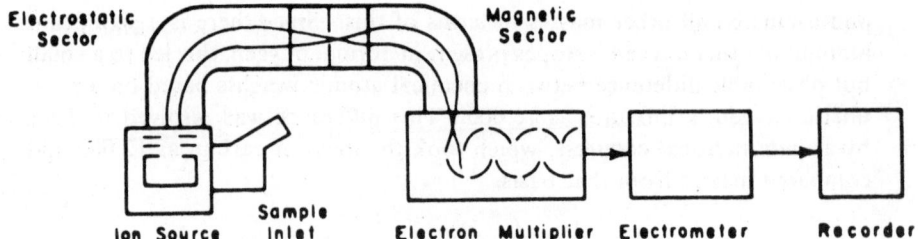

FIGURE 6.5: Modern mass spectrograph (block diagram).

6.2 INSTRUMENTATION

The major function of the mass spectrometer is to provide means to disso-ciate the target molecule and ionize the fragments, to separate the ionized fragments according to their mass, and to provide a detector to measure the relative abundance of each ion type according to its m/e ratio.

A. ION FORMATION

In the ionization chamber, the molecules of the sample are exposed to bombardment with energetic electrons. Below the ionizing potential of the molecule, usually about 10 V, there is no response. At the ionizing potential, an electron is removed from the molecule to form the parent ion, which is the positively charged molecular ion. At still higher applied potentials, enough energy is supplied that one or more bonds in the molecule are broken. The fragments may be sufficiently stable that they proceed through to the detector, or they may react with other fragments or may decompose. The actual mechanism of ion formation is not well understood; much danger of mis-information arises by unwarily accepting the presence of a fragment in a spec-trum on the basis of its mass number.

In some spectra, peaks are present which can best be interpreted as if the molecule had lost *two* electrons. The peak appears at exactly *half* the mass of the fragment.

When the ion fragment decomposes in flight, the resulting metastable ions have less than full kinetic energy, so the apparent mass is less than the true mass. The apparent mass M_a is related to the masses of the original ion M_0 and the daughter ion M_d by the equation

$$M_a \doteq \frac{(M_d)^2}{M_0}$$

One fragment of the decomposed particle retains the negative charge. It is possible to modify instruments to study this negative ion, but intensities are usually low, so most work has been done with the positive ion. Certain halo-genated compounds are satisfactorily studied as negative ions.

The chief limitation of mass spectrometer is the necessity for having the sample in the vapor phase. Heating the injection port is normal procedure, but this frequently brings about thermal degradation of the compound. Large molecules can only be studied if information can be gained from the more volatile breakdown products.

B. ION SEPARATION

The positively charged molecular fragment is subjected to an accelerating potential and passed through collimating slits into a magnetic field, which

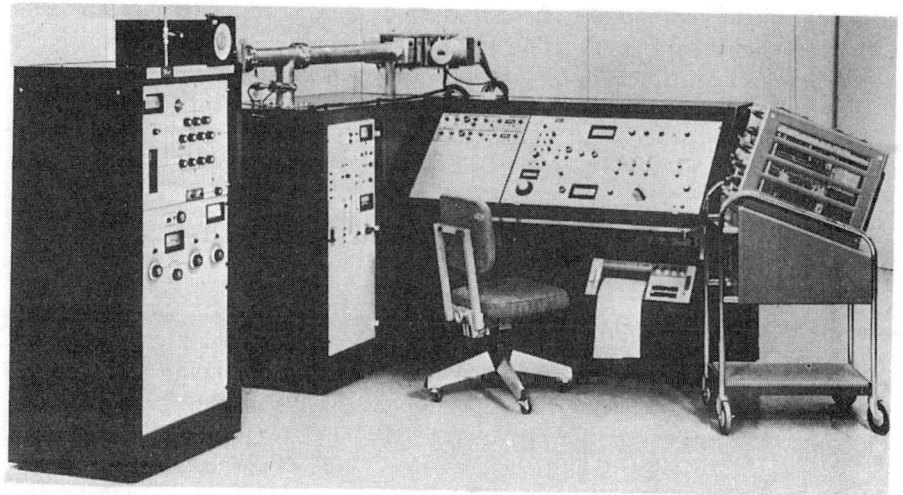

PLATE 6.I: Bendix time-of-flight mass spectrometer. Courtesy Bendix Corporation.

sorts the fragments into groups with the same m/e ratio. With high resolution instruments, species of the same mass, but differing in elemental composition and therefore in fractional mass, can be resolved. A precision of 1 in 100,000 is claimed.

A time-of-flight instrument resolves the ion beam by measuring the flight time of the ions of each mass. A pulse of ions is formed by bombarding the sample molecules with a short burst of ionizing electrons. In rapid sequence the ionizing electrons are turned off and an accelerating pulse pulls the ion into high speed acceleration, which then passes into a field-free "drift" tube. All ions receive the same kinetic energy, and so those with heavier mass move more slowly. The ions are collected by an electron multiplier with the extremely fast response needed to record a complete spectrum in 100 μsec. The simplest instrument to serve as a recording system is an oscilloscope (see Plate 6.I), which is very useful in monitoring the fast-changing effluent from a gas chromatograph.

C. ION DETECTION[25]

Photographic detection was used originally, but development is tedious, and the precision of densitometry is not great. An electrometer or electron multiplier is nearly always used with modern instruments. Electronic amplification of the ion current allows an oscilloscope, a pen and ink recorder, or a digital computer to be used. High resolution instruments again use photographic detection, but the resolution rather than the intensity is the feature of interest.

The instrument must cover a mass range at least beyond the molecular weight of the compound to be investigated, and the resolving power should be such that the species differing by 1 mass unit in the region of the molecular weight can be distinguished. The advantages of mass spectrometry are that a very small sample is needed—not more than a few micrograms—and the accuracy of analyses is within a few tenths of 1 %.

6.3 INTERPRETATION OF SPECTRA

Mass spectrometry can give information at several levels of complexity. The evaluation of rearrangements and bond fissions that have caused each peak in a spectrum is a concentrated effort in advanced research. Such information is then applied to elucidation of the structure of complex molecules. Mass spectral data are used to determine relative quantities of known materials in a mixture. This ability has wide applications in industrial work. The gross pattern can give qualitative identification of known substances, and is an excellent check for absence of a suspected material. Both infrared and ultraviolet examination are often inadequate for this purpose. Gas chromatography is an excellent technique for separating compounds, but a single peak does not mean the compound is pure and so is unreliable for identification of a material. When gas chromatography and mass spectrometry are combined, a powerful analytical tool results.

Now realize that ions in a mass spectrograph can appear through several processes. *First*, a single electron can be removed with no other bond being affected. This gives a positive ion with mass equal to the molecular weight, referred to as the *parent* ion. While some researchers use the symbol *P*, this peak is referred to here as the *M* peak. The strongest peak in the spectrum is referred to as the base peak and intensities of all other peaks are expressed as per cent of the base (Fig. 6.6).

Second, a single bond can be split:

$$R - CH_3 \rightarrow R + CH_3^-$$

This type of cleavage gives rise to most of the principle fragments. More usually when the energy is high enough to disrupt *one* bond, the molecule will split in several places and the pattern must be interpreted in terms of the possible fragments. Relative quantities of these ions indicate the strength of the bonds uniting them to the rest of the molecule.

Third, the incident energy can cause cleavage with rearrangement of a hydrogen atom or larger group.

Fourth, two electrons may be lost, giving rise to an ion of apparent m/e one-half the true value. These four effects can explain most of the major peaks in a spectrum.

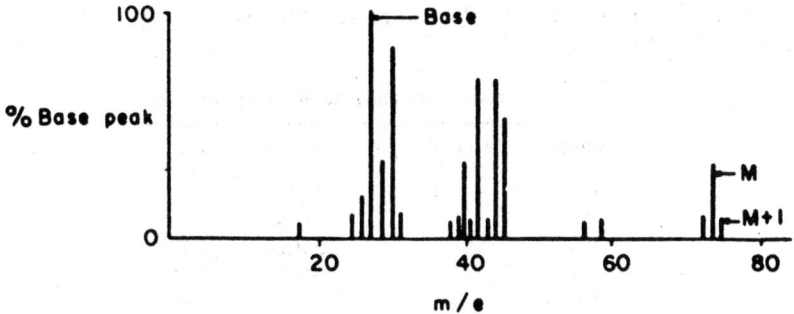

FIGURE 6.6: Idealized mass spectrum.

The greatest aid to qualitative analysis is a large library of mass spectra which have been coded for information retrieval. Most laboratories have collected spectra of the compounds and compound types in which they are interested. Study of the available patterns yields many generalizations regarding the particular compound under study.[17] Libraries of spectra of common materials are also commercially available.

One seldom begins with a sample of completely unknown origin. It may be an impurity isolated from a raw material, a single spot isolated from a thin-layer chromatogram, or a compound isolated by an organic chemist from a reaction. Any of these suggest related compounds, potentially helpful in interpretation of your sample. If this still has not solved the problem, a rough approach is to proceed as follows:

Establish the parent peak ($m/e = M$) using a low voltage spectrum and note the fragment peaks that are produced. Measure the $M + 1$ and $M + 2$ peaks carefully. These are isotopic ion peaks which can give information on the number of carbons, nitrogens, and oxygen atoms. The $M + 1$ peak is due to the 1.1 % of ^{13}C in normal carbon, plus the 0.4 % ^{15}N in normal nitrogen. Silicon and sulfur also contribute significantly to the $M + 1$ peak if they are present. The $M + 2$ peak arises from the ^{18}O isotope of oxygen, and the contribution of silica and sulfur. The halogens chlorine and bromine both have isotopes 2 mass units apart; their relative abundance, 3:1 for chlorine and 1:1 for bromine, give relative intensities for M and $M + 2$ peaks which are quite characteristic and once observed are easily remembered (Table 6.1).

In normal spectra, the parent ion is frequently observed not at all, or in only trace amounts, so that assignment of a parent is frequently difficult. Usually interpretation is begun with a reasonable guess, which must then be justified.

Now using this information, calculate roughly the number of carbons and other atoms in the molecule. Next, examine the first major peaks lower than the parent and using background knowledge relate the difference to loss of a particular fragment. To do this more satisfactorily, some generalizations have been gathered to help.

TABLE 6.1: Natural Abundance of Heavy Isotopes

Isotope	Molecular ion	$M + 1$	$M + 2$
^{12}C	1	0.011	
^{14}N	1	0.004	
^{16}O	1	0.0004	0.002
^{32}S	1	0.008	0.04
^{35}Cl	1		0.33
^{79}Br	1		1

A. ALIPHATIC HYDROCARBONS[4,13,18]

Usually the most intense ion of an aliphatic hydrocarbon occurs in the C_2 to C_6 mass range. These ions appear at masses

29    $(CH_3—CH_2^+)$

43    $(CH_3—CH_2—CH_2^+)(CH_3—CH^+—CH_3)$

57    $CH_3—(CH_2)_3 \left(\begin{array}{c} CH_3 \\ CH—CH_2^+ \\ CH_3 \end{array} \right) \left(\begin{array}{c} CH_3 \\ CH_3—\overset{+}{C}—CH_3 \end{array} \right)$

71    $(CH_3—(CH_2)_3—CH_2^+) \left(\begin{array}{c} CH_3 \\ CH—CH_2—CH_2^+ \\ CH_3 \end{array} \right) \left(\begin{array}{c} CH_3 \\ CH_3—\overset{+}{C}—CH_2^+ \\ CH_3 \end{array} \right)$

Cleavage is easier at bonds around a substituted carbon, for example,

$$R—\overset{\overset{\textstyle H}{|}}{\underset{\underset{\textstyle R_3}{|}}{C}}—R_2 \rightarrow \ ^-\overset{\overset{\textstyle H}{|}}{\underset{\underset{\textstyle R_3}{|}}{C}}—R_2 \quad R_1—\overset{\overset{\textstyle H}{|}}{\underset{\underset{\textstyle R_3}{|}}{C^+}} \quad R_1—\overset{\overset{\textstyle H}{|}}{\underset{\underset{\textstyle R_3}{|}}{\overset{+}{C}}}—R_2$$

B. CYCLOPARAFFINS

Compounds containing cyclohexyl rings give fragments at m/e 83, 82, and 81, corresponding to the ring, and the ring with one and two hydrogens removed. Alkyl substituents are split off and also form a positive ion.

C. MONO OLEFINS

Compounds with one double bond tend to rupture at the carbon-carbon bond beta to the double bond. The fragment with the double bond usually loses the electron to become a positive ion.

$$H_2C{=}CH{-}CH_2{\}}R \rightarrow H_2C{=}CH{-}CH_2^+ + R$$

D. ALKYL BENZENES[13.15]

Meyerson has studied the dissociation in alkylbenzenes. He reported the principle ion usually results from cleavage of bonds beta to the ring. The

mass 91

phenyl-group ion, mass 77, is found in the intensity range 1 to 14%.

E. ALIPHATIC ALCOHOLS[6]

Primary alcohols $R{-}CH_2OH$ cleave to give water ($M - 18$) and CH_2OH^+ ($m/e = 31$). The molecular ion $R{-}CH_2{-}OH^+$ is important in lower molecular weight alcohols.

Secondary alcohols can be divided depending on the location of the hydroxyl on the 2, 3, or 4 carbon atom.

The peak at m/e 31 is also strong.

Tertiary alcohols split off the largest alkyl group to form the base peak.

$$
\underset{\underset{OH}{|}}{\overset{\overset{R_2}{|}}{R_1-C-R_3}} \rightarrow R_1 + \underset{\underset{OH}{|}}{\overset{\overset{R_2^+}{|}}{C-R_3}}
$$

F. ALDEHYDES[7]

Aldehydes give more predictable spectra than alcohols. The parent ion is found even in higher molecular weight compounds. Cleavage is at either the

$$R \overbrace{}^{} CH_2 \overbrace{}^{} CHO$$

α or β bond. β-Bond cleavage gives a fragment m/e 44 which is the base peak in straight-chain aldehydes and a fragment equal to molecular weight less 44. This is characteristic for identification work.

$$
R-CH \Big\langle \begin{matrix} H \\ \\ CH_2CHO \end{matrix}
$$

G. KETONES

A study by Sharkey et al.[19] showed most of the compounds they studied split off the larger of the alkyl groups of the ketone. The fragment ($m/e = 43$)

$$
\overset{\overset{H}{|}}{OC-CH_2}
$$

formed by rearrangement of ions is frequently a base peak or at least a major one.

H. ACIDS[11]

The base peak in almost all acids is attributed to a fragment (m/e 60) formed as follows:

$$
R-CH \Big\langle \begin{matrix} H \\ \\ CH_2-COOH \end{matrix}
$$

The expected COOH fragment (m/e 45) is also common. When strong peaks are observed at m/e 60 and m/e ($M - 45$), an organic acid can be expected.

I. ESTERS[2,20]

Esters tend to dissociate at the valence bonds:

$$
\underset{}{\overset{\overset{O}{\|}}{R_1-OC{\nmid}R_2}} \qquad \underset{}{\overset{\overset{O}{\|}}{R_1-O{\nmid}CR_2}} \qquad \underset{}{\overset{\overset{O}{\|}}{R_1{\nmid}OCR_2}}
$$

and have an intense ion due to the fragments from

$$R\!-\!CH\!\!\underset{CH_2\!-\!COOR_1}{\overset{H}{\Bigg\langle}}$$

J. ETHERS[14]

Symmetrical ethers undergo cleavage alpha to the oxygen atom. Mixed aliphatic ethers also cleave beta to the oxygen, losing the more highly substituted fragment.

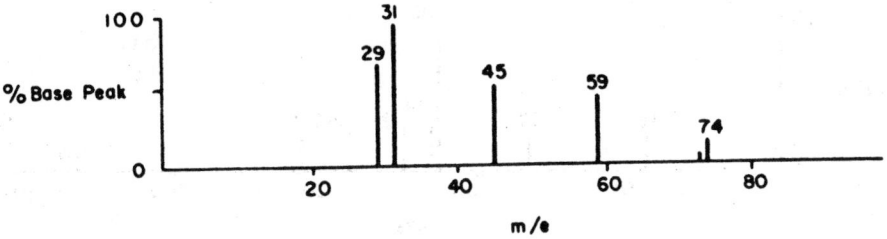

FIGURE 6.7: Major peaks of diethyl ether mass spectrum.

Peaks 74, 59, 45, and 29 in Fig. 6.7 are readily explained:

$$M = CH_3CH_2OCH_2CH_3 = 74$$
$$M - 15 = C_3H_5OCH_2^+ = 59$$
$$M - 29 = C_2H_5O^+ = 45$$
$$29 = C_2H_5^+$$

The 31 peak is most logically formed by rearrangement

$$CH_3\!\!\underset{H\!\!\mid\!\!CH_1}{\overset{\mid}{\Big\langle}}\!\!CH_2\!-\!O\!\!\mid\!\!CH_2 \quad \rightarrow \quad \underset{H}{CH_2}\!-\!O^+ = 31$$

K. MERCAPTANS[13]

These compounds give mass spectra similar to alcohols. The main differences are in the relatively abundant molecular ion peak, and the existence of significant $M + 2$ due to the 4.4% of the sulfur isotope mass 34.

L. ALIPHATIC AMINES[8]

The amino group directs the fragmentation pattern so strongly that the low mass region gives most information about the molecule. The most important process is cleavage of the bond adjacent to the nitrogen.

$$R_1\!\!\mid\!\!C\!-\!NR_2R_3 \rightarrow R_1 + C\!-\!NR_2R_3$$

M. PIPERIDINE RINGS

Piperidine rings are considerably more complicated than aliphatic amines because there are more possibilities for bond cleavage of molecular ions and rearrangement products. Base peak is

(*m/e* 84), but the associated peaks need careful assessment before assignment. A study on Vitamin B₁ by Hesse et al.[12] describes the decomposition pattern

FIGURE 6.8: Major peaks of thiamine mass spectrum.

of a nitrogen-containing compound. The major peaks in the spectrum are given in Fig. 6.8.

The molecule, mass 300, gives major peaks at 264, 233, 157, 143, 122, 113, 112, 85, and 45.

The obvious splits between the two rings provides three of these peaks:

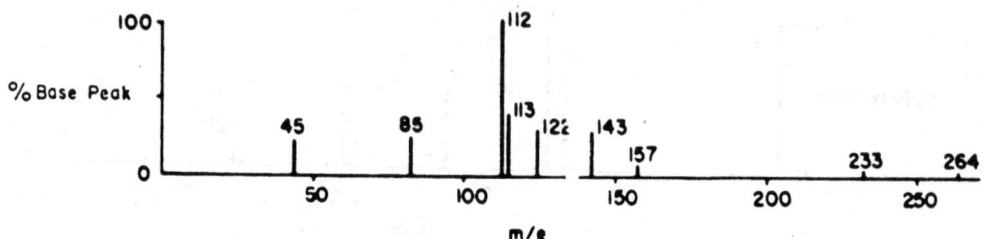

thiamine *m/e* = 300

m/e = 157

m/e = 122

m/e = 143

m/e = 143

m/e = 113

m/e = 112

m/e = 85

Cleavage of the alcohol side chain from one of the rings gives the base peak at 112. This fragment decomposes to give the fragment with m/e 85 and the $S=CH$ fragment, m/e 45.

To explain the peaks at 264 and 233, the authors postulated an initial cyclization to form

$m/e = 264$ $m/e = 233$

N. HALOGEN COMPOUNDS[22]

The presence of chlorine and bromine in a compound may be deduced from the high ratio of the corresponding heavy isotope. For a compound with one atom of chlorine, the $M + 2$ peak will be about one-third of the molecular ion. Further confirmation for chlorine is the presence of a strong doublet at masses 35 and 37, which are also in the $3:1$ ratio. The $M + 2$ peak in a monobrominated compound has an intensity nearly equal to the parent (M) peak. This is because the two isotopes, masses 79 and 81, are present in equal amounts.

In general, be wary. Rearrangements often occur which obscure qualitative interpretation or mislead one into confidence in an obvious but erroneous interpretation. High resolution examination of the major fragments assists greatly in avoiding this last pitfall. Also, attempting to supply a reasonable explanation of all peaks, even minor ones, will often necessitate rejection of a comfortable, obvious solution.

6.4 MASS SPECTROSCOPY IN PHARMACY

In pharmaceutical work, it is frequently necessary to assay one component in a complex mixture. Many procedures are used to effect specificity; separations by classical techniques and chromatography, specific function analysis by colorimetry, kinetic studies, or infrared absorbance. Mass spectrometry is another technique, often very rapid, for performing these specific analyses. One example is a procedure for determination of aspirin, phenacetin, and caffeine in pharmaceutical products.[21] Many materials used in pharmacy are not readily handled by other techniques, so frequently mass spectrometry will give the only useful data. As instrumentation becomes available, more of this type of work will be reported in the pharmaceutical literature.

A. QUANTITATIVE ANALYSIS[24]

Washburn developed a general procedure for quantitative calculations with mass spectrometry where a mixture is being examined to determine quantitatively one component; all possible components must be run separately to observe peaks that are present and to determine the ion current per unit partial pressure. We assume that contributions to each peak are additive for all components present in the mixture and expect we can find at least one peak for each compound which is unique. Where this condition is met, ion currents can be expressed using a four-component mixture as an example:

$$l_1 = i_{1a}P_a + i_{1b}P_b + i_{1c}P_c + i_{1d}P_d$$

$$l_2 = l_{2a}P_a + i_{2b}P_b + i_{2c}P_c + i_{2d}P_d$$

.

.

.

$$l_m = i_{ma}P_a + i_{mb}P_b + i_{mc}P_c + i_{md}P_d$$

TABLE 6.2: Mass Spectral Data for Pure Solvents[a]

m/e	Cyclohexanone	Toluene	Dioxane	Acetone
28	12.5	0.5	100	1.9
43	12.2	1.8	10.9	100
55	100	—	—	0.3
58	0.1	—	23.6	27.8
91	—	100	—	—
Division/micron pressure	29	47	39	48

[a] Per cent of base.

where i_{ma} is current at mass m due to component a, P_a is partial pressure of component a, and I_m is the observed total ion current at mass m. One may calculate the relative peak height for a given component based on a reference peak being 1. A solution for the equations gives component peak heights. A residual peak is the value of a peak intensity after all other known compounds have been mathematically subtracted.

For example, if a solvent mixture of cyclohexanone, toluene, dioxane, and acetone is to be examined, the important peaks in the spectra of the pure solvents are given in Table 6.2. The peak at m/e 28 in a mixture of these four compounds is made up of:

$$\frac{12.5c}{100} + \frac{0.5t}{100} + \frac{100d}{100} + \frac{1.9a}{100}$$

where c, t, d, and a are the contributions of the individual solvents in the mixture. By setting up simultaneous equations equal to the number of components, the contribution of each component to the spectrum may be calculated. The contribution of the component is divided by the sensitivity (divisions/micron) to obtain the partial pressure of that component. The partial pressures, when normalized, give the molar ratios of components.

B. GAS CHROMATOGRAPHY

Joint use of a mass spectrometer and a gas chromatograph has been hailed as the ultimate analytical instrument. The gas chromatograph (see Chapter 18) is a powerful tool for effecting separations of closely related organic compounds. The mass spectrometer enables a qualitative study and identification of the highly purified fractions separated by the gas chromatograph (Plate 6.II).

PLATE 6.II: Combined gas chromatograph-mass spectrometer. Courtesy Perkin–Elmer Corporation.

When pharmaceutical materials are studied, this combination gives both qualitative and quantitative information. A routine gas chromatographic survey demonstrated the presence of an unknown volatile material in a tank carload of specially denatured alcohol. Using mass spectrometry, the unknown was identified as diethyl ether. Knowing its identity, the concentration could then be calculated as about 10 ppm. This enabled a producer to locate a potential contaminating procedure and gave added assurance to the pharmaceutical company that it was using only high quality materials for its processing.

C. PYROLYSIS[10]

Pyrolysis of a sample followed by mass spectrometric examination of the volatile products has been used to identify many polymers and other large nonvolatile molecules. An apparatus has been described by Happ and Maier which enables examination of both volatiles and residue. This identification is important in examination of packaging materials in which pharmaceutical materials are shipped as raw materials and finished products. Infrared examination and physical tests are complementary procedures.

D. ISOTOPE DILUTION[16,27]

In the complex analytical studies needed for determination of the metabolic pathway of a new drug substance, the use of radioactive isotopes has already shown its utility. One disadvantage, however, is that the radioactivity itself may affect the pathway. Another is that the half-life of the isotope may be too short to complete a lengthy study. With a mass spectrograph, it is possible to use stable isotopes and a larger range of isotopes to obtain this information.

The method requires that a sample of the compound to be determined be prepared with one element enriched in its isotope. It is assumed that the test organism cannot distinguish the labeled compound from the normal one. When the labeled compound is added to an unknown mixture, it is diluted by the normal compound. A fractionation is performed to obtain a pure sample of the component; the isotope ratio is determined on the recovered material and from the change in ratio, the amount of the compound originally present is calculated.

The weight W of component in the original mixture is given by

$$W = W_1\left(\frac{A_0 - A}{A}\right)$$

where A_0 is the concentration of the isotope in the added sample *in excess of normal*, A is the concentration of isotope in the *recovered* sample in excess of normal, and W_1 is the weight of isotope-enriched material added.

E. COMPUTER APPLICATIONS[3,25]

Studies using a high resolution, double-focusing mass spectrometer can differentiate among molecules with the same nominal molecular weight Plate 6.III). The exact mass of the elements are sufficiently different that on fragments $^{12}CH_3$ and $^{14}NH_1$ with exact mass, respectively, 15.023 and 15.0108, can be readily separated. Bieman and co-workers[3] have reported a technique where the exact measurements of mass are made for most of the on fragments in a spectrum of a complex material and possible elemental structures are calculated. A computer, which is programmed to sort the

PLATE 6.III   Double focusing mass spectrometer.

exact fragment weight into columns according to content of hetero atoms in ascending order of numbers of carbon and hydrogen, is used for data reduction. A very comprehensive picture of the molecular structure of even complex molecules is obtained this way (Table 6.3).

Computers also find use in the resolution of mixtures when the spectra of pure compounds are inserted. The computer can be programmed to calculate the relative amounts of the individual materials in a mixture. These calculations are otherwise very tedious and time-consuming (6.6). If the spectrum of a pharmaceutical mixture is inserted initially, the computer can watch for changes in the spectrum of subsequent samples which have been subjected to adverse conditions in stability studies.

TABLE 6.3: Exact Mass of Common Nuclides

Nuclide	Mass
^1H	1.007825
^{12}C	12.000000
^{13}C	13.003354
^{14}N	14.003074
^{16}O	15.994914

F. PLANT CONTROL

When production of a material is routine and the market will sustain continuous production, engineering groups are organizing automated plants where the materials going into the plant, and the products, are monitored by instruments such as mass spectrometry and gas chromatography. The data are accumulated and interpreted by massive computer installations which can be programmed to adjust the process to maintain the product within the set limits. This approach must be kept in mind by those working in the pharmaceutical industry.

PROBLEMS

P6.1. If ^{16}O as a base for atomic weight values is taken as 16.0000 and the isotopes ^{17}O and ^{18}O are present to the extent of 0.04 and 0.20%, respectively, calculate the mean atomic weight of normal oxygen on the mass spectrograph scale (16.0044). Calculate the value of ^{16}O based on $^{12}C = 12.0000$ (15.9949).

P6.2. Where major peaks appear in the spectrum at the following mass peaks, list possible structural units which may be present

$$M - 15 \qquad (-CH_3)$$
$$M - 18 \qquad (-OH_2)$$
$$M - 57 \qquad (t\text{-butyl} \quad CH_3-\overset{\overset{\displaystyle O}{\|}}{C}-CH_2 \quad CH_2CH_2CHO)$$
$$M - 45 \qquad (CH_3-CH-OH \quad -COOH)$$

P6.3. Determine the number of carbon atoms in each of the compounds with the following M, $M + 1$, and $M + 2$ ratios

Compound	M	$M + 1$	$M + 2$	
a	48	0.51	—	(1)
b	31	1.02	—	(3)
c	58	1.4	0.26	(2)
d	27	1.5	—	(5)

P6.4. Determine whether the compound with molecular weight (310.3110) is $C_{20}H_{40}NO$ or $C_{20}H_{38}O_2(C_{20}H_{40}NO)$.

P6.5. Djerassi and co-workers considered the following possible formulas for an alkaloid: $C_{45}H_{54}N_4O_8$, $C_{42}H_{48}N_4O_{61}$, $C_{42}H_{50}N_4O_7$, $C_{43}H_{50}N_4O_6$, and $C_{42}H_{46}N_4O_7$. High resolution mass spectrometry indicated the molecular weight was 718.3743. Which formula was accepted? Ans: $(C_{43}H_{50}N_4O_6)$.

P6.6. Using the spectral data for the four pure solvents given in Table 6.2, calculate the contribution of each of the components to the following mass spectrum of a mixture of cyclohexanone, toluene, dioxan, and acetone:

m/e	Intensity
28	368.3
43	300.0
55	500.6
58	126.9
91	400

Gas	c	t	d	a
28	62.5	2.0	300	3.8
43	61.0	7.2	31.8	200
91	—	400	—	—

QUESTIONS

Q6.1. Define
   (a) isotope
   (b) molecular weight
   (c) ionization potential
   (d) ion
   (e) nuclide

Q6.2. Define metastable ion. Ans: If the moving ion can decompose in the region between the electrostatic accelerating and magnetic deflecting fields, the resulting ion is then deflected by the magnetic field. The m/e ratio of this ion, called the "metastable ion," relates the accelerate ion to the deflected ion.

Q6.3. Explain in what ways a chemist can obtain information directly from a mass spectrum.

Q6.4. How would you establish the exact mass of an unknown ion using a high resolution mass spectrometer? Ans: Run first the mass spectrum of a known material of molecular weight close to that of the unknown. Determine the difference between the known and the unknown weight.

Q6.5. Known mass 87.327 reads 39471 on a mass spectrograph. Known mass 95.318 reads 43920. Unknown mass reads 41392; what is the unknown mass?

Ans: 90.605

REFERENCES

1. F. H. Aston, *Science*, **52**, 559 (1920).
2. J. H. Beynon, R. A. Saunders, and A. E. Williams, *Anal. Chem.*, **33**, 221 (1961).
3. K. Bieman, *Tetrahedron Letters*, **1964**, 1725.
4. R. A. Brown, *Anal. Chem.*, **23**, 430 (1951).
5. A. J. Dempster, *Phys. Rev.*, **20**, 631 (1922).
6. R. A. Freidel, J. L. Shultz, and A. G. Sharkey, *Anal. Chem.*, **28**, 926 (1956).
7. J. A. Gilpin and F. W. McLafferty, *Anal. Chem.*, **29**, 990 (1957).

8. R. S. Gohlke and F. W. McLafferty, *Anal. Chem.*, **34**, 1281 (1962).

9. E. Goldstein, *Ber. Preuss. Akad. Wiss.*, **39**, 691 (1886).

10. G. P. Happ and D. P. Maier, *Anal. Chem.*, **36**, 1678 (1964).

11. G. P. Happ and D. W. Stewart, *J. Am. Chem. Soc.*, **74**, 4404 (1952).

12. M. Hesse, N. Bild, and H. Schmid, *Helv. Chim Acta*, **50**, 808 (1967).

13. I. W. Kinnear and G. L. Cook, *Anal. Chem.*, **24**, 1391 (1952).

14. F. W. McLafferty, *Anal. Chem.*, **29**, 1782 (1957).

15. S. Meyerson, *Appl. Spectry.*, **9**, 120 (1955).

16. D. Rittenberg and G. L. Foster, *J. Biol. Chem.*, **133**, 737 (1940).

17. S. M. Rock, *Anal. Chem.*, **23**, 261 (1951).

18. P. N. Rylander, S. Meyerson, and H. M. Grubb, *J. Am. Chem. Soc.*, **79**, 842 (1957).

19. A. G. Sharkey, J. L. Shultz, and R. A. Friedel, *Anal. Chem.*, **28**, 934 (1956).

20. A. G. Sharkey, J. L. Shultz, and R. A. Friedel, *Anal. Chem.*, **31**, 87 (1959).

21. A. Tatematsu and T. Goto, *J. Pharm. Soc. Japan*, **85**, 624 (1965).

22. R. C. Taylor, R. A. Brown, W. S. Young, and C. G. Headington, *Anal. Chem.*, **20**, 398 (1948).

23. J. J. Thomson, *Phil. Mag.*, **13**, 561 (1907).

24. H. W. Washburn, H. F. Wiley, and S. M. Rock, *Ind. Eng. Chem. Anal. Ed.*, **15**, 541 (1943).

25. F. A. White and T. L. Collins, *Appl. Spectry.*, **8**, 169 (1964).

REVIEW ARTICLES

26. K. Biemann, *Ann. Rev. Biochem.*, **32**, 755 (1963).

27. F. W. McLafferty, *Science*, **151**, 641 (1966).

BOOKS

Beynon J. H., *Mass Spectrometry and Its Application to Organic Chemistry*, Van Nostrand, Princeton, N.J., 1960.

Biemann, K., *Mass Spectrometry Organic Chemical Applications*, McGraw-Hill, New York, 1962.

Budzikiewicz, H., C. Djerassi, and D. H. Williams, *Interpretation of Mass Spectra of Organic Compounds*, Holden-Day, San Francisco, 1964.

Freeman, S. K. (ed.), *Interpretive Spectroscopy*, Reinhold, New York, 1965.

Kolthoff, I. M., and P. J. Elving (eds.), *Treatise on Analytical Chemistry*, Part I, Vol. 4, Wiley (Interscience), New York, 1961.

Silverstein, R. M., and G. C. Bassler, *Spectrometric Identification of Organic Compounds*, Wiley, New York, 1963.

Willard, H. H., H. Merritt, and J. A. Dean, *Instrumental Methods of Analysis*, Van Nostrand, Princeton, N.J., 1965.

CHAPTER 7

Nuclear Magnetic Resonance Spectroscopy

Mathias P. Mertes

DEPARTMENT OF MEDICINAL CHEMISTRY
SCHOOL OF PHARMACY
UNIVERSITY OF KANSAS
LAWRENCE, KANSAS

7.1 INTRODUCTION

The intensity of research activity centered on nuclear magnetic resonance as a tool or delicate probe used to determine the fine structure of molecules is incredible when one considers the fact that the first experiment was performed in 1946. Nuclear magnetic resonance was not considered to have broad application until several years later when high resolution studies on ethanol revealed the absorption peaks of the ethanol hydrogens were not alike but were dependent on the environment and the nature of the atom to which they were bonded. The outcome of these initial experiments was the development of stable high resolution instruments capable of probing the fine structure about hydrogen, fluorine, phosphorus, boron, and other isotopes with a non-zero spin number.

(Nuclear magnetic resonance like infrared and ultraviolet is the process whereby energy from an external source is absorbed and causes a change or resonance to an "excited" or high energy state. The energy required for nuclear magnetic resonance is in the low energy or long-wavelength radio-frequency end of the electromagnetic spectrum.) The equivalent of the monochromator in other forms of spectroscopy is an electrically varied radiofrequency or a variable magnetic field. The detector, unlike the usual

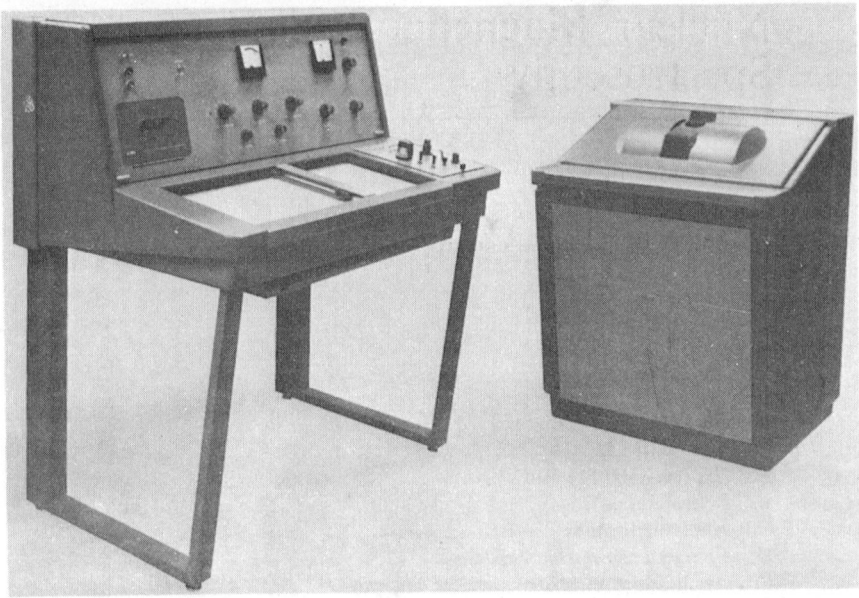

FIGURE 7.1: Varian A-60 A nuclear magnetic resonance spectrometer. Courtesy of Varian Associates, Palo Alto, California.

photomultiplier of ultraviolet or infrared, is a radio receiver. Thus, nuclear magnetic resonance requires a source of energy that resonates or is in tune with the nuclear magnetic and a detector. Figure 7.1 illustrates a nuclear magnetic resonance spectrometer; a schematic diagram of the same instrument is presented in Fig. 7.2.

The prerequisite for application of nuclear magnetic resonance is that the atom under examination has a nuclear spin number I greater than 0. When this condition is satisfied ($I > 0$), as, for example, in ^1H, ^{13}C, ^{15}N, ^{19}F, and ^{31}P, then magnetic resonance can be induced. The peculiar property associated with a spinning nucleus with $I = 1/2$ is that, although the nucleus is a symmetrically charged sphere, a spinning charged body creates a magnetic field. In effect, the spinning nucleus acts as a tiny bar magnet. If this spinning nucleus is placed in a strong external magnetic field, it will align with or

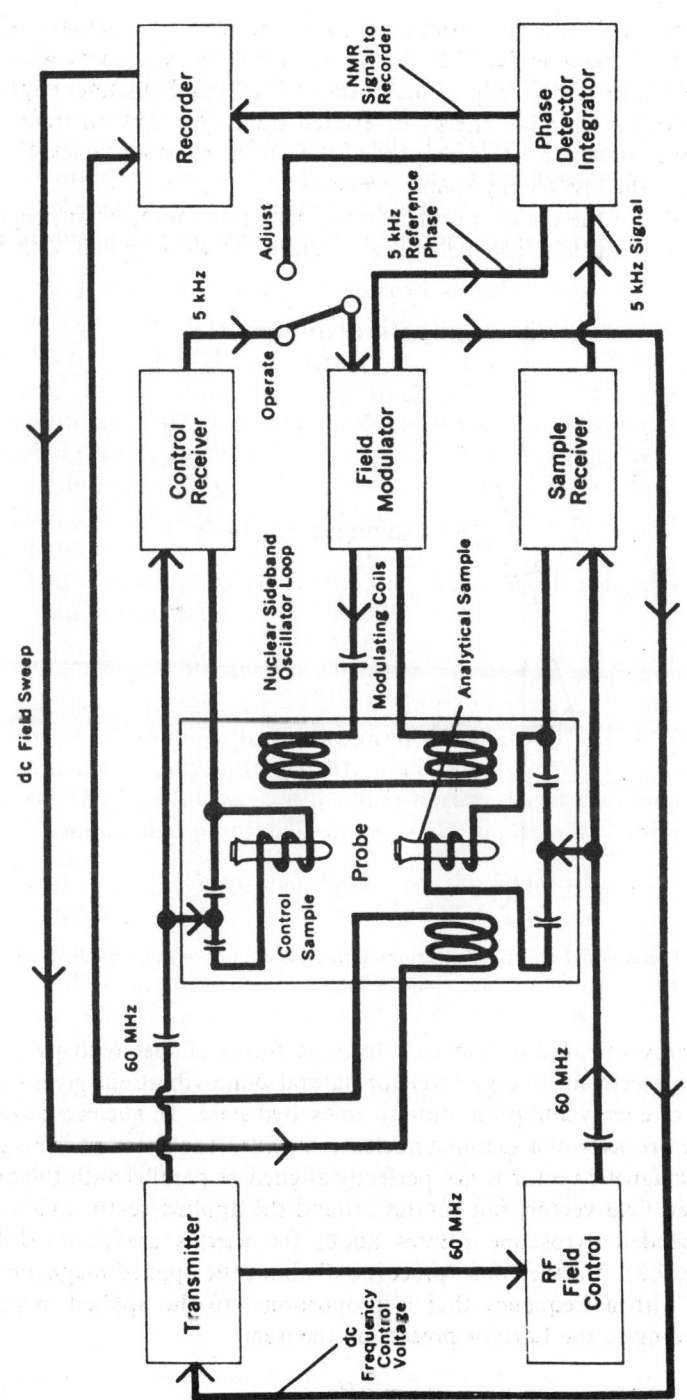

FIGURE 7.2: Schematic block diagram of an A-60 A spectrometer. Courtesy of Varian Associates, Palo Alto, California.

against the external field much the same as the needle on a compass aligns with the earth's magnetic field. It should be noted that alignment with the field represents the stable or low energy state $(-1/2)$ and alignment against the field $(+1/2)$ is the high energy or excited state. An upward transition from the lower state can be accomplished if a discrete energy "packet" can be supplied to the low energy spinning nucleus.

Proceeding into the theory of nuclear magnetic resonance, what property of this spinning nucleus can be utilized and externally supplied with this

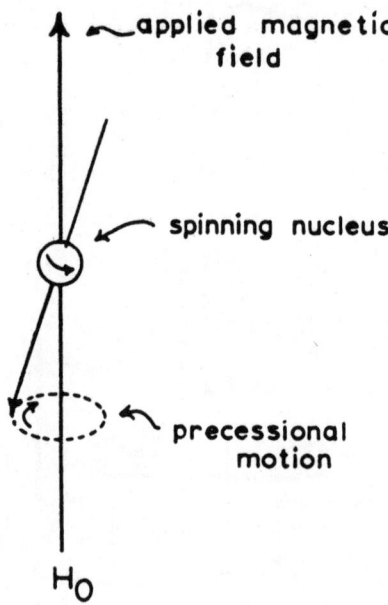

FIGURE 7.3: Precessional motion of a spinning nucleus aligned with an applied magnetic field (H_0).

discrete energy to cause resonance? In other forms of spectroscopy "resonance" with electronic energy levels or natural bond vibrations gives rise to absorption of energy and promotion to an excited state. In nuclear magnetic resonance a property of a spinning nucleus, in contrast to a static bar magnet, is that the magnetic vector is not perfectly aligned or parallel with the externally applied field vector, but rotates around the applied vector much as a freely suspended gyroscope rotates about the earth's gravitational field vector (Fig. 7.3). Further this "precession" about the applied magnetic vector occurs with a frequency that is proportional to the applied magnetic field, according to the Larmor precession theorem.

$$\omega = \gamma H_0 \tag{7.1}$$

The constant of proportionality γ, known as the gyromagnetic constant, is dependent on the nucleus; H_0 is the applied field strength in gauss, and ω, the angular momentum, is 2π times the frequency (ν) of precession. Rearranging to give

$$\gamma = \frac{\omega}{H_0} = \frac{2\pi\nu}{H_0} \tag{7.2}$$

$$\gamma/2\pi = \frac{\nu}{H_0} \tag{7.3}$$

we find that for a given nucleus $(I = 1/2)$ the precessional frequency ν increases as the externally applied field increases. For the hydrogen nucleus the

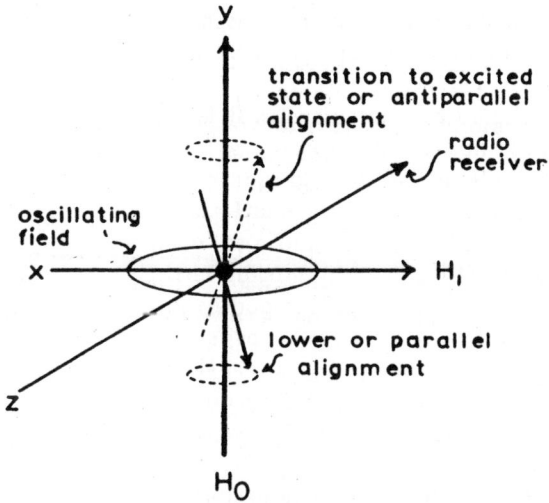

FIGURE 7.4: Alignment of a spinning charged nucleus in the applied field H_0; resonance of the oscillating field H_1 with the precessing nuclear vector induces a transition to the high energy state.

following values for frequency: 20, 40, 60, and 100 MHz correspond to applied fields of 4700, 9400, 10,400, and 17,300 G respectively. Examination of an idealized model in Fig. 7.4 reveals that a hydrogen nucleus aligned with an external field (H_0) of 10,400 G precesses with a certain frequency. If another electrically generated oscillating field (H_1) is now applied perpendicular to H_0, then at the frequency of oscillation in H_1 that corresponds exactly to the frequency of precession of the nuclear vector, the latter will absorb energy and "flip" to the high energy antiparallel or excited state. In the process of flipping, the precessing nuclear magnet induces a current in a radio receiver oriented perpendicular to the other two fields and tuned to the nuclear precessional frequency (60 Mc).

The idealized model does not account for the disturbing effect of thermal motion which effectively reduces the number of protons that align in the applied field. This limitation can be resolved by using stronger applied fields such as the 17,300-G magnet in the 100-mC instrument, which gives a greater proportion of aligned protons. While aligned in this strong field, all of the nuclear magnets theoretically would align parallel or in the low energy state $(-1/2)$. This would be the situation at $0°K$, but at room temperature thermal motions allow only a slight excess aligned in the low energy state. Irradiation of this collection of nuclear magnets with the proper radiofrequency gives equal probability of an upward transition and a downward transition. Fortunately there is a slight excess in the lower state and a signal can be detected from the finite excess of upward transitions.

After irradiation the two states have been equally populated, and unless the lower state can be repopulated in excess, there is no net absorption of energy and the signal fades. In infrared and other spectroscopic methods of analysis the return to ground state is rapidly achieved by emission of the excess energy and by thermal motions. The process of return in nuclear magnetic resonance to an excess population in the lower state requires some mechanism for dissipation of the excess energy. Emission of energy only takes place with an exchange of nuclear states and no net increase in the lower state.

Transition between states that occur by mechanisms other than emission are responsible for the return to an excess in the ground state, an essential feature for nuclear magnetic resonance. These relaxation processes can be divided into spin-spin relaxation and spin-lattice relaxation. In its simplest form the former process, often referred to as "transverse relaxation" (T_2), is a transfer of energy by a mutual exchange of spin states contributing little to the return to equilibrium. Spin-lattice or longitudinal relaxation (T_1) is the principal mode of establishing the initial equilibrium state. Violent thermal motions of nuclei produce random oscillatory magnetic fields some of which have frequency components equal to the precessing, excited nuclear magnet. By a transfer of the excess energy from the precessing excited nucleus to the reasonating random thermal fields dissipation as thermal energy is accomplished and the ground state is repopulated.

The natural line width observed for absorption of energy is proportional to the reciprocal of the relaxation time. If the latter is short, for example, in solids, or viscous liquids, or in the presence of paramagnetic molecules (O_2), the resonance signal is broad. Accordingly, moderately long relaxation times give rise to sharp peaks.

Examination of this simplified view of the theory of nuclear magnetic resonance gives some indication of the instrumental design. The strongest, homogeneous magnet obtainable is desirable to achieve the greatest field alignment and highest resolution in the spinning nuclei. The gyromagnetic ratio of the particular spinning nucleus being examined dictates the oscillating radiofrequency that will resonate with the precessing nuclear magnetic vector.

The detector, situated at a right angle to the oscillating input, is tuned to receive radio frequency generated by the moving nuclear magnet as it undergoes an upward transition. In the center is placed the sample, which responds in part to each of the components just discussed. Alignment of the nuclear spin state with the applied field (H_0) on the y axis is followed by irradiation of the sample by a component of oscillating current on the x axis. When the second field (H_1) oscillates at the same frequency as the precessing nuclear magnet, energy is absorbed in the lower spin to flip to the higher energy spin

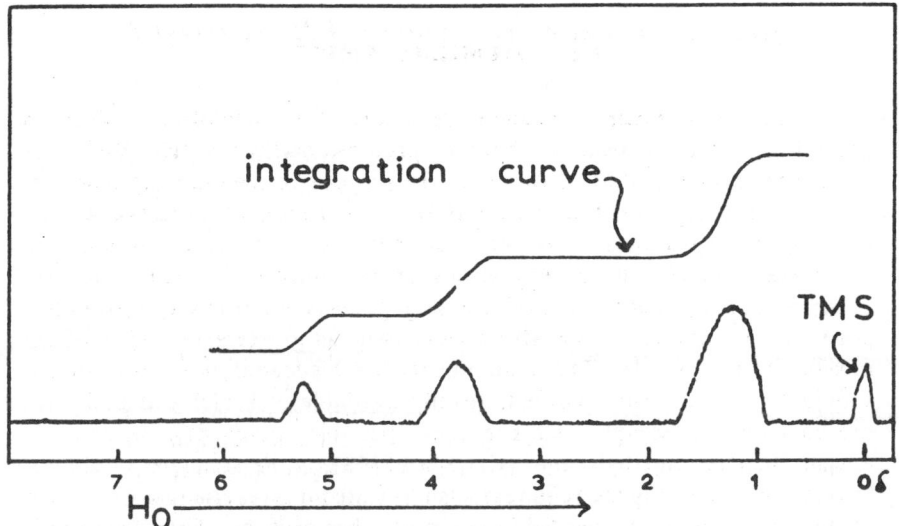

FIGURE 7.5: Low resolution spectrum of ethanol.

state. During the transition a signal is generated by the flipping nuclear magnet that is detected by the receiver installed on the z axis. The entire process of alignment, irradiation, detection, and recording the absorption complete the sequence. The theory of nuclear magnetic resonance predicts that a proton or any nucleus with nonzero spin should have an absorption peak at a frequency of radiation that varies with field strength according to the gyromagnetic ratio. If this is true, then all protons should resonate at a certain frequency in a given magnetic field. Several years after the initial demonstration of nuclear magnetic resonance an experiment utilizing a stronger magnet revealed that under higher resolution not one line for all protons but three resonance lines were observed for ethanol Fig. 7.5. The three lines, with relative areas under the curve of 1-2-3, were separated by only a few milligauss in a strong magnetic field. Almost immediately the significance of this finding was realized—the resonance absorption lines of protons varied over a narrow range because local magnetic fields in the environment of the proton affected the resonance frequency. The three observed peaks for ethanol represented the protons in different environments—on oxygen, C_1, and C_2; the distances between lines were termed the "chemical shift."

High resolution nuclear magnetic resonance thus required not only a homogeneous applied field and a stable applied oscillating radio frequency for irradiation but also a means of varying and measuring either the field or frequency over a very small range with a high degree of accuracy. Commercial instruments presently available utilize both methods—that is, varying radio-frequency in a stable field or varying field at a stable frequency.

7.2 CHEMICAL SHIFT

Application of nuclear magnetic resonance for qualitative analysis is dependent on the chemical shift between protons and the nature of the forces causing the chemical shift. Examining the protons of the classical example, ethanol, there are six protons in three different environments or rather bonded to three different atoms. The chemical shift is the relative position of the resonance signals for the three types of protons. Since absolute measurement of a milligauss in a field of several thousand gauss is not practical, the relative position of an absorption signal is usually assigned in reference to a standard mixed with the sample. The requirements for this standard are such that it should be stable, easily purified, readily available, soluble, and easily removed from the sample. Of the many possible standards considered it appears that tetramethylsilane, $(CH_3)_4Si$ (TMS), is the standard of choice. In addition to meeting the requirements just stated it gives rise to a very sharp resonance signal at a higher field than most other protons. Thus a proton is designated as having a chemical shift at a lower field by some arbitrary value from the TMS absorption peak.

Assuming operation of an instrument with a variable magnetic field and a stable frequency, examination of a solution of ethanol and tetramethylsilane in an inert, nonprotonated solvent (see Fig. 7.5) gives three major peaks in addition to the TMS reference peak. The oxygen-bound proton appears at the lowest magnetic field, followed in order by the protons on C_1 those on C_2, and the methyl protons of TMS, the latter appearing at the highest field. According to theory, the protons absorb radio frequency of a certain value that is dependent on the applied magnetic field.

The protons of ethanol, although subjected to the same applied field, experience the effect of very minute local fields that varies with the immediate chemical environment. For example, the field about the CH_3 protons of ethanol is the summation of applied field and the local field created by the bonding electrons about the proton nucleus. In addition to aligning the nucleus the spinning electrons also align with the applied field, but usually in the opposite direction. The electron motion creates a tiny induced field in the applied field H_0 and changes the field experienced by the proton H_{eff} according to the following:

$$H_{eff} = H_0 - \sigma H_0 \tag{7.4}$$

where σH_0 is the induced field generated in the applied field H_0. This "shielding" alters the effective field in the immediate vicinity of the proton nucleus to some value either higher or lower than the applied field. The protons attached to C_2 experience "shielding" relative to the protons on C_1 and oxygen.

Regarding the values used in assigning relative chemical shifts, certain conventions have been adopted to adjust for the various commercially available instruments. Table 7.1 lists several nuclei with their resonating frequency

TABLE 7.1: Resonating Frequency at Various Field Strengths

| Nucleus | NMR frequency | | Frequency at various fields | | |
	Spin no.	Megacycles/ kilogauss field	9400 G	14,092 G	23,490 G
^1H	1/2	4.26	40.00	60.00	100.00
^2H	1	0.65	6.15	9.21	15.35
^{11}B	3/2	1.37	12.93	19.25	32.08
^{13}C	1/2	1.07	10.06	15.08	25.14
^{14}N	1	0.308	2.89	4.33	7.22
^{17}O	5/2	0.577	5.42	8.13	13.55
^{19}F	1/2	4.00	37.65	56.44	94.07
^{31}P	1/2	1.72	16.19	24.29	40.48

at various fields. Although most routine examinations utilize a 60-Mc instrument with a field of 14,092 G and employ TMS as the reference, a method of comparison to results obtained using instruments at other field strengths is used in assigning chemical shifts.

Three parameters have been used to relate the position of a resonance signal relative to TMS. The recording of a nuclear magnetic resonance spectrum employs an xy recorder. Using the common 60-Mc instrument with a field that can be varied several hundred milligauss on either side of 14,092 G the x axis represents increasing field strength in moving from left to right on the chart. Since the field is varied and the radio frequency is stable, it would be logical to report peak positions (chemical shift) relative to the standard in milligauss. This method of assignment has not been adopted, but rather the resonance peak is recorded in units of cycles per second (hertz) shift from the reference; almost all protons can be observed at 60 Mc to appear over a range of about 700 Hz. If an absorption peak was observed at a $\Delta\nu$ of -100 Hz (the negative sign indicates a lower field) relative to tetramethylsilane at 60 Mc, this would be the same as a $\Delta\nu$ of -67 Hz recorded at 40 Mc or -167 Hz recorded at 100 Mc. Therefore the position of the chemical shift recorded in cycles per second is field dependent. A better method of reporting or assigning resonance signals that is field independent is based on the following:

$$\delta = \frac{H_r - H_s}{H_r} = \frac{\Delta\nu \times 10^6}{\text{Oscillator frequency (Hz)}} \tag{7.5}$$

The chemical shift δ is the difference in field positions of the reference (H_r) and the sample (H_s) divided by the field strength of the reference in gauss. Since the positions are recorded in cycles per second, the ratio of the shift from TMS in cycles per second ($\Delta\nu \times 10^6$) to the oscillator frequency of the instrument also gives δ, the chemical shift, expressed in parts per million (ppm). Applying this to the example of a -100-Hz resonance signal at 60 Mc

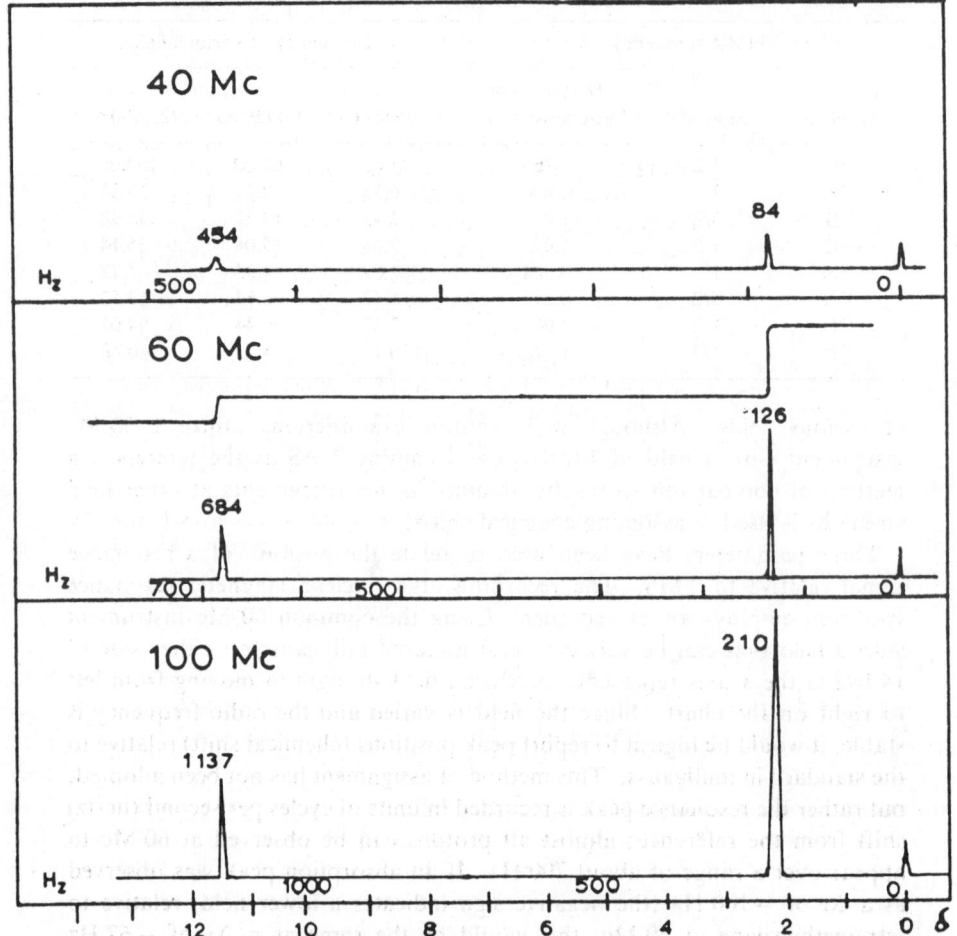

FIGURE 7.6: 40, 60, and 100 Mc nuclear magnetic resonance spectra of acetic acid in $CDCl_3$ with TMS standard.

the δ is 1.66 ppm. Examination will show that δ, independent of field strength and oscillator frequency, is 1.67 ppm in the example just given for all instruments using TMS as a standard. A further modification has been introduced that assumes TMS as the standard:

$$\tau = 10.00 - \delta \qquad (7.6)$$

This simply states that the field position τ is relative to the TMS peak, which is at the highest field, $+10.00$. Most organically bound protons fall at a lower field position ranging from -1.0 to $9.5\ \tau$. In the example shown in Fig. 7.6 acetic acid has absorption signals at -84 and -454 Hz in the top spectra (40 Mc), at -126 and -684 Hz when examined at 60 Mc, and at -210 and -1137 Hz run at 100 Mc. The chemical shift for these peaks is $\delta = 2.10$ and 11.37 ppm or $\tau = 7.80$ and -1.37. The reference TMS peak appears to 0 Hz on the right-hand or high field side of all three spectra.

Another property of nuclear magnetic resonance that has great utility is that the strength of the signal or rather the total area under the resonance peak is proportional to the number of protons resonating at that frequency. All commercial instruments are equipped with an integrator or a device to measure the area under a curve. The horizontal line in Fig. 7.5 and across the middle spectra in Fig. 7.6 is a recording of the area under the respective resonance peaks. The area encompassed by the low field peak (11.37 ppm) is one-third that of the 2.10-ppm signal.

A. EXPERIMENTAL METHOD

Instrument design and the properties of nuclear magnetic resonance, like ultraviolet and infrared, place certain restrictions on sample preparation and examination. Short relaxation times and the resulting broad, poorly resolved resonance signals observed in solids and viscous liquids recommends the use of dilute solutions. Solvents commonly used are carbon tetrachloride, carbon disulfide, deuterochloroform ($CDCl_3$), deuterium oxide, deuteroacetone (CD_3COCD_3), d_6-dimethylsulfoxide (CD_3—SO—CD_3), and other deuterated liquids. Almost any hydrogen-containing solvent can be used if the resonance signals for the solvent protons do not interfere or overlap with the sample signals; the signals due to the high concentration of solvent will be very strong.

The solutions are normally prepared in the range of 15–20% w/v in a minimum volume of 0.5 ml. The reference, tetramethylsilane, is usually prepared as a 1% solution in the solvent to be used. In polar solvents the sodium salt of a trimethylsilylalkane sulfonate is used. The sample is put in a high precision tube designed for the instrument and inserted in the sample holder (probe). After the instrument has been stabilized, scanning the sample and integration usually can be accomplished in 15 min.

B. INTERPRETATION

The first cursory glance at a nuclear magnetic resonance spectrum places it somewhere between an infrared and an ultraviolet spectra in appearance— certainly more complicated than an ultraviolet spectrum, but not quite as formidable as infrared. The first attempts to assign peaks in any of the three forms of spectroscopy can be an awkward experience. Although all three analytical methods are essential in structure determination, it is generally

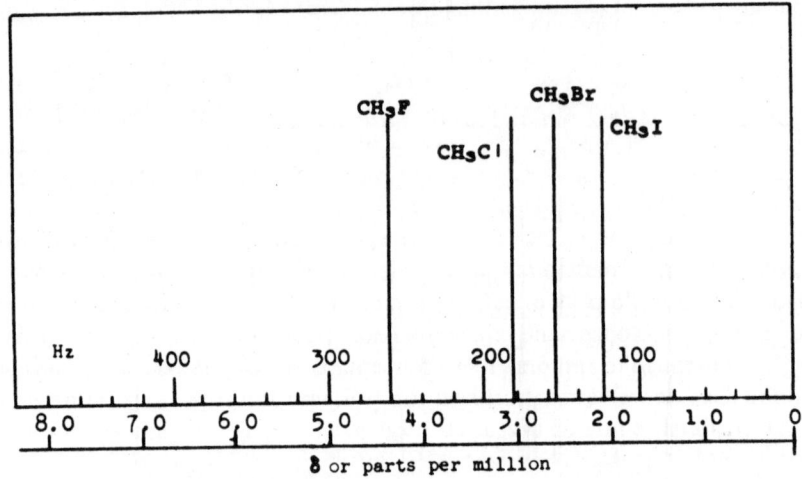

FIGURE 7.7: Scan from low to high fields.

agreed that nuclear magnetic resonance yields the most information about the structure of an organic molecule.

Before proceeding to the theory behind the chemical shift and interpretation of spectra the student should be aware of the terms used in nuclear magnetic resonance. As noted in the introduction, the spectrum in Fig. 7.7 recorded from left to right represents a "scan" or "sweep" from lower to high fields usually terminating at the TMS peak on the extreme right-hand side (high field or upfield), which is arbitrarily set at 0 Hz or 0 ppm (δ). Almost all proton peaks will appear at lower field (downfield or to the left of the TMS peak. The range in cycles per second that usually is examined is either −1000 or −500 Hz for a 60-Mc instrument and should include all but the most unusual proton resonance signals. Various mechanisms give rise to "shielding" the nucleus from the applied field. A diamagnetic shift or shielding effect is a chemical shift to a signal at high field (to the right). A downfield shift to absorption at lower applied fields is indicated by a paramagnetic shift or deshielding. Interpretation of the spectra and assignment of peaks will be greatly simplified if a qualitative understanding of the shielding mechanisms is acquired. With the aid of empirical rules and the published data on nuclear magnetic resonance often complete assignment of the structural environment of protons can be made.

Shielding or deshielding of the proton nucleus is caused by local magnetic fields generated in the applied field by the electrons in the immediate vicinity of the resonating nucleus. Tetramethylsilane gives a single sharp peak at high field because of strong shielding. Silicon, having relatively low electronegativity, permits the generation of strong local fields caused by the circulating silicon electrons. These fields are in opposition to the applied field and in effect partially neutralize the latter. This requires the application of a higher field to overcome the neutralization and induce resonance in the methyl protons, hence the term "shielding." Analogy can be seen in the methyl halides, where progressing to less electronegativity $F \rightarrow Cl \rightarrow Br \rightarrow I$ gives an increase in the shielding; the methyl proton peaks in the respective halides appear as seen in Fig. 7.7 at 4.36, 3.05, 2.68, and 2.16 ppm. An important feature of the opposing field caused by circulating electrons is that the strength of this induced field is proportional to the applied field, therefore the chemical shift in cycles per second for a given proton increases as the applied field is increased: A shift of -100 Hz at 60 Mc moves to -167 Hz at 100 Mc.[1]

Many additional examples of a deshielding effect (downfield shift) with increasing electronegativity can be shown, for example, in a comparison of methane, methyl alcohol, and methylamine. The C—H of methane absorbs at high field (0.3 ppm) compared to the C—H of methylamine (2.5 ppm) and methanol (3.6 ppm).

Further consideration of the effect of the applied field on electrons reveals another mechanism contributing to the chemical shift. In the presence of an applied field, electrons circulate in a perpendicular path and generate a magnetic field opposite to the applied field in the center of the electron "coil" (Fig. 7.8). A proton or any nucleus within the coil is shielded in proportion to the strength of the electron field. In contrast a nucleus outside or on the periphery of the circulating electrons experiences a reverse field. In Fig. 7.8 a nucleus X within the "coil" is shielded from the applied field; a nucleus Y outside of the coil is deshielded. This deshielding is the summation of both the applied and induced fields, and less applied field is required to resonate the nucleus. Deshielding by this mechanism is termed a "paramagnetic effect."

Paramagnetic effects are particularly important in relatively rigid molecules that are oriented in the applied field. As an illustration consider the circulating electrons of Fig. 7.9 to represent the electrons of toluene; both the ring and

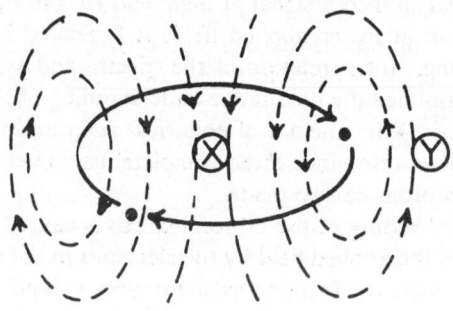

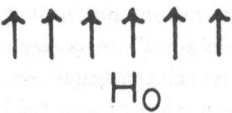

FIGURE 7.8: Effect of an applied field. Dashed lines represent fields generated by circulating electrons. Nucleus X is shielded and at a higher field. Nucleus Y is deshielded and resonates at a low applied field.

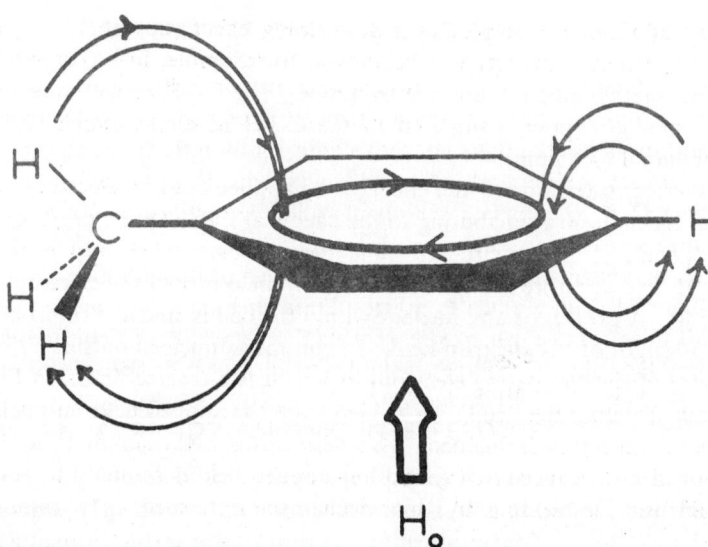

FIGURE 7.9: Paramagnetic effects in relatively rigid molecules. Deshielding in aromatic systems. The methyl group of toluene resonates at 2.32 δ and the ring protons are at 7.17 δ.

methyl protons experience the additive effect of the applied and the para-magnetic field. Subsequently, less applied field is required to resonate both types of protons; the aromatic signals are at 7.2 ppm and the methyl at 2.3 ppm. When this mechanism, specific proton orientation with respect to an induced field, is operative it is termed an "anisotropic effect." The elec-tron current in an oriented molecule such as an aldehyde, being diamagnetic within the coil, has a strong paramagnetic component in the region of the aldehydic proton, sufficient to allow resonance of this proton in a very low applied field, about 9 to 10 ppm. Anisotropic effects can be observed in many types of organic molecules such as acetylene, olefins, ketones, esters, acids, and nitriles, and the magnitude of the effect is dependent on the specific orientation of the proton in the field.

In summary, many factors in addition to electronegativity contribute to the chemical shift; paramount are the diamagnetic (shielding), paramagnetic (deshielding), and anisotropic effects.

Complete assignment of resonance signals to particular protons in a given molecule is relatively easy if some prior knowledge of the structure is available. Examination of an "unknown" will reveal the nature of the groups to which the protons are bonded and integration will give the relative numbers of each type of proton. Particularly in relatively simple molecules, complete structural determination often can be made on the basis of the nuclear mag-netic resonance spectrum.

Many reference tables are available for assistance in correlation of proton peaks and the assignment of structure. The student is referred to the bibli-ography for excellent sources of reference tables and charts.[1-7] Experience gained in assignment peaks can be invaluable in interpretation, and for this reason the student is referred to the *Varian Associates Catalog of Spectra*, volumes I and II, for a diverse range of examples.[7]

Much of the value of nuclear magnetic resonance as a tool for structure determination is because of the relatively narrow region for absorption by a particular proton. It will be apparent throughout spectral interpretation that the shielding or deshielding mechanisms are only operative over a very short distance and the effect is rarely extended beyond two saturated carbon atoms. A further consideration that will be obvious is that protons on flexible or freely rotating groups or molecules will experience an averaging of the local field effects, whereas rigid molecules or conformationally restricted protons, particularly in cyclic systems, will have unique field effects not shared by the flexibly system.

The conventions adopted in Figs. 7.10 and 7.11 and discussion have been to refer to peak positions in δ (ppm) relative to TMS at 0: Conversion to τ values is a simple matter since $\tau = 10 - \delta$. Several examples will be noted, but it is generally safe to assume that progression from a methyl ($CH_3-C\lessgtr$) to a methylene ($\gtrless C-CH_2-C\lessgtr$) and finally a methine $CH-C_3\lessgtr$, all other parameters remaining constant, results in a slight downfield shift. Further,

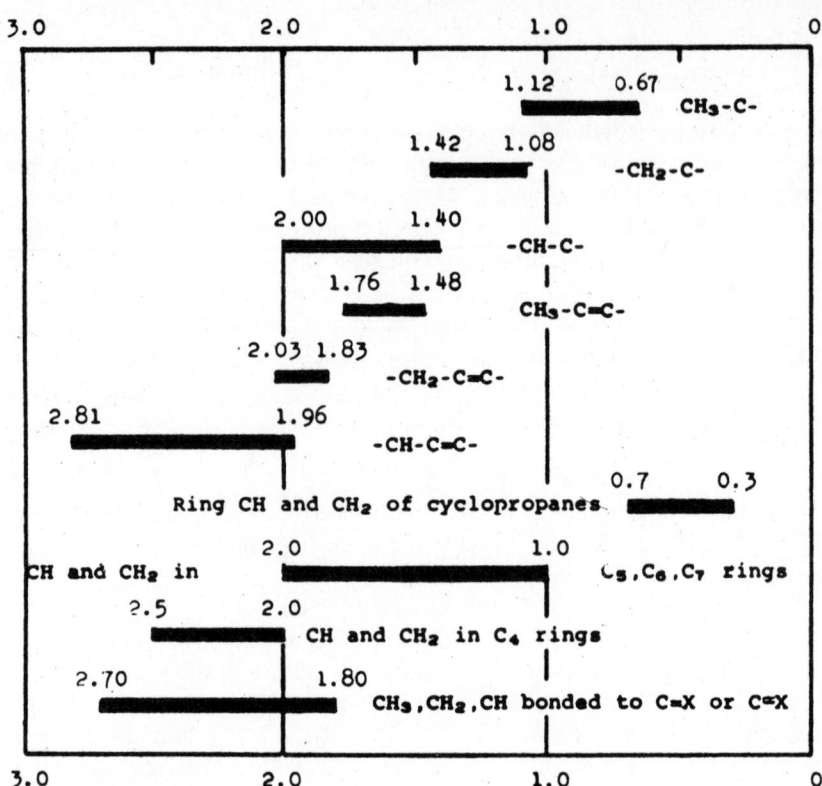

FIGURE 7.10: Resonance peak assignments for CH$_3$—C, CH$_2$—C, and CH—C.

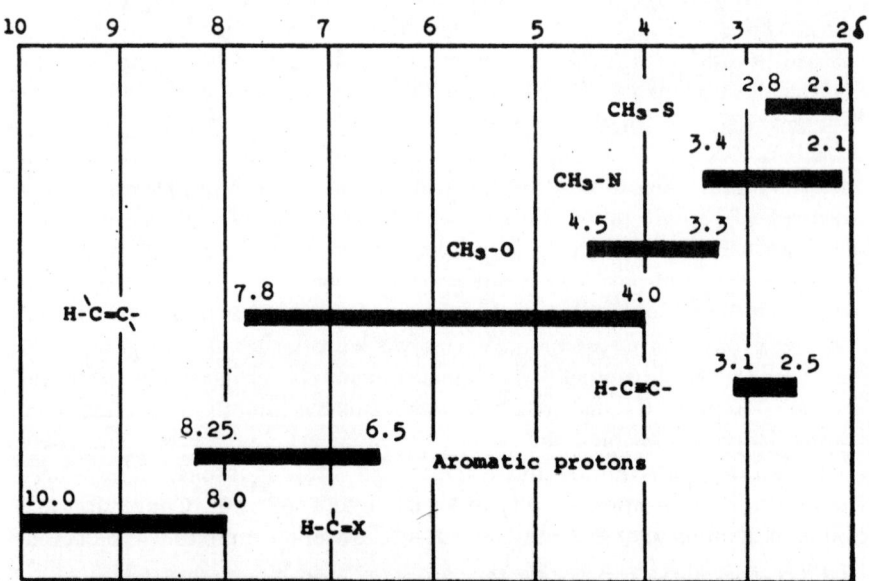

FIGURE 7.11: C—H bonded to atoms other than saturated carbon.

in addition to conformational effects, other factors such as ring currents will alter the positions of methylene and methine protons; therefore, cyclic systems will not be analogous to their paraffinic counterparts.

C. ALIPHATIC PROTONS

Methane protons resonating at 0.3 ppm undergo a downfield shift to 0.8 ppm when one of the protons is replaced by carbon. Successive replacement gives absorption at 1.3 and 1.4 ppm for the methylene and methine protons, respectively (Fig. 7.10). Precise assignment of the methyl can be made, but in general a range from 0.6 to 1.12 ppm will include all paraffinic methyls at least one saturated carbon atom removed from a group other than a saturated carbon. The inclusive range for methylenes is 1.08–1.42 ppm and for methines is 1.40–2.0 ppm.

When these groups are situated one carbon removed from an electronegative group or atom X,

$$CH_3-\overset{\displaystyle |}{\underset{\displaystyle |}{C}}-X$$

where X represents halogenic, olefinic group, oxygen, nitrogen, or a similar group other than a saturated carbon, a slight paramagnetic shift is experienced and in general the respective field positions for the CH_3, CH_2, and CH will be shifted downfield about 0.1 to 0.2 ppm from their paraffinic counterpart.

A methyl attached alpha to an olefin

$$(CH_3-\overset{\displaystyle \diagdown}{C}=\overset{\displaystyle \diagup}{C})$$

appears in the region from 1.48 to 1.76 ppm; corresponding shifts for

$$-CH_2- \quad \text{and} \quad -\overset{\displaystyle |}{CH}-$$

are in Fig. 7.10. Direct substitution on an aromatic ring results in deshielding to between 1.95–3.15 ppm. The toluene methyl appears at 2.44 ppm, the CH_2 of ethylbenzene at 2.62 ppm, and the methine of isopropylbenzene at 2.87 ppm. Methylenes or methine protons in cyclic systems experience deshielding due to the ring currents, and the shift is dependent on ring size. Large shielding effects are noted in cyclopropanes with signals appearing between 0.3 and 0.7 ppm. Protons attached to four-membered cyclic system resonate near 2.5 ppm with C_5 and larger rings at an intermediate position, between 1.0 and 2.0 ppm. Double bonds in the ring enhance the deshielding of methylenes to a range between 2.0 and 2.7 ppm.

Substitution of a methyl alpha to a carbonyl or nitrile (ketone, aldehyde, ester, amide, nitrile, oxime, etc.) creates a relatively strong field in the proton region that gives a large downfield shift to between 1.80 and 2.70 ppm. For example, the methyl protons of acetic acid, acetone, and acetaldehyde resonate at 2.10, 2.17, and 2.20 ppm, respectively; the usual region for methyls alpha to a carbonyl is between 2.0 and 2.6 ppm, for methylenes 2.1–2.4 ppm, and methines 2.4–2.6 ppm.

Protons on carbon bonded to halides, nitrogen, oxygen, sulfur, and other atoms experience strong deshielding (see Fig. 7.11). The role of electro-negativity in deshielding mechanisms cannot be discounted as evidenced by the correlation between the electronegativity of the halide in the halomethanes and the chemical shift (Fig. 7.7). Methyl sulfides or mercaptans resonate between 2.1 to 2.8 ppm. Variations in the substitution on nitrogen may

TABLE 7.2: Shoolery's Effective Shielding Constants

Group	σ_{eff}, ppm[a]
Cl	2.53
Br	2.33
OR	2.36
I	1.82
CR=O	1.70
C≡N	1.70
SR	1.64
NR′R²	1.57
C≡CH	1.44
CR′=CH′R²	1.32
CH₃	0.47

[a] $\delta = 0.233 + \Sigma\sigma_{eff}$.

affect the signal; for example, a simple aminomethylene near 2.5 ppm is deshielded to between 3.2 and 3.1 ppm in N-methylacetamide and sulfon-amides and in quaternary salts moves to 3.4 ppm. Protons on carbons alpha to oxygen normally absorb between 3.3 and 4.5 ppm. The magnitude of the paramagnetic shift is determined in part by the nature of the oxygen; a methylene in a dialkyl ether at 3.4 ppm shifts to 3.6 ppm for the corresponding alcohol and further in a phenolic ether (3.95 ppm), aliphatic ester (4.1 ppm), aromatic ester (4.2 ppm), and finally is found at 4.3 ppm in ethyl trifluoro-acetate.

The relative predictability of the shift associated with alkyl protons alpha to atoms other than saturated carbons has been used in the formulation of constants that are useful in structural assignment.[1-3] Assuming either a methylene or methine proton attached to two or three other groups or atoms, the following equation can be used to calculate the position of absorption of the protons:

$$\delta = 0.233 + \Sigma\sigma_{eff} \tag{7.7}$$

In this equation the shift δ for an alkylproton is the position of methane proton absorption, 0.233 ppm, plus the sum of the effective shielding con-stants (σ_{eff}) shown in Table 7.2. The observed and calculated position of the resonance signal for methylenes usually agree within ±0.05 ppm.

Protons joined to multiple-bonded carbons are found in olefinic, acetylenic, and aromatic compounds, where the multiple bond is to another carbon and in aldehydes, aldoxines, and formic acid and its derivatives, where multiple bonding to oxygen or nitrogen is involved. Simple correlation with electronegativity or inductive effects are not applicable in these compounds because strong anisotropic effects often are responsible for unusually large chemical shifts. The exact position of the chemical shift for an olefinic proton within the expected range of 4.6 to 6.4 ppm is often used in assigning the orientation of the protons. In a monosubstituted vinyl system the terminal methylene normally is near 4.9 ppm with a slightly higher field signal for the terminal proton *cis* to the adjacent proton. The nonterminal proton absorbs near 5.7 ppm. Cyclic olefinic protons are shifted to lower field compared to their acyclic analogs. Likewise conjugation enhances the deshielding to even lower field values extending the total range for all types of olefinic protons to between 4.0 and 7.8 ppm. The acetylenic proton is subjected to a rather large shielding effect due to diamagnetic anisotropic effects and resonates between 2.5 and 3.1 ppm.

Correlations of the chemical shift with structure in aromatic proton systems have been very useful in assigning aromatic substitution patterns. Absorption in benzenoid systems in the range of 6.5 to 8.0 ppm follows the predictable effects of ring substitution. Diamagnetic shielding due to the electron density on the carbon to which the proton is attached has the greatest effect on the chemical shift; electron-donating substituents increase the electron density at the ortho and para carbons and result in a shift in absorption of these protons to higher fields relative to benzene. Low-field shifts result when the electron density at the C—H is lowered by electron-withdrawing substituents. The diamagnetic anisotropic effect, which falls rapidly as the distance between the proton and the oriented group increases, is influential only in causing an upfield shift of the ortho proton signals. The third factor that modifies the resonance absorption of aromatic protons is the paramagnetic deshielding caused by ring currents and the magnitude of the induced field; electron-donating groups increase the electron density, giving larger fields that are parallel to the applied field in the vicinity of the proton and cause a downfield shift. The overall shift therefore is the combined total of these factors and is strongly influenced by ring substituents. As a rule protons on aromatic rings containing an electron-withdrawing substituent are shifted a maximum of 1.0 ppm downfield from the benzene proton peak at 7.27 ppm; the protons ortho to nitro and carbonyl groups experience the strongest shift. Electron donors cause a shielding effect on the aromatic protons and with the exception of iodobenzene and aminobenzenes the shift of the ortho, meta, and para protons is approximately the same.

In contrast, the protons on heterocyclic systems undergo much larger shifts, particularly when the protons are alpha to the heteroatom. Correlation of the nuclear magnetic resonance spectra of heterocyclic molecules is complicated by the effects of both the ring substituents and the hetero atom.

Aldehydes, aldehyde derivatives, formic acid derivatives, and aldoximes offer a special case of protons attached to sp_2 carbon since the double bond is to a hetero atom. The strong paramagnetic shielding caused by anisotropic effects shifts the aldehyde proton to the region 9.4 to 10.0 ppm, a unique shift very useful in structural assignment. Low field absorption is characteristic of all protons attached to systems containing multiple bonding to a hetero atom; the single proton peak for methylformate is observed at 8.08 ppm, that for salicylaldoxime is at 8.18 ppm.

salicylaldoxime

Protons attached to atoms other than carbon are encountered in amines, amides, mercaptans, halogen acids, metal hydrides, alcohols, phenols, carboxylic acids, water, and other molecules such as SiH_4, PH_3, H_2S, NH_3, and H_2. However, the discussion will be limited to the use of nuclear magnetic resonance primarily in structure determinations in organic molecules containing alcohols, phenols, acids, amines, amides, and mercaptans. Two characteristic properties associated with protons bonded to oxygen, nitrogen, and sulfur are (1) their rapid chemical exchange, which is responsible for sharp peaks, and (2) the ability to hydrogen bond. Hydrogen-bonded protons resonate at lower fields than the corresponding unassociated protons. Intramolecular hydrogen-bonded hydroxyl groups are strongly deshielded; for example, the phenolic proton of salicyaldehyde is at 10.9 ppm, and in

salicylaldehyde   acetylacetone

enols such as acetylacetone the enolic proton is at 14.9 ppm. The characteristic properties of intermolecular hydrogen bonds—their concentration and temperature dependence—are readily observed in nuclear magnetic resonance. The resonance signal for protons bonded to oxygen is variable, appearing at higher fields on dilution with inert solvents or as temperature is increased. The predominance of the dimeric form in carboxylic acids and the strength of the hydrogen bonds stabilizes the signal in the region of 10 to 12 ppm, even on dilution with inert solvents.

Amines undergoing rapid chemical exchange in neutral or basic solvents exhibit a sharp absorption peak that changes in appearance to a broad band for the ammonium salt. Although there are exceptions, the normal range for the N—H of aliphatic and aromatic amines is between 1 and 5 ppm with the aromatic amine protons in the low field end of this range (3.5–5.0 ppm). Amide N—H resonates in the region of 5 to 9 ppm as a broad peak.

Aliphatic mercaptan protons absorb in the relatively high field range of 1.2 to 1.6 ppm; propane-1,3-dithiol has the S—H resonance signal at 1.35 ppm. Benzylic mercaptans are found near 2 ppm, i.e., furfuryl mercaptan at 1.90 ppm, while aromatic mercaptans appear at lower fields; the S—H of p-chlorothiophenol is assigned at 3.45 ppm, that of p-methylthiophenol at 3.27 ppm.

The property of rapid chemical exchange associated with O—H, S—H, and N—H groups is employed in the interpretation of nuclear magnetic resonance spectroscopy. Since deuterium does not resonate (absorb) in the proton region, after obtaining the spectrum of such a compound, the addition of several drops of deuterium oxide to the sample tube results in exchange of the O—H, N—H, or S—H protons to give the corresponding O—D, N—D, or S—D function. The latter does not give a signal, and by differences in the two spectra the absorption peak for the exchangeable proton is easily assigned.

D. INTERACTIONS BETWEEN NUCLEI

Prediction of the nuclear magnetic resonance of a simple molecule should not be too difficult after the fundamentals of the mechanisms causing the chemical shift are understood. A halide causes a certain amount of deshielding of the methyl protons that is dependent on the relative electronegativity of the halide; shifts to lower fields (deshielding) accompany increases in electronegativity. Thus, the methyl protons of CH_3I appear in the high field end of the alkyl halide region ($\sim$ 2–4 ppm); it is found at 2.15 ppm. Likewise, in the spectrum of iodoethane, the methylene flanked by a methyl and an iodine should fall at even lower fields and is observed at 3.20 ppm, perhaps somewhat more deshielded than expected. The methyl, adjacent to a CH_2, should be near the region 1–2 ppm; it is at 1.83 ppm. On examination, the spectrum for iodoethane (Fig. 7.12) appears to be complicated by multiple peaks for both the methyl and the methylene, in contrast to the sharp single peak observed for the methyl of CH_3I. Integration of the spectrum for iodoethane results in a 2:3 ratio for the multiplets, the centers of which are 3.20 and 1.83 ppm, respectively. Returning to the theory of nuclear magnetic resonance, each proton must be examined in detail with respect to the field created by a neighboring proton.

Inspection of the two types of protons on iodoethane will be greatly simplified if each proton is designated in turn H_{A_1} and H_{A_2} for the two methylene protons and H_{B_1}, H_{B_2}, and H_{B_3} for the three methyl protons. Each of the five protons oriented in the applied field can be aligned with the field ($\downarrow$), in which case it has a low energy of $-1/2$ spin or antiparallel ($\uparrow$) for a $+1/2$ spin. The three equivalent methyl protons resonate in the applied field by

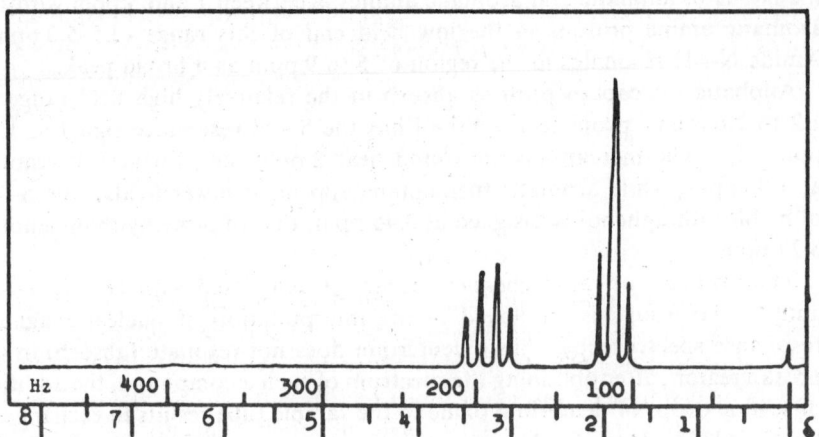

FIGURE 7.12: 60-mC spectrum of iodoethane.

flipping from the $-1/2$ spin to $+1/2$ spin orientation and in iodomethane, absorbed at 2.15 ppm, as a sharp singlet peak. In contrast, the methyl of iodoethane is a triplet ($\sim$ 1.83 ppm) with relative areas 1:2:1 for the three peaks of the triplet. The chemical shift or the field position of the resonance signal is determined by the effective field, which is the summation of the applied field (H_0) and local fields generated by circulating electrons. The additional induced field that is now operating in splitting a peak in this instance into a triplet is that generated by neighboring nonequivalent protons aligned with or against the field. In Table 7.3 all of the possible and equally probable combinations of spin states for H_{A_1} and H_{A_2}, the methylene protons, are given. Therefore the effective field experienced by the methyl protons is that due to the applied field, the induced electron field, and the neighboring nuclear or proton field. Taken in order, state 1 in Table 7.3 has both neighboring protons H_{A_1} and H_{A_2} parallel with and partially neutralizing the applied field; this spin orientation requires a slightly higher applied field to resonate the three methyl protons. States 2 and 3 each total zero in their effective field and therefore do not change the applied field; this methyl peak

TABLE 7.3: Spin States of the Methylene Protons of Iodoethane

	Orientation		Spin state		
	H_{A_1}	H_{A_2}	H_{A_1}	H_{A_2}	Total
1	↓	↓	$-1/2$	$-1/2$	-1
2	↓	↑	$-1/2$	$+1/2$	0
3	↑	↓	$+1/2$	$-1/2$	0
4	↑	↑	$+1/2$	$+1/2$	$+1$

is at intermediate field. In the last orientation, both methylene protons in the high energy state add to the applied field, resulting in slightly less applied field necessary to resonate the methyl protons. Both 1 and 4 are equal but opposite in their effect, and the methyl peaks for these two are equally separated from the central intermediate peak. Integration of the methyl triplet reveals a ratio of 1:2:1; the central peak is twice the area of the two satellite peaks because two states, 2 and 3, contribute to the intermediate peak. The methyl peak has been split into a triplet by a mechanism termed "spin-spin splitting." The magnitude of the splitting or the distance between the lines, in contrast to the chemical shift, is independent of the field strength and subsequently is

TABLE 7.4: Spin States of the Methyl Protons of Iodoethane

	Orientation			Spin state			
	H_{R_1}	H_{B_2}	H_{B_3}	H_{B_1}	H_{B_2}	H_{B_3}	Total
1	↓	↓	↓	$-1/2$	$-1/2$	$-1/2$	$-3/2$
2	↓	↓	↑	$-1/2$	$-1/2$	$+1/2$	$-1/2$
3	↓	↑	↓	$-1/2$	$+1/2$	$-1/2$	$-1/2$
4	↑	↓	↓	$+1/2$	$-1/2$	$-1/2$	$-1/2$
5	↓	↑	↑	$-1/2$	$+1/2$	$+1/2$	$+1/2$
6	↑	↓	↑	$+1/2$	$-1/2$	$+1/2$	$+1/2$
7	↑	↑	↓	$+1/2$	$+1/2$	$-1/2$	$+1/2$
8	↑	↑	↑	$+1/2$	$+1/2$	$+1/2$	$+3/2$

measured in cycles per second (hertz), not parts per million (δ). This measurement, the spin-spin coupling constant (J), for the methyl of iodoethane is 7 Hz.

Analysis of all the possible spin states for the methyl-group protons (Table 7.4) reveals eight possible orientations, theoretically an eight-line multiplet. On inspection of the total spin effect only four field effects are possible: a $-3/2$ effect; 2, 3, and 4 are the same at $-1/2$; 5, 6, and 7 are $+1/2$; and 8 is $+3/2$. The total relative areas of the multiplet therefore are 1:3:3:1 with a coupling constant $J = 7$ Hz. The mutual splitting of H_A and H_B results in the same J for both protons.

Assignment of the chemical shift for the methylene is made in the middle of the quartet, 3.20 δ. The number of lines in the spectrum of the methyl is predictable from the rule $n + 1$, where n is the number of equivalent protons on an adjacent group.

Coupling rarely extends further than between two neighboring proton-bearing groups unless an additional effect such as a double bond is present between the mutually splitting groups. In any event the magnitude of the coupling constant J falls rapidly with distance. The simple rule $n + 1$ for predicting the number of lines in the multiplet applies to those instances where the difference between the chemical shift ($\Delta \nu$) for mutually splitting groups

TABLE 7.5: Common Spin-Spin Coupling Constants

Group	J, Hz
$-\overset{\mid}{C}H-\overset{\mid}{C}H-$	2–9
(geminal CH$_2$ structure)	12–15
(cyclohexane ring, X—Y, H H)	5–10 axial-axial; $\angle = 180°$ 2–4 axial equatorial; $\angle = 60°$ 2–4 equatorial-equatorial; $\angle = 60°$
$\underset{H}{\overset{H}{C=C}}$ (gem vinyl)	0.5–3
$\underset{H}{\overset{H}{C=C}}\overset{H}{}$ (cis)	7–12
$\underset{H}{\overset{H}{C=C}}$ (trans)	13–18
$C=C-\overset{H}{\underset{}{C}}-H$	4–10
$H-C=C-\overset{}{C}-H$	0.5–2.5
$C=CH-CH=C$	9–13
$CH-C\equiv C-H$	2–3
$CH-CH=O$	1–3
$\underset{CHO}{\overset{H}{C=C}}$	6–8

$(\delta_A - \delta_H)$ is much greater than the coupling constant J of these groups. When $\Delta\nu$ approaches J in magnitude (overlapping), a much more complex multiplet is encountered and the multiplicity of lines, δ, J, and intensity for each line must be calculated.

Table 7.5 gives the expected or predicted spin-spin coupling constants for commonly encountered groups. The usual J for adjacent paraffinic protons is 2–9 Hz; when nonequivalent protons are attached to the same group, usually in rigid or cyclic systems, geminal splitting occurs to give $J = 12 - 15$ Hz.

TABLE 7.6: Spin-Spin Coupling Constants

Aromatic	Spin-spin coupling constants		
	ortho 6–9 meta 1–3 para 0–1		
	.$X = O$	$X = NH$	$X = S$
	$\alpha\beta$ 1.6–2.0	2.0–2.6	4.6–5.8
	$\alpha\beta'$ 0.6–1.0	1.5–2.2	1.0–1.8
	$\alpha\alpha'$ 1.3–1.8	1.8–2.3	2.1–3.3
	$\beta\beta'$ 3.2–3.8	2.8–4.0	3.0–4.2
	$\alpha\beta$ 4.9–5.7		
	$\alpha\gamma$ 1.6–2 6		
	$\alpha\beta'$ 0.7–1.1		
	$\alpha\alpha'$ 0.2–0.5		
	$\beta\gamma$ 7.2–8.5		
	$\beta\beta'$ 1.4–1.9		

For substituted cyclohexanes the dihedral angle between the adjacent protons determines the J, and nuclear magnetic resonance is valuable in rigid, cyclic, and olefinic systems in the determination of stereochemistry. It can be readily observed in Table 7.6 that in aromatic systems the magnitude of the splitting constant decreases with increased distance. The magnitude of the J and the possibility of small chemical shift differences often results in very complicated spectra.

Analysis of spin-spin splitting is relatively easy when the chemical shift (δ) is much greater than the coupling constant (J); in these instances the two splitting nuclei are termed an "AX system" since the chemical shift for nuclei A is quite different than nuclei X. Iodoethane is an example of an A_2X_3 system—A_2 represents the methylene protons, X_3 the methyl protons, and the multiplets observed follow the $n + 1$ rule. An AX system would give two doublets, while an A_2X_2 gives two triplets, and finally an A_3X_3 results in two quartets.

When the magnitude of the chemical shift, ν measured in cycles per second (hertz), between the interacting nuclei ($\Delta\nu$) is small and approaches the magnitude of the splitting constant (J), then the splitting becomes complicated.

As an example of an AB system, a closely spaced four line spectrum is predicted. As the ratio $J_{AB}/\Delta\nu_{AB}$ becomes larger, the character of the four lines changes—the inner two lines increase at the expense of the outer lines, and the distances in cycles per second between the inner and outer lines increases (Fig. 7.13). Assignment of the δ and J is accomplished by the following formula:

$$|B_2 - A_2| = |B_1 - A_1| = \sqrt{\Delta\nu_{AB}^2 + J_{AB}^2} \qquad (7.8)$$

where the lines for the protons are labeled with increasing field B_2, B_1, A_2, and A_1.

Three interacting nuclei can be assigned: AX_2, AB_2, AMX, and ABX. In the first instance, where the difference in the chemical shift between the coupling nuclei A and X is large relative to the coupling constant ($\Delta\nu_{AX} > J_{AX}$), a simple triplet-doublet spectra is seen (Fig. 7.14). The A_2B example of Fig. 7.14 is complicated by the fact that the multiplet pattern is dependent on the ratio $J_{AB}/\Delta\nu_{AB}$.

The AMX spectrum is a multiplet of three quartets since in all states the $\Delta\nu \gg J$. In Fig. 7.15 the observed pattern is valid for $J_{AM} > J_{AX} > J_{AM}$. Analysis of ABX and ABC spectra gives rise to complicated spectra that must be analyzed with care; many examples of this type are found in unsaturated and aromatic compounds.

Examples of four (A_2X_2, A_2B_2, A_3B) or five (A_2B_3) interacting nuclei can be simple, easily recognizable patterns when J_{AA}, J_{XX}, and J_{BB} equal zero and the $\Delta\nu$ is much greater than J. The ethyl group represents such an example of an A_2B_3 spectra.

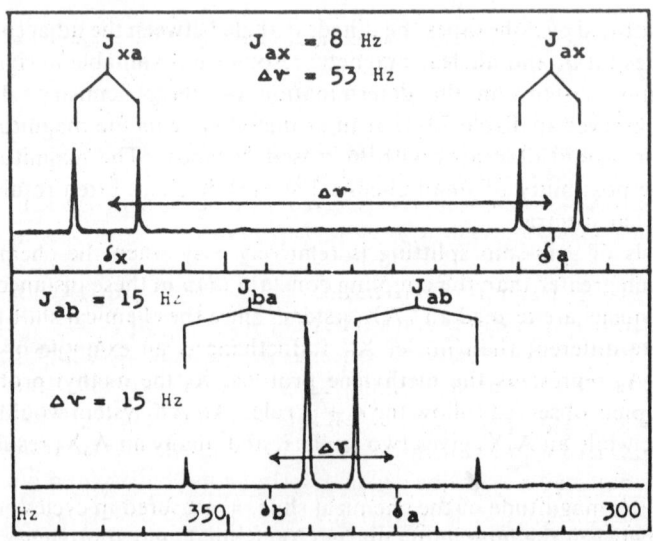

FIGURE 7.13: Coupling between AX and AB protons.

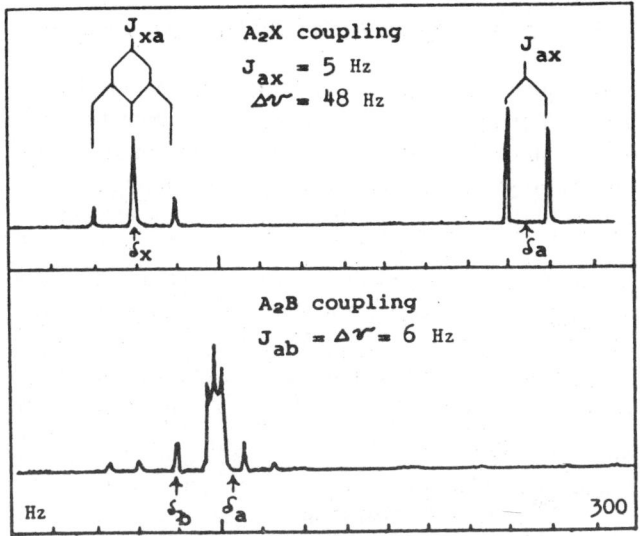

FIGURE 7.14: Coupling between A₂X and A₂B protons.

The analysis and interpretation of complex spectra that are observed in ABX, ABC, and other systems often requires mathematical treatment and a prior knowledge of the structural possibilities for each pattern. Since the magnitude of the coupling constant is independent of the field strength, resolution and assignment of complex multiplet patterns are simplified by examination at higher field strengths where the chemical shift difference ($\Delta\nu_{AB}$) increases while the coupling constant (J_{AB}) remains constant, effectively decreasing the $J/\Delta\nu$ ratio.

The application of spin-decoupling and double-resonance techniques, particularly at higher fields, is routinely employed in analysis of such systems. Excellent discussions of complex splitting patterns and their analysis are available and should be referred to for additional information.[1]

Nuclear magnetic resonance of isotopes other than hydrogen has been of great interest to chemists, particularly the resonance associated with fluorine, phosphorus, boron, and carbon-13.

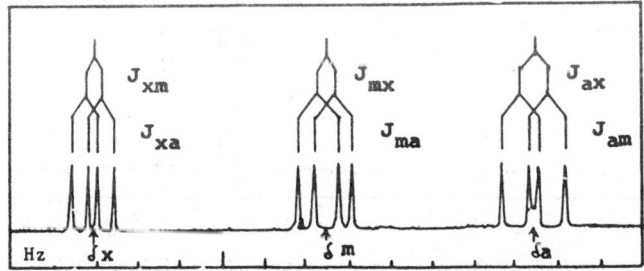

FIGURE 7.15: Coupling between AMX protons: $J_{AM} = 6$ Hz, $J_{AX} = 4$ Hz, $J_{MX} = 2$ Hz.

E. APPLICATIONS IN PHARMACEUTICAL ANALYSIS

The analysis of pharmaceuticals from a qualitative view is concerned with the identification of medicinals and the detection of impurities. Quantitative analysis is applied in determination of the concentration of a drug in dosage form, often in the presence of other drugs or compounding agents. Development of methods in drug analysis employs the entire spectrum of analytical techniques.

Application of modern instrumental methods such as infrared and ultraviolet spectroscopy, polarography, and chromatographic procedures in the analysis of drugs is increasing at a remarkable pace. Although the role of nuclear magnetic resonance in research for structural determination is firmly established, the applications to drug analysis only recently have been explored.[8]

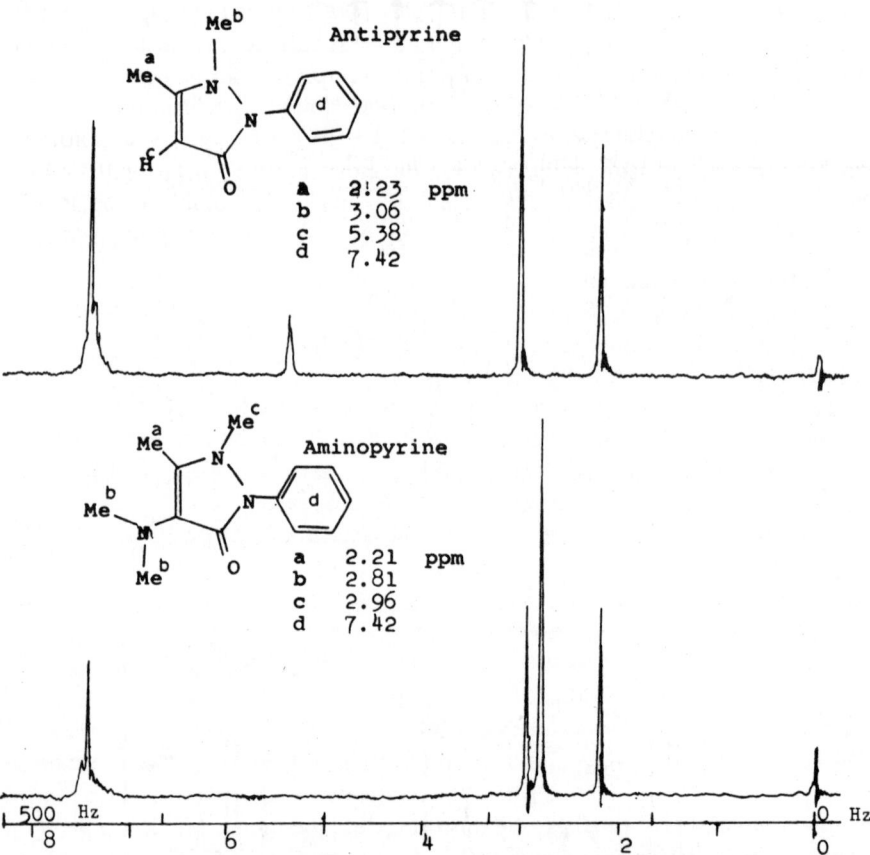

FIGURE 7.16: Nuclear magnetic resonance spectra of antipyrine and aminopyrine.
Courtesy of Varian Associates, Palo Alto, California.

The identification of a particular drug using nuclear magnetic resonance often is not possible due to the close chemical relationship among the various analogs in a particular class of drugs; however, as the first approach, the general structural features can be determined. Thus, limiting the unknown to a particular class of drugs, for example, the phenothiazines, comparison to standards using infrared or other "fingerprinting" methods leads to rapid identification. The inherent limitations of nuclear magnetic resonance stem partially from the fact that relatively large quantities of the sample are required—10 to 40 mg. The detection of impurities in the sample is hampered by the insensitivity of the detection method, requiring more than 5% of the impurity in the sample, and furthermore, the impurity must have a strong resonance signal in a region where the sample does not absorb. These requirements are not often encountered in drugs for two reasons: first the potency of drugs is such that a small quantity of impurity may have strong undesirable pharmacological effects; second the impurity, usually a side product in the synthesis, is closely related structurally to the drug and would exhibit proton resonance in the same regions. Efforts to solve the problem of low sensitivity and relatively low signal-to-noise ratio in nuclear magnetic resonance have centered on computer techniques. Since random noise detection will average to zero over many sweeps of a particular sample and a true resonance signal is additive, by repetitive scans of a sample and storage of the resultant signal in a computer the intensity of a weak signal can be increased manyfold, while the signal due to noise is cancelled to zero. An accessory called a "computer of average transients" (CAT) has been designed for use in increasing the signal-to-noise ratio.

As an example of qualitative analysis of drugs by nuclear magnetic resonance, the spectra of antipyrine and aminopyrine (Fig. 7.16) exhibit only one obvious difference. In antipyrine the sole heterocyclic ring proton, which is replaced by a dimethylamino group in aminopyrine, resonates at 5.38 ppm; the protons of the two methyls on the nitrogen in aminopyrine introduce a peak at 2.81 ppm.

As an aid in characterization of phenothiazines the infrared, ultraviolet, and nuclear magnetic resonance spectra of twenty-three analogs have been published.[9] The fingerprint like spectra in Fig. 7.17 represent three classes of molecules—colchicine, eserine, and procaine. The respective protons have been assigned in Table 7.7.

Quantitative analysis of mixtures has employed ultraviolet and infrared spectroscopy where the Beer-Lambert law or calibration plots are applicable. The counterpart used for quantitative analysis in nuclear magnetic resonance is the proton integration curve where the relative areas under a resonance peak can be compared to standards; by this procedure the ratio of the components in the mixture can be estimated. In analogy to the Beer-Lambert law, where the absorbance is related to the number of molecules, the area under the resonance peak as measured by an integrator gives the relative number of

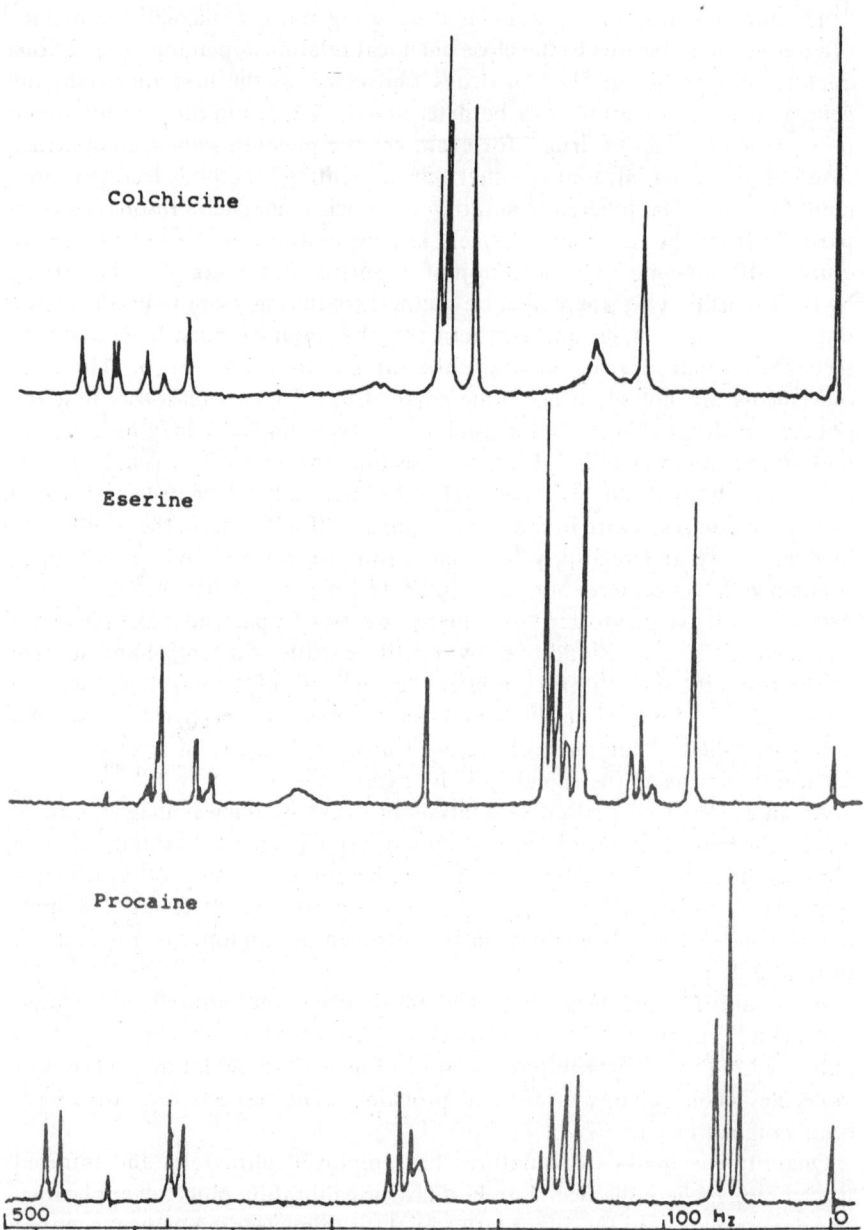

FIGURE 7.17: Nuclear magnetic resonance spectra of colchicine, eserine, and procaine. Courtesy of Varian Associates, Palo Alto, California.

protons in the respective peaks. For example, examination of an equimolar mixture of methanol and acetone gives a spectrum having absorption signals at 2.2 ppm for the six methyl protons of acetone and at 3.5 ppm for the three methyl protons of methanol. Integration shows a ratio of 2:1 for the proton area under the respective peaks, which calculates for a molecular ratio of 1:1. By careful experimentation the error in such an application usually is less than 3%.

TABLE 7.7: Assignment of Proton Values

	Assignments, ppm			
	Colchicine			
	a	1.96	g	6.92
	b	2.43	h	7.37
	c	2.43	i	7.63
	d	3.67	j	8.38
	e	4.62		
	f	6.55		
	Eserine			
	a	1.42	g	4.12
	b	1.95	h	5.33
	c	2.55	i	6.37
	d	2.70	j	6.78
	e	2.82	k	6.87
	f	2.92		
	Procaine			
	a	1.05	g	7.83
	b	2.62		
	c	2.82		
	d	4.13		
	e	4.33		
	f	6.63		

Acetylsalicylic acid-phenacetin-caffeine combinations have been analyzed by nuclear magnetic resonance.[10] In the reported procedure it was found that all three components give unique resonance peaks. Acetylsalicylic acid has a sharp methyl peak at 2.3 ppm, phenacetin has the ethyl peaks centered at 1.3 ppm (CH_3) and 4.0 ppm (CH_2). Caffeine has peaks at 3.35, 3.55, and 4.0 ppm for the three methyl groups. The peaks selected for quantitative analysis were 2.3 ppm for acetylsalicylic acid, 3.4 and 3.6 ppm for caffeine, and 4.0 ppm for phenacetin. The area under the acetylsalicylic acid signal has three protons, representing one molecule of that compound. Since the concentration of caffeine is low relative to the other ingredients, measurement of both the 3.4 and 3.6 ppm signals gives six protons representing one molecule of caffeine. The phenacetin methylene quartet at 4.0 ppm is complicated by the N_7 methyl of caffeine, which also absorbs at 4.0 ppm. Thus measurement of the area under the 4.0-ppm multiplet gives two protons representing each phenacetin molecule and three protons for each caffeine molecule.

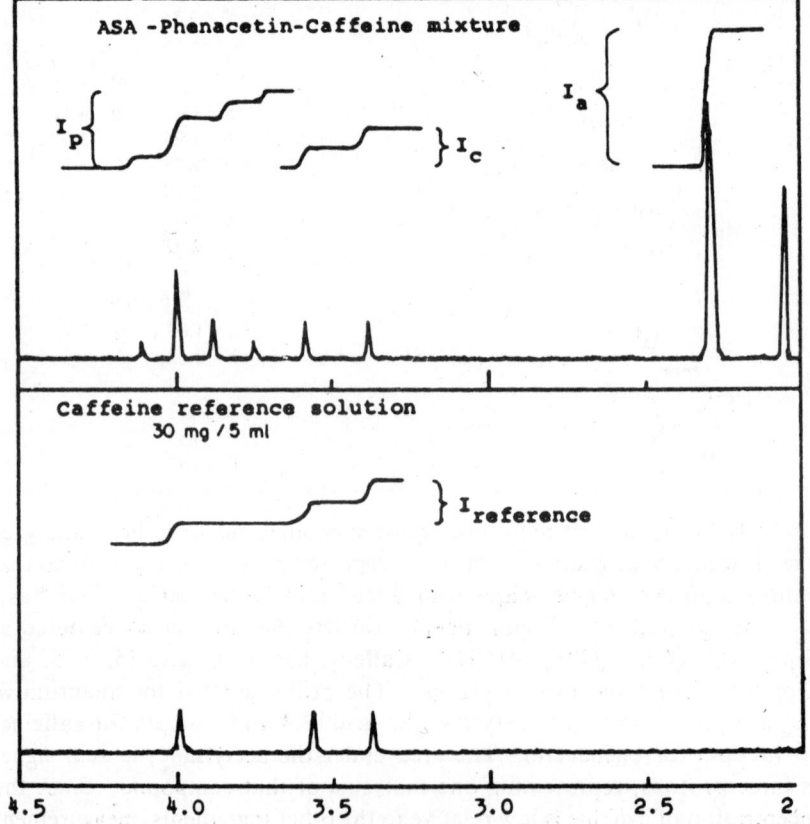

acetylsalicylic acid   phenacetin   caffeine

Since the number of caffeine molecules has been measured independently at the 3.4- and 3.6-ppm regions, the 4.0-ppm integration can be corrected for phenacetin by subtracting one-half of the 3.4 and 3.6-ppm integrated areas. Correcting all ratios to give equimolar equivalents, the molecular ratios of

FIGURE 7.18: Spectrum of ASA–Phenacetin–Caffeine mixture and caffeine reference solution.

each ingredient are given by the following equation where I is the ratio of the integrated area for each peak

Acetylsalicylic acid-caffeine-phenacetin $= (I_A):(\frac{1}{2}I_C):\frac{3}{2}(I_P - \frac{1}{2}I_C)$

In the example given in Fig. 7.18 the areas for each region I_A, I_C, and I_P are 101:28:58; accordingly, the molecular ratios are:

$$A:C:P = 101:14:66 = 7.2:1:4.7$$

The relative concentrations c can be calculated from the molecular weight;

$$c_A:c_C:c_P = 7.2 \times 180:1 \times 194:4.7 \times 184$$
$$= 1295:194:865 = 6.67:1:4.46$$

The absolute concentrations can be calculated from comparison to a standard. A sample containing 30 mg of caffeine in 5 ml of solvent was examined under the same conditions as the unknown and was found to give an integral area (I_R) of 24 mm for the 3.4- and 3.6-ppm signals. The concentration of caffeine unknown (C_S) in milligrams per 5 ml is proportional to the integrated areas for the reference and the sample:

$$C_S = \frac{I_S}{I_R} C_R = \tfrac{28}{32}(30 \text{ mg/5 ml}) = 26 \text{ mg/5 ml}$$

The concentration of the acetylsalicylic acid (C_A) and the phenacetin (C_P) in milligrams per 5 ml then can be determined

$$C_A:C_C:C_P = 7.2:1:4.7 = 187:26:122 \text{ mg/5 ml}$$

The advantages are speed and the nondestructive nature of the method; the 1% error found in the integration curve is the chief disadvantage, particularly when there is a wide difference in concentration of the ingredients in the mixture. By this method and application of a slight correction introduced for the resonance signal of carbon-13 Hollis reported deviations of only 1.1% for acetylsalicylic acid, 2.2% for phenacetin, and 3.2% for caffeine.

Although few applications of nuclear magnetic resonance to analysis of pharmaceutical systems have been reported, the possibilities for the future in both qualitative and quantitative analysis of pharmaceuticals are encouraging. Nuclear magnetic resonance certainly will not replace infrared and ultraviolet, but will add another dimension to pharmaceutical analysis.

REFERENCES

1. L. M. Jackman, *Applications of Nuclear Magnetic Resonance Spectroscopy in Organic Chemistry*, Pergamon Press, New York, 1959.
2. J. D. Roberts, *Nuclear Magnetic Resonance*, McGraw-Hill, New York, 1959.
3. J. R. Dyer, *Applications of Absorption Spectroscopy of Organic Compounds*, Prentice-Hall, Englewood Cliffs, N.J., 1965.

4. J. D. Roberts, *An Introduction to the Analysis of Spin-Spin Splitting in High-Resolution Nuclear Magnetic Resonance Spectra*, Benjamin, New York, 1962.
5. R. H. Bible, *Interpretation of NMR Spectra: An Empirical Approach*, Plenum Press, New York, 1965.
6. A. K. Bose, in *Interpretive Spectroscopy* (S. K. Freeman, ed.), Reinhold, New York, 1965.
7. Varian Associates, *High Resolution NMR Spectra Catalog*, Vol. I (1962) and Vol. II (1963), Palo Alto, Calif.
8. T. G. Alexander and S. A. Koch, J. *Assoc. Offic. Agr. Chemists*, **48**, 618 (1965).
9. R. J. Warren, I. B. Eisdorfer, W. E. Thompson, and J. E. Zarembo, *J. Pharm. Sci.*, **55**, 144 (1966).
10. D. P. Hollis, *Anal. Chem.*, **35**, 1682 (1963).

CHAPTER **8**

Turbidimetry; Nephelometry; Colloidimetry*

Fred T. Semeniuk

SCHOOL OF PHARMACY
UNIVERSITY OF NORTH CAROLINA
CHAPEL HILL, NORTH CAROLINA

* *turba*, cloud (L.); *nephelē* (νεφέλη), cloud, mist (Gk.); *kolla eidos*, glue form (Gk.).

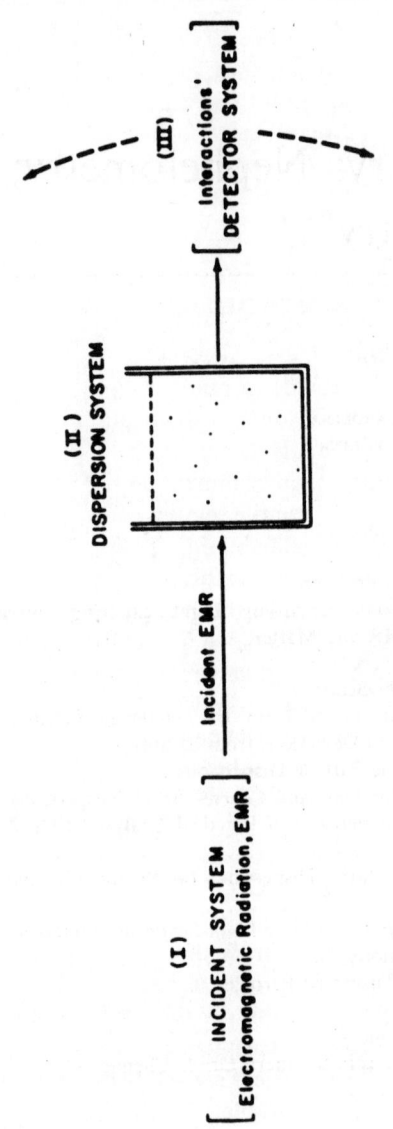

FIGURE 8.1: The systems involved.

8.1 PREFACE

The established applications of these methods have had rather limited direction to pharmacy. As the scope of this chapter, it is the purpose to report upon the *meaning* of the methods because it is felt that present technologies based in them are important resources for pharmacy.

Also, because there are so many facets to the methods, there appears to be less pedagogical purpose to reciting specific procedures than to directing interests into specific areas by citing selected references according to topics (see Appendix).

8.2 INTRODUCTION

The environment of man provides a number of ordinary examples for his relative visual awareness and reasonings about matter whenever it is distributed in fine particulate state by its dispersion medium.

Such phenomena of ordinary environment as are called milkiness, opalescence, haze, mist, fog, smog, smoke, clouds, searchlight beams, etc., are readily sensed and qualitatively understood. Equally sensed but less understood are the deep-seated cause/effect relationships of such phenomena as the blues of tobacco smoke and of sky, the rich spectral hues of the sunrise, sunset and rainbow, the gray of the dawn and twilight.

It is appropriate to state early and generally here that underlying all such phenomena is the fact of interaction between dispersed fine particles and visible electromagnetic radiation (EMR) passing through the dispersion; and that the factors in interaction have been subjected to rigorous theories and experiments.

A. THE MEANING OF THE PROBLEM

Figure 8.1 and Table 8.1 demonstrate the elements of both the primitive and the advanced problems. From them it can be surmised readily that the accuracy to be expected of any analysis for the characteristics of a dispersion via measurement of interactions is dependent upon as many as possible of the factors and their functional interdependencies, in and among the three systems, being controllable and measurable.

The significant literature on the science of the total problem begins with the reports of recognitions by Richter[1] (using a gold sol) and. Tyndall[2] (using an atmosphere of butyl nitrite and HCl) that the colors from or the path of a beam of light directed through the dispersion is revealed by visual observation from the side.

TABLE 8.1: Factors Involved in Interactions between EMR and Dispersions

I	II	III
Intensity I_0	Dispersion, as to components':	Interactions:[a]
Wavelength λ	Number	Scatterings:
Bandwidth	Composition	Diffractions
	Phase condition	Refractions
	Optical properties	Dispersions (spectral)
	Scattering particles as to:	Reflections
	Size	Polarizations
	Size distribution	Interferences
	Shape	Absorptions
	Concentration	Kinetics
	Dielectrics	Thermodynamics
	Length of EMR path	

[a] It is understood that EMR undergoes no change in wavelength as a result of interaction.

Such phenomena, due to light-scattering, are popularly termed the "Tyndall effect"; the path as the "Tyndall cone." It may be demonstrated more readily by the following simple experiment: Prepare a 6″-square sheet of card-paper with a slit (2″ × 1/16″) cut at its center. Into a 400 ml. beaker of clear water stir a few drops of milk or a pinch of powdered silica gel. With the room darkened and with a flashlight beam directed through the horizontal slit and the solution, attempt to observe the Tyndall effect and cone from the various angles relative to the direction of the beam path. Observe also the change in direction of the path as the flashlight or the slit is elevated or lowered; the path from its transmitted direction; the effect of turning the slit slowly through a 90° angle; the relative intensities of incident, transmitted and scattered EMR.

Demonstrations with the same purpose but involving more elaborate apparatus are described elsewhere.[3,4]

It is generally understood that observations of the Tyndall effect and cone are made from a horizontal position at a 90° angle. If observation be from the vertical position through the meniscus, the considerable distortion which results can obliterate the otherwise clear path of the cone.

Richards' use[5] of the Tyndall effect, through his design of the first *nephelometer* to enable his determination of the atomic weight of Sr, constituted the classical beginning for practical measurement of dispersions on the basis of their EMR/particle light-scattering interaction. Cruder visual judgments of relative concentration of dispersed particles had been attempted prior to Richards' work.

Since the size factor was recognized early by theorists to be important to interaction characteristics, this discourse concerning the principles involved is greatly assisted by consistent meaning for prevalent terms as to sizes, as follows·

1. *Fine* particles are:
   a. *Small* (or Rayleigh-size), when $L < \lambda/10$

    b. *Large* (or Mie-size), when $\lambda/10 < L < 2.5\ \mu$; measurable by electron microscopy or ultramicroscopy

2. *Coarse* particles: when $L > 2.5\ \mu$

where L is the longest dimension, disregarding shape of particle, e.g., diameter of a sphere, or length of a rod.

Because of the small dimensions which in part characterize EMR wavelengths and particle sizes, the following approximate L-values are helpful for the perspective they have when compared with EMR-wavelengths . . . : H_2: 0.1 mμ; O_2: 0.16 mμ; Sucrose: 0.7 mμ; Starch: 8 mμ; Colloidal range: 1–250 mμ; *B. coli*: 1500 mμ; *Anthrax bacillus*: 6 μ; RBC: 8 μ.

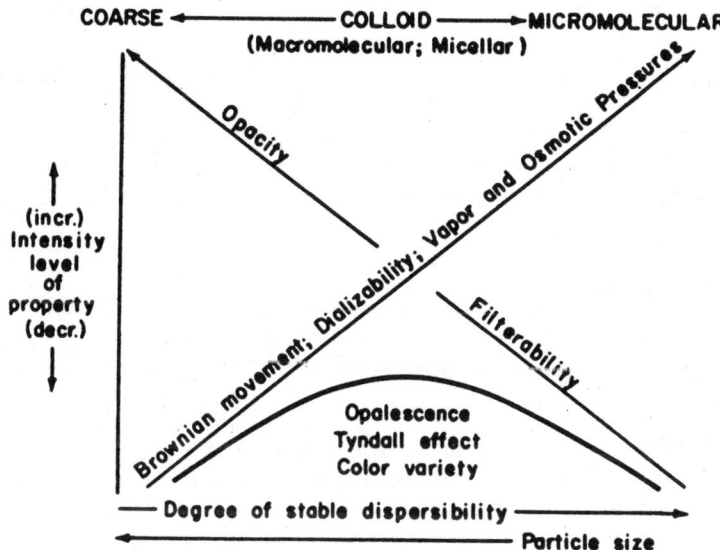

FIGURE 8.2: Effect of particles upon properties of their dispersions. Adaptation, after McBain.[*]

It should be emphasized that light-scattering methods are but one type among many other (such as microscopic, centrifugal, osmotic, X-ray, diffusion, and gas absorption) methods for determining dispersion characteristics.

A correlation may be made among various magnitudes of dispersed particles and their physical properties. Figure 8.2 is a representation of the general effects of three broad categories of particle sizes upon some of the most familiar of these properties. The figure is analogous to being a plot of dispersed particle sizes (X axis) vs. consequent dispersion properties (Y axis). Thereby, the concept is clarified that the gradual change in intensity of any such property and the gradual change in particle size are interdependent, at least for a given species of matter. Specific data would be required to detail the nature of the interdependence.

B. SOME GENERALIZATIONS REGARDING LIGHT-SCATTERING INTERACTIONS BETWEEN VISIBLE EMR AND MATTER

Absorption of EMR by matter is not included as a light-scattering phenomenon. In practical considerations, however, its coincidence with scattering must be evaluated for significance as a factor in the attenuation of incident EMR.

Light scattering principles for dispersions must be considered under the following conditions. It should be emphasized at the beginning that a dispersion system must be such that the particles have the conditioned freedom for mobile random distribution. Thus, the point-sources for scattering interactions are random particles.

1. Matter as a Pure Gas

The dispersed particles may be contained or uncontained. In either case, each interacts individually and without interference by neighboring particles. Thus, the total interaction is the sum of the interactions by the individual particles.

2. Matter as a Pure Liquid

The individual particles are no longer independent of neighboring particles, for the dispersion is much more dense so that in relation to EMR wavelength, the EMR-scattering effects by adjacent particles are out of phase.

If the number of particles in an interacting volume of matter as a liquid is n times the number of particles in an interacting equal volume of it as a gas (both at atmospheric pressure), the resultant interaction will show to be less for the liquid form; for, the particles not being independent, their scattering effects will not be independent but rather interferent (cf. Section 8.3A.2).

3. Solutions and Suspensions as Dispersion Systems

Inasmuch as such dispersions mean composition by more than one component, the generalizations are various according to analytic purpose for considering EMR/matter scattering interactions. Each purpose, therefore, determines the necessity for analytic data concerning:

1. The resultant interaction due to the dispersion
2. The resultant interactions due to each:
   a. Dispersing medium
   b. Dispersed phase

These imply a range of possible needs for data which derive *relative* to *absolute* values for interactions with EMR by the dispersion and its components. Thus, in a practical relative analysis, the coincident use of a primary-standard dispersion or of a blank dispersion may provide comparable data of sufficient significance to permit cancellation, without measurement, of interaction due to dispersion medium when the dispersed phase is the primary object of the analysis; and in analyses based on absolute values for interactions, the theoretical and applied significance for such values is exemplified in Section 8.3C.2.e.

The following are the general effective factors in dispersion systems subjected to interaction with EMR. According to what is available in the literature, several of the factors have not yet been evaluated.

1. Number of components in the system:
   a. Dispersed phase (1, 2, 3, . . . components)
   b. Dispersing phase; i.e., the medium (1, 2, 3, . . . components)
2. All components, as to:
   a. Material differences
   b. Optical differences. Ideally, a dispersion which is perfectly transparent will scatter no light. The degree of opalescence (see Fig. 8.2) of a dispersion is due in part to the magnitude of difference between the refractive indices of particles of the dispersed and dispersing phases.

   The property of refraction is not an independent factor since it is intimately related to the material (part 2.a., just listed), to other optical properties (cf. Fig. 8.3), and to particle size and shape.
   c. Gas-, liquid-, and solid-phase states. The necessary qualification that a dispersion must be such that the particles have freedom (though conditioned) for mobile random distribution in order to possess the property of light-scattering restricts them to being either gas-, liquid-, or solid-in-gas or gas-, liquid, or solid-in-liquid.

   Gas-, liquid-, or solid-in-solid systems constitute immobile cases, and are not considered to be in the general realm of dispersions. They do not conform in theory or practice to conventional principles because the point-sources for scattering interaction are fixed in position, resulting in complete interference among scattered wavelets.

Encounter of visible EMR with dispersions may involve its passage along, among, into, and from the particles, depending upon the optical properties. The interactions possible are of several appreciable forms which were cited but not defined in Table 8.1. It is helpful to realize that the relative complexity of a dispersion system contributes, with seeming directness, to the number of such interactions which have a role; and that the interactions involved may somehow combine, interfere with, or be independent of each other.

Figure 8.3 provides a review of the basic meanings of the various types of scattering-interactions phenomena. In each case it is seen that matter in its path is represented to effect some change in EMR. Such changes are directional (q.v. diffraction, refraction, reflection), electromagnetically differentiated (q.v. polarization), and λ_{band} component separated (q.v. spectral dispersion).

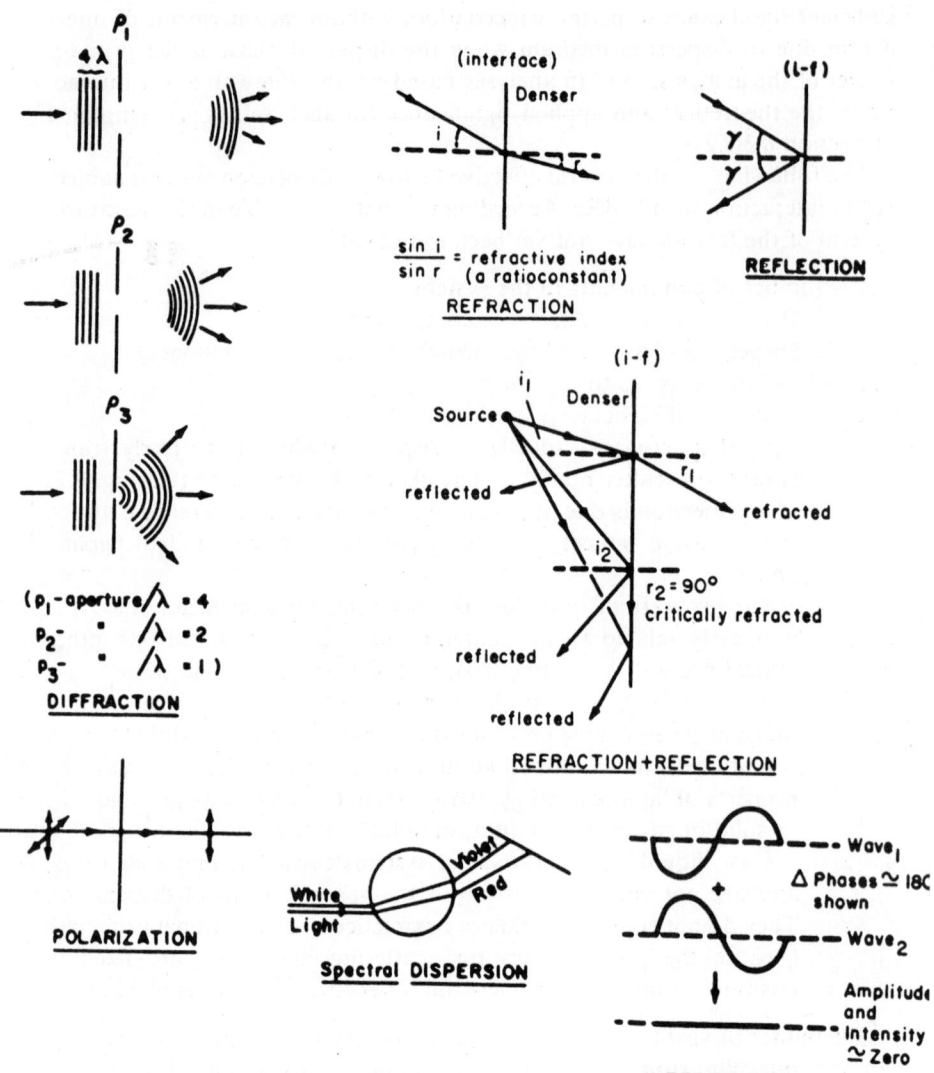

FIGURE 8.3: Basic meaning of EMR/particle interactions.

Projection of Fig. 8.3 evokes Fig. 8.4 to represent the possible angular effects of one particle upon incident EMR. Figure 8.4, in turn, evokes Fig. 8.5 to introduce the significance of interference (q.v. Fig. 8.3). The implications from further projections of meanings of Figs. 8.3–8.5 to dispersion *systems* will be considered in sections which follow. Before continuing however, it is essential to review the meanings of basic terms of spectrometry, such as (true) absorption, transmission, and optical density (see Chapter 1),

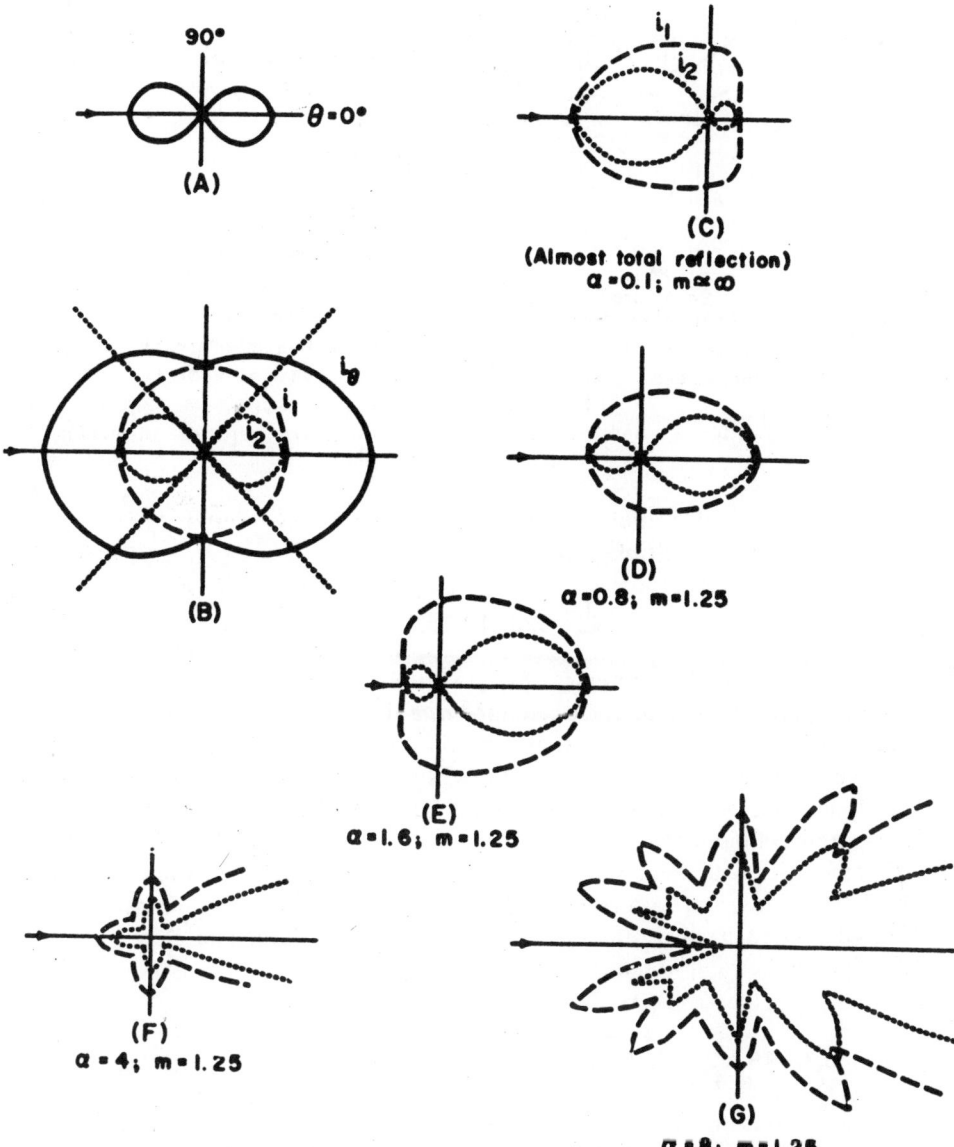

FIGURE 8.4: Theoretical light-scattering envelopes. Particle is at 0/0/0 of the XYZ coordinates. Adaptation, after Blumer.[11]

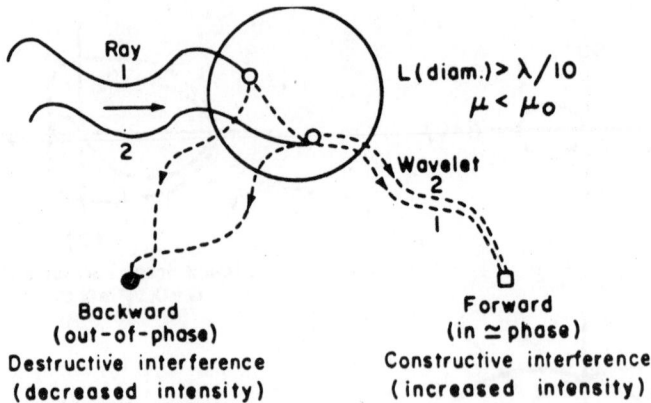

FIGURE 8.5: The forward and backward scatterings from a *large* particle. Adaptation, after Zimm et al.[23]

necessary for the definition of the meanings of such new terms as "apparent absorption" and "multiple scattering."

C. SPECTROMETRY

Concerning all of the interactions between EMR and dispersions, the most extensive science and literature belong to *true* absorption; in fact, much of the knowledge gained from spectral-absorption studies have been basic for studies extended to light-scattering interactions, so that many of the concepts and analytic terms of spectrometry apply equivalently to turbidimetry, nephelometry, and colloidimetry. For adequate background in addition to Chapter 1, familiarity with "Spectrometric Nomenclature"[7] (for conventional terms and definitions) and with references[8-10] [for their historical documentation of the attributed names: Bouguer (1729), Lambert (1760), and Beer (1852) for spectrometric laws] are recommended.*

8.3 PRACTICAL THEORY ON LIGHT-SCATTERING

Throughout the pertinent literature concerning light-scattering interactions in dispersions, there have been a few minor differences among reporters in symbols attributed for scattering factors and terms. The following list collects, for brief definitions, most of the symbols of this section; it includes those which reconcile differences for the purpose of this text.

* The conglomerate meaning of the Bouguer, Lambert, and Beer laws has been expressed variously by names in modern spectrometric literature, most prevalently by Beer's law and the Beer-Lambert law. This text selects "Beer's law" as the term for referring to that meaning.

I_0 = incident EMR ⎱ intensity (energy flux, ergs per square centimeter
I = transmitted EMR ⎰ per second)
λ = wavelength (in vacuo) ⎱
λ' = wavelength (in medium) ⎰ $(\lambda' = \lambda/\mu_0)$
n = wavelength exponent (of λ^n; $n = 4$ in Rayleigh theory)
θ = scattering angle; reference: transmission angle $(\theta = 0°)$
i_{sc} = scattered-EMR intensity at $\theta°$
i_{sc}/I_0 = Tyndall ratio
r = radius of a spherical particle
$L = 2r$ or D (diameter) of a sphere; longest dimension, nonsphere
α = size-parameter for *fine* particles; $\alpha = 2\pi r/\lambda'$
μ = refractive index of particle; of solution in Debye theory
μ_0 = refractive index of medium
m = refractive parameter; $m = \mu/\mu_0$
d = distance of detector from particle (or dispersion volume)
ρ = polarization effect of particle
v' = volume of particle $(4/3\pi r^3)$
n = number of particles per unit volume of dispersion
nv' = volume of particles per unit volume of dispersion
c = concentration of dispersion (grams per centimeter cubed)
l = length of transmission (or scatter) path traversed
M = molecular weight of solute
N = Avogadro's number; $N \simeq 6.03 \times 10^{23}$
K_{sa} = "scattering area" coefficient (of Mie theory)
τ = turbidity

A. FORMULATIONS OF FACTORS AND TERMS FOR FINE-PARTICLE DISPERSIONS

Though the more fundamental theory is based upon consideration of light-scattering by a single particle, applied theory projects single-particle theory to the obvious condition that many particles are present.

The assumption of an *ideal case* consists of the conditions listed in Table 8.2. Complications due to deviations from ideality are discussed in Section 8.3B.

This ideal case enables theoretical considerations to light-scattering phenomena involving *fine* particles to be classified into principal types. Each of these types is based on two parameters of the scattering particles, namely their size (expressed as $\alpha = 2\pi r/\lambda'$) and their refractive index relative to that of the medium (expressed as $m = \mu$ (of particle)$/\mu_0$ (of medium)); it is also based on the scattering-angle parameter θ.

Figure 8.4 summarizes the principal types of light-scattering patterns ("envelopes") from single particles. Of these types A and C have not received extensive study. Incident EMR is unpolarized, except in type A. The solid lines (where shown) represent "finite" limits of vectors (i.e., directed magnitudes) of total scattered intensities. The dashed and dotted lines refer to the scattered polarized components of the incident EMR; thus, the i_1 or $i_\perp$ (dash) and the i_2 and $i_\parallel$ (dot) are the perpendicular and horizontal components, respectively; and thus, the vector sum $i_\theta = i_1 + i_2$ is the resultant intensity at angle θ. Recognition that the vectors are all-directioned evokes the realization that a theoretical single-vector intensity is an "infinitesimal" fraction of the incident-EMR intensity.

TABLE 8.2: An Ideal Case for Theoretical Study
of Interactions[a]

1. Incident-EMR intensity I_0 is constant
2. Incident-EMR wavelength λ is monochromatic
3. Isotropicity of ⎫
4. Nonabsorption by ⎬ scattering particles
5. Monodispersity of ⎭
6. Two-component (dispersing + dispersed) system
7. Adequately dilute dispersion for independent EMR/particle interactions

[a] Presumed are: Stability of the dispersion system; variability of the detector position.

Blumer[11] derived the types C, D, E, and F envelopes by calculations from Mie equations. Concurrently, Gans[12] described a correlation between particle size and wavelength (as in $\alpha = \pi D/\lambda'$) by means of scattering envelopes.

I. Rayleigh-Type Scattering[13-17]

Rayleigh-type scattering involves *small* particles, $L < 0.05\,\mu$ (see Fig. 8.4B). The longest particle-dimension (diameter in this instance) does not exceed $\lambda/10$. Under this condition each particle interacts as a single-point dipole oscillator. At $\theta = 90°$ the scattered light is completely polarized. Since the envelope is completely symmetrical, scattering is believed to be due almost entirely to diffraction, practically not at all to refraction and/or reflection.

Rayleigh scattering theory has received expression principally in two equation forms, depending upon the considered type of dispersion and factors involved. Thus, the scattered/incident intensity ratio i_{sc}/I_0:

a. For a gas in a vacuum (e.g., atmosphere; the refractive index relationship between μ_0 and μ need not be considered because the medium is not material) is given by

$$\frac{i_{sc}}{I_0} = \frac{8\pi^4 n\rho^2}{\lambda^4 d^2}(1 + \cos^2\theta) \quad \text{(Refs. 18,19)} \quad (8.1)$$

b. For a very dilute solution (the medium is material) is given by

$$\frac{i_{sc}}{I_0} = \frac{n\nu'^2}{\lambda'^4 d^2}\left(\frac{\mu^2 - \mu_0^2}{\mu^2}\right)(1 + \cos^2\theta) \quad \text{(Refs. 20,21)} \quad (8.2)$$

The following points emphasize some of the scattering principles which are conveyed by Eqs. (8.1) and (8.2):

$i_{sc} \propto I_0$ implies that experimental proof of a scattered-intensity envelope, via evaluation of its vectors, would require great precision within and co-ordination of both the incident and detector systems.

$I \propto I_0 - \Sigma i_{sc}$, in the hypothetical and absolute sense, means that attenuation of incident-EMR intensity is in all directions without loss of EMEnergy because the dispersion is a perfect vacuum having particles which upon encounter undergo only scattering interactions.

$i_{sc} \propto (1 + \cos^2 \theta)$, calculated for all values of θ, describes the geometric shape of the scattered-intensity envelope (i_{sc}) and of the polarized components (i_1 and i_2) of i_{sc}. For a given θ-vector, $i_\theta = i_1 + i_2$; and for a 90° vector, $\cos^2 \theta = 0$ and $i_{sc} = i_1$. The ratio $(1 + \cos^2 \theta)/2$ is termed the "depolarization factor."

$i_{sc} \propto 1/\lambda^4$ is a consistent relationship for *small*-particle dispersions. The relationship is manifest beautifully in nature by atmosphere as the dispersion. Since $\lambda_{blue} < \lambda_{red}$ of the visible-EMR range, the former is scattered more and transmitted less than the latter. Thus, the *spectral* dispersions at sunrise and sunset are "riotous" and the blues of the sky, the ocean, and tobacco smoke are less so.

The rainbow as a *spectral* dispersion warrants a different explanation because the raindrop particles are *coarse* in size, prismatic in function, and falling.

$i_{sc} \propto 1/d^2$ expresses that EMR-scattering is in accord with the principle of conservation of energy. Here, EMEnergy (expressed in terms of intensity) is scattered as vectors from a point-source. To a detector in position at a cross-sectional area of vectors, the scattered intensity varies inversely as the square of the distance of the detector from the point-source; in other words, doubling the distance quarters the intensity.

$i_{sc} \propto v'^2$ expresses the role of particle size as a scattering factor. Because of the relationship of v' with r, this role becomes even more impressive when expressed in the equivalent terms of $i_{sc} \propto r^6$; hence the importance of excluding alien particles from dispersions to be subjected to analysis by light-scattering or by spectral-absorption methods.

2. Rayleigh-Gans-Type Scattering[12,16,22,23]

Rayleigh-Gans-type scattering appears to be easiest to understand when it is positioned somewhat "intermediate" in meaning between Rayleigh-type (Section 8.3A.1) and Mie-type (Section 8.3A.3) scatterings. To this end, Fig. 8.5 represents a particle of diameter comparable with λ to have at least two point-sources for scattering. It is to be noted that the forward-scattered wavelets are favorable and the backward-scattered wavelets are unfavorable contributors to scattered-intensity vectors (i_θ) for Rayleigh-type scattering. The figure does not attempt to distinguish precisely among the possible scattering mechanisms (see Fig. 8.3).

The two principal and limiting conditions for so-called R-G-type scattering have it that:

a. The refractive-index ratio m is near 1 in value. (If it were 1, there would be no distortion of the path of incident EMR traversing through the particle, and hence no scattering.)

b. The particle size may be greater than that of Rayleigh particles only to the extent that, because of condition a, each volume element of the particle (corresponding to each of its point-sources) interacts as an independent Rayleigh-type scatterer.

Under condition b, the different positions in space of volume elements (and point-sources) result in interference between backward vectors of the independent Rayleigh-type scattering elements.

Figure 8.4D and E exemplify the general intermediate characteristics of R-G-type scattering compared with Rayleigh-type (B) and Mie-type (F, G) scatterings.

3. Mie-Type Scattering[24,25]

Mie-type scattering involves *large* particles, $0.05\mu < L < 2.5\mu$ and refractive-index ratio m not near 1 in value. See Fig. 8.4F and G and Fig. 8.5 in the perspective of R-G-type scattering. Refraction and some diffraction are principal mechanisms for the Mie-scattering phenomena. Particle size (expressed as $\alpha = 2\pi r/\lambda$) and the refractive-index ratio (expressed as $m = \mu/\mu_0$) together constitute the so-called m-α domain, and together with θ are the factors which have received most of the experimental attention to verification of Mie-scattering theory.

The theory of the involvement of the foregoing and other factors was established by Mie[24] and shortly thereafter by Debye,[25] independently. Reviews of and extensions upon their basic theory were conducted subsequently by many others.

Within a clarifying summary of the meaning of the complex Mie theory in relation to light-scattering methods for determining particle characteristics (size, molecular weight, shape), Bender[26] presented the following component equations to explain the involvements of the various factors and terms: when

$$m \gg 1.33$$

$$L \gg \lambda'/6$$

$$\frac{i_1}{I_0} = \frac{\lambda'^2 n}{8\pi^2 d^2}\left[\frac{a_1}{2} + \frac{a_1 + p_1}{2}\cos\theta\right]^2 \tag{8.3}$$

$$\frac{i_2}{I_0} = \frac{\lambda'^2 n}{8\pi^2 d^2}\left[\frac{a_1}{2}\cos\theta + \frac{a_2}{2}\cos 2\theta + p_1\right]^2 \tag{8.4}$$

whereof (electric dipole moment):

$$a_1 = 2\alpha^3 \frac{(m^2 - 1)}{(m^2 + 2)} \tag{8.5}$$

(electric quadripole moment):

$$a_2 = \frac{-\alpha^5}{6} \frac{(m^2 - 2)}{(m^2 + 3/2)} \tag{8.6}$$

(magnetic dipole moment):

$$p_1 \simeq \frac{-\alpha^5}{15} (m^2 - 1) \tag{8.7}$$

It is because of the *large*-particle size (see Fig. 8.5) that the a_2 and p_1 effects come into play. Moreover, particles having larger m and L magnitudes than those just given contribute more terms to the [ ]² factor. *Small* particles have negligible a_2 and p_1 values, so that Eqs. (8.3) and (8.4) become Eq. (8.2) in meaning (Rayleigh-type scattering).

B. COMPLICATIONS DUE TO DEVIATIONS* FROM IDEALITY

Figure 8.4B implies that Rayleigh-type scattering has the characteristics for being the most ideal within the ideal case (see Table 8.2), inasmuch as the scattering envelope for it is the most symmetrical in reference to intensity vectors from the scattering particle.

Factors within interacting EMR and dispersion systems which are not in accord with the assumed criteria for factors of the ideal case are realistic because they are *the* case. Several of them have received theoretical and experimental attention.

While considering the deviation from ideality of each factor in turn, it is a helpful simplification to consider all other factors as remaining ideal. In fact, appreciation for the known and unknown complexities of the interaction phenomena of the most primitive problem (Fig. 8.1) dictates this simplification.

1. ΔI_0 (Change in Incident-EMR Intensity)

Since $i_{sc} \propto I_0$, it is reasonable to expect a change of magnitude, but not of shape, of the scattered-intensity envelope. Furthermore, fluctuating or insufficient I_0 results in problems for the detector system.

2. Polychromaticity, cf. Monochromaticity

When λ_{band} is incident and the particles of the dispersion system are *large* and have the appropriate optical properties, relative possibility exists for interaction by the mechanism of *spectral* dispersion. This is because the

* Bibliography to theoretical and experimental studies of these deviations are cited via the topics index of the Appendix.

exponent n of λ^n has two important dependencies in the *large*-particle realm, namely, $n \propto 1/\lambda'$ and $n \propto 1/$particle size (as α), both of which are implied in Mie theoretical equations.

Spectral dispersion as a prominent scattering mechanism has been demonstrated experimentally, under conditions approaching ideality, by a variety of prepared sols and using λ (visible range). Under the best attained conditions, the interaction results in a "riot" of colors by dispersion of incident λ_{band} into its spectrum. The several θ angles at which the most easily detected components (red and green) appear are termed "orders" (1, 2, 3, . . .) of which those of "red hue" are the more prominent.

The principal application of this phenomenon (high order Tyndall spectrum, HOTS) is to particle-size analysis of colloidal systems.

3. Anisotropy, cf. Isotropy

The basic formulations of the Rayleigh and Mie theories assume that the light-scattering particles are spherical. Anticipation of a possible "-tropic" factor in interaction phenomena begins with the realization that:

    a. An *isotropic* particle is uniformly shaped in that it will effect the same type of scattered-intensity envelope regardless of its orientation to I_0.

    b. An *anisotropic* particle is not uniformly shaped in that its scattered-intensity envelope is changed by at least certain of its orientations to I_0.

The following are some of the principal implications from anisotropy compared with isotropy:

    a. Of a *small* particle.

        A small amount of dissymmetry of the scattering envelope occurs, favoring the forward direction. Even though the particle is still interacting as a single-point dipole oscillator, vector norms (most noticeably ca. $\theta = 90°$) are distorted; that is, polarization is less than complete at that angle, resulting in a quantitative circumstance which is termed "depolarization ratio" and expressed as i_2/i_1.

        The extent of this dissymmetry is of course relative and small; but even so it may be of sufficient significance to need correction by a factor ("Cabannes factor") in analyses requiring special precision (e.g., molecular-weight determinations). Although polarization measurements at $\theta = 90°$ are difficult because the intensity at that angle is low and small ratios are difficult to detect, approximate measurements have been attempted from a detector position *near* $\theta = 90°$ using a so-called polarization photometer, which is capable of reading for i_1 and i_2 through a bipartite (polarized) disk (one-half of which is for i_1, one-half for i_2).

    b. Of a *large* particle.

        Figure 8.4D, E, F, and G for spheres show the dissymmetry (reference, $\theta = 90°$ or $270°$) between corresponding forward/backward i_{sc} vectors, due to particle size (although μ also is a factor). Gans[27] theorized to some extent regarding the influence upon scattering by deviation from spherical shape.

It is readily grasped that if the particle is anisotropic (e.g., rod or coil), the number of factors contributing to scattering increases over the number due to particle size alone, so that approximations become necessary. Bender[26] has summarized the meanings of some of the approximations and of the application of the concept of "dissymmetry ratio" (expressed, for example, as i_{45}/i_{135}) which, via appropriate equations, can be related to L of the particle if the shape is known. Thus for each combination of size and anisotropic and refractive properties a *large* particle has in theory its characteristic scattered intensity envelope, depolarization ratios, and dissymmetry ratios. From this it is interesting to anticipate the changing effects upon the envelope and ratios by a theoretical change of a spherical *large* particle elongating to a rod to a thread (v' and μ remaining constant); and to anticipate the changing considerations required for evaluating these changing effects from the changing relative particle dimensions within this type of anisotropy.

4. Absorption, cf. Nonabsorption

In experimental light-scattering studies, it is unrealistic to expect to attain the absolute, that is, a dispersion system completely devoid of interactions with EMR other than scattering. Studies reported have been of conditions known or assumed to be practically devoid of *true* absorption, which is an interaction property of relative magnitude and significance, depending upon the chemical composition of dispersion components. Evaluation of the effect of true absorption coincident to any extent with *apparent* absorption has been a neglected aspect of light-scattering analytic methods, for the problems such coincidence presents are formidable and generally have been avoided by design in experimental studies to test validity of scattering-interaction theories. The problems emerge, when true absorption is coincident, because of its dependence as a function of λ'; furthermore, the wavelength exponent of λ'^n is a function of both $1/\lambda'$ and $1/\alpha$ for *large* particles and is 4 for *small* particles, when scattering alone is the interaction.

Spectrometry uses the transmission angle ($\theta = 0°$) to evaluate true absorption by correlating I with I_0 through expression as absorbance (A) or as transmittance ($\%T$), instrumentally. To test whether a dispersion of *small* particles has sufficient true absorption property to affect the practical validity of data concerning its light-scattering properties, the transmission angle is chosen. Since the dispersion itself constitutes a constant (i.e., is invariable for the test), $I_0 - I$ will vary as $1/\lambda'^4$ for measurements over a range of wavelengths when no true absorption is coincident.

When true absorption is unavoidable, the prevalent inclination is to seek a wavelength at which it is minimal and at which light-scattering data are valid in terms of Beer's law or are derived under calibrated conditions.

When the three systems of Fig. 8.1 are properly coordinated and there is no stray EMR, $I = I_0 - [\Sigma i_{sc} + I \text{ (absorbed)}]$.

5. Polydispersity, cf. Monodispersity

It is helpful to begin with the assumption that a polydispersion is an integration of constituent monodispersions (of Section 8.3B.6). The following are principal considerations which polydispersity evokes:

    a. Several values for the expression $\alpha = 2\pi r/\lambda'$ are coincident. Thus, in reference to Fig. 8.4 (assuming m constant):

        1. If the polydispersity is entirely within the limit for *small* particle sizes, the depolarization and dissymmetry ratios for the dispersion system are scarcely deviant from their respective values of zero and 1 in the ideal case.

        2. If the polydispersity (a) spans coincidentally into both *small* and *large* particle realms, or (b) is entirely within the size limits for *large* particles, and because such a dispersion is relatively stable, the determination of size range and size frequencies are formidable theoretical and experimental problems for light-scattering methods. Practical analyses for these size characteristics are directed to obtaining average values which represent concentration of dispersed particles in terms of numbers-average and/or weight-average.

    b. The properties of the components and the conditions in a given dispersion system determine the relative monodispersity . . . polydispersity which prevails. Statement of this implicates a great variety of possible phenomena in dispersions. Particular reference here is to the properties of particles and conditions which can result in changes of their sizes and numbers, or in inherent stability or preservation of their sizes and numbers, for a given weight concentration.

        Thus implied are possibilities for such phenomena as aggregation, coagulation and flocculation, micelle formation, polymerization, growth of microorganisms; and for reverse phenomena of subdivision. Inseparable from considerations to changes in particle size are considerations to coincident changes in particle shape.

    c. Relatively ideal monodispersions are attainable by preparation, but not without thorough knowledge of the properties of components and dispersions and of the required conditions.

6. Multicomponent Systems, cf. Two-component Systems

    e.g.: One medium + two particle-species
          Two media + one particle-species

It is helpful to begin with the assumption that a three-component system is an integration of constituent one- and two-component systems (cf. Section 8.3B.5). The following are principal considerations which multicomponency evokes:

a. The optical property of refraction differs (μ_0' and μ_0'') as between two media, and differs (μ' and μ'') as between two particle-species. Thus, there are different m ($= \mu/\mu_0$) ratios which have roles.

b. The ideal case would have it that the two particle-species are of the same L; therefore, in regard to their number of point-sources for scattering they are equivalent. However,

1. If L is *small*, there is little if any distortion of scattering behavior insofar as Rayleigh-type scattering is principally diffractive [Eq. (8.2)].

2. If L is *large*, refractions become important factors of the integrated scattering behavior, and their evaluation constitutes a complex problem according to Mie theory [Eq. (8.3–8.7)].

It is not resolved whether or not two miscible media of different μ_0 values are correctly combined into a single μ_0 value for evaluation of scattering properties, although assumption may suggest that this be done for practical purposes. Factors μ and μ_0 do appear as factors and terms (other than as m) of theoretical equations which have to do with extended concepts about scattering phenomena (see Sections 8.3C.2.b, c, and f).

7. Concentrated, cf. Dilute Systems

Rayleigh and Mie equations for scattered intensities by *fine*-particle dispersions are based upon the ideal conditions that the total scattered-intensity from all particles interacting with EMR is the sum of the effective scattered-intensities from each particle. This means that no particle stands in the way of another along any EMR path in their system, and that *infinite dilution* ($c \rightarrow 0$) most satisfies this ideal. Projection of the meanings of Figs. 8.4–8.5 for an individual-particle system to their meaning for a community-of-particles system emphasizes the importance of the dilution condition.

The experimental approach is by *adequately* dilute dispersion and *adequately* short paths. In this way the secondary, tertiary, ... multiple scatterings due to interactions by neighboring particles are avoided as much as possible. Data which correlate the scattering effects of a dispersion with concentration of scattering particles reveal the validity of adequate dilution when they are extrapolated to zero concentration to enable comparison with corresponding theoretical data.

The "fluctuation theory" referred to in Section 8.3C.2.d involves considerations which lead to an analytic first-approximation due to secondary scattering. This places it in close relation with the condition of "adequately dilute dispersion" of the ideal case (see Table 8.2).

C. THE TURBIDITY OF *FINE*-PARTICLE DISPERSIONS

The empirical meaning of *turbidity* is the visual-sensory one of *cloudiness*. However, in the perspective of the full meanings of light-scattering phenomena and of photoelectric measurement of them, its meaning is profound to the extent that any dispersion of which the scattering particles are in mobile random distribution has a turbidity (cf. Section 8.2B).

The *small* and *large* particles (which, if they are spherical, pertain precisely to Rayleigh, Rayleigh-Gans, and Mie light-scattering principles) together constitute ranges of sizes, shapes, and optical properties which involve interactions principally by diffraction and refraction, and by *spectral* dispersion, under specialized conditions. Within such ranges are particles which are micromolecular to macromolecular, polymeric, aggregate, coagulate, micellar, and microbiological. Of such particles, the determinations of unit and aggregate weights, sizes and size distributions, shapes, and concentrations are at the heart of the purposes of turbidimetry, nephelometry, and colloidimetry.

1. Basic Concept: Turbidity as a Coefficient

Rayleigh and Mie theories indicate that the $\theta = 0°$ detector position (see Fig. 8.4) is a favorable one for measuring an emergent vector of high intensity to characterize a *fine*-particle dispersion by the attenuation it exacts upon the intensity of incident EMR. Thus, in the absence of true absorption $I = I_0 - \Sigma i_{sc}$.

This places the attenuation due to scattering (*apparent* absorption) into analogous relationship with the *true* absorption Beer's law, which is expressed by:

$$I = I_0 e^{-\epsilon cl} \qquad (8.8)*$$

so that

$$\log_{10} \frac{I_0}{I} = \epsilon cl/2.303 \qquad (8.9)$$

where $e = 2.7184$ (the base of natural logarithms; whence, $2.7184^{2.303} = 10$) and ϵ is the extinction coefficient, characteristic of λ and the absorbing property of the solute species.

Equations (8.8) and (8.9) mathematically define the amount of attenuation of I_0 as it traverses an adequately dilute solution of the species. The following equations are analogous for the expression of turbidity τ as an extinction phenomenon:

$$I = I_0 e^{-\tau cl} \qquad \text{(Ref. 28a)} \qquad (8.10)†$$

---

* USP and NF use P and P_0 in place of I and I_0; BP uses I and I_0. Equivalence is factorial, since power = intensity × time.

† Cf. Rayleigh's expression[13] for extinction: $I = I_0 \exp(-k\lambda^{-4} x)$

so that

$$\log_{10} \frac{I_0}{I} = \tau cl/2.303 \qquad (8.11)$$

and

$$I = I_0 e^{-\tau l} \qquad \text{(Refs. 18,19,29,30)} \qquad (8.12)$$

so that

$$\log_{10} \frac{I_0}{I} = \tau l/2.303 \qquad (8.13)$$

where τ (dimension, cm^{-1}) is the extinction coefficient, characteristic of λ and of the scattering property of the dispersed particles. Scattering property depends on the number of point-sources and on size of particles; i.e., on effective surface area (n/cm^2).

Thus, turbidity is defined as the fractional decrease, due to scattering, of incident intensity through unit thickness of dispersion traversed. The inclusions of factor c in Eqs. (8.10) and (8.11) clarify the cm^{-1} dimension mathematically, since n/cm^2 of the particles is a function of c. In Eqs. (8.12) and (8.13), the mathematical value of c is *included with* that of τ because of the direct dependence of τ on c.

Equations (8.10)–(8.13) apply the photometric data for the calculation of τ.

The analogy as extinction concepts between spectrometry and turbidimetry may be given further expression as a parallelism, as follows:

Spectrometry (absorption)	Turbidimetry (scattering)
$T = \dfrac{I}{I_0}$
(Transmittance)
(I_0 is of monochromatic λ)
(Absorbance, *true*) $A = \log 1/(I/I_0)$ $= \log I_0/I$ $= \epsilon cl$	(Absorbance, *apparent*) $A = \log 1/(I/I_0)$ $= \log I_0/I$ $= \tau cl*$
(A vs. c plot, slope ϵl)	Straight line origin 0/0 for Beer's law	(A vs. c plot, slope τl)
$A = 2 - \log \% T$
$\epsilon = 2.303 \log_{10} I_0/I$	$\tau = 2.303 \log_{10} I_0/I$

* The measurement of A or of $\% T$ as the indication of relative τ is prevalent analytic practice.

Notable are other comparisons, though of lesser equivalence:

a. The I/I_0 (transmittance) ratio is somewhat analogous to the i_{sc}/I_0 (Tyndall) ratio of Eqs. (8.1) and (8.2), and to the component i_1/I_0 and i_2/I_0 (Mie) ratios of Eqs. (8.3) and (8.4).

b. Regarding Beer's law, reasons for deviations by dispersions intended for spectrometry are analogous in principle but not the same in specifics as reasons for deviations by dispersions intended for turbidimetry, the EMR/particle interactions being different. The specifics of the reasons for the former have been documented extensively as involving instability in the number and function of the absorbing species, i.e., involving changes of a chemical nature (dissociation, isomerization, ...); and the specifics of the reasons for the latter involve instability in the number and function of the scattering species, i.e., involving changes of a physical nature (effective surface area for scattering interactions, mechanism of scattering, ...). Related to both categories regarding specifics is the factor of concentration of species. Thus the analytic controls for avoiding or minimizing deviations by dispersions regarding Beer's law are a common purpose of both spectrometry and turbidimetry. .

2. Extended Concepts

Beyond the basic concept, there are several concepts which require the involvement of other physical considerations and which are variously less familiar.

The following is a selection from the literature of some various other equations containing the turbidity symbol τ.

a.

$$\tau = \frac{kcld^4}{d^4 + \alpha\lambda^4} \tag{8.14}$$

where k is a constant of dispersion and method, α is a constant of method, and d is diameter of particle. Basing the concept in both Beer's spectrometric law [Eq. (8.8) and Rayleigh theory, Wells[20] and Yoe[21] showed the derivation of the useful approximation which Eq. (8.14) represents. The equation is assumed valid for both *small*- and *coarse*-particle dispersions, but not for the intermediate (*large*-particle) range, and is in reasonable agreement with the dilution principle of Beer's law.

The USP[31] and NF[32] engage this concept for their concept of *turbidance S* as:

$$S = \log_{10} P_0/P = \frac{kbcd^3}{d^4 + \alpha\lambda^4} \tag{8.15}$$

where b is thickness of dispersion and d is average diameter of particles. The application of these concepts for analyses is in the derivation of S vs. c curves for unknown dispersions, for comparison with such a curve for a standard dispersion.

Furthermore, the compendia use an extension of the concepts for nephelometric analysis by measurement of that portion of incident power which is scattered in a normal $(\theta = 90°)$ direction. The expression of this is given the form:

$$P_s = P_0 - P = P_0(1 - 10^{-S})$$
(8.16)

where P_s is that vector of scattered power. When the conditions for the analysis of a given dispersion are fixed, the k, b, d, α, and λ (monochromatic) are a set of constants coverable by the collective constant K. Whence:

$$P_s = P_0(1 - 10^{-Kc})$$
(8.17)

when $S = Kc$.

b.
$$\tau = \frac{8\pi}{3}\left(\frac{2\pi}{\lambda}\right)^4 n\rho^4$$
(8.18)

This is turbidity derived by Debye[30,33] from Rayleigh scattering theory for an *ideal gas* (n molecules per cubic centimeter) of low density, referring to the transmission angle. If the index of refraction μ is introduced as a factor, ρ can be eliminated on the basis that $\mu - 1 = 2\pi n\rho$, deriving (for complete polarization at $\theta = 90°$ where $1 + \cos^2 \theta = 1$):

$$\tau = \frac{32\pi^3}{3}\frac{(\mu - 1)^2}{\lambda^4}\frac{1}{n}$$
(8.19)

Further derivations by Debye obtained:

$$\tau = \frac{32\pi^3}{3}\frac{1}{\lambda^4}\frac{kT}{\kappa}\left(\mu\frac{\partial\mu}{\partial p}\right)^2$$
(8.20)

where κ is the compressibility, p is hydrostatic pressure, and T is absolute temperature, as the analogous turbidity for a liquid, wherein refraction property is changed from that for a gas in Eq. (8.19) by greater proximity of particles; and obtained:

$$\tau = \frac{32\pi^3}{3}\frac{\mu(\mu - \mu_0)^2}{\lambda^4}\frac{1}{n}$$
(8.21)

as the analogous turbidity for a solution.

c.
$$\tau = HcM$$
(8.22)

where
$$H = \frac{32\pi^3}{3}\frac{\mu_0^2}{N\lambda^4}\left(\frac{\mu - \mu_0}{c}\right)^4$$

is the "refraction constant" of the dispersion, is Debye's equation[33] for the determination of molecular weight by light-scattering to $\theta = 90°$ in very dilute solutions.

The value of n of Eq. (8.21) is impractical to determine, whereas c is convenient because it can be prepared accurately from pure solute. This is evident from the expression:

$$\frac{\text{No. of solute particles/cc}}{\text{G. of solute/cc}} = \frac{\text{no. of solute particles/mole}}{\text{g. of solute/mole}}$$

which is $n/c = N/M$ and $n = cN/M$ [introduced into Eq. (8.21)].

For derivation of M, μ and μ_0 (by differential refractometry), and τ must be determined by measurement and c must be known by preparation; τ is measured from a pertinent range of concentrations to obtain a linear plot* of c/τ vs. c, which is suitable for extrapolation to zero concentration. M is the reciprocal of the intercepted c/τ value at $c = 0$.

Note that τ/c means "specific turbidity" and c/τ means reciprocal "specific turbidity."

d.
$$H \frac{c}{\tau} = \frac{1}{M} + 2Bc \tag{8.23}$$

where B is an "interaction constant" depending upon the solvent, defines the turbidity due to fluctuations in concentration.

The Debye equation (8.22), and its transposed form ($Hc/\tau = 1/M$) is valid within an upper limit of adequately dilute solution. Inaccuracies from use of it emerge, in solutions which are more concentrated than those which conform to van't Hoff's osmotic pressure law for ideal solutions (wherein osmotic pressure $\propto n$, and N), from incidence of secondary scattering, which causes deviation from the ideal plot-linearity of c/τ vs. c. Also incident are refractive-index and osmotic-pressure changes from density and concentration fluctuations, resulting from thermodynamic interactions; local inhomogeneities occur.

The thermodynamic factors in density and concentration fluctuations have been coordinated within the "fluctuation" theories of Smoluchowski[34] and Einstein[35], and within subsequent elaborations and discussions by others.[6,26,36-42] Application of the correction which the "$2Bc$" term of Eq. (8.23) represents restores the linear relationship of c/τ vs. c, and implies that the equation reduces to Eq. (8.22) in meaning as c reduces toward ideality.

e.
$$\tau = \frac{8\pi}{3} \frac{Ir^2}{I_0} = \frac{8\pi}{3} R_0 \tag{8.24}$$

$$\tau = \frac{16\pi}{3} \frac{i_{90}r^2}{I_0} = \frac{16\pi}{3} R_{90} \tag{8.25}$$

whereof $R_\theta = i_\theta r^2/I_0$ is termed the "Rayleigh ratio," which descriptively means the reduced-intensity vector of I_0 scattered by a small-particle

* In such plots there is a λ' of maximum turbidity.

dispersion; and r means the radius of a spherical detector-distance d from the dispersion. In the perspective of Eqs. (8.1) and (8.2), R_θ is seen to contain the "Tyndall ratio" as a factor.

So as to establish a theoretical relationship between turbidity (as an extinction coefficient) and i_{sc} (according to Rayleigh theory), Doty and Steiner[43] derived the concept for Eqs. (8.24) and (8.25) by integrating Eq. (8.2) to give it the meaning of scattering to the surface of the sphere (just discussed) of radius r, as by:

$$\int_0^\pi i_\theta 2\pi r^2 \sin\theta \, d\theta = \frac{8\pi}{3} i_0 r^2$$

Equations (8.24) and (8.25) have notable agreements with the dimensions of Fig. 8.4B, as revealed by:

1. Equating τ in Eq. (8.24) with τ in Eq. (8.25),

$$\frac{8\pi}{3} \frac{Ir^2}{I_0} = \frac{16\pi}{3} \frac{i_{90}r^2}{I_0}$$

whence, $I = 2i_{90}$.

2. The "depolarization factor," $(1 + \cos^2\theta)/2$, in that it has the value $1/2$ at $\theta = 90°$; that is, $i_\theta = I(1 + \cos^2\theta)/2$.

R_θ calculations have primary usefulness in their application for determining *absolute turbidity* of pure liquids used as dispersion media, and for calibration of instruments. Furthermore, the correction which absolute turbidity implies is applicable to necessary accuracy in determinations of M by light-scattering methods, according to Eqs. (8.22) and (8.23).

f.
$$\tau = K_{sa}\pi r^2 n l \qquad (8.26)$$

where K_{sa} is the "scattering area" coefficient. It is a function of $\alpha(= 2\pi r/\lambda')$ and $m(= \mu/\mu_0)$. When τ has the meaning of absorbance (apparent), $K\pi r^2 n l$ is expressed as "extinction coefficient" in a form analogous to transmission-type Eqs. (8.8) and (8.10), as:

$$I = I_0 e^{-K\pi r^2 n l} \qquad (8.27)$$

so that

$$\log_{10}\frac{I_0}{I} = K\pi r^2 n l/2.303 \qquad (8.28)$$

K_{sa} is coefficient for *large* spherical particles, being related to Eqs. (8.3) and (8.4) via Eqs. (8.5)–(8.7) Its values for incremental values of α and m have been derived by calculations from the Mie equations and have been published as tables of scattering functions. Thus enabled are direct and intrapolated values for application with measured data in the analysis of such dispersions.

Equations (8.26)–(8.28) should be recognized in the literature to have a number of revised and extended forms.

8.4 CORRELATION OF THEORY FOR *FINE*- AND *COARSE*-PARTICLE DISPERSIONS (cf. Fig. 8.2)

What are known of the light-scattering properties of *fine*-particle dispersions form much of the basis for the recognition and solution of equally if not more difficult problems attendant in *coarse*-particle dispersions. As the quantitative considerations to scattering interactions traverse from the *small*- through *large*-particle sizes, there is encountered a relatively "gray" range of sizes where interactions cannot be evaluated by Mie theoretical equations except by approximations. Through and beyond the "gray" range the considerations become somewhat empirical.

The following points describe generally the reasons for empiricisms and limited precisions:

1. The effective scattering-area decreases with increasing size for a given mass-concentration of particles. (Cf. Table 8.3).

TABLE 8.3: Effect of Cubed Subdivision upon Total Surface Area

Edge of cube, cm	No. of cubes	Total volume, cm³	Total surface area, cm²
1	1	1	$6 \times 10^0 \times 10^0 = 6$
0.1	10^3	1	$6 \times 10^3 \times 10^{-2} = 60$
0.01	10^6	1	$6 \times 10^6 \times 10^{-4} = 600$
0.001	10^9	1	$6 \times 10^9 \times 10^{-6} = 6000$
0.0001 (1 μ)	10^{12}	1	$6 \times 10^{12} \times 10^{-8} = 60,000$

2. As particle size increases from *small* to *coarse*, the principal mechanism for interaction changes from diffraction through refraction to reflection. (Cf. Fig. 8.3.)

   In the absence of absorptive property, relative reflectivity of matter appears as a relative opacity which is a property which may be due to the thermodynamic factors having roles in the formation, growth, composition, and structure of *coarse* particles.

3. Characteristics inherent in *coarse* particles are that their effective surfaces are not likely to be spherical and smooth (anisotropy), and that the sizes are not likely to be uniform (polydispersity). Furthermore, any *changes* in effective surface from changes in particle size due to growth, aggregation, or fracture result in changes in reflectivity, intensity-vectors, and polarizations, and in anisotropy and polydispersity; and furthermore, the dimensions of *coarse* anisotropy compared to those of visible EMR may contribute to destructive interference. (Cf. Fig. 8.5.)

4. The factors L, λ', μ, and μ_0 have not been determined as to their precise functions in EMR/*coarse*-particle interactions. Lack of evidence of

significant functions is suggestive that reflective interaction is not sensitive to these factors. (Cf. Fig. 8.4C.)

5. As particle size increases from *small* through *large* into *coarse*, there is a coincident increase in proneness to decrease in concentration of their dispersions through gravitational forces. (Cf. Fig. 8.2.)

While sedimentation is a detrimental factor in some analyses, it is a useful factor in such others as size and size-distribution analyses by light-scattering methods.

Of the foregoing list, the implications of items 3, 4, and 5 indicate that characteristics of *coarse*-particle dispersions are most practically determined by measurements via the transmission angle. To this end, scattering characteristic, expressed as an extinction due to apparent absorption (as absorbance) is the concept preferred for most analyses. Thus, Rose and Lloyd[44] reported upon meaning and application of the formula:

$$\log_{10} \frac{I_0}{I} = KclA_p/2.303 \qquad \text{(cf. Ref. 29)} \qquad (8.29)$$

where K is the total-scattering coefficient, as the ratio: scattering cross-section per geometric cross-section; values are generally near 2, about which it is an oscillating function; c is the concentration of particles (grams per milliliter); l is the length of transmission path; A_p is the projected average particle-area; $A_p = S$ (the so-called Cauchy relation), where S is the *specific surface* of dispersed particles (surface area per unit weight). They applied their formula for the study of a variety of prepared dispersions. These were a series of narrow size distributions. All particles were within the size limits, ca. 2 μ to 60 μ; and were assumed to be spherical. Further reference to the application of their formula is made in Section 8.6C.2.

It is seen that Eq. (8.29) is, in effect, a formula for calculation of *turbidity* and that, within the limits of the dilution principal of Beer's law, it resembles and may be taken to represent an extension of Eq. (8.28), which is based on Mie theory. It is further seen that it retains the analogy with Eqs. (8.10)–(8.13).

8.5 NEPHELOMETRY, cf. TURBIDIMETRY: FOR PRACTICAL ANALYSIS OF DISPERSIONS

Equations (8.10)–(8.13) and (8.26)–(8.29) are considered to be the most useful of the formulas of established theory to coordinate available data for the direct expression of the property of *turbidity* as a function of concentration. With the detector position along the 0° vector, the data sought by measurement basically concerns the transmission ratio (transmittance) I/I_0 when the dispersion is within the limits for accord with the dilution principle of Beer's law.

Nephelometry is a blend of similarities to and differences from turbidimetry. These are expressed within the following items:

1. The principal purpose is the derivation of concentration.
2. The data sought by measurement basically concerns the Tyndall ratio i_{90}/I_0.
3. The detector position is along the 90° vector, against a dark background.
4. The dispersion must be within the limits for accord with the dilution principle of Beer's law.
5. For a given dispersion, the concentration limits for the accord are lower.
6. In general and in the absence of proved data for commenting specifically, its optimal analytic precision derives from dispersions of particles which are intermediate within the broad *small*-to *coarse*-range of sizes.

On the basis of the foregoing comparisons, nephelometry and turbidimetry are not practical alternatives for the analysis of dispersions. The specific reasons are enmeshed with the fact that, for a given dispersion, the data from measurement of the scattered 90° vector is not related to concentration in the same way as is the data from measurement of the 0° vector, which is dependent (by difference) upon *all scattered*-intensity vectors. That is, different segments of the same community of dispersion properties and conditions are the objects of scrutiny by nephelometry and turbidimetry.

The correlation of items 5 and 6 with nephelometry suggests that a short incident wavelength is advantageous when a dispersion of relatively small particles is analyzed. This is because the 90° angle receives a greater proportion of the shorter wavelengths scattered than at lesser angles, and because it is important to derive as much intensity as possible at that angle.

With the support of theory and experiment, Wells[45] gave early (1922) recognition to the advantages of the nephelometric angle over the transmission angle for the analysis of very dilute dispersions, and in fact concluded that the "depth ratios by reflection and by transmission are equal" . . . "in an intermediate range of concentration." Yoe[21] in 1929 stated in introduction: ". . . the nephelometer is limited to the measurement of substances in *low concentration*, usually not stronger than 100 milligrams per liter."

A. HISTORY

Definite pedagogic values emerge from selected reference to the early development of practical light-scattering methods for the analysis of dispersions.

I. General

The review by Wells[20] in 1927 titled "The Present Status of Turbidity," and the book by Yoe[21] in 1929 titled "Nephelometry" are of permanent value for their coverage by text and by bibliography of the classical era of concepts, methods, apparatus, and instruments.

Recognized early during the evolution of these were the many problems inherent in the dispersions themselves, preparatory to their measurement. Prominent among these problems were those which demanded considerations to the particulate properties and dispersion stabilities and reproducibilities; some indication as to the considerations to some of these is given in the reports of Tolman et al.[46-49]

2. The Early Methods

Wells[20] classified the concepts for the photometric measurement of light-scattering dispersions into three main types, which may be described as follows:

a. **Extinction Index.** The extinction index is undoubtedly the oldest concept and is based on empirical theory that there are relationships (though in fact complex* and mathematically unresolved) between the relative turbidity of a dispersion and the thickness of it required to just extinguish visual detection of the source of incident light. The concept is exemplified at one extreme by the more than century-old method in oceanography for determining depths, and at the other extreme by the "Parr Turbidimeter".[50]

b. **Density (Optical).** Like the first concept, that of optical density involves the transmission angle; but unlike the first, it permits concentrations which are sufficiently low so that the dispersion system is in accord with the dilution principle of Beer's law and τ vs. c or A vs. c plot-linearity.

c. **Tyndall Ratio.** The Tyndall ratio involves the nephelometric angle and requires that the dispersion system be in accord with the dilution principle of Beer's law, as previously mentioned. Not previously mentioned are that (1) nephelometry represents an indirect approach, and (2) the requirement of accord is relative, so that near accord may be a first approximation. From these there follow that (1) the indirect approach is enabled by a comparative analysis (correlation of nephelometric data from the dispersion analyzed with that from a prepared standard dispersion of known concentration), and (2) the dilution principle is reexpressed as the inverse proportion principle. This reexpression has the form of an equation which is suitable for comparative analyses:

$$C_u \times l_u = C_s \times l_s \qquad (8.30)$$

* It is expected that secondary-, tertiary-, . . . multiple-scatterings contribute much to the complexity.

where C_u and l_u are the concentrations of, and C_s and l_s are the lengths of paths through the unknown and standard dispersions.

Yoe,[21] in his Chapters 2 and 3, described the evolution of the ten visual-photometric models of nephelometers (involving scale readings) spanning the interval from that of the original of Richards[5] in 1894 to that of Kleinmann in 1927. Of the ten, Kober's[51,52] models (1912, 1917, and 1921) initiated the

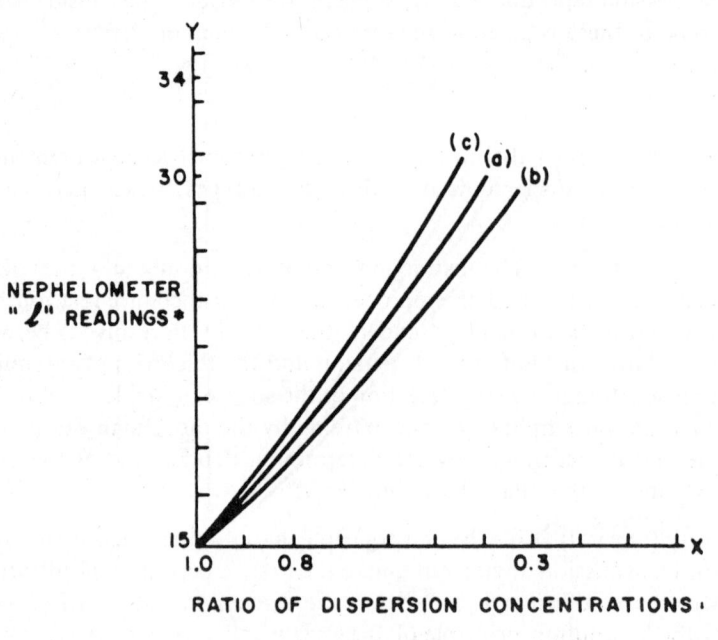

RATIO OF DISPERSION CONCENTRATIONS '

(b) c(dil.)/c(undil.) forms of std.; (c) c(unkn.)/c(std.
(for evaluation of K)

FIGURE 8.6: Nephelometer curves for standardization and analysis. (a) Hypothetical curve: $X = S/Y$; (b) experimental curve from dilutions of std; (c) analytical curve [relative position, left of (a) if K is $-ve$]: $Y = S/X - (1 - X)SK/X^2$. *"l" for the standard is stationary. Adaptation, after Kober.[51,52]

adaptation to nephelometry of the dual-comparator/adjustable-height (l) colorimeter model of Duboscq.

Strictly, accord with Eq. (8.30) by dual dispersions requires the support of optically sound photometry. Kober's nephelometers admittedly had the imperfection that they performed with *near* accord; that is, smooth and nearly straight l vs. c plots were obtained. To calibrate for this, for improved accuracy, he developed theoretical equations:

$$Y = \frac{S}{X} - \frac{(1 - X)SK}{X^2} \tag{8.31}$$

a standard-dilution formula (standard solutions being, relatively: $X = 1.0$, 0.9, 0.8, 0.7, . . .)

$$X = \frac{S + SK - \sqrt{(S + SK)^2 - 4SKY}}{2Y} \tag{8.32}$$

$$K = \frac{X(S - XY)}{S(1 - X)}, \quad K = kS \tag{8.33}$$

where Y is height (scale reading) of unknown, S is height (scale reading, set position) of standard, X is ratio of dispersions (C_u/C_s), K is the "nephelometric constant" (correlates photometric performance for a given unknown with a given standard and its dilutions, and calculable when Y, S, and X are known); k is a related constant. The expression $X = S/Y$, from Eq. (8.31), is equivalent but not equal to Eq. (8.30), though their factors agree in meaning. Figure 8.6 illustrates this, and that related experimental and analytical plots are also nonlinear. The second term of Eq. (8.31) serves to calibrate when Eq. (8.30) does not apply.

Kober's contribution points to some of the early and perpetual difficulties in nephelometry, viz., the controls over stray light, dispersion characteristics, and photometric design which are essentials to linearity of the l vs. c relationship.

The visual-photometric detector of Kleinmann[21] and subsequently the electrophotometric detectors (e.g., Klett-Summerson, Fisher nephluorophotometer) were dual-comparator instruments designed for built-in accord with Eq. (8.30). The greatest disadvantages of visual photometric methods have been the subjective factors (e.g., capability, fatigue) involved in matching or equal light intensities.

3. The Modern Era

Fifty years after Richards[5] introduced the first nephelometer in 1894, Debye's paper[33] of 1944 on "Light Scattering in Solutions" initiated concepts which were to so stimulate subsequent investigations that practical distinctions became possible from analysis of dispersions of most particle sizes. In step with advanced light-scattering theories have been the advanced designs of instruments to test them. Since the 1930's, the increasing availability of photoelectric detector systems and their application to light-scattering methods have left the subjective factors of the analysis of dispersions only in the *preparation* of the dispersions. Within recent years and for a few specialized analyses, automation for the preparation of the dispersions is receiving experimental attention.

B. QUALIFICATIONS OF PREPARED DISPERSIONS FOR PRACTICAL TURBIDIMETRY OR NEPHELOMETRY

The following summarizes the general conditions which may require controls. Underlying all conditions leading to measurement of dispersions are

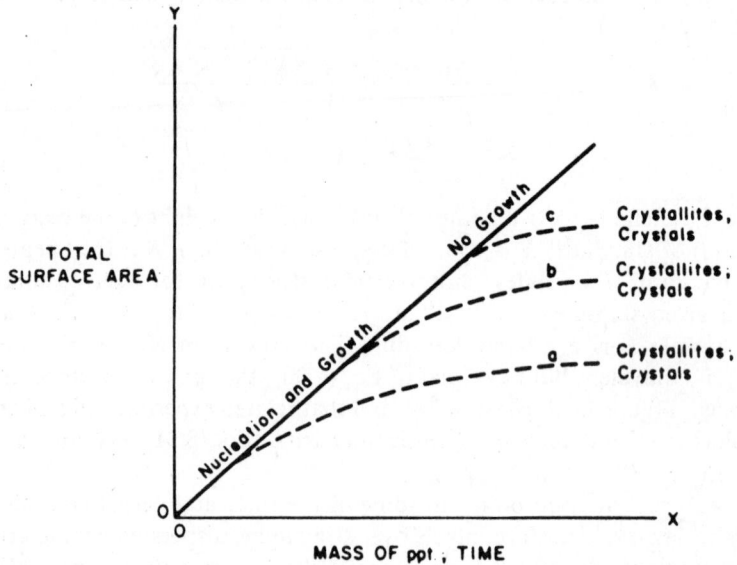

FIGURE 8.7: Nucleation and growths (a,b,c after increasing numbers of nucleates are formed).

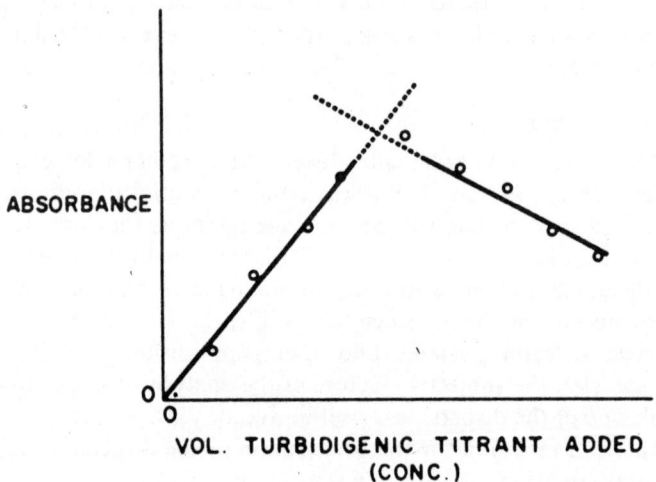

FIGURE 8.8: A turbidimetric titration plot, including extrapolated endpoint.

the environmental factors: time, temperature, interfering coincident substances, ...

1. The dispersed phase:
   a. Is of known identity
   b. When generated by designed chemical reaction: (1) is formed by mixing, in the best order and rate, very dilute solutions of reactants; (2) is formed quantitatively
   c. Is insoluble in the dispersion medium
   d. Is stable as to: (1) identity; (2) particle size and shape
2. The dispersion:
   a. Is stable during the required interval for measurements to be completed; if a stabilizing agent is added, its effect on data must be known
   b. Undergoes no change except in concentration when dilutions are made
   c. Is reproducible
   d. Is free of alien debris and air bubbles

8.6 SPECIAL DISPERSION ANALYSES INVOLVING LIGHT-SCATTERING METHODS

A. TURBIDIMETRIC TITRATIONS

Considerations to the photometry of light-scattering interactions in dispersions of which the suspended particles are by design formed by precipitation are given enhanced meaning by the involved phenomena and mechanisms of nucleation and growth of such particles. Figure 8.7 illustrates a general correlation regarding effective surface area for interaction during the preparation of suspended precipitates in time. The knowledge and, if necessary, control of such phenomena within an analytic procedure can be directed to improved significance of photometric data (cf. Section 8.4).

"Photometric titrations" (originated in concept by Tingle[53] in 1918) is a term with a broad base, including physical-chemical reactions, of stoichiometric importance, of which the end points are obtained from plotted data by measurement of increments in interactions between EMR and a forming dispersion. Susceptible to end-point determinations by this means are such reactions as neutralization, oxidation-reduction, precipitation and complexation. Figure 8.8 illustrates one of the prevalent general types of plots.

Any precipitation reaction whose stoichiometric end point can be designed to appear via a dispersion in which light-scattering interactions are in accord with Beer's law (e.g., $A \propto c$, as a linear plot of apparent absorbance vs. concentration in terms of volumes of standard titrant added) constitutes sound basis for analysis by *turbidimetric* titration. Ringbom[54] initiated this concept in 1941.

The method has significant advantages, principal among which are that:

1. Precise reproducibility of the dispersion is not a critical matter to quantitative validity; this is so:
   a. As long as the dispersed particles remain sufficiently *fine* for stable suspension and for minimum reflectivity
   b. As long as the dispersion system, though changing, maintains its accord with Beer's law
   c. As long as the lines that are plotted to intersect for end point are sufficiently different in slope
2. Though the experimental points in the region of the end point may not define the point, the definition is obtainable by extrapolating the lines to their point of intersection
3. A standard dispersion is not required for reference

Two other matters warrant comment: the plot of different slope, after the end point, represents decrease in apparent absorbance due to dilution of the dispersion by the added volume of unreacting titrant; also, since stirring is an important factor in titrimetry, its effect upon nucleation and growth of particles may indicate it to be a factor requiring control.

B. MICROORGANISMS AS DISPERSED PARTICLES

The measurement of turbidities, which can be related quantitatively to concentration of microorganisms, has evolved, since initiated by McFarland[55] in 1907, to become a substantial field of applied photometry. Turbidity and concentration so related depend upon the absence of immeasurable by-products affecting the dispersion.

Dispersions of microorganisms have many characteristics that are sufficiently different from those of other dispersion systems to warrant description of important features and implications. These are enumerated as follows:

1. The most favorable dispersions for photometric analysis are of those microorganisms whose longest cell (particle) dimensions are within the approximate 1 to 2 μ range, the upper limit of the *large*-particle range. Microorganisms of this range are notable in that they include species which are particulately isotropic (the cocci) and anisotropic (the bacilli); however, cells of a single species may not be of uniform size, even in a pure culture.

Also used are dispersions of a considerable variety of other microorganisms, sizes, and shapes; for, in general, any organism which will give adequately uniform dispersion throughout the steps of a procedure may be used with analytic purpose. Bacteria, protozoa, fungi, and yeasts have been engaged.

2. Because of the *living* property of microorganisms and of all which this embraces, it is to be expected that their dispersions differ profoundly from dispersions of particles without this property. From the standpoint of the

"living" dispersions, their differences appear to be in regard to reasons and mechanisms for changes in particle size, shape, and associations, responses to other materials and their "stoichiometric" relationships, mobility, interaction with EMR, etc.

Furthermore, inasmuch as the environmental factors (such as time, temperature) have special significance to "living" dispersions, their controls within analytic procedures are critical.

3. In the perspective of light-scattering methods, the practical purposes for prepared dispersions of microorganisms are the assay of agents (vitamins and other essential biochemicals; antibiotics) to which specific microorganisms show demonstrable vital response (favorable; unfavorable). The determination of response is by measurement of change in turbidity in time, translated to terms of potency units or chemical equivalents of agent. An essential requirement of the methods is that the analytic procedure is parallel to and its data are comparable with procedure and data based upon a reference standard agent having statutory or other conventional directives for this purpose. The procedures are various and specific.

The foregoing distinctions and implications indicate that light-scattering phenomena in "living" dispersions probably are more complex and more fraught with unknown factors than are the dispersions of lower order.

Their practical turbidimetry assumes their accords with basic concepts (q.v. Section 8.3C.1) by use of the following equations for representing relative turbidity:

$$OD = 2 - \log G \qquad (8.34)$$

(in "optical density" units as indicated by galvanometer readings) and,

$$A = 2 - \log\% T \qquad (8.35)$$

(in *apparent* spectrometric units, as by absorbance through per cent transmittance), and

$$D = sC \qquad \text{(Ref. 56, p. 143)} \qquad (8.36)$$

(where D is optical density, s is a constant (including dispersion thickness l), and c is number of particles per milliliter). Figure 8.9 gives the general meaning of growth-response curves as determined by photometry via the transmission angle.

The practical nephelometry of such dispersions assumes their accord with the concepts in Eq. (8.30), the inverse proportion principle. The indirectness of nephelometry as a method for dispersion analysis dictates the need for nephelometric standards, a series of temporary, semipermanent or permanent comparable dispersions of graded concentrations. The standards enable *relative* dispersion-concentration to be expressed in terms of nephelometer-scale readings. Permanent nephelometric standards (e.g., Coleman Nephelos system), made possible with durable components, are used extensively.

As implied in Section 8.5, the concentration of microorganisms in a given dispersion determines which of the transmission and nephelometric angles is selected for deriving the photometric data.

Analyses via growth-response curves derived turbidimetrically (Fig. 8.9) or nephelometrically evoke considerations to possible differentiation between chemical and biological stoichiometries. The relationship between growth response and dose is seen to be linear from the origin for growth-promoting agents, but to be nonlinear for growth-inhibiting agents; that is, there is less than complete reciprocal relationship between the two types of responses.

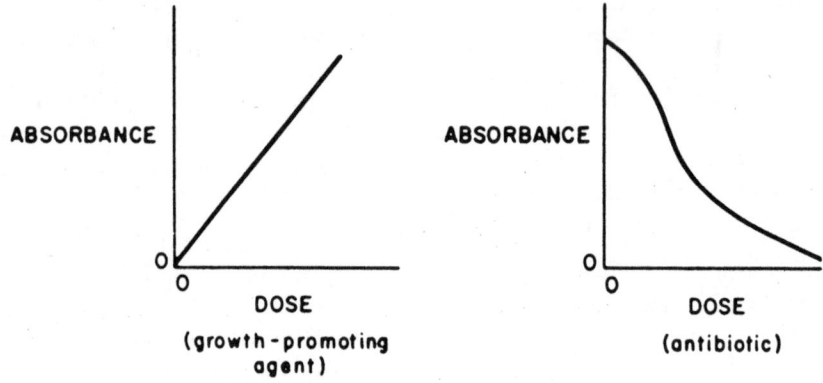

FIGURE 8.9: Growth-response curves.

As the response to a growth-promoting agent, it would appear, each chemical equivalent (and number of molecules) of it results in a fixed increase in the number of cells, whereas the response to a growth-inhibiting agent is less regular because of some growth, some inhibition, and some annihilation. In practice, annihilation is not sought or is not attainable, and may be undesirable because of the unknown factors this condition may introduce.

C. "UNSTABLE" DISPERSIONS

Reference here is to two types of dispersions which individually present special problems for their analysis by light-scattering methods. Their inclusion is for their principles and because they have received considerable study and are certain to receive a great deal of future attention because of the technologies they invite.

I. Aerosols (Literally: Solutions or Suspensions in Air)*

Aerosols are *fine*-particle dispersions of liquid or solid matter in gas media. The apparent examples in nature are within what are accepted to be "atmospheric conditions"—clouds, mist and fog, dust, smog and smoke—all of which

* It is also defined, in modern technology, as a "pressurized package."

are of varying position and concentration because they are relatively free (uncontained). In laboratory studies, the containment necessary ("cloud chamber") for natural samples and for prepared samples is alone sufficient to render the best of such samples poor imitations.

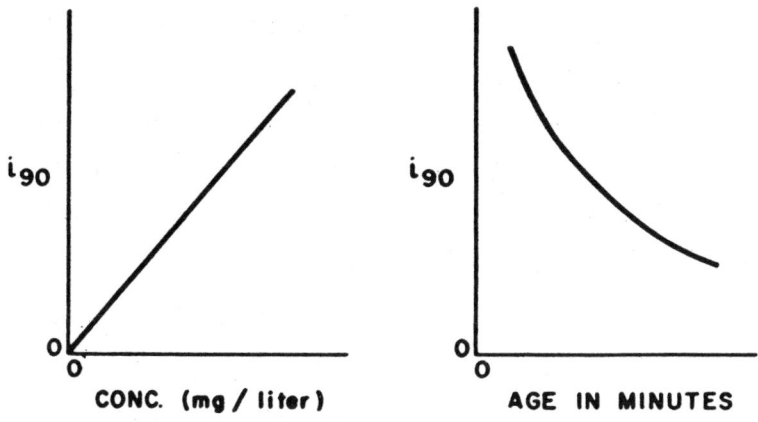

FIGURE 8.10: Relationships with tyndallmeter readings (i_{90}) in smoke aerosols.

Particle size and concentration contribute the most critical factors to the practical "aerosol problem," and data concerning them are at the heart of solutions to the atmospheric pollution problem. In fact, it was for the latter purpose (in the context of chemical warfare science) that the basic work (apparatus, theory, experiments) of Tolman et al.[46-49] in 1919 was motivated.

They submitted as limiting forms for the relation between tyndallmeter reading (T, intensity in footcandles) and size of particles: when concentration $\propto$ number of particles $\times$ diameter cubed,

$$T = knd^6 \quad \text{and} \quad T = kcd^3 \tag{8.37}$$

(according to Rayleigh law for *small* particles); and

$$T = k'nd'^2 \quad \text{and} \quad T = k'c/d' \tag{8.38}$$

(for larger particles, when reflecting area $\propto d'^2$) where k and k' are constants, n is the number of particles per cubic centimeter, and d and d' are particle diameters (determined microscopically). To test the theory they used their designed tyndallmeter, smokes of rosin, tobacco and NH_4Cl (for *small* particles), and liquid suspension of silica (for larger particles). They derived data for the types of correlations illustrated by Fig. 8.10. The linear relationship in the first graph is in accord with the dilution principle (i.e., tyndallmeter intensity $i_{90} \propto c$). The curve in the second illustration is of the effect on i_{90}

of progressive aggregation, settling and adherence to the walls of the container by the particles as a smoke aerosol disappears in time. Furthermore, several distinctions were derived: within certain concentration limits and for the same mass-concentration, the smaller the particle size, the higher the i_{90} because of larger total reflecting surface; for the same mass-concentration, the smaller the particle size the greater the aggregating and aerosol disappearing rates; for different mass-concentrations of the same particle size, the higher the concentration the greater the aerosol disappearing rate.

Subsequently, spectrometric-type equations were introduced for the evaluation of aerosols. For example, Gumprecht and Sliepcevich[57] in 1953 combined corrected forms of such equations with Stokes' law of settling for the study of aerosol polydispersions of kerosene. Thus, the corrected form of Eq. (8.27) is

$$I = I_0 e^{-RK\pi r^2 n l} \qquad (8.39)$$

where R is a correction factor for K, for the subtended angle. This, in the integrating form:

$$\log_n \frac{I}{I_0} = -\pi l \int_0^\infty RKr^2 n \, dr \qquad (8.40)$$

is for a summation based on the concept that a polydispersion is composed of monodispersions or narrow polydispersions.

Orr and Dallavalle[58] have detailed the involvements in the concept and the use of these equations for size-distribution analysis, with $I \propto$ settling intervals for n and r.

It is noteworthy that photoelectric methods for counting number of particles and oscillatory amplitude methods for measuring particle size in aerosols are dependent upon light-scattering interactions.

2. Coarse-Particle Dispersions for the Technology of Powders

In reference to their work with Eq. (8.29), Rose and Lloyd[44] stated: "The mathematical laws of light transmission through a suspension are exact," and to the effect that deviations are due to some transparency of particles and to apparatus imperfections and inadequate experimental design. Thus, if the apparatus can subtend a very small solid angle at the center of l, there is a linear relationship between the attenuation of I, i.e., log I/I_0, and c, even for dispersions of particles of μ sizes.

They listed the following assumptions for ideality in the derivation of their equation:

a. The particles are completely opaque
b. The amount of I_0 reflected from transmissibility is proportional to the total projected area of the particles
c. There are no multiple reflections between particles and between particles and container surfaces

d. l and c are not so great that more than one particle at one time is in a single line of the transmission path

For size-distribution analysis, further derivations gave:

a. That for all sizes (d_x), Eq. (8.29) takes the form

$$\log_n \frac{I_0}{I} = kCl \sum_0^{d_x} K_x n \, d_x^2 \quad \text{(Ref. 44)} \tag{8.41}$$

b. That for a d_1 to d_2 size distribution, Eq. (8.41) takes the form

$$\log_n \frac{I_0}{I_2} - \log_n \frac{I_0}{I_1} = kCl \sum_{d_1}^{d_2} K_x n \, d_x^2 \quad \text{(Ref. 59)} \tag{8.42}$$

where

$k = $ a particle-shape factor
$K = $ the opacity factor, a proportionality constant of transmission data
$C = $ concentration of powder
$n = $ number of particles per gram of powder
$d = $ particle diameter
and where particles are assumed to be spherical.

Equations (8.41) and (8.42) and other published equations of related meanings to particle-size distribution are a basis of the technology of powders for product-quality control in a variety of industries (e.g., pharmaceutic, cosmetic, ceramic). In this perspective, the light-scattering method which is termed "photoextinction" and "photosedimentation"[28,44,59] is prominent, and is of pedagogic value.

D. ANALYSES FOR BIOCHEMICALS

The chemical entities in biological systems are, in general terms, of the following types: micromolecular to macromolecular in range; inorganic or organic; animal, plant, or microorganism as to host.

As is to be expected when any such constituent is the object of analysis by light-scattering methods, its appropriate separation from its biological medium and its transposition into a prepared dispersion medium can be fraught with difficulties which are numerous and technical. To the extent that all difficulties cannot be overcome, which is characteristic of biological systems, resort within procedure is prevalent to use of: replicate sampling, reference standards, control or blank determinations, and null reading. Though analytic results from these resorts are relative, refinements minimize error from any source.

One of the most important problems involved in biochemical analyses is the selection of its most appropriate stoichiometric derivative to represent it in the prepared dispersion, for the derivative must have acceptable formative and particulate as well as insoluble characteristics.

The analysis of enzymes is, as in microbiological growth-promoting or inhibiting methods (Section 8.7B), based upon the measurement of response to progressive action on a suitable substrate. Depending upon the solubility or insolubility of the substrate or products of action, the response is measured as an increased or decreased turbidity with reference to the turbidity at the beginning of action. Time and temperature are important environmental factors.

Biological systems are neither constants nor are they comfortably functional outside of the limits set for them by nature. They are susceptible to abnormalizing conditions which may be recognizable through changes in amounts or identities of biochemicals essential to physiological balance. The cause/effect relationships underlying such conditions and changes are much more obscure than clear.

In the human area of clinical diagnosis of diseased conditions, the scope of biochemical analysis includes considerations to environment, nutrition, microorganisms, drugs, and poisons as well as to the inherent factors of the cause/effect relationships.

E. ANALYSES FOR IMPURITIES IN STANDARD-GRADE CHEMICALS

Near-absolute purity of chemicals may be impractical to achieve or be unnecessary to have. The extent to which impurities may be "overlooked" or be allowed to remain depends upon method of preparation, refinement, and purposes. For the more particular purposes, the labeling of production-control results on containers of analytical and reagent-grade chemicals and the specifications for conformity and limit tests in drug compendia are standard practices.

Analytic procedures, in accord with the principles of turbidimetry and nephelometry (Section 8.5), are available or may be designed for determining *exact* amounts of impurities; all of them require selection of a suitable insoluble derivative in a suitable dispersion of it. When the analytic purpose is merely to determine conformity of a chemical with specified *limit* for amount of impurity, as for Cl^- and SO_4^{2-} in the official compendia, instrumental[31,32] or visual[351] methods for comparing dispersions are used.

8.7 COMMERCIAL INSTRUMENTS

Since Kober's[51] first adaptation of a colorimeter (Duboscq) for the analysis of suspensions, the apparatus and instruments for light-scattering measurements have been those of colorimetry and spectrophotometry with adaptations when necessary. Furthermore, fluorometers are being used for nephelometry since the photometric angle is the same.

These fortunate overlaps, the abundance of manufacturers and suppliers, catalogs and bulletins, diagrams of optical systems and descriptions of accessory apparatus, operating, care, and service instructions, and the rate of obsolescence make textual inclusions about these unnecessary; the following outline seems sufficient here.

1. *Models using visual photometry:* Parr turbidimeter,[50] extinction principle; Jackson turbidimeter, optical-density principle; Hellige turbidimeter,[71] combines transmission and Tyndall principles; St. Louis turbidimeter, Tyndall principle with Nessler tubes.

2. *Models using electrophotometry:* Coleman nepho-colorimeter (and certified Nephelos standards), for nephelometry and turbidimetry; Klett-Summerson photoelectric colorimeter, for turbidimetry (micro-macro); Bausch & Lomb Spectronic-20 colorimeter, for turbidimetry; spectrophotometers; fluorometers.

3. *Models of extraordinary precision and applications:* Aminco-Chance dual-wavelength spectrophotometer, for recording small changes in optical transmission through turbid media; Hewlett-Packard light-scattering photometer and Brice-Phoenix light-scattering photometer and Aminco light-scattering photometer, for angular scattering, dissymmetry and depolarization and Rayleigh ratios, absolute turbidity, M.

When refractive-index values are required data, the following instrumental models for their determination are used with characteristic precisions: Pulfrich, Abbé, dipping and differential.

Periodical information about optical instruments is published in: *Instruments and Control Systems* (and its *Buyers' Guide* to manufacturers), *Review of Scientific Instruments*, *Journal of Scientific Instruments*, *Applied Optics*, *Journal of the Optical Society of America*.

8.8 EXPERIMENTAL PROCEDURE

DETERMINATION OF CHLORIDE (Cl⁻) AND SULFATE (SO₄²⁻) LIMITS

DETERMINATION OF CHLORIDE (Cl^-) AND SULFATE (SO_4^{2-}) LIMITS

(Cf. References USP, XVII, p. 870; NF, XII, p. 438; BP, 1963, p. 1052).

Equipment. 1. Individually labeled "dispensing" burettes are convenient if set up for community sources of approximate and exact volumes of specified solutions and reagents.

2. 50-ml volumetric flasks; three per determination.

3. Spectronic-20 colorimeter; specified "test tubes".

Notes. 1. The instrument is conveniently preset if it is to be used in community; thence, the adjustment knobs need not be touched except to check the adjustment periodically. With the instrument set for frequent readings, a null reading made according to:

a. The wavelength (ca. 350 mμ) for maximum absorbance

Cl⁻ AND SO₄²⁻ IN CALCIUM GLUCONATE

Procedure and Data: (Date:

Locker No: Sample No: Surname (print):

Cl^- Determination
Standard Turbidity

1. Measure 1.00 ml of 0.02 N HCl into a clean[a] 50-ml volumetric flask
2. Add 35 ml distilled water
3. Add 1 ml concentrated HNO_3
4. Add 1 ml $AgNO_3$-T.S.[b]
5. Add distilled water, q.s. to 50.00 ml
6. Stopper and invert several times[c]
7. Allow to stand for 5 min, protected from direct sunlight
8. Fill a Spectronic-20 test tube to the mark with the suspension
9. Wipe test tube clean, insert, read

$$A_{std}^{Cl^-} =$$

Sample Turbidity

1–2. Weigh about 1 g of sample exactly (Wt. =   ) and transfer via solutions in 35 ml distilled water into a clean[a] 50-ml volumetric flask.

3–9. As for Standard.

$$A_{sample}^{Cl^-} =$$

Calculations

USP Limit. A 1-g sample of calcium gluconate shows no more Cl⁻ than corresponds to 1 ml of 0.02 N HCl (ca. 700 ppm).

$$\text{ionic wt. } Cl^- = 35.453$$

$$\frac{0.02}{1000} \times 35.453 = 0.00070906 \text{ g } Cl^- \text{ in}$$

1.0000 g calcium gluconate is 709.06 ppm

$$Wt._{sample}^{Cl^-} = \frac{A_{sample}^{Cl^-}}{A_{std}^{Cl^-}} \times Wt._{std}^{Cl^-}$$

$$= \frac{(\qquad)}{(\qquad)} \times (\qquad)$$

$$= (\qquad) \text{ g}$$

$$\frac{(\qquad \text{g})}{Wt. \text{ sample} (\qquad \text{g})} \text{ represents} \qquad {}^{d} \text{ ppm}$$

SO_4^{2-} Determination
Standard Turbidity

1. Measure 1.00 ml of 0.02 N H_2SO_4 into a clean[a] 50-ml volumetric flask
2. Add 35 ml distilled water
3. Add 1 ml dilute HCl
4. Add 3 ml $BaCl_2$-T.S.[b]
5. Add distilled water, q.s. to 50.00 ml
6. Stopper and invert several times[c]
7. Allow to stand for 10 min
8. Fill a Spectronic-20 test tube to the mark with the suspension
9. Wipe test tube clean, insert, read

$$A_{std}^{SO_4^{2-}} =$$

Sample Turbidity

1–2. Weigh about 2 g of sample exactly (Wt. =   ) and transfer via solution in 35 ml distilled water into a clean[a] 50-ml volumetric flask

3–9. As for Standard.

$$A_{sample}^{SO_4^{2-}} =$$

Calculations

USP Limit. A 2-g sample of calcium gluconate shows no more SO_4^{2-} than corresponds to 1 ml of 0.02 N H_2SO_4 (ca. 500 ppm).

$$\text{ionic wt. } SO_4^{2-} = 96.0616$$

$$\frac{0.02}{1000} \times \frac{96.0616}{2} = 0.000960616 \text{ g } SO_4^{2-}$$

in 2.0000 g calcium gluconate is 480.308 ppm

$$Wt._{sample}^{SO_4^{2-}} = \frac{A_{sample}^{SO_4^{2-}}}{A_{std}^{SO_4^{2-}}} \times Wt._{std}^{SO_4^{2-}}$$

$$= \frac{(\qquad)}{(\qquad)} \times (\qquad)$$

$$= (\qquad) \text{ g}$$

$$\frac{(\qquad \text{g})}{Wt. \text{ sample} (\qquad \text{g})} \text{ represents} \qquad {}^{d} \text{ ppm}$$

---

[a] May be wet.
[b] Omitted in blank.
[c] Do not agitate vigorously.
[d] For undetectable absorbance reading (A_{sample}) report undetectable *trace*. As an added exercise, plot correlation (A vs. t in minutes) as "standard turbidity" tubes are saved for extended intervals.

Signature:

b. 100% transmittance for the *respective blanks*; i.e., the *blanks* are prepared exactly as for "standard" turbidities, except that they contain distilled water in place of the volumes of precipitant specified in step 4. Thus, these respective *blanks* represent zero absorbance and no turbidity, relatively.

2. The readings for standard and sample turbidities are made from the absorbances scale (as *A*, for apparent absorbance, turbidity) so as to make applicable the formula corresponding to the colorimetric function of the instrument. It should be emphasized, however, that a reading must be made without undue delay after the specified waiting period (step 7), since suspended particles may eventually aggregate or/and settle to change the status or/and the homogeneity of the turbidity.

3. Significance of results (e.g., accuracy, reproducibility) can be improved by any means that can effect uniformity in size and stability of suspended particles.

4. It is axiomatic that solvent and diluent water used in the procedure must be *turbidimetrically blank* as to suspended particles and as to common and interfering ions.

QUESTIONS

Q8.1. Cite the types of interactions possibly involved upon encounter of EMR with matter. In a general way, diagram the meaning in principle of each type.

Q8.2. What are the principal correlations between particle sizes and types of light-scattering interactions, assuming sphericity for all of the sizes?

Q8.3. Cite the qualifications within the ideal case for the theoretical study of fine-particle dispersions.

Q8.4. List the symbols for the principal factors and terms of the scattering theories for *fine*-particle dispersions. Explain the role of each factor.

Q8.5. Title and describe the complications due to deviations from the ideal case (Q8.3).

Q8.6. What special characteristics do *coarse*-particle dispersions have which set them apart from *fine*-particle dispersions as to analytic approaches to them?

Q8.7. Describe practical turbidimetry and nephelometry, wherein they are similar and wherein they are different.

Q8.8. Cite the qualifications of prepared dispersions for practical turbidimetry and nephelometry.

Q8.9. Speculate as to factors most difficult to control so as to minimize margins of error in the preparation of dispersions for instrumental analysis.

Q8.10. Table 8.3 deals with cubed subdivision. Speculate, and if possible solve, an analogous problem dealing with sphered subdivision (beginning with a diameter of 1 cm).

Q8.11. Reasoning from the substance of the exemplary experiment in Section 8.8, and citing comparable quantities of its appropriate materials for use, design an experiment (and enumerate the steps of its procedure) which will yield data that turbid dispersions are in relative accord/discord with the principles of Beer's law.

APPENDIX: TOPICS INDEX TO REFERENCES AND EXTENDED BIBLIOGRAPHY RELATING TO DISPERSIONS AND LIGHT-SCATTERING PRINCIPLES[a]

General discussions: 6, 18, 20, 21, 26, 29, 81, 125, 195, 236, 247, 278, 298, 299, 300, 352, 353, 394, 418

Colloids, general: 6, 195, 278

Book coverages: 6, 21, 58, 59, 63, 82, 125, 195, 236, 247, 261, 277, 278, 352, 370, 394, 418, 442, 470

Reviews: 18, 19, 20, 21, 26, 29, 30, 56, 58, 72, 75, 80, 81, 99, 134, 140, 141, 142, 151, 153, 154, 197, 198, 201, 247, 262, 299, 303, 304, 344, 352, 353, 361, 372, 416, 421, 436, 440, 443, 444, 448, 449, 451, 470, 482, 484, 491, 493

Symposia: 308, 370, 437

Theoretical discussions: 1, 2, 11, 12, 13, 14, 15, 16, 17, 19, 22, 23, 24, 25, 26, 27, 30, 33, 38, 39, 40, 41, 42, 43, 44, 45, 48, 51, 52, 54, 57, 63, 82, 85, 95, 99, 100, 103, 111, 130, 135, 136, 146, 147, 168, 170, 172, 192, 205, 208, 231, 234, 247, 252, 253, 254, 262, 268, 272, 278, 284, 285, 286, 308, 309, 311, 316, 321, 323, 325, 326, 327, 328, 334, 336, 340, 341, 342, 344, 347, 359, 364, 370, 378, 379, 380, 384, 408, 414, 417, 426, 428, 436, 439, 442, 446, 455, 456, 458, 467, 469, 470, 477, 478, 481, 484, 485, 491, 493, 495, 496

Fluctuation theory: 18, 23, 26, 30, 34, 35, 36, 37, 38, 39, 40, 41, 42, 43, 110, 111, 206, 235, 278, 376, 377, 389, 392, 409, 410, 471, 476, 487, 495 496,

K (scattering coeff.):   121, 170, 180, 183, 192, 261[b], 277[b], 288, 312, 316, 323, 336, 362, 373, 403, 461, 484, 486

Absolute turbidity (τ): 82, 101, 161, 178, 179, 182, 189, 190, 191, 195, 202, 204, 212, 227, 242, 244, 299, 425

R_θ (Rayleigh ratio): 19, 43, 150, 158, 161, 178, 179, 180, 184, 189, 190, 191, 194, 211, 212, 227, 229, 242, 244, 271, 299, 334, 344, 364, 377, 385, 409, 410, 425, 428, 433, 456, 465, 475, 477, 492, 495

π (Osmotic pressure): 6, 38, 123, 127, 135, 149, 169, 205, 207, 228, 245, 272, 278, 336, 348, 422, 424, 468, 474, 479, 485

η (Viscosity):   6, 123, 131, 143, 150, 173, 207, 235, 238, 258, 272, 281, 282, 291, 321, 348, 349, 350, 360, 406, 417, 466, 468, 475, 479, 490

m, μ (Refractive index): 11, 12, 22, 23, 38, 84, 101, 102, 110, 111, 147, 149, 150, 160, 171, 173, 208, 213, 214, 226, 252, 253, 268, 273, 278, 281, 284, 287, 309, 316, 318, 323, 325, 327, 336, 340, 343, 347, 356, 360, 379, 380, 387, 388, 415, 426, 461, 463, 475, 480, 482, 485, 486, 487, 495

    [a] Many of the citations are shown assignable after more than one topic.   [*] (asterisked) citations are to official compendia: U.S.P., N.F. and B.P. V within citations is for specifying volume numbers.
    [b] Tables correlating Mie-scattering functions.

Oligomer: 414

Sols of:

Polystyrene: 38, 41, 43, 95, 104, 105, 131, 137, 158, 161, 171, 200, 203, 211, 214, 233, 257, 269, 274, 288, 330, 343, 359, 362, 370, 388, 427, 439, 464, 465, 477, 480, 485, 488, 491, 492, 493, 496

Silica: 47, 48, 49, 144, 173, 179, 182, 183, 211, 212, 244, 332, 333

Silver halides: 122, 297, 312, 316, 331, 335, 373, 375, 412, 435, 459, 486, 489

Starch: 483

Sulfur: 106, 107, 114, 119, 120, 133, 134, 157, 180, 430

Aerosols: 6, 46, 47, 48, 49, 57, 60, 113, 116, 132, 140, 141, 142, 148, 152, 153, 162, 219, 308, 394, 412, 418, 444, 452, 457, 458, 459, 460, 463

Colored sols: 296

Nucleation and related
growth: 20, 122, 133, 187, 199, 223, 230, 248, 276, 280, 297, 301, 302V3, 306, 314, 335, 367, 373, 375, 391, 435, 441, 459, 477, 489

Turbidimetric titrations: 53, 54, 155, 197, 201, 237, 280, 301, 302V3, 307, 345, 365, 373, 421, 439, 445, 464

Aggregation, coagulation,
flocculation: 6, 30, 85, 119, 123, 129, 144, 173, 174, 191, 205, 216, 260, 262, 304, 332, 333, 358, 429, 430, 435, 445, 447, 458, 460, 489, 490

Binding, adsorption,
replication: 262, 287, 318, 319, 320, 321, 322, 377, 426, 445, 464, 483, 485

Micelles: 129, 149, 174, 191, 205, 210, 229, 256, 258, 271, 289, 290, 371, 389, 433

Emulsions: 6, 75, 323, 484

Surface-active agents,
quaternaries, detergents: 77, 124, 129, 149, 174, 191, 205, 229, 256, 258, 263, 271, 280, 302, 313, 371, 389, 433, 438, 446, 471

Anisotropy: 18, 26, 27, 36, 63, 82, 99, 101, 110, 115, 130, 131, 135, 136, 143, 145, 146, 150, 151, 168, 169, 172, 174, 191, 209, 225, 244, 259, 262, 272, 292, 348, 350, 370, 410, 417, 428, 456, 467, 469, 477, 496

Angular scattering: 11, 18, 43, 101, 130, 131, 135, 143, 150, 151, 158, 230, 231, 232, 272, 284, 285, 309, 321, 330, 357, 370, 374, 378, 385, 388, 410, 413, 417, 425, 426, 431, 437, 458, 465, 471, 486, 490, 492, 494, 496

Dissymmetry ratio,
$i_\theta/i_{180^\circ - \theta}$:   18, 26, 101, 123, 135, 143, 151, 188, 191, 202, 225, 244, 278, 282, 312, 330, 370, 411, 423, 425

ᵃ Limit tests in official compendia.

Quantitative analysis for
biochemicals (procedures
and/or references to
procedures):

Acetone: 125V3; *Alkaloids:* 125V4; *Amino acids:* 56, 267, 303; *Ammonia:* 125V2; *Amylase:* 125V4; *Antibiotics:* 31, 56, 83, 92, 93, 97, 125V4, 159, 167, 303, 393, 396, 398, 454; *Caffeine:* 125V4; *Calcium:* 125V2; *Carbohydrates:* 222, 240, 313, 368; *Carbon dioxide:* 125V2; *Casein:* 21, 52; *Cathepsin:* 125V4; *Chloride:* 125V2; *Cholesterol:* 125V4, 181, 249, 315, 346; *Diastase:* 125V4; *Enzymes:* 21, 51, 125V4, 238, 265, 275, 293, 382, 402; *Essential oils:* 21, 126; *Fats and oils:* 125V3; *Fibrinogen:* 125V4; *Folic acid:* 125V4; *Globulin:* 125V4; *Glycogen:* 125V3, 222; *Guanidine:* 125V4; *Heparin:* 294; *Hyaluronic acid:* 313; *Hydrogen cyanide:* 70, 177; *Inositol:* 125V3; *Lecithin:* 125V4; *Lipids:* 21, 125V3, 279; *Magnesium:* 125V2; *Morphine:* 125V4; *Mucopolysaccharides:* 240, 368; *Nicotine:* 73, 78, 125V4, 196, 221; *Nucleic acids:* 21, 238, 400; *Penicillin:* 56, 125V4; *Pepsin:* 125V4; *Polysaccharides:* 222; *Potassium:* 125V2, 302V1; *Proteins:* 21, 52, 125V4, 163, 185, 186, 196, 338, 369, 400; *Purines:* 21; *Quinine:* 125V4; *Sodium:* 125V2; *Sulfate (biol.):* 61, 79, 125V2, 125V4, 164, 187, 305, 368; *Sulfide (biol.):* 125V2; *"Sulfur" (biol.):* 61, 67, 125V2, 187; *Terramycin:* 56, 125V4; *Trypsin:* 125V4; *Vitamins:* 31*, 56, 74, 87, 90, 94, 117, 118, 125V3, 128, 154, 165, 239, 303, 384, 390, 393

Dispersion analysis for
microorganisms:

55, 72, 86, 110, 115, 119, 137, 142, 184, 215, 216, 241, 244, 245, 246, 295, 337, 339, 355, 397, 399, 400, 401, 402, 438, 452

Thymol turbidity analysis: 89, 96, 186, 337

Parenterals: 250, 363, 404, 419, 448, 449, 450, 451, 453

Apparatus: 5, 28b, 44, 46, 58, 59, 131, 139, 140, 141, 142, 148, 150, 156, 166, 189, 270, 271, 278, 299, 300, 357, 364, 366, 386, 388, 405, 416, 443, 444, 452, 453, 460, 493

Instruments: 20, 21, 45, 46, 50, 51, 52, 53, 55, 62, 65, 66, 70, 71, 72, 76, 80, 81, 88, 101, 102, 112, 113, 116, 125V1, 140, 152, 155, 166, 179, 187, 212, 215, 220, 255, 259, 264, 299, 300, 323, 345, 357, 361, 372, 383, 386, 399, 421, 444, 475, 482, 484, 491, 492, 494, 496

Automation: 295, 345, 355, 396, 397, 398, 399, 400, 401, 402, 434, 452

Miscellaneous topics and
studies (arbitrary order
and simplified
classification):

Inverse proportion principle: 21, 45, 51, 52, 81, 125, 195
Rayleigh law limits: 347, 408
Opalescence: 6, 34, 35, 432, 487
Cloud-point curves: 491
Dissolution-temperature curves: 387

Haze by water in hydrocarbons: 432
Turbidity indicators: 395
Precipitating agents: 263
Kinetics: 34, 133, 199, 223, 246, 271, 335, 358, 435, 489
Thermodynamics: 23, 392, 409, 416, 427, 487
High-temperature effects: 383
Gel structure: 260, 469
Water structure: 377
Liquid interfaces: 471
Surface tension: 471
Particles in electric field: 467
pH: 216, 358
Binary mixtures: 392, 432, 487
Chelates and complexes: 251, 296, 302
Agitation and stirring effects: 341, 388
Polarimetry of turbid dispersions: 64, 354
Spectrophotometry of turbid dispersions: 175, 288, 296, 403, 420
(cf. Topics: H.O.T.S; colored, absorbing and fluorescing disps.)
Photo-extinction/-sedimentation methods: 28b, 44, 57, 59, 443
Metabolic probings: 246, 339, 454
Chloroplasts: 407
Mitochondria: 454
Phosphorylation: 407

REFERENCES

1. J. B. Richter, *Über die Neuern Gegenstande der Chemie*, **11,** 81 (1802); through J. R. Partington, *Ann. Sci.*, **7,** 196 (1951).
2. J. Tyndall, *Phil. Mag.*, **37,** 384 (1869); *Proc. Roy. Soc. (London)*, **17,** 223 (1869).
3. H. J. Abrahams and A. Dubner, *J. Chem. Educ.*, **20,** 61 (1943).
4. P. B. Ackerson, *J. Chem. Educ.*, **36,** A309 (1959).
5. T. W. Richards, *Proc. Am. Acad. Arts Sci.*, **30,** 385 (1894).
6. J. W. McBain, *Colloid Science*, Heath, Boston, 1950, p. 91, etc.
7. Spectrometry Nomenclature, *Anal. Chem.* (reprinted at end of each volume).
8. H. G. Pfeiffer and H. A. Liebhafsky, *J. Chem. Educ.*, **28,** 123 (1951).
9. D. R. Malinin and J. H. Yoe, *J. Chem. Educ.*, **38,** 129 (1961).
10. F. C. Strong, *Anal. Chem.*, **24,** 338, 2013 (1952).
11. H. Blumer, *Z. Physik*, **32,** 119 (1925).
12. R. Gans, *Ann. Physik*, **76,** 29 (1925).
13. Lord Rayleigh (J. W. Strutt), *Phil. Mag.*, **41,** 107 (1871).
14. Lord Rayleigh, *Phil. Mag.*, **41,** 274 (1871).
15. Lord Rayleigh, *Phil. Mag.*, **41,** 447 (1871).
16. Lord Rayleigh, *Phil. Mag.*, **12,** 81 (1881).
17. Lord Rayleigh, *Phil. Mag.*, **47,** 375 (1899).
18. G. Oster, *Chem. Rev.*, **43,** 319 (1948).
19. P. Doty and J. T. Edsall, *Advan. Protein Chem.*, **6,** 35 (1951).
20. P. V. Wells, *Chem. Rev.*, **3,** 331 (1927).
21. J. H. Yoe, *Photometric Chemical Analysis*, Vol. II, Wiley, New York, 1929.
22. Lord Rayleigh, *Proc. Roy. Soc. (London)*, **A90,** 219 (1914).
23. B. H. Zimm, R. S. Stein, and P. M. Doty, *Polymer Bull.*, **1,** 90 (1945).

24. G. Mie, *Ann. Physik*, **25**, 377 (1908).
25. P. Debye, *Ann. Physik*, **30**, 57 (1909).
26. M. Bender, *J. Chem. Educ.*, **29**, 18 (1952).
27. R. Gans, *Ann. Physik*, **37**, 881 (1912).
28. *Standard Methods of Chemical Analysis* (F. J. Welcher, ed.), Part A, Van Nostrand, Princeton, N.J., 6th ed., 1966: a. R. B. Fischer, p. 277; b. B. H. Kaye, *ibid.*, p. 809.
29. G. L. Beyer, in *Physical Methods of Organic Chemistry* (A. Weissberger, ed.), Vol. I, Part I, Wiley (Interscience), New York, 3rd ed., 1959, pp. 191, 231.
30. P. Debye, *J. Phys. Colloid Chem.*, **51**, 18 (1947).
31. *United States Pharmacopeia*, Mack Printing, Easton, Pa., 17th rev. ed., 1965, pp. 806, 833, 870, 929, 933.
32. *The National Formulary, Twelfth Edition*, Mack Printing, Easton, Pa., 1965, pp. 438, 495.
33. P. Debye, *J. Appl. Phys.*, **15**, 338 (1944).
34. M. von Smoluchowski, *Ann. Physik*, **25**, 205 (1908).
35. A. Einstein, *Ann. Physik*, **33**, 1275 (1910).
36. C. V. Raman and R. Ramanathan, *Phil. Mag.*, **45**, 213 (1923).
37. P. M. Doty, B. H. Zimm, and H. Mark, *J. Chem. Phys.*, **12**, 144 (1944).
38. P. M. Doty, B. H. Zimm, and H. Mark, *J. Chem. Phys.*, **13**, 159 (1945).
39. B. H. Zimm, *J. Chem. Phys.*, **13**, 141 (1945).
40. S. N. Timasheff, M. Bier, and F. F. Nord, *J. Phys. Chem.*, **53**, 1134 (1949).
41. J. G. Kirkwood and R. J. Goldberg, *J. Chem. Phys.*, **18**, 54 (1950).
42. M. Fixman, *J. Chem. Phys.*, **23**, 2074 (1955).
43. P. Doty and R. F. Steiner, *J. Chem. Phys.*, **18**, 1211 (1950).
44. H. E. Rose and H. B. Lloyd, *J. Soc. Chem. Eng.*, **65**, 52, 65 (1946).
45. P. V. Wells, *J. Am. Chem. Soc.*, **44**, 267 (1922).
46. R. C. Tolman and E. B. Vliet, *J. Am. Chem. Soc.*, **41**, 297 (1919).
47. R. C. Tolman, L. H. Reyerson, E. B. Vliet, R. H. Gerke, and A. P. Brooks, *J. Am. Chem. Soc.*, **41**, 300 (1919).
48. R. C. Tolman, E. B. Vliet, W. M. Peirce, and R. H. Dougherty, *J. Am. Chem. Soc.*, **41**, 304 (1919).
49. R. C. Tolman, R. H. Gerke, A. P. Brooks, A. G. Herman, R. S. Milliken, and H. D. Smith, *J. Am. Chem. Soc.*, **41**, 575 (1919).
50. S. W. Parr and W. D. Staley, *Ind. Eng. Chem. Anal. Ed.*, **3**, 66 (1931).
51. P. A. Kober, *J. Biol. Chem.*, **13**, 485 (1913)
52. P. A. Kober, *J. Biol. Chem.*, **29**, 155 (1917).
53. A. Tingle, *J. Am. Chem. Soc.*, **40**, 873 (1918).
54. A. Ringbom, *Z. Anal. Chem.*, **122**, 263 (1941).
55. J. McFarland, *J. Am. Med. Assoc.*, **49**, 1176 (1907).
56. F. Kavanagh (ed.), *Analytical Microbiology*, Academic Press, New York, 1963, p. 141. (Also, for: Antibiotics, pp. 249, 278; Antifungal Agents, p. 391; Vitamins and Amino Acids, p. 411).
57. R. O. Gumprecht and C. M. Sliepcevich, *J. Phys. Chem.*, **57**, 95 (1953).
58. C. Orr and J. M. Dallavalle, *Fine Particle Measurement*, Macmillan, New York, 1959, p. 114.
59. H. E. Rose, *The Measurement of Particle Size in Very Fine Powders*, Chemical Publishing, New York, pp. 58, 66.

EXTENDED BIBLIOGRAPHY*

60. P. V. Wells and R. H. Gerke, *J. Am. Chem. Soc.*, **41**, 312 (1919).
61. W. Denis and L. Reed, *J. Biol. Chem.*, **71**, 191, 205 (1926).

* Cited in the Appendix, with foregoing references, according to principal topics.

62. J. W. Baylis, *Ind. Eng. Chem.*, **18**, 311 (1926).
63. J. Cabannes (with Y. Rocard), *La Diffusion Moléculaire de la Lumière*, Les Presses Universitàires de France, 1929.
64. H. K. Miller and J. G. Andrews, *Ind. Eng. Chem. Anal. Ed.*, **2**, 283 (1930).
65. S. D. Gehman and J. S. Ward, *Ind. Eng. Chem. Anal. Ed.*, **3**, 300 (1931).
66. E. J. Dunn, *Ind. Eng. Chem. Anal. Ed.*, **4**, 191 (1932).
67. K. S. Klemmerer and P. W. Bowtwell, *Ind. Eng. Chem. Anal. Ed.*, **4**, 423 (1932).
68. I. M. Kolthoff and H. Yutzy, *J. Am. Chem. Soc.*, **55**, 1915 (1933).
69. A. R. Osborn, *Assoc. Offic. Agr. Chemists*, **17**, 135 (1934).
70. E. T. Bartholomew and E. C. Raby, *Ind. Eng. Chem. Anal. Ed.*, **7**, 68 (1935).
71. R. T. Sheen, H. L. Kahler, E. M. Ross, W. H. Betz, and L. D. Betz, *Ind. Eng. Chem. Anal. Ed.*, **7**, 262 (1935).
72. H. Mestre, *J. Bacteriol.*, **30**, 335 (1935).
73. R. Hofmann, *Mikrochemie*, **18**, 24 (1935).
74. P. M. West and P. W. Wilson, *Science*, **88**, 334 (1938).
75. L. H. Cohan and N. Hackerman, *Ind. Eng. Chem. Anal. Ed.*, **12**, 210 (1940).
76. H. H. Willard and G. H. Ayers, *Ind. Eng. Chem. Anal. Ed.*, **12**, 287 (1940).
77. C. M. Alter and D. S. Thomas, *Ind. Eng. Chem. Anal. Ed.*, **12**, 525 (1940).
78. L. N. Markwood, *J. Assoc. Offic. Agr. Chemists*, **23**, 800 (1940).
79. G. Medes and E. Stavers, *J. Lab. Clin. Med.*, **25**, 624 (1940).
80. R. H. Muller, *Ind. Eng. Chem. Anal. Ed.*, **13**, 715 (1941).
81. T. R. P. Gibb, *Optical Methods of Chemical Analysis*, McGraw-Hill, New York, 1942, p. 160.
82. S. Bhagavantam, *Scattering of Light and the Raman Effect*, Chemical Publishing, New York, 1942.
83. J. W. Foster, *J. Biol. Chem.*, **144**, 285 (1942).
84. C. E. Barnett, *J. Phys. Chem.*, **46**, 69 (1942).
85. R. Landshoff, *J. Phys. Chem.*, **46**, 778 (1942).
86. F. J. Halliman, *Am. J. Public Health*, **33**, 137 (1943).
87. L. A. Atkin, A. S. Schultz, W. L. Williams, and C. N. Frey, *Ind. Eng. Chem. Anal. Ed.*, **15**, 141 (1943).
88. J. G. Baier, *Ind. Eng. Chem. Anal. Ed.*, **15**, 144 (1943).
89. N. F. MacLagan, *Brit. J. Exptl. Pathol.*, **25**, 234 (1944).
90. P. R. Burkholder, I. McVeigh, and D. Moyer, *J. Bacteriol.*, **48**, 385 (1944).
91. L. L. Merritt, *Ind. Eng. Chem. Anal. Ed.*, **16**, 758 (1944).
92. J. B. McMahan, *J. Biol. Chem.*, **153**, 249 (1944).
93. S. W. Lee, E. J. Foley, J. Epstein, and J. H. Wallace, *J. Biol. Chem.*, **152**, 485 (1944).
94. H. P. Sarett and V. H. Cheldelin, *J. Biol. Chem.*, **155**, 153 (1944).
95. B. H. Zimm and P. M. Doty, *J. Chem. Phys.*, **12**, 203 (1944).
96. N. F. MacLagan, *Nature*, **154**, 670 (1944).
97. D. A. Joslyn, *Science*, **99**, 21 (1944).
98. E. Canals, A. Charrâ, and G. Riety, *Bull. Soc. Chim.*, **12**, 1055 (1945).
99. V. K. LaMer and I. Johnson, *J. Am. Chem. Soc.*, **67**, 2055 (1945)
100. R. H. Ewart, *Advan. Colloid Sci.*, **2**, 226 (1946).
101. R. S. Stein and P. Doty, *J. Am. Chem. Soc.*, **68**, 159 (1946).
102. P. P. Debye, *J. Appl. Phys.*, **17**, 392 (1946).
103. W. Heller and E. Vassy, *J. Chem. Phys.*, **14**, 565 (1946).
104. W. Heller, H. B. Klevens, and H. Oppenheimer, *J. Chem. Phys.*, **14**, 566 (1946).
105. R. H. Ewart, C. P. Roe, P. Debye, and J. R. McCartney, *J. Chem. Phys.*, **14**, 687 (1946).
106. V. K. LaMer and M. D. Barnes, *J. Colloid Sci.*, **1**, 71 (1946).
107. M. D. Barnes and V. K. LaMer, *J. Colloid Sci.*, **1**, 79 (1946).
108. F. K. Lindsay, D. G. Braithwaite, and J. S. D'Amico, *Ind. Eng. Chem. Anal. Ed.*, **18**, 101 (1946).

109. E. D. Bailey, *Ind. Eng. Chem. Anal. Ed.*, **18**, 365 (1946).
110. G. Oster, *Science*, **103**, 306 (1946).
111. J. Bardwell and C. Sivertz, *Can. J. Res.*, **25B**, 255 (1947).
112. H. P. Kortschak, *Ind. Eng. Chem. Anal. Ed.*, **19**, 692 (1947).
113. F. T. Gucker, H. B. Pickard, and C. T. O'Konski, *J. Am. Chem. Soc.*, **69**, 429 (1947).
114. I. Johnson and V. K. LaMer, *J. Am. Chem. Soc.*, **69**, 1184 (1947).
115. G. Oster, P. M. Doty, and B. H. Zimm, *J. Am. Chem. Soc.*, **69**, 1193 (1947).
116. F. T. Gucker, C. T. O'Konski, H. B. Pickard, and J. N. Pitts, *J. Am. Chem. Soc.*, **69**, 2422 (1947).
117. J. C. Rabinowitz and E. E. Snell, *J. Biol. Chem.*, **169**, 631 (1947).
118. J. C. Rabinowitz and E. E. Snell, *J. Biol. Chem.*, **169**, 643 (1947).
119. G. Oster, *J. Colloid Sci.*, **2**, 291 (1947).
120. M. D. Barnes, A. S. Kenyon, E. M. Zaiser, and V. K. LaMer, *J. Colloid Sci.*, **2**, 349 (1947).
121. V. K. LaMer and M. D. Barnes, *J. Colloid Sci.*, **2**, 361 (1947).
122. R. Ginell, A. M. Ginell, and P. E. Spoerri, *J. Colloid Sci.*, **2**, 521 (1947).
123. P. Doty, H. Wagner, and S. Singer, *J. Phys. Colloid Chem.*, **51**, 32 (1947).
124. J. F. Gain and C. A. Lawrence, *Science*, **106**, 525 (1947).
125. F. D. Snell and C. T. Snell, *Colorimetric Methods of Analysis* (including some *Turbidimetric* and *Nephelometric* Methods), Vols. I–IV, Van Nostrand, Princeton, N.J., 3rd ed., 1948–1954.
126. E. M. Burdick and J. S. Allen, *Anal. Chem.*, **20**, 539 (1948).
127. M. Halwer, *J. Am. Chem. Soc.*, **70**, 3985 (1948).
128. J. C. Rabinowitz, N. I. Mondy, and E. E. Snell, *J. Biol. Chem.*, **175**, 147 (1948).
129. R. F. Stamm, T. Mariner, and J. K. Dixon, *J. Chem. Phys.*, **16**, 423 (1948).
130. B. H. Zimm, *J. Chem. Phys.*, **16**, 1093 (1948).
131. B. H. Zimm, *J. Chem. Phys.*, **16**, 1099 (1948).
132. L. White and D. G. Hill, *J. Colloid Sci.*, **3**, 251 (1948).
133. E. M. Zaiser and V. K. LaMer, *J. Colloid Sci.*, **3**, 571 (1948).
134. V. K. LaMer, *J. Phys. Colloid Chem.*, **52**, 65 (1948).
135. B. H. Zimm, *J. Phys. Colloid Chem.*, **52**, 260 (1948).
136. P. Doty, *J. Polymer Sci.*, **3**, 750 (1948).
137. P. Doty and S. J. Stein, *J. Polymer Sci.*, **3**, 763 (1948).
138. H. E. Rose and C. C. J. French, *J. Soc. Chem. Ind.*, **67**, 283 (1948).
139. T. R. Gillette, P. F. Meads, and A. L. Holven, *Anal. Chem.*, **21**, 1228 (1949).
140. D. Sinclair and V. K. LaMer, *Chem. Rev.*, **44**, 245 (1949).
141. F. T. Gucker and C. T. O'Konski, *Chem. Rev.*, **44**, 373 (1949).
142. R. M. Ferry, L. E. Farr, and M. G. Hartman, *Chem. Rev.*, **44**, 389 (1949).
143. A. M. Bueche, *J. Am. Chem. Soc.*, **71**, 1452 (1949).
144. P. Debye and R. V. Nauman, *J. Chem. Phys.*, **17**, 664 (1949).
145. P. Doty and R. F. Steiner, *J. Chem. Phys.*, **17**, 743 (1949).
146. B. H. Zimm and W. H. Stockmayer, *J. Chem. Phys.*, **17**, 1301 (1949).
147. H. C. van de Hulst, *J. Colloid Sci.*, **4**, 79 (1949).
148. F. T. Gucker and C. T. O'Konski, *J. Colloid Sci.*, **4**, 541 (1949).
149. P. Debye, *J. Phys. Chem.*, **53**, 1 (1949).
150. a. R. H. Blaker, R. M. Badger, and T. S. Gilmann, *J. Phys. Chem.*, **53**, 794 (1949); b. R. M. Badger and R. H. Blaker, *ibid.*, **53**, 1056 (1949).
151. G. Oster, *Rec. Trav. Chim.*, **68**, 1123 (1949).
152. F. T. Gucker, *Science*, **110**, 372 (1949).
153. F. T. Gucker, *Sci. Monthly*, **68**, 373 (1949).
154. E. E. Snell, in *Vitamin Methods*, Vol. I (P. György, ed.), Academic Press, New York, 1950, p. 327.
155. M. L. Nichols and B. H. Kindt, *Anal. Chem.*, **22**, 785 (1950).

156. T. W. Sarge, *Anal. Chem.*, **22**, 1435 (1950).
157. M. Kerker and V. K. LaMer, *J. Am. Chem. Soc.*, **72**, 3516 (1950).
158. W. B. Dandliker, *J. Am. Chem. Soc.*, **72**, 5110 (1950).
159. R. C. Kersey, *J. Am. Pharm. Assoc.*, **39**, 252 (1950).
160. W. H. Stockmayer, *J. Chem. Phys.*, **18**, 58 (1950).
161. C. I. Carr and B. H. Zimm, *J. Chem. Phys.*, **18**, 1616 (1950).
162. V. K. LaMer, E. C. Y. Inn, and I. B. Wilson, *J. Colloid Sci.*, **5**, 471 (1950).
163. J. de la Huerga and H. Popper, *J. Lab. Clin. Med.*, **35**, 459 (1950).
164. L. A. Nalefski and F. Takano, *J. Lab. Clin. Med.*, **36**, 468 (1950).
165. Association of Vitamin Chemists, in *Methods of Vitamin Assay*, Wiley (Interscience), New York, 2nd ed., 1951, p. 129.
166. A. C. Mason, *Analyst*, **76**, 172 (1951).
167. R. C. Kersey, *Antibiot. Chemother.*, **1**, 173 (1951).
168. D. P. Riley and G. Oster, *Discussions Faraday Soc.*, **11**, 107 (1951).
169. M. Halwer, G. C. Nutting, and B. A. Brice, *J. Am. Chem. Soc.*, **73**, 2786 (1951).
170. G. F. Lothian and F. P. Chappel, *J. Appl. Chem.*, **1**, 475 (1951).
171. P. Debye and W. M. Cashin, *J. Chem. Phys.*, **19**, 510 (1951).
172. F. Bueche, P. Debye, and W. M. Cashin, *J. Chem. Phys.*, **19**, 803 (1951).
173. R. V. Nauman and P. Debye, *J. Phys. Chem.*, **55**, 1 (1951).
174. P. Debye and E. W. Anacker, *J. Phys. Colloid Chem.*, **55**, 644 (1951).
175. J. Fog, *Analyst*, **77**, 454 (1952).
176. M. Kaufman, *Anal. Chem.*, **24**, 683 (1952).
177. F. Brown and J. Graham, *Anal. Chem.*, **24**, 1032 (1952).
178. F. Tietze and H. Neurath, *J. Biol. Chem.*, **194**, 1 (1952).
179. W. F. H. M. Mommaerts, *J. Colloid Sci.*, **7**, 71 (1952).
180. R. H. Dinegar and R. H. Smellie, *J. Colloid Sci.*, **7**, 370 (1952).
181. O. J. Pollack and B. Wadler, *J. Lab. Clin. Med.*, **39**, 791 (1952).
182. G. Oster, *J. Polymer Sci.*, **9**, 525 (1952).
183. H. E. Rose, *Nature*, **169**, 287 (1952).
184. D. A. Goring and P. Johnson, *Trans. Faraday Soc.*, **48**, 367 (1952).
185. A. Hiller, *Practical Clinical Chemistry*, Thomas, Springfield, Ill., 2nd ed., 1957, pp. 73, 143, 149, 153.
186. American Association of Clinical Chemists, *Standard Methods of Clinical Chemistry* (S. Meites, ed.), Academic Press, New York, 1953, p. 113, Vol. I; p. 231, Vol. V.
187. G. Toennies and B. Bakay, *Anal. Chem.*, **25**, 160 (1953).
188. B. A. Brice, G. C. Nutting, and N. Halwer, *J. Am. Chem. Soc.*, **75**, 824 (1953).
189. R. F. Stamm and P. A. Button, *J. Chem. Phys.*, **21**, 304 (1953); **23**, 2456 (1955).
190. M. Halwer, G. C. Nutting, and B. A. Brice, *J. Chem. Phys.*, **21**, 1425 (1953).
191. E. W. Anacker, *J. Colloid Sci.*, **8**, 402 (1953).
192. R. O. Gumprecht and C. M. Sliepcevich, *J. Phys. Chem.*, **57**, 90 (1953).
193. G. B. Alexander and R. K. Iler, *J. Phys. Chem.*, **57**, 932 (1953).
194. B. H. Zimm, *J. Polymer Sci.*, **10**, 351 (1953).
195. B. Jurgensons and M. E. Straumanis, *A Short Textbook of Colloid Chemistry*, Wiley, New York, 1954, pp. 100, 178, 393.
196. G. W. Leonard, D. E. Sellers, and L. E. Swim, *Anal. Chem.*, **26**, 1621 (1954).
197. R. F. Goddu and D. N. Hume, *Anal. Chem.*, **26**, 1740 (1954).
198. H. R. Lang (ed.), *Brit. J. Appl. Phys. Suppl.*, S64 (1954).
199. R. A. Johnson and J. D. O'Rourke, *J. Am. Chem. Soc.*, **76**, 2124 (1954).
200. F. W. Billmeyer, *J. Am. Chem. Soc.*, **76**, 4636 (1954).
201. A. L. Underwood, *J. Chem. Educ.*, **31**, 394 (1954).
202. D. A. I. Goring and P. G. Napier, *J. Chem. Phys.*, **22**, 147 (1954).
203. W. Heller, J. N. Epel, and R. M. Tabibian, *J. Chem. Phys.*, **22**, 1777 (1954).
204. R. W. Fessenden and R. S. Stein, *J. Chem. Phys.*, **22**, 1778 (1954).

205. E. Hutchinson, *J. Colloid Sci.*, **9**, 191 (1954).

206. K. J. Mysels, *J. Phys. Chem.*, **58**, 303 (1954).

207. H. Benoit, A. M. Holtzer, and P. Doty, *J. Phys. Chem.*, **58**, 635 (1954).

208. B. H. Zimm and W. B. Dandliker, *J. Phys. Chem.*, **58**, 644 (1954).

209. H. J. L. Trapand and J. J. Hermans, *J. Phys. Chem.*, **58**, 757 (1954).

210. M. Kerker, G. L. Jones, J. B. Reed, C. N. P. Yang, and M. D. Schoenberg, *J. Phys. Chem.*, **58**, 1147 (1954).

211. S. H. Maron and R. L. H. Lou, *J. Polymer Sci.*, **14**, 29 (1954).

212. S. H. Maron and R. L. H. Lou, *J. Polymer Sci.*, **14**, 273 (1954).

213. E. S. Cohn and E. M. Schuele, *J. Polymer Sci.*, **14**, 309 (1954).

214. R. Tremblay, Y. Sicotte, and M. Renfret, *J. Polymer Sci.*, **14**, 310 (1954).

215. E. O. Powell, *J. Sci. Instr.*, **31**, 360 (1954).

216. G. Oster, in *Physical Techniques in Biological Research*, Vol. I (G. Oster and A. W. Pollister, eds.), Academic Press, New York, 1955, p. 51.

217. J. G. Reinhold, *Anal. Chem.*, **27**, 239 (1955).

218. H. J. Keily and L. B. Rogers, *Anal. Chem.*, **27**, 459 (1955).

219. G. D. Clayton and P. M. Giever, *Anal. Chem.*, **27**, 708 (1955).

220. H. J. Keily and L. B. Rogers, *Anal. Chem.*, **27**, 759 (1955).

221. R. H. Robinson, *Anal. Chem.*, **27**, 1351 (1955).

222. J. A. Cifonelli and F. Smith, *Anal. Chem.*, **27**, 1639 (1955).

223. J. D. O'Rourke and R. A. Johnson, *Anal. Chem.*, **27**, 1699 (1955).

224. P. J. De La Rubia and L. R. F. Blasco, *Chemist-Analyst*, **44**, 58 (1955).

225. F. W. Billmeyer, J. R. de Than, and C. B. de Than, *J. Am. Chem. Soc.*, **77**, 4763 (1955).

226. W. Heller, *J. Chem. Phys.*, **23**, 342 (1955).

227. J. Kraut and W. B. Dandliker, *J. Chem. Phys.*, **23**, 1544 (1955).

228. S. H. Maron and R. L. H. Lou, *J. Phys. Chem.*, **59**, 231 (1955).

229. J. N. Phillips and K. J. Mysels, *J. Phys. Chem.*, **59**, 325 (1955).

230. C. K. Sloan, *J. Phys. Chem.*, **59**, 834 (1955).

231. J. H. Chin, C. M. Sliepcevich, and M. Tribus, *J. Phys. Chem.*, **59**, 841 (1955).

232. J. H. Chin, C. M. Sliepcevich, and M. Tribus, *J. Phys. Chem.*, **59**, 845 (1955).

233. P. H. Scott, G. C. Clark, and C. M. Sliepcevich, *J. Phys. Chem.*, **59**, 849 (1955).

234. C. M. Chu and S. W. Churchill, *J. Phys. Chem.*, **59**, 855 (1955).

235. S. N. Chinai, P. L. Scherer, and D. W. Levi, *J. Polymer Sci.*, **17**, 117 (1955).

236. K. A. Stacey, *Light Scattering in Physical Chemistry*, Academic Press, New York, 1956.

237. H. J. Keily and L. B. Rogers, *Anal. Chim. Acta.*, **14**, 356 (1956).

238. C. A. Thomas, *J. Am. Chem. Soc.*, **78**, 1861 (1956).

239. H. W. Loy, W. P. Parrish, and S. S. Schiaffino, *J. Assoc. Offic. Agr. Chemists*, **39**, 172 (1956).

240. N. Di Ferrante, *J. Biol. Chem.*, **220**, 303 (1956).

241. D. H. Taysum, *J. Chem. Phys.*, **25**, 183 (1956).

242. D. K. Carpenter and W. R. Krigbaum, *J. Chem. Phys.*, **24**, 1041 (1956).

243. R. M. Tabibian, W. Heller, and J. N. Epel, *J. Colloid Sci.*, **11**, 195 (1956).

244. W. J. Hughes, P. Johnson, and R. H. Ottewill, *J. Colloid Sci.*, **11**, 340 (1956).

245. J. Mager, M. Kuczynski, and G. Schatzberg, *J. Gen. Microbiol.*, **14**, 69, (1956).

246. Y. Avi-Dor, M. Kuczynski, G. Schatzberg, and J. Mager, *J. Gen. Microbiol.*, **14**, 76 (1956).

247. H. C. van der Hulst, *Light Scattering by Small Particles*, Wiley, New York, 1957.

248. A. E. Nielsen, *Acta Chem. Scand.*, **11**, 1512 (1957).

249. R. L. Costello and G. L. Curran, *Am. J. Clin. Pathol.*, **27**, 108 (1957).

250. E. J. Morrisey and W. J. Hartop, *Drug. Std.*, **25**, 1 (1957).

251. J. E. Land, *J. Chem. Educ.*, **34**, 38 (1957).

252. W. Heller and W. J. Pangonis, *J. Chem. Phys.*, **26**, 498 (1957).
253. W. Heller, *J. Chem. Phys.*, **26**, 920 (1957).
254. W. Heller, *J. Chem. Phys.*, **26**, 1258 (1957).
255. W. Heller and R. M. Tabibian, *J. Colloid Sci.*, **12**, 25 (1957).
256. L. M. Kushner, W. D. Hubbard, and A. S. Doan, *J. Phys. Chem.*, **61**, 371 (1957).
257. V. K. LaMer and I. W. Plesner, *J. Polymer Sci.*, **24**, 147 (1957).
258. L. M. Kushner, W. D. Hubbard, and R. A. Parker, *J. Res. Natl. Bur. Std.*, **59**, 113 (1957).
259. W. F. Wright, *Rev. Sci. Instr.*, **28**, 129 (1957).
260. M. B. McEwen and M. I. Pratt, *Trans. Faraday Soc.*, **53**, 535 (1957).
261. W. J. Pangonis, W. Heller, and A. Jacobson, *Tables of Light Scattering Functions for Spherical Particles*, Wayne State Univ. Press, Detroit, 1957.
262. E. P. Geiduschek and A. Holtzer, *Advan. Biol. Med. Phys.*, **6**, 431 (1958).
263. R. A. Post and O. H. Miller, *Am. J. Hosp. Pharm.*, **15**, 150 (1958).
264. C. T. O'Konski, C. T. Biton, and W. I. Higuchi, *Light Scattering Instrumentation for Particle-Size Distribution Measurement*, A.S.T.M. Publ. No. 234, 1958.
265. J. C. Houck, *Arch. Biochem. Biophys.*, **73**, 384 (1958).
266. L. E. Cohen, *Chemist-Analyst*, **47**, 65 (1958).
267. S. S. Schiaffino, J. J. McGuire, and H. W. Loy, *J. Am. Assoc. Offic. Agr. Chemists*, **41**, 420 (1958).
268. W. Heller and H. J. McCarty, *J. Chem. Phys.*, **29**, 78 (1958).
269. R. M. Tabibian and W. Heller, *J. Colloid Sci.*, **13**, 6 (1958).
270. L. Packer, *J. Opt. Soc. Am.*, **48**, 503 (1958).
271. R. H. Ottewill and H. C. Parreira, *J. Phys. Chem.*, **62**, 912 (1958).
272. R. L. Cleland, *J. Polymer Sci.*, **27**, 349 (1958).
273. E. J. Meehan, *J. Polymer Sci.*, **27**, 590 (1958).
274. M. Kerker, *J. Polymer Sci.*, **28**, 429 (1958).
275. J. K. Donahue, J. C. Houck, and R. J. Coffey, *Surgery*, **44**, 1070 (1958).
276. D. H. Klein and L. Gordon, *Talanta*, **1**, 334 (1958).
277. R. H. Boll, J. A. Leacock, G. C. Clark, and S. W. Churchill, *Light Scattering Functions*, Univ. of Mich. Press, Ann Arbor, 1958.
278. K. J. Mysels, *Introduction to Colloid Chemistry*, Wiley (Interscience), New York, 1959, pp. 415ff.
279. A. C. Arcus, *Anal. Chem.*, **31**, 1618 (1959).
280. R. B. Fischer, M. L. Yates, and M. M. Batts, *Anal. Chim. Acta*, **20**, 501 (1959).
281. A. Holtzer and S. Lowey, *J. Am. Chem. Soc.*, **81**, 1370 (1959).
282. W. W. Everett and J. F. Foster, *J. Am. Chem. Soc.*, **81**, 3464 (1959).
283. W. W. Everett and J. F. Foster, *J. Am. Chem. Soc.*, **81**, 4359 (1959).
284. W. Heller, M. Nakagaki, and M. L. Wallach, *J. Chem. Phys.*, **30**, 444 (1959).
285. M. Nakagaki and W. Heller, *J. Chem. Phys.*, **30**, 783 (1959).
286. W. Heller and M. Nakagaki, *J. Chem. Phys.*, **31**, 1188 (1959).
287. P. Debye and L. K. H. van Beek, *J. Chem. Phys.*, **31**, 1595 (1959).
288. J. B. Bateman, E. J. Weneck, and D. C. Eshler, *J. Colloid Sci.*, **14**, 308 (1959).
289. M. J. Kronman and S. N. Timasheff, *J. Phys. Chem.*, **63**, 629 (1959).
290. L. H. Princen and K. J. Mysels, *J. Phys. Chem.*, **63**, 1781 (1959).
291. R. Chiang, *J. Polymer Sci.*, **36**, 91 (1959).
292. Q. A. Trementozzi, *J. Polymer Sci.*, **36**, 113 (1959).
293. E. D. Korn, *Methods Biochem. Anal.*, **7**, 176 (1959).
294. L. B. Jaques and H. J. Bell, *Methods Biochem. Anal.*, **7**, 291, 305 (1959).
295. J. S. Harrison, *Rept. Progr. Appl. Chem.*, **44**, 401 (1959).
296. Z. P. Zagorski and M. Cyrnkowska, *Talanta*, **2**, 380 (1959); cf. *Anal. Chem.*, **27**, 120 (1955).
297. D. H. Klein, L. Gordon, and T. H. Walnut, *Talanta*, **3**, 177, 187 (1959).

298. G. W. Ewing, *Instrumental Methods of Chemical Analysis*, 2nd ed., McGraw-Hill, New York, 1960, pp. 15, 60, 64, 410.
299. G. Oster, "Light Scattering," in *Technique of Organic Chemistry* (A. Weissberger, ed.), 3rd ed., Vol. I, Part III, Wiley (Interscience), New York, 1960, p. 2107.
300. H. A. Strobel, *Chemical Instrumentation*, Addison-Wesley, Reading, Mass., 1960, p. 217.
301. M. Bobtelsky, *Heterometry*, Elsevier, New York, 1960, pp. 1, 6.
302. H. Flaschka and A. J. Barnard, "Tetraphenylboron (TPB) as an Analytical Reagent," in *Advances in Analytical Chemistry and Instrumentation* (C. N. Reilly, ed.), Wiley (Interscience), New York; Vol. 1, 1960, pp. 1, 64; A. L. Underwood, "Photometric Titrations," *ibid.*, Vol. 3, 1964, p. 93.
303. F. Kavanagh, *Advan. Appl. Microbiol.*, **2**, 65 (1960).
304. J. G. Reinhold, *Advan. Clin. Chem.*, **3**, 83 (1960).
305. G. Zdybek, D. S. McCann, and A. J. Boyle, *Anal. Chem.*, **32**, 558 (1960).
306. R. B. Fischer, *Anal. Chim. Acta*, **22**, 501 (1960).
307. U. Niwa and E. P. Parry, *Chemist-Analyst*, **49**, 102 (1960).
308. "The Physical Chemistry of Aerosols," coll., *Discussions Faraday Soc.*, **30**, 1 (1960).
309. M. Nakagaki and W. Heller, *J. Chem. Phys.*, **32**, 835 (1960).
310. J. F. Boyd and J. W. Sommerville, *J. Clin. Pathol.*, **13**, 85 (1960).
311. D. Stigter, *J. Phys. Chem.*, **64**, 114 (1960).
312. E. J. Meehan and W. H. Beattie, *J. Phys. Chem.*, **64**, 1006 (1960).
313. J. E. Scott, *Methods Biochem. Anal.*, **8**, 145 (1960).
314. A. E. Nielsen, *Acta Chem. Scand.*, **15**, 441 (1961).
315. G. R. Kingsley and O. Robnett, *Anal. Chem.*, **33**, 561 (1961).
316. E. J. Meehan and W. H. Beattie, *Anal. Chem.*, **33**, 632 (1961).
317. K. K. Georgieff, *Anal. Chem.*, **33**, 1432 (1961).
318. J. Brahms and J. Brezner, *Arch. Biochem. Biophys.*, **95**, 219 (1961).
319. L. F. Cavalieri, J. F. Deutsch, and B. H. Rosenberg, *Biophys. J.*, **1**, 301 (1961).
320. L. F. Cavalieri and B. H. Rosenberg, *Biophys. J.*, **1**, 317 (1961).
321. L. F. Cavalieri and B. H. Rosenberg, *Biophys. J.*, **1**, 323 (1961).
322. L. F. Cavalieri and B. H. Rosenberg, *Biophys. J.*, **1**, 337 (1961).
323. J. D. S. Goulden, *Brit. J. Appl. Phys.*, **13**, 456 (1961).
324. G. Perkins, J. W. Wimberley, J. F. Lamb, and L. E. Maurer, *J. Chem. Educ.*, **38**, 358 (1961).
325. W. Pangonis, W. Heller, and N. A. Economou, *J. Chem. Phys.*, **34**, 960 (1961).
326. W. Heller, W. J. Pangonis, and N. A. Economou, *J. Chem. Phys.*, **34**, 971 (1961).
327. A. F. Stevenson, W. Heller, and M. L. Wallach, *J. Chem. Phys.*, **34**, 1789 (1961).
328. M. L. Wallach, W. Heller, and A. F. Stevenson, *J. Chem. Phys.*, **34**, 1796 (1961).
329. E. C. Peterson and A. H. Tincher, *J. Colloid Sci.*, **16**, 87 (1961).
330. G. D. Deželić and J. P. Kratohvil, *J. Colloid Sci.*, **16**, 561 (1961).
331. R. H. Ottewill and R. F. Woodbridge, *J. Colloid Sci.*, **16**, 581 (1961).
332. P. Debye and R. V. Nauman, *J. Phys. Chem.*, **65**, 5 (1961).
333. P. Debye and R. V. Nauman, *J. Phys. Chem.*, **65**, 10 (1961).
334. W. J. Prins, *J. Phys. Chem.*, **65**, 369 (1961).
335. E. J. Meehan and W. H. Beattie, *J. Phys. Chem.*, **65**, 1522 (1961).
336. M. Kerker, J. P. Kratohvil, and E. Matijević, *J. Phys. Chem.*, **65**, 1713 (1961).
337. J. D. Bauer, G. Toro, and P. G. Ackerman, *Bray's Clinical Laboratory Methods*, 6th ed., C. V. Mosby, St. Louis, 1962, pp. 288, 290, 405.
338. A. Cantarow and M. Trumper, *Clinical Biochemistry*, 6th ed., Saunders, Philadelphia, 1962, pp. 164, 506.
339. M. K. Coultas and D. G. Hutchison, *J. Bacteriol.*, **84**, 393 (1962).
340. W. Heller, H. L. Bhatnager, and M. Nakagaki, *J. Chem. Phys.*, **36**, 1163 (1962).
341. T. Gillespie, *J. Colloid Sci.*, **17**, 290 (1962).

342. R. J. Gledhill, *J. Phys. Chem.*, **66**, 458 (1962).
343. W. Heller and R. Tabibian, *J. Phys. Chem.*, **66**, 2059 (1962).
344. J. P. Kratohvil, G. Deželić, and E. Matijević, *J. Polymer Sci.*, **57**, 59 (1962).
345. J. M. Stearne and J. R. Urwin, *Makromol. Chem.*, **56**, 76 (1962).
346. J. J. Kabara, *Methods Biochem. Anal.*, **10**, 291 (1962).
347. M. J. Jaycock and G. D. Parfitt, *Nature*, **194**, 77 (1962).
348. T. G. Fox, J. B. Kinsinger, H. F. Mason, and E. M. Schuele, *Polymer*, **3**, 71 (1962).
349. E. Cohn-Ginsberg, T. G. Fox, and H. F. Mason, *Polymer*, **3**, 97 (1962).
350. T. G. Fox, *Polymer*, **3**, 111 (1962).
351. *British Pharmacopoeia*, 1963, The Pharmaceutical Press, London, p. 1052.
352. R. R. Irani and C. F. Callis, *Particle Size Measurement, Interpretation and Application*, Wiley, New York, 1963.
353. The Particle Size Analysis Sub-Committee (Analytical Methods Committee, Society for Analytical Chemistry (Britain)), *Analyst*, **88**, 156 (1963).
354. A. L. Rouy, B. Carroll, and T. J. Quigley, *Anal. Chem.*, **35**, 627 (1963).
355. T. A. Haney, J. R. Gerke, and J. F. Pagano, "Automation of Microbiological Assays: B. Photometric Methods," in *Analytical Microbiology* (F. Kavanagh, ed.), Academic Press, New York, 1963, p. 227.
356. M. F. Gellert and S. W. Englander, *Biochemistry*, **2**, 39 (1963).
357. D. Froelich, C. Strazielle, G. Bernardi, and H. Benoît, *Biophys. J.*, **3**, 115 (1963).
358. A. T. Ansevin and M. A. Lauffer, *Biophys. J.*, **3**, 239 (1963).
359. L. H. Sperling, *J. Appl. Polymer Sci.*, **7**, 1891 (1963).
360. E. F. Woods, S. Himmelfarb, and W. F. Harrington, *J. Biol. Chem.*, **238**, 2374 (1963).
361. S. Z. Lewin, *J. Chem. Educ.*, **40**, A423, 5 (1963).
362. G. Deželić, N. Deželić and B. Tezak, *J. Colloid Sci.*, **18**, 888 (1963).
363. P. C. Eisman, E. Ebersold, J. Weerts, and L. Lachman, *J. Pharm. Sci.*, **52**, 183 (1963).
364. Y. Tomimatsu and K. J. Palmer, *J. Phys. Chem.*, **67**, 1720 (1963).
365. G. J. Howard, *J. Polymer Sci.*, **A1**, 2667 (1963).
366. R. H. Shipman and E. Farber, *J. Polymer Sci.*, **B1**, 65 (1963).
367. A. G. Walton and T. Hlabse, *Talanta*, **10**, 601 (1963).
368. G. D. Zdybek, D. S. McCann, and A. J. Boyle, *Talanta*, **10**, 879 (1963).
369. R. J. Henry, *Clinical Chemistry: Principles and Technics*, Harper and Row, New York, 1964, pp. 45, 186.
370. D. McIntyre and F. Gornick, ed., *Light Scattering from Dilute Polymer Solutions*, (International Science Review Series), Gordon and Breach, New York, 1964.
371. B. A. Mulley, *Advan. Pharm. Sci.*, **1**, 97 (1964).
372. J. P. Kratohvil, *Anal. Chem.*, **36**, 458R (1964).
373. E. J. Meehan and G. Chiu, *Anal. Chem.*, **36**, 536 (1964).
374. Y. Tomimatsu, *Biopolymers*, **2**, 275 (1964).
375. E. J. Meehan and G. Chiu, *J. Am. Chem. Soc.*, **86**, 1443 (1964).
376. R. Zwanzig, *J. Am. Chem. Soc.*, **86**, 3489 (1964).
377. K. J. Mysels, *J. Am. Chem. Soc.*, **86**, 3503 (1964).
378. R. Bersohn, *J. Am. Chem. Soc.*, **86**, 3505 (1964).
379. H. L. Bhatnager and W. Heller, *J. Chem. Phys.*, **40**, 480 (1964).
380. W. Heller, *J. Chem. Phys.*, **40**, 2700 (1964).
381. W. W. Graessley and J. H. Zufall, *J. Colloid Sci.*, **19**, 516 (1964).
382. F. P. Siegel and B. Ecanow, *J. Pharm. Sci.*, **53**, 831 (1964).
383. C. J. Stacy and R. L. Arnett, *J. Polymer Sci.*, **A2**, 167 (1964).
384. M. Kerker, J. P. Kratohvil, and E. Matijević, *J. Polymer Sci.*, **A2**, 303 (1964).
385. B. R. Jennings and H. G. Jerrard, *J. Polymer Sci.*, **A2**, 2025 (1964).
386. W. C. Taylor and J. P. Graham, *J. Polymer Sci.*, **B2**, 169 (1964).
387. A. Peterlin and G. Meinel, *J. Polymer Sci.*, **B2**, 751 (1964).
388. E. E. Lindsey, D. C. Chappelear, D. M. Sullivan, and V. A. Augstkalns, *J. Polymer Sci.*, **C5**, 55 (1964).

389. S. P. Wasik and W. D. Hubbard, *J. Res. Nat. Bur. Std. A*, **68**, 359 (1964).
390. C. A. Storvick, E. M. Benson, M. A. Edwards, and M. J. Woodring, *Methods Biochem. Anal.*, **12**, 229, 237 (1964).
391. D. H. Klein and B. Fontal, *Talanta*, **11**, 1231 (1964).
392. D. J. Coumou and E. L. Mackor, *Trans. Faraday Soc.*, **60**, 1726 (1964).
393. Remington's *Pharmaceutical Sciences* (E. W. Martin, ed.), Mack Publishing Co., Easton, Pa., 1965, pp. 1604, 1606.
394. R. D. Cadle, *Particle Size*, Reinhold, New York, 1965, pp. 51, 341.
395. T. Higuchi and K. A. Connors, *Advan. Anal. Chem. Instr.*, **4**, 127 (1965).
396. T. B. Platt, J. Gentile, and M. J. George, *Ann. N.Y. Acad. Sci.*, **130**, 664 (1965).
397. J. R. McMahan, *Ann. N.Y. Acad. Sci.*, **130**, 680 (1965).
398. R. Dewart, F. Naudts, and W. Lhoest, *Ann. N.Y. Acad. Sci.*, **130**, 686 (1965).
399. A. Ferrari, J. R. Gerke, R. W. Watson, and W. W. Umbreit, *Ann. N.Y. Acad. Sci.*, **130**, 704 (1965).
400. J. R. Gerke, *Ann. N.Y. Acad. Sci.*, **130**, 722 (1965).
401. R. W. Watson, *Ann. N.Y. Acad. Sci.*, **130**, 733 (1965).
402. J. D. Dealy and W. W. Umbreit, *Ann. N.Y. Acad. Sci.*, **130**, 745 (1965).
403. K. D. Straub and W. S. Lynn, *Biochim. Biophys. Acta*, **94**, 304 (1965).
404. A. R. Matz, *Bull. Parenteral Drug Assoc.*, **19**, 72 (1965).
405. C. C. H. Macpherson, *Chem. Ind.* (*London*), **45**, 1934 (1965).
406. P. V. McKinney, *J. Appl. Polymer Sci.*, **9**, 583 (1965).
407. G. Hind and A. T. Jagendorf, *J. Biol. Chem.*, **240**, 3195 (1965).
408. W. Heller, *J. Chem. Phys.*, **42**, 1609 (1965).
409. J. P. Kratohvil, M. Kerker, and L. E. Oppenheimer, *J. Chem. Phys.*, **43**, 914 (1965).
410. G. Cohen and H. Eisenberg, *J. Chem. Phys.*, **43**, 3881 (1965).
411. B. R. Jennings and H. G. Jerrard, *J. Colloid Sci.*, **20**, 448 (1965).
412. W. F. Espenscheid, E. Willis, E. Matijević, and M. Kerker, *J. Colloid Sci.*, **20**, 501 (1965).
413. J. P. Kratohvil and C. Smart, *J. Colloid Sci.*, **20**, 875 (1965).
414. W. Heller, *J. Polymer Sci.*, **A3**, 2367 (1965).
415. Z. Tuzar and P. Kratochvil, *J. Polymer Sci.*, **B3**, 17 (1965).
416. F. W. Billmeyer, *J. Polymer Sci.*, **C8**, 161 (1965).
417. J. H. Bradley, *J. Polymer Sci.*, **C8**, 305 (1965).
418. C. N. Davies, ed., *Aerosol Science*, Academic Press, New York, 1966.
419. I. Vessey and C. E. Kendall, *Analyst*, **91**, 273 (1966).
420. W. B. Elliott and G. F. Doebbler, *Anal. Biochem.*, **15**, 463 (1966).
421. J. P. Kratohvil, *Anal. Chem.*, **38**, 517R (1966).
422. K. Banerjee and M. A. Lauffer, *Biochemistry*, **5**, 1957 (1966).
423. E. Chiancone, M. R. Bruzzesi, and E. Antonini, *Biochemistry*, **5**, 2823 (1966).
424. H. Fujita, A. Teramoto, K. Okita, T. Yamashita, and S. Ikeda, *Biopolymers*, **4**, 769 (1966).
425. A. Bello and G. M. Guzman, *European Polymer J.*, **2**, 79 (1966).
426. A. G. Ogston and B. N. Preston, *J. Biol. Chem.*, **241**, 17 (1966).
427. G. C. Berry, *J. Chem. Phys.*, **44**, 4550 (1966).
428. G. Deželić, *J. Chem. Phys.*, **45**, 185 (1966).
429. K. P. M. Heirwegh, *J. Colloid Interfacial Sci.*, **21**, 1 (1966).
430. S. N. Timasheff, *J. Colloid Interfacial Sci.*, **21**, 489 (1966).
431. J. P. Kratohvil, *J. Colloid Interfacial Sci.*, **21**, 498 (1966).
432. P. H. J. Hermanie and M. van der Waarden, *J. Colloid Interfacial Sci.*, **21**, 513 (1966).
433. W. P. J. Ford, R. H. Ottewill, and H. C. Parreira, *J. Colloid Interfacial Sci.*, **21**, 522 (1966).
434. W. F. Beyer, *J. Pharm. Sci.*, **55**, 622 (1966).
435. E. J. Meehan and G. Chiu, *J. Phys. Chem.*, **70**, 1384 (1966).

436. J. J. Hermans, *J. Polymer Sci.*, C12, 51 (1966).
437. R. H. Marchessault, ed., *J. Polymer Sci.*, C13, 1 (1966).
438. M. R. W. Brown, *J. Soc. Cosmetic Chemists*, 17, 185 (1966).
439. C. F. Cornet and H. van Ballegooijen, *Polymer*, 7, 293 (1966).
440. P. Debye, *Pure Appl. Chem.*, 12, 23 (1966).
441. D. Mealor and A. Townshend, *Talanta*, 13, 1191 (1966).
442. R. G. Newton, *Scattering Theory of Waves and Particles*, McGraw-Hill, New York, 1967.
443. I. C. Edmundson, *Advan. Pharm. Sci.*, 2, 146 (1967).
444. A. P. Altshuller, *Anal. Chem.*, 39, 10R (1967).
445. M. B. Abramson, R. Katzman, R. Curci, and C. E. Wilson, *Biochemistry*, 6, 295 (1967).
446. C. E. Smith and M. A. Lauffer, *Biochemistry*, 6, 2457 (1967).
447. R. A. F. Shalaby and M. A. Lauffer, *Biochemistry*, 6, 2465 (1967).
448. R. W. Lines, *Bull. Parenteral Drug Assoc.*, 21, 113 (1967).
449. N. F. Ho, *Drug Intelligence*, 1, 7 (1967).
450. M. A. Gross, *Drug Intelligence*, 1, 12 (1967).
451. Anon., *Drug Intelligence*, 1, 21 (1967).
452. L. Wahlström and S. Öberg, *Drug Intelligence*, 1, 158 (1967).
453. N. F. H. Ho, R. L. Church, and H. Lee, *Drug Intelligence*, 1, 356 (1967).
454. C. E. Wenner, E. J. Harris, and B. C. Pressman, *J. Biol. Chem.*, 242, 3454 (1967).
455. H. Yamakawa, *J. Chem. Phys.*, 46, 973 (1967).
456. S. Kielich, *J. Chem. Phys.*, 46, 4090 (1967).
457. K. Takahashi and A. Iwai, *J. Colloid Interfacial Sci.*, 23, 113 (1967).
458. E. Willis, M. Kerker, and E. Matijević, *J. Colloid Interfacial Sci.*, 23, 182 (1967).
459. S. Kitani and S. Ouchi, *J. Colloid Interfacial Sci.*, 23, 200 (1967).
460. M. J. Matteson and W. Stöber, *J. Colloid Interfacial Sci.*, 23, 203 (1967).
461. B. A. Seiber and P. Latimer, *J. Colloid Interfacial Sci.*, 23, 509 (1967).
462. M. Rebhun and H. Sperber, *J. Colloid Interfacial Sci.*, 24, 131 (1967).
463. K. Takahashi, *J. Colloid Interfacial Sci.*, 24, 159 (1967).
464. K. W. Suh and D. H. Clarke, *J. Polymer Sci. A-1*, 5, 1671 (1967).
465. T. P. Wallace and J. P. Kratohvil, *J. Polymer Sci.*, B5, 1139 (1967).
466. Z. Tuzar, P. Kratohvil and M. Bohdanecky, *J. Polymer Sci.*, C16, 633 (1967).
467. S. P. Stoylov, *J. Polymer Sci.*, C16, 2435 (1967).
468. J. M. Barrales-Rienda and D. C. Pepper, *Polymer*, 8, 337 (1967).
469. M. C. A. Donkersloot, J. H. Gouda, J. J. van Aartsen, and W. Prins, *Rec. Trav. Chim.*, 86, 321 (1967).
470. I. L. Fabelinskii, *Molecular Scattering of Light* (Russian transl. R. T. Beyer), Plenum Press, New York, 1968.
471. A. Vrij, *Advan. Colloid Interfacial Sci.*, 2, 39 (1968).
472. M. A. Lauffer and C. L. Stevens, *Advan. Virus Res.*, 13, 1 (1968).
473. L. G. Hargis, *Anal. Letters*, 1, 471 (1968).
474. R. Jaenicke, D. Schmid, and S. Knof, *Biochemistry*, 7, 919 (1968).
475. J. A. Harpst, A. I. Krasna, and B. H. Zimm, *Biopolymers*, 6, 585 (1968).
476. J. V. Sengers and A. L. Sengers, *Chem. Eng. News*, 46, 114 (1968).
477. A. E. M. Keijzers, J. J. van Aartsen, and W. Prins, *J. Am. Chem. Soc.*, 90, 3107 (1968).
478. P. J. Flory and R. L. Jernigan, *J. Am. Chem. Soc.*, 90, 3128 (1968).
479. M. D. Baijal and L. P. Blanchard, *J. Appl. Polymer Sci.*, 12, 169 (1968).
480. E. M. Barrall, M. J. R. Cantow, and J. F. Johnson, *J. Appl. Polymer Sci.*, 12, 1373 (1968).
481. B. Chu, *J. Chem. Educ.*, 45, 224 (1968).
482. L. E. Maley, *J. Chem. Educ.*, 45, A467 (1968).
483. Z. Mansour and E. P. Guth, *J. Pharm. Sci.*, 57, 404 (1968).

484. M. J. Groves and D. C. Freshwater, *J. Pharm. Sci.*, **57**, 1273 (1968).
485. K. Okita, A. Teramoto, K. Kawahara, and H. Fujita, *J. Phys. Chem.*, **72**, 278 (1968).
486. E. J. Meehan and J. K. Miller, *J. Phys. Chem.*, **72**, 1523 (1968).
487. R. L. Schmidt and H. L. Clever, *J. Phys. Chem.*, **72**, 1529 (1968).
488. T. Kato, K. Miyaso, and M. Nagasawa, *J. Phys. Chem.*, **72**, 2161 (1968).
489. E. J. Meehan and J. K. Miller, *J. Phys. Chem.*, **72**, 2168 (1968).
490. J. Vavra, J. Lapčik, and J. Sabidoš, *J. Polymer Sci. A-2*, **5**, 1305 (1968).
491. R. Koningsveld and A. J. Staverman, *J. Polymer Sci. A-2*, **6**, 305, 349 (1968).
492. G. Meyerhoff, U. Moritz, and R. L. Darskus, *J. Polymer Sci.*, **B6**, 207 (1968).
493. C. F. Cornet, *Polymer*, **9**, 7 (1968).
494. D. Cooke and M. Kerker, *Rev. Sci. Instr.*, **39**, 320 (1968).
495. G. D. Parfitt and J. A. Wood, *Trans. Faraday Soc.*, **64**, 805 (1968).
496. M. J. French, J. C. Angus, and A. G. Walton, *Science*, **163**, 345 (1969).

CHAPTER **9**

Optical Crystallography

John A. Biles

SCHOOL OF PHARMACY
UNIVERSITY OF SOUTHERN CALIFORNIA
LOS ANGELES, CALIFORNIA

9.1 INTRODUCTION

About 1885 the petrographic microscope was introduced by geologists to give more adequate information regarding thin slices of rocks. In 1916 Wright[1] suggested that the petrographic microscope be used for the identification of chemical compounds. Winchell's *The Optical Properties of Organic Compounds* was first published in 1945 and has since been revised.[2] This text included the optical crystallographic data on all of the organic compounds reported in journals to which he had access at that time. At the present time there are several texts available to those interested in the use of optical crystallography.[3-7] Today the petrographic microscope has assumed importance

337

in the various fields of chemistry including pharmaceutical chemistry, medicine, police and military intelligence, and the various scientific technologies. The use of optical crystallography has been emphasized in identification and determination of molecular orientations in the crystalline state. Its use often permits the evaluation of properties that cannot be determined by other means.

The use of optical crystallography developed with the theory of the interaction of light with the electrons of the molecules present in the crystal. The theory used in the explanation of optical crystallography is derived from the electromagnetic theory of light. Fletcher introduced the optical indicatrix to explain the effect of light passing through a nonopaque crystal.[8] Early in the twentieth century, Barlow and Pope published an article on "The development of the atomic theory which correlates chemical and crystalline structure and leads to a demonstration of the nature of valency."[9] Lewis[10] stated that this was an extremely bold but carefully elaborated theory of the nature of valence and chemical structure based on the crystallographic study of chemical compounds. The theory leads to the idea that elements having the same valence occupy the same fraction of the total volume. Jaeger contributed to the theory of Barlow and Pope using saccharin and phthalimide as models.[11] Other investigators who contributed to the theory of Barlow and Pope were Rodd,[12] Drugman,[13] and Colgate and Rodd.[14]

In 1918 Wherry published his work on the correlation of the axial ratios and refractive index ratios for compounds belonging to the tetragonal crystal system.[15] Déverin investigated the relationship of symmetrically substituted aromatic compounds to threefold and sixfold symmetry.[16] After Lorentz[17] and Lorenz[18] developed the formula for molecular refraction, calculations were made for the structural units in the molecule. Brühl found that conjugated unsaturated groups caused an increase of the molecular refractive and dispersive power.[19] This optical exhaltation was quantitative and dependent on the nature and number of unsaturated atomic groups. It has been shown that the mathematical relationship of the total induced polarization which is caused by the distortion of the electron shells in the molecule is equivalent to the molar refraction. Optical crystallographers have included the determination of the molar refraction of the molecules in the solid state in their publications.

Krishnan and other workers have studied the magnetic susceptibilities of solids.[20] The results of those studies showed that there was a correlation of magnetic and optical anisotropy. The investigators stated that

Optical measurements in theory have the same value, but owing to strong mutual influence of the optical dipoles induced in neighboring molecules in the crystal, even in those cases where the molecules happen to be oriented in the same manner, the birefringence of the crystal, as a whole, does not directly give that of the individual molecules. In order to correlate the optical constants of the crystal, the relative *positions* of the latter in the crystal have also to be considered; thus the problem is more complicated than in the corresponding magnetic case.

Lonsdale and Krishnan[21] formulated mathematical relationships between crystal and molecular diamagnetic susceptibilities and molecular orientations for the different crystal systems. The abnormal susceptibilities of aromatic molecules led Pauling to the conclusion that three bond orbitals of the carbon atom in the benzene ring are coplanar and the fourth orbital possesses lobes above and below the benzene ring.[22] The just mentioned studies led to the use of optical anisotropy in predicting molecular orientations. Wells[23] and Evans[24] have summarized these relationships.

Another area of microscopy was developed by the Kofler group and the McCrone group. The Kofler group has been responsible for the development of "Thermo-Mikro-Methoden" and the McCrone group has been responsible for the development of fusion methods in the United States. The work of McCrone may be appreciated by referring to his text.[25] Fusion methods have been utilized in the study of solids of pharmaceutical and medical importance, identification of explosives, study of insecticides, and the investigation of polymorphic crystal forms.

Most recently in the pharmaceutical field there has been considerable interest in the relationship of polymorphic forms as well as crystalline forms of solvated molecules to drug availability. The use of optical crystallography provides a convenient tool for the rapid identification of these various crystalline forms. Since a microscopic examination can be done with a minimum amount of material, positive and rapid identification and definite conclusions can be made and drawn utilizing optical crystallography.

Having introduced the reader to the historical and applied aspects of optical crystallography, a discussion of the tool, presentation of theory, and application of the knowledge are in order.

9.2 PETROGRAPHIC MICROSCOPE

The petrographic microscope, a polarizing microscope, differs from an ordinary compound microscope in that it has a rotating stage, a polarizer located in the condenser below the stage, an analyzer located in the body tube, and an Amici-Bertrand lens located in the body tube. Of importance also is the convergent lens and iris diaphragm located in the condenser. Just above the objective is a slot into which a quartz wedge, first-order test plate, or quarter-wavelength test plate may be inserted. A cross section of the petrographic microscope is shown in Fig. 9.1.

The polarizer and analyzer are either Nicol prisms or Polaroid lens. The polarizer is used to resolve all light into one plane of vibration. The analyzer is placed in the body tube so that its plane of vibration is perpendicular to that of the polarizer. When the planes are perpendicular, the Nicol prisms are crossed. The planes of vibration of the prisms coincide with the cross hairs in the ocular. The analyzer can be pushed in and out of the body tube. When the analyzer is inserted into the tube, no light will pass through. The result

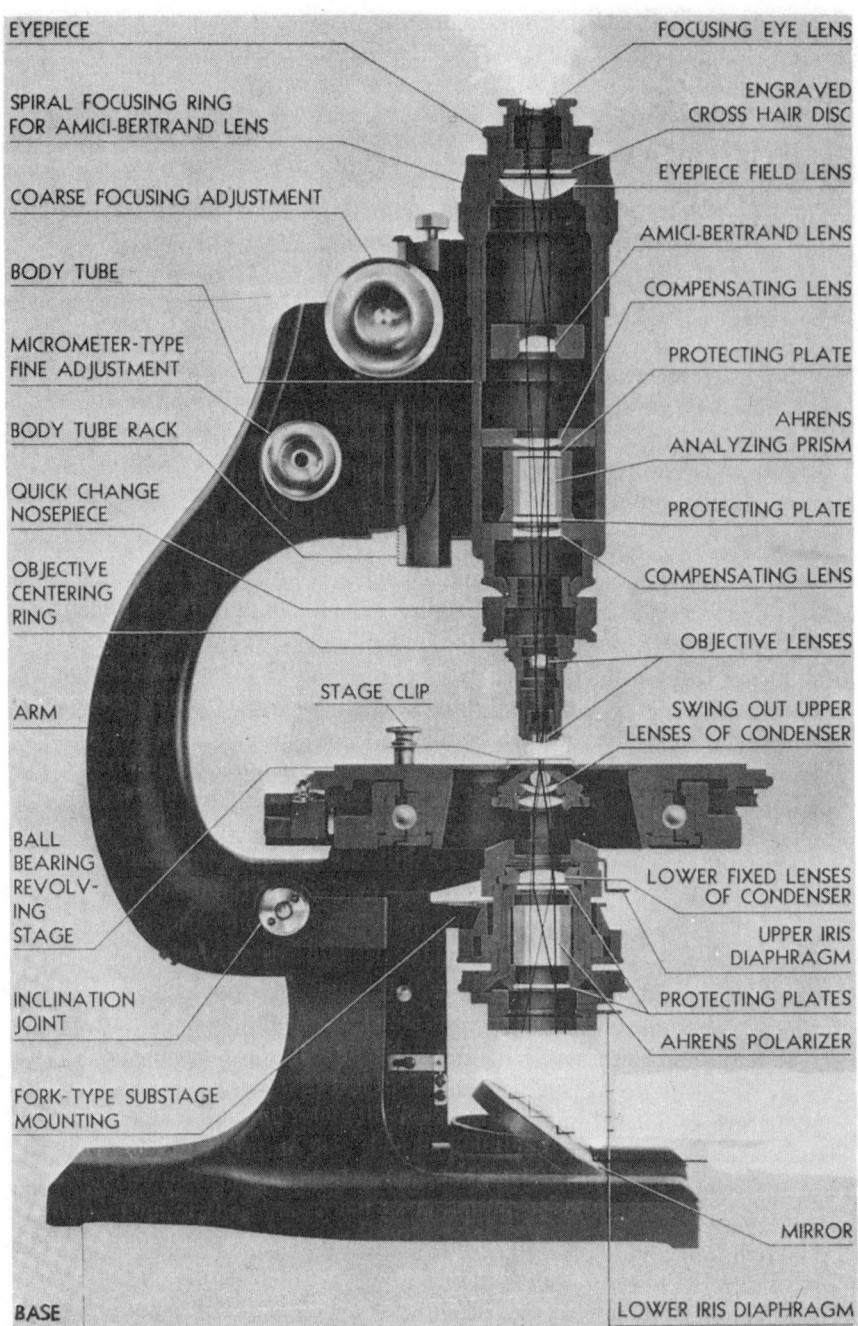

FIGURE 9.1: The Petrographic microscope. Courtesy of American Optical Company, Instrument Division.

is termed "extinction." If an anisotropic material is placed on the stage between the crossed nicols, light will pass through the analyzer provided the planes of vibration of the crystal are not parallel to the planes of vibration of the Nicol prisms. The light passing through may be colored, and these are called polarization colors. The intensity of the light in the optical system may be controlled by altering the diameter of the iris diaphragm.

The polarization colors may be made a great deal more intense by rotating the convergent lens into position above the condenser. If then the Amici-Bertrand lens is inserted into the body tube, the image will be focused at infinity. The resultant shadow or combination of colors and shadows is called an "interference figure." This figure is of great importance in determining almost all of the optical crystallographic properties of a given crystalline substance. Complete determination of the properties is made by inserting the test plates into the slot above the objective and observing the resulting effects on either the interference figure or crystal. A more detailed description of the petrographic microscope and the passage of light through the optical system is found in any suitable textbook on optical crystallography[4,7] or chemical microscopy.[5]

9.3 CRYSTALS

An ideal crystal is a regular polyhedral solid bounded by plane faces which represents an extended array of atoms arranged in a definite order in all directions. Such a crystal contains a unit cell or a unit of structure, repetition of which in three dimensions produces the crystal. Each unit cell for a specific crystal is the same size and contains the same number of atoms similarly arranged. Even though size and shape of such a crystal may vary, the angles between the faces will remain constant. At times many of the crystal faces may not be evident, but there still remains an orderly internal arrangement of atoms.

Crystals vary in their angular relationships and symmetry. On this basis, a crystal is classified into one of six crystal systems. The systems, depending on the symmetry, are subdivided into 32 classes. Certain forms are characteristic of each class. A measurement of the interfacial angles reveals a number of faces of the same type. Faces of the same type are referred to as a "form." In some instances one form is all that is needed to define the three-dimensional solid. In other instances, and at times demanded, two or more forms are present.

Each system is defined in terms of the crystallographic axes, which are imaginary lines used to describe the position of the plane faces in space. The systems are defined as follows:

1. *Isometric System.* This system includes those crystals which are referred to three mutually perpendicular equal axes. The axes are identified as A_1, A_2, and A_3.

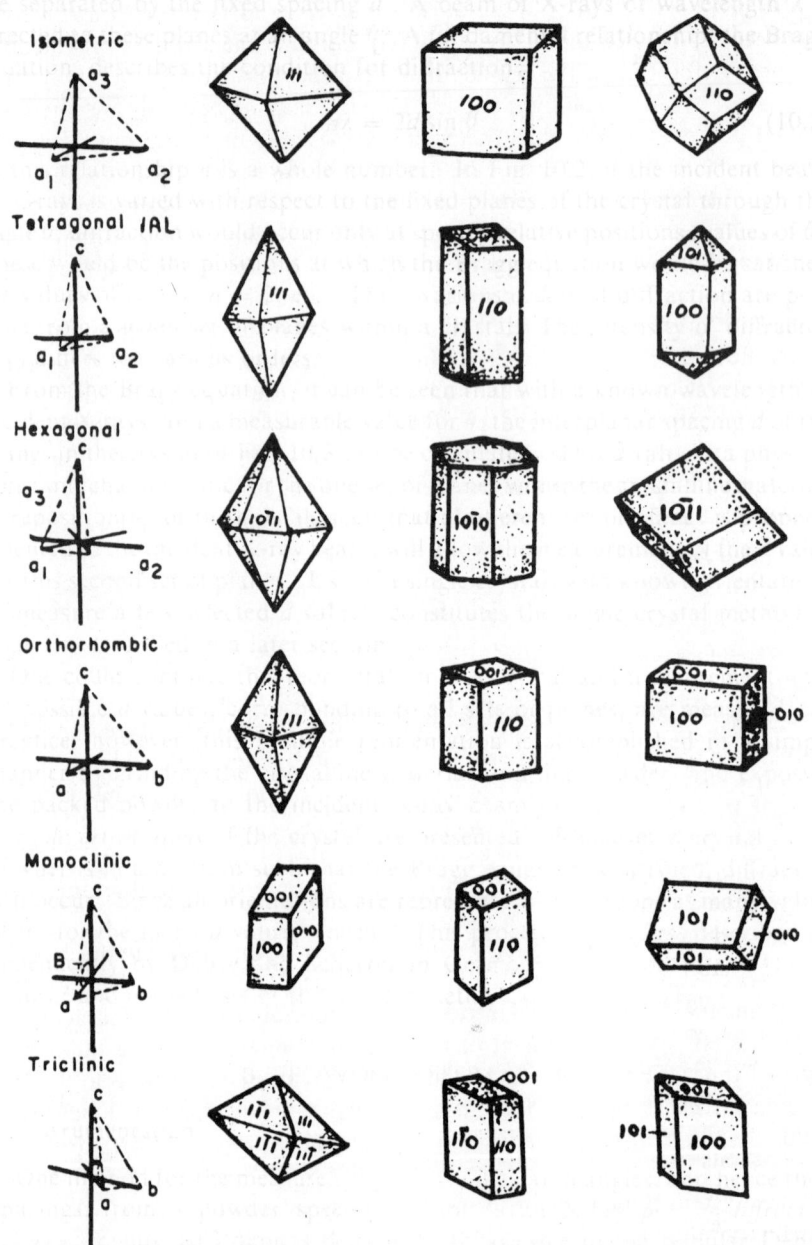

FIGURE 9.2: The crystal systems. Courtesy of John Wiley & Sons.

2. *Tetragonal System.* This system includes those crystals which are referred to three mutually perpendicular axes, two equal in length, and one longer or shorter. The equal axes are A_1 and A_2. The third axis is C.

3. *Hexagonal System.* This system includes those crystals which are referred to three equal axes lying in one plane and intersecting at angles of sixty degrees, and a fourth axis, either longer or shorter and perpendicular to the plane of the other three. The equal axes are A_1, A_2, and A_3. The perpendicular axis is C.

4. *Orthorhombic System.* This system includes those crystals which are referred to three mutually perpendicular axes of unequal length. The axes are a, b, and c.

5. *Monoclinic System.* This system includes those crystals which are referred to three unequal axes, two axes perpendicular to the third axis but not to each other. The two axes are a and c, the third axis is b.

6. *Triclinic System.* This system includes those crystals which are referred to three unequal axes intersecting at oblique and obtuse angles. The axes are a, b, and c.

The axes, simple forms, and combinations in the six crystal systems are illustrated in Fig. 9.2.

Crystals of a given substance may vary in size, relative development of a given face, and the number and kind of faces or forms present. Crystals showing such variations have different crystal habits. The habit acquired depends upon the solvent used, the temperature, pressure, concentration, impurities, and the rate of precipitation. If the crystal faces in the three dimensions are equidimensional, or nearly so, the habit is termed "equant." If a pair of opposite and parallel faces are larger, the crystal is flattened and is said to be "tabular." If this is carried to the extreme, the crystal is said to be "lamellar." If three or more faces all parallel to a line are larger and elongated, the crystal is said to be "columnar"; the extreme tendency of this type is termed "acicular."

The habit of a crystal can also vary because the crystal is easily cleaved. Cleavage occurs in the plane where the intramolecular forces are the weakest. A crystal may possess several cleavages which are parallel to common crystal faces. The cleavage may range from perfect to poor. Cleavage is of great help in suggesting intermolecular arrangement, particularly with aromatic organic substances.

Crystals of a given substance may also vary in crystal form. This means that the unit cell will vary, and the crystal exhibits polymorphism. Such substances are termed dimorphic, trimorphic, etc., according to whether they possess two, three, or more forms. On standing one of the forms may convert to another form. This form is termed "metastable." If the form conversion is reversible, the polymorphism is enantiotropic; if the form conversion is irreversible, the polymorphism is monotropic. Polymorphic conversions are dependent on pressure and temperature. In addition various polymorphic forms may be isolated using different solvents. When different crystalline forms of a single substance are isolated, one must also consider the possibility of isolating solvated forms. Also a single substance may exist in two different species in the solid state. For example, two different species can exist because of tautomerism. This difference is called "dynamic isomerism" and should not be confused with polymorphism.

9.4 CRYSTALLIZATION

Macroscopic and microscopic procedures are valuable in obtaining crystals for microscopic study. Generally speaking well-formed crystals are obtained using microscopic methods. Such crystals should be used for detailed study. Pharmaceutical chemists are well aware of the macroscopic procedures. A clean microscope slide is used for microscopic procedures. About 2 drops of the solvent is added to the slide and a small amount of the solid material to be crystallized. A supersaturated solution is obtained at room temperature. To dissolve the small amount of excess solid, the slide is heated using a micro-burner. On heating the solvent will evaporate and solid will crystallize at the edge of the solution. A glass rod is used to push the crystals to the center of the drop of solution. As this crystallization begins and proceeds, the slide is removed from the microburner. To prevent further evaporation the solution containing crystallized solid is covered with a clean cover slip.

Macroscopic procedures are advantageous because the rate of crystallization may be carefully controlled. When necessary, large volumes of solvent may be used for crystallizing slightly soluble solute. Concentrations of more soluble solids may be carefully controlled. Temperature is more readily controlled. Crystallization on a microscopical scale has great advantages in that crystallization is rapid, the crystal growth may be followed directly, and physical-chemical changes which occur can be readily identified. Crystals from microscopic procedures may also be smaller and more nearly perfect. Application may be done by doing experiments 9.1–9.4.

9.5 POLARIZATION COLORS—ANISOTROPY

Substances in which the velocity of the transmission of light differs in different directions are called "anisotropic." Tetragonal, hexagonal, ortho-rhombic, monoclinic, and triclinic crystals are anisotropic. When white light, which has many planes of vibration, passes through an anisotropic crystal, it is resolved into two planes of vibration. The velocity of light in each plane of vibration is different. Since the refractive index is a measure of the velocity of light, it can be said that the anistropic media shows double refraction. The numerical difference between the two indices of refraction is a measure of the birefringence of the crystal.

The plane of vibration of the crystal may be located under the microscope with the aid of the polarizer and the analyzer. Should the plane of vibration of the crystal be parallel to the plane of vibration of the polarizer, light will pass through the crystal with no change in direction of vibration. The light vibrating in the plane of the polarizer will then be totally reflected in the analyzer. As a result, between crossed Nicol prisms the crystal will be extinct. If the crystal is rotated by moving the stage, the crystal will then take on a color. The intensity of the color reaches a maximum at 45° from the position of extinction. The color may be any in the visible spectrum. These are called polarization colors. With continued rotation of the stage, the crystal will then become extinct again. The extinction position is 90° from the previous extinction position. Thus, two different planes of vibration have been located. The refractive index of the crystal is determined when the crystal is in the position of extinction. The principal planes of vibration of the tetragonal, hexagonal, and orthorhombic systems parallel the crystallographic axes.

The principal indices of refraction are obtained by locating the principal planes of vibration. A principal index of refraction parallels the b axis of the monoclinic system. Principal indices of refraction may or may not parallel one of the crystallographic axes of the triclinic system.

If an observer looks down the c axis of a tetragonal or hexagonal crystal between crossed Nicol prisms, the crystal section will show complete extinction. If the observer looks at a section including the a and c crystallographic axes of the tetragonal or hexagonal system, extinction will occur when the long edge of the crystal is parallel to the cross hairs. This occurs because the plane of vibration is parallel to the crystallographic axis, which in turn is parallel to the long edge of the crystal. This extinction is called "parallel extinction." Orthorhombic crystals will show parallel extinction. However no complete extinction is observed, as is shown by tetragonal and hexagonal crystals. On close examination of the principal planes of vibration through the determination of the principal indices of refraction, one finds that orthorhombic crystals possess three different planes of vibration, all mutually perpendicular. Monoclinic and triclinic crystals possess three planes of vibration and therefore three different principal indices of refraction. When one looks at a section of a monoclinic crystal which includes the b axis, the extinction is parallel. When one looks down the b axis of this monoclinic crystal, extinction between crossed Nicol prisms occurs when the long edge of the crystal is at an angle to the cross hairs. This extinction is called "inclined" or "oblique." Triclinic crystals show oblique extinction. This discussion is summarized in Table 9.1. Experiments 9.5–9.9 allow one to observe the extinction, polarization colors, and habit of various crystalline substances.

TABLE 9.1: Effect of Crystals on Light

System	Velocity of transmission	Extinction	Principal indices of refraction
Isometric (cubic)	Isotropic	Complete	One
Tetragonal	Anisotropic	Parallel and complete	Two
Hexagonal	Anisotropic	Parallel and complete	Two
Orthorhombic	Anisotropic	Parallel	Three
Monoclinic	Anisotropic	Parallel and inclined	Three
Triclinic	Anisotropic	Inclined	Three

9.6 DETERMINATION OF REFRACTIVE INDEX

The refractive index governs the visibility of all transparent objects. It can be defined as the ratio of the velocity of light traveling in a vacuum to the light traveling in a given medium. The refractive index has also been defined as the ratio of the sine of the angle of incidence to the angle of refraction when incident light is refracted into a medium of less density to a medium of greater density. These ratios result because of the interaction of light with the electron shells of the lesser and denser media. The interaction of light with the electron cloud causes a distortion of the cloud. A measure of the ease of distortion is called the "polarizability." The refractive index is directly proportional to the polarizability. The distortion causes an optical dipole to be induced. The total polarization, which is a measure of the dielectric constant, is a function of the induced dipole and the permanent dipole. For light of long wavelength Maxwell showed that the dielectric constant was equal to the square of the refractive index. For this reason the molar polarization can be calculated knowing the refractive index.

The refractive index of liquids can be easily determined with the use of the Abbé refractometer. Knowing the refractive index of liquids, the refractive index of solids may be determined, since a nonopaque solid will not be seen in a liquid provided both solid and liquid have the same refractive index.

To determine the refractive index of solids, it is necessary to prepare a series of liquids having specific refractive-index values. On mixing mineral oil and α-bromnaphthalene in various proportions a series of liquids can be prepared having refractive-index values varying from 1.48 to 1.658. Usually the oils are prepared in which each oil varies from the next oil by 0.005. Equal volumes of mineral oil and α-bromnaphthalene provide a liquid having an index midway between 1.48 and 1.658. The series of oils may be continued by mixing α-bromnaphthalene with methylene iodide (1.74). Another oil may be prepared by saturating methylene iodide with sulfur, iodoform, and iodides. This provides a liquid with a refractive index of 1.86. Mixing this latter oil with methylene iodide will provide various liquids having refractive indices between 1.74 and 1.86. Most organic solids are not sufficiently soluble in these oils. Thus, they may be used to determine the refractive index of the solids. The refractive index of the liquids will vary about 0.0004 units/1°C.

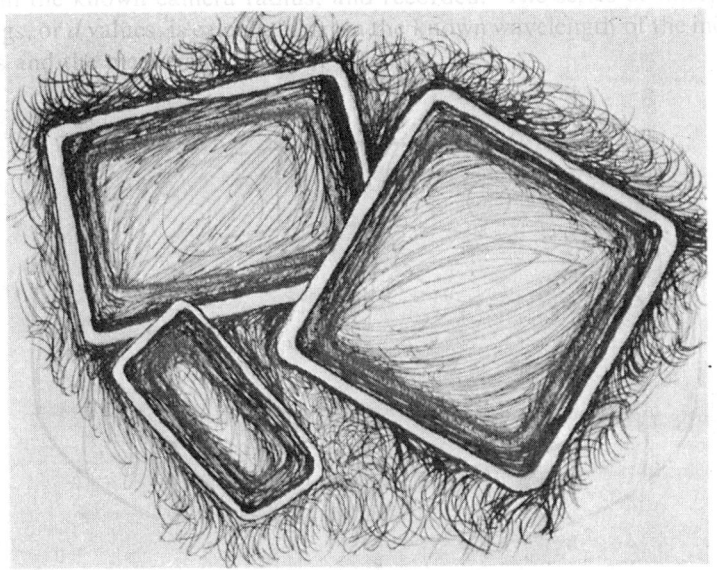

FIGURE 9.3: The Becke line is the bright band of light surrounding the crystal.

To determine the refractive index of an isometric crystal such as sodium chloride, freshly crystallized material is isolated and dried. A few crystals are placed in a microscope slide and immersed with a couple of drops of refractive-index oil. The slide is mounted on the revolving stage of the microscope and focused sharply. The Becke line is used to determine if the crystal has a higher or lower refractive index than the surrounding medium. The Becke line is a bright band of light which moves to the medium of higher index of refraction as the body tube is focused upward. The bright band of light will move to the medium or lower index of refraction as the body tube is focused downward. The Becke line is difficult to observe if the field is too well illuminated. Therefore, the iris diaphragm must be adjusted so that the field is not too bright; the field should have a slight touch of gray. The Becke line is illustrated in Fig. 9.3. The movement of the Becke line in opposite directions when the body tube is moved upward or downward may be explained with the use of Fig. 9.4. As the figure suggests the majority of rays from the light source is reflected when they hit the interface between the media of high and low refractive index. The reader may do experiments 9.10 and 9.11 to familiarize himself with the determination of the refractive index of solids.

When a crystal is mounted in a chosen immersion oil, the Becke line is observed. If the Becke line moves toward the center of the crystal as the focus is raised, then the surrounding medium has the lower refractive index. In order to determine the refractive index of the crystal, a higher mounting oil

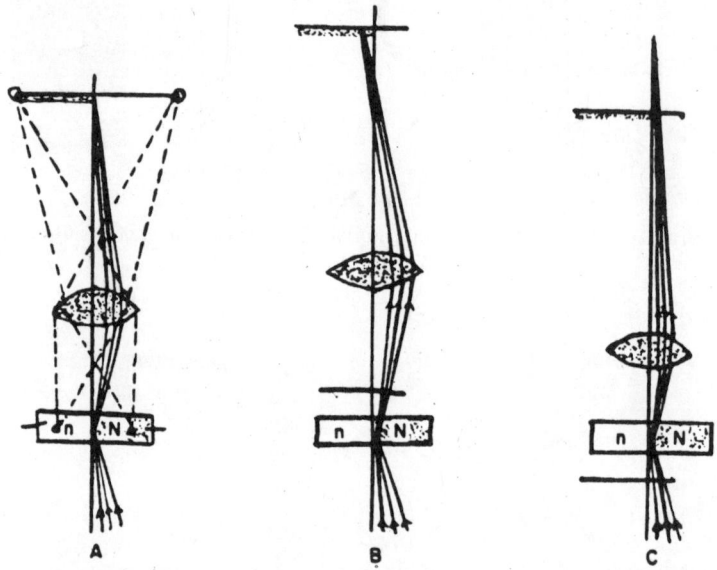

FIGURE 9.4: Movement of the Becke line. (A) Crystal in sharp focus. (B) Becke line within crystal as the body tube is focused upward. (C) Becke line outside crystal as the body tube is focused downward. Courtesy of John Wiley & Sons.

is chosen. After mounting the crystal in an oil of higher refractive index, it may be observed that the Becke line moves outward from the edge of the crystal as the focus is raised. With proper choice of oils, the crystal is found to disappear when mounted in one of the oils, or a mixture of two oils adjacent in refractive index. When crystals do disappear, it is concluded that the solid has the same refractive index as the liquid. Should the crystal be colored, then a filter of the same color may be placed between the condenser and the light source so that the field will have a similar color. In this manner the refractive index of the colored crystal may be easily determined.

When an isometric crystal is mounted in an oil having the same refractive index, the crystal will not be observed, even when the stage is rotated through 360°. If the difference between the refractive index of the liquid and solid is low, the contrast is low; when the difference is great, the contrast is high. When one mounts an anisotropic crystal in a liquid and focus on the solid sharply under the microscope, it can be observed that the contrast will vary as the stage is rotated through 360°. This indicates that the refractive index of the crystal changes as the stage is rotated. This incidentally will occur only when the polarizer is in position in the condenser and when the analyzer is not inserted into the body tube.

In determining the refractive indices of an anisotropic crystal, the orientation of the crystal must be known. Thus the experience gained in observing the crystal as it is rolled in Canada balsam is most beneficial in determining the principal indices of refraction. Consider a columnar crystal which belongs to the tetragonal system. When looking at a face of this crystal (rectangular shaped) two principal indices of refraction can be determined. One index will be determined when the long edge is parallel to the vertical cross hair. A second index will be determined when the long edge is parallel to the horizontal cross hair. These positions represent extinction positions of the crystal.

To determine the refractive indices of the tetragonal or hexagonal crystal, a crystal showing the maximum polarization colors between crossed Nicol prisms is chosen for study. This crystal is rotated to a position of extinction. It is then noted whether the long edge of the crystal is parallel to the vertical or horizontal cross hair. The analyzer is then removed and the movement of the Becke line is noted. The crystals are mounted in the various immersion oils using the previously described procedure to determine the refractive index which corresponds to the chosen plane of vibration (as determined from the extinction position). Having determined the first refractive index, then the refractive index corresponding the plane of vibration at 90° is determined. This may be observed by performing experiment 9.12.

When Table 9.1 is examined, it will be observed that crystals belonging to the orthorhombic, monoclinic, and triclinic crystal systems possess three principal indices of refraction. These three indices are difficult to determine without the knowledge or use of the interference figures to determine the crystal orientation. Therefore, the procedure is somewhat "hit-and-miss." Generally speaking the orientation showing the highest polarization colors will yield the highest and lowest indices of refraction. If then the crystal is rolled 90°, the intermediate index may be obtained. These determinations, however, may not yield the three principal indices of refraction. If one is studying a crystal possessing a tabular habit, he may roll the crystal in Canada balsam to determine the planes of vibration and relate the polarization colors to each position studied. The planes of vibration may then be identified with the crystallographic axes. Then two orientations need be chosen to determine the three indices of refraction. One of the indices will be common to both orientations. This is illustrated in Fig. 9.5. The refractive indices for an orthorhombic crystal may be obtained by performing experiment 9.13.

If the higher index of refraction parallels the long edge of the crystal, then the elongation is positive. If the opposite occurs, then the elongation is negative. Elongation is not too significant since the habit of the crystal is quite susceptible to change.

The two indices of refraction in tetragonal and hexagonal crystals are identified as epsilon and omega. Omega parallels the *a* crystallographic axis and epsilon parallels the *c* crystallographic axis. The three indices of refraction in the orthorhombic, monoclinic, and triclinic crystals are referred to as alpha, beta, and gamma. Alpha is the lowest index value and gamma is the highest index value, beta therefore being the intermediate value.

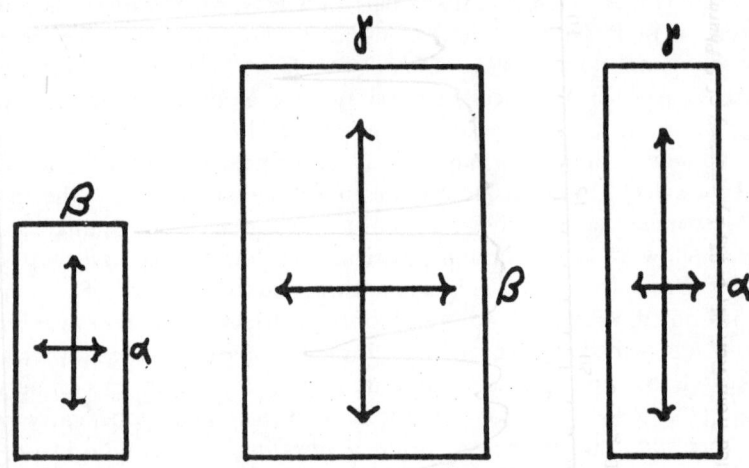

FIGURE 9.5: The three principal planes of vibration as observed from the three faces of the orthorhombic crystal. The refractive indices are alpha, beta, and gamma. The crystal shows positive elongation.

It is possible that omega can be a higher or lower value when compared to epsilon. When omega is the higher refractive index, the crystal by definition is optically negative. When epsilon is the higher refractive index, the crystal is optically positive. When the values for alpha, beta, and gamma are compared, one can conclude that beta, the intermediate value, may have an index value closer to alpha than gamma. If this occurs, one states the orthorhombic, monoclinic, or triclinic crystal is optically positive. The crystal is optically negative when beta has an index value closer to gamma than to alpha.

9.7 INDICATRIX

To understand the optic sign of anisotropic crystals more adequately and comprehend the effect of crystals on light, the indicatrix is used. By definition the indicatrix is a "three dimensional geometric figure showing the variation of the indices of refraction of a crystal for light waves in their directions of vibration. Each radius vector represents a vibration direction whose length measures the index of refraction of the crystal for waves vibrating parallel to the direction." Consider a pinpoint source of light in the center of an isometric crystal. Allow the light waves to propagate in all directions for an instant. Then draw wave normals perpendicular to the direction of propagation. The summation of all wave normals for the isometric crystal would be a sphere. This sphere by definition is an isotropic indicatrix. This is illustrated in Fig. 9.6.

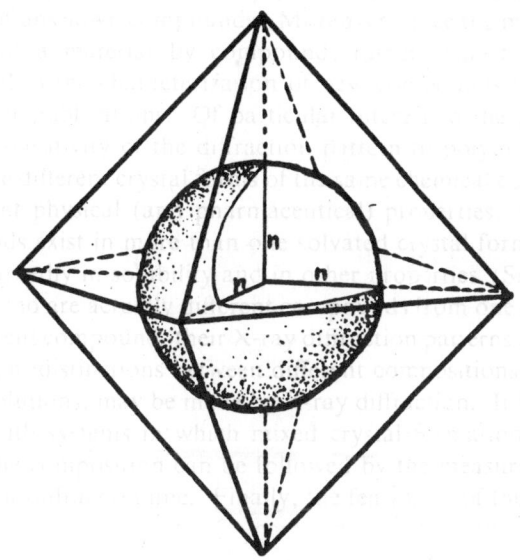

FIGURE 9.6: The isotropic indicatrix. Courtesy of John Wiley & Sons.

The radius of the sphere is equal to the refractive index of the crystal. Therefore, the refractive index of the isometric crystal is constant.

When considering a tetragonal or hexagonal crystal, an indicatrix shaped like an oblate or prolate spheroid is formed. When light travels in the direction of the c crystallographic axis, it behaves as though the crystal is isotropic. Therefore, two radii of the spheroid are equal. When light travels in a direction perpendicular to the c axis, the velocity of light varies depending on the position of the section of the crystal. Therefore, the radii will vary. This direction, coinciding with the c axis, is called the "optic axis." Consequently in the three-dimensional section, the indicatrices appear as shown in Fig. 9.7.

The radius of the longer axis of the prolate spheroid corresponds to the refractive index for epsilon. As defined earlier, this corresponds to a uniaxial positive indicatrix. When the refractive index corresponding to epsilon coincides to the radius of the shorter axis of the oblate spheroid, the uniaxial indicatrix is negative. If one considers any section of the positive or negative indicatrix, one axis always corresponds to the refractive index for omega and the other axis will vary in length up to a maximum which corresponds to

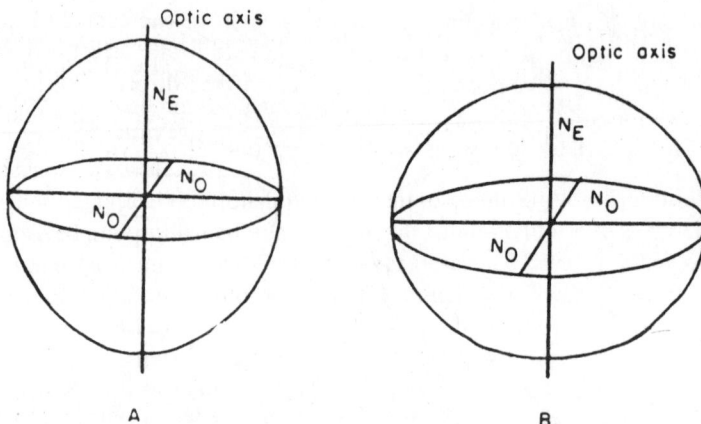

FIGURE 9.7: The uniaxial indicatrix for (A) a positive crystal and (B) a negative crystal.

epsilon. Consequently the value for omega will always be constant irregardless of the position of the crystal (or position of the indicatrix). However, only the highest value for epsilon for a positive crystal and the lowest value for epsilon for a negative crystal should be reported. These values are called "principal indices of refraction." The exact positions of the indicatrices for

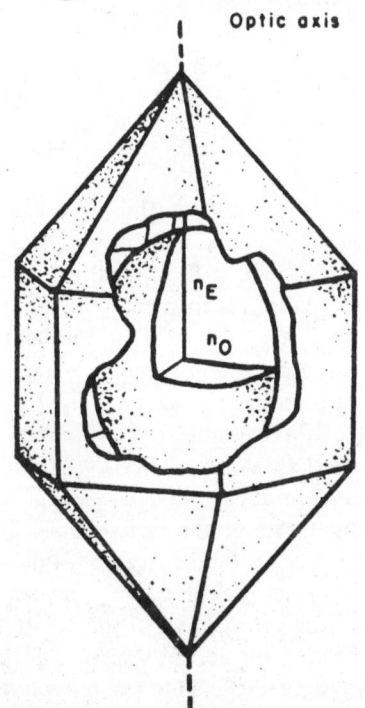

FIGURE 9.8: Uniaxial positive crystal (quartz). Courtesy of John Wiley & Sons.

quartz (positive crystal) and sodium nitrate (negative crystal) are illustrated in Figs. 9.8 and 9.9.

In the quartz and $NaNO_3$ crystal the section of the indicatrix perpendicular to the optic axis is circular. All other sections are elliptical. When looking at a section perpendicular to the optic axis, the crystal shows complete extinction between crossed Nicol prisms. This always holds true and is a convenient way

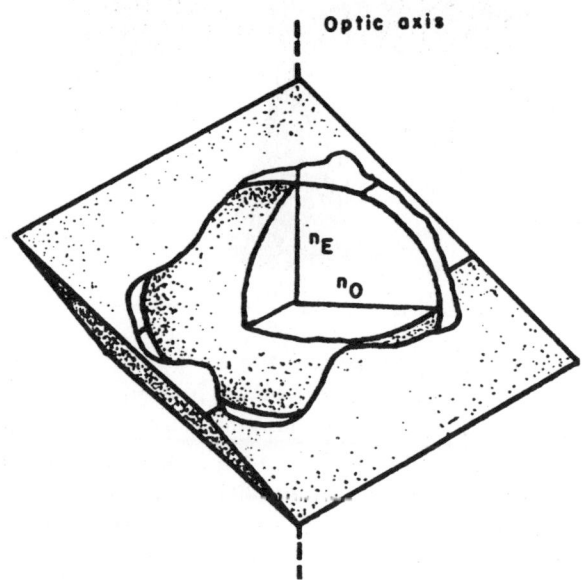

Optic axis

FIGURE 9.9: Uniaxial negative crystal ($NaNO_3$). Courtesy of John Wiley & Sons.

to locate the optic axis. All other observed sections of the crystal correspond to elliptical sections of the indicatrix and will show parallel extinction. That is, when looking at the crystal planes corresponding to the elliptical sections, one sees the crystal becoming alternately light and dark between crossed Nicol prisms. The crystal is extinct when one of the axes of the indicatrix is parallel to the plane of the polarizer. The greatest difference in the indices of refraction, i.e., the maximum birefringence, is obtained when a section is viewed parallel to the plane of the optic axis. This orientation gives the two principal indices of refraction. When a section is viewed perpendicular to the optic axis, the crystal shows complete extinction between crossed Nicol prisms. With this orientation only omega is obtained.

The orientation of tetragonal and hexagonal crystals and the optic sign are determined from interference figures. An interference figure is a telescopic image of convergent light passing through an anisotropic crystal between crossed Nicol prisms. The definition of the interference figure also represents the procedure by which the figure may be obtained. A crystal is focused sharply and placed at the intersection of the cross hairs in the ocular.

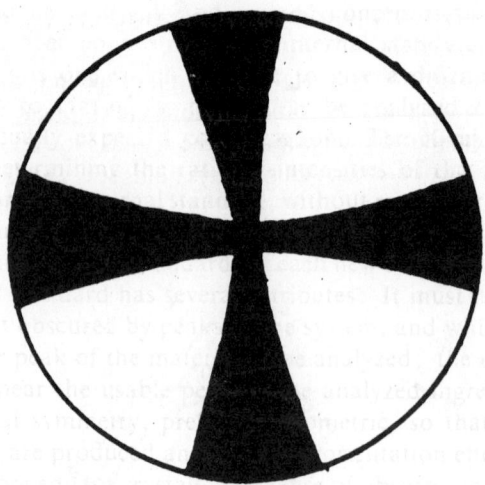

FIGURE 9.10: Centered uniaxial optic-axis interference figure.

The analyzer is then placed in position, the convergent lens is inserted into position and the Amici-Bertrand lens is placed in position so that the power of the ocular will be nullified. If one is looking at a section of a tetragonal or hexagonal crystal perpendicular to the *c* axis (i.e., looking down the *c* axis or looking down the optic axis) one sees an interference as shown in Fig. 9.10. This figure is called a centered optic-axis uniaxial interference figure. If one examines Figs. 9.8 or 9.9, one concludes that the crystal must rest on its apex to get the centered figure (i.e., in order to look down the optic axis). This is impossible unless the crystal is mounted in a viscous liquid such as Canada

FIGURE 9.11: Off-centered uniaxial optic-axis interference figure.

balsam. It is more likely that the crystal will rest on one of its faces. If the direction of propagation of light is at an angle to the optic axis, then the figure will not be centered, as illustrated in Fig. 9.11.

The crystal may be lying on a predominant face and the optic axis is outside the field of the microscope. Then the cross will not be observed, only one of its brushes. This is illustrated in Fig. 9.12.

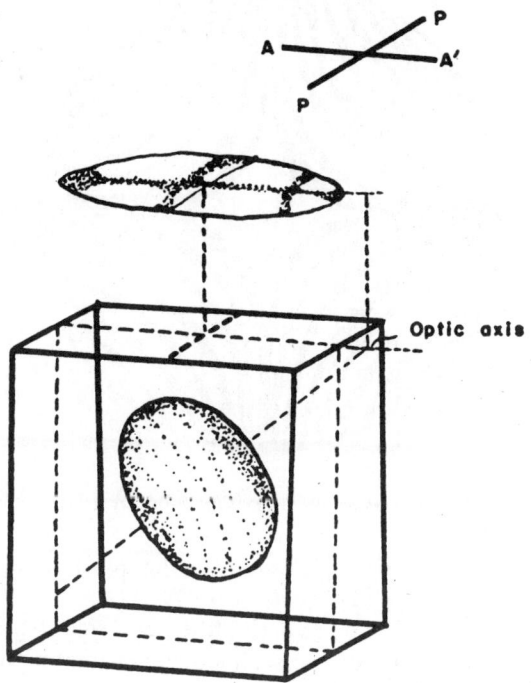

FIGURE 9.12: Off-centered optic-axis figure. Point of emergence of optic axis is outside the field of the microscope. Courtesy of John Wiley & Sons.

Should the crystal be lying on a face which represents a section including the *a* and *c* crystallographic axis and therefore including the optic axis, one will observe a flash figure. This interference figure is illustrated in Fig. 9.13.

The optic sign may be determined using the optic-axis interference. Before this procedure is described, a review of the interaction of light waves (electromagnetic wave theory) is necessary. Without any crystal under the microscope and with the polarizer and analyzer in position, different Newton colors are observed when the quartz wedge is inserted in the slot located above the objective. As the wedge is pushed in, some colors will reappear. On first appearance, the color is first order; on second appearance it is second order, etc. The colors of the first order include, in order: gray, gray blue, white, yellowish white, yellow, and red. Following the red color, include, in order, the second-order colors, namely: violet, blue, green, yellow, orange, red. The third-order colors include violet, green, yellow, red. In reference to red color, it can be said that a color of lower order would be yellow or orange and a color of higher order would be blue or violet. The quartz wedge is so constructed that the fast ray vibrates in the direction of the long edge and the slow

FIGURE 9.13: The uniaxial flash figure. Optic axis in the diagonal position.

ray vibrates in the direction of the width of the quartz wedge. The quartz wedge is illustrated in Fig. 9.14.

Another test plate used in determining the optic sign is the gypsum plate or selenite plate, which is ground to such a thickness that it gives a first-order red color. This test plate is shown in Fig. 9.15 and the planes of vibration of the fast and slow rays are identified.

Consider the anisotropic crystal to be of such thickness that its polarization or interference color is gray. If this anisotropic crystal is now placed under the microscope in its extinction position and the first-order red test plate is inserted, the field will appear red. When rotated 45°, the crystal may appear blue. If the stage is now rotated an additional 90°, the crystal will appear yellow or orange. These observations are explained as follows: when the slow and fast rays of the test plate and crystal are vibrating, parallel reinforcement occurs. Thus, the color (red) will be reinforced to give a color of a higher order (blue). If the slow and fast rays of the test plate and crystal are vibrating perpendicularly, then compensation occurs and the result is a color of a lower order (i.e., color changes from red to yellow or orange). Reinforcement is called "constructive interference" and compensation is called "destructive interference."

When Fig. 9.10 is examined, one observes a cross, termed "isogyres," which represents positions of complete extinction. The light areas in the four quadrants represent slow and fast rays of the crystal vibrating mutually perpendicular of such wavelength difference that destructive interference is not complete. Thus in the four quadrants rays are vibrating mutually perpendicular, one ray representing the radial ray, the other representing the tangent ray. The velocity of one of the rays does not change and is termed the "ordinary ray," the other ray velocity changes and is called the "extraordinary

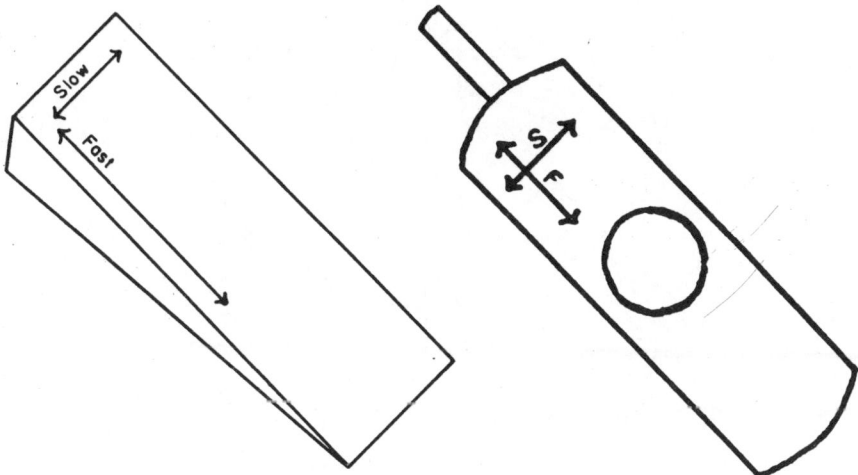

FIGURE 9.14: The quartz wedge.   FIGURE 9.15: The first-order red test plate.

ray." The former ray is the tangent ray and the latter ray is the radial ray. The traces of the planes of vibration of the ordinary and extraordinary components in the optic-axis uniaxial interference figure are illustrated in Fig. 9.16. The optic axis interference figure is dashed in. If the first-order red test plate is inserted into the slot, the isogyres will appear red. The slow ray of the test plate (see Fig. 9.15) will vibrate parallel to the *o* ray in quadrants 2 and 4 and will vibrate parallel to the *e* ray in quadrants 1 and 3. Let us assume that the *e* ray is faster than the *o* ray. Then the optic sign may be determined from the interaction of the rays of the crystal with the rays of the test plate. Figure 9.17 illustrates the resulting effect. The isogyres will be red. In quadrants 1 and 3 adjacent to the isogyre a yellow color will appear. In quadrants 2 and 4 adjacent to the isogyre a blue color will appear. Using a first-order red test plate, blue appears when interference is constructive. This results when like rays (i.e., slow rays) vibrate parallel. Thus, in quadrants 2 and 4, the slow ray of the test plate is vibrating with the slow ray of the crystal. The *o* ray is the slow ray because its plane of vibration is parallel to that of the slow ray of the test plate. By definition, when the *o* ray is the slow

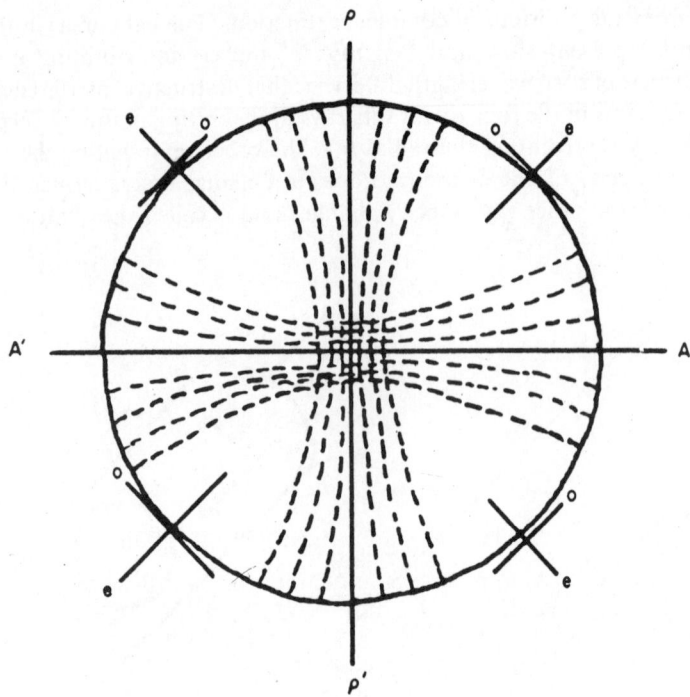

FIGURE 9.16: The uniaxial interference figure.

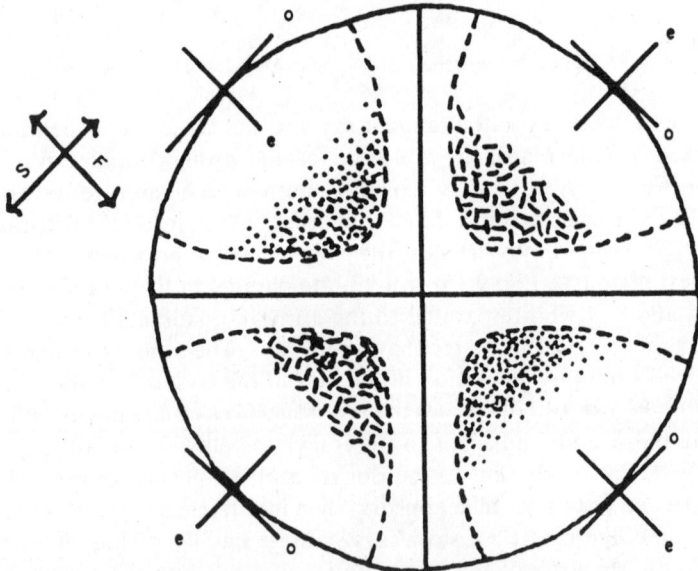

FIGURE 9.17: The uniaxial interference figure. When the first-order red plate is inserted into the body tube, blue color will appear, as represented by the dotted areas in quadrants 2 and 4. The similar areas in quadrants 1 and 3 will appear yellow.

ray and the *e* ray is the fast ray, the crystal is optically negative. Should the crystal have been optically positive (i.e., the optic sign is positive) blue would have appeared in quadrants 1 and 3, whereas yellow or orange would have appeared in quadrants 2 and 4. For a more detailed discussion of the determination of the optic sign of uniaxial crystals, using the various test plates, the reader is referred to texts in optical crystallography.

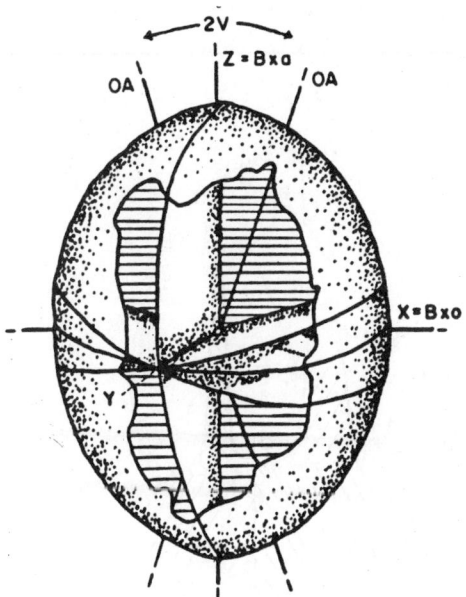

FIGURE 9.18: The positive biaxial indicatrix. *OA* is the optic axis. Courtesy of John Wiley & Sons.

Since the refractive index is inversely proportional to the velocity of the light, the refractive index for the ordinary ray (omega) is higher for a negative crystal than for a positive crystal.

The examination of the interference figures, the determination of the optic sign may be done by performing experiments 9.13 and 9.14.

The biaxial indicatrix is common to the orthorhombic, monoclinic, and triclinic crystal system. Biaxial indicates that this three-dimensional form contains two optic axes. This indicatrix corresponds to a triaxial ellipsoid. Just as the section perpendicular to the optic axis of a uniaxial indicatrix is circular, so must the section perpendicular to the optic axis of the biaxial indicatrix. The radius of each of the three axes of the biaxial indicatrix corresponds to the numerical value of the refractive index. Thus the orthorhombic, monoclinic, and triclinic crystals possess three indices of refraction. The values are identified in increasing numerical numbers as alpha, beta, and gamma. The positive and negative biaxial indicatrices are illustrated in Figs. 9.18 and 9.19.

The three axes of the triaxial ellipsoid are XX' (shortest axis), YY' (intermediate axis), and ZZ' (longest axis). The optic axes lie in the XZ plane, and XX' or ZZ' will bisect the acute angle made between the optic axes. If XX' bisects the acute angle, the indicatrix is negative. If ZZ' bisects the acute angle, the indicatrix is positive. Respectively then the ZZ' or XX' will be the obtuse bisectrix. Perpendicular to the optical axes and to the XZ plane is YY'. This intermediate axis is termed the "optic normal." Perpendicular to each optic axis is a circular section, the diameter being equivalent to YY'.

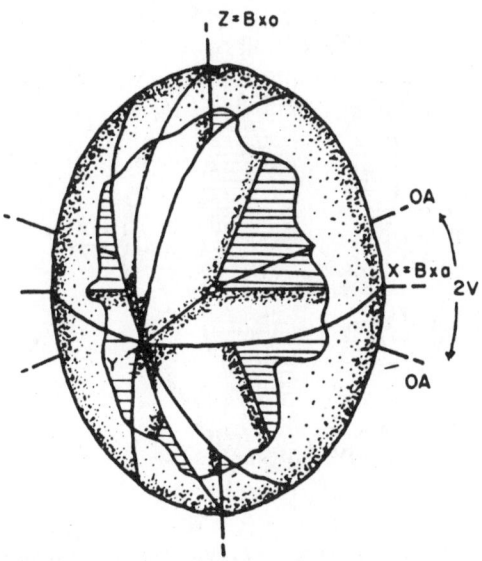

FIGURE 9.19: The negative biaxial indicatrix. OA is the optic axis. Courtesy of John Wiley & Sons.

Whereas the biaxial indicatrix as such is not observed under the microscope, observations are seen which can be correlated with the indicatrix. If the observer looks at a section parallel to the XZ plane, he is looking down the Y axis. An interference figure called the "optic normal figure" is seen. This figure is somewhat analogous to the flash figure observed with a uniaxial figure. When the crystal with the optic normal orientation is placed in the position of extinction, the interference figure is a cross. When the stage is rotated, the isogyres (black brushes) will rapidly leave the field in either the first and third quadrant or the second and fourth quadrant depending on what direction the stage is rotated. On continued rotation the isogyres will come back into the field in the alternate quadrants and form the cross at the position of extinction.

If the observer looks at a section perpendicular to the axis bisecting the acute angle made by the two optic axes (termed the "acute bisectrix" and abbreviated as Bxa) in the position of extinction, a cross is observed. On rotating the stage, the isogyres separate and come together again during a 90° rotation. If the observer is looking at a section perpendicular to the obtuse bisectrix when the crystal is in the extinction position, a cross is observed and will separate on rotation of the stage. The obtuse bisectrix figure separates more rapidly when the stage is rotated than does the acute bisectrix figure. The optic normal separates the most rapidly on rotation of the stage. If the acute angle is small enough, the isogyres will not leave the field when the stage is rotated. With most microscopes, the isogyres will leave the field when the obtuse bisectrix figure is rotated. If one examines a crystal so oriented that the observed section corresponds to the circular section of the biaxial indicatrix, a cross is not observed—only a single brush. The interference figures corresponding to the various sections of the triaxial indicatrix are shown in Figs. 9.20 and 9.21.

When examining Fig. 9.20A–C and 9.21A–B, one observes concentric rings. The center of the concentric rings is the point of emergence of the optic axis. Therefore, the distance between the point of emergence of the two optic axes in Fig. 9.20A–C is a measure of the angle between the two optic axes (termed "$2V$").

In Fig. 9.22 the axes of the triaxial ellipsoid are located in the crystal. In Fig. 9.22A the crystal is located in the extinction position and in Fig. 9.22B the crystal is located in the 45° position. When the crystal is between crossed Nicol prisms, the convergent lens and Amici-Bertrand lens inserted, the interference figure is seen. In 9.22A the optic normal (YY') coincides with the vertical cross hair and the obtuse bisectrix coincides with the horizontal cross hair. When the crystal is rotated 45° so that the isogyres are located in quadrants 2 and 4, the optic normal (YY') is in the diagonal position in quadrants 1 and 3. The obtuse bisectrix is in the diagonal position in quadrants 2 and 4. The observer is looking down the acute bisectrix. Figure 9.22B corresponds to Fig. 9.20B or C.

By definition, the length of the semiaxis of the triaxial ellipsoid corresponding to the refractive index, XX', YY', and ZZ', corresponding to alpha, beta, and gamma. It has been previously stated that the refractive index is inversely proportional to the velocity of the light. Therefore, the velocity of light vibrating in the XX' plane is the fastest, the velocity of light vibrating in the ZZ' plane the slowest, and the velocity of light vibrating in the YY' plane the intermediate. In Fig. 9.22B, the light vibrating in the diagonal plane in quadrants 1 and 3 represents rays parallel to the optic normal and the light vibrating in the diagonal plane in quadrants 2 and 4 represents the rays parallel to the Bxo.

This condition holds for that section between the convex sides of the isogyres.

With this knowledge, the optic sign of the crystal may be determined. Consider that a crystal fragment is so oriented that a centered acute bisectrix figure is obtained. The observer places the crystal such that the isogyres are in quadrants 1 and 3 in the 45° position (see Fig. 9.20C). The plane of vibration of YY' is diagonal in quadrants 2 and 4. The plane of vibration of ZZ' is diagonal in quadrants 1 and 3. The planes of vibration of the fast and slow rays are therefore known. The first-order red test plate is then inserted in the slot above the objective. Yellow or orange spots are seen next to the optic axis on the convex side of the isogyres. As recalled, yellow or orange results

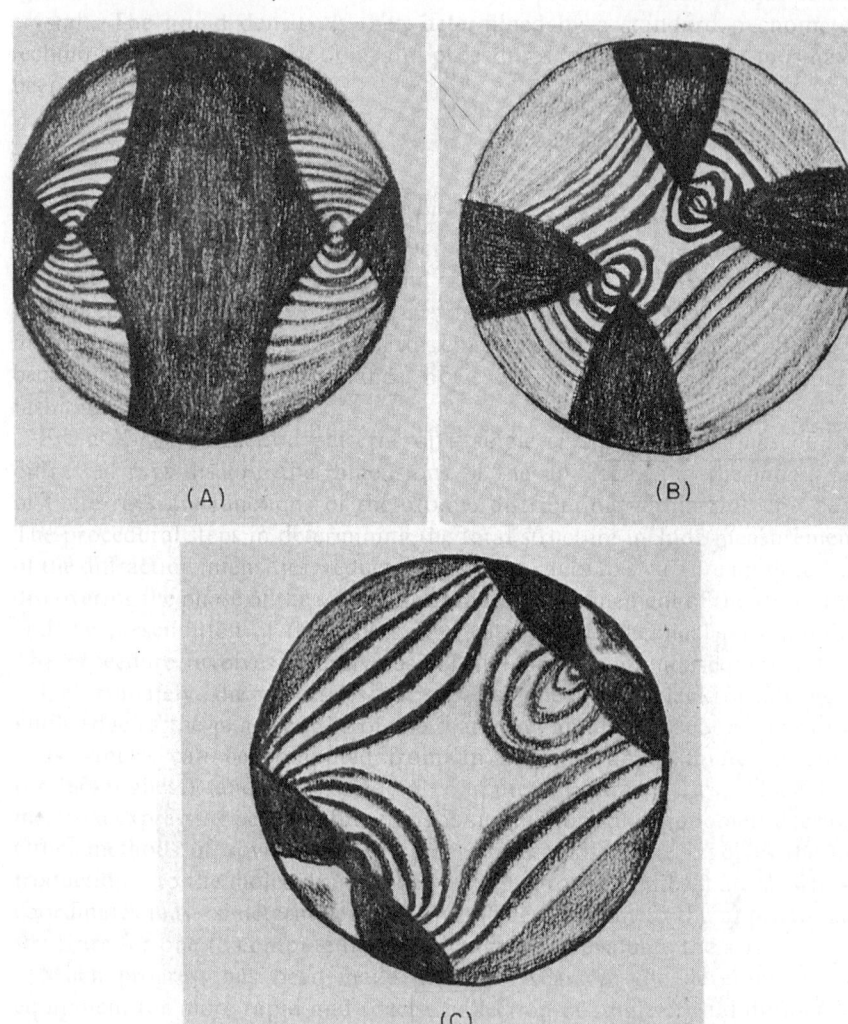

FIGURE 9.20: The acute bisectrix interference figure. (A) Crystal in extinction position. (B) Acute bisectrix with small $2V$. (C) Acute bisectrix with moderate $2V$.

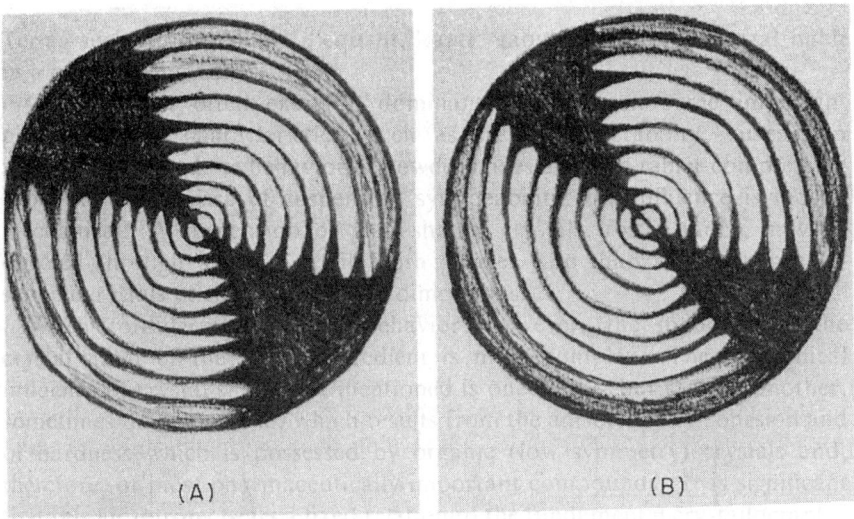

FIGURE 9.21: The biaxial optic axis interference figure. (A) Symmetrical isogyre indicating 2V to be 90°. (B) Nonsymmetrical isogyre indicating 2V to be less than 90°.

because the plane of vibration of the slow ray of the first-order red plate is vibrating parallel to the fast ray of the crystal fragment. Knowing the direction of vibration of the slow ray of the test plate, the direction of vibration of the fast ray of the crystal fragment is thus determined. If the crystal fragment is then rotated 90° so that the isogyres are in quadrants 2 and 4, one will observe a blue patch next to the point of emergence of the optic axis on the convex side of the isogyre.

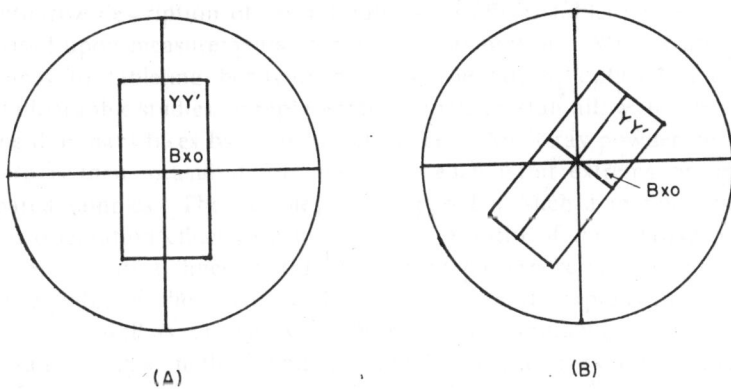

FIGURE 9.22: A section of an orthorhombic crystal including the optic normal and the obtuse bisectrix. (A) The section in the extinction position and (B) in the 45° position.

9.8 DISPERSION

In some instances, which is true of many organic compounds, there is an abnormal change of refractive index with different wavelengths of light. This condition is called the "dispersion of the indices of refraction." Dispersion is essentially the scattering of the components of white light along different paths. Dispersion of the refractive indices is possible in any of the six crystal systems. Rarely, if at all, is the dispersion of the indices in the anisotropic crystals regular.

In biaxial crystals there can be dispersion of the optic axes. In uniaxial crystals such dispersion does not occur because the optic axis for different wavelengths of light parallels the c axis. In orthorhombic crystals dispersion of the optic axes is termed "axial dispersion." The optic axes for different wavelengths of light lie in the XZ plane, but $2V$ changes with different wavelengths of light.

Axial dispersion is expressed by a dispersion formula which states that $2V$ for red light is less than or greater than $2V$ for violet light. Red and violet are chosen as these colors are most generally seen. When white light is used, violet light is cut out along the optic axis for this monochromatic light. The color at the opposite end of the spectrum, red, appears. At the point of emergence of the optic axis for red light, violet light will appear. If $2V$ for red light is greater than $2V$ for violet light, the dispersion formula is $r > v$. Some crystals will show axial dispersion, $v > r$. Another type of dispersion which may occur in orthorhombic crystals is crossed axial plane dispersion. In this instance the optic planes for red and violet light are perpendicular. At some intermediate wavelength of light, the crystal becomes uniaxial.

Three different types of dispersion may arise in monoclinic crystals, depending on the orientation of the triaxial ellipsoid. One of the planes of symmetry of the indicatrix is parallel to the plane of symmetry of the crystal. One of the axes of the indicatrix coincides with the b axis of the crystal.

When the acute bisectrix parallels the b axis, the obtuse bisectrix and the optic normal lie in the plane of symmetry. Dispersion takes place by the rotation of the optic plane around the b axis. The optic normal and the obtuse bisectrix lie in the plane of symmetry. In this instance the optic planes for different wavelengths of light are crossed. This condition is termed "crossed dispersion."

When the obtuse bisectrix parallels the b axis, the acute bisectrix and the optic normal lie in the plane of symmetry of the crystal. Dispersion takes place by rotation of the optic plane around the b axis. In the direction of the acute bisectrix the optic planes for different wavelengths of light are parallel to each other. This condition is termed "horizontal dispersion."

It is to be noted that when the B_{xa} shows crossed dispersion, the B_{xo} shows horizontal dispersion. Conversely, if the B_{xa} shows horizontal dispersion, the B_{xo} shows crossed dispersion.

A third type of dispersion is seen if the optic normal coincides with the *b* axis. The acute and obtuse bisectrices lie in the plane of symmetry. The dispersion arises from the rotation of the optic plane on the optic normal. The optic plane for different wavelengths of light are in the same plane. However, the optic axes for the different wavelengths of light are inclined to each other. This condition is termed "inclined dispersion".

In the three types of dispersion in monoclinic crystals, which are independent of the dispersion of the indices of refraction, there is always dispersion of the optic axes—namely, axial dispersion. Crossed axial plane dispersion may occur in monoclinic crystals.

The triclinic system has no plane of symmetry as the crystallographic axes are inclined to each other. None of the ellipsoidal axes corresponds to the crystallographic axes, so dispersion of the refractive indices, the optic axes, and the bisectrices are irregular.

Dispersion of the optic axes and the ellipsoid is seen when looking at an interference figure. When the crystal is in the extinction position, axial or inclined dispersion is not seen as the optic plane, for different wavelengths of light, is in the same plane as white light. When the crystal is in the extinction position, crossed or horizontal dispersion can be seen. With horizontal dispersion red light will appear in quadrants 1 or 2 or. 3 and 4. Violet light will appear in quadrants 3 and 4 or 1 and 2, respectively. With crossed dispersion red light will appear in quadrants 1 and 3 or 2 and 4. Violet light will appear in 2 and 4 or 1 and 3, respectively.

Dispersion of the indices of refraction is observed when the source of light is white light and the image of the crystal is seen. In such instances the crystal may not disappear in immersion oils. However, when the index of refraction in question is obtained, a band of blue light and a band of yellow light around the edge of the crystal will move in opposite directions as the focus is raised.

9.9 MOLECULAR ORIENTATIONS IN THE SOLID STATE

Optical and magnetic studies led to the use of anisotropy in predicting molecular orientations. Wells and Evans have summarized these relationships.

In crystals containing planar atoms or groups with all their planes parallel, a large negative birefringence is expected with two high and comparable refractive indices for light polarized in the plane of the groups and one much lower index for vibrations at right angles to this plane. When planar groups are all parallel to a line but not parallel to each other, strong positive birefringence is expected, since the single vibration direction parallel to all the groups will be associated with a much smaller velocity than in any other direction. When planar groups are inclined in all directions, no large birefringence is to be expected.

Vibrations along the length of rod-shaped molecules or atom groups correspond to much greater polarization than vibrations in any other direction.
A large positive birefringence is expected when the rod-shaped groups are all
parallel to one direction. When the groups are all parallel to a plane but not
to each other, a large negative birefringence is expected. If the rod-shaped
groups are inclined in all directions, the resulting crystal is isotropic or approximately so.

EXPERIMENTAL PROCEDURES

E9.1. Recrystallize sodium chloride from an aqueous solution using the microscopic procedure described previously. After placing the cover slip on the
recrystallized sodium chloride in the solution, place the cooled slide on the
stage of the microscope. Examine the crystals using white light. Examine
the crystals between crossed Nicol prisms. It will be noted that all of the
crystals examined show complete extinction between the crossed Nicol
prisms. Sodium chloride is isomatic.

E9.2. Recrystallize sodium chloride from an aqueous solution of urea and compare the habit obtained in this manner with the habit obtained by crystallization from distilled water. Cubic forms are obtained when crystallized from
distilled water and octahedral forms are obtained when crystallized from a
solution of urea.

Habit variations may also be observed by first crystallizing adipic acid
from distilled water. Crystallize a second batch from 0.1% benzalkonium
chloride. Finally crystallize a third batch from 0.1% sodium lauryl sulfate.

E9.3. Recrystallize methylprednisolone (Medrol) from a oiling solution of n-butyl
alcohol. Examine the crystals under the microscope obtained when
crystallized above 117°C. Recrystallize the steroid at a temperature below
117°C. Also heat the solution above 120°C and allow to cool slowly. After
standing for a few minutes examine the crystals under the microscope. Two
forms should be evident. Allow the solution to stand for 24 hours and
reexamine the crystals. The low temperature form should be present.
The high temperature form is therefore metastable at room temperature
and also more soluble than the low temperature form. Methylprednisolone
is dimorphic.

Recrystallize diethylbarbituric acid (Barbital) from hot water. How
many different forms can you identify? Repeat the microscopic study by
crystallizing from concentrated ammonium solution and also after sublima-
tion. Barbital is polymorphic.

E9.4. Recrystallize ouabain from water at 4, 20, and 50°C. Examine the crystals
under the microscope. They represent respectively the nonahydrate,
octahydrate, and dihydrate of ouabain. The tetrahydrate may be obtained
by recrystallizing from 97% methanol. The $4\frac{1}{2}$ hydrate is obtained by
recrystallizing from 95% ethanol.

The monoethanol solvate form of hydrocortisone t-butylactate may be
obtained by crystallization from 95% ethanol. When crystallized from
20% ethanol, the anhydrous form of the ester is obtained. Examine the
two crystalline forms under the microscope.

E9.5. Examination of tetragonal crystal with and without the use of crossed Nicol prisms. Recrystallize urea from water using the microscopic technique. Note the columnar crystals. The extinction is parallel. Between crossed Nicol prisms different crystals will show different polarization colors. The colors vary because the thickness of the crystals is different.

Obtain some fluorocortisone acetate and crystallize from approximately 50% ethanol. Examine under the microscope with and without the analyzer in the body tube. Complete extinction should be observed with most of the crystals studied. Crystallize the fluorocortisone acetate using the macroscopic technique. Filter and dry the crystals. Place a few crystals on a microscope slide and suspend them in Canada balsam. Place a cover slip on the balsam. Focus on a crystal under the microscope. If the cover is gently moved, the crystal, if not too large, will roll. After balancing the fluorocortisone acetate crystal on its side, one will observe parallel extinction between crossed Nicol prisms. Danthron and potassium dihydrogen phosphate are tetragonal.

E9.6. Examination of hexagonal crystals under the microscope. Repeat experiment 9.5 by (a) crystallizing lead iodide from water, (b) iodoform from water, (c) thymol from water-ethanol solution, (d) tartar emetic from water, (e) sodium nitrate from water. Examine crystals between crossed Nicol prisms, rolling the crystals in Canada balsam.

E9.7. Examination of orthorhombic crystals under the microscope. Repeat experiment 9.5 by (a) crystallizing sulfacetamide from water-ethanol (b) sulfanilamide from water, (c) potassium nitrate or potassium sulfate from water, (d) p-aminohippuric acid from water, (e) codeine sulfate from water-ethanol. Examine between crossed Nicol prisms, rolling the crystals in Canada balsam.

E9.8. Examination of monoclinic crystals under the microscope. Repeat experiment 9.5 by (a) crystallizing antipyrine, aprobarbital, ephedrine HCl, pheniramine maleate from water. Other monoclinic crystals include salicylamide, quinine dihydrochloride, p-aminobenzoic acid. Many organic medicinals are monoclinic substances.

E9.9. Examination of triclinic crystals under the microscope. Repeat experiment 9.5 by (a) recrystallizing copper sulfate from water, (b) phenolphthalein from water or water-ethanol, (c) naphazoline hydrochloride from water.

E9.10. Recrystallize sodium chloride so that suitable crystals are obtained. Dry the crystals well. Mount the crystals on a microscope slide using the following immersion oils (a) mineral oil, (b) Canada balsam or nitrobenzene and (c) α-bromnaphthalene or methylene iodide. Focus sharply on the crystals and then note the movement of the Becke line as focused upward and downward.

E9.11. Using the refractive index oils, determine the refractive index of (a) NaCl, (b) NaBr, (c) NaI. Relating the refractive index to the polarizability, and knowing that the polarizability of the halides increases with molecular weight, is it surprising that the relationship of the refractive index to the sodium halides exist as found in this experiment?

E9.12. Determine the two principal indices of refraction for urea. To gain experience, also determine the two principal indices of refraction for fluorocortisone acetate and lead iodide. With the latter two compounds, the habit is such that the common orientation is a section perpendicular to the c crystallographic axis. This orientation will show complete extinction. The crystals must then be crushed so that various random orientations will be obtained. Choose the orientation showing the greatest polarization colors.

E9.13. Place a few crystals of sodium nitrate on a microscope slide. Cover the crystals with a cover slip. Heat the slide over a microburner until the crystals have liquified. Cool the slide and examine the fused material under the microscope. Insert the convergent lens, analyzer, and Amici-Bertrand lens. Move the slide around until an optic-axis uniaxial interference figure is obtained. With the aid of the first-order test plate, determine the optic sign. Also observe the movement of the isochromatic curves appearing in the four different quadrants when the quartz wedge is inserted.

E9.14. Obtain some fluorocortisone acetate, crystallize from 50% ethanol, and determine the crystal system, optic sign, and the refractive indices. Repeat the procedure for iodoform, urea, and tartar emetic. The common orientation for the first two compounds is a centered optic-axis figure. The common orientation for the latter two compounds is the flash figure. You will observe that the two indices of refraction will be more difficult to determine when the common orientation is the optic-axis figure.

E9.15. Obtain a small sheet of mica. Separate the sheet into thinner sections. Mount one of these sections on the petrographic stage. Examine the interference figure. This is an acute bisectric. Rotate the stage until the isogyred are in quadrants 2 and 4. Determine the optic sign using the first-order red test plate. Repeat the examination with the isogyres in quatrants 1 and 3.

E9.16. Obtain some ouabain and crystallize from water at room temperature. The octahydrate is isolated. Mount isolated crystals in Canada balsam and examine the crystal. Determine what interference figure is common to this section. Then roll the crystal $90°$ so that you are looking at another section of the crystal. Determine the interference figure common to this new section. Having determined the orientation of the biaxial indicatrix by rolling the crystal, then determine the optic sign, dispersion, and principal indices of refraction for ouabain octahydrate.

QUESTIONS

Q9.1. Compare and/or differentiate:
    (a) isoaxial, uniaxial, and biaxial
    (b) dynamic isomerism and dimorphism
    (c) monoclinic and triclinic
    (d) dispersion in tetragonal and orthorhombic crystals

Q9.2. Compare the refractive indices (handbook values) of NaCl, NaBr, and NaI. Explain the values. (*Hint:* Check the values with the specific refraction values of the halogens.)

Q9.3. Crystals of aromatic molecules such as purines will cleave readily one in plane, but not in other planes of the crystals. The cleavage is parallel to the plane of the purine nucleus. Explain why this characteristic cleavage occurs.

Q9.4. (a) Will the refractive index for isometric crystals change when different monochromatic light sources are used?

   (b) How would one determine the refractive index of a colored isometric crystal using the refractive index oils described?

Q9.5. Under what conditions will optical crystallography be valuable to a structural crystallographer?

Q9.6. (a) Assuming the orthorhombic crystal illustrated in Fig. 9.22 to be a negative crystal, what kind of an interference figure would be seen if:

   (i) the investigator looked at a section including the XX' and ZZ' axes?

   (ii) the investigator looked down the XX' axis?

   (b) Does the crystal illustrated in Fig. 9.22B show $(+)$ or $(-)$ elongation?

REFERENCES

1. F. E. Wright, *J. Am. Chem. Soc.*, **38**, 1647 (1916).
2. A. N. Winchell, *The Optical Properties of Organic Compounds*, Academic Press, New York, 2nd ed., 1954.
3. A. Johannsen, *Manual of Petrographic Methods*, McGraw-Hill, New York, 1918.
4. A. N. Winchell, *Elements of Optical Mineralogy*, Part I, Wiley, New York, 1937.
5. E. M. Chamot and C. W. Mason, *Handbook of Chemical Microscopy*, Vol. I, Wiley, New York, 2nd ed., 1938.
6. N. H. Hartshorne and A. Stuart, *Crystals and the Polarising Microscope*, Arnold, London, 1934.
7. E. E. Wahlstrom, *Optical Crystallography*, Wiley, New York, 1960.
8. L. Fletcher, *The Optical Indicatrix*, Frowde, London, 1892; quoted by E. E. Jelley, in *Physical Methods of Organic Chemistry*, (A. Weissberger, ed.), Wiley (Interscience), New York, 1945, p. 458.
9. W. Barlow and W. J. Pope, *J. Chem. Soc.*, **89**, 1675 (1906).
10. G. N. Lewis, *CA*, **1**, 1809 (1907).
11. F. M. Jaeger, *Z. Krist.*, **44**, 61 (n.d.); see *CA*, **2**, 498 (1908).
12. E. H. Rodd, *Proc. Roy. Soc. (London)*, **A292**, 313 (1913–1914).
13. J. Drugman, *Z. Kryst. Mineral*, **53**, 240 (n.d.); see *CA*, **8**, 1111 (1914).
14. R. T. Colgate and E. H. Rodd, *J. Chem. Soc.*, **97–98**, 1585 (1910).
15. E. T. Wherry, *J. Wash. Acad. Sci.*, **8**, 277, 319 (1918).
16. L. Déverin, *Chem. Zentr.*, **109**, 847 (1938, II).
17. H. A. Lorentz, *Ann. Physik.*, **9**, 641 (1880).
18. L. Lorentz, *Ann. Physik*, **11**, 70 (1880).
19. J. W. Brühl, *J. Chem. Soc.*, **91**, 115 (1907).
20. K. S. Krishna, B. C. Guha, and S. Banerjee, *Trans. Roy. Soc. (London)*, **A231**, 235 (1933).
21. K. Lonsdale and K. S. Krishnan, *Proc. Roy. Soc. (London)*, **A156**, 597 (1936).
22. L. Pauling, *J. Chem. Phys.*, **4**, 673 (1936).
23. A. F. Wells, *Structural Inorganic Chemistry*, Oxford, New York, 1945, p. 230.
24. R. C. Evans, *An Introduction to Crystal Chemistry*, Cambridge, New York, 2nd ed., 1964, p. 116.
25. W. C. McCrone, *Fusion Methods in Chemical Microscopy*, Wiley (Interscience), New York, 1957.

CHAPTER **10**

X-Ray Analysis

John W. Shell

DIRECTOR OF RESEARCH
ALLERGAN PHARMACEUTICALS
SANTA ANA, CALIFORNIA

10.1 PRELIMINARY REMARKS.

There are a number of X-ray methods of analysis which are useful to the pharmaceutical scientist. Some of these bear little relationship to the others, except for the common employment of X-radiation as an energy source. For the reader who is unfamiliar with these analytical methods, some preliminary remarks are pertinent.

X-Ray *diffraction* procedures apply only to *crystalline* materials, and the results disclose information about the material as a *compound*. Analyses by both X-ray *emission* (fluorescence) and X-ray *absorption* techniques may be applied to material in any physical state, solid, liquid, or gas. These methods provide analysis by chemical *elements*, rather than by compound, and are largely insensitive to the combination state of the element. They are applicable to all elements except those of low atomic number. Emission analysis is more sensitive than absorption analysis at trace levels, although at higher element concentration there are advantages to the absorption method.

The purpose of this chapter is to acquaint the pharmaceutical scientist with the various X-ray methods which he may find useful. No description has been made of the extensive array of instruments and accessories now available, nor of the mechanical or electronic principles involved in their operation. In general, the manufacturers' literature is the best source of this information.

Some of the figures used in this chapter have appeared before. The author is grateful to the editors of the *Journal of Pharmaceutical Sciences* for permission to reuse Fig. 10.6 and Figs. 10.8 through 10.14, which first appeared in that journal as part of original publications of the author.

10.2   X-RAY SAFETY

Two hazards are involved in working with X-rays. One stems from the high voltages associated with X-ray generators (20 to 60 kV). Such potentials can be lethal, but fortunately the manufacturers of modern-day equipment have built in many protective features, and thus risks from high voltages are virtually nonexistent with proper use of the equipment.

An additional hazard stems from the X-rays themselves. Although most X-ray equipment provides features to protect against radiation damage to personnel, the routine use of the equipment allows many opportunities for oversight on the part of the operator which result in radiation exposure. This can be from "scattered" radiation throughout the room as well as from the direct X-ray beam itself.

From sufficient exposure, somatic injuries (those to the individual operator) include leukemia and other malignant diseases, ocular lens opacities, impaired fertility, and shortening of life. Genetic injuries (those to the offspring of the irradiated individual) may not become apparent for many generations. Constant awareness and concern of the operator for the safety of himself and others is essential.

Convenient radiation monitoring devices in the form of "r-meters" are available commercially, and should be used to monitor the laboratory routinely. It is also advisable for operating personnel to wear a "film badge," which is a small plastic device, clipped to the clothing, and which contains a strip of unexposed film. This film, usually a standard "dental pack," is replaced and developed regularly, providing evidence of accumulated exposure. The subject of the maximum permissible dose for a variety of circumstances has been thoroughly studied, and recommendations have been published.*)

10.3 PRODUCTION AND PROPERTIES OF X-RAYS

X-Ray tubes for analytical use are devices for bringing about, in a controlled fashion, the bombardment with fast-moving electrons of a pure target metal, which in turn emits the X-radiation. Several processes occur when the electrons impinge on the target metal. One is a quantum process in which an inner-shell electron of the target atom is displaced by collision, thereby ionizing the atom. The vacancy may then be filled by an outer-shell electron, a process accompanied by a release in energy. This energy is in the form of X-rays, of wavelength characteristic of the atom. A second quantum process occurring as a result of the electron bombardment of the target metal is simple decreasing of the velocity of the impinging electrons, due to the influence of the electric field near the atomic nucleus. The decrease in energy ΔE of the electron is compensated by an energy release in the form of an X-ray of frequency ν, according to the Einstein equation

$$h\nu = \Delta E \qquad (10.1)$$

where h is Planck's constant. The X-rays thus produced are not characteristic of the bombarded atom, but consist of a wide range of continuously varying wavelengths, limited in range by the energy of the bombarding electrons.

Superposition of the two types of emitted X-rays gives an overall pattern similar to that shown in Fig. 10.1, which is an idealized drawing of the intensity of emitted radiation as a function of its wavelength, for sufficiently intense electron bombardment of a target atom. The two peaks represent X-radiation characteristic of the target element. The $K\alpha$ peak represents the energy emitted by an L-shell electron filling a K-shell vacancy; the $K\beta$ peak represents energy emitted by an M electron filling a K-shell vacancy. Similarly, the filling of L-shell vacancies by outer-shell electrons gives rise to an L series of peaks, and so forth. A relationship known as Moseley's law relates the wavelength of the characteristic radiation to the atomic number of the target element.

Monochromatic X-radiation is produced by effectively removing all radiation of wavelength lower than the $K\alpha$ radiation. This is accomplished by

* National Bureau of Standards Handbooks No. 76 and 93, Superintendent of Documents, U.S. Government Printing Office, Washington, D.C. 20402.

passing the radiation beam through a material whose mass absorption properties allow passage of wavelengths immediately greater than some specific
value and effect absorption of wavelengths immediately below this value.
The value is called the "absorption edge "for the filtering element; it is the
wavelength at which a discrete discontinuity exists in a plot of the mass
absorption coefficient vs. wavelength for the element. 'This discontinuity

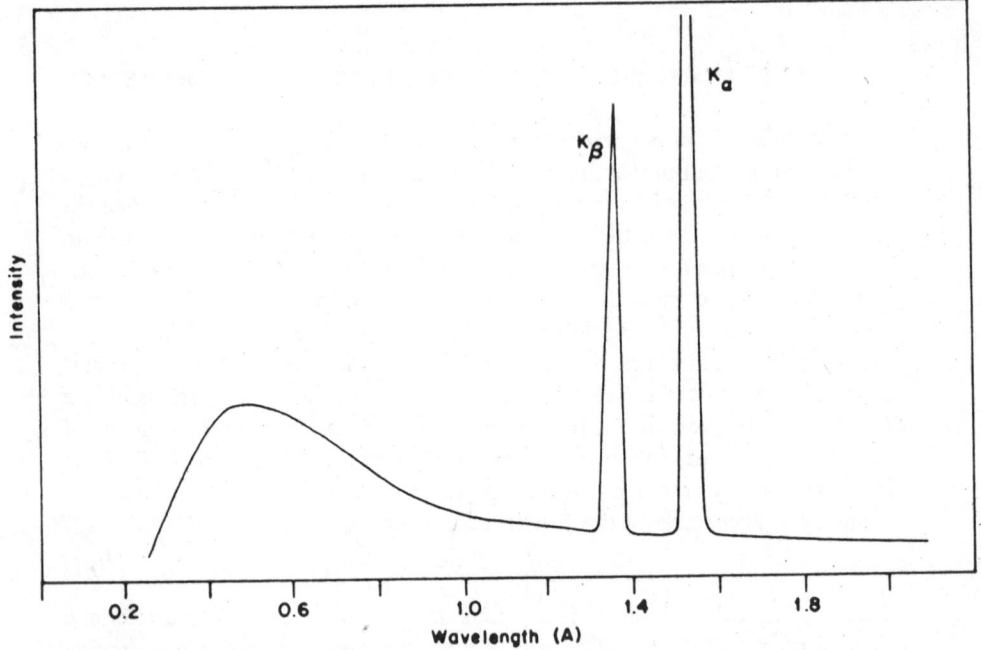

FIGURE 10.1: Intensity distribution in an X-ray beam from a copper target.

for nickel, for instance, is at a wavelength which is between the wavelength
values for the $K\alpha$ and $K\beta$ radiation from a copper target. A thin film of
nickel therefore selectively filters $K\beta$ and lower "white" wavelengths from
copper radiation to produce fairly monochromatic X-rays of known wavelength. (Copper $K\alpha = 1.5418$ Å).

10.4 X-RAY DIFFRACTION

A. PRINCIPLES OF DIFFRACTION

As noted in preliminary remarks to this chapter, the analysis of materials
by X-ray diffraction is limited to those materials which are crystalline. Ideal
crystals are regular polyhedral forms bounded by smooth surfaces (faces)

which reflect an orderly internal arrangement of atoms or molecules. In the pharmaceutical field one seldom encounters large, ideal crystals with faces present, but the same internal arrangement persists through grinding or micronizing of the crystalline material, and it is this inner structure of crystals—a three-dimensional repeat pattern called a "space lattice"—which is responsible for the diffraction of X-rays.

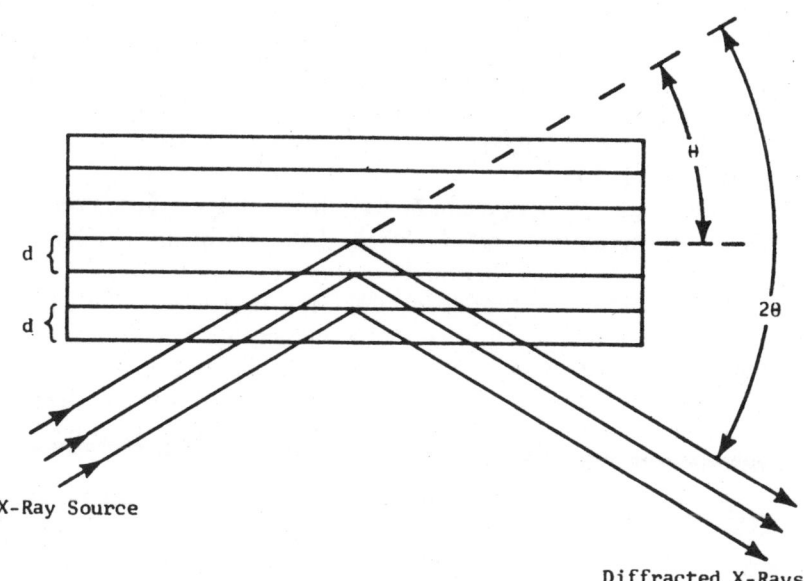

FIGURE 10.2: Diffraction of X-rays from an oriented crystal.

It is significant that the units of the space lattice are separated by distances of the same order of magnitude as the wavelength of X-rays. Upon impingement, the X-rays are scattered by the electrons within the atoms making up the space lattice. As the atoms are regularly arranged in a repeat pattern, the wave fronts emerging from each scattering center form a pattern. It is the reinforcement in specific directions of scattered X-rays which comprises the diffracted beam.

The atoms or molecules making up the internal structure of the crystal lie at intersections of the space lattice. This gives rise to a number of sets of planes, each set bearing a different angular relationship to the other sets. The planes within any one set are mutually parallel and are separated from each other by a fixed distance, called the interplanar spacing d. In general, each set of planes has a different d value from the other sets. (The sets of planes most densely populated with lattice points are most likely to have natural external crystal faces parallel to them.)

Consider a crystal whose edge view is shown in Fig. 10.2. The lines shown within the crystal are traces of planes which are parallel to one another and are separated by the fixed spacing d. A beam of X-rays of wavelength λ is directed to these planes at an angle θ. A fundamental relationship, the Bragg equation, describes the condition for diffraction:

$$n\lambda = 2d \sin \theta \qquad (10.2)$$

In this relationship n is a whole number. In Fig. 10.2, if the incident beam of X-rays is varied with respect to the fixed planes of the crystal through the angle θ, diffraction would occur only at specific relative positions (values of θ). These would be the positions at which the Bragg equation would be satisfied for values of $n = 1$, $n = 2$, etc. Thus, various *orders* of diffraction are possible from a given set of planes within a crystal. The intensity of diffracted rays differs for various orders.

From the Bragg equation, it can be seen that with a known wavelength of incident X-rays, and a measurable value for θ, the interplanar spacing d of the planes in the crystal of Fig. 10.2 can be calculated. This d value is a physical constant, characteristic for the one set of planes within the crystalline material. A repositioning of the crystal, such that a different set of planes is properly oriented to the incident X-ray beam, will allow the measurement of the d value for this second set of planes. Use of a single crystal, with known orientations to measure a few selected d values, constitutes the single crystal method of analysis, described in a later section.

One could continue this reorientation of a crystal and measurement until all possible d values, corresponding to all sets of planes, are measured. In practice, however, this multiple reorientation is accomplished in a simple manner by grinding the crystalline material to a fine powder, and exposing the packed powder to the incident X-ray beam in such a manner that *all possible orientations* of the crystal are presented. Whenever a crystal in the powder is in a position such that the Bragg equation is satisfied, diffraction will occur. Since all orientations are represented, diffractions at many values of θ—for the many d values—occur. This procedure was first described independently by Debye and Scherrer in Germany[1] and Hull in the United States,[2] and is the basis of the powder method, described next.

B. POWDER DIFFRACTION

I. Instrumentation

One method for the measurement of the diffraction angles, and hence the d spacings, from a powder specimen employs an X-ray *powder-diffraction camera*. Figure 10.3 depicts the use of this device in the popular Debye-Scherrer method. The camera is a light-tight container which holds a strip of X-ray film in the shape of a cylinder of known radius. Incident X-rays of known wavelength are collimated by a beam tunnel, and directed at the powdered sample contained in a thin-walled capillary. The specimen is automatically rotated at a slow rate during the exposure time (2 to 4 hr. depending upon film characteristics and the nature of the sample), allowing a large number of crystallites within the sample to fall into a position relative to the incident beam such that the Bragg equation is satisfied. The diffraction from each set of planes occurs as a diverging cone from the sample, exposing the film strip in a series of arcs. Following development of the film, the distances between corresponding arcs are measured, converted to values of 2θ from the known camera radius, and recorded. The series of interplanar spacings, or d values, is calculated from the known wavelength of the incident X-rays and the measured values for 2θ.

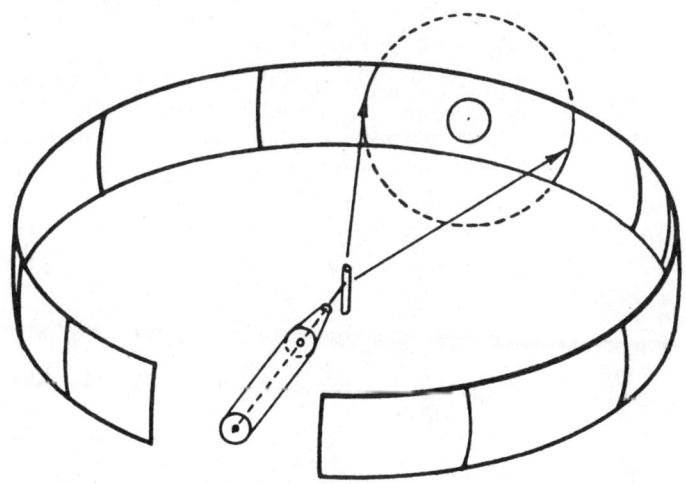

FIGURE 10.3: X-ray powder diffraction in a film camera.

Because of its high degree of versatility, the X-ray *diffractometer* (spectrometer) has considerable value to the pharmaceutical scientist. Among other features, the diffractometer permits measurement of the 2θ values from a powder sample without the use of film recording and the long film exposure and development times. X-Ray diffraction equipment produced by two of the several manufacturers who supply such instrumentation are pictured in Figs. 10.4 and 10.5. That portion of the diffractometer designed to measure angles is referred to as a "goniometer." Its operation is depicted in Fig. 10.6, where S represents an edge view of a shallow tray containing the powdered sample with a planar surface.

A proportional or other electronic counter C rotates about S at an angle of 2θ, while the sample turns at θ, relative to the fixed X-ray source. X-ray beam slitt (sl) define the beam parallel to the sample surface. While the scanning operation is in progress, a strip-chart recorder operates to plot the diffracted intensity (detected by the counter) as a function of the angle 2θ.

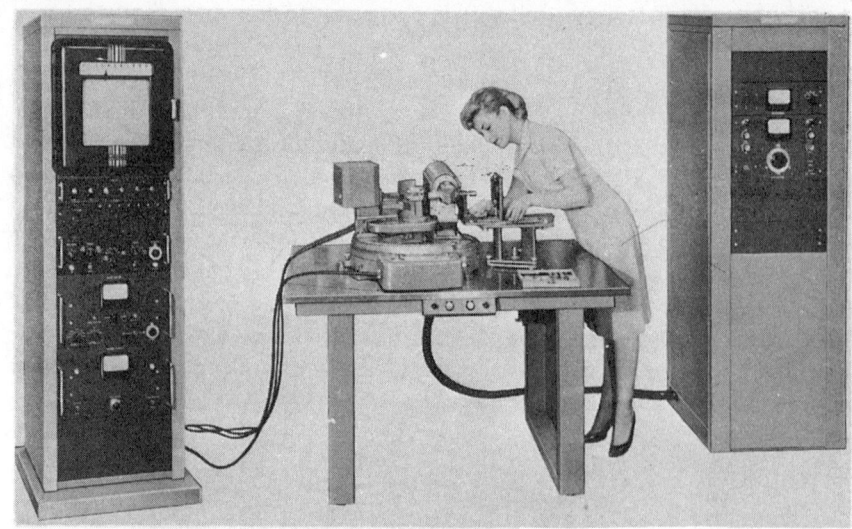

FIGURE 10.4: General Electric diffraction unit, model XRD-5. Courtesy General
Electric Co.

FIGURE 10.5: Norelco diffraction system. Courtesy Philips Electronic Instruments.

The intensity is recorded as counting rate, in counts per second (cps). Additional major components of commercially available equipment are the X-ray power source, the X-ray tube, and an electronic scaler and timer for quantitative measurement of diffracted intensity, in counts per second.

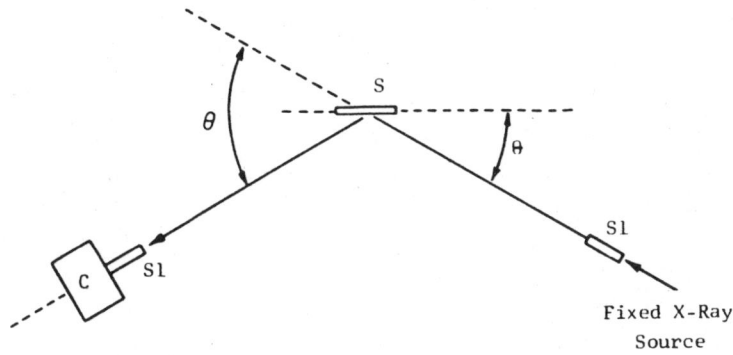

FIGURE 10.6: The sample (S) rotates through θ while the counter (C) scans through 2θ. An edge view of the planar surface sample is shown. X-ray beam slits (sl) define the beam parallel to the sample surface. Reprinted from Ref. (11), p. 25, by courtesy of *J. Pharm. Sci.*

2. Qualitative Analysis

Qualitative analysis by X-ray powder diffraction techniques is straightforward and fairly simple. Essentially it involves the measurement of a series of d spacings, the interplanar spacings, from the positions of the diffraction peaks (from the diffractometer) or the diffraction lines (from the powder camera film), as described in the foregoing section. As it is most practical from an instrumental point of view to record the diffraction angle in terms of 2θ, rather than θ, tables have been prepared from which all values of 2θ are readily converted to d values, expressed in Angstrom units (Å), for a given wavelength of X-rays. These tables are available from a variety of sources, including the equipment manufacturers.

At the time of recording the values of 2θ, the relative intensity of each line (peak) is noted. If the diffractometer is used, the intensities are measured as peak height above background, and expressed as percentages of the strongest line.

For purposes of identification, the table of measured d values, together with the relative intensities, is compared with similar listings for known compounds. A library consisting of a collection of data cards for known compounds, called the ASTM X-Ray Powder Data File, is commercially available.* This file contains powder diffraction data for essentially all metals,

* The American Society for Testing Materials, 1916 Race Street, Philadelphia, Pennsylvania.

minerals, and inorganic compounds, and an ever-increasing number of the more common organic compounds. Use of the file is made by entering the *d* values of the three strongest lines from the unknown. This usually produces a few cards representing possibilities. As each card also lists the complete table of *d* values and relative intensities, the positive identification readily follows.

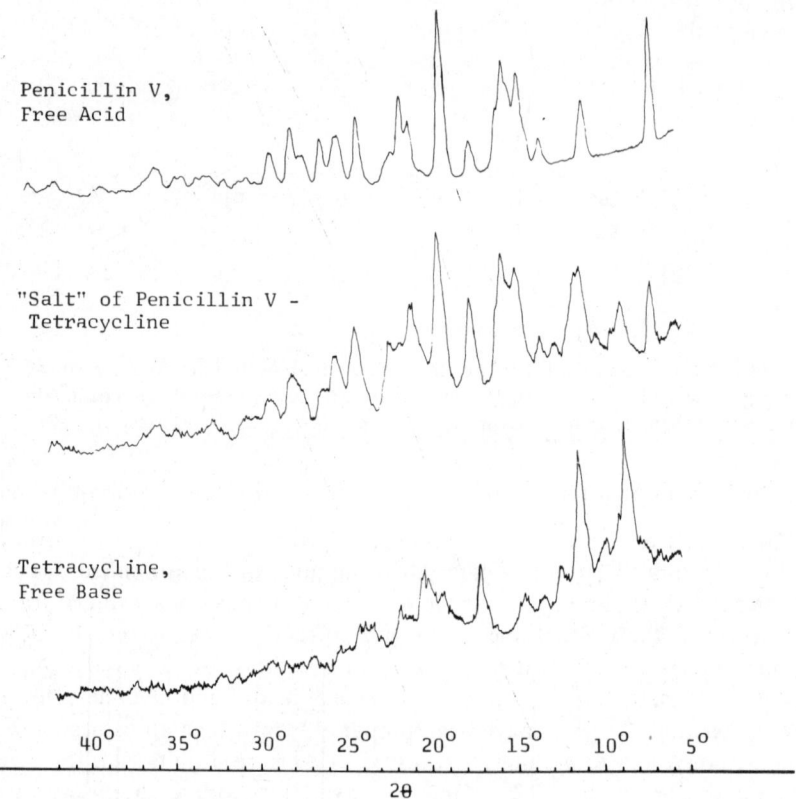

FIGURE 10.7: X-ray powder diffraction patterns of antibiotic materials.

Many investigators using the powder diffraction technique maintain a file of original diffraction patterns (or films) representing compounds of interest to them, and make identifications of unknowns by direct comparison of the patterns.

A major source of error in determining the relative intensities of diffraction lines is due to *preferred orientation* of the particles in the powder sample. If the material to be examined is made up of flat, micaceous crystals, and these are permitted to orient preferentially so that most of them lie flat in the sample holder, most of the resulting diffraction will be from the planes within the crystals which are parallel to this flat face. This diffraction line will appear very strong, and other lines, requiring other crystal orientations for emergence, will be either missing or, at best, low in intensity. For a diffraction pattern reflecting true relative intensities, great care must be exercised to ensure random orientation of the crystals.

The most widespread use of X-ray powder-diffraction analysis is in the identification of unknown compounds. Moreover, since the method provides identification of a material by compound, rather than by its elements, it is very useful in the characterization of new compounds for purposes of patents or other publications. Of particular interest to the pharmaceutical chemist is the sensitivity of the diffraction pattern to polymorphic changes. Polymorphs are different crystal forms of the same chemical compound which possess different physical (and pharmaceutical) properties. Further, many drug compounds exist in more than one solvated crystal form; these forms often differ markedly in solubility and in other properties. (Since solvates of a given compound are actually different compounds from one another as well as from the parent compound, their X-ray diffraction patterns are quite different. In addition, distinctions between different compositions of mixed crystals, or solid solutions, may be made by X-ray diffraction. It is interesting to observe that with systems in which mixed crystal formation obtains in all proportions, the composition can be followed by the measurable shift in 2θ value of a major diffraction line. Finally, the feasibility of the use of diffraction to identify the components of a complex mixture without the necessity of separation is worthy of note. Figure 10.7 shows the diffraction patterns of penicillin V (free acid), tetracycline (free base), and a material described in patent literature as the salt of these two antibiotics. A comparison of the three diffraction patterns allows the positive conclusion that the material represents a simple mixture of the reaction starting materials, rather than a true salt. Although this is a trivial example in one sense, the importance of the method becomes evident when one searches for alternative methods for proving—or disproving—the existence of a salt of large, complex molecules.

3. Quantitative Analysis

Among quantitative methods, those based on X-ray diffraction are unique in that they combine the absolute specificity usually found only with bioassays with the high precision of typical chemical assays. This fact is of particular interest to the pharmaceutical chemist who must develop assays to guide stability studies which are specific for the intact molecule. Further, such methods may often be applied directly to complex mixtures without separation or knowledge of the other ingredients.

Some mention of the possibilities of quantitative diffraction was first made by Hull as early as 1919,[3] but the first work reported was by Clark and Reynolds on the analysis of mine dust in 1936.[4] This work, and other work following,[5] was based upon microphotometric density measurements of X-ray film following exposure. This method of measuring the intensity of diffracted X-rays was highly inaccurate, and it was not until the advent of the Geiger counter diffractometer (spectrometer)[6] that truly quantitative diffraction became possible.

The mathematical relationships pertinent to quantitative diffraction analysis were derived and published in an important fundamental paper by Alexander and Klug appearing in 1948.[7] This paper described conditions under which standard curves alone could be used, and under which standard curves based on internal standards were required, depending upon absorption effects. Relationships permitting, in certain instances, quantitative analysis of differential absorption systems without the use of internal standards were described in 1953.[8] In 1958, Copeland and Bragg[9] described special conditions under which calibration curves could sometimes be eliminated and multiple components determined.

The papers just mentioned refer to investigations of inorganic systems, which, due to higher mass absorption coefficients of the ingredients present special problems not so often encountered with organic systems. Only a few papers have appeared reporting applications to organic systems; some of these of pharmaceutical interest concerned applications to the determination of sodium penicillin G, published in 1948,[10] general applications to typical pharmaceutical systems, published in 1963,[11] and the analysis of intact tablets, published in 1964.[12]

The feasibility of quantitative X-ray diffraction stems from the fact that the intensity of a diffracted beam of X-rays is a function of the amount of diffracting material. The linearity of this response, for a fixed set of experimental conditions, depends upon the difference in the amount of absorption of X-ray energy between the compound of interest and its surrounding matrix. The energy absorption by any material depends upon the mass absorption coefficients of its constitutive atoms, which in turn are generally a function of atomic number for a given wavelength of radiant energy. Thus, as a compound differs chemically from its matrix, so will it differ in absorption for the radiant energy. The result of a large difference is a very nonlinear relationship between the amount of diffracting material and the diffracted intensity.

It is significant that in inorganic systems the probability of wide variation in X-ray absorption between the compound of interest and its surrounding matrix is high, whereas with organic systems the variation is usually low. This feature of organic systems results in an advantageous gain in linearity of response and allows, in many instances, the use of a simple calibration curve for the quantitative analysis of complex organic systems.

When diffraction from a powder sample occurs at a given 2θ value, it does so because a sufficient number of crystals have the same set of planes, whose d spacings correspond to the 2θ value, properly oriented with respect to the X-ray beam. The intensity of the diffracted ray is a function of the amount of material so oriented. If truly random orientation is assured,* and except for absorption effects, the diffracted intensity becomes proportional to what may be termed the "specific lattice volume." It is highly significant that when the intensity of a single diffraction peak is measured at a fixed 2θ value, both additive and constitutive effects are being measured. This unique fact is the basis for the specificity of quantitative diffraction analysis.

In developing a method for a component of a given mixture, the procedure of choice will vary according to the complexity of the system. The complexity here refers more to the absorptive qualities of the matrix and the constancy of the matrix than to the number of ingredients present. If a diffraction assay is feasible at all, one of the procedures which follow should apply. These procedures assume the use of a modern X-ray diffractometer, equipped with counter, scaler, and timer circuits.

a. Use of Simple Calibration Curves. No errors due to changing absorption effects occur with simple, two-component mixtures, or with multicomponent mixtures whose composition, except for the component being analyzed, remains constant. The determination of a crystalline ingredient in such systems may be based on calibration curves prepared from synthetic mixtures.

The procedure begins with an examination of the diffraction pattern of the ingredient to be assayed. A major diffraction peak is first found in an area free from interferences from other components of the mixture (the matrix); synthetic mixtures are then prepared containing known amounts of the compound of interest, over the range of concentration of interest, and the intensity of the selected diffraction peak measured for each mixture.

The intensity measurement is made by setting the diffractometer to the 2θ position of maximum peak height, and engaging the timer and scaler to measure the peak height in counts per second. Using the diffraction pattern of the pure material for reference, the instrument is then set at a 2θ position for the measurement of the background intensity, usually near the base of the selected peak. Again, the intensity in counts per second is measured by use of the scaler and timer circuits. The background intensity for each sample is subtracted from the corresponding peak intensity, and this difference is plotted as a function of sample composition, to form the standard curve for use in subsequent assays.

Illustrating an assay based on a simple calibration curve is the determination of pamoic acid in a matrix of the pamoic acid salt of a basic antibiotic.[11] The method was found useful in testing for completeness of the salt-forming

---

* In all quantitative diffraction studies the effects of preferred orientation of the crystals must be minimized. Simple grinding, followed by careful packing in the sample tray, usually accomplishes this for organic compounds. (In instances where an extreme of one crystal habit is represented, the advantageous use of 2% carbon black added to the mixture has been reported.[10]

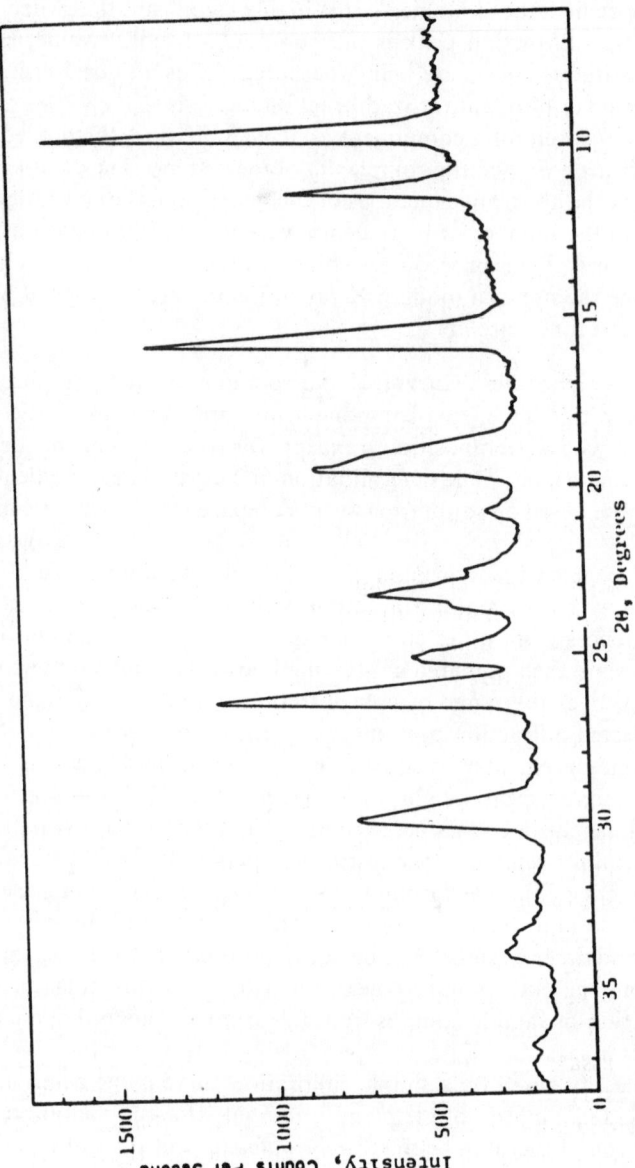

FIGURE 10.8: The pamoic acid diffraction pattern. Reprinted from Ref. 11, p. 27, by courtesy of *J. Pharm. Sci.*

reaction, and particularly in support of the long-term stability studies of a formulation of the antibiotic salt.

The diffraction pattern of pamoic acid is shown in Fig. 10.8. Although major diffraction peaks occur at 2θ values of 9.50, 11.20, 15.60, and 19.40°, all of these peaks were at least partially obscured when superposed on the pattern of the matrix materials. For this reason the peak at 26.25° 2θ was chosen as a basis for the determination.*

For the preparation of the standard curve, use was made of synthetic mixtures containing known concentrations of pamoic acid in the antibiotic salt. Following thorough mixing and grinding in a mortar, each sample was poured into a standard 2-in. sample tray, which required about 200 mg of material. In order to assure random crystal orientation, the excess powder was carefully removed by use of the edge of a glass microscope slide, and the sample surface finally packed slightly by use of the surface of a rough, low-grade blotting paper. This procedure eliminates large errors due to preferred orientation of the crystals.

The diffraction peak intensity of each sample was determined in the following manner. After placing the sample in the diffractometer, the goniometer was set for 26.25° 2θ and the instrument set to record the time required for the accumulation of 20,000 counts. This operation was followed by a setting of the goniometer at 26.80° 2θ and a second recording of the time required to accumulate 20,000 counts. The diffraction intensity at 26.25° (peak intensity) in counts per second was calculated, and the intensity at 26.80° (background intensity) in counts per second was also calculated.† The background intensity was subtracted from the peak intensity and the resulting value for each of the standard samples representing synthetic mixtures was plotted vs. the known concentration. The results are shown in Fig. 10.9. The total counting time for each point was approximately 3 min.

Since organic molecules are large, organic crystals have large unit cells. Most interplanar spacings are, therefore, large. Such spacings give rise to diffraction peaks at small values of 2θ, which is the region of maximum response to scattering of the X-rays by the powdered sample. Thus it sometimes happens that a diffraction peak of interest occurs in an area of maximum background. This is particularly true when a sample contains a large amount of amorphous, or scattering material. Such a situation is illustrated by Fig. 10.10.[11] Under these conditions it has been found advantageous to

* One has a choice of several possibilities upon which to base a proposed analysis. If the matrix is not obtainable in pure form, and there is doubt as to the freedom from interference of diffraction peaks, two simultaneous methods may be developed, each using a separate peak. Freedom from interfering peaks is assured when results of the determinations agree.

† Since the statistical accuracy of a counting process depends upon the total number of counts taken, and not upon the time it takes to make the count, a preset count, rather than preset time, is used to determine the intensity in counts per second. This procedure allows all values, when compared later, to be of the same statistical accuracy.

determine the net peak height by a different procedure in order to maintain accuracy. Using the scaler and synchronous timer circuits, the intensity P (Fig. 10.10) is measured at 7.25° 2θ. The intensities at R_1 and R_2 are measured at 2θ values of 7.0 and 7.80°. The background B is calculated by interpolation:

$$B = R_2 + \frac{7.80 - 7.25}{7.80 - 7.00}(R_1 - R_2) = R_2 + 0.688(R_1 - R_2) \quad (10.3)$$

The net peak height ($P - B$) can then be calculated for each sample.

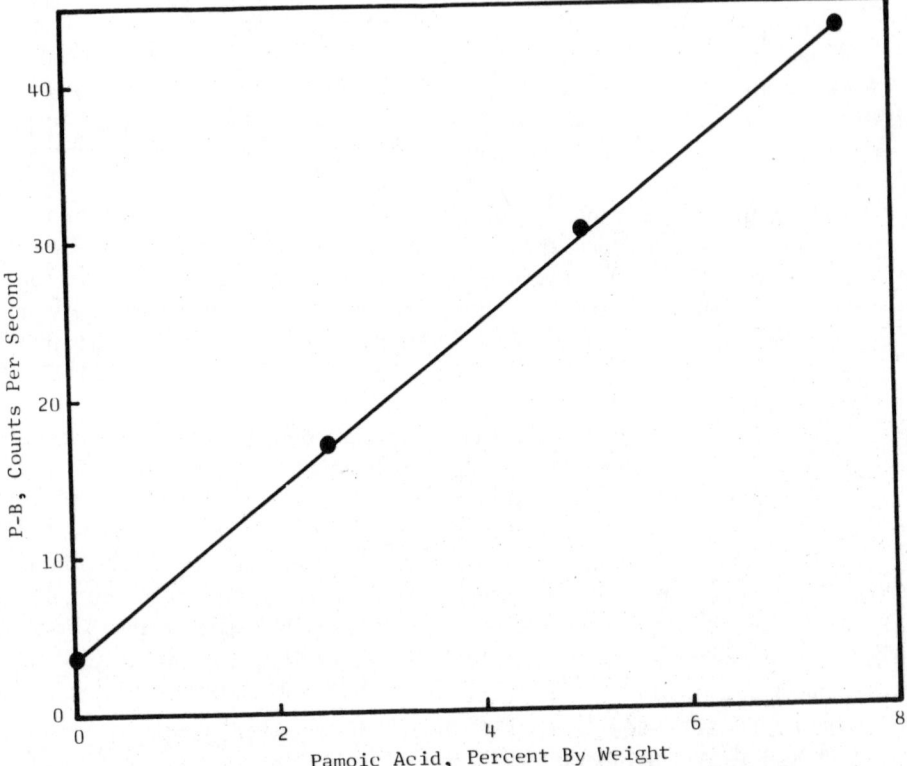

FIGURE 10.9: Intensity difference, peak minus background, as a function of pamoic acid concentration. Reprinted from Ref. 11, p. 27, by courtesy of *J. Pharm. Sci.*

If all variables such as instrumental drift and the density of sample packing could be held constant, the background correction could be eliminated. Its use, however, permits day-to-day validity of the standard curve with good accuracy.

In the two previous examples, peak intensities were corrected by subtracting background intensities. It is just as valid to use a peak-to-background intensity ratio in place of the arithmetical difference. Moreover, the choice of position for the background measurement is not limited, an obvious advantage in instances of multiple peak overlapping.

b. Use of an Internal Standard. The most universally applicable method for quantitative diffraction involves the use of an internal standard. This method is free from all matrix effects as well as errors due to variations in sample packing density and variations in instrumental conditions. It is ideal for drug systems containing unknown mixtures, such as those containing possible degradation products.

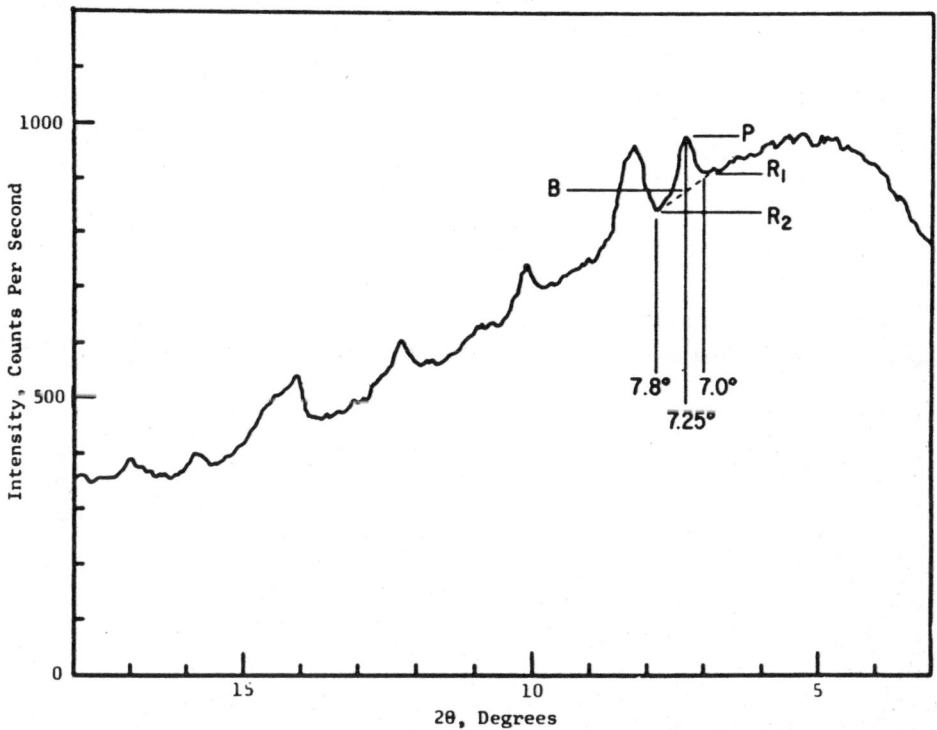

FIGURE 10.10: Partial diffraction pattern showing peak of interest over large background. Reprinted from Ref. (11), p. 27, by courtesy of *J. Pharm. Sci.*

The procedure involves examination of the diffraction patterns of the material to be assayed, and of the matrix in which it is found, and selecting a peak produced by the material which is free of interference from neighboring peaks. A suitable inert crystalline material, the internal standard, must then be selected which has a peak in a clear region with respect to the system to be analyzed.

When a suitable internal standard has been found, a standard curve is prepared from samples containing known but different concentrations of the compound to be analyzed and a fixed concentration of the internal standard. The proper concentration of internal standard to establish for use in all samples is one which is found to give a diffraction peak about equal in intensity to that of the material to be analyzed when the latter is present in the usually expected concentration. The X-ray procedure consists simply of determining the ratio of intensities of the diffraction peaks of the unknown and the internal standard, without regard for the background. The ratio is linearly proportional to the concentration of the unknown.

Finding a suitable internal standard for each new system is often a problem. An ideal internal standard has several attributes: It must have a diffraction peak which is not obscured by peaks of the system, and which will not interfere with a major peak of the material to be analyzed; the internal standard peak should be near the usable peak of the analyzed ingredient; it should be of high crystal symmetry, preferably isometric, so that strong but few diffraction peaks are produced and preferred orientation effects will be minimized (by the more equant crystals); because of absorption effects, it should contain only elements of low atomic number; it should have a density not too far removed from those of the system ingredients (for aid in maintaining homogeneity in mixing); and it should be chemically stable in the presence of the system. No one compound can qualify universally on all counts, but a material which has been found to have many of these attributes is hexamethylenetetramine, one of the few organic compounds of isometric (cubic) symmetry.

It is worth mentioning that in instances in which the system is of many variable components, requiring therefore an internal standard, but for which no suitable standard can be found, an alternative procedure exists. This consists of adding known amounts of the ingredient to be analyzed to the unknown system, plotting the peak intensity values as a function of the amount added, and extrapolating the curve. The intercept value gives a measure of the original amount of ingredient present.

c. Use of Integrated Diffraction Peak Areas. Some organic compounds, when produced in successive batches over a period of time, tend to vary in the degree of crystallinity possessed by any one batch. A decrease in the degree of crystallinity of a compound is accompanied by a drop in the apparent diffraction peak intensity as measured by the peak height. Such a drop in peak height, however, is accompanied by a peak broadening. It is significant that the area of the diffraction peak is relatively constant for a wide variation in crystallinity. It is, therefore, sometimes of value to base a diffraction assay on peak areas rather than on peak heights. Further, the use of peak areas is advantageous when the particle size of the crystalline ingredient is very small due, again, to line broadening and a significant drop in peak height. Measurable line broadening occurs in the particle size range below $0.2\,\mu$. Finally, the peak-area method offers an advantage in being free from errors due to apparent shifts in peak maxima. A standard curve for an assay based on the use of an internal standard and on integrated peak areas is always linear and intersects the origin.

To illustrate the method, a portion of a diffraction pattern is presented in Fig. 10.11, which shows a doublet at 10° from a sulfonamide from aqueous suspension, and a peak at 11.5° from an added internal standard, $CaSO_4 \cdot 2H_2O$.[11]

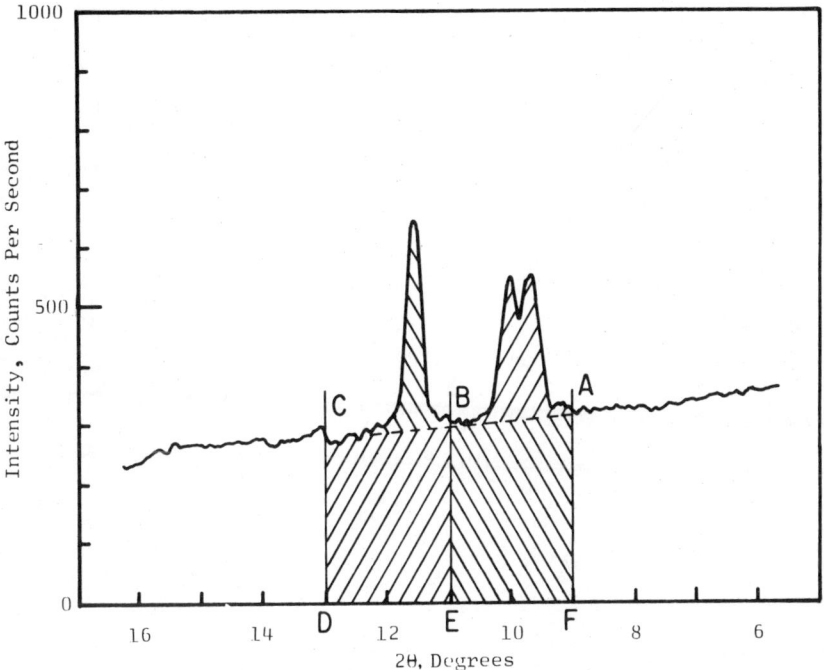

FIGURE 10.11: Partial diffraction pattern showing sulfonamide peaks (doublet at 10°) and internal standard peak ($11\frac{1}{2}°$). Reprinted from Ref. (11), p. 26, by courtesy of *J. Pharm. Sci.*

Integrated intensity values were determined from this system by the following procedure. With the scaler set for a preset time of 100 sec, the counting rate at 9° was determined in counts per second. This was repeated for positions at 11 and 13°. Diagrammatically these three values of background counting rates are represented in Fig. 10.11 by the lengths of the lines *AF*, *BE*, and *CD*. With the goniometer set at somewhat less than 9°, an automatic scan was begun at 2°/min. As the scan reached 9°, the counting switch was turned on, simultaneously starting the scaler and timer. As the scan crossed 11°, this switch was turned off. The total count and total elapsed time (approximately 60 sec) were recorded. The elapsed time (in seconds) is represented by *EF* on the diagram, and when multiplied by the

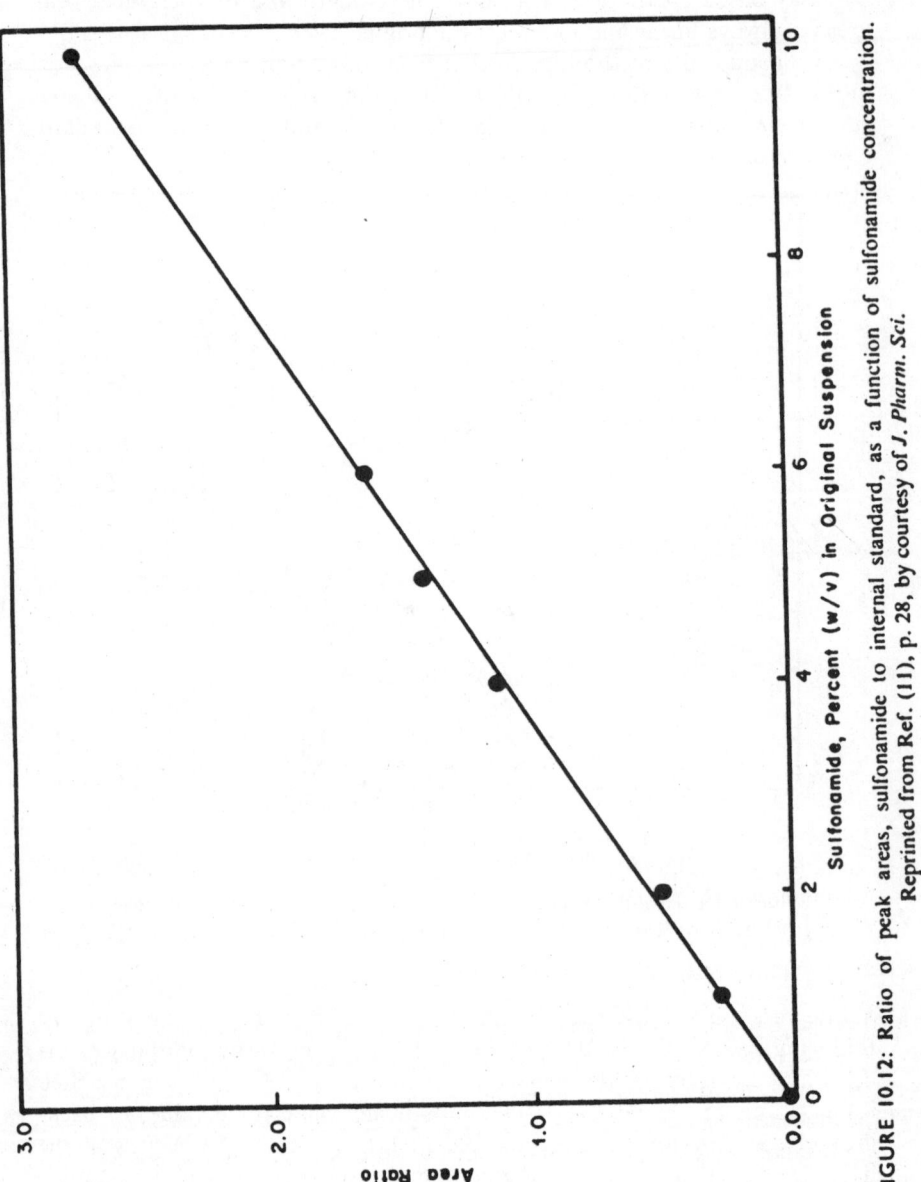

FIGURE 10.12: Ratio of peak areas, sulfonamide to internal standard, as a function of sulfonamide concentration. Reprinted from Ref. (11), p. 28, by courtesy of *J. Pharm. Sci.*

average of background counting rates measured at 9 and 11° (in counts per second) gives the area of background under the curve (*ABEF*) expressed in total count. This value was subtracted from the total count accumulated in the scan operation. The difference was the area of only the net sulfonamide diffraction peak, expressed in total count. The operation was repeated between 11 and 13° to determine the area of the internal standard peak, and the ratio of sulfonamide peak area to standard peak ares was then computed.

Figure 10.12 shows a least squares plot of values determined by this procedure for the ratios of sulfonamide peak areas to internal standard peak areas as a function of sulfonamide concentration in aqueous suspension.[11] Following sample preparation, X-ray analysis by this method required about 15 min/sample. The values indicate an accuracy of ±0.15% of the amount present. The method is specific for the intact sulfonamide molecule in the crystalline state and, therefore, sensitive to any product degradation within the aqueous suspension.

4. Particle-size Measurement

An X-ray powder-diffraction technique may be used to measure the mean dimensions of the crystallites composing a finely divided powder, provided the dimensions are small enough. The method is generally applicable to particle sizes from 20 to 2000 Å.

Most powders and suspensions of pharmaceutical interest contain particles in the micron-size range. Extensive research in the biopharmaceutical field has shown enhancement of drug dissolution rates and general drug availability from extremely fine particles, however, and with the promise of significant enhancement of these factors, some commercial formulations have emerged with particle sizes in the submicron range. Size-measuring methods which employ visible light are limited by the wavelength of the light in this range and are, therefore, not applicable.

The Bragg relationship predicts diffraction from an ideal set of planes within a crystal at a discrete angle θ. Truly ideal conditions do not exist, however, and in practice one finds diffraction from a set of planes occurring over a small range of angles near θ, resulting in an apparent broadening of the diffraction peak. The factors which contribute to this line broadening are mosaic structures within the crystal, nonuniform strain on the crystals, and small crystallite size. Fortunately, contributions to line broadening by the first two factors is often minimal from organic crystals, permitting a measurement of the broadening to reflect mean particle size.

The fundamental relationship was described by Scherrer[13]:

$$D = \frac{K\lambda}{\beta \cos \theta} \qquad (10.4)$$

where

$$D = \text{crystallite dimension}$$
$$K = \text{crystallite shape constant}$$
$$\lambda = \text{X-ray wavelength}$$
$$\beta = \text{corrected line breadth}$$
$$\theta = \text{Bragg angle}$$

In practice, one carefully records the X-ray powder-diffraction pattern of the crystalline powder of interest, and measures the observed peak breadths at half maximum intensity, in terms of angular degrees, of several diffraction peaks. The accuracy with which Eq. (10.4) can be applied is limited by the uncertainties within the K factor and the accuracy with which β can be deduced; for precise work corrections must be made.

Reference may be made to Jones[14] for the correction of the observed line width for unresolved K_α radiation. The resulting value must then be corrected for instrumental line broadening by reference to an experimentally determined plot of actual line breadth vs. diffracted angle, measured from a diffraction pattern produced by a powder sample of particles too large to produce line broadening. Finally, this value for β may be further refined by reference to a set of curves produced by Alexander[15] for use with diffractometers with narrow sources.

The value of the shape factor K approaches unity, and this value has been used for approximations. When nothing is known of the shape of the crystallites or of the indices of the prominent planes, the value of 0.9 is recommended for use for this factor.[16,17]

As noted previously, the applicability of X-ray line broadening to the determination of particle size is limited to the range of crystallite size from 20 to about 2000 Å. The sizes measured are average sizes, weighted toward the larger sizes present in a mixture. The accuracy of the determination is low; it diminishes rapidly with increasing size range.

A more detailed presentation of this method has been presented by Klug and Alexander.[18]

5. Crystal Habit Quantitation

The symmetry of a crystal is fixed by the crystal system and class to which it belongs. Its relative dimensions, however, are independent of its symmetry. As a crystal grows from solution, a variety of factors, notably crystallization rate and the presence of impurities, tend to influence the relative amount of growth of the possible faces. Extremes of the possible conditions result in acicular, or needle-shaped, crystals as a consequence of unidimensional growth (bidimensional retardation) and tabular, or plate-shaped, crystals, as a consequence of bidimensional growth (unidimensional retardation).

Terms such as "acicular," "equant," and "tabular" describe crystal habit in a qualitative manner.

Crystal habit often exerts a dominant influence on some important pharmaceutical characteristics, such as suspension stability, suspension syringeability, and the behavior of powder mixes during a tablet-compressing process. In the case of suspension syringeability, the influence is mostly mechanical. A suspension of plate-shaped crystals, for instance, may be injected through a small needle with greater ease than one with needle-shaped crystals of the same overall dimensions.

When considering tableting behavior, however, the influence of the crystal habit of the active ingredient is more complex. The mechanical influence of crystal shape just mentioned is one factor, but there is another, sometimes dominant one, which results from the anisotropy of cohesion and of hardness which is possessed by organic (low symmetry) crystals and, therefore, of most pharmaceutically important compounds. It is significant that this anisotropy bears a fixed relation to the fundamental crystallographic directions. Therefore, as crystal habit varies, the dominant faces may vary in their relation to this anisotropy, and it is the influence of the dominant faces which tends to orient the crystals during a packing or compression process. Thus, major habit variations of an active ingredient can influence greatly the ease or the difficulty of making satisfactory compressed tablets. This is particularly true when the active ingredient makes up a large portion of the total tablet mass.

To evaluate tableting behavior as influenced by crystal habit, the habit must be expressed in quantitative terms which reflect some relationship between the dominant faces and the principal crystallographic directions. Qualitative terms describing shape are, in some instances, not sufficient.

For a typical pharmaceutical composition, it has been found that a quantitative description of crystal habit as it affects tableting behavior can be based upon measurements of preferred orientation. After relating habit extremes to tableting behavior by experimentation, optical* and X-ray crystallographic studies on representative single crystals allow the designation of the dominant faces by their Miller indices. An X-ray powder-diffraction pattern is then measured for crystals of each habit extreme on specially prepared samples. The samples are prepared in such a manner that preferred orientation effects are maximized. A ratio of the relative peak intensities of critical lines in this diffraction pattern serves to indicate the average habit of the crystals. The ratio is useful in predicting tableting behavior, as well as serving, when desired, as a manufacturing specification.

Details of this method and a general illustrative example have been published.[19]

* Methods for the measurement of optical properties of crystals are described by J. A. Biles in Chapter 9.

C. SINGLE-CRYSTAL DIFFRACTION

I. Determination of Molecular Weights

For the precise determination of molecular weights, the X-ray method, where applicable, has long been the method of choice. Applied with reasonable care, the accuracy of this method need be limited only by errors due to crystal imperfections, which are often of the order of 0.04% if due to void spaces and much less if due in part to inclusions.

The feasibility of the X-ray method for molecular weight determinations derives from the fact that an exact and determinable number of molecules occupies a *unit cell* which, by three-dimensional repetition, generates the space lattice of the crystal. The procedure involves the measurement of the unit cell dimensions by X-ray diffraction, and the calculation of the unit cell volume. Knowing the volume of the cell, its weight is then calculated from an experimentally determined value for the crystal density The molecular weight M is then found by

$$M = \frac{Vd}{1.6604\,N} \tag{10.5}$$

where V is the cell volume in Å^3, d is the density, 1.6604 is a constant used to refer the cell weight to oxygen at 16, and N is the number of molecules per unit cell. The value for N can be determined directly if an approximate value for M is known; N is always some whole number, usually small, and limited to a few possibilities by the crystal symmetry.

The determination of the unit cell dimensions is most easily accomplished by the use of measurements from oriented, single crystals. Although the more easily acquired powder-diffraction pattern may be used to measure unit cell constants, this is readily accomplished only if the crystal is of high symmetry. Procedures for the indexing of powder patterns for this purpose have been described in detail.[20] Unfortunately, most compounds of pharmaceutical interest are organic, and these compounds most often crystallize in lower symmetry systems, requiring the single-crystal approach.

A suitable crystal, usually about 0.5 mm on an edge, is selected and cemented to the end of a fine glass fiber with the aid of a microscope. If a film camera is to be used, the crystal is oriented and then rotated in the X-ray beam about one of its crystallographic axes while the diffraction pattern is recorded on the film, arranged cylindrically about the rotation axis. Following development, "layer lines" appear on the film, the distances between which relate to and permit the calculation of the dimension of the unit cell along this rotation axis. The other dimensions of the cell are measured in similar fashion, following crystal reorientation and rotation about each of the other axes. If the crystal is of low symmetry (monoclinic or triclinic), the angular relationships of the crystal (and hence the cell)

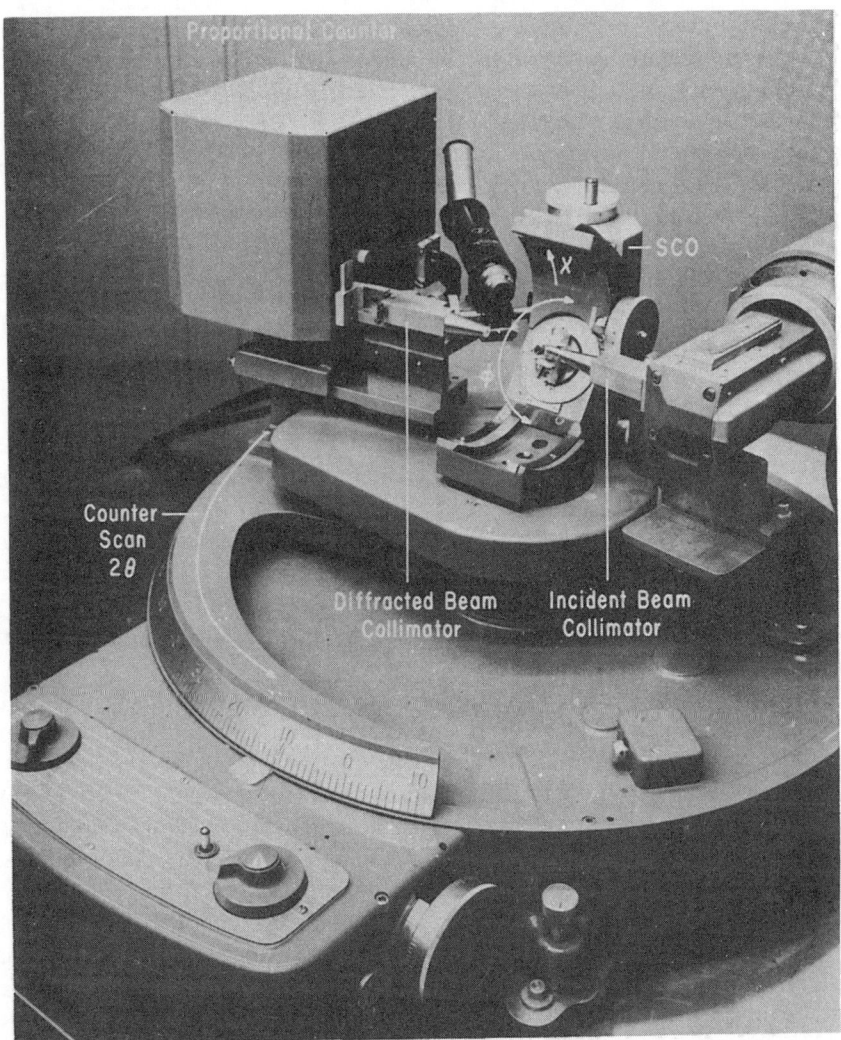

FIGURE 10.13: X-ray diffractometer and single-crystal orienter (SCO). Crystal rotation directions ϕ and χ are shown (radiation shields removed). Reprinted from Ref. (21), p. 155, by courtesy of *J. Pharm. Sci.*

must be determined; this may often be accomplished by use of the calibrated stage of a chemical microscope.

Several instrument manufacturers make available single-crystal diffractometers employing Geiger or proportional counters for rapid measurement of lattice parameters and relative diffraction intensities. One of these instruments is the single-crystal orienter, manufactured by General Electric Co., depicted in Fig. 10.13. In this photograph, the adjustments ϕ and χ

permit all possible orientations of the crystal from a single mounting. This allows rapid determination of all the dimensions of the unit cell (from 2θ measurements), as well as the angular relationships of the cell axes (from the ϕ and χ settings). Detailed procedures have been published[21] covering the use of such an instrument for the measurement of unit cell constants of crystals of all symmetries, particularly as applied to organic crystals.

Crystal densities are most accurately determined by flotation methods, wherein a liquid is generated whose density is precisely equal to that of the crystal. The liquid density is then determined by a standard pycnometer technique. Detailed descriptions and procedures for two such methods have been reported.[21]

2. Total Structure Determination

By far the most sophisticated X-ray methods are those whose objective is the elucidation of the total structure of complex molecules. From a suitable single crystal of a compound it is possible to discover and describe with precision the relative spacial arrangement of all of the atoms of the molecule. It is beyond the scope of this chapter to cover this area of endeavor; because of its immense importance, however, mention in at least summary fashion is pertinent.

From X-ray diffraction patterns of a single crystal, the *positions* of the diffracted rays disclose the dimensions of the unit cell, and the *intensities* of these rays are functions of the atomic distribution within this unit cell. The procedural steps in determining the total structure include measurement of the diffraction intensities, reduction of these values to structure amplitudes, discovering the phase of the scattering amplitudes, refinement of the structure, and the presentation of the results in terms of stereochemical information. The procedure involves extensive use of high-speed computer equipment.

Unfortunately, the process is not a direct one, due to lack of advanced knowledge of the phase angles of the scattering amplitudes. Sometimes the phase angles can be computed from an assumed approximate structure (or lucky guess), and these used to compute a Fourier series, which is a means of expressing periodic functions as summations of trigonometric terms. Other methods of obviating the phase problem exist; one involves the introduction into the molecules of a heavy (high atomic number) atom, whose coordinates may be determined, giving sufficient insight to the approximate structure for one to compute the phase angles and evaluate the series.

Much progress has been made in recent years in the development of equipment for more rapid and precise collection of single-crystal diffraction data, and for the automatic handling of these data by special computers. It is possible that the phase problem may be solved in time by a direct approach the possibility of discovering the phase of the amplitude by simultaneous recording of the scattering amplitudes from two independent X-ray sources and correlated detectors has been reported.[22] Possibly another approach may involve the use of extrapolated values from various wavelengths of X-rays.

10.5 X-RAY EMISSION (FLUORESCENCE)

A. QUALITATIVE ANALYSIS

In the section of this chapter relating to the production and properties of X-rays, a description of the origin of X-ray spectra was presented. It was noted that several processes occur when matter is bombarded by intense radiation energy. One of these is a quantum process wherein atomic electrons are displaced, with the vacancies thus caused filled by electrons from an outer shell. Such a process must be accompanied by a release of energy, and it is significant that this energy is in the form of X-rays. As there are several energy levels associated with the various electron shells, quantum theory predicts a number of possible transitions, some more likely than others. Thus an element, when bombarded by sufficiently intense radiant energy, will emit several X-ray "lines," each at a specific wavelength, characteristic of the element. Qualitative analysis of a sample for the elements present is based upon the measurement of the wavelength of the emission lines.

In practice, the radiant energy directed to the sample is supplied by an X-ray tube. The determination of wavelength of the emitted X-ray lines is accomplished by diffraction using a large, single analyzing crystal of known orientation and known interplanar spacing; the wavelength of each emission line is then given by solution of the Bragg equation. The commercially available spectrometers make this process easy by producing an automatic plot of the diffracted intensity against the Bragg angle 2θ. The "peaks" from this spectrum can then be read off in terms of 2θ, and converted not only to wavelength, but directly to the element by use of prepared tables. These tables are available from the various equipment manufacturers, and provide the conversion for each of several standard analyzing crystals.

The same commercially available equipment used for X-ray diffraction (see Fig. 10.4, and 10.5) may also be used for emission analysis, simply by the use of added accessories. The conversion to a spectrometer (spectro-goniometer) consists of replacement of the lower rated diffraction tube by an X-ray tube of higher intensity output (a tungsten tube is commonly used); the placement of a sample holder in the primary X-ray beam; and the positioning of an analyzing crystal in the path of the collimated rays from the sample holder. The conversion may involve changing the electronic counter, depending upon the analytical wavelength of interest.

Several advantages of the X-ray emission method make it especially adaptable for pharmaceutical problems. Due to the low atomic number of carbon, hydrogen, and oxygen constituting organic systems, matrix component interferences, encountered in X-ray emission applications to other systems, are often absent. Moreover, the method is applicable directly to samples, either liquid or solid, and usually without need for separations, permitting assays on final product formulations. Further advantages are speed, specificity, the nondestructive nature of the method, the inherent simplicity of X-ray spectra, and sensitivity for trace amounts of material.

The probability that an excited atom will emit radiation depends upon many factors; it is related to what has been termed the "fluorescent yield," a quantity which rises precipitously with increasing atomic number of the element excited. Presently, X-ray emission analysis is limited in its application to elements of higher atomic number: elements of atomic number less than 12 are not detectable with present-day equipment; those elements of atomic number from 12 to 22 are detectable by use of vacuum paths or helium-filled paths, and a detector (such as a gas-flow proportional counter) for long wavelength emission. Elements of atomic number greater than 22 are detected with no special conditions being required. The nonapplicability to low atomic number elements is a disadvantage when considering X-ray emission as a general method, but is a decided advantage when the method is used for the detection or determination of heavy elements in a light (organic) matrix.

B. QUANTITATIVE ANALYSIS

Whereas the emission of X-ray lines at specific wavelengths from an irradiated element serves as a basis for qualitative analysis, the fact that the intensity of any selected line is proportional to, among other factors, the number of emitting atoms is the basis for a quantitative method. The method generally requires independent calibration from standards of known concentration.

As noted in the preceding section, samples for X-ray emission analysis may be in any physical state. For quantitative work, it is only essential that samples for the development of the required standard curve and for the analyses be prepared and analyzed under identical conditions. For solid samples, powdering and briquetting has been found to be an ideal method for maintaining reproducibility. A variety of holders for liquid samples are available; most of these employ a thin Mylar window to pass the incident and emitted radiation.

In practice, the spectrometer is set to measure the emission intensity of a selected "line" from previous knowledge of the 2θ value of the line. The signal is maximized by slight adjustments of the spectrometer, and the intensity measured by use of the counter and scaler circuits. The peak intensity, in counts per second, is corrected by either subtracting the background counting rate, as determined by counting at a convenient 2θ setting near the base of the peak of interest, or computing the peak-to-background ratio. In systems comprising a light matrix (low atomic number elements) containing

a heavy element, a valid analysis for the heavy element may be based simply
on the net peak height (or the peak-to-background ratio) thus determined.

Figure 10.14 shows a typical plot of the ratio quantity obtained by counting
0.5 g, 0.5 in. diameter briquettes of a steroid at the 2θ position for the
selenium K_α line for several concentrations of selenium.[23] The detector was
set to record the time required to accumulate 200,000 counts, and the

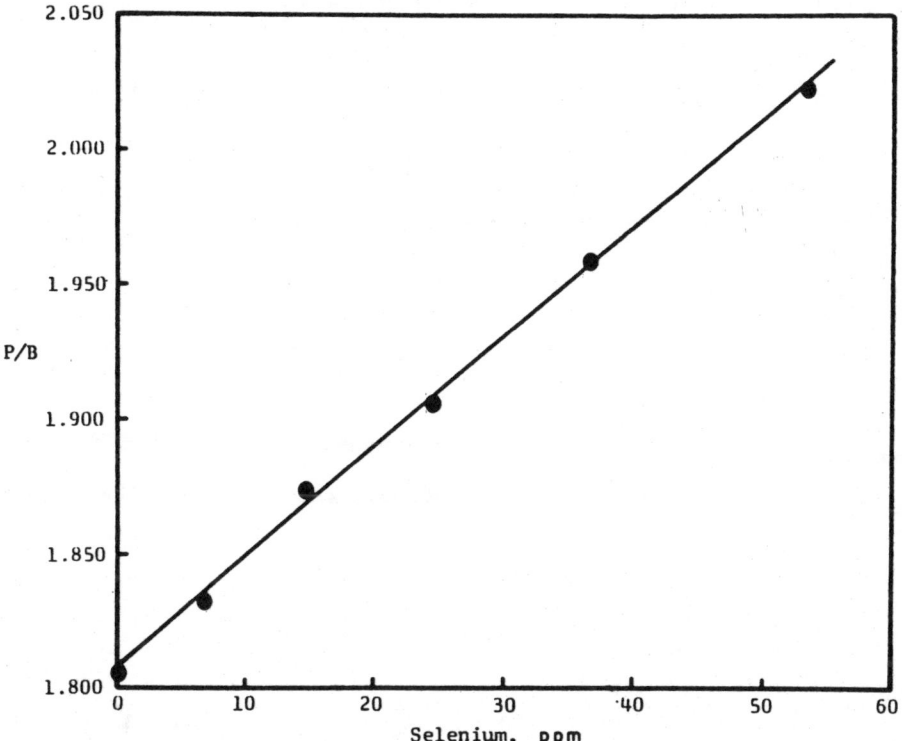

FIGURE 10.14: Selenium, peak to background ratio vs. concentration. Reprinted from
Ref. (23), p. 733, by courtesy of *J. Pharm. Sci.*

counting rate was calculated from this quantity. This required, for the
highest concentration sample, 153 sec. This particular assay was of interest
due to the use of this toxic material as a dehydrogenation catalyst in the
steroid synthesis.

If the element of interest is contained in a matrix which is not made
exclusively of light elements, a variety of effects, collectively referred to as
"absorption effects," occur. Their overall effect destroys the linearity of the
response elicited by simpler systems. The most practical way to overcome
these effects is through use of an internal standard, selected for an appropriate emission wavelength of one of its lines, and added in a fixed amount

to all standard curve samples and unknown samples. The ratio of the peak height (counts per second) of the element of interest to the peak height (counts per second) of the internal standard element, without background corrections, is then plotted against concentration to form the standard calibration curve, and measured on the unknown samples for their analysis.

In instances where high background-counting rates, or interferences from an emission line of another element present problems, use may be made of an electronic circuit called a "pulse-height selector," which is available for most commercial spectrometers. This device permits electronic discrimination against undesired wavelengths, with an overall gain in peak-to-background ratio.

Application of quantitative X-ray emission analyses appearing in the literature are numerous, and their number is steadily increasing. Of particular interest to the pharmaceutical chemist is the application of the method to the determination of selected elements in biological materials,[24] and the application to trace elements in general pharmaceutical systems.[23] Extension of the range of applicability to the gamma concentration range in solutions through use of ion exchange membranes has been reported.[25] Finally, two reference texts are selected[26,27] which cover the fundamentals of X-ray emission analysis in detail.

10.6   X-RAY ABSORPTION

When the intensity of X-rays which have passed through a pure element is measured as a function of the X-ray wavelength, several abrupt discontinuities are observed at specific wavelength values. These critical absorption wavelengths are called "absorption edges" for the element, and occur because the mass absorption coefficient of the element suffers abrupt discontinuities at the excitation potentials of the various spectra (K series, L series, etc.).

Several facts regarding the absorption edge make it of interest to the analytical chemist. The position of the edge is characteristic for only the one element producing it; the change in transmitted intensity at the edge is related to the amount of the element present; and the change in transmitted intensity is not influenced by other elements which might be present.

For a monochromatic X-ray beam transmitted through a sample containing an element of interest, it has been shown[28] that

$$\ln I''/I' = (\mu'_M - \mu''_M)WG \tag{10.6}$$

where I'' and I' are transmitted intensities on either side of the absorption edge of the element, μ'_M and μ''_M are the mass absorption coefficients at these positions, W is the weight fraction of the element, and G is mass thickness in grams per square centimeter of the sample (thickness in centimeters X density).

In practice, one may use a standard X-ray diffractometer, and produce continuous monochromatic radiation by placing a standard analyzing crystal in the sample space. Samples for absorption analysis may be solids in the form of briquettes or liquids in special cells,[29] placed in the X-ray beam. Wavelength of the radiation is determined from the 2θ settings of the diffractometer, related by the Bragg equation to the known spacings of the particular analyzing crystal being used.

As the quantitative feature is strictly valid only at the absorption edge, it is practical to determine the transmitted intensities on either side of the edge,

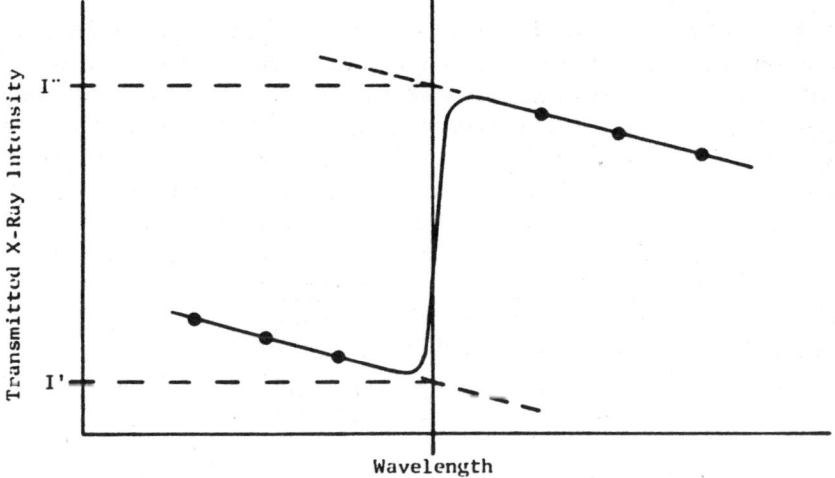

FIGURE 10.15: X-ray absorption near the absorption edge for the element.

and extrapolate to the edge. The values I'' and I' are thus determined in Fig. 10.15 by extrapolation from three experimentally determined points on either side of the absorption edge. The wavelengths of the absorption edges of a given element are handbook values, but they may be determined experimentally as well.

In setting up a working equation for routine use, the value in Eq. (10.6) of $(\mu'_M - \mu''_M)$ may be determined experimentally from a known concentration of the element of interest. For example, Rose and Flick[30] developed a working formula for the determination of iodine in thyroid extract:

$$\% \text{ iodine} = \frac{\log I''/I'}{\text{wt. sample}} \times 24.8 \qquad (10.7)$$

in which the constant 24.8 represented, in addition to conversion factors, the term $(\mu'_M - \mu''_M)$ as determined experimentally from potassium iodide solutions.

Like X-ray emission methods, X-ray absorption analysis is limited to elements of higher atomic number. It is not so sensitive as the emission method for low concentrations of the element of interest, but when applied to higher concentrations, it may be preferred because of its freedom from interference from other elements present.

10.7 LOW-ANGLE X-RAY ANALYSIS

A. MEASUREMENT OF EXCEPTIONAL PERIODICITIES

A consideration of the Bragg equation $n\lambda = 2d \sin \theta$ [Eq. (10.2)] reveals that very large spacings d must correspond to extremely small values of θ. It will only be mentioned here that modifications of diffraction equipment is possible for accurate measurement in the areas of $2\theta < 2°$ for large periodic spacings, from 50 to 1000 Å. This is the range of spacings found in certain muscle tissues, high polymers, and fibered proteins. This method for the detection and characterization of viruses holds great promise.

For the accurate recording of diffraction from large periodicities, several approaches are useful. They involve increasing the degree of divergence at the low angle by increasing the specimen-to-detector distance, using fine beam slits for maintaining resolution; employing longer wavelengths to increase the diffraction angle; the use of staining of the tissue with selective agents; and the use of longer exposures.

B. MEASUREMENT OF LARGE MOLECULES IN SOLUTION: RADIUS OF GYRATION

With the simple addition of special elongated beam slits, commercially available for the purpose, and a readily fashioned liquid sample holder, a standard X-ray diffractometer may be adapted for the determination of the radius of gyration of large molecules in solution by low-angle X-ray scattering.

The radius of gyration is defined as the root mean square of the distance from all of the electrons in the molecule to its center of gravity. While not too definitive in itself, it is of interest because it is a readily measured value which is sensitive to change as the molecular configuration changes. It is a parameter by which complex molecular changes may be followed, when the changes cannot be followed by other methods. Hence, the pH at which abrupt configurational changes occur, or the ionic strength at which such changes occur with certain proteins, may readily be found. Of particular interest to the pharmaceutical chemist is the sensitivity of the method to changes in macromolecules in the presence of drugs, due to binding. It is a useful method for evaluating the effects of agents on protein binding of drugs.

Use is made of the theory as developed by Guinier[31] according to which the intensity of X-rays scattered by a group of identical particles at sufficiently small angles is

$$I_{(\phi)} = Nn^2I_e \exp(-4\pi^2R^2\phi^2/3\lambda^2) \tag{10.8}$$

where

N = total number of particles irradiated
n = number of electrons per particle
I_e = Thomson scattering at zero angle for one electron
R = radius of gyration of the particle
ϕ = scattering angle in radians
λ = wavelength of X-rays used

R is defined with respect to the electron density, rather than the mass density. It may be obtained readily from the slope of a plot of log $I_{(\phi)}$ vs. the square of the scattering angle:

From Eq. 10.8, let Nn^2I_e = a constant A.

Then

$$I_{(\phi)} = Ae^{-(4\pi^2R^2\phi^2/3\lambda^2)} \tag{10.9}$$

Let

I_1 = intensity at low angle of plot
I_2 = intensity at high angle of plot
ϕ_1 = angle at lower angle of plot
ϕ_2 = angle at higher angle of plot

(all from the slope of the line, log I vs. ϕ^2), then

$$\ln I_1 = \ln A - \frac{4\pi^2R^2\phi_1^2}{3\lambda^2} \tag{10.10}$$

$$\ln I_2 = \ln A - \frac{4\pi^2R^2\phi_2^2}{3\lambda^2} \tag{10.11}$$

Combining (10.10) and (10.11),

$$\ln I_1 - \ln I_2 \left[= \ln\frac{I_1}{I_2} \right] = \frac{4\pi^2R^2}{3\lambda^2}[\phi_2^2 - \phi_1^2] \tag{10.12}$$

where ϕ is in radians, or

$$2.303 \log\frac{I_1}{I_2} = \frac{4\pi^2R^2}{3\lambda^2}(\phi_2^2 - \phi_1^2) \times \frac{\pi^2}{(180)^2} \tag{10.13}$$

where ϕ is in degrees, then

$$R^2 = \frac{2.303 \log\frac{I_1}{I_2} \times 3\lambda^2 \times (180)^2}{4\pi^2(\phi_2^2 - \phi_1^2)\pi^2} \tag{10.14}$$

For copper K_α radiation ($\lambda = 1.539$ Å),

$$R = 36.89\sqrt{\frac{\log\frac{I_1}{I_2}}{\phi_2^2 - \phi_1^2}} \text{ Å} \tag{10.15}$$

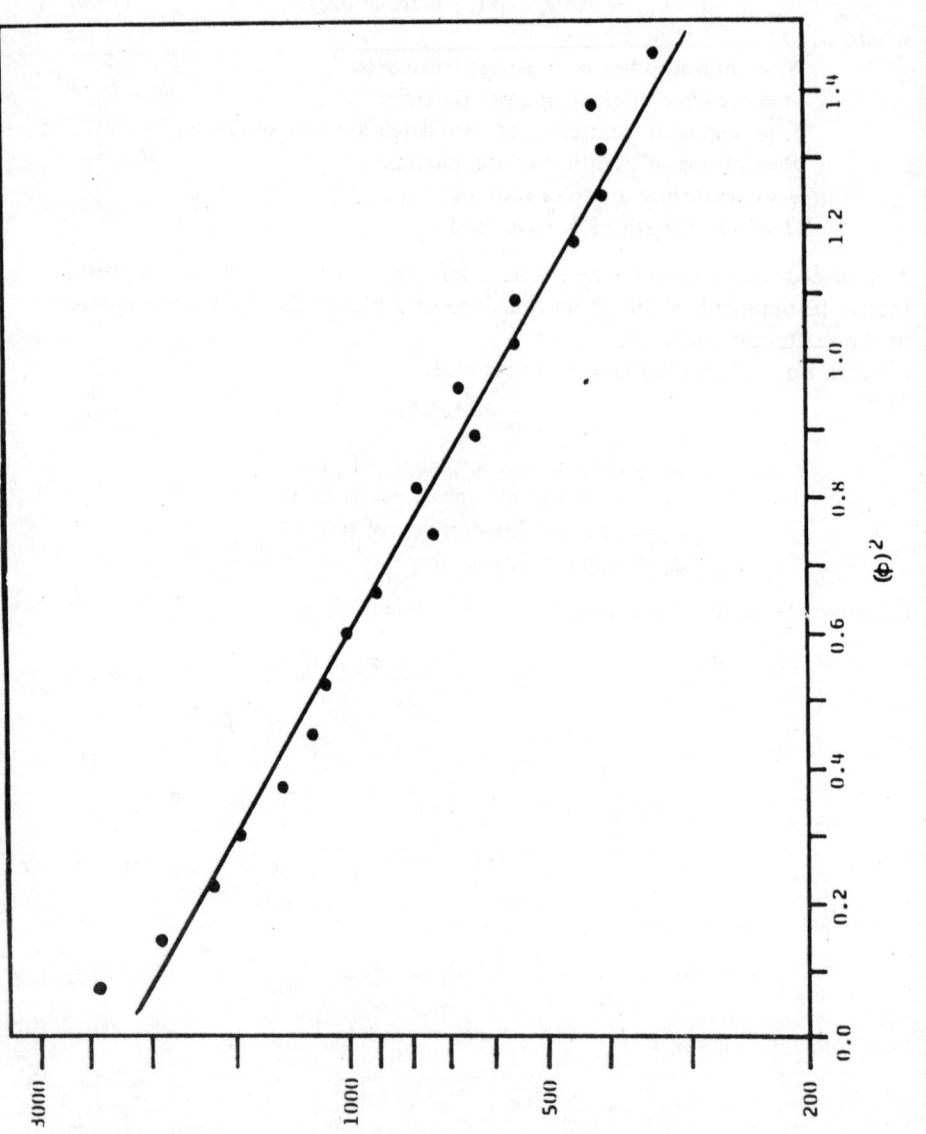

FIGURE 10.16: A plot of log *I* (scattered intensity) as a function of ϕ^2 (scattering angle) for bovine serum albumin.

In practice, use is conveniently made of plastic, cylindrical test tubes for liquid sample holders. Rectangular cells with Mylar windows may be used as well, as the path length of the cell does not effectively change with the small angular changes involved. Concentrations must be kept below about 20 to 30%, as the plot of log I vs. ϕ^2 produces an S-shaped curve, rather than a straight line, at higher concentrations. Unlike solutions for light-scattering experiments, solutions for X-ray scattering do not have to be so scrupulously free from all dust particles, which is a major advantage of the X-ray method.

Intensity determinations are made by counting for approximately 200 sec at a number of 2θ settings, from 2θ values of 0.25° to 1.2°. Some 15 to 20 intermediate settings should be made for counting, spaced closer together at larger 2θ settings so that the points for the ϕ^2 plot will be equidistant. A goniometer zero correction should be applied to the 2θ values to produce values for ϕ. The pure solvent is run in the same cell to determine a blank value in counts per second (cps), which is subtracted from each measured solution intensity value. The log I values are then plotted against the ϕ^2 values.

Figure 10.16 shows a plot of these values as determined from a 5% solution of bovine serium albumin using copper radiation.[32] From Eq. (10.15), taking values from Fig. 10.16,

$$R = 36.89 \sqrt{\frac{\log \dfrac{2380}{348}}{1.40}} \text{ Å}$$

$$= 28.3 \text{ Å}$$

QUESTIONS

Q10.1. In what physical state must a sample be in order that X-ray diffraction procedures may be applied? X-ray emission? X-ray absorption?

Q10.2. Provide a definition for the X-ray powder-diffraction pattern. Let your answer disclose what a single "line" or "peak" represents. Make use of the Bragg relationship.

Q10.3. Describe the principle of the commonly used method for obtaining monochromatic energy for use in X-ray diffraction.

Q10.4. State the particle size range over which X-ray line broadening occurs, sufficient for size analysis.

Q10.5. Give an equation showing the relationship of constants and measurable parameters relating to the determination of molecular weights by X-ray diffraction. Define each term in the equation.

Q10.6. Write a short paragraph describing the origin of X-ray spectra.

Q10.7. What is the major contributing factor responsible for the nonlinearity of the calibration curve used for the X-ray emission analysis of complex mixtures?

Q10.8. State the principle advantages of the X-ray absorption method of analysis.

REFERENCES

1. P. Debye and P. Scherrer, *Physik. Z.*, **17**, 277 (1916); **18**, 291 (1917).
2. A. W. Hull, *Phys. Rev.*, **9**, 84 (1917); **10**, 661 (1917).
3. A. W. Hull, *J. Am. Chem. Soc.*, **41**, 1168 (1919).
4. G. L. Clark and D. H. Reynolds, *Ind. Eng. Chem. Anal. Ed.*, **8**, 36 (1936).
5. S. T. Gross and D. E. Marten, *Ind. Eng. Chem. Anal. Ed.*, **16**, 95 (1944).
6. H. Friedman, *Electronics*, **18**, 132 (1945).
7. L. Alexander and H. P. Klug, *Anal. Chem.*, **20**, 886 (1948).
8. J. Leroux, D. H. Lennox, and K. Kay, *Anal. Chem.*, **25**, 740 (1953).
9. L. E. Copeland and R. H. Bragg, *Anal. Chem.*, **30**, 196 (1958).
10. C. L. Christ, R. B. Barnes, and E. F. Williams, *Anal. Chem.*, **20**, 789 (1948).
11. J. W. Shell, *J. Pharm. Sci.*, **52**, 24 (1963).
12. G. J. Papariello, H. Letterman, and R. E. Huettemann, *J. Pharm. Chem.*, **53**, 663 (1964).
13. P. Scherrer, *Göttinger Nachrichten*, **2**, 98 (1918).
14. F. W. Jones, *Proc. Roy. Soc. (London)*, **A166**, 16 (1938).
15. L. Alexander, *J. Appl. Phys.*, **25**, 155 (1954).
16. P. Scherrer, *Göttinger Nachrichten*, **2**, 98 (1918).
17. L. Bragg, *The Crystalline State*, Bell, London, 1948, p. 189.
18. H. P. Klug and L. E. Alexander, *X-Ray Diffraction Procedures*, Wiley, New York, 1954, Chap. 9, p. 491.
19. J. W. Shell, *J. Pharm. Sci.*, **52**, 100 (1963).
20. L. V. Azároff and M. J. Buerger, *The Powder Method*, McGraw-Hill, New York, 1958.
21. J. W. Shell, *J. Pharm. Sci.*, **52**, 153 (1963).
22. M. L. Goldberg, H. W. Lewis, and K. M. Watson, *Phys. Rev.*, **132**, 2764 (1963).
23. J. W. Shell, *J. Pharm. Sci.*, **51**, 731 (1962).
24. G. V. Alexander, *Anal. Chem.*, **34**, 951 (1962).
25. P. D. Zemany, W. W. Welbon, and G. L. Gaines, Jr., *Anal. Chem.*, **30**, 299 (1958).
26. L. S. Birks, *X-Ray Spectrochemical Analysis*, Wiley (Interscience), New York, 1959.
27. H. A. Liebhafsky, H. G. Pfeiffer, E. H. Winslow, and P. D. Zemany, *X-Ray Absorption and Emission in Analytical Chemistry*, Wiley, New York, 1960.
28. R. E. Barieau, *Anal. Chem.*, **29**, 348 (1957).
29. J. Leroux, P. A. Maffett, and J. L. Monkman, *Anal. Chem.*, **29**, 1089 (1957).
30. H. A. Rose and D. E. Flick, *J. Pharm. Sci.*, **53**, 1153 (1964).
31. A. Guinier, *Ann. Physik*, **12**, 161 (1939).
32. J. W. Shell, unpublished work, 1962.

CHAPTER **11**

Refractometry

Robert A. Locock

FACULTY OF PHARMACY
UNIVERSITY OF ALBERTA
EDMONTON, ALBERTA

11.1   INTRODUCTION

The refractive index is a dimensionless constant determined by the character and state of the medium under specified conditions of temperature and wavelength of light. Refractive indices may be employed as criteria for the identification of materials, as an aid in the structural elucidation of molecules, to deduce molecular parameters and energies, and to determine concentration or changes in composition of a system.

When light from an optically isotropic medium (a medium in which the optical properties are the same in all directions) enters another optically isotropic medium, a change in the velocity of light occurs. This velocity change is accompanied by a change in direction unless the light has entered perpendicularly to the boundary of the two media. The relationship between the angle of incidence of a light ray i and the angle of refraction r is given by Snell's law of refraction;

$$\frac{\sin i}{\sin r} = \frac{\eta_r}{\eta_i} \tag{11.1}$$

where η is the refractive index of the first medium and η_r is the refractive index of the second medium. The refraction of a light ray for isotropic media is shown diagrammatically in Fig. 11.1.

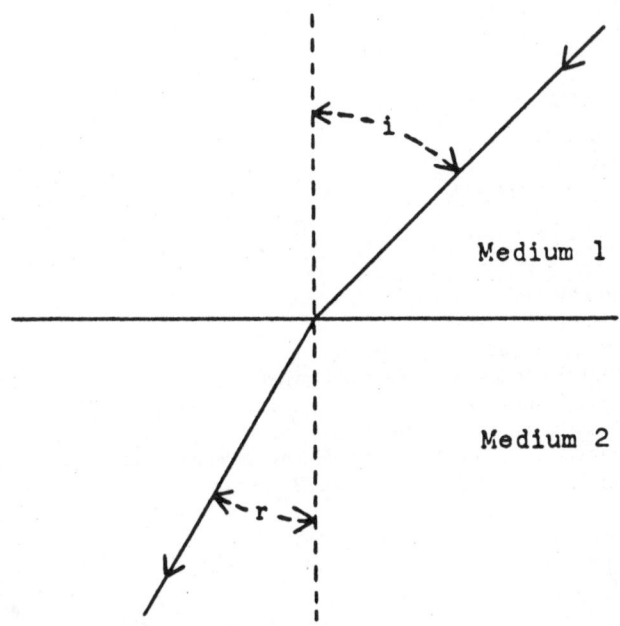

FIGURE 11.1: The fundamental geometric relationship of an isotropic light ray from medium 1 entering an optically denser medium 2. Both media are optically isotropic.

The refractive index of a single medium η is the ratio of the velocity of light in this medium to the velocity of light in a reference medium. Absolute refractive indices are those which are measured with reference to the velocity of light in a vacuum, which is approximately 3×10^{10} cm sec^{-1} for all electromagnetic radiation. Therefore,

$$\eta_{abs} = \frac{v_{vac}}{v_{med}} = \frac{\sin i_{vac}}{\sin r_{med}} \tag{11.2}$$

where η_{abs} is the absolute refractive index, v_{vac} is the velocity of light in a vacuum ($\approx 3 \times 10^{10}$ cm sec^{-1}), v_{med} is the velocity of light in a given medium, sin i_{vac} is the sine of the angle of incidence with the normal on the plane interface between the two media, and sin r_{med} is the sine of the angle of refraction with the normal.

It is common practice to refer the index of refraction of liquids and solids to a reference medium of air so that Eq. (11.2) becomes

$$\eta_{med} = \frac{v_{air}}{v_{med}} = \frac{\sin i_{air}}{\sin i_{med}} \qquad (11.3)$$

where η_{med} is the refractive index of the medium. The refractive index of dry air at 0°C, 1 atm, and a wavelength of 5893 Å is 1.000277. Thus the absolute refractive index may be estimated by multiplying a relative refractive index such as η_{med} by 1.000277.

$$\eta_{abs} = 1.000277\, \eta_{rel} \qquad (11.4)$$

In reporting refractive indices the wavelength is commonly designated by a subscript and the temperature by a superscript. For example, η_D^{25}, 25 refers to the temperature of the measurement, 25°C, D refers to the wavelength of the light used for the measurement, the D line of sodium, 5893 Å (actually a doublet $D_1 = 5890$ Å and $D_2 = 5896$ Å). Other commonly used wavelengths of measurement are the hydrogen lines, $H_\alpha(C) = 6563$ Å, $H_\beta(F) = 4861$ Å, and $H_\gamma(G) = 4340$ Å, and the mercury lines at 5790, 5461, 4358, and 4047 Å.

11.2 ORIGIN OF REFRACTIVE INDEX

The index of refraction of a substance has its origin in its tendency to undergo distortion in an electric field. This tendency, termed "optical polarizability," is related to the number, charge, and mass of the vibrating particles in the material, which in turn is determined by the atoms in the structure and by the type of electronic bonding. The refractive index is principally related to the motion of electrons in molecules and therefore both refraction and absorption of light may be considered to be two different aspects of the same phenomenon—the interaction of electromagnetic radiation with valence electrons.

The relationship between refractive index and absorption is illustrated in Fig. 11.2. If we select a hypothetical compound which has only one important absorption band as the wavelength of light changes in the visible and ultraviolet region of the spectrum, then the refractive index decreases as the wavelength of the radiation increases in a region of "normal" dispersion, and the refractive index increases as the wavelength increases in a region of "anomalous" dispersion. The curve obtained by plotting η vs. wavelength is called a "dispersion curve" since natural or "white" light of many wavelengths is dispersed or refracted to different angles by a given medium.

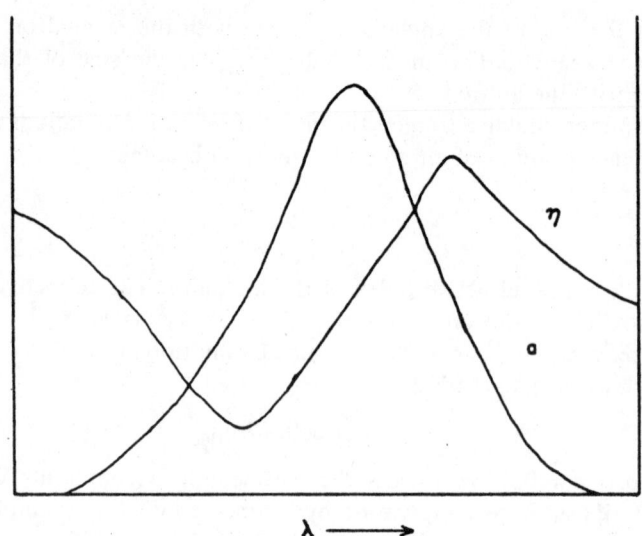

FIGURE 11.2: Theoretical absorption and dispersion curves; η = refractive index, a = absorbancy index, λ = wavelength increasing from left to right from the ultraviolet to the visible region of the spectrum.

One of the first attempts to empirically correlate wavelength and refractive index is the Cauchy formula

$$\eta = A + \frac{B}{\lambda^2} + \frac{C}{\lambda^4} \tag{11.5}$$

or

$$\eta = A + \frac{B}{\lambda^2} \tag{11.6}$$

where A and B are constants evaluated from data η_1 and η_2 at λ_1 and λ_2. Thus

$$A = \frac{\eta_1 \lambda_1^2 - \eta_2 \lambda_2^2}{\lambda_1^2 - \lambda_2^2} \tag{11.7}$$

and

$$B = (\eta_1 - A)\lambda_1^2 \tag{11.8}$$

In a region of normal dispersion the Cauchy formula may be used for a rough interpolation of refractive index, however serious errors will occur as the absorption band is reached. The Cauchy formula is without theoretical basis.

A more satisfactory relationship for normal dispersion curves, which is based on classical electromagnetic theory, is the Sellmeier equation

$$\eta^2 - 1 = \sum_i \frac{K_i \lambda^2}{\lambda^2 - \lambda_i^2} \tag{11.9}$$

where η is the refractive index, λ_i represents the wavelength of the absorption band of the substance, λ is the wavelength of measurement of η, and K_i is a constant arising from electromagnetic theory, which contains factors involved with transition probabilities of the absorption band. Another form of this expression, the Lorenz form, is:

$$\frac{\eta^2 - 1}{\eta^2 + 2} = \sum_i \frac{K_i \lambda^2}{\lambda^2 - \lambda_i^2} \tag{11.10}$$

Because of the complexity of Eq. (11.9) and (11.10) graphical methods are usually used to show the quantitative variation of refractive index with wavelength in regions of normal dispersion.

Regions of anomalous dispersion in the dispersion curve are explained by interaction of the scattered light (oscillating electrons) of the absorption phenomenon with the original incident light. It is possible to develop theoretical relationships between absorption curves and dispersion curves and to discuss them both qualitatively and quantitatively using classical electromagnetic theory and modern quantum theory. Such a discussion is beyond the scope of this present chapter.

11.3 REFRACTIVE INDEX AND TEMPERATURE

From the foregoing section it may be understood that the spectral characteristics of a molecule may be related to the refractive index measurement. This is a factor in the explanation of the temperature dependence of refractive index. An increase in temperature tends to shift ultraviolet absorption bands to longer wavelengths, which will lead to an increased refractive index. Conversely infrared absorption is shifted to shorter wavelengths by an increase in temperature, which tends to decrease the refractive index.

Other factors which cause the refractive index to be temperature dependent are changes in density of the sample medium and reference medium as the temperature changes. If the temperature increases, the light beam will encounter fewer molecules per unit volume in passing through a given medium. Thus a decrease in density of the sample will decrease the absolute refractive index, whereas a decrease in density of air will increase the relative refractive index.

In general the refractive index of organic liquids will decrease by a factor of 4.5×10^{-4} for every centigrade degree of increase in temperature. Therefore it is necessary to specify the temperature of a refractive index measurement and to control the temperature at which the measurement is taken to within $\pm 0.2°C$ for a precision of η of 1×10^{-4}. For even rough qualitative measurement the temperature should be controlled to within $3°C$ for an accuracy of η of ± 0.002.

11.4 REFRACTIVE INDEX AND DIELECTRIC CONSTANT

At low frequencies the refractive index may be related to the dielectric constant by electromagnetic theory. If the measurement of refractive index is made with infrared radiation in a region where the radiation is not absorbed, then it may be demonstrated that the refractive index η is related to the dielectric constant ϵ by the equation:

$$\eta^2 = \epsilon \tag{11.11}$$

This relationship is derived with the assumption that the substance contains no dipoles or that contribution to η of the permanent dipoles of the substance is negligible since permanent dipoles of nuclei cannot keep up with a very rapidly alternating electric field such as that associated with light waves and only electrons in molecules can adjust themselves.

At low frequencies, long wavelengths, the dielectric constant is a measure of the total moment of the substance in an electric field. The equation relating the dielectric constant and the polarization of a substance is:

$$P_m = \frac{\epsilon - 1}{\epsilon + 2} \frac{m}{d} \tag{11.12}$$

where P_m is the polarization per mole, m is the molecular weight, and d is the density. The total molar polarization may also be expressed by the following equation:

$$P_m = 4/3\pi N \left(\alpha + \frac{\mu^2}{3kT} \right) \tag{11.13}$$

where N is Avogadro's number, α is the electrical polarizability of the molecules, μ is the moment of the permanent dipoles, k is the Boltzman constant, and T is the absolute temperature.

The term $4/3\pi N$ gives the contribution of the induced dipoles to molar polarization, which is proportional to α, and the term $\mu^2/3kT$ gives the contribution corresponding to permanent dipoles.

The contribution of the induced dipoles to the dielectric constant may be evaluated by a calculation of the molar refraction R_m:

$$R_m = \frac{\eta^2 - 1}{\eta^2 + 2} \frac{m}{d} = 4/3\pi N_0 \alpha \tag{11.14}$$

A discussion of the significance and application of the calculation of molar refraction follows in the next section.

11.5 MOLAR REFRACTION

It may be shown that the refractive index is related to the polarizability of a substance through the molar refraction R_m. Thus

$$R_m = \frac{\eta^2 - 1}{\eta^2 + 2} \frac{m}{d} = 4/3 \pi N_0 \alpha = \frac{N e^2}{3 \pi m_e} \sum_i \frac{f_i}{v_i^2 - v^2} \qquad (11.15)$$

where N_0 is the number of atoms per cubic centimeter, e is the electronic charge, m_e is the electron mass, f_i is the oscillator strength of an individual absorption band, a unitless measure of absorption intensity v_i is the frequency of the absorption band, and v is the frequency of the radiation used for measurement. The equations used in expression (11.15) apply for the steady state condition in which the electron moves in phase with the electromagnetic radiation and contributions to molar refraction from any permanent dipoles are ignored (see previous section).

As a consequence of the final expression in Eq. (11.15), it may be seen that weak absorption bands which have small oscillator strengths will make small contributions to the refractive index at all frequencies not close to the frequency of the weak band. Nearly all the molar refraction, calculated from measurement at the sodium D line, of organic compounds that do not have significant molar extinction coefficients above 200 mμ arises from absorption bands in the far ultraviolet. For example, the molar extinction coefficient of the carbonyl band of simple aliphatic ketones, ϵ_{max} 15–20 near 290 mμ, has an f_i value of 3×10^{-4} and contributes very little to refractive index even at measurements made near the absorption maximum. Oscillator strengths close to unity are found for these substances with molar extinction coefficients of the order of 10^4 and 10^5 at wavelengths below 200 mμ.

As discussed previously, molar refraction is a calculation of the contribution of the induced dipoles in an electric field to the dielectric constant and is therefore a calculation of the "polarizability" of the substance. The molar refraction may be found from the measurement of refractive index using the Lorenz-Lorentz equation

$$R_m = \frac{\eta^2 - 1}{\eta^2 + 2} \frac{m}{d} \qquad (11.16)$$

or it may be estimated by adding up the contributions of individual atoms or groups of atoms in the molecule. The latter method, the summing of individual atomic refractivities to give an estimation of molar refractivity, does not always yield results which agree with the Lorenz-Lorentz calculation. If the electronic structure of the substance under consideration is altered by bringing two or more atoms together the mean value of the frequency of absorption of all effective bands is changed. An example of this is the introduction of a carbon-carbon double bond to a system of conjugated double bonds or to a system which creates conjugation and thus delocalization of electrons. In such instances, the calculated molar refractivity becomes the sum of atomic refractivities plus "exhaltations" due to unsaturation, ring formation, or special groups which may be considered chromophores in the ultraviolet region of the spectrum.

TABLE 11.1: Atomic Refractivities of Some Elements and Structural Units at 20°C (D line)

C	2.418	C=C (double bond)	1.733
H	1.100	C≡C (triple bond)	2.398
O (hydroxyl)	1.525	N in amines	
O (carbonyl)	2.211	primary	2.322
O (ester)	1.64	secondary	2.499
O (ether)	1.643	tertiary	2.840
—CH₂—	4.618	S (mercaptan)	7.69
Cl	5.967	C≡N	5.459
Br	8.865	C——C \ / C	0.71
		C—C | | C—C	0.48

Table 11.1 presents values for the atomic refractivities of some elements and structural units.

The molar refraction of an organic compound may be calculated from the observed refractive index using the Lorenz-Lorentz equation. For example in order to calculate the molar refraction of eucalyptol (cineol), $C_{10}H_{18}O$, the values of the molecular weight, 154.24, the density (d_4^{20}), 0.9267, and the observed refractive index η_D^{20}, 1.4584, are substituted in the Lorenz-Lorentz equation:

$$R_{m(obs)} = \left[\frac{(1.4584)^2 - 1}{(1.4584)^2 + 2}\right]\frac{154.24}{0.9267}$$

$$= 45.45$$

The molar refraction may also be calculated from the values of the atomic refractivities listed in Table 11.1. Using the previous example, eucalyptol, the $R_{m(calc)} = 45.62$.

10 carbon atoms = 10 × 2.418 = 24.78
18 hydrogen atoms = 18 × 1.100 = 19.8
1 oxygen atom (ether) = 1 × 1.643 = 1.643
$R_{m(calc)}$ = 45.62

The values in Table 11.1 may also be used to estimate the value of refractive index for a given structure. For example in order to calculate the refractive index of α-pinene at 20°C using the sodium D line, the calculated molar refraction is found using the values from Table 11.1:

α-pinene

$$10 \times C = 10 \times 2.418 = 24.18$$
$$16 \times H = 16 \times 1.100 = 17.6$$
$$1 \times C{=}C = 1 \times 1.733 = 1.733$$
$$1 \times C{-}C = 1 \times 0.48 = 0.48$$
$$\begin{matrix} | & | \\ C & - & C \end{matrix}$$

$$R_{m(calc)} = 43.99$$

The value of the calculated molar refraction for α-pinene, 43.99, the molecular weight, 136.23, and the density (d_4^{20}), 0.8585, are then placed in the Lorenz-Lorentz equation and the refractive index calculated:

$$43.99 = \frac{\eta^2 - 1}{\eta^2 + 2} \cdot \frac{136.25}{0.8585}$$

$$\frac{\eta^2 - 1}{\eta^2 + 2} = 0.27721$$

$$\eta^2 - 1 = 0.2772\eta^2 + 0.5544$$

$$\eta^2 = 2.1505$$

$$\eta = 1.466$$

The observed refractive index (found experimentally) is 1.4663. The molar refraction, calculated from the observed refractive index using the Lorenz-Lorentz equation, is 43.97.

Since molar refraction is an additive property of substances, the refractive index of mixtures may be calculated using equations of the following form:

$$R_{1,2} = X_1 R_1 + X_2 R_2 \tag{11.17}$$

and

$$R_{1,2} = \frac{\eta^2 - 1}{\eta^2 + 2} \frac{X_1 m_1 + X_2 m_2}{d_{1,2}} \tag{11.18}$$

where X is the mole fraction, η is the refractive index of the mixture, and d is the density of the mixture.

If the volume does not change when two substances are mixed, it may be possible to calculate the approximate refractive index using the following equation:

$$\eta = \frac{\eta_1 \nu_1 + \eta_2 \nu_2}{\nu_1 + \nu_2} \tag{11.19}$$

where v is volume and η is the refractive index. The refractive index, however, will more often vary linearly with the mole fraction of the components.

The specific refraction r defined by Lorenz and Lorentz is

$$r = \frac{\eta^2 - 1}{\eta^2 + 2}\left(\frac{1}{d}\right) \tag{11.20}$$

Specific refraction like molar refraction is an additive property of substances and therefore it is possible to calculate the refractivity of a mixture from the specific refractivity of the components with an equation of the type

$$C_{rm} = Ar_a + Br_b \tag{11.21}$$

where C is a mixture of A grams of a and B grams of b with specific refractivities r_a and r_b, respectively.

11.6 MEASUREMENT OF REFRACTIVE INDEX

When light passes from a less-dense medium to a more-dense medium, the angle of refraction is always less than the angle of incidence. As the angle of incident light is increased, the angle of refracted light will also increase. The angle of refracted light will reach a maximum as the angle of incidence approaches 90°. If the angle of refraction is increased to a value at which the angle of incidence is 90°, light no longer enters the denser medium, no refracted light is possible, and light will be reflected through the less-dense medium. The value of the angle of refraction where the incident light approaches 90° incidence (also termed "grazing incidence") and which separates refracted and reflected light is called the "critical angle of refraction." In this limiting instance, since sin 90° = 1, Snell's law may be simplified to

$$\frac{\eta_r}{\eta_i} = \frac{1}{\sin r_c} \tag{11.22}$$

where r_c is the critical angle in the denser medium at the wavelength used, η_r is the refractive index of the denser medium, and η_i is the refractive index of the less-dense medium. The critical angle is characteristic of each substance and the critical-angle principle is used to measure refractive indices by means of so-called critical-angle refractometers. If either η_r or η_i are known and the critical angle r_c is observed, then the refractive index may be found. In most critical-angle refractometers η_i is the refractive index of the unknown sample and η_r is the known refractive index of a prism.

A typical design of a critical-angle refractometer (Abbe) is schematically shown in Fig. 11.3¹ and a commercial instrument, the Bausch & Lomb "Abbe 3L" refractometer, is pictured in Fig. 11.4. A thin layer of liquid is placed between the illuminating and refracting prisms, which may be equipped with water jackets to provide temperature control. Light enters from below by means of a reflecting mirror. The angle of emergence of the critical ray is measured by scanning the upper edge of the refracting prism with a telescope. Since each wavelength of light has a critical angle, wavelengths other than that of the desired wavelength (usually the D line of sodium) are dispersed by Amici prisms—small direct vision spectroscopes. These compensating prisms are rotated to obtain a sharp colorless light-dark boundary and the sharp edge of the boundary is centered on the cross hairs of the telescope. The refractive index of the liquid in terms of monochromatic sodium light can be obtained directly from a calibrated glass or metal scale. The accuracy that can be obtained with an instrument of this type is ±0.0001 in a range of η from 1.30 to 1.70. Substances which have refractive indices greater than that of the refracting prism of the Abbe refractometer cannot be measured since total reflection would be the result.

The dispersion of the sample may be estimated with some instruments of the Abbe type. Dispersion ν is calculated from the equation

$$\nu = \frac{\eta_D - 1}{\eta_F - \eta_C} \qquad (11.23)$$

where η_D is the refractive index for the sodium D line, η_F is the refractive index for blue hydrogen line (4681 Å), and η_C is the refractive index for red hydrogen line (6563 Å). The values for partial dispersion $\eta_F - \eta_C$ may be obtained from the reading ("Z" value) of the Amici compensator drum and reference to a table furnished with the instrument. The partial dispersion and dispersion are sometimes useful for characterization purposes. Another parameter that has characteristic values for certain classes of compounds is the specific dispersion δ.

$$\delta = \frac{\eta_F - \eta_C}{d} \times 10^4 \qquad (11.24)$$

Values found for dispersion, partial dispersion, and specific dispersion may vary depending on the actual instrument and must be used with caution.

Another type of critical-angle refractometer is the dipping or immersion type. See Fig. 11.5. This instrument differs from the Abbe type in that it does not have an illuminating prism and the single refracting prism which is immersed in the liquid sample is rigidly mounted in the telescope containing the compensator and the eyepiece. A scale below the eyepiece is moved by means of a micrometer screw at the top of the instrument until the nearest division falls on the light-dark boundary. Both the scale and the micrometer are read and the refractive index found by reference to a table. The immersion type of critical-angle refractometer yields a greater precision than the standard Abbe refractometer. By the use of ten interchangeable prisms the range of η, 1.325 to 1.647, can be read with a precision of ±3 × 10⁻⁵. However this type of refractometer requires a sample size of 5 to 30 ml, whereas the Abbe

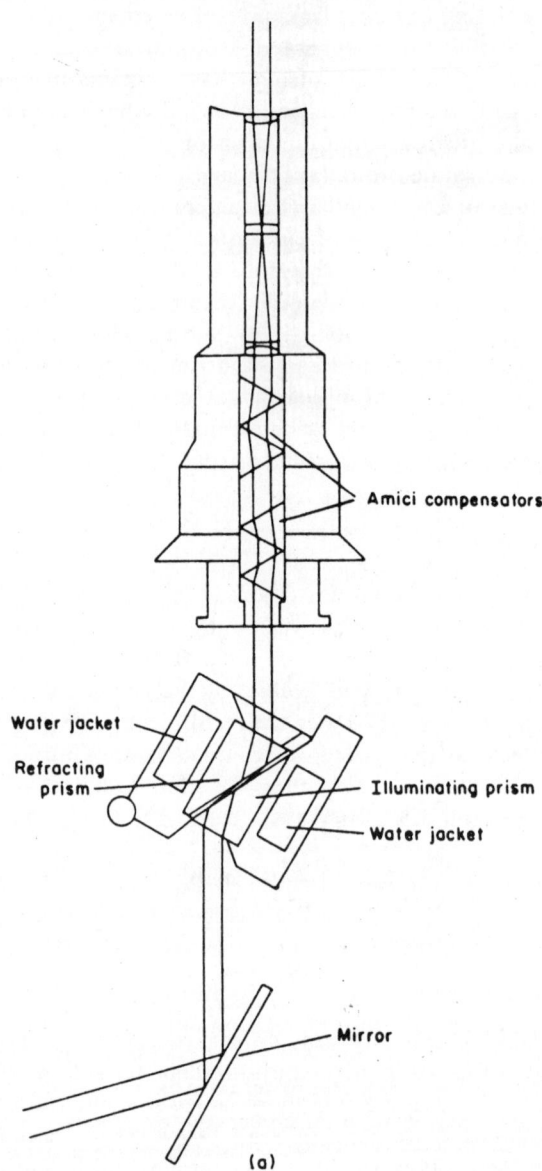

Amici compensators

Water jacket

Refracting prism

Illuminating prism

Water jacket

Mirror

(a)

FIGURE 11.3: (a) Schematic diagram of a typical critical ray refractometer; (b) enlarged view of prisms of refractometer (r_c = critical angle, α = angle of emergence of critical ray); (c) enlarged view of Amici compensator.

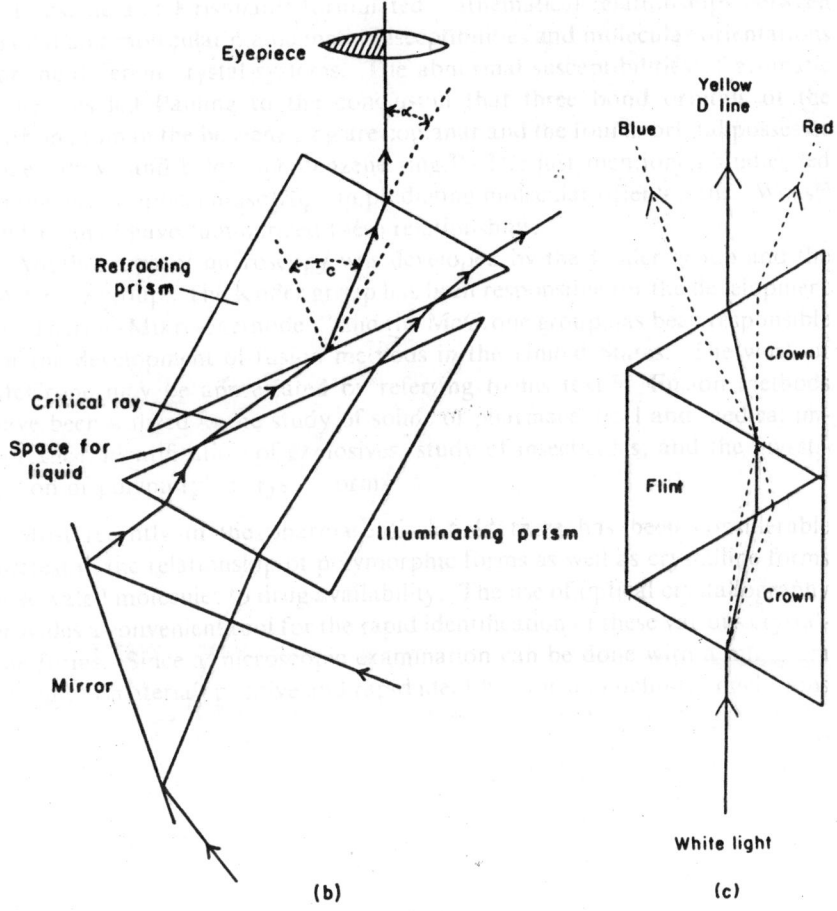

FIGURE 11.3 (continued)

instrument may require from 0.05 to 0.10 ml. The immersion refractometer is especially useful for the routine precision analysis of the concentration of aqueous and alcoholic solutions.

The most precise measurements of refractive index are made with instruments using the principle of interference of light waves. In this type of refractometer parallel rays of white or monochromatic light pass through matched cells containing the sample substance and a standard of known refractive index. The light then passes to a slit at the end of each cell. If the two light beams are of equal length, they will arrive in phase on a plane surface perpendicular to the light path and result in a series of bright bands. At places on the plane surface the two light beams are not the same length (out of phase) and dark bands are produced. Thus a pattern of interference

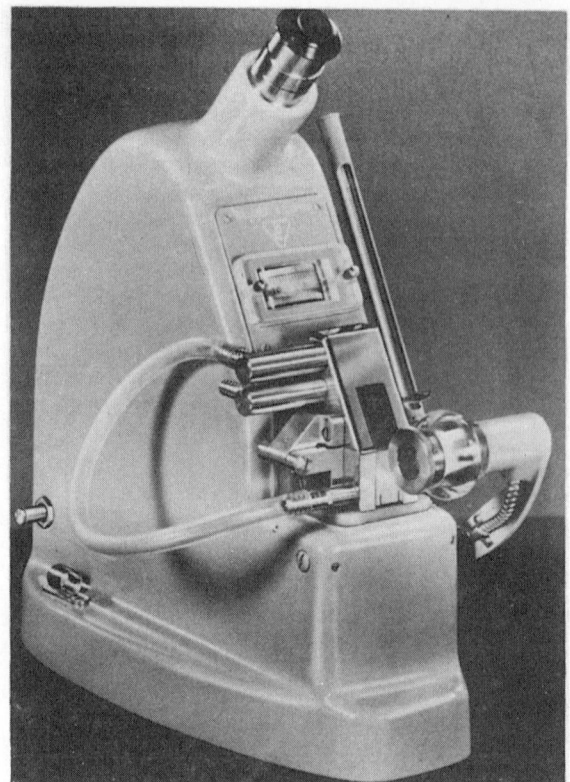

FIGURE 11.4: Bausch & Lomb Abbe 3L refractometer.

bands is produced. The relative position of the interference bands will depend on the optical length of the light paths. If the sample has a slightly greater refractive index than the standard, the optical path of light through the sample will be increased since the velocity of light through the sample is decreased. The change in the interference pattern produced by unequal optical paths is noted visually as a shift in the position of the interference pattern. The changed interference pattern may be matched with a fixed band pattern by means of a variable glass compensator in the sample path. A differential reading of refractive index $\Delta\eta$ is obtained from the scale of the compensator micrometer screw. A schematic diagram of a typical Rayleigh (interference) refractometer is shown in Fig. 11.6.

Interference refractometers are usually used for the differential measurement of refractive indices over a narrow range. The maximum difference $\Delta\eta$ that may be obtained with liquids is 0.05 units of η. These measurements may be made with an ultimate precision of $\pm 2 \times 10^{-7}$ units of η for liquids and $\pm 3 \times 10^{-8}$ units of η for gases. The interferometer may be used to

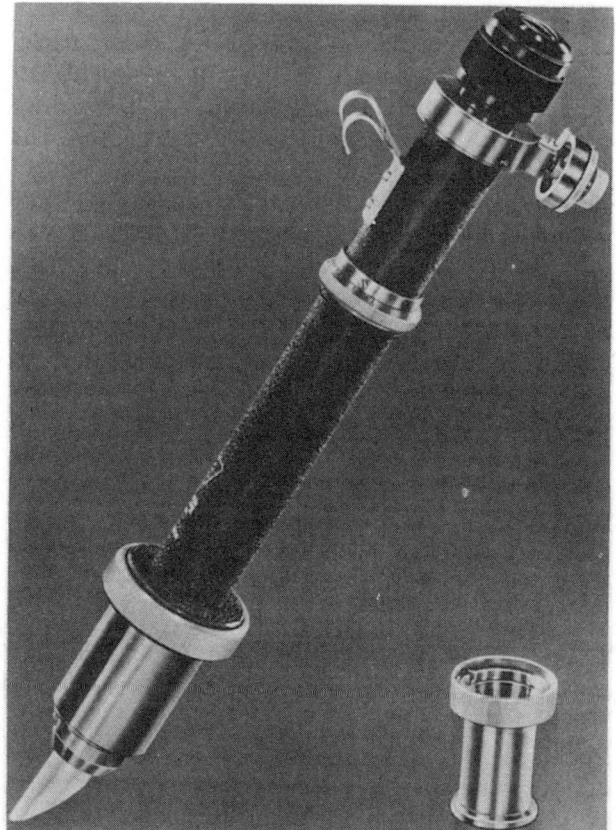

FIGURE 11.5: Bausch & Lomb dipping refractometer.

analyze gases, *e.g.*, CO_2 in air, in the analysis of dilute solutions and small changes in the composition of mixtures.

Precision spectrometers may be used to measure refractive index. In these instruments the material to be measured is formed into a prism, which is placed at the center of a large, precisely ruled circle of metal. Monochromatic energy is focused through a slit and passed through the sample prism. The

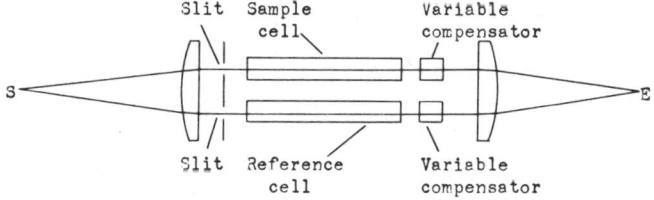

FIGURE 11.6: Schematic diagram of a typical Rayleigh refractometer. S = light source; E = eyepiece.

angular setting of the displacement of the image of the slit is a measurement of the index of refraction of the substance at the particular wavelength and temperature. This type of refractometer can yield good precision, $\pm 1 \times 10^{-6}$ units of η, and is most versatile. Wavelengths in the ultraviolet and infrared regions of the spectrum may be used and there is no limit on the value of η, since any angle of refraction can be determined. However careful angular measurement and close temperature control are necessary to obtain precise measurements and these instruments are not generally suitable for routine laboratory analyses.

A type of image-displacement refractometer which may be used for a very rough measurement of refractive index is the *Jelley-Fisher* refractometer. As with the prism-spectrometer type, the sample liquid is formed into a prism by the beveled edge of a glass slide. The operator looks through this liquid sample prism at a graduated illuminated scale containing a slit. The refractive index is read by observing the refracted slit image formed by the eye focused in the plane of the scale. Using the most yellow part of the dispersed white light of the slit image, it is possible to estimate a refractive index within ± 0.002 units of η.

11.7 APPLICATIONS OF REFRACTIVE INDEX MEASUREMENTS

Refractive index measurements may be used as physical constants for the identification of substances. For example, in pharmaceutical analysis refractive index may be used as a means for the identification and detection of impurities in volatile oils. The standard Abbe refractometer usually has sufficient range and precision for the comparison of pharmacopoeial materials.

Refractometry may be employed as a convenient method of measurement and evaluation of the separation of complex materials by physical means such as chromatography, distillation, electrophoresis, and extraction.

Since the more modern development of ultraviolet, infrared, and nuclear magnetic resonance spectroscopy and mass spectrometry, refractive index measurements are not as important as they once were in qualitative organic analysis. However refractometric data may still be of aid, as a nondestructive method, in the determination of molecular structure.

The precision and accuracy of quantitative refractometric measurements depend on the type of refractometer used and the temperature control. Often it is necessary to prepare accurate calibration curves using pure components since refractive index composition curves are not always linear over the range of concentrations to be determined. Quantitative refractometry has been used for serum, sugar, fat, oil, and petroleum analyses.

EXPERIMENTS

E11.1. Qualitative and quantitative analysis of liquids using the Abbe refractometer. (a) Read and note the laboratory directions for the instrument. Determine the refractive index of water at a constant temperature of 20 or 25°C. Take five readings and determine the average. The refractive index of water at 25°C, η_D^{25}, = 1.3325 and at 20°C, η_D^{20}, = 1.3330.

(b) Determine the refractive index of a pure liquid obtained from the instructor. Compare your value with the literature value. Take five readings in either direction on the Amici compensator. With the value for the refractive index and the compensator value, calculate the dispersion of the substance using the dispersion table supplied with the instrument. Using the reported density of the substance, calculate specific dispersion. Use your observed value of refractive index to calculate molar refraction from the Lorenz-Lorentz equation. Calculate the molar refraction from a table of atomic and group refractivities and compare the result with the Lorenz-Lorentz equation.

(c) Determine the percentage of sucrose in samples of syrup supplied by the instructor. (If the instrument is not supplied with a sugar scale, a calibration curve must be constructed with standard solutions of sucrose.)

E11.2. Determination of alcohol content using the Pulfrich immersion refractometer. Determine the percentage volume to volume of ethyl alcohol in the preparation supplied by the instructor according to the procedure of the *British Pharmacopoeia*, 1968, Appendix XII H, p. 1278.

QUESTIONS

Q11.1. Define the following terms and describe how they are applied to the measurement of refractive index:
(a) dispersion
(b) isotropic light
(c) grazing incidence.

Q11.2. How may measurements of refractive index be used in quantitative analyses?

Q11.3. Briefly and qualitatively describe the relationship between the refractive index and absorption coefficient of a substance.

Q11.4. Atomic refractivities of some elements and structural units may be used to calculate the refractive index of a particular structure. In some instances the calculated refractive index may be quite different from the observed refractive index. Explain.

PROBLEMS

P11.1. Calculate the molar refraction of triethanolamine; η_D^{20} = 1.4852, d_4^{20} = 1.1242. Compare the results with the theoretical molar refraction calculated by a summation of atomic and group refractivities.

P11.2. Calculate the molar refraction of nicotine; η_D^{20} = 1.5820, d_4^{20} = 1.0097. Compare the result with the theoretical molar refraction calculated by a summation of atomic and group refractivities.

P11.3. The molar refraction of limonene (dipentene) $C_{10}H_{16}$ is 45.25. Calculate the refractive index if its density is 0.840, using the Lorenz-Lorentz equation. Compare the result with the literature value.

P11.4. What is the composition of a mixture of n-hexane ($\eta_D^{20} = 1.3751$) and cyclohexane ($\eta_D^{20} = 1.4264$), which has a refractive index of 1.3800 at 20 C? Assume a linear relationship between mole fraction and refractive index.

P11.5. Calculate the weight per cent of 50 ml of a mixture of morpholine and pyrrole if the density of the mixture is 0.9880 and $\eta_D^{20} = 1.4920$. Pure pyrrole has a d_4^{20} of 0.9691 and η_D^{20} of 1.5085. Morpholine has a density of 0.9994 and η_D^{20} of 1.4545.

BIBLIOGRAPHY

Bauer, N., K. Fajans, and S. Z. Lewin, in *Physical Methods of Organic Chemistry* (A. Weissberger, ed.), Part II, Vol. 1, Wiley (Interscience), New York, 3rd ed., 1960.

Berl, W. G. *Physical Methods in Chemical Analysis*, Vol. 1, Academic Press, New York, 1950.

Kauzmann, W., *Quantum Chemistry*, Academic Press, New York, 1957.

Meloan, C. E., and R. W. Kiser, *Problems and Experiments in Instrumental Analysis*, Merrill, Columbus, Ohio, 1963.

Strobel, H. A., *Chemical Instrumentation*, Addison-Wesley, Reading, Mass., 1960.

CHAPTER **12**

Polarimetry

Robert A. Locock

FACULTY OF PHARMACY
UNIVERSITY OF ALBERTA
EDMONTON, ALBERTA

12.1 INTRODUCTION

Polarimetry, which may be defined as the measurement of the rotation of polarized light, is a classical method of quantitative analysis and was the first optical property of organic compounds to be interpreted in terms of molecular structure. The measurement of rotation, accomplished with a polarimeter, is a measurement of the change in direction of linearly polarized light after its passage through an anisotropic medium. If the change in direction is caused by anisotropic refraction (circular birefringence), i.e., the refractive indices of right- and left-hand components of linearly polarized light are different, then the change is called "optical rotation" and the medium is said to be "optically active" or, more correctly, to have "optical rotatory power." If the change in rotation is caused by anisotropic absorption or

425

scattering of linearly polarized light, then the medium is said to exhibit "circular dichroism."

(Polarimetry is employed as a method of quantitative analysis of organic compounds that are optically active; notably the sugars, alkaloids, terpenes, and steroids. Optical rotation is used widely as a criterion in the establishment of identity of substances, the characterization of stereoisomers, the elucidation of configuration, and to solve problems involving stereochemical reaction mechanisms and conformational and configurational changes of molecules in solution.)

Although the phenomenon of the rotation of polarized light has been known for approximately 150 years, it is only comparatively recently, with the development of commercial spectropolarimeters, that measurements have been extended to wavelengths other than those of the sodium D doublet. The measurement of optical rotatory dispersion i.e., the variation in rotation with wavelength, has become an important tool for the characterization of organic structures. An even more recent development is the availability of instruments which provide measurement of the absorption of circularly polarized light with change in wavelength. Such circular diochroism spectra may have certain advantages for studying optical activity relative to optical rotatory dispersion spectra. These closely related phenomena have become exceedingly important as aids in the solution of complex qualitative problems involving asymmetric molecules.

12.2 ORIGIN OF OPTICAL ROTATION

(Light may be described as having a transverse wave motion. The vibrations of the wave motion are at right angles to the direction of propagation and

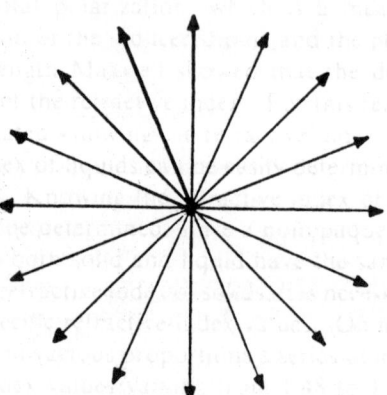

FIGURE 12.1: End view of unpolarized light. The unresolved electric vectors are in a plane perpendicular to the direction of propagation and the equal probability of all directions of vibration is schematically illustrated.

FIGURE 12.2: End view of plane- or linearly polarized light. The resolved electric vector is confined to one plane perpendicular to the direction of propagation.

thus there may or may not be perfect symmetry around the direction of travel. If the light does not have perfect symmetry then, it may be said to be "polarized." Figure 12.1 schematically illustrates an end-on view of unpolarized light. Such a beam of ordinary light may be assumed to consist of millions of waves, each with its own plane of vibration. Since there are waves vibrating in all planes with equal probability, the light beam is said to have "perfect symmetry." If the waves of ordinary light are restricted by some means to vibration in planes parallel to each other, then the light is said to be "plane polarized." Plane or linearly polarized light is a form of electromagnetic radiation in which the energy or electric vector is restricted to vibration in one plane in a sinusoidal manner. See Fig. 12.2 and 12.3.)

(A plane-polarized light beam may be resolved theoretically and experimentally into two circularly polarized components. The two circular beams

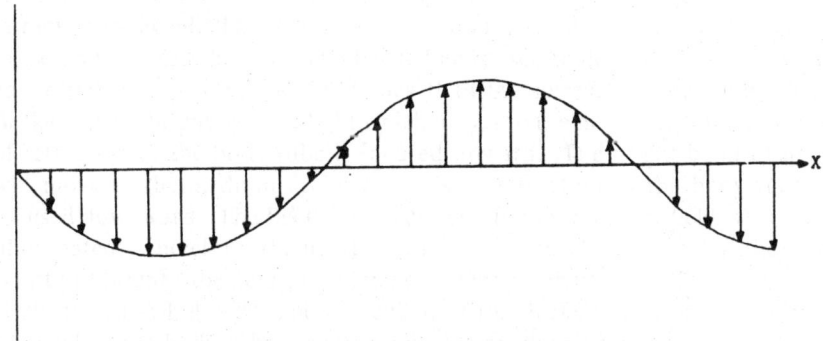

FIGURE 12.3: The electric vector of plane-polarized light at constant time travels in the X direction and vibrates in the plane of the paper.

are of equal magnitude and have opposite senses of rotation. The sinusoidal vibration of the electric vector of right circularly polarized light will describe a right-handed helix in the direction of propagation and similarly a left-handed helix for left circularly polarized light. These two components of plane-polarized light are illustrated in Figs. 12.4 and 12.5.¹

⟨If in passing through a medium the two circular components of plane-polarized light are propagated with unequal velocities, there is a phase shift between the component beams and the resultant beam of plane-polarized light is rotated in its plane of polarization as it emerges from the medium.

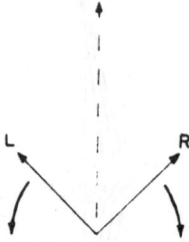

FIGURE 12.4: End view of the resolution of the electric vector of plane-polarized light into right and left circularly polarized light.

The extent to which the resultant beam is rotated is quantitatively given by Fresnel's law:

$$\alpha = \frac{\pi}{\lambda}(\eta_L - \eta_R) \qquad (12.1)$$

where α is the angle of rotation in radians per centimeter, λ is the wavelength of the incident plane-polarized light beam in centimeters, η_L is the refractive index of the left circular polarized vector, and η_R is the refractive index of the right circularly polarized vector of plane-polarized light. It may be recalled that the refractive index is the ratio of the velocity of light in a vacuum relative to its velocity in a medium, $\eta = c/v$, thus refractive indices serve as a convenient gauge for measuring the unequal velocities of left and right circularly polarized light. The angle of rotation may be converted to the more usual units of degree reciprocal decimeters by multiplying by $1800/\pi$:

$$\alpha(\text{deg dm}^{-1}) = \frac{1800}{\lambda}(\eta_L - \eta_R) \qquad (12.2)$$

⟨As can be seen from Eqs. (12.1) and (12.2), the rotation α depends upon a velocity difference (represented by $\eta_L - \eta_R$ or $\Delta\eta$), and the wavelength of the plane-polarized light beam. The angle of rotation is also proportional to the number of molecules in the light beam, which may be given by the product of the density or concentration and the path length. Thus

$$\alpha = [\alpha]_\lambda^t(l\rho) \quad \text{or} \quad \alpha = [\alpha]_\lambda^t(lc) \qquad (12.3)$$

where $[\alpha]_\lambda^t$ is a proportionality constant called the "specific rotation," l is the light-path length through the optically active sample in decimeters, and ρ is the density of the pure liquid or c is the concentration of the solute in grams per milliliter.) A compound may be reported as having $[\alpha]_D^{25} = +62.5°$ ($c = 26$, H_2O), which means that the compound has a dextro-rotatory specific rotation (the polarization plane is rotated clockwise as one faces the

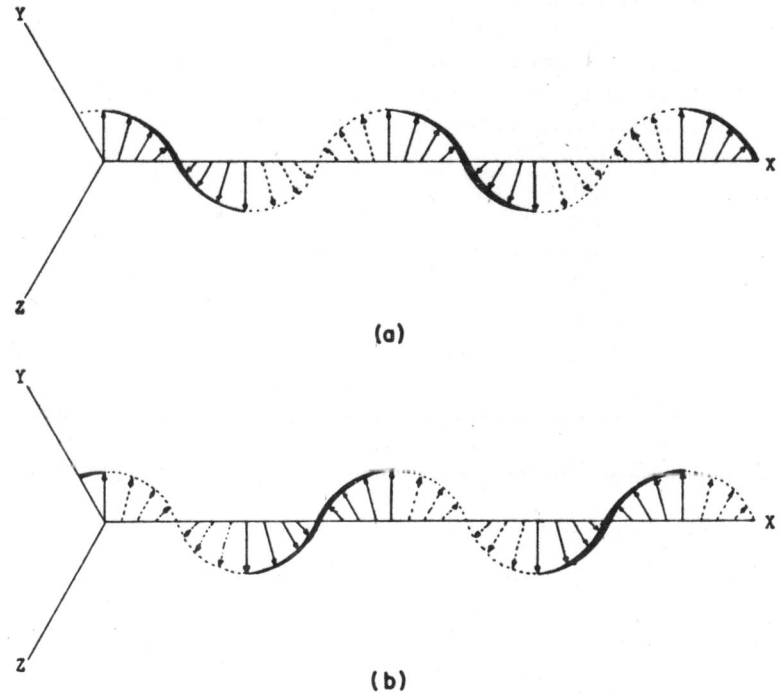

(a)

(b)

FIGURE 12.5: (a) The electric vector of the right circularly polarized component of plane-polarized light, at constant time. The vector travels in the X direction as a right-handed helix and vibrates above and below the plane on the page. (b) The electric vector of left circularly polarized light.

emerging beam) of 62.5° at a temperature of 25°C, the rotation being measured with sodium D light in a tube 1 dm long at a concentration of 26 g/100 ml of aqueous solution. (When substances of different molecular weight or differing rotation are compared, the most suitable experimental quantity is the molecular rotation ϕ:

$$[\phi]_\lambda^t = \frac{[\alpha]_\lambda^t \cdot M}{100} \tag{12.4}$$

where M is the molecular weight of the optically active substance.

Both α and ϕ are, in principle, constants which are independent of concentration or density and cell path length, however, in some instances,

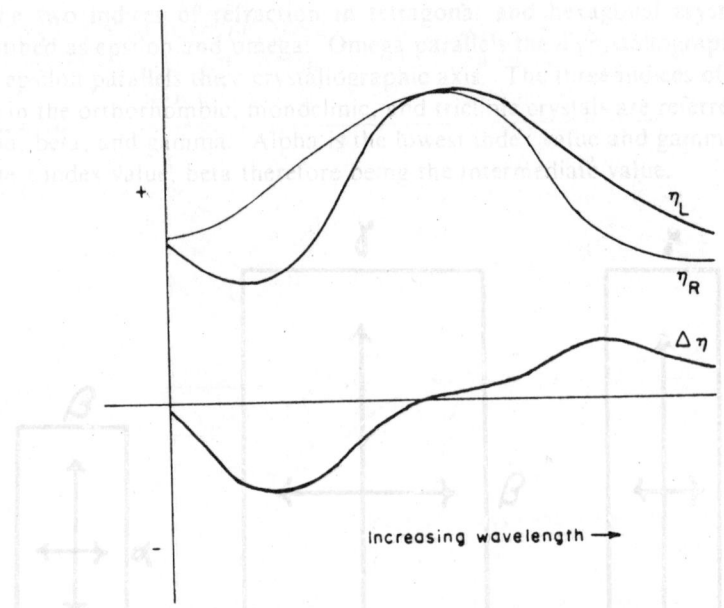

+

η_L

η_R

$\Delta\eta$

Increasing wavelength ⟶

−

FIGURE 12.6: The change in refractive index associated with rotatory dispersion. $\Delta\eta$ represents the difference in dispersion of left and right circularly polarized light.

when the solute may be involved in some type of associative equilibrium such as hydrogen-bonding with the solvent, these constants will depend on concentration.)

(The term "rotatory dispersion" arises from the fact that different wavelengths of plane-polarized light are rotated by different amounts. Figure 12.6 illustrates the change in rotation and refractive index associated with rotatory dispersion. Although diagrammatically in Fig. 12.6 a significant curve is shown for the difference $\Delta\eta$, actually it is extremely small. For example at a wavelength of 5893 Å (sodium D doublet) and a cell path length of 1 dm a rotation of 10° corresponds to a $\Delta\eta$ value of approximately 3.27×10^{-8}:

$$\eta_L - \eta_R = \Delta\eta = \frac{\lambda\alpha}{1800}$$

$$\Delta\eta = \frac{10 \times 5893 \times 10^{-9}}{1800} = 3.27 \times 10^{-8}$$

Since ordinary refractive indices may be considered to be

$$\frac{\eta_L + \eta_R}{2}$$

and are commonly in the range of 1.3 to 1.7, the difference $\Delta\eta$ is at least one ten-millionth of the refractive index.)

'An empirical equation developed by Drude quantitatively expresses the variation of α with wavelength for regions of the spectrum where the optically active material does not absorb light:

$$\alpha = \frac{k_1}{\lambda^2 - \lambda_1^2} + \frac{k_2}{\lambda^2 - \lambda_2^2} \cdots \sum \frac{k_i}{\lambda^2 - \lambda_i^2} \tag{12.5}$$

In this equation k_i is a constant found by experiment and λ and λ_i are the wavelengths of the incident linearly polarized light and of characteristic active absorption, respectively. Implicit in this equation is the inference that optical rotation is related to an absorption process. The Drude formula may be simplified to a form

$$[\alpha] = \frac{A}{\lambda^2 - \lambda_0^2} \tag{12.6}$$

where $[\alpha]$ is the specific rotation, A is the rotation constant, λ is the measuring wavelength, and λ_0 is the closest wavelength at which the material exhibits

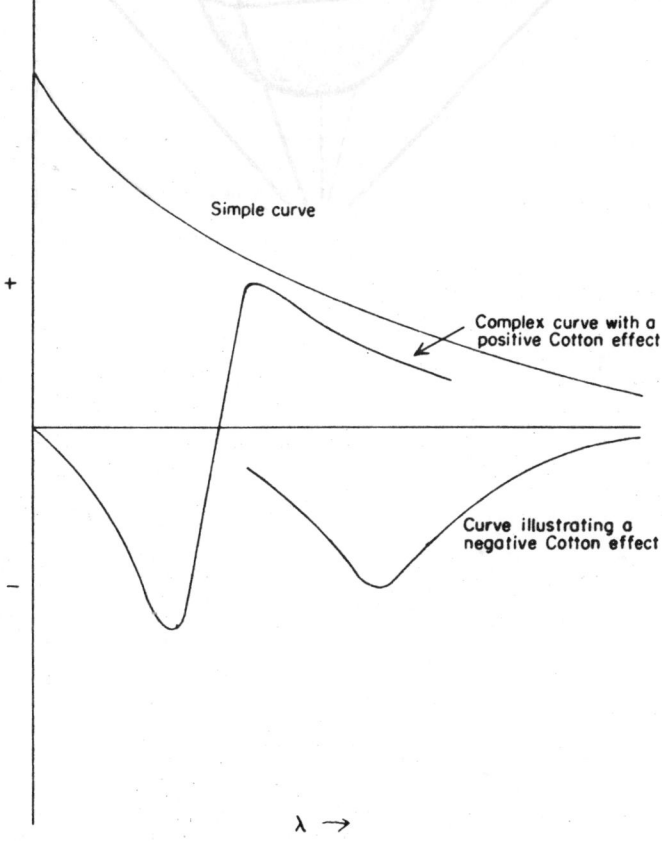

FIGURE 12.7: Types of curves that may be found in optical rotatory dispersion data.

optically active absorption i.e., the nearest absorption band of the molecule contributing to optical activity. The Drude equation is generally applicable to plain optical rotatory dispersion curves (also called "simple curves"), which do not have a chromophore within the range of wavelengths under investigation and thus do not show any maximum and minimum. Other

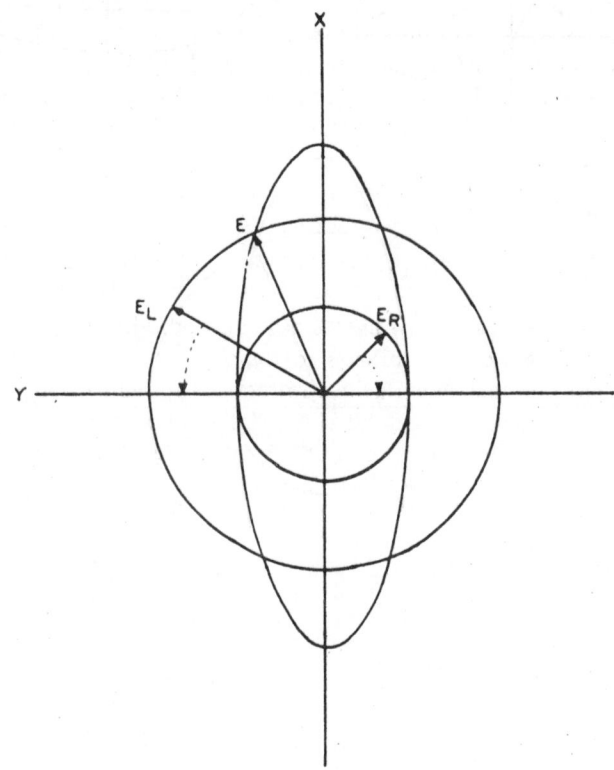

FIGURE 12.8: Elliptically polarized light, which results from unequal absorption of the components of plane or linearly polarized light. E_L represents the electric vector of the left circularly polarized component rotating as a helix counterclockwise away from the viewer. E_R represents the electric vector of the right circularly polarized component. E is the resultant vector $E_L - E_R$ represented as an elliptically shaped helix.

forms of the Drude equation with two or more terms are required for complex rotatory dispersion curves.

In the wavelength region of an optically active absorption maximum, the optical rotation increases, then reverses and becomes zero at or near the wavelength of maximum absorption and then reaches another maximum of opposite sign. This characteristic anomaly, which is associated with circular dichroism in the absorption region, is called the "Cotton effect," after its discoverer. Figure 12.7 illustrates the types of curves that may be found in optical rotatory dispersion data. The Cotton effect arises from the fact that in the vicinity of an absorption maximum, optically active molecules will absorb the circular components of linearly polarized light to different extents. The origin of the Cotton effect of an absorption band is the vibrating electrical momentum of the band itself. If this vibrating momentum is isotropic i.e., unidirectionally orientated, the molecules will not exhibit circular dichroism. If the vibrating electrical momentum is associated with

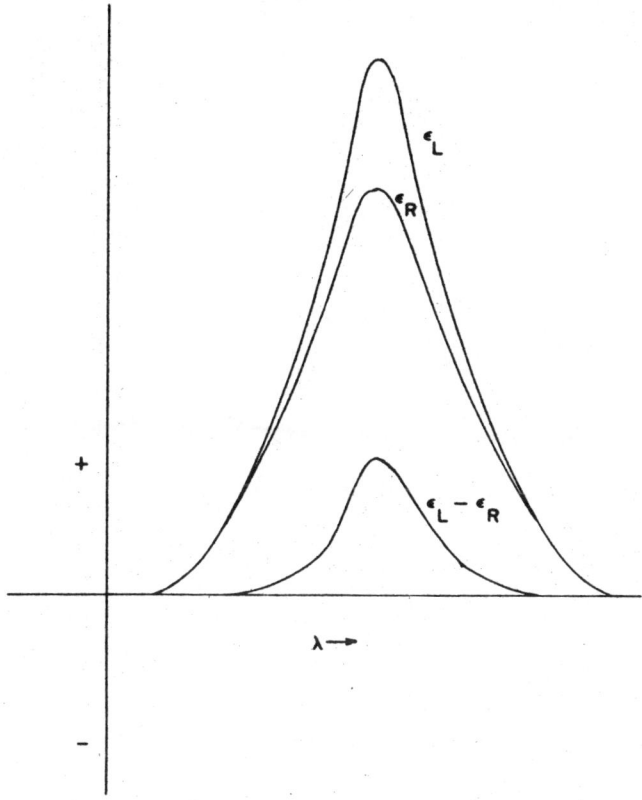

FIGURE 12.9: Schematic diagram showing the origin of a circular dichroism curve $(\epsilon_L - \epsilon_R)$.

an electronic polarizability which is not isotropically orientated, then the molecules will respond differently to left and right circularly polarized light and will show circular dichroism and optical rotation. In the absorption process, the light is not only rotated in its plane of polarization but also becomes elliptically polarized. See Fig. 12.8. The resulting angle of ellipticity may be expressed as the molecular ellipticity $[\theta]_\lambda'$ which is proportional to the molecular absorption coefficients, ϵ_L and ϵ_R for the left- and right-hand components of linearly polarized light.

$$[\theta]_\lambda' = 3300(\epsilon_L - \epsilon_R) \qquad (12.7)$$

\Just as the optical rotatory dispersion of an optically active material is obtained by measurement of the rotation of linearly polarized light, the circular dichroism of the optically active material is obtained by the measurement of the preferential absorption i.e., the difference in absorption between right and left-hand circularly polarized light. Figure 12.9 illustrates a circular dichroism curve resulting from the difference, $\epsilon_L - \epsilon_R$, of absorption

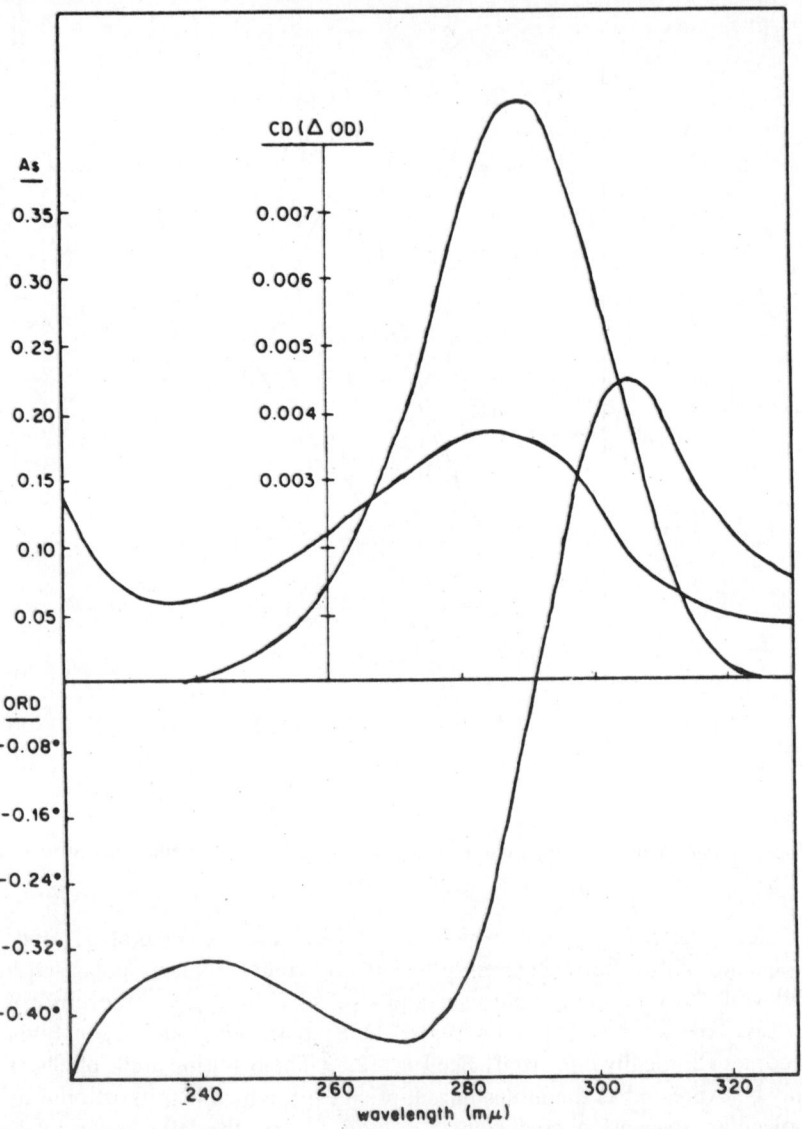

FIGURE 12.10: Absorption, rotatory dispersion, and circular dichroism spectra of a solution of *d*-camphor sulfonic acid in water.

coefficients of left and right circularly polarized light. The relationship between optical rotatory dispersion, circular dichroism and absorption is illustrated in Fig. 12.10 by the spectra of a solution of *d*-camphor sulfonic acid in water.'

12.3 MOLECULAR REQUIREMENTS FOR OPTICAL ROTATORY POWER

Substances which refract and/or absorb right and left circularly polarized light to different extents are "optically active" or show optical rotatory power. Such substances usually lack a plane or center of symmetry and thus are

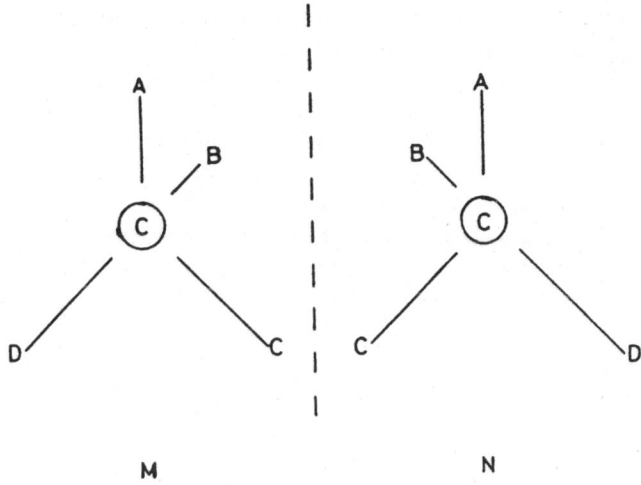

FIGURE 12.11: Simple asymmetric molecules *M* and *N* which are related to one another as object and mirror image.

termed "asymmetric." In general the absence of symmetry is a necessary criterion of optical activity.

A substance will demonstrate optical activity if its mirror image is not superimposable upon the original. This is a fundamental stereochemical requirement which was first elaborated by Pasteur and which led to the concept of the tetrahedral carbon atom. Carbon with four bonds attached to different groups extending to the corners of a regular tetrahedron is asymmetric since it has no plane of symmetry and it cannot be superimposed on its mirror image. Figure 12.11 illustrates this simple case of asymmetry.

Molecules such as M and N in Fig. 12.11, which are related to one another as objects and nonsuperimposable mirror image, are called "enantiomers" or "optical antipodes." For every optical active asymmetric molecule there exists one and one only enantiomer, however there may exist several asymmetric spatial arrangements which are not enantiometic but diastereomeric. The term "diastereoisomerism" is applied to all molecules which may be optically active and are not related to one another as object and mirror image. In contrast to enantiomers, diastereoisomers have different physical properties e.g., spectral properties such as ultraviolet, infrared, nuclear magnetic resonance, and mass spectra; melting points, densities, solubilities, boiling points, refractive indices, viscosities, etc. Enantiomers of the same substance (object and mirror image) have identical physical and chemical properties and differ only in the sign of their optical rotations. A mixture composed of equal molar quantities of two pure enantiomers is called a "racemic modification." Such a modification is optically inactive since it is composed of equal numbers of dextrorotatory (+) and levorotatory (−) molecules and therefore the average rotation is zero.

Frequently, racemic modifications have crystal structures which differ from those of the pure enantiomers. In such instances, a racemic compound or racemate is formed which has the properties of a eutectic mixture and therefore the melting point, density, solubility, and solid-state infrared spectrum differ from the corresponding properties of the pure enantiomer, either (+) or (−). True racemization is the irreversible process of formation of (+)(−) pairs by the reversible interconversion of enantiomers.

Other processes involving asymmetric atoms which are closely related to racemization are mutarotation and epimerization. Mutarotation is the change, with time, in optical rotation of a solution of a pure optically active substance. Such a solution eventually reaches an equilibrium value of optical rotation which is usually not zero. Epimerization is a process involving the change of configuration of one asymmetric atom in a molecule which has more than one asymmetric atom. Epimerization usually involves the interconversion of diastereoisomers, and the diastereoisomers that result differ in configuration at only one asymmetric atom and are called "epimers."

An example of epimerization and mutarotation is the process of mutarotation of $\alpha(+)$-glucose, which involves epimerization (a change in configuration) at the number 1 carbon atom—the so-called anomeric carbon atom. A solution of $\alpha(+)$-glucose, $[\alpha]_D^{20} = +113°$, with time undergoes spontaneous mutarotation to an equilibrium value of $[\alpha]_D^{20} = +52.5°$. This change involves the hemiacetal formation at C-1, which opens to the aldehyde form and then may close with a different configuration at C-1—the hemiacetal (anomeric) carbon. Thus the mutarotation of glucose is a spontaneous epimerization since it involves only a change in configuration at the number 1 carbon atom. The epimers that are formed are diastereoisomers and in the special nomenclature of carbohydrate chemistry are called "anomers," since they differ only by the configuration at the anomeric carbon atom C-1.

The presence of four different substituents on carbon, creating an asymmetric center, may not be a sufficient condition for the demonstration of optical rotatory power. For example, tartaric acid $COOH \cdot CHOH \cdot CHOH \cdot COOH$, in which the two central carbon atoms each have four different substituents, exists as $(+)$-tartaric acid, $(-)$-tartaric acid, $(\pm)$-tartaric acid,

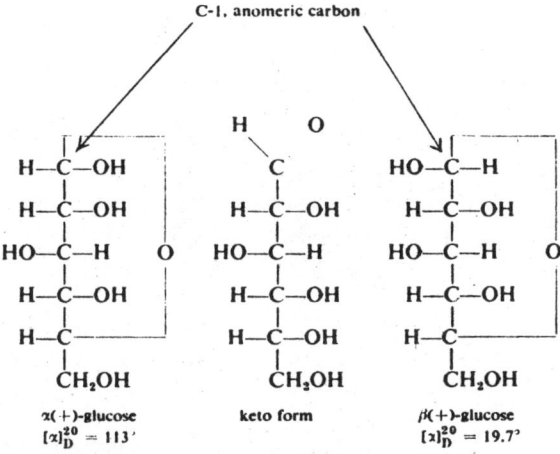

FIGURE 12.12: Formulas for tartaric acid.

and *meso*-tartaric acid. The last two forms of tartaric acid are optically inactive. The $(\pm)$-tartaric acid is a racemate and the *meso*-tartaric acid is optically inactive due to internal compensation in the molecule, which produces a plane of symmetry. These forms of tartaric acid are illustrated in Fig. 12.12.

Optical activity may exist in carbon compounds which do not have asymmetrically substituted carbon tetrahedra. Typical examples of these types of compounds are shown in Fig. 12.13. Thus optical activity which is often but not always associated with asymmetrically substituted carbon, which may confer a molecular dissymmetry on the molecule, may also be present in certain spiranes, allenes, diaryls, and helical conjugated compounds (e.g., hexahelicine) in which substituents prevent coplanarity. Optical activity is also exhibited by the helical structures of proteins and may be exhibited by compounds containing the elements arsenic, phosphorus, nitrogen, antimony, and sulfur, which like carbon may be asymmetric. In general, molecular dissymmetry is a more basic criterion than asymmetric carbon for optical activity, however, it is probably most accurate to state that molecular symmetry is a condition which does not produce optical activity, rather than that molecular asymmetry is a condition for optical activity. Optical activity, in a broad sense, involves a diastereoisomeric interaction, since a dissymmetric sample interacts with the enantiomeric circularly polarized components of plane-polarized light.

All substances, regardless of their molecular symmetry or lack of it, may be induced to have optical activity either by an electrical field (the Kerr effect) or by a magnetic field (the Faraday effect). Induced optical activity by these methods has found little direct application to pharmaceutical analysis.

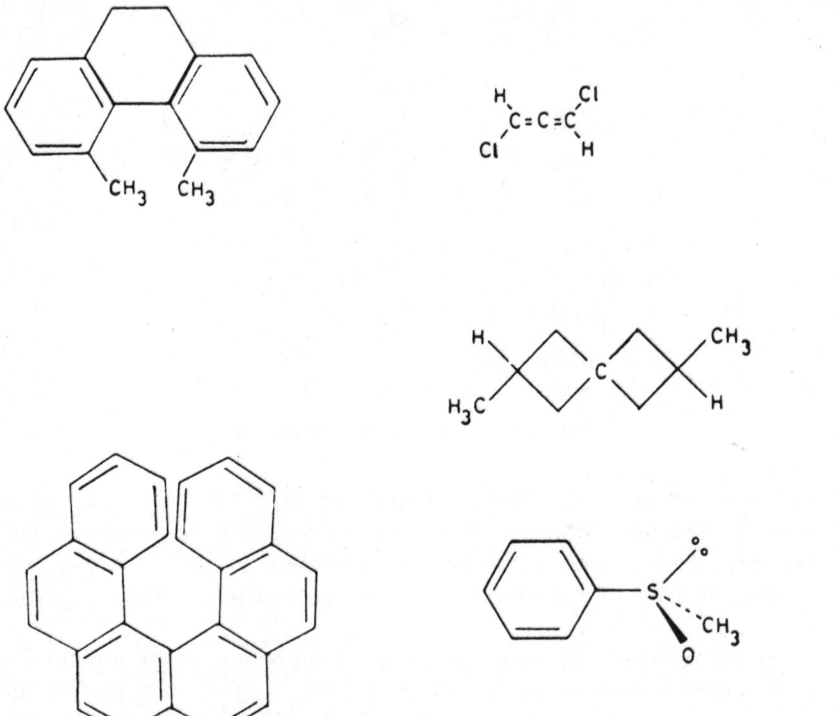

FIGURE 12.13: Some structures which have optical rotatory power, but do not have asymmetrically substituted carbon atoms.

12.4 INSTRUMENTATION

The measurement of optical rotation is the estimation of the angular rotation of a beam of plane-polarized light. The sample under study may be a solid, liquid, or solution, but must be transparent to the incident light. Since many optical materials disperse white light, a monochromatic light source must be used. A simple polarimeter consists of a polarizer, a sample tube, and an analyzer mounted in a graduated circle.

A simple polarimeter such as the instrument schematically illustrated in Fig. 12.14 may be used to estimate the strengths of sugar solutions. Such an instrument, called a "saccharimeter," consists of a polarizer, which produces plane-polarized light, a sample tube or cell for containing the optically active solution, and an analyzer for measuring the rotation, together with the appropriate lenses necessary to allow a parallel beam of monochromatic light (usually the sodium D line) to pass through the solution and enter the

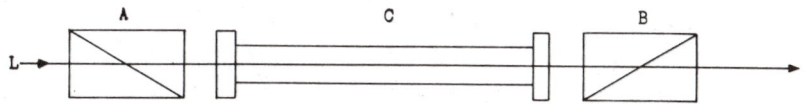

FIGURE 12.14: Schematic diagram of a simple polarimeter: *L*, light source; *A*, polarizing prism; *B*, analyzing prism; *C*, sample tube.

observer's eye. The cell is filled with water and the analyzer is turned until no light passes. The cell is then filled with the sugar solution under test and the analyzer is rotated until the field is dark again. The angle through which the analyzer has been turned is read on a circular scale. Under most practical conditions the angle of rotation ψ is never greater than π and thus $\psi = \alpha$ or $\psi = 180 - \alpha$. To distinguish between these two possibilities the solution is diluted to half-strength and a new value of α is measured.

In its simplest form, the measurement of optical rotation is a differential measurement since the difference produced by the sample is the angular difference between the setting on the circular scale with and without the sample in position. The measurement of optical rotation using the simple polarimeter thus far described involves the detection of a totally dark field where the rate of change of illumination is at a minimum and therefore the accurate location of the angle of zero illumination is quite difficult. Various "half-shadow" devices have been incorporated into visual polarimeters to make the measurement easier by allowing the observer to estimate small differences in the intensity between two halves of a bipartite field. These half-shadow devices alter the plane of polarization of one-half of the beam emerging from the polarizer, creating a split field. A typical example is a small Nicol prism covering one half of the field from the polarizer. Another design incorporates a pair of Nicol prisms, one of the pair covering one-third of the field emerging from the polarizer and the other prism covering another one-third of the field, thus creating a tripartite field in which the intensity of minimum illumination is matched to make the measurement. The small Nicol prisms, which may be adjusted in some polarimeters, rotate the polarized light beam small amounts in opposite directions and thus introduce a small phase difference usually between

1 and 7°. Another half-shadow device is the Laurent quarter-wave plate—a quartz plate cut so that for a given wavelength of light (usually the sodium D line) the radiation is altered in phase by half a wavelength. In visual polarimeters, the zero position of the instrument is that angle at which the two or three divisions of the observed field are equally dim.

Plane-polarized light may be produced by reflection, by transmission through a pile of plates, or by absorption of the monochromatic light through Polaroid film [an iodine poly(vinyl alcohol) complex which is made

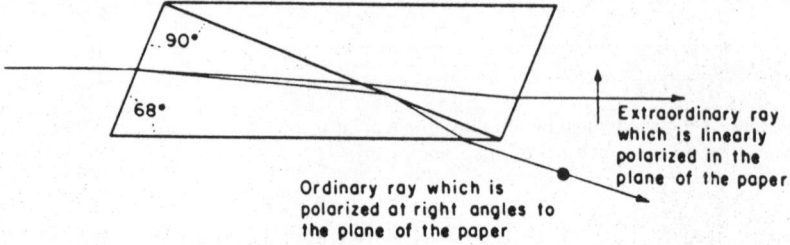

FIGURE 12.15: The Nicol prism.

dichroic by stretching the film]. These methods either involve great loss of energy or do not give complete polarization and most polarimeters use prisms for the production of polarized light.

Prisms of the Rochon and Wollaston type, which are usually made of either quartz or calcite, produce two beams of light polarized in mutually perpendicular planes and traveling in different directions. One of the two beams is termed the "ordinary ray" and the other the "extraordinary ray." If one ray is removed by inserting a stop, the other may be used as the incident plane-polarized light of the polarizer. In the Nicol, Foucault, and Glan-Thompson prism designs the difference in the angles of refraction of the ordinary and extraordinary rays of plane-polarized light allow separation of one light ray by total reflection. See Fig. 12.15.

In most visual polarimeters a pair of Nicol prisms is used. One prism serves as the polarizing prism, which produces linearly polarized light, and the other Nicol prism, located after the sample cell, is used as an analyzing prism which is rotated to the angle of extinction. If the two prisms are oriented identically with respect to their optic axes, the radiation is "uncrossed" and maximum light is transmitted. If the analyzing prism is situated so that its optic axis is at an angle of 90° (crossed position) to the optic axis of the polarizing prism the intensity of the light is a minimum. According to the law of Malus, the energy of the linearly polarized light beam transmitted by the analyzer varies as the square of the cosine of the angle between the optic axes of the polarizing and analyzing prism. This is written

$$E = E_0 \cos^2 \theta \qquad (12.8)$$

where E_0 is the energy of the beam as it leaves the polarizing prism, E is the energy of the beam as it leaves the analyzing prism, and θ is the angle between the optic axis of the polarizer and the optic axis of the analyzer. An optically active substance placed in the sample cell will change the angle of extinction to which the analyzer must be set to obtain minimum intensity. The law of Malus then becomes

$$E = E_0 e^{-Klc} \cos^2 (\theta + \alpha) \tag{12.9}$$

where K is the absorption coefficient of the sample, l is the path length of the sample cell, c is the concentration of the sample, and α is the optical rotation introduced by the sample. If the sample does not absorb light and a minimum of reflected and unpolarized light is produced, then Eq. (12.9) becomes

$$E = E_0 \cos^2 (\theta + \alpha) \tag{12.10}$$

The change introduced by the optically active sample (α) is observed on a large circular scale which is graduated in angular degrees and usually may be read with a precision of approximately 0.005–$0.010°$.

Since visual polarimeters use the human eye as the sensing device, most accurate determinations must be made with the polarimeter in a darkened room where the operator's eye is dark adapted. Repeated routine measurements of an almost extinct field may become less precise due to fatigue and thus polarimeters employing a photoelectric sensing element have been designed. The substitution of photoelectric devices for the end-point determination has allowed the use of visual instruments for automatic recording and has facilitated the design of spectropolarimeters.

A simple photoelectric polarimeter may employ a chopper, which allows alternative transmission of the two areas of a conventional bipartite field of a visual polarimeter to a photomultiplier detector. The electrical signal of the photomultiplier which is generated by oscillation of the plane of polarization when the two halves of the bipartite field differ in intensity is amplified and fed to a servomotor mechanism, which rotates the analyzing prism to the balance point and thus automatically compensates for intensity differences and the angular rotation may be read on a mechanically coupled digital indicator.

Several modifications in the design of the simple photoelectric polarimeter just described have been designed. These instruments eliminate the conventional half-shadow device of the visual polarimeter and differ mainly in the method by which the plane of polarization is caused to oscillate. This oscillation may be achieved by mechanically rotating the polarizing prism over a small adjustable "symmetrical" angle. Balance is obtained by manually rotating the analyzing prism until a galvanometer attached to the photomultiplier tube gives equal readings for the two positions of the oscillating prism and the angle of rotation may be read conventionally. Another modulator which produces oscillation of the plane-polarized light is a rotating "half-shadow" disk, which consists of two semicircular pieces of D and L quartz. The plane of polarization may also be caused to oscillate by means of a Faraday cell, which consists of an electromagnet surrounding a rod of transparent material such as dense flint glass. When current flows in the solenoid magnet, the rod becomes optically active and the optical rotation of the rod varies in direct proportion to the magnetic field. Reversal of current to the magnet causes reversal of rotation and thus alternating current will cause an alternating rotation of optical activity of the rod. The oscillating polarized light beam which is produced by the polarizing device and the alternating-current Faraday cell pass through to an analyzer, which consists of a polarizer in crossed orientation to the initial polarizer. At balance, a uniform light wave strikes the photomultiplier. An optically active

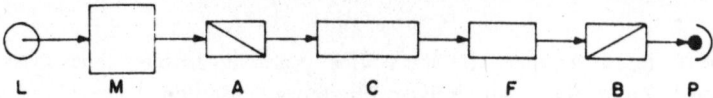

FIGURE 12.16: Schematic diagram of photoelectric polarimeter: A, polarizing prism; B, analyzing prism; C, sample tube; F, Faraday cell or half-shadow prism or half-shadow disk; L, light source; M, monochromator; P, photomultiplier tube.

sample will distort this wave pattern and the distortion may be analyzed by a phase-sensitive detector which then supplies direct current to another Faraday cell that acts as a null device and compensates for the optical rotation of the sample. The current supplied to the second Faraday cell is proportional to the optical rotation of the sample and may be measured by a galvanometer or by various read-out devices such as a chart or digital meter. A schematic diagram of the photoelectric polarimeters just described is given in Fig. 12.16. Most designs of the photoelectric polarimeter may be altered to produce spectropolarimeters (which measure the change in optical rotation with change in wavelength of plane-polarized light) simply by incorporating a monochromator between the energy source and the polarizing device.

Spectrophotometers may be converted into photoelectric polarimeters and spectropolarimeters by inserting polarizing and analyzing prisms into the light path of these instruments. In single beam spectrophotometers, the prisms are arranged so that the analyzer can be moved from $+\theta$ to $-\theta$. A solution which is not optically active transmits the same amount of light at both settings, whereas an optically active solution produces two different intensities for each position of the analyzer. The ratio of these two intensities is then related to the angular rotation of the sample α by the following equation

$$R = \frac{E_2}{E_1} = \frac{\cos^2(-\theta + \alpha)}{\cos^2(\theta + \alpha)} \tag{12.11}$$

where R is the ratio of the two intensities, E_1 at position $+\theta$ of the analyzer, and E_2 at position $-\theta$ of the analyzer. R values are converted to optical rotation through the use of tables or may be approximated to angular rotation α by the following equation

$$\tan \alpha = \frac{(1 - \sqrt{R})}{(1 + \sqrt{R})} \cot \theta \qquad (12.12)$$

Double-beam spectrophotometers compare the energy of two identical units, each consisting of a polarizer and an analyzer and constructed to fit the sample and reference compartments of the spectrophotometer. The polarizer and the analyzer are crossed in each of the two compartments except for the small angle of $+\theta$ and $-\theta$ in the two units, respectively. When an optically active solution is substituted for the solvent, $+\theta$ and $-\theta$ become $\theta \pm \alpha$ and $-\theta \pm \alpha$, depending on the sign of the optical rotation of the solution. The intensities of the transmitted plane-polarized light in the two beams are no longer equal, and as before with the single-beam spectrophotometer the ratio of the two intensities R, which is directly determined, is related to the optical rotation.

The major disadvantage of this method for determination of optical rotation and optical rotatory dispersion is that the recorded curve is in terms of R values or transmittance values rather than optical rotation, and the values of the ratio is not always directly proportional to the angle of rotation, particularly when $|\alpha| \geq |\theta|$ and/or the solute exhibits circular dichroism at the wavelengths used.

To record the circular dichroism of an optically active solution, linearly polarized light must be resolved into two circularly polarized components. Passage of the monochromatic polarized beam through an appropriate quarter-wave plate, a Babinet-Soleil compensator, the use of a Fresnel rhomb, and a Billings' cell (Pockels' effect modulator) are four methods currently available for providing this resolution in commercial circular dichroism instruments. Such instruments directly measure the difference in absorption for the two circular components, and this difference is converted to an extinction coefficient, which is related to the molecular ellipticity by Eq. (12.7). The theory and design of circular dichroism instruments is beyond the scope of this present chapter.

12.5 APPLICATIONS

Measurement of the optical rotation of a compound may be used as one of the physical constants of a pure liquid or solute. Optical activity is the only physical property that will distinguish $(+)$ and $(-)$ isomers. The specific rotation is a physical constant, but, as noted previously, the concentration, temperature, solvent, and wavelength of plane-polarized light must always be stated when the specific rotation is reported.

Optical rotation may be used as a quantitative method for the determination of optically active compounds. An optically active substance may be determined in the presence of inactive compounds. It may be possible to analyze two component mixtures at a single wavelength in which both substances are optically active since optical activities may be additive, and therefore there is only a single composition possible for any value of rotation between that of the pure compounds. In most instances, the optical rotation is not strictly linear with concentration and a calibration curve must be constructed in which the observed rotation is plotted as a function of concentration with the other specified variables such as cell path length, temperature, and wavelength of plane-polarized light kept constant.

The determination of optical rotation finds its greatest application in the qualitative indentification of organic compounds. Several methods have been developed which empirically correlate rotational data usually obtained at the sodium D line with the structure of organic compounds. Two examples follow. A method of molecular rotational differences based upon the empirical correlation of differences in molecular rotation of a series of known compounds may be applied to an unknown compound by carrying out a transformation on the unknown and assigning structural requirements to this unknown based on a correlation of the differences in rotation with known molecules. This approach to structure determination has been used successfully in steroid structure determination. Optical rotation is also used in carbohydrate structure determination in which empirical rules such as Hudson's rules may be widely applied to carbohydrate molecules.

Optical rotatory dispersion data have found most use in the determination of conformational and configurational information about known organic sturctures. Measurements of optical rotatory dispersion provide a means of distinguishing between random-coil and regular helical arrangements in polypetides and proteins. Such optical rotatory dispersion spectra are dependent on conformation and disappear when the macromolecule assumes a random-coil arrangement.

The optical rotatory dispersion method for low molecular weight compounds involves study of the Cotton-effect curves which result when there is absorption of plane-polarized light by a chromophore, usually the carbonyl group, which has some sort of asymmetric environment. The underlying principle of the method is the octant rule, which qualitatively predicts the contribution of the substituents to rotatory dispersion when a substituent is introduced into a carbonyl-containing compound at a specific site. To apply the rule, a structure such as cyclohexanone in the chair configuration is divided into eight octants by three mutually perpendicular planes. One plane passes through the carbonyl carbon atom and carbons 2 and 6; a second plane passes through oxygen of the carbonyl and carbon 4; the third plane bisects the carbonyl group and is perpendicular to the other two planes. See Fig. 12.17. If a substituent is placed at any position other than on one of these planes it destroys the symmetry of the system and the orbitals involved in the $\eta \rightarrow \pi^*$ transition of the carbonyl group may interact with the substituent. Therefore axial or equatorial substituents at carbon 4 do not contribute to the Cotton effect. Equatorial substituents at carbon 2 or carbon 6 have almost no effect on the sign of the Cotton effect. Axial or equatorial substituents at carbon 5 and axial substituents at carbon 2 have

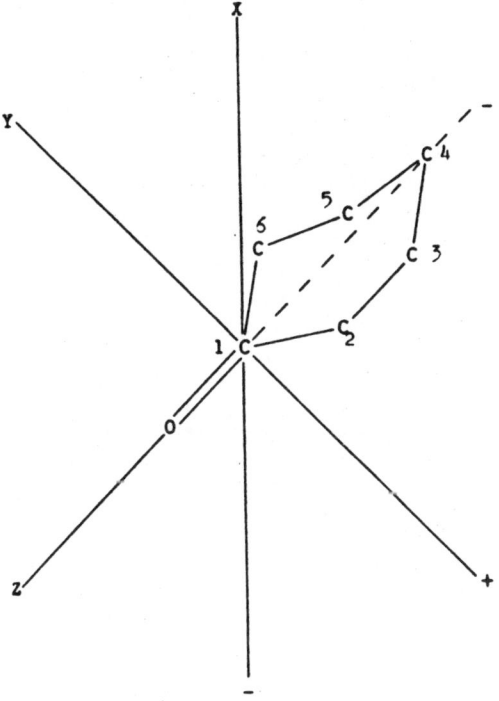

FIGURE 12.17: The basis for the octant rule.

a positive contribution to the Cotton effect and substituents at all other positions of the back four octants contribute negatively. In the example of cyclohexanone, the front four octants may be ignored. The octant rule allows prediction of the sign of the $\eta - \pi^*$ transition of the asymmetrically perturbed carbonyl group and allows certain conclusions about configuration and conformation of its environment to be made which have been experimentally substantiated for a large number of simple ketones.

Circular dichroism curves, for the most part, yield the same type of qualitative information as optical rotatory dispersion curves, and theoretically at least one type of curve may be converted to the other. Circular dichroism curves may be more useful for quantitative measurement of the Cotton effect, since the transitions which are observed are much sharper, like ultraviolet absorption spectra, and do not tail off as do optical rotatory dispersion curves. (See Fig. 12.10.) Circular dichroism curves, therefore, may facilitate quantitative study of isolated asymmetrically perturbed chromophores which, in optical rotatory dispersion data, would have a continuous background due to chromophores which absorb at more distant wavelengths superimposed on the Cotton effect of a chromophore.

EXPERIMENTS

E12.1. Specific rotation of camphor. Fill a polarimeter tube with ethanol and determine the zero reading of the polarimeter using a sodium-arc light source. Note the temperature at which the measurement is made. The half-shade angle of the instrument may be adjusted to obtain maximum sensitivity of the zero setting.

Accurately make-up a solution containing approximately 10% w/v camphor in ethanol. Fill the prepared polarimeter tube, note the temperature of the solution, and measure its rotation. Calculate the specific rotation $[\alpha]_D^t$ of camphor, $[\alpha]_D^t = \alpha/lc$ where α is the observed rotation in degrees, l is the polarimeter tube length in decimeters and c is the concentration in grams per milliliter.

Determine the rotation of a camphor solution supplied by the instructor and using your calculated specific rotation calculate the concentration of camphor in this unknown solution. Is the unknown sample material natural or synthetic camphor?

Explain any deviation between the concentration determined by the measurement of optical rotation and the actual concentration of the unknown. What improvements may be suggested to make your determination more accurate?

E12.2. Determine the fructose content of fructose and sodium chloride injection NF supplied by the instructor, according to the procecure of the *National Formulary*, twelfth edition, p. 171.

E12.3. Determine the epinephrine content of epinephrine solution USP according to the procedure of the *U.S. Pharmacopeia*, XVII, p. 226.

E12.4. Acid hydrolysis of sucrose.* The rate of hydrolysis of sucrose may be followed by measuring the decrease of optical rotation with time. The complete hydrolysis of sucrose, $[\alpha]_D^{20} = +66.5$ (c = 26, H_2O), yields equimolar quantities of glucose $[\alpha]_D^{20} = +52.5-53.0$ (c = 10, H_2O), and fructose $[\alpha]_D^{20} = -92$ (c = 2, H_2O). During the process of hydrolysis the sign of the observed rotation decreases because of the high negative rotation of fructose, which more than balances the positive rotatory contribution of glucose. This hydrolysis reaction has been called "inversion" because of the change in sign of the rotation, and the reaction products are known as "invert sugar."

Prepare six solutions containing 10.0 g of sucrose per 100 ml of solution of the following molarities of hydrochloric acid; 0.25, 0.50, 1.00, 1.50, 2.00, and 3.00 M. Measure the optical rotation α_0 immediately after mixing

---

* Dawber, J. G., D. R. Brown, and R. A. Reed, *J. Chem. Educ.*, 43, 34 (1966).

the solutions, i.e., at time zero, and repeat the observations of the rotation of each solution at suitable time intervals α_t to observe the initial stages of the hydrolysis. Measure the optical rotation after approximately 1 week, α_∞, to estimate the rotation of the completed reaction. Close control of temperature is necessary in this experiment. The solutions should be stored and the measurements should be made at a constant temperature.

The rate constant of the reaction k is given by the following equation:

$$k = \frac{2.303}{t} \log_{10} \left[\frac{\alpha_0 - \alpha_\infty}{\alpha_t - \alpha_\infty} \right]$$

For each solution plot a graph of t against $\log_{10} [\alpha_t - \alpha_\infty]$ and evaluate k (−slope). Plot the rate constant values vs. the molarity of the acid solutions and also $\log_{10} k$ values against the Hammett acidity functions H_0 for the various molarities of HCl.

H_0 values for the HCl solutions are; $0.25\,M = 0.55$, $0.5\,M = 0.20$, $1.00\,M = -0.18$, $1.50\,M = -0.47$, $2.0\,M = -0.67$, $3.0\,M = -1.05$, where

$$H_0 = pK_{BH^+} - \log_{10} \frac{[BH^+]}{[B]}$$

which is derived from the protonation of a neutral base B.

$$B + H^+ \rightleftharpoons BH^+$$

If the acid-catalyzed hydrolysis of sucrose is a unimolecular process, a graph of $\log_{10} k$ against H_0 should be linear and have a slope of 1.0. If the graph of k vs. acid molarity is linear, then the reaction mechanism is bimolecular.

QUESTIONS

Q12.1. Describe the relationship between refractive index and optical rotatory power.

Q12.2. Explain the following terms:
(a) enantiomer
(b) diastereoisomer
(c) racemate
(d) mutarotation
(e) inversion of configuration
(f) elliptically polarized light

Q12.3. Discuss the following statements:
(a) Optical rotation is not a universal property of all substances.
(b) The measurement of optical activity is inherently a measurement of molecular dissymmetry.
(c) The optical rotation of a solid, liquid, or pure compound in solution serves as a characteristic physical property of the substance and often will give a more specific identification than will the determination of other physical constants such as melting or boiling points.
(d) The octant rule relates the sign of the rotatory power of an optically active ketone in the 300-mμ region to the substituents creating dissymmetry in the molecule.

Q12.4. Which of the following structures are optically active? Draw all isomers showing enantiomeric relationships.

(a)

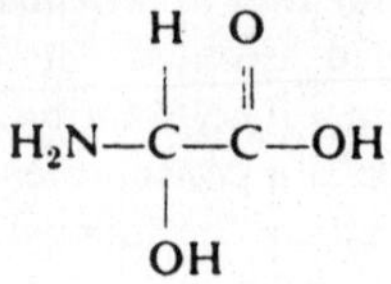

$$H_2N-\overset{\overset{\displaystyle H}{|}}{\underset{\underset{\displaystyle OH}{|}}{C}}-\overset{\overset{\displaystyle O}{\|}}{C}-OH$$

(b)

$$\begin{array}{c} CH_2OH \\ | \\ H-C-OH \\ | \\ H-C-OH \\ | \\ \overset{C}{\underset{O}{\diagup}}{\diagdown}OH \end{array}$$

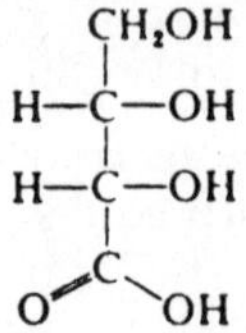

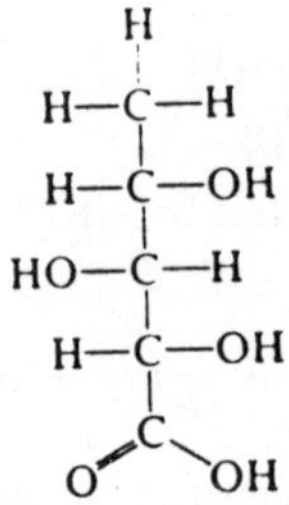

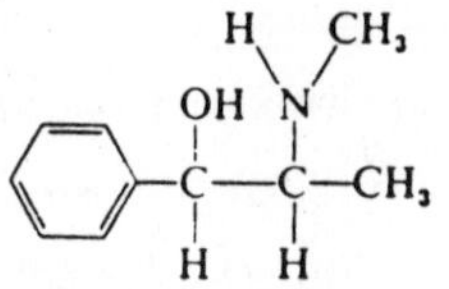

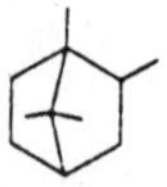

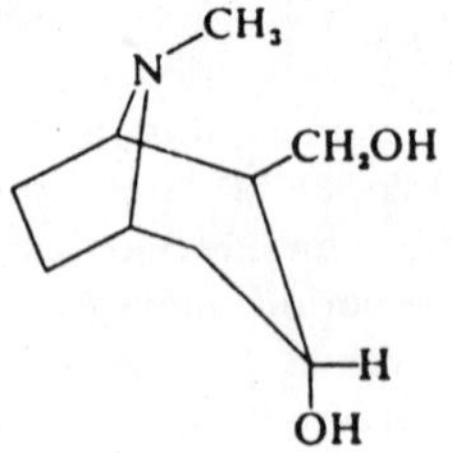

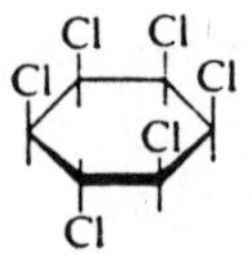

(f)

(c)

$$\begin{array}{c} H \\ | \\ H-C-H \\ | \\ H-C-OH \\ | \\ HO-C-H \\ | \\ H-C-OH \\ | \\ \overset{C}{\underset{O}{\diagup}}{\diagdown}OH \end{array}$$

(g)

(d)

(h)

(e)

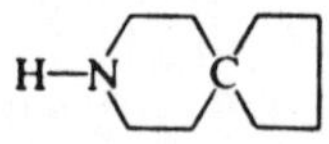

(i)

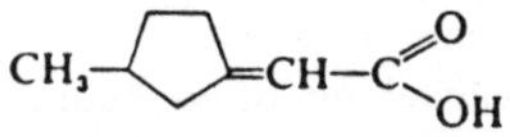

PROBLEMS

P12.1. The specific rotation, $[\alpha]_D^{20}$, of scopolamine is $-28°$. If a 1-dm tube is used and an optical rotation of $-0.69°$ is observed, what is the concentration of scopolamine in this solution?

P12.2. The specific rotation of quinidine sulfate, $[\alpha]_D^{25}$, in ethanol is $212°$. What is the concentration of a solution of this salt which gives an observed rotation of $25.05°$ in a 2-dm tube?

P12.3. A solution of chloramphenicol containing 1 g/100 ml of ethanol was placed in a 20-cm tube and had an optical rotation of $+0.38°$ at 25°C. Calculate the specific rotation of chloramphenicol.

P12.4. Chlortetracycline has a specific rotation of $-275.0°$. Calculate the expected optical rotation of a methanol solution containing 250 mg in 50 ml using a 0.5-dm tube.

P12.5. The specific rotation, $[\alpha]_D^{20}$, of α-D-glucose is $+113°$ and the specific rotation of the corresponding β epimer, β-D-glucose, is $+19.7°$. A solution of α-D-glucose undergoes spontaneous epimerization to an equilibrium mixture of the α and β forms (mutarotation) which has a specific rotation of $+52.5°$. Calculate the percentage of the α and β forms in the equilibrium mixture.

BIBLIOGRAPHY

Crabbe, P., *Optical Rotatory Dispersion and Circular Dichroism in Organic Chemistry*, Holden-Day, San Francisco, 1965.

Djerassi, C., *Optical Rotatory Dispersion*, McGraw-Hill, New York, 1960.

Heller, W., in *Physical Methods of Organic Chemistry* (A. Weissberger, ed.), Part II, Vol. 1, Wiley (Interscience), New York, 3rd ed., 1960.

Lowry, T. M., *Optical Rotatory Power*, Longmans, Green, London, 1935.

Mason, S. F., *Quart. Rev. (London)*, **17**, 20 (1963).

Mislow, K., *Introduction to Stereochemistry*, Benjamin, New York, 1966.

Snatzke, G. (ed.), *Optical Rotatory Dispersion and Circular Dichroism in Organic Chemistry*, Heyden, London, 1967.

Struck, W. A., and E. O. Olson, in *Treatise on Analytical Chemistry* (I. M. Kolthoff and P. Elving, eds.), Part 1, Vol. 6, Wiley (Interscience), New York, 1965, pp. 3933-4008.

Velluz, L., M. Legrand, and M. Grosjean, *Optical Circular Dichroism*, Academic Press, New York, 1965.

CHAPTER **13**

Potentiometric Titrations

J. George Jeffrey

COLLEGE OF PHARMACY
UNIVERSITY OF SASKATCHEWAN
SASKATOON, SASKATCHEWAN

13.1 INTRODUCTION

Potentiometry, or the measurement of potential, is important in chemical analysis because the potential developed by a particular electrode immersed in a solution is dependent on the presence of some entity in the solution and on its concentration in the solution.

During a titration procedure the active ingredient being determined decreases in concentration; this decrease in concentration may be followed by measuring the potential developed by an electrode immersed in the

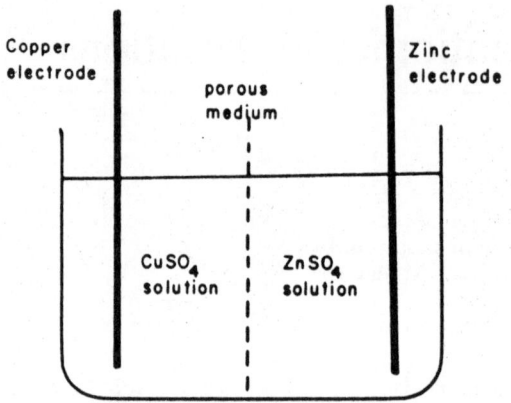

FIGURE 13.1: Daniell cell.

solution. Or, the titrant used in a titration may cause production of a potential at the electrode immersed in the reaction mixture; at the end point of the titration, when the titrant is no longer being used up, the sudden increase in concentration of titrant may result in a sudden increase of potential developed at the electrode. The course of the titration reaction may be followed by following the potential developed at the immersed electrode.

An electrochemical cell is made up of two half-cells. Each of these two half-cells develops a particular potential or electromotive force. When two half-cells are connected together (internally and externally), the electromotive force developed causes a current to flow in the external circuit.

The classical example of two half-cells is the Daniell cell, where a copper electrode is immersed in a solution of copper sulfate and a zinc electrode is immersed in a solution of zinc sulfate, with the two solutions being separated by a porous material which permits passage of ions but prevents the mixing of the solutions. A Daniell cell is represented in Fig. 13.1.

The two half-cells are connected electrically internally, even though the porous barrier prevents mixing of the two solutions. If the copper electrode and the zinc electrode are connected by an external wire, current will flow in the wire. If the two electrodes are connected to a meter, we can measure the potential developed across the two half-cells. As current is allowed to flow around the external circuit, it is noted that the potential falls slightly and continues to fall. Consequently, if we wish to know the potential developed, it should be measured without drawing any current; this is done by using an instrument called a "potentiometer."

A potentiometer in simple form consists of two electrical circuits in combination. The first circuit consists of a length of resistance wire of uniform resistance throughout its length, and a working battery to drive a small current through the length of the resistance wire. The second circuit is made

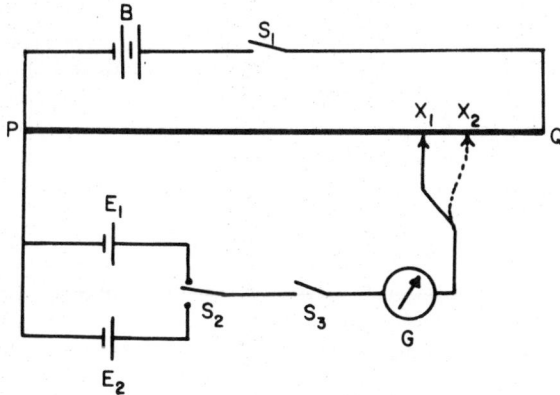

FIGURE 13.2: Potentiometer circuit.

up of two cells (one of accurately known potential, the other whose potential is to be determined), a switch to include one or other of the cells in the circuit, a galvanometer to show when current is flowing in the second circuit, and a sliding contact on the length of resistance wire. These components can be arranged as shown in the diagram labeled Fig. 13.2. When switch S_1 is closed, the battery B sends electricity flowing from the battery B to the end P of the resistance wire PQ. At point P the current from battery B can divide, part of it flowing through the resistance wire PQ and then back to the battery B; the remainder tends to flow through the second part of the circuit: the small cell (either E_1 or E_2), the switches S_2 and S_3 and, via the sliding contact X, through the XQ part of the resistance wire and then back to the battery B.

The tendency for electricity from battery B to flow in the second circuit is opposed by the tendency for electricity to flow from cell E_1 and E_2 up to P and through the part PX of the resistance wire and back to E_1 or E_2 through the galvanometer G. The galvanometer G then will indicate whether current is flowing in the second part of the circuit and in which direction. By moving the sliding contact X, the overall resistance of the second circuit can be adjusted until the tendency of the battery B to send current through this circuit is just balanced by the tendency to oppose this flow as represented by the potential of cell E_1 (if switch S_2 is set to include cell E_1 in the circuit); at this point of balance, represented by position X_1 of the sliding contact, there will be no current flowing in the second circuit and this will be shown by the galvanometer registering zero deflection. To prevent drawing appreciable current from the cell E_1, this second circuit is protected by a normally open tapping key; this key is depressed to close the circuit only momentarily while observing the movement of the galvanometer needle and released immediately.

Let us assume that position X_1 represents the position of the sliding contact for zero deflection of the galvanometer when the cell of known potential, E_1, is included in the circuit by switch S_2; and let us further assume that position X_2 represents the position of the sliding contact for no current to flow (zero deflection of galvanometer) when the cell E_2 of unknown potential is included in the circuit.

The potentials developed by the cells E_1 and E_2 can be related by comparison of the resistances represented by the lengths of the wire of uniform resistance included in the corresponding circuits.

That is, the potential of the cell of known potential E_1 is to the potential of the cell of unknown potential E_2 as the resistance of length PX_1 is to the resistance of length PX_2. Since the resistance wire is of uniform resistance, the lengths PX_1 and PX_2 can be substituted for the corresponding resistances, i.e.,

$$\frac{E_1 \text{ (in volts)}}{E_2 \text{ (in volts)}} = \frac{PX_1 \text{ (in units of length)}}{PX_2 \text{ (in units of length)}} \tag{13.1}$$

Crossmultiplying,

$$E_2 \text{ (in volts)} = \frac{PX_2 \times E_1 \text{ (in volts)}}{PX_1} \tag{13.2}$$

Thus, the potential of the unknown cell is that fraction of the potential of the known cell represented by the ratio of the respective lengths of the resistance wire required to be included in the circuit when no current is flowing.

13.2 ELECTROCHEMICAL CELLS AND HALF-CELLS

It was stated previously that an electrochemical cell is made up of two half-cells, with each half-cell developing a particular potential.

A. CLASSIFICATION OF HALF-CELLS

I. Metal in Equilibrium with its Ions

Each half of the Daniell cell is an example of the type of half-cell consisting of a metal electrode immersed in a solution of ions of that metal.

A system of symbols is used to indicate in brief form the composition of the half-cell; for example, a half-cell made up of a copper electrode immersed in a solution of cupric ions is represented:

$$Cu \mid Cu^{2+}_{a_1}$$

where the activity of the Cu^{2+} ions is a_1.

Similarly, a half-cell made up of a zinc electrode immersed in a solution of zinc ions is represented:

$$Zn \mid Zn^{2+}_{a_2}$$

where the activity of the Zn^{2+} ions is a_2.

If these two half-cells are connected together, they can be represented:

$$Cu \mid Cu^{2+}_{a_1} \mid \mid Zn^{2+}_{a_2} \mid Zn$$

$$Zn \mid Zn^{2+}_{a_2} \mid \mid Cu^{2+}_{a_1} \mid Cu$$

In these examples the single vertical line represents a junction or interface of metal and solution; and the double vertical lines represent a junction between two solutions where electrical contact is made, but where any electrical potential set up at the junction is so small that it can be neglected. This electrical potential set up at the interface or junction of two liquids is referred to as "liquid-liquid junction potential" or simply as "liquid junction potential."

The "activity," referred to previously, means "effective concentration." In a very dilute solution, activity and concentration are practically synonymous; but in more concentrated solutions, the activity differs from the concentration by an activity coefficient. That is,

$$a = fc \tag{13.3}$$

where a is the activity, c is the concentration, and f is the activity coefficient.

The chemical reaction which occurs at an electrode consisting of a metal and its cation can occur as a reduction reaction or an oxidation reaction, i.e.,

$$Cu^{2+} + 2e = Cu \tag{13.4}$$

or

$$Cu = Cu^{2+} + 2e \tag{13.5}$$

The reduction reaction [Eq. (13.4)] states that a cupric ion can take on two electrons to produce one copper atom. The oxidation reaction [Eq. (13.5)] states that a copper atom can produce a cupric ion and two electrons.

Accompanying these chemical reactions is the production of electrical potential. The electrical potential depends on two factors: (a) the particular metal and cation involved, and (b) the activity of the cation in the solution. If the activity of the cation is unity (i.e., 1 gram ion/liter), the electrode potential developed is the "standard electrode potential" for that metal. Standard electrode potentials are expressed by comparison with the standard hydrogen electrode, of which the potential is arbitrarily taken as zero. This is not to say that a standard hydrogen electrode produces no electrical potential, but it is taken as the reference standard and is arbitrarily said to be zero; that is, all other electrodes are considered positive or negative with respect to the standard hydrogen electrode.

Table 13.1: Standard Reduction Potentials For Selected Metals[a]

Electrode reaction	E^v, volts
$Li^+ + e = Li$	−3.045
$K^+ + e = K$	−2.925
$Ba^{2+} + 2e = Ba$	−2.90
$Ca^{2+} + 2e = Ca$	−2.87
$Na^+ + e = Na$	−2.714
$Ce^{3+} + 3e = Ce$	−2.48
$Mg^{2+} + 2e = Mg$	−2.37
$Al^{3+} + 3e = Al$	−1.66
$Ti^{2+} + 2e = Ti$	−1.63
$Mn^{2+} + 2e = Mn$	−1.18
$Zn^{2+} + 2e = Zn$	−0.763
$Cr^{3+} + 3e = Cr$	−0.74
$Fe^{2+} + 2e = Fe$	−0.440
$Cd^{2+} + 2e = Cd$	−0.403
$Co^{2+} + 2e = Co$	−0.277
$Ni^{2+} + 2e = Ni$	−0.250
$Sn^{2+} + 2e = Sn$	−0.136
$Pb^{2+} + 2e = Pb$	−0.126
$2H^+ + 2e = H_2$	0.000
$Cu^{2+} + 2e = Cu$	+0.337
$Cu^+ + e = Cu$	+0.521
$Hg_2^{2+} + 2e = 2Hg$	+0.789
$Ag^+ + e = Ag$	+0.7991
$Pt^{2+} + 2e = Pt$	+1.2
$Au^{3+} + 3e = Au$	+1.5

[a] Wendell M. Latimer, *The Oxidation States of the Elements and Their Aqueous Solutions*, 2nd ed., ç 1952. Adapted by permission of Prentice-Hall, Inc., Englewood Cliffs, New Jersey.

Whether the reaction for the copper electrode immersed in the solution of cupric ions is written as a reduction reaction [Eq. (13.4)] or an oxidation reaction [Eq. (13.5)] will make a difference in the electrode potential produced. The numerical value will be the same, but the sign will differ. The standard electrode potential for Eq. (13.4) is 0.337 V and is referred to as the "standard reduction potential"; the standard electrode potential for Eq. (13.5) is −0.337 V and is referred to as the "standard oxidation potential." Standard electrode potentials are readily available in tables. These are usually expressed as standard reduction potentials in texts on chemical analysis and in texts which make use of the European convention; however, physical chemists use the American convention, which expresses the standard electrode potentials as standard oxidation potentials. The reversed signs are more appropriate for the use of physical chemists in their calculations from thermodynamic data.

The difference in sign of the two conventions is evident from the following:

$$E_S - E_E = -(E_E - E_S) \tag{13.6}$$

where E_S is the potential of the solution, and E_E is the potential of the metal electrode.

We have said that the potential developed by a metal electrode immersed in a solution of cations of that same metal at an activity of one is the standard electrode potential and is readily available in tables. A number of standard electrode potentials are shown in Table 13.1. These are standard reduction potentials since the reaction is written as a reduction reaction.

When a metal electrode is immersed in a solution of cations of the same metal, the standard electrode potential occurs only if the activity of the cations is unity (1 gram ion/liter). If the activity of the cations is at some other value, then the potential developed at the electrode will have some value other than the standard electrode potential. What it actually is can be calculated from the Nernst equation. The Nernst equation will be considered in greater detail later, but for a metal-metal ion half-cell, it becomes simplified.

For a half-cell where a metal M is in equilibrium with positively charged ions of the same metal at an activity a and which can be represented diagrammatically

$$M \mid M_a^+$$

the reduction reaction can be written

$$M^+ + ne = M \tag{13.7}$$

That is, a positively charged ion of the metal M can be reduced by taking on a number of electrons n to produce one atom of the metal.

For this half-cell the Nernst equation becomes simplified to:

$$E = E° + \frac{RT}{nF} \ln a_{M^+} \tag{13.8}$$

where

E = measured potential
$E°$ = standard reduction potential for the electrode
R = 8.314 J/deg/mole (the gas constant)
T = absolute temperature
n = number of electrons transferred in the reduction reaction
F = 96,500 coulombs (the Faraday)
$\ln$ = natural logarithm (base e)
a_{M^+} = activity of solution of cations M^+

Since $\ln a_{M^+} = 2.303 \log a_{M^+}$, where "log" means common logarithm (base 10), and since a common room temperature is 25°C (298° K), if we substitute these values and values for R and F, Eq. (13.8) simplifies further to:

$$E = E° + \frac{0.0591}{n} \log a_{M^+} \tag{13.9}$$

Thus, the potential developed at a metal electrode is dependent on the activity (or effective concentration) of the solution of cations.

Equation (13.9) then shows how the standard reduction potential of a particular metal electrode is modified by the activity of the solution of cations.

It was stated previously that a standard hydrogen electrode is used as a standard and is considered to have zero potential. This electrode for the present can be compared with a metal-metal ion electrode if we loosely consider hydrogen gas as a metal and hydrogen ion as a metal ion:

$$2H^+_{(a=1)} + 2e = H_2 \, gas_{(p=1 \, atm)} \tag{13.10}$$

If the activity (effective concentration) of the hydrogen ion is 1, then we have a reference electrode with zero potential; if the activity differs from 1, then the reduction reaction is

$$2H^+_a + 2e = H_2 \, gas_{(p=1 \, atm)} \tag{13.11}$$

and the potential developed will be:

$$E = E° + \frac{0.0591}{2} \log a^2_{H^+} \tag{13.12}$$

$$E = 0 + 0.02955 \log a^2_{H^+}$$

If Eq. (13.11) be regarded as the same as

$$H^+_a + e = H \, gas_{(p=1 \, atm)} \tag{13.13}$$

then the E can be found as follows:

$$E = E° + \frac{0.0591}{1} \log a_{H^+}$$

or

$$E = 0 + 0.0591 \log a_{H^+} \tag{13.14}$$

Equation (13.14) is essentially the same as Eq. (13.12), since the log of a number is the same as one-half the log of the number squared. Most tables show the reduction reaction for hydrogen as in Eq. (13.11).

a. Calculation of Potential of a Half-Cell. Example: Calculate the potential developed at an electrode consisting of a cadmium electrode immersed in a solution of cadmium ions at an activity of 0.05 gram-ion/liter.

The half-cell can be represented

$$Cd \mid Cd^{2+}_{(a=0.05)}$$

and the reduction reaction is:

$$Cd^{2+}_{(a=0.05)} + 2e = Cd$$

From Table 13.1 the standard reduction potential $E°$ is seen to be -0.403 V. Substituting in Eq. (13.9),

$$E = E° + \frac{0.0591}{n} \log a_{M^+} \tag{13.9}$$

$$= -0.403 + \frac{0.0591}{2} \log (0.05)$$

$$= -0.403 + 0.02955(\bar{2}.6990)$$

$$= -0.403 + 0.02955(-1.3010)$$

$$-0.403 - 0.03844$$

$$= -0.441$$

i.e., the measured potential of the electrode should be -0.441 V

b. Calculation of Potential of Two Half-Cells. Consider an example of two half-cells of different metals. Calculate the potential developed in the following system:

$$Sn \mid Sn^{2+}_{(a=4\times10^{-2})} \mid \mid Cu^{2+}_{(a=2\times10^{-3})} \mid Cu$$

At the tin electrode

$$E_{Sn} = E°_{Sn} + \frac{0.0591}{n} \log a_{Sn^{2+}}$$

$$= -0.136 + \frac{0.0591}{2} \log (4 \times 10^{-2})$$

$$= -0.136 + 0.02955(0.6021 - 2)$$

$$= -0.136 + 0.02955(-1.3979)$$

$$= -0.136 - 0.0413$$

$$= -0.177 \text{ V}$$

At the copper electrode

$$E_{Cu} = E°_{Cu} + \frac{0.0591}{n} \log a_{Cu^{2+}}$$

$$= +0.337 + \frac{0.0591}{2} \log (2 \times 10^{-3})$$

$$= +0.337 + 0.02955(0.3010 - 3)$$

$$= +0.337 + 0.02955(-2.6990)$$

$$= 0.337 - 0.07975$$

$$= 0.257 \text{ V}$$

In this example, the tin electrode has a negative potential of 0.177 V and the copper electrode a positive potential of 0.257 V. The overall potential of the two half-cells is the algebraic difference 0.257 V − (−0.177 V) or +0.434 V. Of the two half-cells, the copper electrode is the positive one.

c. Calculation of Potential of a Concentration Cell. If two half-cells are set up, using the same metal immersed in solutions of the same cation, but with the activities of the cations at differing values, such a combination is spoken of as a "concentration cell." For example calculate the overall potential in the following:

$$\text{Zn} \mid \text{Zn}^{2+}_{(a=3\times10^{-4})} \mid\mid \text{Zn}^{2+}_{(a=1\times10^{-2})} \mid \text{Zn}$$

For the half-cell on the left

$$E = E^\circ + \frac{0.0591}{n} \log a_{\text{Zn}^{2+}}$$

$$= -0.763 + \frac{0.0591}{2} \log (3 \times 10^{-4})$$

$$= -0.763 + 0.02955(0.4771 - 4)$$

$$= -0.763 + 0.02955(-3.5229)$$

$$= -0.763 - 0.1041$$

$$= -0.867 \text{ V}$$

For the half-cell on the right

$$E = E^\circ + \frac{0.0591}{n} \log a_{\text{Zn}^{2+}}$$

$$= -0.763 + \frac{0.0591}{2} \log (1 \times 10^{-2})$$

$$= -0.763 + 0.02955(0 - 2)$$

$$= -0.763 - 0.0591$$

$$= -0.822 \text{ V}$$

The overall potential is the potential of the left half-cell minus the potential of the right half-cell, or

$$-0.867 \text{ V} - (-0.822 \text{ V})$$

$$= -0.867 \text{ V} + 0.822 \text{ V}$$

$$= -0.045 \text{ V}$$

If the subtraction were done in the opposite direction, that is, the potential of the right half-cell minus the potential of the left half-cell, the answer would be positive. So the difference in potential between left and right electrodes is 0.045 V. Of the two electrodes, the one on the right is the less negative of the two so the right electrode as written is the positive electrode.

2. A Metal in Equilibrium with a Saturated Solution of a Slightly Soluble Salt

Half-cells of this type are widely used as reference electrodes. Included in this group are the silver-silver chloride electrode and the various calomel electrodes.

A silver-silver chloride electrode is prepared in such a way that silver and silver chloride are in intimate contact with each other and with chloride ion. It can be represented as

$$Ag\,|AgCl|\,Cl^-$$

If the activity of Cl^- ion in this electrode is one, then the potential of the half-cell is $+0.222$ V in comparison with a standard hydrogen electrode. The electrode reaction can be written:

$$AgCl + e = Ag + Cl^- \qquad (13.15)$$

A Calomel electrode is composed of mercury, calomel, and chloride ion. It can be represented as

$$Hg\,|Hg_2Cl_2|\,Cl_a^-$$

The electrode reaction can be written.

$$Hg_2Cl_2 + 2e = 2Hg + 2Cl^- \qquad (13.16)$$

3. Two Soluble Species in Equilibrium at an Inert Electrode

Another kind of half-cell is that in which a solution contains two soluble species which can convert to each other by an oxidation-reduction reaction. In order that the electrical potential developed may be led to a potentiometer, it is necessary that some metallic electrode be inserted into the solution. This electrode should not itself enter into reaction with the solution, so should be made of some inert metal, usually platinum.

A general diagram of a half-cell containing two soluble species is:

$$Pt\,|\,O_{a_1} + R_{a_2}$$

The reduction reaction for such a half-cell could be written:

$$xO + ne = yR \qquad (13.17)$$

where O is the oxidized form, or the species being reduced, R is the reduced form, or the reduction product, n is the number of electrons involved in the reaction, x and y are stoichiometric coefficients, and a_1 and a_2 are the activities of the species O and R, respectively.

The Nernst equation for half-cells of this type at 25°C is:

$$E = E° + \frac{0.0591}{n} \log \frac{a_O^x}{a_R^y} \qquad (13.18)$$

An example of a half-cell containing two soluble species in equilibrium is:

$$Pt\,|\,Fe_{a_1}^{3+} + Fe_{a_2}^{2+}$$

The reduction reaction for this electrode is:

$$Fe^{3+} + e = Fe^{2+} \tag{13.19}$$

A number of examples of standard reduction potentials are listed in Table 13.2.

TABLE 13.2: Standard Reduction Potentials for Selected Soluble Species[a]

Electrode reaction			$E°$, volts
Cr^{3+}	$+ e = Cr^{2+}$		-0.41
Ti^{3+}	$+ e = Ti^{2+}$		-0.37
$2H^+$	$+ 2e = H_2$		0.00
$S_4O_6^{2-}$	$+ 2e = 2S_2O_3$		0.08
Sn^{4+}	$+ 2e = Sn^{2+}$		0.15
Cu^{2+}	$+ e = Cu^+$		0.153
$Fe(CN)_6^{3-}$	$+ e = Fe(CN)_6^{4-}$		0.36
I_2	$+ 2e = 2I^-$		0.536
Fe^{3+}	$+ e = Fe^{2+}$		0.771
$2Hg^{2+}$	$+ 2e = Hg_2^{2+}$		0.920
Br_2	$+ 2e = 2Br^-$		1.065
Cl_2	$+ 2e = 2Cl^-$		1.360
Ce^{4+}	$+ e = Ce^{3+}$		1.61
Co^{3+}	$+ e = Co^{2+}$		1.82
Ag^{2+}	$+ e = Ag^+$		1.98
F_2	$+ 2e = 2F^-$		2.65

[a] Wendell M. Latimer, *The Oxidation States of the Elements and Their Aqueous Solutions*, 2nd ed., copyright © 1952. Adapted by permission of Prentice-Hall Inc., Englewood Cliffs, New Jersey.

4. General Case

There are instances where substances, other than the substance actually undergoing reduction, enter the reaction. These instances can be considered in the general case:

$$xO + pW + ne = yR + qZ \tag{13.20}$$

where O is the substance being reduced, R is the main reduction product, W is some other species entering the reduction reaction, Z is some product other than the main reduction product, n is the number of electrons involved in the reaction, and x, y, p, and q are stoichiometric coefficients.

The Nernst equation for the general reaction shown in Eq. (13.20) is:

$$E = E° + \frac{RT}{nF} \ln \frac{a_O^x a_W^p}{a_R^y a_Z^q} \tag{13.21}$$

At 25°C, this simplifies to:

$$E = E° + \frac{0.0591}{n} \log \frac{a_O^x a_W^p}{a_R^y a_Z^q} \tag{13.22}$$

A very common example of species represented by W and Z are hydrogen ions and water molecules, respectively, as in the reduction of permanganate ion to manganous ion:

$$MnO_4^- + 8H^+ + 5e = Mn^{2+} + 4H_2O \tag{13.23}$$

Equation (13.22) for the reaction where W and Z represent hydrogen ions and water becomes

$$\dot{E} = E° + \frac{0.0591}{n} \log \frac{a_O^z a_{H^+}^p}{a_R^y a_{H_2O}^q} \tag{13.24}$$

For dilute solutions, the activity of water is taken as unity so Eq. (13.24) simplifies to

$$E = E° + \frac{0.0591}{n} \log \frac{a_O^z a_{H^+}^p}{a_R^y} \tag{13.25}$$

Taking out the factor involving activity of hydrogen ion,

$$E = E° + \frac{0.0591}{n} \log \frac{a_O^z}{a_R^y} + \frac{0.0591}{n} p \log a_{H^+} \tag{13.26}$$

Since

$$pH = -\log a_{H^+} \tag{13.27}$$

then

$$E = E° + \frac{0.0591}{n} \log \frac{a_O^z}{a_R^y} - 0.0591 \frac{p}{n} \cdot pH \tag{13.28}$$

Equation (13.28) relates the electrode potential to the pH of the medium. Equation 13.28, if applied, must be used with discrimination, since obviously some species cannot exist in the same form if the pH is changed too markedly.

Some examples of standard reduction potentials for the general type of electrode reaction are given in Table 13.3.

TABLE 13.3: Standard Reduction Potentials for Selected Species[a]

Electrode reaction	$E°$, volts
$2H^+ + 2e = H_2$	0.00
$S + 2H^+ + 2e = H_2S$	0.141
$Sb_2O_3 + 6H^+ + 6e = 2Sb + 3H_2O$	0.152
$BiO^+ + 2H^+ + 3e = Bi + H_2O$	0.32
$H_3AsO_4 + 2H^+ + 2e = HAsO_2 + 2H_2O$	0.559
$O_2 + 2H^+ + 2e = H_2O_2$	0.682
$NO_3^- + 3H^+ + 2e = HNO_2 + H_2O$	0.94
$NO_3^- + 4H^+ + 4e = NO + 2H_2O$	0.96
$O_2 + 4H^+ + 4e = 2H_2O$	1.229
$MnO_2 + 4H^+ + 2e = Mn^{2+} + 2H_2O$	1.23
$Cr_2O_7^{2-} + 14H^+ + 6e = 2Cr^{3+} + 7H_2O$	1.33
$ClO_3^- + 6H^+ + 5e = Cl_2 + 3H_2O$	1.47
$MnO_4^- + 8H^+ + 5e = Mn^{2+} + 4H_2O$	1.51
$BrO_3^- + 6H^+ + 5e = Br_2 + 3H_2O$	1.52

[a] Wendell M. Latimer, *The Oxidation States of the Elements and Their Aqueous Solutions*, 2nd ed., copyright © 1952. Adapted by permission of Prentice-Hall, Inc., Englewood Cliffs, New Jersey.

Example: Calculate the potential of a platinum electrode immersed in a solution which is 4×10^{-3} M in $Cr_2O_7^{2-}$ and 3×10^{-2} M in Cr^{3+} and has a pH of 3.

For the electrode reaction,

$$Cr_2O_7^{2-} + 14H^- + 6e = 2Cr^{3-} + 7H_2O$$

the standard potential is 1.33 V.

$$E = E^\cdot + \frac{0.0591}{n} \log \frac{a_o^r}{a_R^v} - 0.0591 \frac{p}{n} (\text{pH}) \tag{13.28}$$

$$= 1.33 + \frac{0.0591}{6} \log \frac{(4 \times 10^{-3})^1}{(3 \times 10^{-2})^2} - 0.0591 \frac{14}{6} \times 3$$

$$= 1.33 + 0.00985 \log (4.444) - 0.0591 \times 7$$

$$= 1.33 + 0.00985(0.6478) - 0.4137$$

$$= 1.33 + 0.00638 - 0.4137$$

$$= 0.92 \text{ V}$$

13.3 ELECTRODES

A determination of potential involves two half-cells or at least two electrodes immersed in a system. One of these electrodes is a reference electrode, that is, one which develops a definite potential to which the potential of the other electrode can be compared. The other electrode is usually referred to as the "indicating electrode" and is such that it develops a potential which is dependent on the quantity of the substance in the solution which we wish to measure.

More correctly speaking, the overall measured potential of the cell (or two half-cells) will be made up of the potential developed in the half-cell containing the reference electrode, plus the potential developed in the half-cell containing the indicating electrode, plus the electrical potential developed at the junction of the two half-cells. This measured overall potential of the cell can be expressed:

$$E_{cell} = E_{reference} + E_{indicator} + E_{junction} \tag{13.29}$$

The potential developed at the junction between the two half-cells, the so-called junction potential or liquid-junction potential [$E_{junction}$ in Eq. (13.29)] must be taken into consideration unless steps are taken to render it negligible. In actual practice, the two half-cells are usually connected by a salt bridge of saturated potassium chloride solution; this has the effect of reducing the junction potential to such a small value that it can be neglected. The form of the salt bridge may vary. In some instances the two electrodes are immersed in solutions contained in separate vessels and the two solutions connected by a glass tube of inverted U-shape containing the potassium chloride solution (see Fig. 13.3a). In other situations one electrode may be contained in a tube containing also the potassium chloride with the tube having a constricted opening by which electrical contact with the other half-cell is maintained

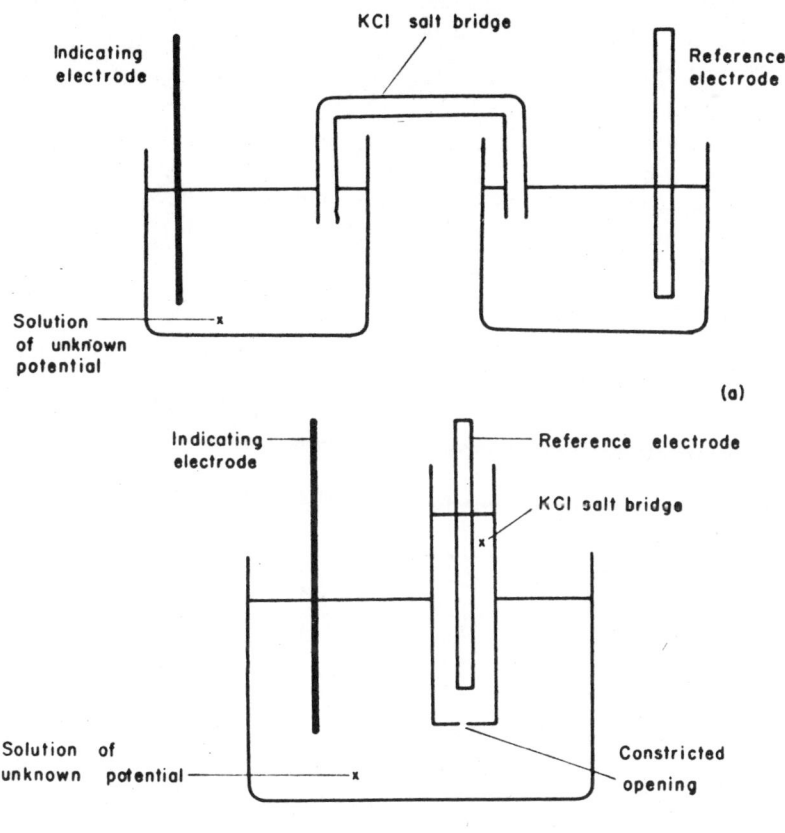

FIGURE 13.3: Two forms of salt bridge.

and through which the potassium chloride solution is allowed to seep slowly; this slow seeping maintains electrical contact but prevents undue contamination of the solution in which the electrode-containing tube is immersed (see Fig. 13.3b).

For those situations where chloride ions cannot be used in a salt bridge (such as where precipitation of silver chloride would occur), the most commonly used solution for a salt bridge is ammonium nitrate solution.

A.   REFERENCE ELECTRODES

I. Standard Hydrogen Electrode

The standard hydrogen electrode is the reference against which potentials are compared. The standard hydrogen electrode develops a potential which is arbitrarily considered to be zero and all other electrodes develop a potential which may be positive or negative compared with the standard hydrogen electrode.

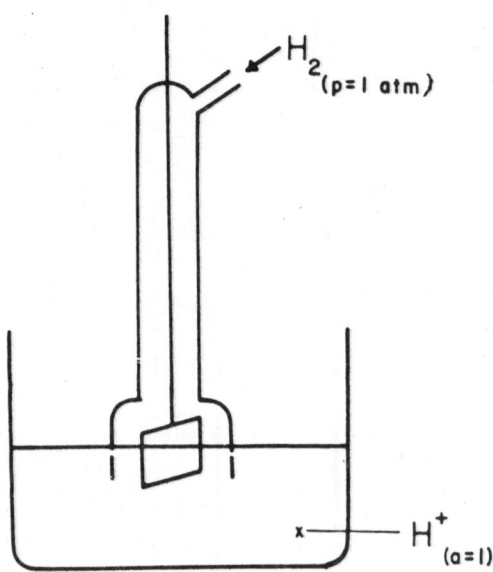

FIGURE 13.4: Standard hydrogen electrode.

The standard hydrogen electrode consists of a surface of platinum covered with finely divided platinum black and kept in intimate contact with hydrogen gas (at a pressure of 1 atm) and hydrogen ions (at an activity of 1). It frequently has the form of a square of platinum sealed into a glass tube into which hydrogen gas can be led. The square of platinum is connected to a wire for electrical connection. The surface of the platinum square must be prepared freshly with a coating of finely divided platinum called "platinum black" or "platinized platinum." The glass tube is inserted into a solution containing hydrogen ions at unit activity (approximately 1.18 normal) so that the square of platinum is partly immersed. A number of holes around the glass tube just below the surface of the solution permit the hydrogen gas to bubble out against atmospheric pressure. The hydrogen gas bubbling out disturbs the surface of the liquid so that the platinum surface is alternately covered with hydrogen gas and solution containing hydrogen ions. A diagram of a standard hydrogen electrode appears in Fig. 13.4.

The standard hydrogen electrode is written diagrammatically:

$$Pt\,|H_2\underset{(p=1\,atm)}{}\quad|\,H^+\underset{(a=1)}{}$$

and the reduction reaction is:

$$2H^+_{(a=1)} + 2e = H_2\underset{(p=1\,atm)}{} \tag{13.10}$$

The standard hydrogen electrode is the ultimate reference against which all other electrode potentials are compared. It has a number of advantages, e.g., it has no error in strongly alkaline solutions. However there are many disadvantages to its use. Some of the more important disadvantages are:

1. It requires that the platinum black surface of the electrode be renewed daily.
2. It requires a long time to come to equilibrium. It requires that hydrogen be swept through the electrode for an hour or more to remove all oxygen from the system.
3. Dissolved gases such as ammonia, carbon dioxide, etc., interfere with its use.
4. The surface of the platinum is easily poisoned with colloidal material or a variety of other substances which might be present.

Because of these and other disadvantages the standard hydrogen electrode is almost never used as a routine reference electrode.

2. Calomel Electrode

The calomel electrode is one which lends itself readily to use as a reference electrode. It is relatively cheap, can be readily prepared, and gives a potential which is reproducible in relation to the standard hydrogen electrode.

The calomel electrode consists of a paste of mercury and calomel in contact with a layer of metallic mercury and with a solution of potassium chloride saturated with the calomel. Electrical contact is made by an inert wire dipping into the layer of mercury.

Figure 13.5 illustrates some forms of the calomel electrode, showing means of providing electrical contact with the solution through a salt bridge of saturated solution of potassium chloride. The junction is cleansed in 13.5a by running in additional potassium chloride from the reservoir, thus displacing solution through the constricted tip on the side tube; in 13.5b by gradual seeping out of the potassium chloride solution in the tube through the fiber tip (usually of asbestos), which fills the tiny opening in the outer tube; in 13.5c by loosening the ground-glass joint to permit solution to flow out of the tube.

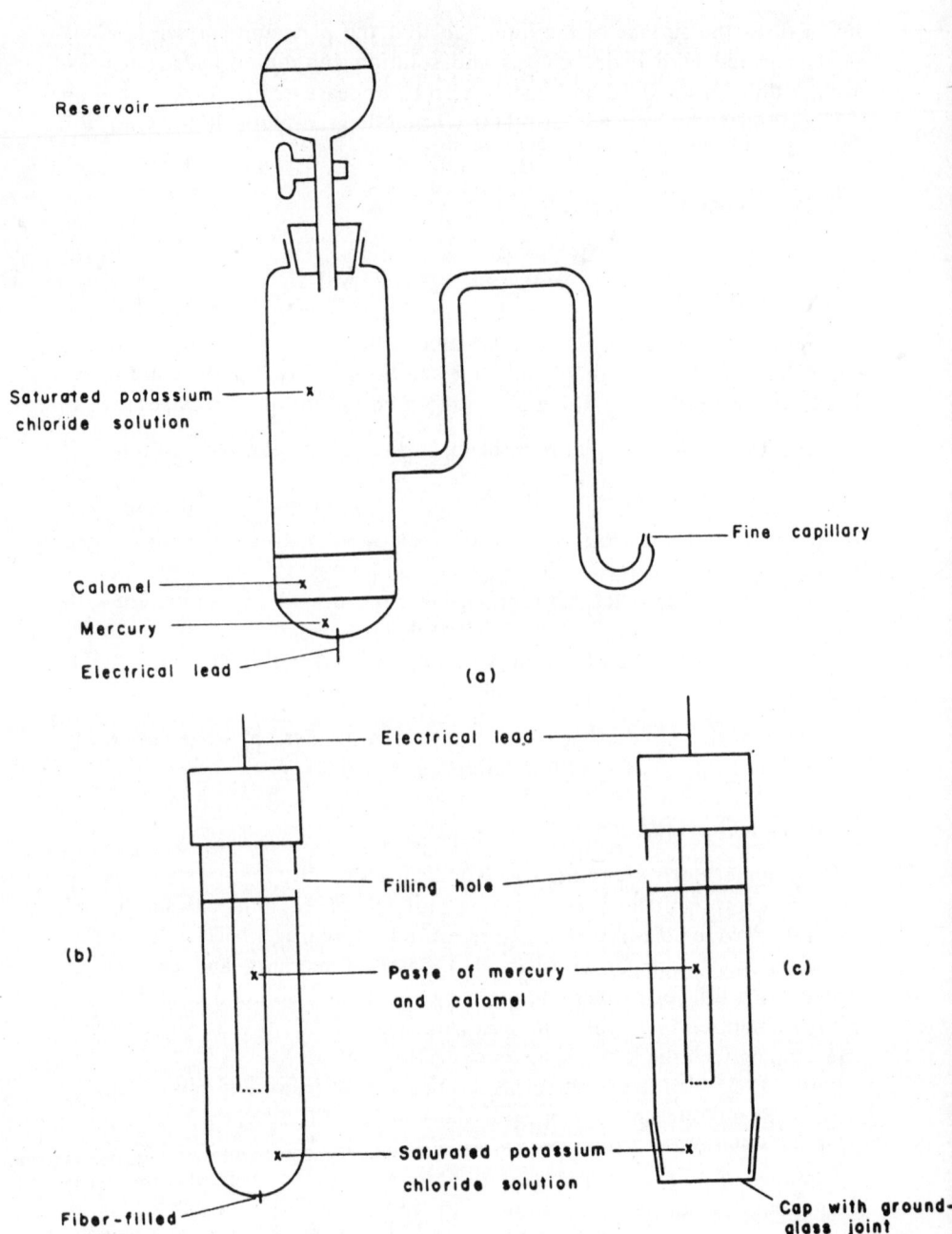

Reservoir

Saturated potassium
chloride solution

Fine capillary

Calomel

Mercury

Electrical lead

(a)

Electrical lead

(b)

Filling hole

Paste of mercury
and calomel

(c)

Saturated potassium
chloride solution

Fiber-filled
opening

Cap with ground-
glass joint

FIGURE 13.5: Some forms of calomel electrode.

The calomel electrode is of the type where a metal exists in equilibrium with a saturated solution of a slightly soluble salt.

$$Hg \,|Hg_2Cl_2|\, Cl^-$$

The reduction reaction is:

$$Hg_2Cl_2 + 2e = 2Hg + 2Cl^- \tag{13.16}$$

The Nernst equation for the calomel electrode becomes simplified to:

$$E = E^\circ + \frac{0.0591}{2} \log \frac{1}{a_{Cl^-}^2} \tag{13.30}$$

Thus the potential developed by the calomel electrode depends on the concentration of the potassium chloride solution, The most common concentration employed is a saturated solution. This is probably because of the ease of maintaining the solution in this concentration in practice; the solution will not become more concentrated when left open to the air and thus change the potential produced.

The potential of a saturated calomel electrode containing saturated potassium chloride solution is +0.244 V with respect to the standard hydrogen electrode. The potential of a normal calomel electrode (and an $N/10$ calomel electrode) containing a normal (or $N/10$) solution of potassium chloride is +0.281 V (or 0.336 V).

3. Saturated Potassium Sulfate Electrode

This electrode is used as a reference electrode where absence of chloride ion is necessary such as in the determination of halide ion in precipitation reactions. The potential of the standard potassium sulfate electrode is 0.615 V in comparison to the standard hydrogen electrode.

4. Silver-Silver Chloride Electrode

This is used commonly as an inner reference electrode in glass electrodes and occasionally as an ordinary reference electrode. It consists of a surface of silver which is coated with silver chloride and can be represented diagrammatically:

$$Ag \,|AgCl|\, Cl^-_{(a=1)}$$

The cell reaction is

$$AgCl + e = Ag + Cl^- \tag{13.15}$$

The voltage of this electrode at 25°C is 0.222 V.

5. Weston Cell

A Weston cell consists of two half-cells in a sealed glass container comprising a single unit. It is easily prepared and remains stable for long periods.

Its most serious disadvantage is that it shows changes in voltage produced with variations in temperature. The maximum voltage produced by the saturated cell is 1.01902 V near 3°C, and the voltage produced decreases from this value with increase or decrease in temperature. The saturated cell can be used from about −20°C to about 43°C. At 20°C the voltage produced

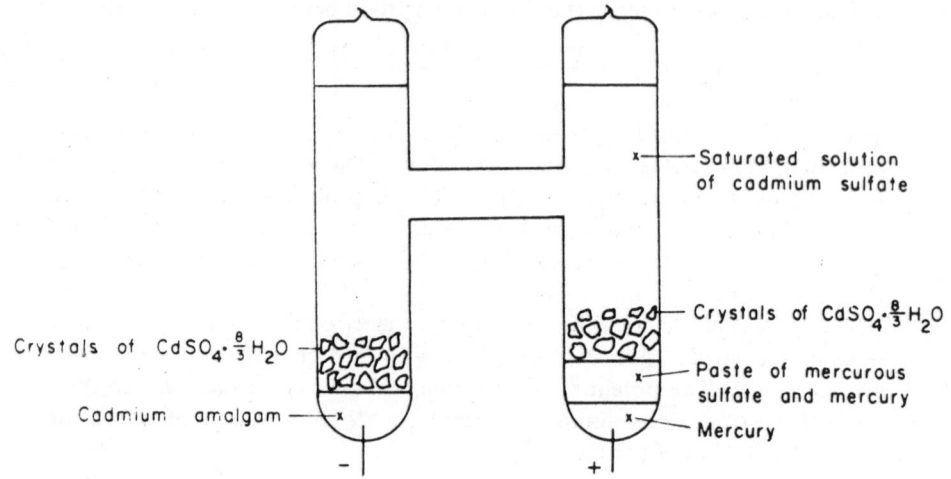

Saturated solution of cadmium sulfate

Crystals of $CdSO_4 \cdot \frac{8}{3}H_2O$

Paste of mercurous sulfate and mercury

Mercury

Crystals of $CdSO_4 \cdot \frac{8}{3}H_2O$

Cadmium amalgam

−

+

FIGURE 13.6: Weston cell.

is 1.01864 and varies according to the formula, where E_t is the potential in volts at the temperature t (in °C)

$$E_t = 1.01864 - 4.06 \times 10^{-5}(t - 20) - 9.5 \times 10^{-7}(t - 20)^2$$
$$+ 1 \times 10^{-8}(t - 20)^3 \quad (13.31)$$

Between 25 and 35°C, the potential produced decreases by about 0.00005 V per degree rise in temperature. Consequently, the control of temperature is quite important for accurate work; this can be accomplished by a jacketed chamber with a thermostat.

The Weston cell is prepared in a glass tube of H-shape. The positive terminal is a mercury electrode covered with a paste of cadmium sulfate and mercury; the negative electrode is a cadmium amalgam (of about 10% cadmium). Filling the bulk of the H-tube and providing electrical contact between the electrodes is a saturated solution of cadmium sulfate; to ensure saturation at all temperatures excess crystals of $CdSO_4 \cdot \frac{8}{3}H_2O$ are present in both arms of the H-tube. The cell can be represented schematically:

$$Cd_{(10\% \text{ amalgam})} \mid CdSO_4 \cdot \frac{8}{3}H_2O, CdSO_{4(\text{sat'd sol'n})}, Hg_2SO_4 \mid Hg$$

A diagram of a Weston cell is shown in Fig. 13.6.

The cell reactions at the electrodes are

$$Cd^{2+} + Hg + 2e = Cd(Hg) \tag{13.32}$$

$$Hg_2^{2+} + 2e = 2Hg \tag{13.33}$$

The Weston cell is so stable and so readily reproducible that it has become the accepted type of reference cell to be built into various instruments (such as pH meters) as a stable source of known emf to which an unknown emf can be compared.

In an effort to decrease the variation of emf with changes in temperature, modifications of the Weston cell have been made. These include the use of a solution of cadmium sulfate saturated at 4°C, but unsaturated at higher temperatures, and the addition of bismuth to cadmium amalgam.

B. INDICATING ELECTRODES

1. Noble Metal

The simplest type of indicating electrode is merely a wire or flat plate or cylinder of a metal which will not enter into reaction with the solution in which it is immersed. The most commonly used metal is platinum; in some instances, silver or some other metal may be used.

The platinum electrode finds frequent use in oxidation-reduction reactions, where it is employed in combination with a calomel electrode as a reference electrode.

2. Glass Electrode

A glass electrode is used in conjunction with a reference electrode for the determination of hydrogen ion concentration, Concentration of hydrogen ion can be shown in terms of electrical potential or in terms of the pH scale.

A "glass electrode" consists of a bulb or covering of a thin pH-sensitive glass membrane, within which is mounted a reference electrode. The contained reference electrode is usually a silver-silver chloride electrode or a calomel electrode in hydrochloric acid or buffered chloride solution. A diagram of a glass electrode appears in Fig. 13.7.

When a glass electrode is immersed in a solution containing hydrogen ions, a potential is set up between the inside and outside solutions separated by the pH-sensitive glass membrane. This potential set up is dependent upon the pH of the solution outside the glass membrane, This potential can be measured, between the reference electrode sealed inside the glass electrode and a calomel reference electrode immersed in the solution.

Such a system can be represented:

Ag |AgCl| HCl |glass membrane| test solution ||KCl$_{sat'd}$| Hg$_2$Cl$_2$| Hg

glass electrode   saturated calomel
                                                           electrode

The type of glass membrane used in most glass electrodes shows reasonable response over the pH range from 3 to 9.; outside this range, the glass membrane is subject to error, tending to give readings too high below a pH of

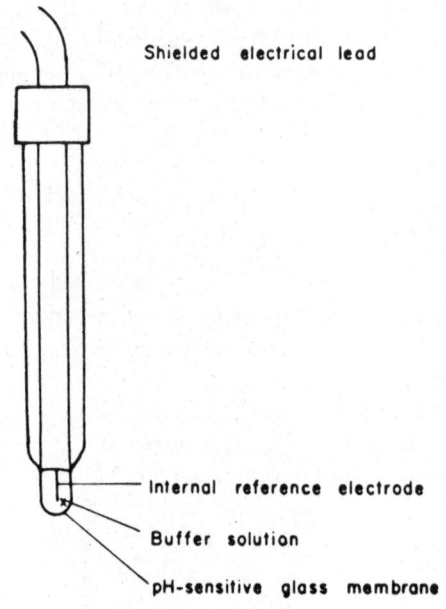

Shielded electrical lead

Internal reference electrode

Buffer solution

pH-sensitive glass membrane

FIGURE 13.7: A form of glass electrode.

about 3 and too low above a pH of about 9. The errors are dependent to a considerable extent upon other ions which may be present. Special types of glass electrodes are available which show considerably less error in the alkaline range.

The usual type of glass electrode is subject to errors in the alkaline range. This error is pronounced in the presence of sodium ions; in fact, at high pH ranges such an electrode can function as an electrode which produces a potential dependent on the sodium ion concentration. This property can be enhanced by using special glass, resulting in an electrode which is sensitive to sodium ion concentration even at relatively low pH (as low as pH 5). Other types of glass are available which are sensitive to concentration of potassium, lithium, rubidium, or caesium. While there is some sensitivity to the other ions, special glasses can be prepared which have a favorable ratio in being sensitive to one ion; electrodes sensitive to sodium or potassium are available commercially. Electrodes sensitive to cations other than hydrogen ion are referred to as "cation-specific electrodes."

Because of the high electrical resistance of the glass membrane it is not possible to measure accurately the potential developed between a glass electrode and calomel reference electrode by using an ordinary potentiometer equipped with a galvanometer. It is possible to measure the potential reasonably accurately using a vacuum tube voltmeter, and a form of vacuum tube voltmeter is included in the usual "pH meter" which employs the glass electrode.

3. Quinhydrone Electrode

· The quinhydrone electrode is composed of a platinum wire (or other inert wire) dipped into a solution to which quinhydrone crystals have been added. Quinhydrone is a solid compound consisting of 1 mole of hydroquinone to 1 mole of quinone. It is only sparingly soluble in water. The solution can be oxidized or reduced as indicated by the reaction:

$$+ 2H^+ + 2e \rightleftharpoons \qquad (13.34)$$

The Nernst equation (13.22) for this reaction becomes:

$$E = E^\circ + \frac{0.0591}{2} \log \frac{[\text{quinone}][H^+]^2}{[\text{hydroquinone}]} \qquad (13.35)$$

Since in quinhydrone the molar proportions of quinone and hydroquinone are equal, Eq. (13.35) simplifies to :

$$E = E^\circ + \frac{0.0591}{2} \log [H^+]^2$$

or

$$E = E^\circ + 0.0591 \log [H^+] \qquad (13.36)$$

Since $pH = -\log [H^+]$,

$$E = E^\circ - 0.0591 \, pH \qquad (13.37)$$

Thus, we see that the measured potential at a quinhydrone electrode is the standard reduction potential of the quinhydrone electrode modified by a factor which is dependent upon pH. The quinhydrone electrode therefore can be used to indicate the pH of a solution; it is used in conjunction with a reference electrode such as the calomel electrode.

The quinhydrone electrode is simple to use and comes to equilibrium fairly rapidly. Among the disadvantages of the quinhydrone electrode are:

(1) At a pH of 8 or over it reacts with alkali, since it is a weak acid, to dissociate into the corresponding anion. This anion is readily subject to air oxidation.

(2) It is also subject to salt errors in the presence of high concentrations of electrolytes, since these change the activity coefficients of quinone and hydroquinone unequally.

4. Antimony Electrode

An antimony electrode consists of a rod of pure antimony which is superficially oxidized. The potential is probably developed by the antimony and the coating of antimonous oxide:

$$Sb_2O_3 + 6H^+ + 6e \rightleftharpoons 2Sb + 3H_2O \qquad (13.38)$$

Since solid antimony and solid antimonous oxide can be considered to have unit activity, the Nernst equation [Eq. (13.22)] becomes simplified to:

$$E = E^\circ + 0.0591 \log a_{H^+} \qquad (13.39)$$

That is, the antimony electrode is dependent upon pH. Deviations arise at low and high pH values and there are a number of interfering substances.

Advantages of the antimony electrode are that it is very rugged and can give useful results when certain substances are present. It is used with a saturated calomel electrode as a reference electrode.

13.4 MEASUREMENT OF POTENTIAL

A. INSTRUMENTS

It has been indicated in Section 13.1 that, when current from an electrochemical cell is allowed to flow, the potential of the cell falls and continues to fall. For this reason, when we wish to know the potential developed by a cell we should measure the potential without drawing any current. It was further indicated that this measurement could be done using a potentiometer. A simple form of potentiometer was described.

A potentiometer can be used to measure the potential produced by a number of electrical cells, including certain electrical cells for measuring hydrogen ion concentration. For example, if the two half-cells consist of a normal hydrogen electrode as a reference electrode and an indicating hydrogen electrode, or if the two half-cells consist of a saturated calomel electrode as reference electrode and an antimony electrode as indicating electrode, these systems can have their electrical potential determined by the type of potentiometer which has been described.

However, if a glass electrode is used as the indicating electrode for the determination of hydrogen ion concentration the potentiometer cannot be used directly to determine the potential produced. This is so because of the very high resistance of the glass membrane (perhaps as great as $10^8 \, \Omega$), thus the current that flows is too small to cause significant deflections of the galvanometer with small changes in potential. To have the small potential produced in a cell (of the order of 1 V) cause sufficient current to flow to cause noticeable deflection of a galvanometer that potential must be amplified. In the usual type of instrument used to measure concentration of hydrogen ion with a glass electrode some amplification is necessary; this type of instrument is called a pH meter. Because of the convenience of use of the usual pH meter and its wide application, the pH meter is very frequently used in all types of potentiometric determinations if suitable electrodes are employed. Thus, pH meters usually have two scales, one reading in pH units and the other reading in millivolts. In order to give more exact readings from a scale of limited length, some pH meters, instead of using the whole scale for the pH range of 0–14, use a "folded scale," where the whole scale is used for the pH range 0–8 and by changing a switch the same scale covers the pH range 6–14; still other pH meters use an "expanded scale," where, in addition to using the whole scale for the complete pH range, they also use the whole scale for a small pH range such as 1 or 2 pH units. Similarly, the millivolt scale can be varied by the use of range switches to cover a portion of a large millivolt range, for example, from -1400 to $+1400$ mV.

B. CLASSIFICATION OF INSTRUMENTS

pH meters including instruments for measuring potential can be classified in a number of ways:

1. By type of meter
   a. Null reading, where the meter is used to indicate when no current is flowing in the potentiometer type of circuit
   b. Direct reading, where the meter is used to provide the reading directly in pH units or millivolts
2. By precision of measurement
   a. High precision, for use in research and other applications where a high degree of accuracy is required
   b. Medium precision, for use where a lesser degree of accuracy is satisfactory
   c. Low precision, where a still lesser degree of accuracy is required. This may often be accompanied by a more rugged type of construction suitable for use in field trips and in industrial plants

3. By power supply
   a. Battery operated, where all the power is supplied by batteries. This renders the instrument self-contained so that it can be readily transported and used in localities far from a regular power source
   b. Line operated, where the power is obtained from an ac line and periodic replacement of batteries is not required

C. CONSTRUCTION OF INSTRUMENTS

If a meter, whose needle or pointer is actuated by a moving coil, is used to measure voltage, some current is used up in flowing through the coil. This current is sufficient to cause the potential to drop. It is possible to use the potential which we wish to measure merely as a control for the

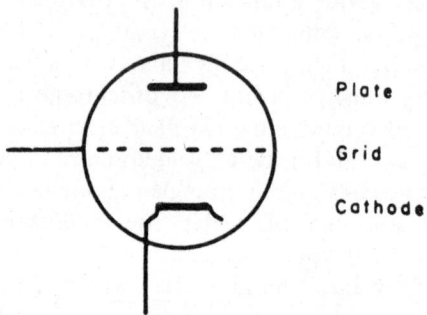

FIGURE 13.8: The three electrodes of a triode.

current provided by another source and which actually moves the coil and thereby deflects the meter needle. That is, the meter needle is actuated by a current which is controlled by the potential we are measuring; small variations in potential can cause a considerable difference in current to flow. This considerable difference of current can cause the meter needle to move without drawing appreciable current from the original voltage source. We can speak of small differences of potential being amplified to considerable differences of current. If the amplification is done by vacuum tubes, we have a "vacuum tube voltmeter" (frequently abbreviated VTVM). A vacuum tube voltmeter then is a convenient way of measuring potential without drawing appreciable current.

A vacuum tube to be used as an amplifier requires at least three electrodes or connections, i.e., must be a "triode." (There may be one or, more commonly, two additional connections to supply current to heat the filament, but these connections are not included among the three connections just mentioned.) Other types of vacuum tubes may have four electrodes (a "tetrode") or five electrodes (a "pentode"). A triode can be represented diagrammatically in Fig. 13.8 to show the essential parts: (a) cathode, (b) grid, and (c) plate, all of which are placed within a glass or metal envelope from which the air has been evacuated. The cathode is heated, which renders it a good source of electrons. (The heating may be done by causing electricity to flow through the cathode, i.e., the cathode takes the form of an electrically heated filament; or the cathode may be heated by a separate filament placed near to the cathode.) A source of potential is placed across the plate and cathode with the plate being positive. (This potential is frequently referred to as B⁺.) The positive potential at the plate attracts electrons given off by the cathode and a stream of electrons passes across the space between the cathode and plate. This stream of electrons constitutes a current called the "plate current." If a smaller potential is placed across the grid and the cathode with the negative potential at the grid, this potential can control the plate current. To illustrate: if the grid has a particular small negative voltage applied to it, the negative potential serves to repel the electrons given off by the cathode and fewer of these electrons will manage to pass the grid and find their way to the positively charged plate. Thus, the negative potential on the grid cuts down the plate current. If the grid is made more negative still, the plate current will be still smaller. If the grid is made less negative, there will be less repulsion of the electrons, so more electrons from the cathode will pass through the grid and reach the plate; consequently, the plate current will be increased. Thus the variations in negative potential applied to the grid control the plate current. Small variations in grid potential result in considerable variations in plate current, i.e., amplification of potential applied to the grid results. There are practical limits to the amplification—if the grid potential is made sufficiently small in a negative way, the plate current eventually reaches a maximum value called "saturation value"; if the grid potential is made sufficiently large, in a negative direction, the plate current becomes so small that it is "cut off" altogether. This value of potential is spoken of as the "cut off grid voltage."

The particular values of plate current depend not only on the grid potential (or grid voltage or grid bias) but also on the potential applied at the plate. If the plate current is plotted against grid voltage, a curve results. Typical curves are shown in Fig. 13.9 for two particular values of plate voltage (E_b).

The essential parts of a vacuum tube voltmeter might be indicated by Fig. 13.10. It will be seen that the resistance R can be adjusted (and thereby the grid bias varied) so that, when no potential is applied at the input terminals, there would be no current flowing in the plate circuit and the meter needle would read zero. The meter could be suitably calibrated to read in volts to correspond to a potential applied at the input terminals.

The vacuum tube voltmeter forms the basis of direct-reading instruments for determination of potential and pH. The direct-reading pH meter has advantages to its use and is incorporated into many instruments available commercially. One of the greatest advantages is that it can be readily adapted

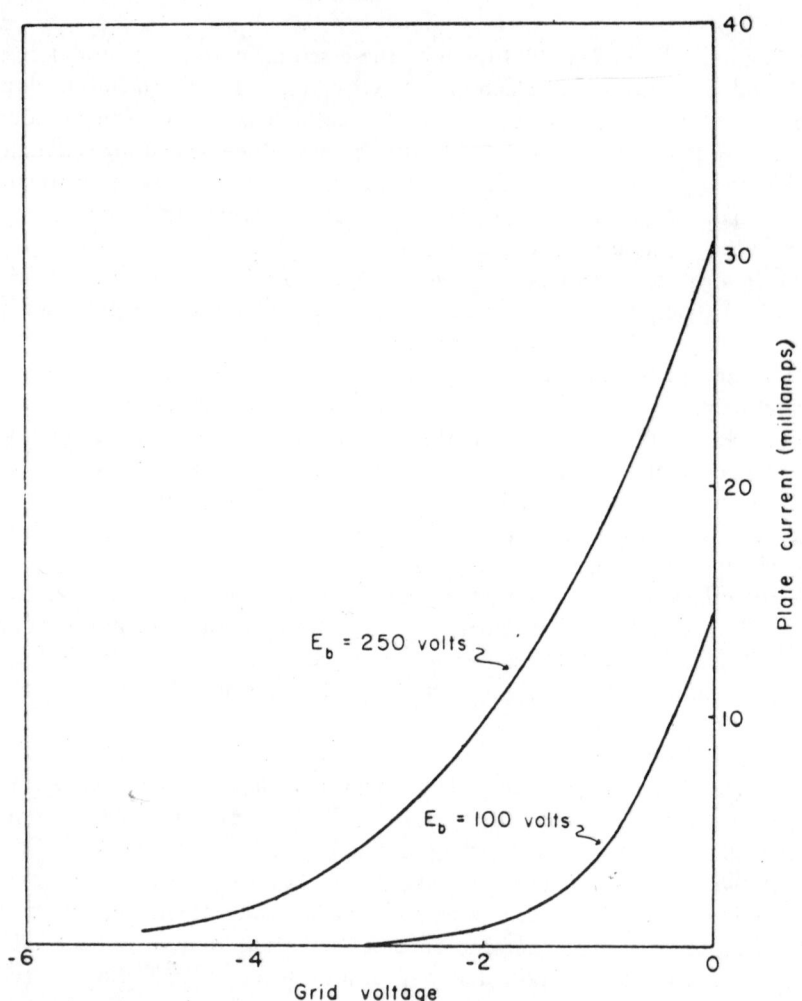

FIGURE 13.9: Variations of plate current with grid voltage for two values of plate voltage E_b applied to a 12AT7 tube.

for use with a recorder for automatically recording the potential or pH as a function of time, or as a function of volume of titrant added, etc.

To provide the advantage of greater accuracy usually built into a null-reading instrument, a potentiometer circuit with a standard reference cell (such as the Weston-cell) must be added to the amplifier circuit. The essential parts are shown in Fig. 13.11. The potentiometer circuit behaves in a similar way to the potentiometer discussed previously except that variations in potential at different settings are amplified by the vacuum-tube amplifier circuit before showing up in the galvanometer. With no voltage applied to the grid, the "zero adjust" resistor is adjusted to yield zero deflection of the meter. The standard cell (such as a Weston cell) is then inserted into the

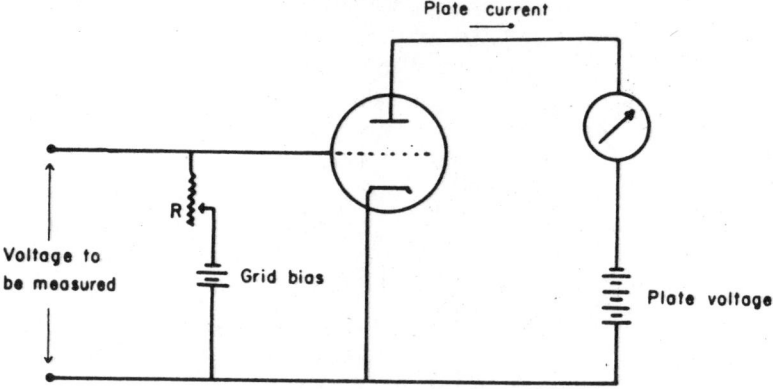

FIGURE 13.10: Essential parts of a vacuum tube voltmeter.

grid circuit using the "function switch" and, with the potentiometer scale set to the known potential of this standard cell, the "current adjust" resistor is varied to give zero deflection of the meter. Following this the unknown potential can be inserted into the grid circuit using the function switch; then, after adjusting the dial of the potentiometer so that the meter deflection is zero, the potential of the unknown can be read directly from the potentiometer dial.

In addition to the essential parts just mentioned for an amplifying potentiometer (either direct-reading or null-reading) there are two other controlled

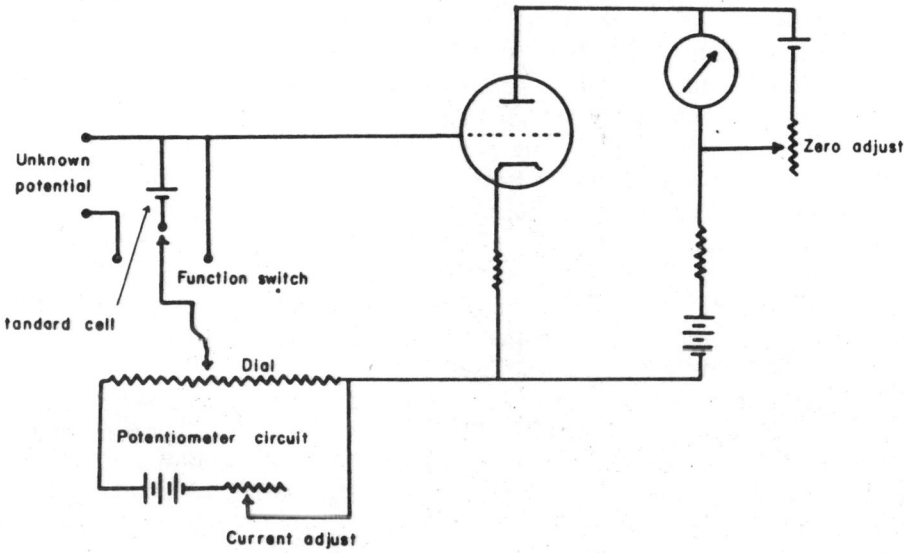

FIGURE 13.11: Essential parts of a null-reading pH meter.

resistances required if the instrument is to be used as a pH meter. These are the asymmetry potential control and the temperature control.

Because all glass electrodes are not alike but vary slightly in the potential set up on the two sides of the glass membrane there exists a variable described as the "asymmetry potential." To compensate for this an additional variable resistor is included in the circuit; this resistor is adjusted while the two electrodes (glass and reference) are immersed in a buffer solution of known pH. In practice, it is preferable to use a buffer solution whose pH is fairly close to the pH of the unknown solution; better still, the use of two buffer solutions will check the condition of the electrodes, expecially if the pH's of the two buffer solutions fall above and below the pH of the unknown solution.

The glass electrode is sensitive to temperature changes; consequently, most pH meters have an adjustment for the temperature to be used. This is usually a knob moving over a dial calibrated in degrees centigrade, with the knob being connected to a variable resistance. For some types of work it is convenient to use a heat-sensing probe immersed in the solution whose pH is to be measured; this probe inserts into the circuit a variable resistance appropriate to the temperature of the circuit. The common type of glass electrode may be used over a reasonable temperature range (5 to 80°C); a special type of glass is available in an electrode suitable for use up to 100°C.

Battery operation of pH meters is used for very expensive, very accurate, research-type meters and for the relatively inexpensive, much less sensitive, portable pH meters. Batteries provide a stability of operation highly desirable in an accurate pH meter for use in research. The use of batteries in portable pH meters for field use means they can be used readily, even where no electrical outlet is available. However, batteries must be changed periodically; this is bothersome and expensive, so most pH meters used routinely in laboratories are of the line-operated type.

Line-operated pH meters must incorporate a power supply to provide the high dc plate voltage as well as the low dc grid voltage and also the working dc voltage for the potentiometer circuit in the null-reading type of instrument. Line-operated instruments frequently employ a Zener diode in place of a Weston cell as a means of providing a definite known voltage for purposes of standardization. Although a line-operated pH meter is very handy to use, this type of instrument is subject to zero-drift and to variations in gain (or amplification of the electrical signal).

Changes in battery voltage or fluctuations in the line voltage and quality of the electronic tubes result in what is termed "zero drift." Among the means used to reduce zero drift are included the use of line-voltage stabilizers, the use of battery-operated transistors causing very little operating current, differential or balanced amplifiers, and a zero-correcting circuit (Beckman Zeromatic pH meter) which automatically applies a zero correction once each second without disturbing the reading.

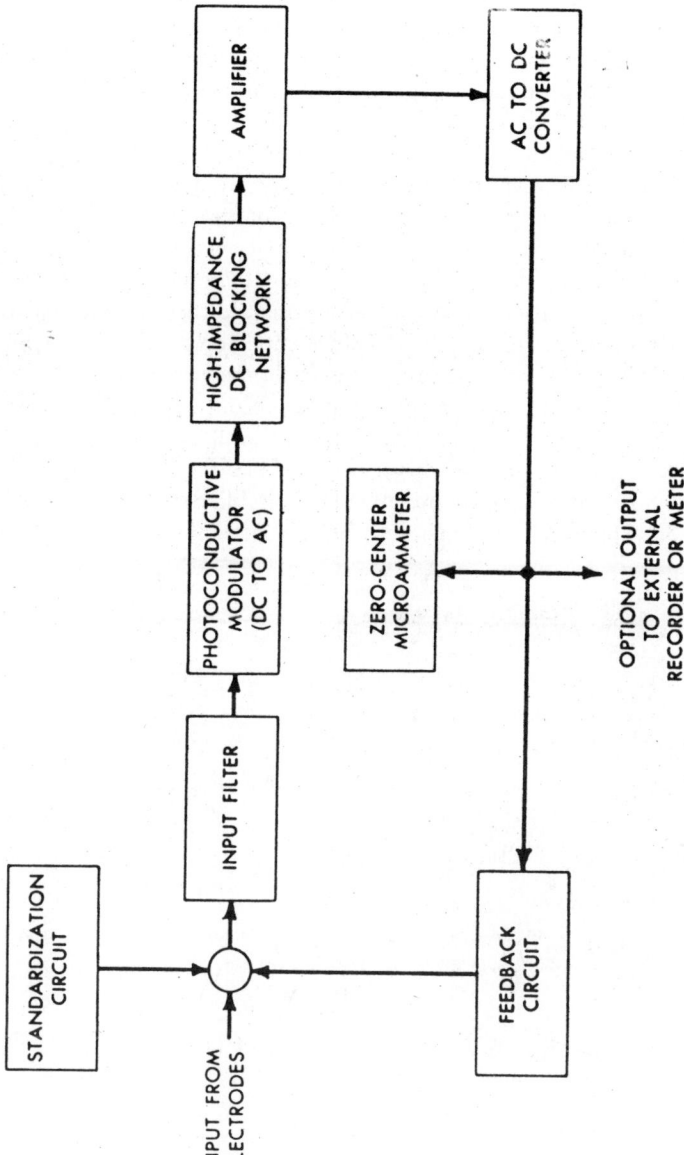

FIGURE 13.12: Simplified circuit of Leeds & Northrup stabilized pH indicator No 7401.

Variations in gain cause proportional errors. This variation in gain can be due to nonlinearity of the electron tube characteristics. The most useful means of reducing proportional errors is to apply negative feedback in the circuit. If a sufficient number of stages of amplification are used, i.e., amplification of the signal repeated several times, then sufficient feedback can be applied to produce independence of nonlinearity of the tube characteristics.

Many of the commercially available pH meters use amplifiers of dc signal as described here. However, a number of currently available commercial instruments use an ac amplifier. To make use of this, the dc signal from the electrodes must first be converted to an ac signal by a chopper or vibrator, then the ac signal is amplified in a number of stages and then converted back to a dc signal. The final dc signal is used to operate the meter and also to provide the feedback. This type of instrument shows excellent stability. An example of such an instrument is the Radiometer titrator no. TTT1, which is sufficiently stable that it can be used to control pH over a period of days, and the Leeds & Northrup No. 7401, a simplified circuit of which is shown in Fig. 13.12.

Illustrations of a number of commercially available pH meters are shown in Fig. 13.13. These are line-operated instruments for routine laboratory

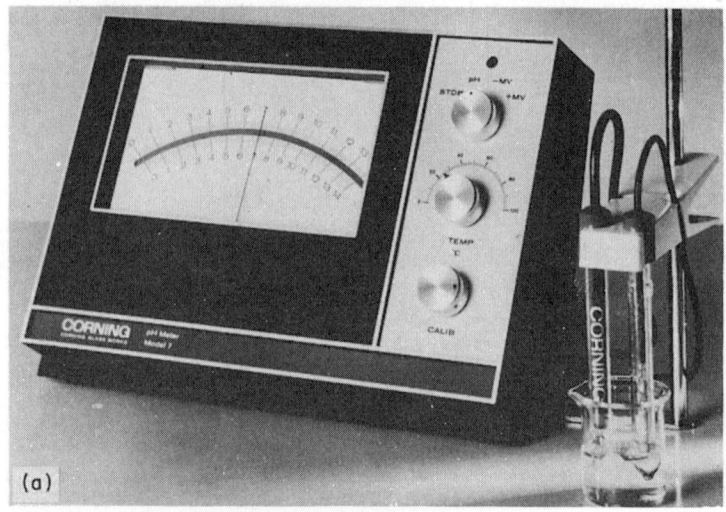

FIGURE 13.13: Line-operated laboratory pH meters: (a) Corning model 7 pH meter (photo courtesy of Corning Glass Works, Corning, N.Y.); (b) Fisher Accumet pH meter model 210 (photo courtesy of Fisher Scientific Company, Pittsburgh, Pa.); (c) Beckman Zeromatic pH meter (photo courtesy of Beckman Instruments, Inc., Fullerton, Calif.); (d) Radiometer model 28 (photo courtesy of Radiometer A/S., Copenhagen, Denmark); and (e) Leeds & Northrup pH meter No. 7401 (photo courtesy of Leeds & Northrup Company).

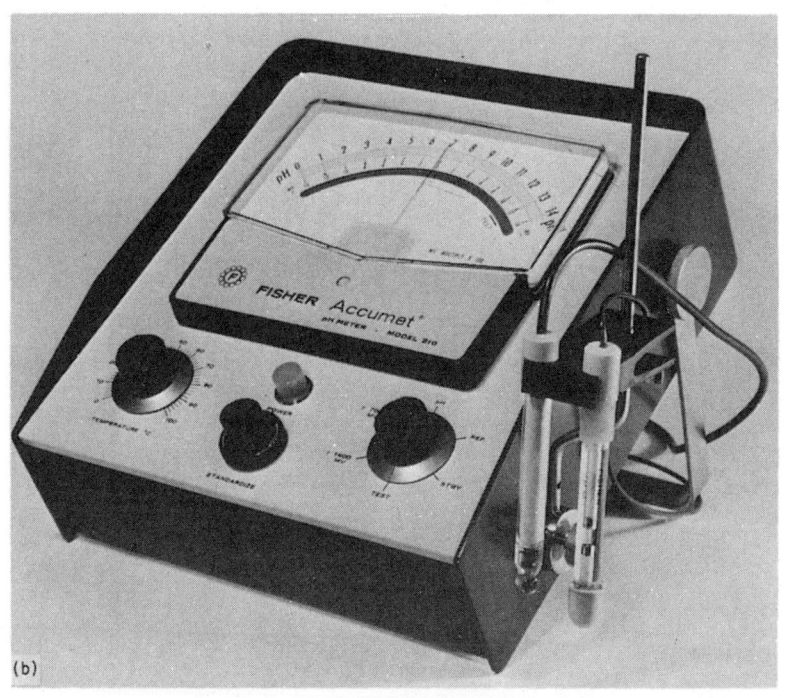

(b)

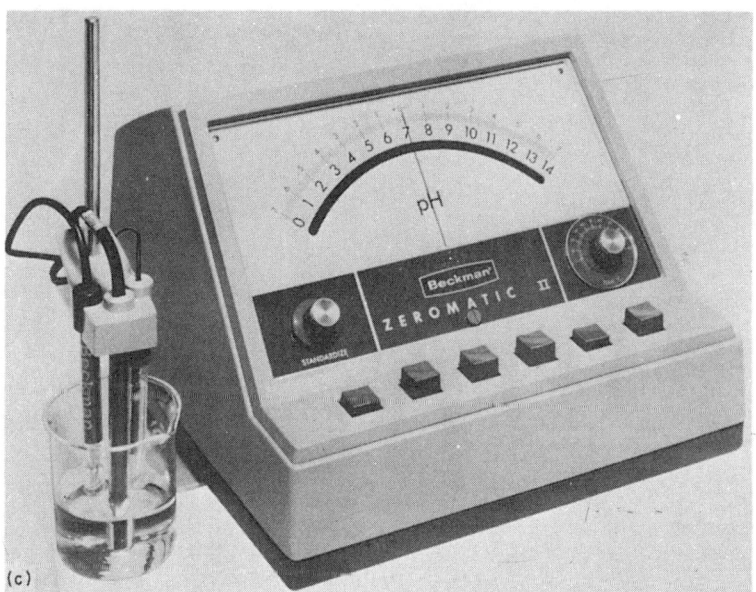

(c)

FIGURE 13.13 (continued)

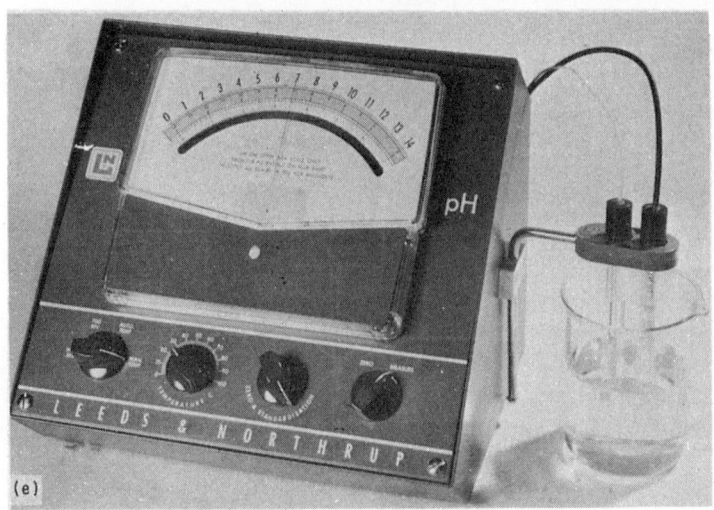

FIGURE 13.13 (continued)

use and include instruments manufactured by Beckman (Zeromatic), Fisher (Accumet model 210), Corning (model 7), Radiometer (model 28), and Leeds & Northrup (model 7410).

Research-type pH meters are illustrated in Fig. 13.14, including Beckman and Radiometer model 4.

A battery-operated pH meter is illustrated in Fig. 13.15 (Beckman model N-2).

D. RELATION OF pH TO POTENTIAL

Most practical pH determinations are based on the measurement of the electrical potential developed in a cell of the type

$$H^+\text{-sensitive electrode} \parallel \text{reference electrode}$$

where the H^+-sensitive electrode may be a hydrogen electrode or more commonly a glass electrode (or a quinhydrone or antimony electrode) and the reference electrode is usually a saturated calomel electrode.

It has been stated that there are a number of disadvantages to the routine use of a hydrogen electrode so that it is commonly replaced by the other types of H^+-sensitive electrodes mentioned; however these all have somewhat imperfect response to activity of hydrogen ion. If this response were not imperfect, we would expect the measured potential to be

$$E = E_{\text{reference}} + E_{\text{indicator}} + E_{\text{junction}} \qquad (13.29)$$

and using the same reference electrode and taking steps to render the junction potential negligible, we could expect this equation to simplify to:

$$E = E_{\text{reference}} + 0.0591 \log a_{H^+} \qquad (13.40)$$

and since $pH = -\log a_{H^+}$

$$E = E_{\text{reference}} - 0.0591 \, pH \qquad (13.41)$$

From Eq. (13.41), the

$$pH = \frac{E_{\text{reference}} - E_{\text{measured}}}{0.0591} \qquad (13.42)$$

However, these electrodes are not responsive uniformly to the activity of hydrogen ion. Consequently, it has become practice to relate pH to an accepted pH standard rather than to the absolute values of potential (measured and reference) suggested by Eq. (13.42). The National Bureau of Standards has described specifications for preparing solutions of various pH. In practice a pH meter is standardized with one of these solutions of known pH, preferably with the pH of the known solution close to the pH of the unknown solution. Having standardized a pH meter against a known pH standard, the meter may be used with a wide range of unknowns, but it should be borne in mind that there may be considerable error in the measurements made at some distance from the pH at which the meter was standardized.

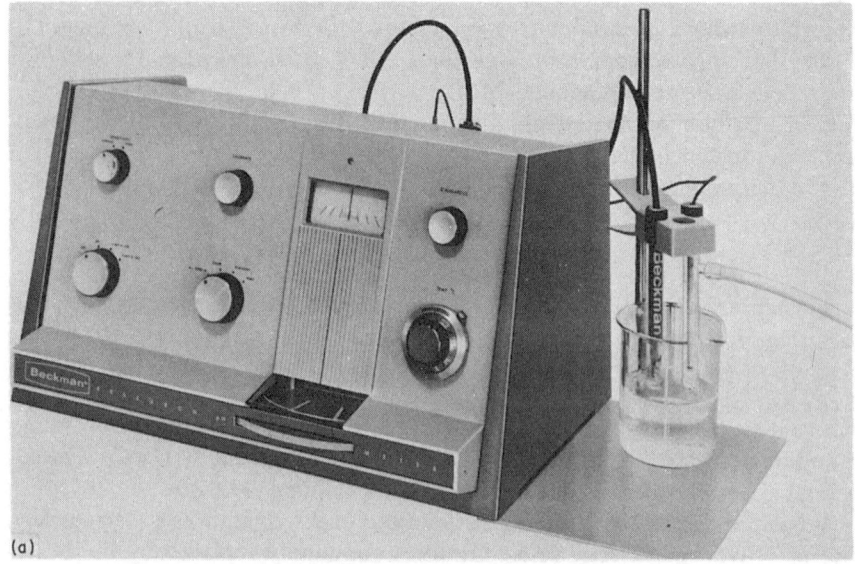

FIGURE 13.14: Research-type pH meters: (a) Beckman research pH meter (photo courtesy of Beckman Instruments, Inc., Fullerton, Calif.); (b) Radiometer model 4 (photo
courtesy of Radiometer A/S. Copenhagen, Denmark).

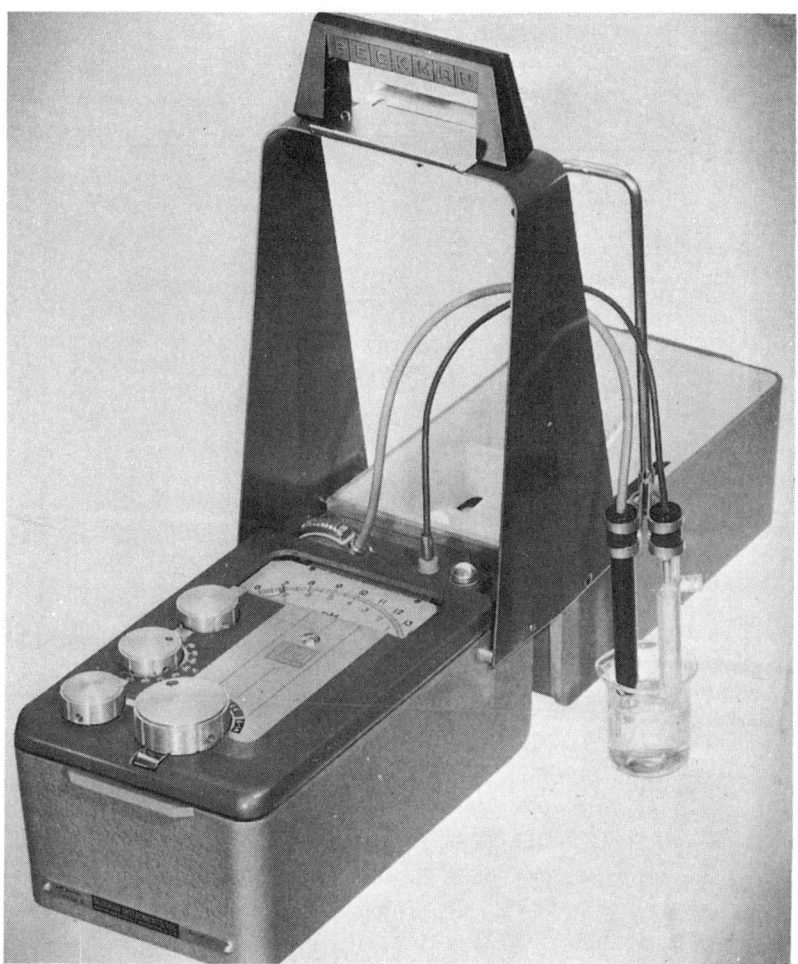

FIGURE 13.15: Battery-operated, portable pH meter, Beckman model N-2. Photo courtesy of Beckman Instruments, Inc., Fullerton, Calif.

The Fig. 13.16 shows an example of a titration curve where both pH value and potential in a particular set of circumstances is plotted against volume of titrant.

E. TITRATION CURVES

A potentiometric titration is a quantitative determination where the end point is derived from the change in potential. If the ingredient being determined produces an electrical potential with a suitable indicating electrode, the course of the titration may be followed by noting the fall in potential as the ingredient is used up in the titration reaction following the addition

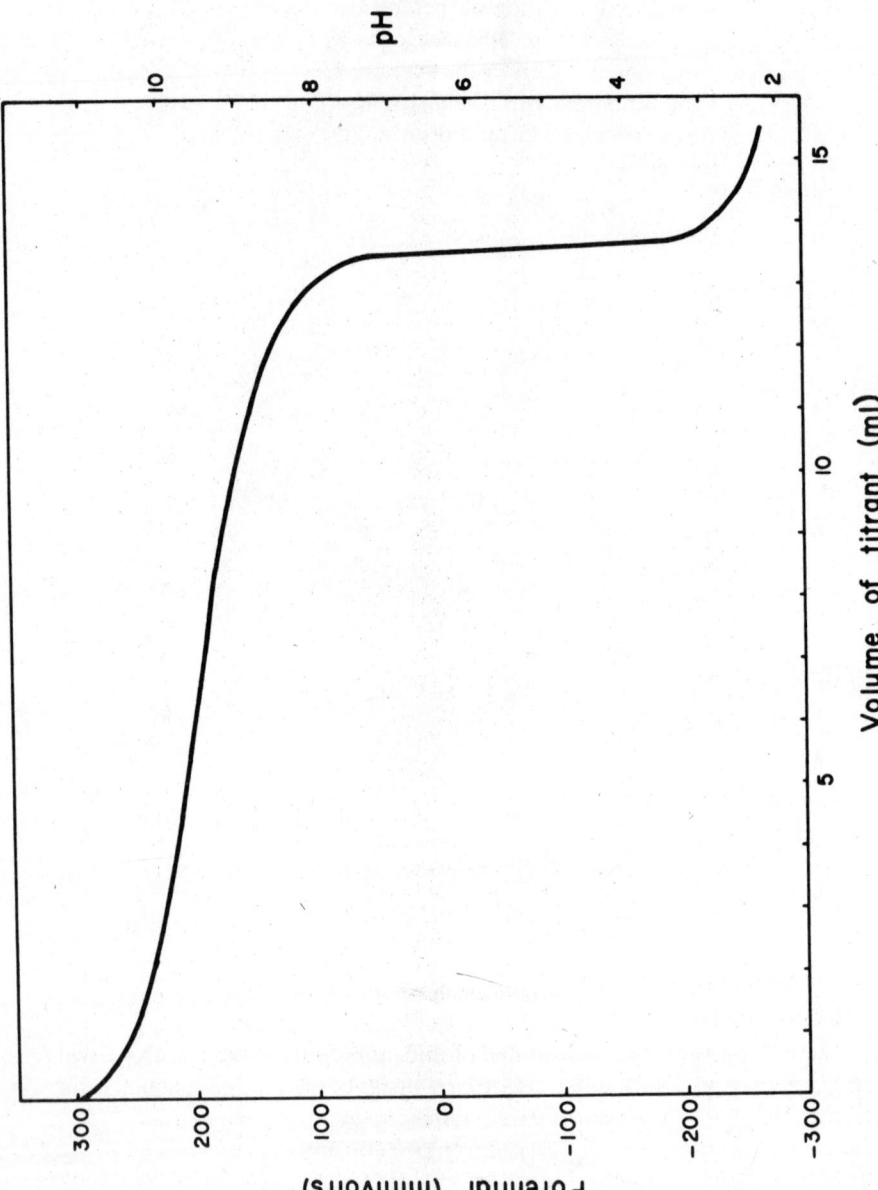

FIGURE 13.16: Titration curve showing both pH and potential plotted against volume.

of increments of titrant. It usually happens that initially there is a relatively small fall of potential with addition of increment of titrant; however, as the titration proceeds there is a somewhat larger fall in potential with addition of the same size increment; then in the region of the endpoint there will be a much larger fall in potential with a very small increment of titrant. Following the end point of the reaction the fall in potential decreases again for an

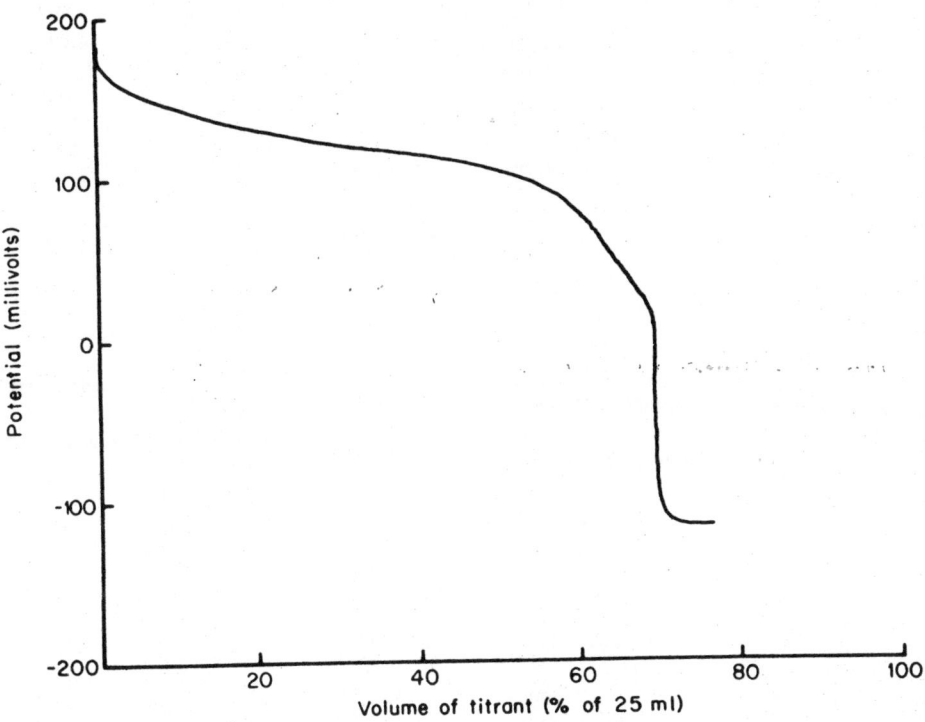

FIGURE 13.17: Titration curve of determination of sulfa drug by sodium nitrite solution.

increment of titrant. If the potential of the sample being determined is plotted against the volume of titrant added to produce a graph, such a graph is termed a "titration curve." An example of a titration curve is shown in Fig. 13.17, where the potential of a solution of a sulfa drug is plotted against the number of milliliters of $M/10$ solution of sodium nitrite.

In this particular titration curve, it is to be noted that there is a sharp fall in potential at the end point or the "equivalence point," as it is also called. In a titration of this kind, we are interested in the number of milliliters of titrant required rather than in what the actual value of the potential is. That is, we are not particularly interested in the numerical value of the potential at

the beginning or at the end of the titration but only in the number of milliliters of titrant required to reach the end point, where the potential changes
sharply from a high value to a low value. Consequently, when performing
a titration of this kind it is not necessary to spend time in determining the
exact potential; instead, the potential may be set arbitrarily at a high
value and the titration continued until the potential changes sharply to a
low value.

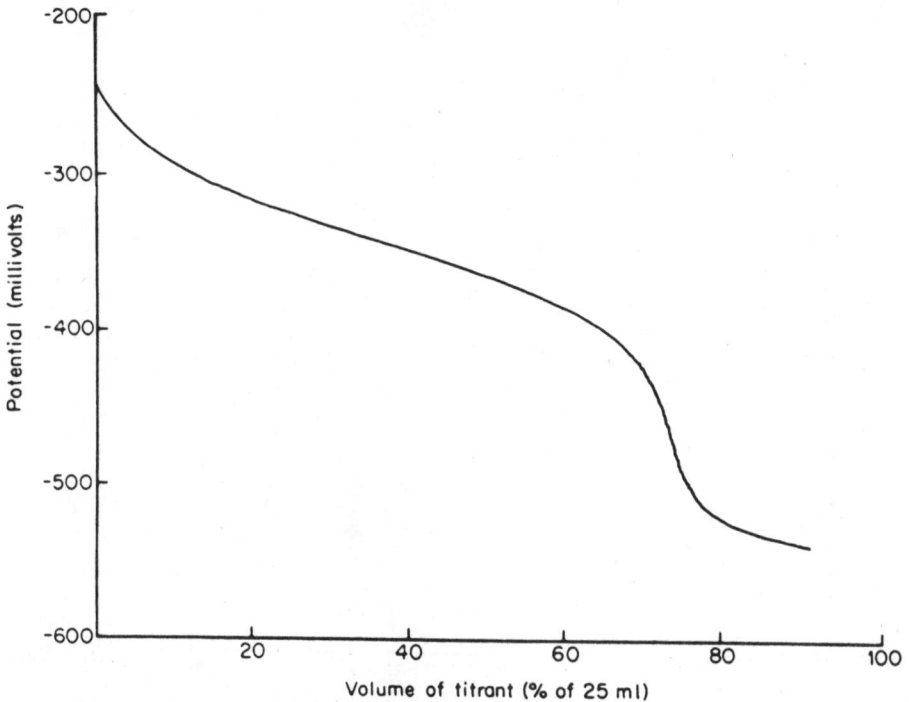

FIGURE 13.18: Titration curve of determination of ferrous ion by dichromate solution.

Another example of a titration curve is shown in Fig. 13.18. This curve
shows the fall in potential which occurs when ferrous ion is titrated with
dichromate solution.

In some titrations the potential may rise sharply, instead of falling, at
the end point. This would be so where the substance being determined
would produce only a low potential with a particular indicating electrode,
but where the titrant used would produce a higher potential with the electrode
system. At the beginning of the titration there would be only a low potential
value measured. Only upon reaching the end point when the substance was
no longer present to react with the titrant would the titrant be present and

so produce the potential. Thus, we would have a sharp rise in potential at the end point of such a titration.

In titrations involving acids and bases, it is customary to prepare titration curves by plotting pH against milliliters of titrant added. Examples of such

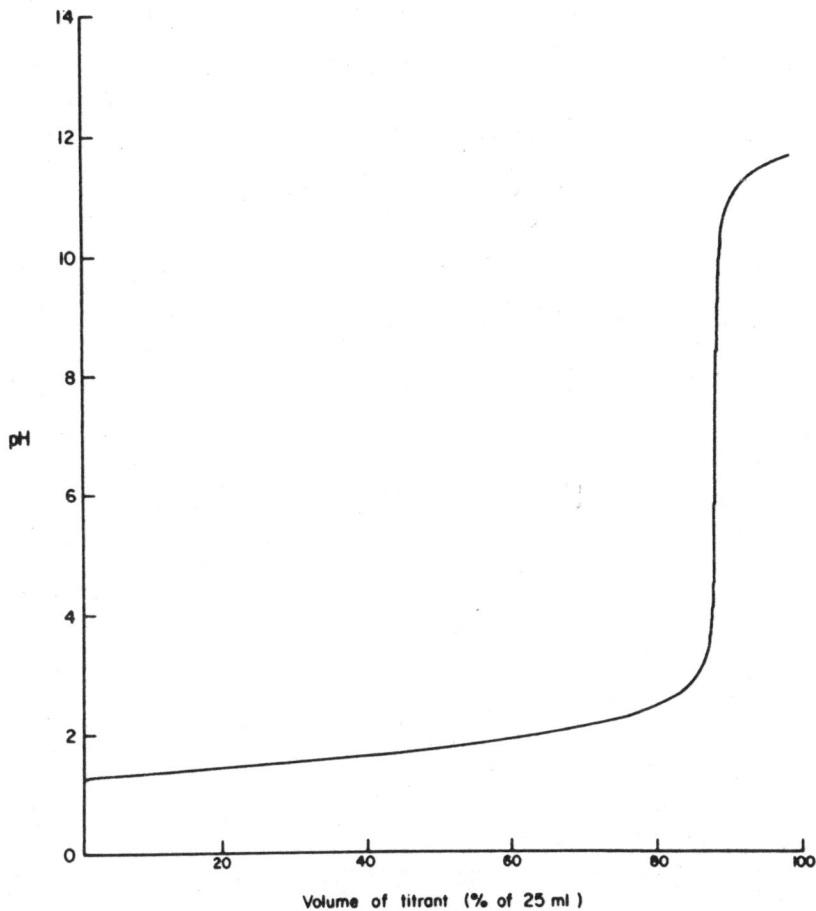

FIGURE 13.19: Titration curve of determination of hydrochloric acid by $N/10$ sodium hydroxide.

titration curves are shown in Fig. 13.19 to 13.22. Figure 13.19 illustrates a titration curve for a strong acid (HCl) titrated by a strong base (NaOH); Fig. 13.20 depicts a titration curve for a strong base (NaOH) titrated by a strong acid (HCl). In both these situations there is a sharp change in pH over several pH divisions at the end point. Figure 13.21 shows a titration curve for a weak acid (acetic acid) titrated by a strong base (NaOH); Fig. 13.22 illustrates a titration curve for a weak base (ammonium hydroxide) titrated with a strong acid (HCl). In both these situations there is also a sharp change in pH, but the change is over only a few pH divisions. (See also Volume 1, pp. 120–121, "Selection of an Indicator.") An example of a titration curve for a polybasic acid (carbonic acid) is shown in Fig. 13.23; the two sharp changes in the titration curve correspond to the two stages

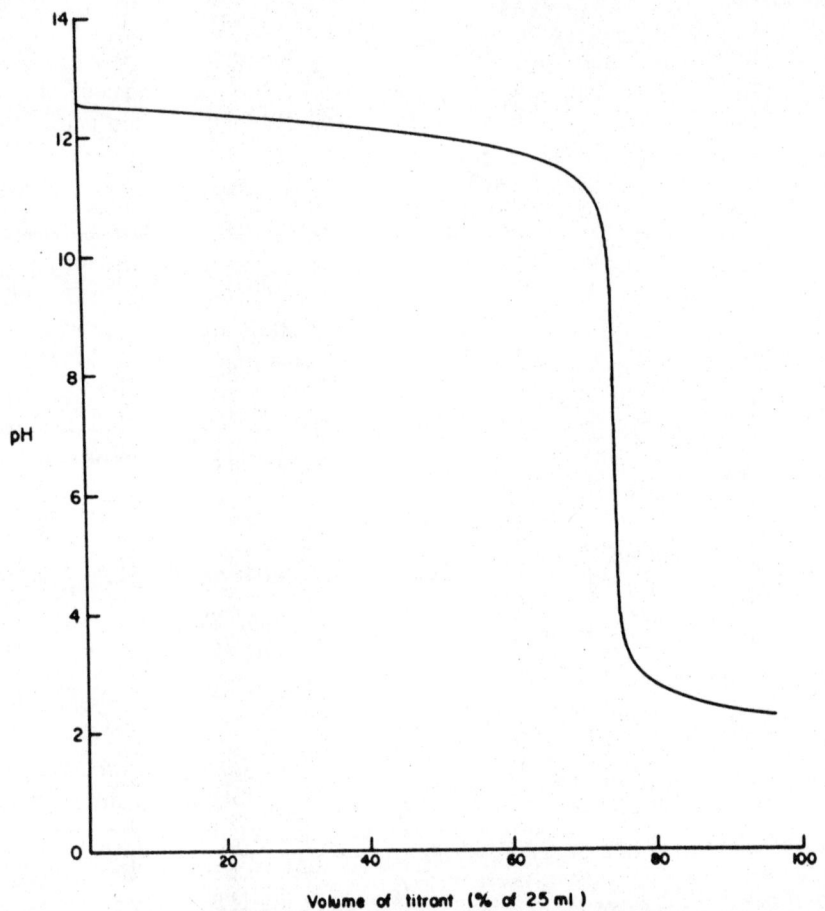

FIGURE 13.20: Titration curve of determination of sodium hydroxide solution by $N/20$ hydrochloric acid.

of ionization of the relatively poorly ionized acid. The less sharp break corresponds to the neutralization reaction:

$$H^+ + CO_3^{2-} \rightarrow HCO_3^- \qquad (13.43)$$

The second break in the curve corresponds to the neutralization reaction

$$H^+ + HCO_3^- \rightarrow H_2CO_3 \rightarrow H_2O + CO_2\uparrow \qquad (13.44)$$

F. DETERMINATION OF END POINTS

It has been indicated in the preceding section on titration curves that the end point in a potentiometric titration is the sharp rise or fall which occurs in the potential produced at an indicating electrode immersed in the titration

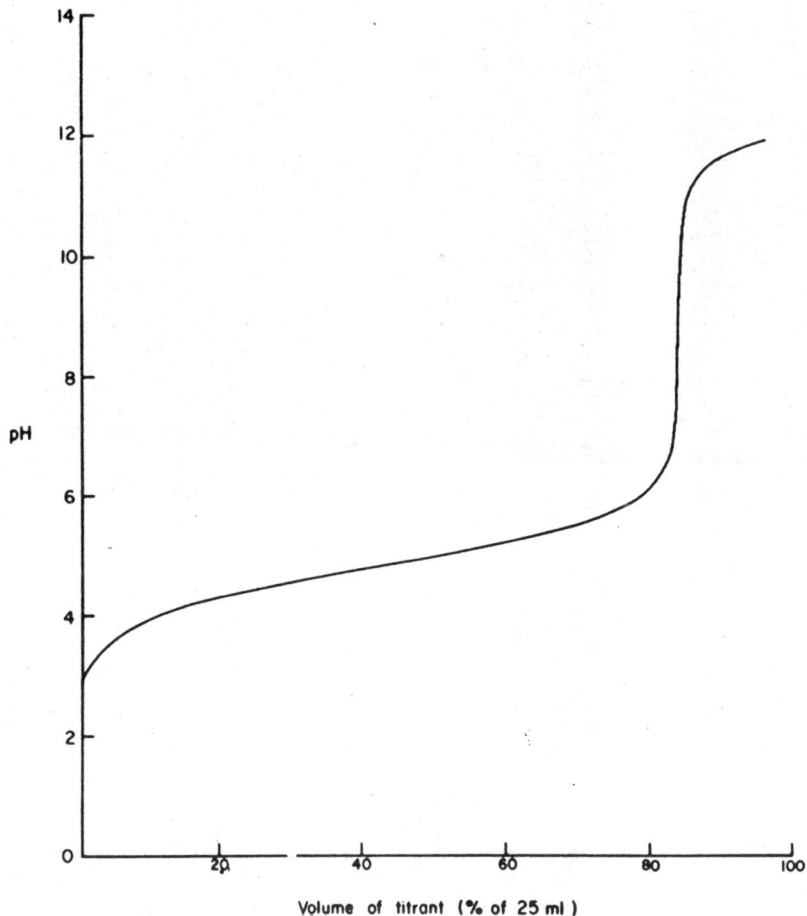

Volume of titrant (% of 25 ml)

FIGURE 13.21: Titration curve of determination of acetic acid solution by $N/10$ sodium hydroxide.

vessel. If the sharp rise or fall is sufficiently steep, it is relatively easy to select the midpoint in the steep part of the titration curve and call this the end point (or equivalence point). However, if the rise or fall through the equivalence point is more gradual it may be difficult to select the end point with any degree of certainty. In such a circumstance it is useful to resort to a first-derivative curve or even to a second-derivative curve to fix the end point. To obtain a first-derivative curve, change in potential per small increment of titrant added is plotted against volume of titrant added. Using symbols, $\Delta E/\Delta V$ is plotted against V, where E is potential and ΔE is the change in potential which occurs following addition of an increment of

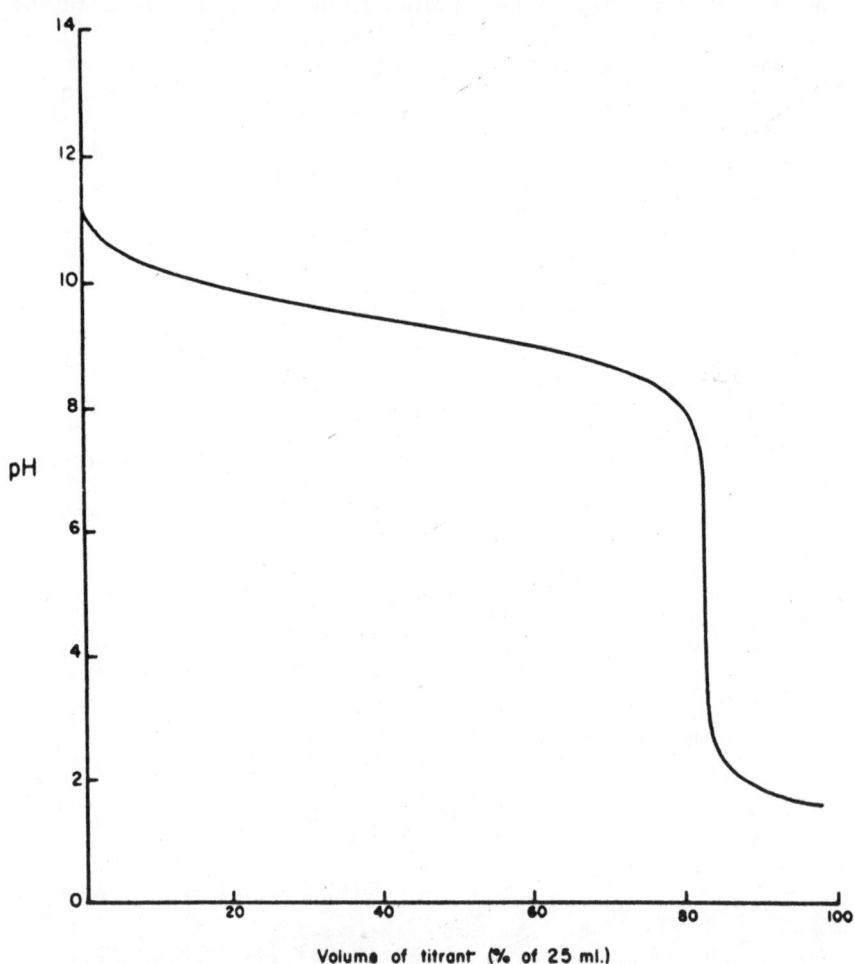

FIGURE 13.22: Titration curve of determination of ammonia solution by $N/2$ hydrochloric acid.

titrant; V is volume of titrant and ΔV is the increment of titrant. Mathematically speaking, dE/dV is plotted vs. V or, the first derivative of potential with respect to volume is plotted as the ordinate against the volume as the abscissa. The titration curve shown in Fig. 13.24 was prepared from the data shown in the first two columns of Table 13.4. In the normal titration curve Fig. 13.24, the potential is falling most rapidly as it passes the end point.

That is, the rate of change in potential is greatest at the end point; so, if we plot the rate of change of potential against volume, we should have a curve with a maximum corresponding to the volume at the end point. Such is the situation as illustrated in Fig. 13.25. This plotting of the first derivative

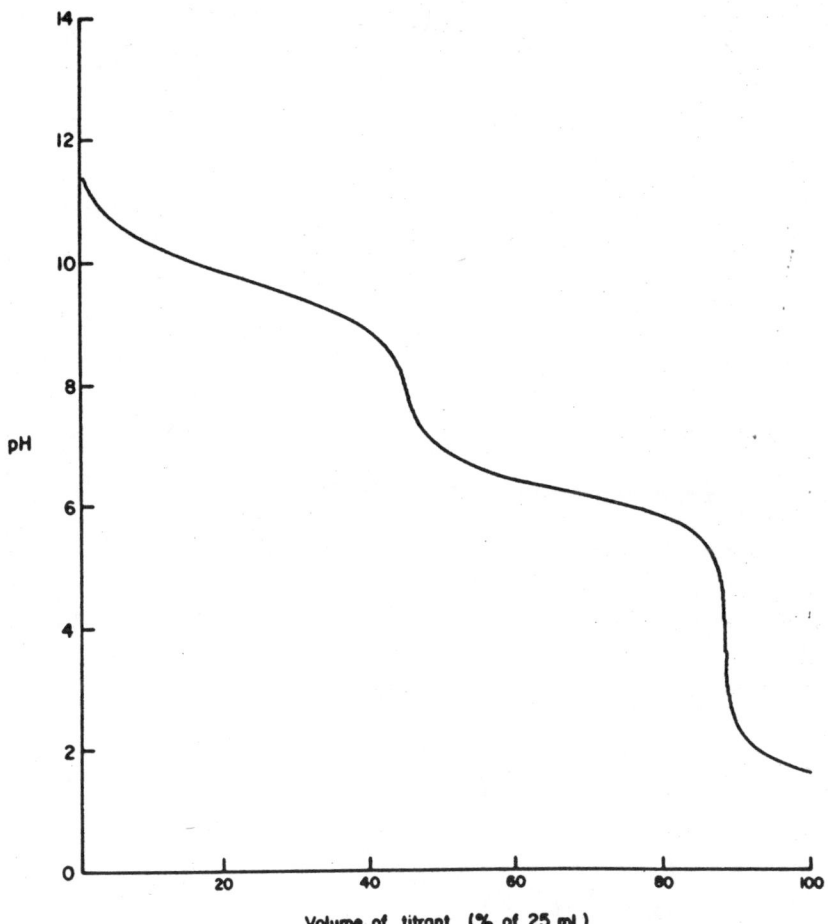

FIGURE 13.23: Titration curve of sodium carbonate solution with $N/2$ hydrochloric acid.

can be done manually as Fig. 13.25 was prepared or it can be traced out automatically on certain automatic titrators. An example of such a titrator is the Metrohm potentiograph, which can produce either a normal titration curve or a first-derivative curve as shown in Fig. 13.26.

It is sometimes of value in determining the end point to prepare the second-derivative curve of the titration. The first-derivative curve is a peak that rises more or less sharply to a maximum and then falls off. The tangent of this first-derivative curve (i.e., the second derivative of the titration curve)

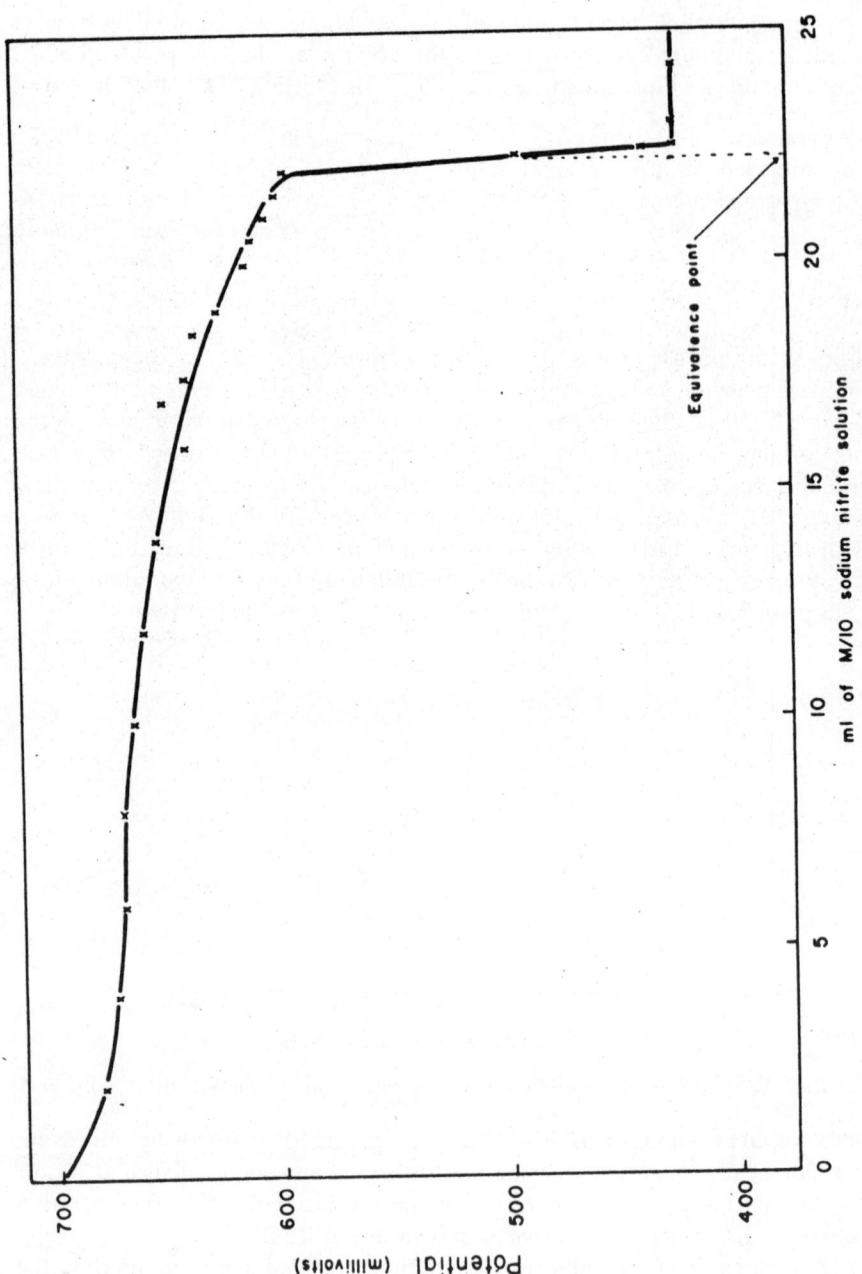

TABLE 13.4: Potentiometric Titration of Sulfanilamide

Volume, milliliters	Potential, mV	ΔV	ΔE	$\dfrac{\Delta E}{\Delta V}$	V'	$\dfrac{\Delta^2 E}{\Delta V^2}$	V''
0.00	700	$(V_2 - V_1)$					
2.10	680						
4.07	674		$(E_2 - E_1)$				
6.06	670						
8.04	670			$\dfrac{(E_2 - E_1)}{(V_2 - V_1)}$			
10.05	665						
12.01	661				$\dfrac{(V_1 + V_2)}{2}$		
14.03	653						
16.03	640					$\dfrac{\Delta(\Delta E/\Delta V)}{\Delta V}$	
17.05	650						
17.50	640						
18.50	636						$\dfrac{(V_1' + V_2')}{2}$
19.07	625						
20.02	613	0.95	12	12.7	19.54		
20.53	610	0.5⊢	3.	5.9	20.27	−6.7	19.90
21.08	604	0.55	6	10.9	20.80	−5.0	20.53
21.50	600	0.42	4	9.5	21.29	−2.4	21.04
22.03	596	0.53	4	7.5	21.76	−1.0	21.52
22.31	493	0.28	103	368	22.17	360	21.96
22.41	439	0.10	54	540	22.36	172	22.26
22.49	425	0.09	14	155	22.45	−385	22.40
23.05	425	0.56	0	0	22.77	−155	22.61
24.21	425	1.16	0	0	23.63	0	23.20
25.02	425	0.81	0	0	24.61	0	24.12

changes from sloping up to the right (positive value) through a horizontal position (slope = 0) to sloping down to the right (negative value) as the volume increases. That is, the second derivative of the titration curve will have the general shape of the curve shown in Fig. 13.27 and the end point will be the volume where the curve passes through zero in changing from a positive value to a negative value.

The second-derivative curve can be prepared manually as was Fig. 13.27, however there are available commercially a number of titrators which record automatically a second-derivative from which the end point can be read quite conveniently; an example of such an instrument is the Sargent-Malmstadt titrator.

G. DEAD-STOP TITRATIONS

One name applied to a type of titration which goes by a variety of other names is "dead-stop." A small potential is applied across two platinum electrodes in series with a variable resistance and a microammeter. If the platinum electrodes are immersed in certain solutions a small constant current will flow. As titration of the solution is carried on, the current

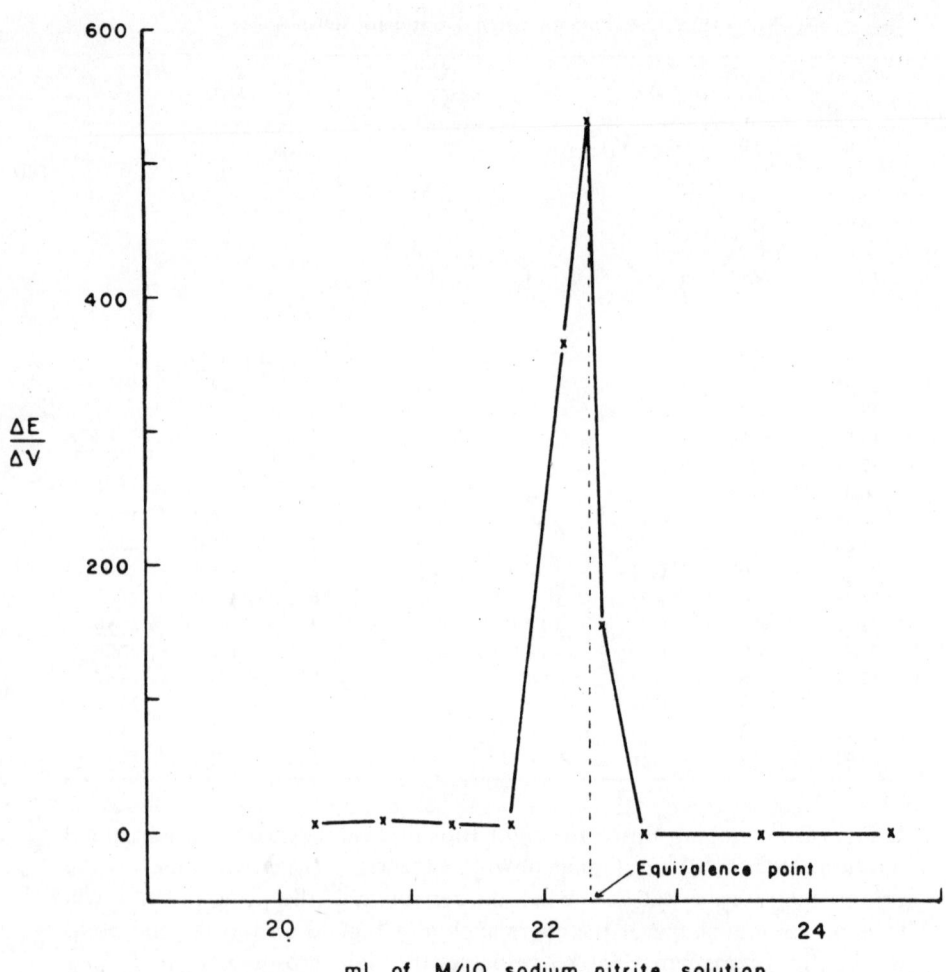

FIGURE 13.25: First-derivative curve, manually prepared.

remains constant until the end point is reached, at which time there is a sudden increase of current. ˙Dead-stop titrations are used by the British Pharmacopoeia for the determination of sulfa drugs and in the Karl Fischer determination of water. Further discussion of dead-stop titrations is given in Chapter 9 of Volume 1 by J. E. Sinsheimer.

13.5 APPLICATIONS OF POTENTIOMETRIC TITRATIONS

A. USES

Any titration can be performed by potentiometric means provided that the change in potential can be followed by a suitable indicating electrode when used in conjunction with an appropriate reference electrode. This includes the usual oxidation-reduction titrations and, in addition, precipitation reactions, and titrations involving complex formation; the traditional methods of analysis for these groups have been dealt with in Chapter 4 of Volume 1 by J. A. Zapotocky.

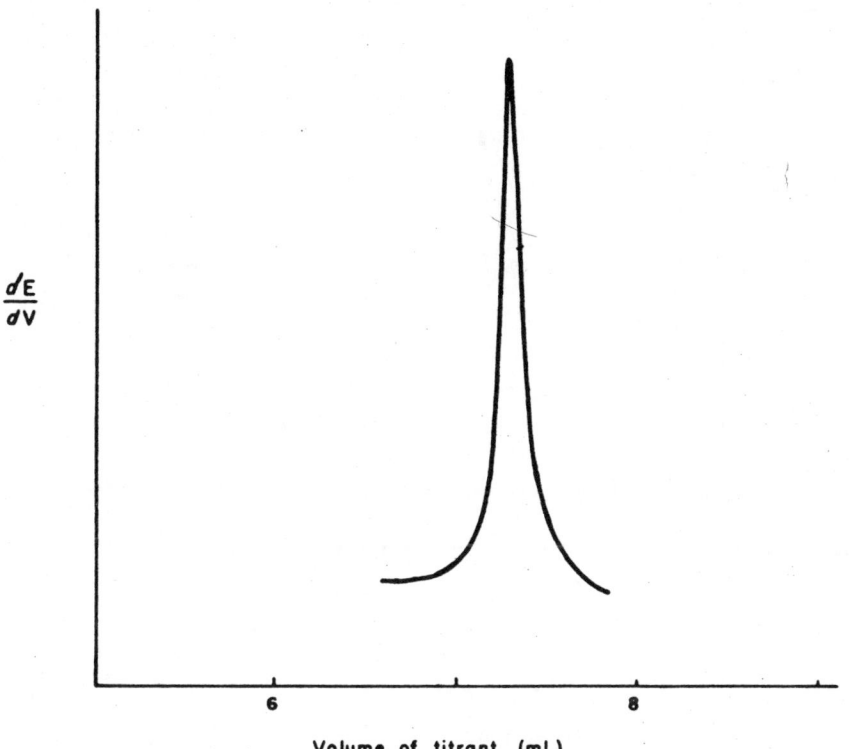

FIGURE 13.26: First-derivative curve, automatically prepared.

Because of the relation between electrical potential and pH, titrations involving acid and base lend themselves to determination of the end point by potentiometric means. The traditional methods of acidimetry and alkalimetry have been dealt with in Chapters 3 and 5 of Volume 1 by J. W. Steele and M. I. Blake; acid-base titrations in nonaqueous media have been dealt with in Chapter 6 of Volume 1 by L. G. Chatten.

B. SELECTION OF ELECTRODES

It is necessary that one electrode of the pair required maintain a potential which does not vary during the titration so that it may serve as the reference electrode. The other electrode, to be the indicating electrode, should develop a potential dependent on the concentration of the ingredient to be measured or the titrant to be added. A number of commonly used electrodes

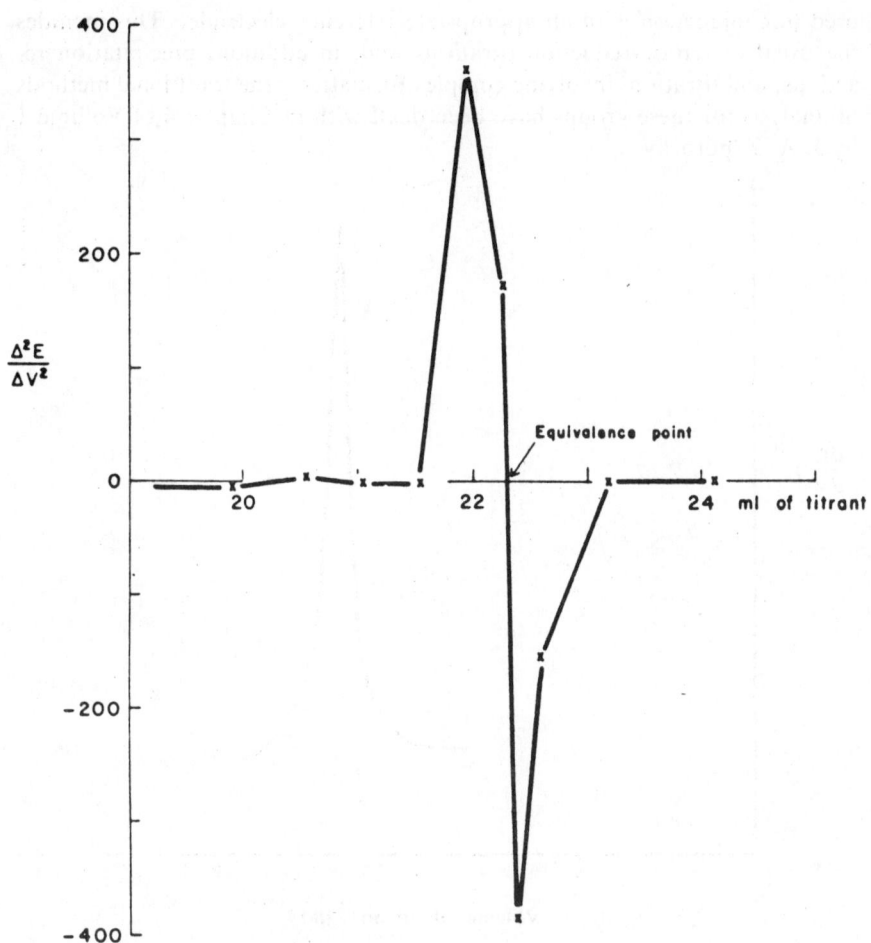

FIGURE 13.27: Second-derivative curve.

are listed in Table 13.5 together with the type of titration in which they find common use.

Specially constructed microelectrodes are available commercially for use in microdeterminations where the volumes are quite restricted.

C. ADVANTAGES IN DETERMINING END POINT

The potentiometric method of titration offers a number of advantages in the determination of the end point:

1. Can be used in colored or turbid solutions where ordinarily the use of a colored visual indicator would be useless.

2. Can be used where there is no satisfactory internal indicator to mark the end point.

3. Can be used where the change in potential (or pH) is rather small at the end point—too small to cause the complete change of color of the indicator. Under these circumstances it is customary to titrate to a particular potential (or pH).

TABLE 13.5: Common Electrode Applications

| | Commonly used electrodes | |
Titration mixture	Indicator	Reference
Acid-base, aqueous	Glass	Calomel
Acid-base	Quinhydrone	Calomel
Acid-base	Antimony	Calomel
Acid-base, nonaqueous	Glass	Calomel
Oxidation-reduction	Platinum	Calomel
Iodimetric	Platinum	Tungsten
Precipitation, silver	Silver	Mercurous sulfate
Precipitation, halide	Silver	Mercurous sulfate

4. A magnetic valve may be fitted to a potentiometer so that when a certain potential or pH is reached the supply of titrant from a burette may be shut off. This permits automatic titrations to be performed. An example of such an instrument is the Fisher titrimeter, model 36, illustrated in Fig. 13.28.

5. Can be used readily with recorders. This permits the preparation of titration curves very easily. Most pH meters manufactured at present have a jack to which a potentiometric recorder can be attached. With some titration apparatus the recorder is an integral part of the instrument. An example of this is the Radiometer titrator with recorder and automatic burette. An illustration of this instrument is shown in Fig. 13.29. The originals of the titration curves in Figs. 13.17–13.23 were prepared on an instrument of this kind. Another example of a recording titrimeter is the Metrohm potentiograph, illustrated in Fig. 13.30.

6. If the recording instrument is fitted with a a differentiating circuit (this usually involves a relatively simple circuit including one electronic tube), it can record the first-derivative curve, rendering the end point quite easy to ascertain. The Metrohm potentiograph shown in Fig. 13.30 is such an instrument.

7. If the instrument is fitted with two differentiating circuits it can make

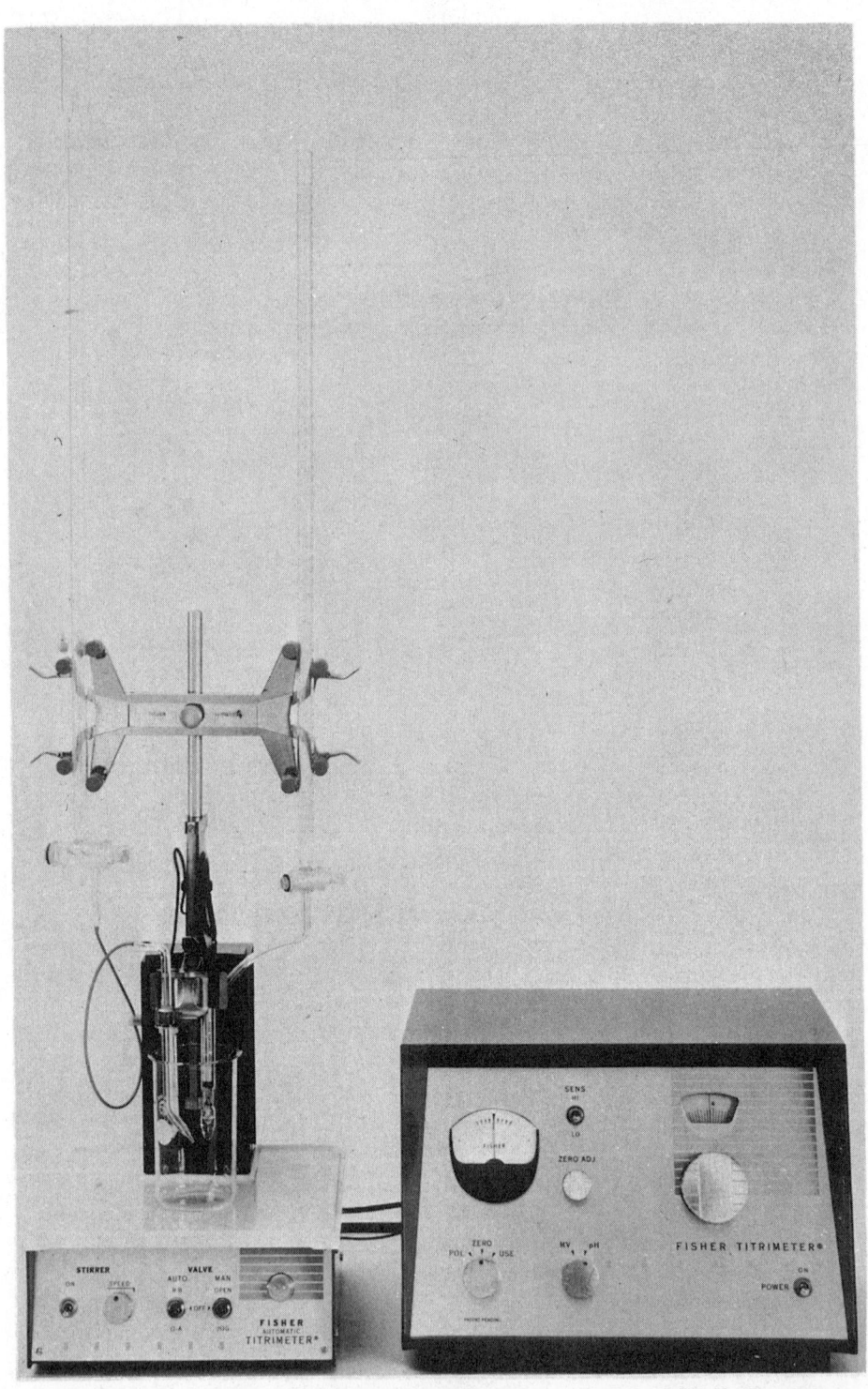

FIGURE 13.28: Fisher titrimeter, model 36. (Photo courtesy of Fisher Scientific Company, Pittsburgh, Pa.)

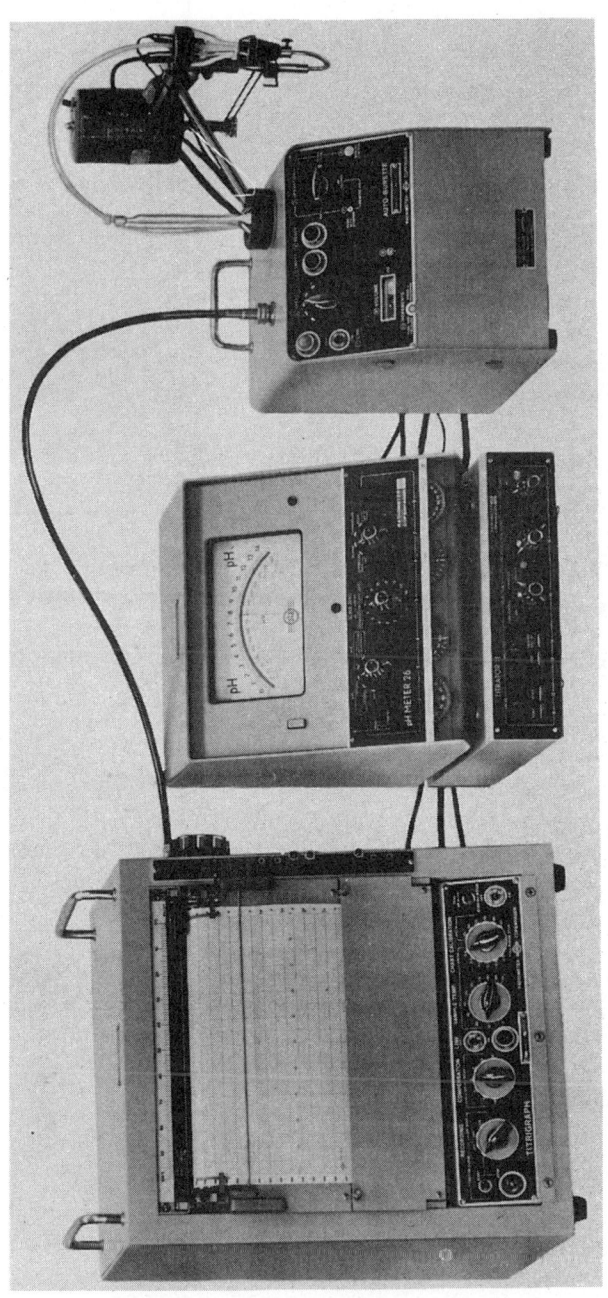

FIGURE 13.29: Radiometer titrator, recorder, and automatic burette. Photo courtesy of Radiometer A/S, Copenhagen, Denmark.

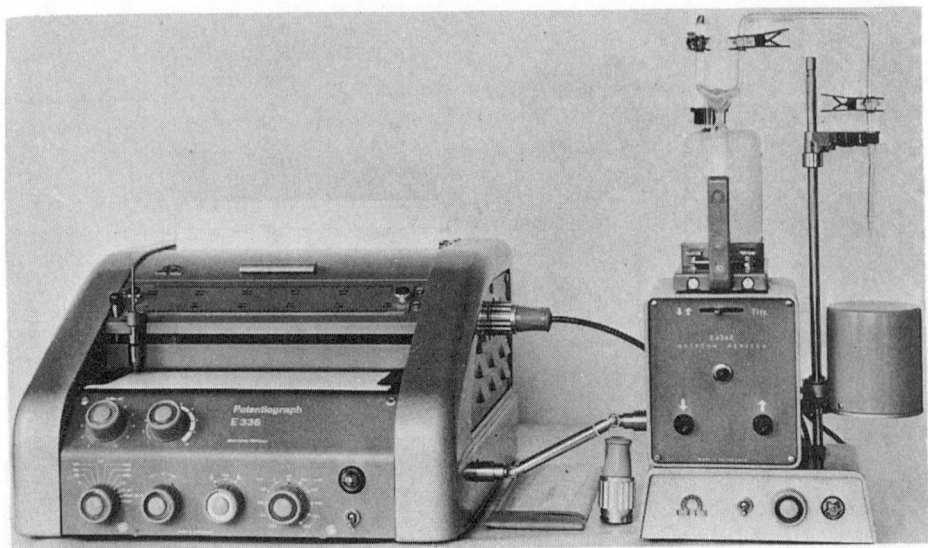

FIGURE 13.30: Metrohm potentiograph model E 336 A. Photo courtesy of Metrohm Ltd., Herisau, Switzerland.

use of the second-derivative curve to ascertain the end point. Such an instrument is the Sargent-Malmstadt titrator shown in Fig. 13.31.

13.6 ANALYTICAL PROCEDURES

A. STANDARDIZATION OF N/2 HYDROCHLORIC ACID AGAINST ANHYDROUS SODIUM CARBONATE, REAGENT GRADE

Apparatus:
Beckman Zeromatic pH meter or other suitable pH meter which can be fitted with a magnetic stirrer under a 250-ml beaker as the titration vessel
Glass electrode for the pH meter
Calomel electrode for the pH meter
Magnetic stirrer
Burette, glass or Teflon stopcock, 50 ml

Reagents:
Sodium carbonate, reagent grade, anhydrous, previously dried at 150°C for 3 hr, then cooled in a desiccator
Hydrochloric acid solution, approximately N/2
Buffer solution of known pH (such as pH 7.0)

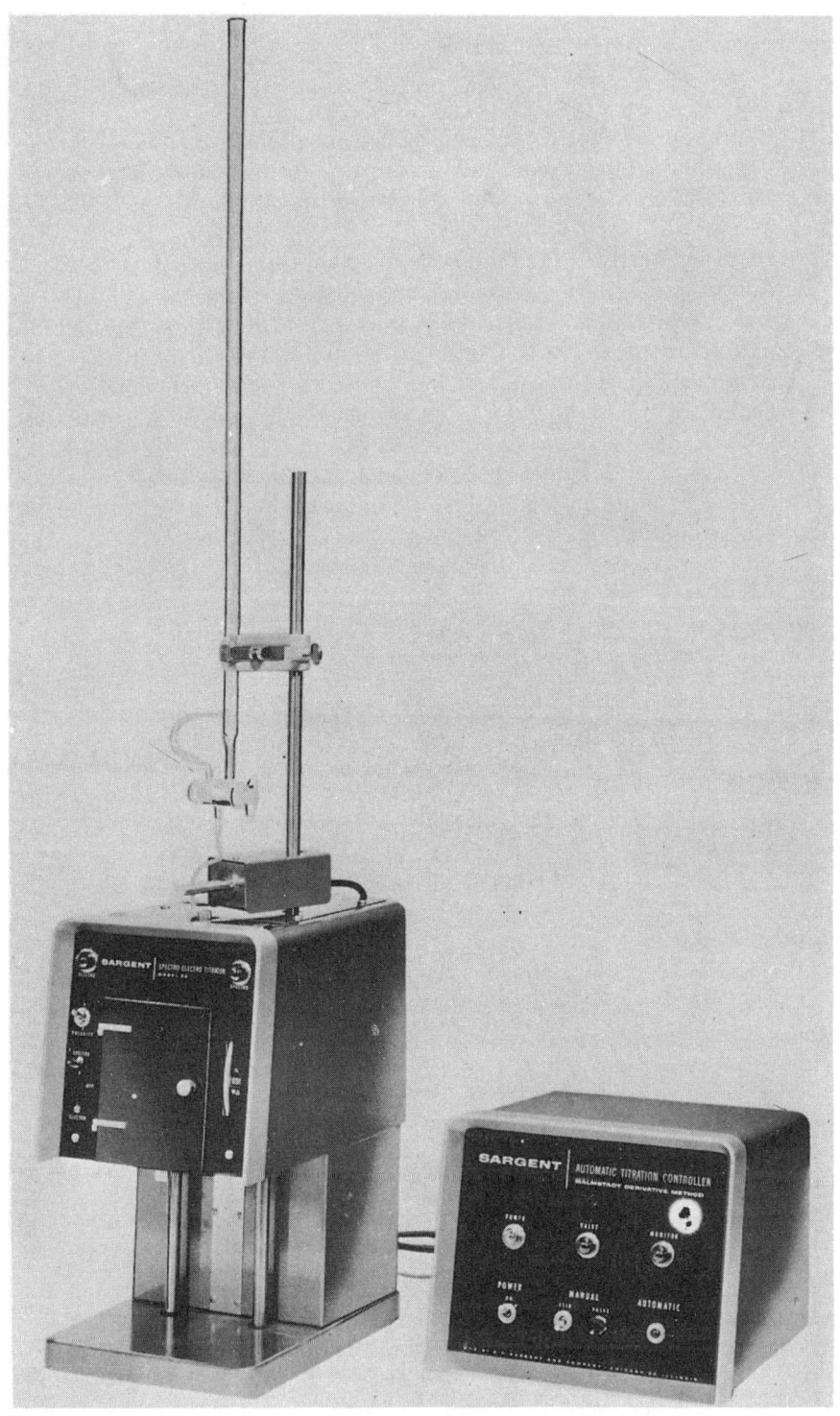

FIGURE 13.31: Sargent model SE spectrophotometric-electrometric titrator. Photo courtesy of E. H. Sargent and Co., Chicago.

Procedure:

1. Use the specific instructions for the individual pH meter,*†

2. Weigh accurately about 1.25 g of anhydrous sodium carbonate. Add 50 ml of distilled water and dissolve in the titration vessel on the magnetic stirrer.

3. Determine the pH of the solution of sodium carbonate.

4. Add by means of the burette some of the approximately $N/2$ hydrochloric acid. After the stirring has been continued long enough to allow the titration mixture to reach equilibrium, determine the pH of the mixture.

5. Continue adding the $N/2$ acid in increments, with stirring, and determine the pH after each addition. Record the volume added and the pH in tabular form. At the beginning, the increments of acid may be 2 ml in size, but as the end point is approached the increments should be made considerably smaller, as little as 0.1 ml. Sufficient of the titrant should eventually be added to go considerably past the end point of the reaction.

6. From the data in the table prepared in step 5, draw a titration curve. This should have the general shape of the curve in Fig. 13.23.

7. From the titration curve indicate the end point and the volume of titrant required to react with all the sodium carbonate.

8. From the Eq. (13.43) and (13.44) it will be seen that 1 molecular weight of sodium carbonate is equivalent to 2 equivalent weights of hydrogen chloride. Accordingly 106 g of sodium carbonate are equivalent to 4000 ml of $N/2$ hydrochloric acid. From this factor calculate the normality of the sample of hydrochloric acid.

B. ASSAY OF SULFANILAMIDE TABLETS USING $M/10$ SODIUM NITRITE SOLUTION

Apparatus:

Fisher titrimeter, model 36, or other suitable potentiometric titrator‡

---

* The instructions for a pH meter will vary with the instrument, but will almost certainly include the following steps:

(i) The instrument is allowed to warm up, with the electrodes attached to the instrument.

(ii) The meter is zeroed. (This may be at midscale, at pH 7.0, or at some other position, depending on the instrument.)

(iii) The asymmetry potential is adjusted. This is done with the electrodes immersed in a solution of known pH and with the temperature control set to room temperature or the temperature of the solution.

(iv) The known buffer solution is replaced by the solution whose pH is to be found. This is done without disturbing the asymmetry or temperature controls.

† Since the electrodes should not be handled when connected electrically to the pH meter, the instrument has a control, often labeled "standby," which should be used whenever changing the solution or rinsing or drying the electrodes.

‡ The apparatus could be a pH meter arranged to be read in millivolts and fitted with calomel and platinum electrodes. A magnetic stirrer and burette for addition of titrant are also required.

Calomel electrode for the titrimeter
Platinum electrode for the titrimeter

Reagents:
Sample of sulfanilamide tablets
USP reference standard sulfanilamide
Sodium nitrite solution, approximately $M/10$
Hydrochloric acid

Procedure:
a. Calibration of Instrument for millivolt range: After being allowed to warm up the instrument should be calibrated for an appropriate millivolt range according to the manufacturer's instructions. The Fisher model 36 titrimeter can be set to the 0- to 1400-mV range as follows: set the mV-pH switch to the mV position; set the pol-zero-use switch to ZERO*; turn potential dial to 14; turn the zero adjust control until the null meter indicates a zero reading.

b. Standardization of $M/10$ sodium nitrite solution: For this part of the procedure the model 36 titrimeter is used on "manual" operation.

1. Fill the burette on the titrimeter stand with the sodium nitrite solution provided (approximately $M/10$).

2. Weigh accurately about 0.5 g of USP reference standard sulfanilamide and place this in a 250-ml beaker together with a polyethylene-coated stirring magnet and a mixture of 50 ml of distilled water and 10 ml of concentrated hydrochloric acid.

3. Dissolve the sulfanilamide by stirring magnetically.

4. Determine the potential (in millivolts) of the solution before adding any titrant. With the model 36 titrimeter the electrodes are immersed in the solution and, while the pol-zero-use switch is in the use position, the potential dial is adjusted until the null meter reads zero.

5. Add a few milliliters of titrant ($M/10$ sodium nitrite solution) from the burette. Allow stirring to continue for a few minutes following the addition and then determine the potential of the solution. Record both the burette reading and the potential measured.

6. Continue adding increments of titrant and determining the potential after a suitable period of stirring. The increments at the beginning of the titration may be rather large (e.g., about 5 ml); as the equivalence point is approached the increments should become smaller (e.g., about 1 ml and then 0.1 ml). After each addition, both burette reading and potential should be recorded in tabular form. Continue the titration until well past the equivalence point.

---

* Since the electrodes should not be handled when connected electrically to the titrimeter, the pol-zero-use switch should be set to zero whenever changing the solution or rinsing or drying the electrodes. Other types of instruments have a similar switch labeled "standby" or some other appropriate term.

7. From the table of burette readings and potential developed, plot the titration curve for this standardization. Plot potential on the ordinate and milliliters of titrant on the abscissa. This equivalence point is the middle of the steepest portion of the curve, or the "point of inflection" (i.e., where the curve stops curving convexly and starts to curve concavely). A vertical dropped from the point of inflection to the abscissa shows the number of milliliters of titrant required to react with the weighed amount of reference standard sulfanilamide.

8. Calculate the strength of the sodium nitrite solution from the factor: 17.22 mg of $C_6H_8N_2O_2S$ are equivalent to 1 ml of exactly $M/10$ $NaNO_2$.

9. Repeat the standardization procedure until duplicate results in close agreement are obtained.

c. Assay of assigned sulfanilamide tablets: For this part of the procedure the model 36 titrimeter is used on "automatic" operation.

1. Refill the burette with the sodium nitrite solution just standardized.

2. Weigh accurately 20 tablets of the individually assigned sulfanilamide tablet sample and powder them in a mortar, taking care to avoid loss.

3. Weigh accurately a quantity of the powder equivalent to about 0.5 g of sulfanilamide. Place this in a 250-ml beaker together with a polyethylene-coated stirring magnet and a mixture of 50 ml of distilled water and 10 ml of concentrated hydrochloric acid.

4. Dissolve as completely as possible by stirring magnetically.

5. Immerse the electrodes in the solution.

6. Adjust the end-point setting of the instrument (potential dial of the model 36 titrimeter) to the potential developed at the equivalence point in step 7 of procedure b, standardization of $M/10$ sodium nitrite solution.

7. Set the instrument to add titrant (on the model 36 titrimeter: auto valve switch to 0-A, pol-zero-use switch to the use position). The titrimeter will add titrant until the equivalence point of the titration is reached.

8. Record the burette reading.

9. From the volume of titrant required, the molarity of the titrant, and the factor for the sulfanilamide, calculate the content of sulfanilamide in the weighed amount of sample used for the titration. From this, calculate the total amount of sulfanilamide in the original 20 tablets. Express the results of the assay both as (a) average weight of sulfanilamide per tablet and (b) per cent of labeled strength.

10. Repeat the assay procedure until results are obtained which agree closely.

QUESTIONS

Q13.1. What is the potential developed at 25 C in each of the following half-cells?

(a) An iron electrode immersed in a solution of ferrous sulfate where the activity of ferrous ion is 3×10^{-4}.

(b) A silver electrode immersed in a solution of silver nitrate where the activity of silver ion is 5×10^{-3}.

Q13.2. What is the overall potential developed at 25°C in the following concentration cell? Which is the positive electrode?

$$Zn \,|Zn^{2+} \underset{(a=4\times10^{-7})}{} \,|\,|\, Zn^{2+} \underset{(a=6\times10^{-3})}{} \,|\, Zn$$

Q13.3. What is the reduction potential developed in the following half-cell at 25°C?

$$Pt\,|\,I_2 \underset{(a=2\times10^{-3})}{} \qquad I^- \underset{(a=1\times10^{-2})}{}$$

Q13.4. What is the potential developed at 25°C in the following half-cell where the pH is buffered at 1.8?

$$Pt\,|\,MnO_4^- \underset{(a=1)}{} + Mn^{2+} \underset{(a=1)}{}$$

Q13.5. In the following system: mercury electrode, solid mercurous chloride, saturated solution of potassium chloride, salt bridge, unknown solution, platinum electrode,
(a) Which is the reference electrode?
(b) Which is the indicating electrode?
(c) What would serve as a salt bridge?

BIBLIOGRAPHY

Bates, R. G., *Determination of pH*, Wiley, New York, 1964.

Beckett, A. H., and J. B. Stenlake, *Practical Pharmaceutical Chemistry*, Athlone Press, London, 1962.

Biffen, F. M., and W. Seaman, *Modern Instruments in Chemical Analysis*, McGraw-Hill, New York, 1956.

Connors, K. A., *A Textbook of Pharmaceutical Analysis*, Wiley, New York, 1967.

Davis, D. G., in *Standard Methods of Chemical Analysis*, Vol. 3, (F. J. Welcher, ed.), Van Nostrand, Princeton, N.J., 6th ed., 1966.

Delahay, P., *Instrumental Analysis*, Macmillan, New York, 1957.

Eisenman, G., R. Bates, G. Mattock, and S. M. Friedman, *The Glass Electrode*, Wiley (Interscience), New York, 1962.

Ewing, G. W., *Instrumental Methods of Chemical Analysis*, McGraw-Hill, New York, 2nd ed., 1960.

Fischer, R. B., *Quantitative Chemical Analysis*, Saunders, Philadelphia, 2nd ed., 1961.

Fisher Scientific Company, *Potentiometric Titration with the Fisher Titrimeter*, Philadelphia, 1961.

Harley, J. H., and S. E. Wiberley, *Instrumental Analysis*, Wiley, New York, 1954.

Jenkins, G. L., A. M. Knevel, and F. E. DiGangi, *Quantitative Pharmaceutical Chemistry*, McGraw-Hill, New York, 6th ed., 1967.

Kortüm, G., *Treatise on Electrochemistry*, Elsevier, Amsterdam, 2nd English ed., 1965.

Leveson, L. L., *Introduction to Electroanalysis*, Butterworth, London, 1964.

Malmstadt, H. V., and C. G. Enke, *Electronics for Scientists*, Benjamin, New York, 1963.

Martin, A. N., *Physical Pharmacy*, Lea & Febiger, Philadelphia, 1960.

Mattock, G., *pH Measurement and Titration*, Macmillan, New York, 1961.

Phillips, J. P., *Automatic Titrators*, Academic Press, New York, 1959.

Walton, H. F., *Principles and Methods of Chemical Analysis*, Prentice-Hall, Englewood Cliffs, N.J., 2nd ed., 1964.

CHAPTER 14

Current Flow Methods

Stuart Eriksen

ALLERGAN PHARMACEUTICALS
SANTA ANA, CALIFORNIA

The property of certain solutes to enable a solvent to conduct electricity when dissolved in it was observed very early in the history of science and was used to distinguish between those classes of solutes called "salts" and those called "nonsalts." The development of our current concept of ions and

511

dissociation stems directly from efforts to explain the conduction of electricity by solutions of salts. At the culmination of a century of experimental efforts and theoretical interpretations as to the nature of salts in solution, our present concept of ionic dissociation was proposed by Svante Arrhenius in 1887[1] when he suggested that ions are present at all times in solutions of salts, the current the solutions carry being the result of simple passive transfer of the ions from one electrode to the other under the impulse of impressed voltage. Future work has indicated many of the details of Arrhenius' original theory were in error, but the general concept of ionization and ionic conduction is still accepted as correct.

The electrical properties of solutions, particularly those properties involving the flow of current, are indications of the ions present in solution, their conditions, and their concentrations, all in response to an externally applied stress, voltage. The measure of the capability of a solution to carry current is the electrical resistance of the solution, better expressed for our purposes as the conductance, the reciprocal of the resistance.

14.1 CONDUCTION OF CURRENT

The process by which current is transported through matter varies greatly as one considers the various states of matter from solids to solutions. Solid conductors in general have the lower resistances, while temperature has a greater effect on the resistance of solutions. The effect of temperature on the conductance of solids and solutions best illustrates the basic differences in the manner and mechanism of electrical conductance in these two materials.

Raising the temperature of a solid normally increases its resistance. Because the charge carriers are electrons and higher temperatures increase the state of kinetic motion of all the atomic elements of the solid, the passage of these electrons is made more difficult, so that the resistance increases. Solutions, however, show very marked decreases in resistance with increasing temperature. The charge carriers are often molecular in size, though always much larger than subatomic elements, and their passage depends heavily on the viscosity properties of the bulk solvent. As the viscosity of most solvent systems decreases markedly with temperature, such a change produces a corresponding drop in solution resistance.

The conductance of a solid therefore is then considered to be a property of the material (although "doping" or alloying very pure metals will produce effects characteristic of the "solution"), while the resistance of a solution is to a large extent a property of the dissolved ions. It is this basic idea that makes conductance useful as an analytical procedure, and it is the methods and ideas underlying the use of conductance as a to tool study these dissolved ions we wish to develop.

14.2 SOME BASIC ELECTRICAL CONCEPTS AND DEFINITIONS

Although our purpose is to consider the electrical properties of solutions, the definitions of terms and the systems used in the measurement of these properties are perfectly general for all conducting solids, and may be best introduced in terms of solid conductors. Only the briefest of reviews will be presented. The student is referred to any elementary physics text for general discussion.[2]

A. OHM'S LAW

Ohm's law is the basic equation describing the relationship between the three common electrical parameters, current, voltage, and resistance. It is an equation having the identical structure of the equations of motion in mechanical systems,

$$\frac{\text{Force applied}}{\text{Resistance to motion}} = \text{resulting motion}$$

$$\frac{\text{Potential difference}}{\text{Resistance}} = \text{current}$$

(Ohm's law)

$$\frac{\text{PD}}{R} = i \tag{14.1}$$

"Potential difference" (PD) is, in precise terms, a force term related to the work required to move a unit charge under its influence, but for the purposes of measurement, a less precise, more accessible potential difference unit is used, voltage. Voltage is defined as the potential difference required to produce a flow of 1 standard ampere through a standard 1 ohm resistance.

A useful analogy to Ohm's law and the relationship it describes involves a water system with hydrodynamic pressure at the input, flow resistance in the form of pipes in the system, and a flow meter to indicate water delivered per unit time (terms corresponding to PD, R, and i). The relationship between these three physical factors is precisely that described by the equations, and the intuitive effect of changes in hydrodynamic pressure or pipe constriction on the flow is precisely that predicted by the equation. Analogous intuitive reasoning with regard to PD, R, and i in the electrical system produces similarly correct predictions.

B. RESISTANCE MEASUREMENT

It is apparent that the resistance is a constant of proportionality between voltage and current leaving units of volts per ampere or ohms (Ω). To put the measured resistance of materials on some standard basis, a correction for the size and/or shape of the sample measured is made to produce a standardized measure, the specific resistance. By definition, the resistance of a cube 1 cm long with 1-cm² faces, the specific resistance (R_s) is related to the measured resistance (R) by

$$R_s = \frac{A}{l} R \tag{14.2}$$

where A and l are the measured sample's cross-sectional area and length, respectively.

For the purpose of describing the flow of current in a system, the inverse of the resistance and specific resistance, the conductance L and specific conductance L_s are used. These terms are directly related to current flowing and can be related to voltage, current, specific resistance, and resistance in simple ways.

$$L = 1/R \qquad L_s = 1/R \tag{14.3}$$
$$i = VL \tag{14.4}$$
$$L_s = \frac{l}{A} L \tag{14.5}$$

Several basic ideas must be kept in mind in accurately measuring values of resistance. Because the resistance is normally measured from the effect it produces on the flow of current, the measured value is the sum of all such effects present. If the connecting leads or the points of attachment possess significant resistances, these will be included in the measured value; for low resistance measurements this phenomenon is of considerable importance. As the resistance increases, these contributions become negligible; except for very poor design, the resistances measured in solutions are too high to be affected by this factor.

Measurements of the resistance of solutions possess a unique and particularly troublesome difficulty, the effect of the electrode reactions on the solution properties. Aside from the changes flowing current produces in the solution by ions moving, the largest single problem is the collection of electrode reaction products (notably hydrogen or oxygen) on the electrode surface itself. These products then oppose the desired electrode reaction with a potential difference of their own, markedly diminishing or totally cancelling the effect of the externally applied voltage. This effect, referred to as "polarization," is of particular concern in direct current methods, where because the current always passes in the same direction, the effects of polarization continuously increase until measurement becomes impossible. This phenomenon is overcome to some extent by coating the electrodes with "platinum black," but the use of alternating current is much more successful. By reversing the field rapidly, the products are not allowed to accumulate and true polarization-free readings are obtained.

C.   MEASUREMENT CIRCUITS

The use of Ohm's law to measure resistance involves a so-called series circuit as shown in Fig. 14.1. By applying a known voltage as the battery, a measured current would permit calculation of the resistance using Eq. (14.1). The problem that the precision of such measurements depends on the calibration of the meter and the battery, as well as their constancy, has led to the use of the bridge principle rather than this series circuit for such measurements, so that no such calibrations or constancy are required. In general terms, the principle involves balancing the current's effects on the unknown resistor against the same current's effects on a calibrated, known

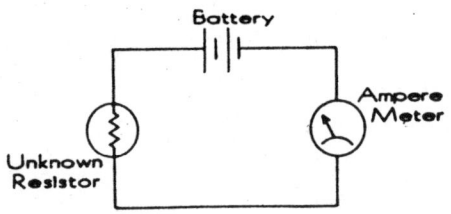

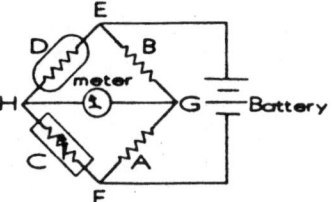

FIGURE 14.1: A simple series circuit suitable for measuring resistance using Ohm's law.

FIGURE 14.2: The Wheatstone bridge for measuring the unknown resistance D by comparing it with the calibration resistance C.

resistor, resistance standards being much easier to maintain than voltage standards. The design normally used is the Wheatstone bridge (Fig. 14.2). The unknown resistor (or solution D) is being balanced against the adjustable standard C using the "ratio arms" A and B to determine the relationship between C and D. When "balanced" so that no current is indicated by the ammeter between G and H, the resistance indicated on C is uniquely related to D. The theory is straightforward and instructive.

In practice, the unknown is placed at D and the resistance of C then adjusted until no current flow is indicated in the meter. At this point, the voltage at H and G are equal (no current is flowing in the ammeter). Designating voltage at a point by V with that point's letter as a subscript, the voltage at $F(V_F)$ is distributed over the parallel circuits A, G, B, and C, H, D, dropping to zero at E, and the voltages for each segment can be calculated using Ohm's law,

$$V_F - V_G = i_A R_A$$
$$V_F - V_H = i_C R_C$$
$$V_H - V_E = i_D R_D = V_H \qquad (14.6)$$
$$V_G - V_E = i_B R_B = V_G$$
$$V_E = 0$$

where i_A is the current through the resistor A, whose resistance is R_A. Because $V_G = V_H$ at balance, the same current passes through both resistors in any one arm, $i_A = i_B$ and $i_C = i_D$; with this in mind,

$$i_A R_A = i_C R_C$$
$$i_A R_B = i_C R_D \tag{14.7}$$

or, eliminating i_A,

$$\frac{R_C}{R_A} = \frac{R_D}{R_B} \qquad R_D = \left[\frac{R_B}{R_A}\right] R_C \tag{14.8}$$

The value in the Wheatstone bridge lies in the fact that neither the actual value of the current flowing in either arm i_A or i_C nor the voltage of the

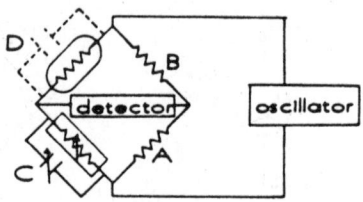

FIGURE 14.3: The simplest form of the low frequency alternating-current bridge. The oscillator is normally audiofrequency, about 1000 Hz, and the detector is often an earphone.

battery V_F is required for the calculation. Although it is not explicit in the equations, variations in V_F play no part either, these having the same effect on both arms. A variable, calibrated resistor R_C can be used to measure very wide resistance ranges by using variable pairs of ratio arms R_A and R_B. Accurate measurements of their resistances are not required, only their ratio R_B/R_A.

As described before, overcoming polarization requires the use of alternating current for the best results, and alternating current presents special problems in measurement and circuitry. There are two general frequency ranges of alternating current devices in use, so-called low frequency (i.e., up to 1–2000 Hz, and high frequency, usually in the range of 1–10 mHz).

Because alternating current is affected by capacitance as well as by resistance, alternating current bridges require adjustment for these capacitance links in various parts of the circuit in order to balance the bridge. Thus both resistance and capacitance balances must be made. In its simplest form low frequency alternating current measurement requires the use of a circuit as shown (Fig. 14.3). The adjustable resistance at C still measures the resistive component of the unknown resistance D the adjustable, non-calibrated, variable capacitor now balances the capacitance present in the electrodes and leads of the unknown resistor, or solution, D. Because it is alternating current power we are using, an electronic oscillator is used as the power source, and an alternating current receiver such as an earphone as a detector. To eliminate as much extraneous alternating current as possible, the oscillator and detector are normally tuned to generate and detect only a particular frequency, usually 1000 or 2000 Hz. Thus noise from the overhead lights and other electric devices does not interfere. Considerably more complex, but also more useful, circuits are available to balance the capacitance of the measuring system to the earth. The best known of those is the Wagner Earthing circuit,[3] as illustrated in Fig. 14.4. .

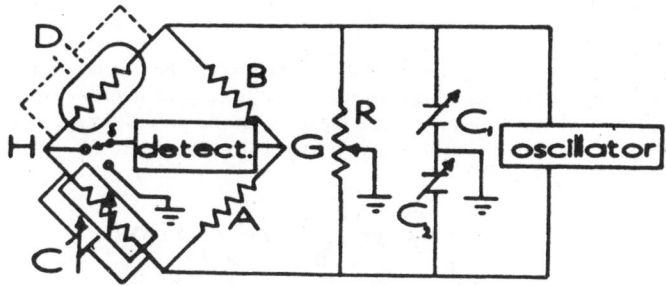

FIGURE 14.4: The Wagner Earthing circuit for eliminating extraneous currents caused by capacitance links to the earth.

The additional "balancing" resistor and capacitors (R, C_1, and C_2) are used to bring point G to ground potential when the detector is switched to ground. At balance then, not only are the resistance and capacitance factors of unknown and standard (C and D) matched, but the detector inputs (G and H) are grounded, preventing stray ground capacitance effects from producing detector currents.

High frequency alternating current measurement (1–10 MHz) requires quite a different system. The unknown resistance, or solution is placed into an oscillating circuit, where its inductance (a function of its resistance and capacitance) influences the frequency at which the circuit involving it oscillates. Several of the circuits used in this way are shown in Fig. 14.5.

A measurement of the change in oscillation frequency or power drawn from the oscillator system by the cell and solution, or a measurement of the adjustment required to bring the system back to its original state are the measurement factors. In the former circuit, changes in values of plate or grid current measured on a sensitive meter are made.

Although several commercial instruments are available, a number of relatively simple devices have been described in the literature. In Fig. 14.5a and 14.5b are two of these showing one with the cell in the feedback loop and one with the cell in the plate circuit. Both are used by reading plate current directly on the meters as functions of added titrant current. The block schematic in Fig. 14.5c represents the concept behind the substitution

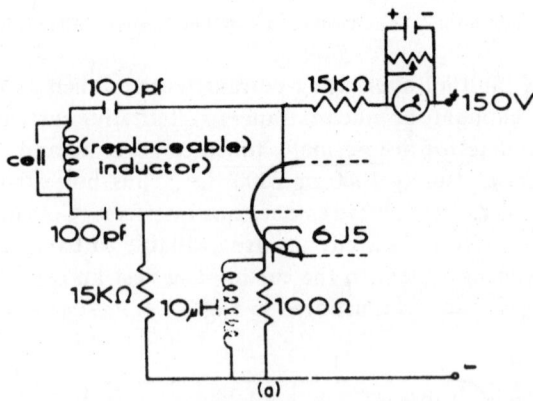

FIGURE 14.5a: High frequency triode oscillation circuit where the conductance cell is part of the feedback loop. The replaceable inductor is used to vary the frequency of oscillation and the meter shunt circuit serves to "zero" the meter at the start of titration. Reprinted from Ref. 4, p. 99, through the courtesy of the publisher.

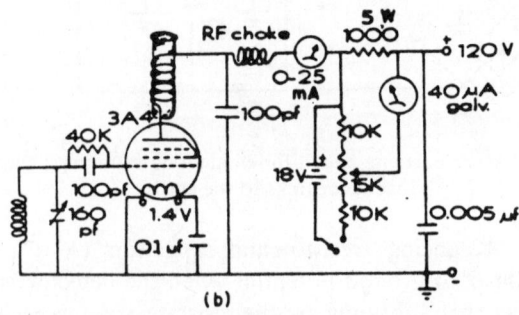

FIGURE 14.5b: A battery-operated high frequency oscillator circuit where the conductance is part of the plate circuit. The adjustable grid capacitor is used to set the starting current, read on the 0–25 mA meter. The galvanometer is biased to allow precise readings of the plate current during titration. Reprinted from Ref. 5, p. 492, through the courtesy of the publisher.

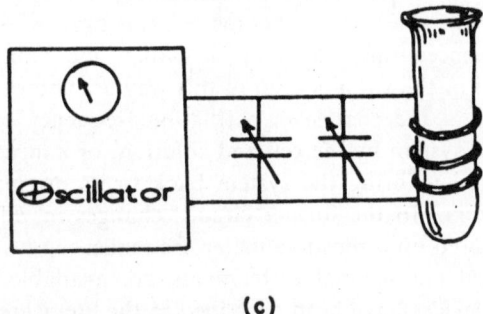

FIGURE 14.5c: Block diagram of the substitution method for high frequency titration. During titration, the adjustment in the two calibrated variable capacitors, required to return the system to its prior state, is noted.

method. After the oscillator frequency is noted, titrant is added and the change in the variable capacitors required to return the oscillator to its former condition is noted as a function of the added titrant.

Systems involving high frequency have not reached the high precision of low frequency measurements and are customarily used only for measuring changes, such as detecting an end point during titration, so that meter readings alone are normally used in making the required graphs.

14.3 PROPERTIES OF SOLUTIONS AS THEY AFFECT ELECTRICAL RESISTANCE MEASUREMENT

Solutions, the prime concern in pharmaceutical analysis, present some very real differences and problems in comparison to solids when the meaning and measurement of their conductance is considered.

A. ELECTRODE REACTIONS

As mentioned earlier, when current flows in a solid no real change occurs in the circuit through which the current flows; in a solution, however, an important but not always appreciated phenomenon occurs; a reaction occurs at *both* electrodes whenever current flows.* As a corollary, some reaction must occur (albeit small) for a voltage to be generated also, but the initial concept is the one of concern in conductimetry. The importance of this idea lies in the fact that no current flows without producing some change in the solution, and in some cases in the electrodes, through which it flows. It behooves the analytical chemist, therefore, to keep such change-producing current flow negligible or in some way cancelled, to prevent his measuring tool from interfering with the measurement he wishes to make.

B. CHARGE CARRIERS

Unlike current flow in a solid, current in a solution is carried by a variety of carriers (ions) derived from the solvent or from entities dissolved in the solvent; these carriers have either positive or negative and single or multiple changes. In addition, because we are most often concerned analytically with dissolved entities, the additional influence of their number and physical state will influence their ability to carry current. The latter factors, concerned with the effects of concentration and physical state, make the analytical prospects of solution conductance great.

* As earlier, the general discussion involved dc aspects. The high frequency units and systems described do *not* require reactions at each electrode as in truth they measure inductance, a combination of factors including resistance, not resistance alone and with proper frequencies, the electrodes do not contact the solution. For our purposes, as dc (and low frequency ac) systems will be considered here, the current flow-reaction idea is correct.

C. MEASURING CELL

The solution being measured must be contained in a conductance "cell," a chamber made of high resistance and heat conducting material in which the "electrodes" are relatively rigidly placed, roughly as shown in Fig. 14.6.

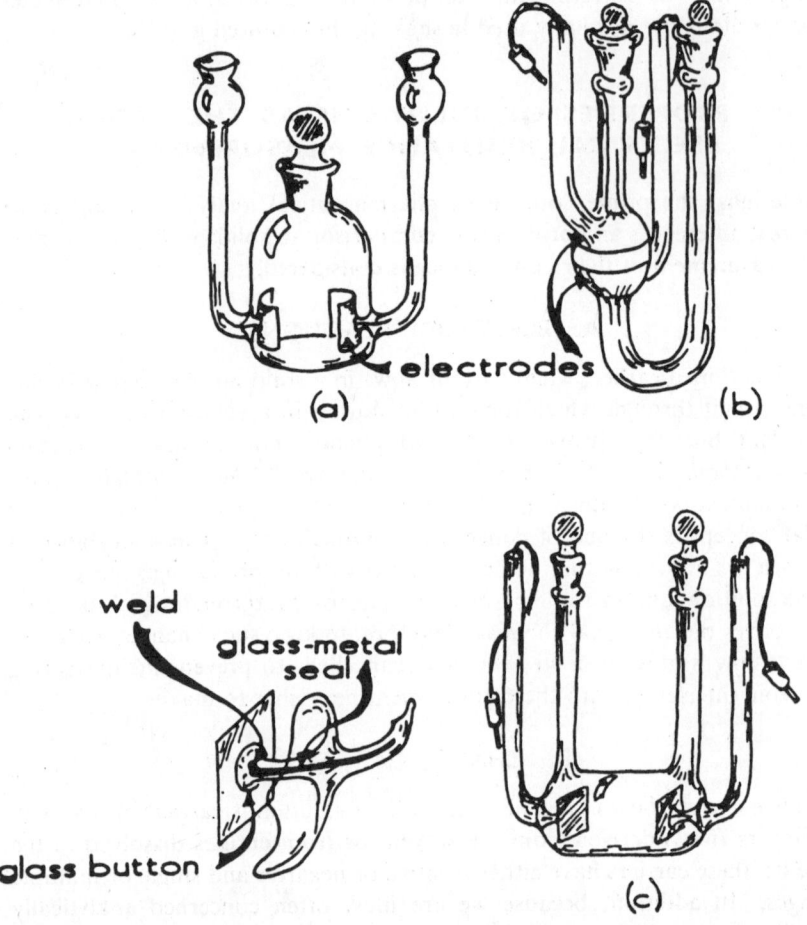

FIGURE 14.6: Several representative conductance cells for titration (a), small volume (b), and high-precision work (c). The detail of a rigid mounting for an electrode is also shown.

The precise dimensions and the design used will depend greatly on the type of analytical work being done and the conductance of the solutions being measured, but some general points may be made.

Glass is the usual external container material because of its easy workability, high electrical resistance at temperatures satisfactory for solution measurements, structural rigidity, and ease of cleaning. The placement of filling tubes depends on the purpose of the experiment. For titrations and experiments where frequent additions are to be made to the cell, one with a large single opening as shown in Fig. 14.6a is often used. Smaller volumes of solution are more simply filled if the tubes are in a vertical plane as in Fig. 14.6b. Platinum is the customary electrode material; because of its inertness and high oxidation potential, the conductance of most solutions can be studied without affecting the electrodes. By proper design of the glass-to-metal joint and the use of heavy gauge lead-in wires, the cell electrodes may be made rigid (see Fig. 14.6) as is required for precise, absolute measurements. The placement of the lead insulation tubes is critical only for ac measurements; because the liquid of the temperature bath intervenes between the leads, the closer these leads are together, the higher the capacitance that must be allowed for in balancing. While rigidity is actually all that is required, measurements are much simpler to make if the leads are kept as far apart as possible. External connections are made either through a drop of mercury in the insulation tubes or by soft-soldering the leads permanently, directly in the insulation tubes, with a small torch.

The platinum electrode plates are usually coated with platinum black after assembly to make the bridge system easier to balance. This may be accomplished by filling the cell with a dilute solution of platinum chloride and passing current through the cell for several minutes in both directions. In cases where platinum black will create catalytic problems, it may be omitted, but an increase in bridge balancing difficulties may be expected.

Cleaning and aging cells after manufacture is of concern in all accurate work. The cell constant will be observed to change slowly for several months after manufacture, after which it will remain stable indefinitely. This process can be hastened at the time of initial cleaning of a new cell by boiling for an hour in concentrated HCl then, after rinsing, boiling in distilled water.

D. SPECIFIC CONDUCTANCE

A solution in a cell has a specific resistance and conductance, defined precisely as discussed previously; it is not normally found by exactly measuring the resistance of a precisely measured gap between the electrodes as is usually done with solids, but by measuring the resistance of the cell filled with a carefully prepared standard solution whose specific resistance and conductance (L_s) is precisely known (Demal solutions, Jones and Bradshaw[6]). From this measured resistance R, the cell constant K is, when required, calculated from the equation

$$K = RL_s \tag{14.9}$$

The constant K is roughly related to the dimensions of the electrode gap by the equation

$$K = \frac{\text{length}}{\text{electrode area}} \tag{14.10}$$

This constant is then used to convert the measured R for an unknown solution to specific conductance L_s, again using Eq. (14.9).

E. EQUIVALENT CONDUCTANCE

To completely describe the electrical properties of a dissolved salt, the conductance of 1 equivalent of the salt at the concentration in question (Λ at concentration c, equivalents per liter) is defined.

In a diagrammatic sense this is the conductance of a cell with electrodes 1 cm apart, but large enough to hold 1 equivalent of the solution. The volume of this cell is $(1000/c)$ ml, and the conductance of this cell per square surface of electrode area is the equivalent conductance Λ,

$$\Lambda = \frac{1000L_s}{c}$$ (14.11)

Most analytical conductimetry data may be used without conversion to these more basic terms, resistance alone being adequate for the plotting required.

F. DISSOCIATION AND ASSOCIATION

The development shown in the last section, of the parameter dependent on concentration, implies in light of our earlier discussion of the relationships between ions and conductance, that all of any dissolved material dissociate upon dissolution into its constituent ions. If this does not occur, the value of specific conductance L_s and thus the equivalent conductance Λ will depend on the ions actually present, i.e., on the completeness of this dissociation. In truth, as discussed when the background of solution conductance was being presented, the development of the present theory of dissociation began with efforts to explain the data obtained from conductance measurements. The equivalent conductance of a completely dissociated solute in solution is normally referred to as the equivalent conductance at infinite dilution Λ_0. As the solution becomes more concentrated, the equivalent conductance is observed to decrease so that the degree of dissociation, or ionization, α was early defined by Arrhenius[1] as

$$\alpha = \frac{\Lambda}{\Lambda_0}$$ (14.12)

This equation appears intuitively correct, but ignores the differences in the solute ions at the two concentrations implied by Λ and Λ_0.

At infinite dilution, it is apparent that all molecules that can dissociate will have done so and the ions so produced will have become as independent of each other as it is possible for them to be; at this concentration each ion will contribute its own intrinsic equivalent conductance to the measured equivalent conductance; these are designated as λ_0^+ or λ_0^-, for cations and anions. These "limiting ionic conductances" can be intuitively seen to depend on the particular speed with which the ion in question moves through the solvent used under a standard potential difference, called "mobility" (μ), and the number of current units (coulombs) the ion carries in 1 equivalent ξ,

$$\lambda_0^+ = \xi\mu_+ \tag{14.13}$$

Once determined, these individual limiting ionic conductances could be combined in any stoichiometric way to deduce (when required) the equivalent ionic conductance at infinite dilution of any salt. As a first approximation, at other concentrations, similar relationships are assumed to hold for that fraction of the salt that is ionized, so that the measured equivalent conductance at infinite and finite dilution is the sum of the contributions of all ions present (see any physical chemistry text for a table of limiting ionic conductances, i.e., Ref. 7). As an example, at infinite dilution, for a salt A_nB_m,

$$\Lambda_0 = n\lambda_0^+ + m\lambda_0^- = \xi(n\mu_+ + m\mu_-) \tag{14.14}$$

$$\Lambda = \alpha\xi(n\mu_+ + m\mu_-) \tag{14.15}$$

where α is the fraction of molecules dissociated. Some of the assumptions inherent in Arrhenius' intuitive equation [Eq. (14.12)] can then be seen,

$$\alpha = \frac{\Lambda(n\mu_+ + m\mu_-) \text{ at infinite dilution}}{\Lambda_0(n\mu_+ + m\mu_-) \text{ at concentration } c} \tag{14.16}$$

and $\alpha = \Lambda/\Lambda_0$ only when μ_+ and μ_- are the same at all concentrations. This factor is often assumed true in analytical work.

G. DEBYE, HUCKEL, AND ONSAGER

The final assumption of the last section was the culprit at whose door a number of strange anomalies were laid as the ionic theory was being developed. Several "strong salts," sodium chloride, etc., even when relatively dilute, seemed to possess a finite α, suggesting that they were not completely ionized in solution (despite other evidence to the contrary). Through the work of Debye, Huckel,[8] and Onsager,[9] these anomalies were corrected on both a theoretical and practical level by theorizing and correcting for the interactions between the ionized molecules in solution, that affect the effect of the external field on those same molecules. At the concentrations normally involved in analytical work these equations permit nearly complete explanation of the anomalies in conductance caused by concentration that are observed in solution of completely ionized salts. The corrections take the form of a simple equation relating measured equivalent conductance and concentration:

$$\Lambda = \Lambda_0 - (\theta\Lambda_0 + \sigma)\sqrt{c} \tag{14.17}$$

indicating that for solutions of finite concentration plots of equivalent conductance vs. the square root of concentration are straight lines. Of passing interest to us here is the fact that the constants θ and σ are theoretically calculable from nonconductance data. Great studies have been made in increasing our knowledge of solution conductance and the serious student is referred to the work of Fuoss and co-workers.[10]

14.4 ANALYTICAL ASPECTS OF SOLUTION RESISTANCE

To the foregoing general discussions, the special problems of the various types of conductimetric analyses must be appended. The more precise problems of the use of conductimetry as a tool for the probing of the nature of ions in solution are rather complex subjects in and of themselves and will be dealt with only briefly. We are concerned mainly with the concentration aspects of conductimetry and the use of that tool in measuring changes in the solution concentrations of ions.

A. CELL AND TEMPERATURE CONTROL

As mentioned before, the conductivity cell is usually made of glass, with platinum electrodes. In high-precision direct current or low frequency work, seeking Λ_0 for instance, exact knowledge of the cell constant is required and rigidly positioned electrodes are essential so that the cell constant, once determined, remains constant. The dimensions of such cells are made such that measured resistance of the solution-containing cell is in the most sensitive region of the bridge system, normally, 1000–10,000 Ω.

As viscosity of the solvent is a prime effector of the velocity of the ions and viscosity is very sensitive to temperature, the temperature of cells whose resistance is being measured must be accurately known and controlled. The viscosity of water varies at a rate of about 2%/deg near room temperature, so that to make measurements of aqueous solutions with a precision of the order of 0.1 % requires that the cell temperature be maintained constant to at least $\pm 0.05°C$.

The problems associated with ac measurements have been discussed in general, but an additional point arises in these systems when the measurement cell is placed in a bath liquid. The capacitance contributed to the system by the electrode leads into the cell depends, of course, on the lead insulator positions and the dielectric constant or the intervening material. For precise work a low dielectric constant bath oil is preferred, but with proper and rigid design, precise low frequency ac measurements can be obtained using cells of the general shape shown in Fig. 14.6c, in a water bath where the leads are separated as much as possible.

Titrations and other low accuracy measurements can be made with much simpler cells: often simply two pieces of wire mesh or foil clamped into a

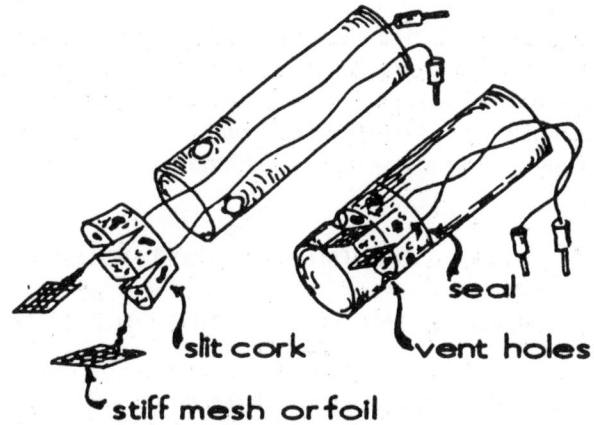

FIGURE 14.7: A "dipping" conductance cell, made of cork and glass tubing. The dimensions may be adjusted to provide any suitable cell constant.

beaker will suffice. The homemade "dip" cell shown in Fig. 14.7 functions well, and can be conveniently varied for any resistance solution by adjusting the area of mesh showing. It is fashioned from a stopper sliced, reassembled with electrodes in the slices, then pushed completely into the glass tubing. A bit of de Kohtinsky cement on top of the stopper completes the seal.

The only requirement for titration cells is that sufficient electrode area be accessible to the solution to permit the bridge system used to be read in its most accurate region and that the electrodes remain in their relative position throughout the one titration.

Higher frequency titrations require (or permit) quite different electrode placement methods. The electrodes are usually *outside* the vessel and rigidly mounted directly to the vessel walls. High frequency conductance cells are normally variations of types shown in Fig. 14.8.

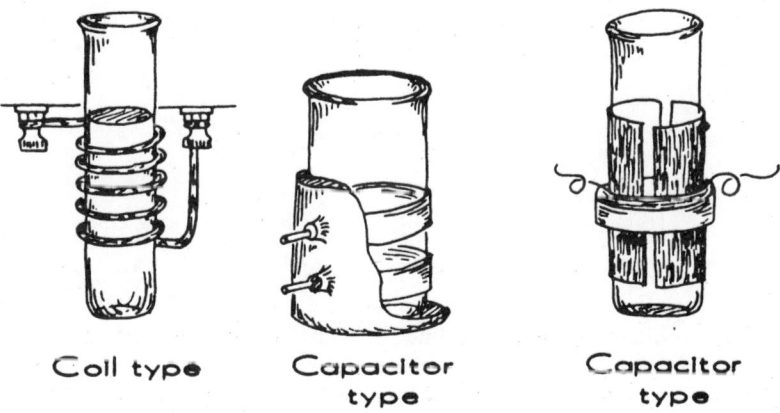

FIGURE 14.8: Three general types of high frequency conductance cell.

In the capacitative cell, the electrodes are cylindrical metal foil rings or plates with the glass solution chamber between. Because glass is a dielectric in itself, the glass wall is kept thin, to contribute as little as possible to the overall capacitance. The coil-type cell usually uses a test-tube-shaped cell made to fit snugly into heavy gauge, self-supporting coils.

14.5 EXPERIMENTAL METHODS

A. PRECISE MEASUREMENT OF LIMITING IONIC CONDUCTANCE

Studies of the physical properties of ions in solution most often require estimates of Λ_0 and/or measurements of the limiting slope of the Onsager plot of Λ vs. $\sqrt{c}$.

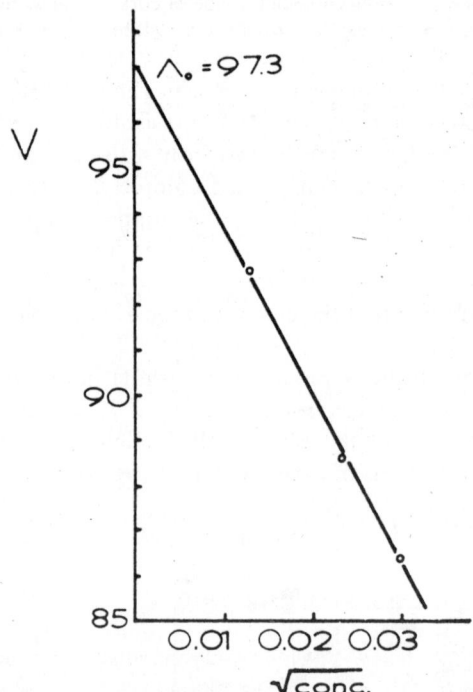

FIGURE 14.9: The plot of the data for Example 1 using Eq. (14.17).

The cell constant for the cell used must be precisely determined using an appropriate one of the standard solutions of Jones and Bradshaw[6]; then, using all the precautions outlined earlier, solutions of decreasing concentration are measured. The contribution of the solvent to the conductance is subtracted from the measured conductance using the equation

$$L_{s(\text{solute})} = L_{s(\text{solution})} - L_{s(\text{solvent})} \tag{14.18}$$

and plots of the Onsager equation or some variation of it used to determine Λ_0 and the slope from the solute data so obtained. An example of such data and its handling are shown below:

EXAMPLE 1: Conductance of tetra-*n*-hexylammonium iodide in methanol at 25°:

a. Cell constant
$L_{25°}$ (0.001 Demal KCl) = 0.0001479 mho (corrected for solvent)
R (0.001 Demal solution) = 5557 Ω, 5524 Ω, 5552 Ω
b. Methanol solutions ($L_{\text{methanol (25°)}}$ = 3.4 × 10⁻⁷ mho)

Concentration, × 10⁻⁴ M	(Mean R value), × 10⁴	L_s, × 10⁻³ mho	L_s (corrected for solvent), × 10⁻⁵ mho	Λ
8.991	1.051	7.79	7.76	86.4
5.464	1.685	4.861	4.827	88.6
1.633	5.295	1.548	1.514	92.7

c. Cell constant $k = 5.544 \times 1.479 \times 10^{-3} = 0.8199\ M$
d. A plot of Λ vs. $\sqrt{c}$ yields Fig. 14.9, suggesting $\Lambda_0 = 97.3$

References to any of the newer texts on electrochemistry,[11] or even one of the older ones,[12] should be made for specifics on high precision work and in particular for aid in the interpretation of the data.

B. LOW FREQUENCY CONDUCTANCE TITRATION

Conductance measurements are most often used for end-point detection in titrations; in this instance, a cell of the simplest type should be used, often only a beaker with clamped electrodes or a dipping cell such as shown in Fig. 14.7. Most titration-curve shapes may be theoretically deduced, and even a brief discussion of the curve shapes for several common titrations and the methods used for deducing them will be sufficiently instructive to permit the student to deduce nearly any others of concern.

The basic assumption underlying conductance titrations is that the conductance will vary linearly with concentration, so that as the ion titrated decreases in concentration to zero during titration, the conductance changes in a similar fashion. It is comforting that although the conductance of an ion in solution by itself decreases linearly with the *square root* of the concentration, in the usual conductance system used for titration, the relatively constant ionic strength of the solution permits strong electrolytes to change conductance quite linearly with concentration. The association-dissociation phenomena exhibited by weaker electrolytes are then included as corrections on this basic assumption.

Titrations of strong acids and with strong bases are the simplest systems; all ions concerned can be accounted for easily, and their individual conductances allowed for in assembling the theoretical curve. The steps involved in theorizing this basic curve shape are shown in Fig. 14.10. All ions present in the initial solution sample are accounted for in ratio to their limiting ionic conductances (note, we actually desire the accounting to be in ratio to their ionic concentrations and their limiting ionic conductances $m\alpha\lambda_0$, where m is

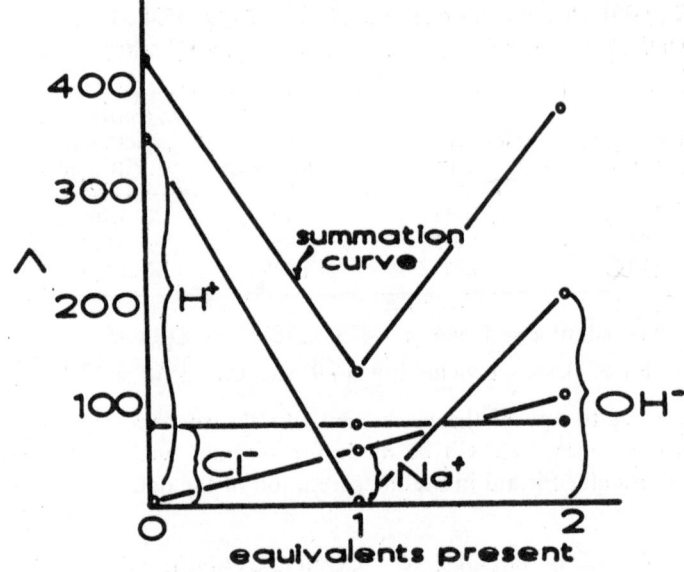

FIGURE 14.10: A hypothetical titration of HCl with NaOH from the start to 1 equivalent beyond the end point. It is assumed that each ion present exerts its own limiting ionic conductance. At each point, the total conductance is represented by the "summation curve."

total equivalents present and α the degree of ionization, but in drawing theoretical curves λ_0 alone will suffice). The same accounting should be done at the end point and at 1 equivalent beyond the end point, assuming ideality. These individual points are shown in Fig. 14.10.

When beyond the end point, it is of real concern to note we are using $n\lambda_0$ for the titrant, where n is the equivalents of titrant present. One assumes, in Fig. 14.10, that all of the H^+ is used with all of the OH^- to reach the end point; the Na^+ added rises with the slope $n\lambda^+_{(Na)}$ and the Cl^- is constant throughout. Connecting all these basic points with straight lines produces a rather good approximation of the conductance curve for this titration as it is obtained. For weaker acids, corrections for variations in the degree of ionization α, occurring during the titration, can be estimated from knowledge of the acid being titrated or the base used.

In effect two alterations are being made when correcting for dissociation; at and near the end point the occurrence of hydrolysis of the salt titration product produces an "apparent non-end point" in that more titrant is present than one would expect, producing a pronounced rounding of the sharp break expected at the end point. For example, at the end point of acetic and boric acid titrations, solvent equilibrium produces the situations,

$$Ac^- + H_2O \rightleftharpoons HAc + OH^-$$
$$Borate^- + H_2O \rightleftharpoons HB + OH^-$$

so that a sharp end point is not achieved. The same result is produced by titrating a weak base with a strong acid.

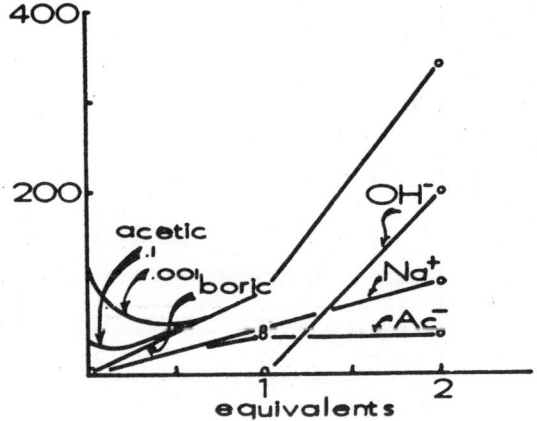

FIGURE 14.11: Hypothetical titration curves for boric and acetic acid. Boric acid, being very weak, contributes no conductance until the borate ion is produced as titration proceeds. Acetic acid shows the effect of initial weakness followed by common-ion suppression of ionization. The degree of ionization is a function of initial concentration.

The amount of curvature to be included in deducing the curve for a weak titrant is in direct proportion to the weakness of the acid or base being titrated. This problem can be diminished somewhat by adding a miscible solvent with a low dielectric constant, such as ethanol or acetone, to decrease hydrolysis. Most often the end point can be satisfactorily determined by extrapolating the straight line portions before and after the end point to their intersection at the actual end point.

The effects of the degree of dissociation are allowed for at the point where the weak solute is in highest concentration; at the beginning of the titration for a weak titrant. The extremes at the end for a weak titrant are shown with hydrochloric acid (for our purposes completely dissociated), Fig. 14.10, and with boric acid (for our purposes completely undissociated), Fig. 14.11. The first shows the completely additive effects of all constituent ions; the second shows the effect of the very low dissociation of the acid, a slow, proportionate rise in all ions present; the borate ion is produced as the hydrogen ion is titrated to the end point.

The titration of acid such as acetic, intermediate between these two extremes, shows a low but finite dissociation and conductance initially, Fig. 14.11 (the dissociation dependent on the concentration being titrated) followed by a suppression of their initial dissociation by the common ion produced in the titration as evidenced by a decrease in conductance, which then rises slowly to the end point. All curves are essentially identical after the end point, in essence, being due to addition of the titrant alone.

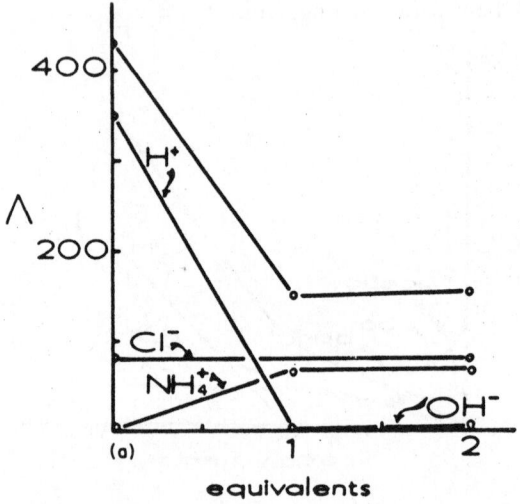

FIGURE 14.12a: A hypothetical titration curve of a strong acid titrated with a weak base. The overall measured curve is shown at the top.

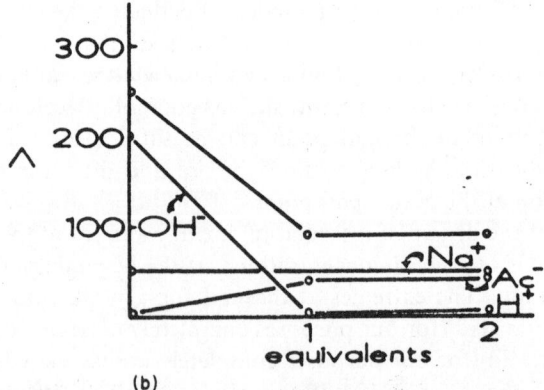

FIGURE 14.12b: A hypothetical titration curve of a strong base titrated with a weak acid. The overall measured curve is shown at the top.

The effects of titration on a strong titrant with a weak base or a weak acid are handled similarly, but produce one interesting variation, as shown in Figs. 14.12a and 14.12b. In passing the end point, little increase in constituent ions is produced by weak titrants both because of their weakness *and* because ionization is further suppressed by the high concentration of counterion present as the product of the titration; the curve flattens almost immediately after the end point.

As an example of the handling of titration data, consider the following titration of a mixture of acids with a strong base.

EXAMPLE 2: Titration of sulfuric and acetic acids: 3 ml of sulfuric and 5 ml of acetic acid solutions of unknown strengths added to 200 ml of water and titrated with 0.2541 N NaOH, produced the following results:

Titrant, ml	$L_{(read)}$, $\times 10^{-5}$ mho
0	17.7
1	14.9
2	11.7
3	9.26
4	8.40
5	8.71
6	9.15
7	9.50
8	9.94
9	11.20
10	12.50

Note that the stronger acid is titrated first and provided the strengths differ by at least 2–3 pK units, the two end points can be easily detected. (See Fig. 14.13.)

Titrations involving other types of systems such as oxidation-reductions and precipitations are handled in a similar manner, i.e., considering the ions present at the three stages of the titration and correcting for the completeness of the reactions involved; curves of generally the same shape are obtained.

The theory just developed has made some very basic assumptions that should be considered; that the conductance system being used is capable of accurate measurements of the solutions involved, that an increase in concentration can be obtained without increasing the volume (the abscissa on these plots is concentration in equivalents, not milliliters added). The first of these can usually be solved by proper concentrations and/or electrode placement, spacing, and bridge design. The second is solved by using titrant at least 10 times as concentrated as the titrand, and/or by correcting for volume changes using the equation,

$$L_{corr} = L_{measured} \left(\frac{V + v}{V} \right) \qquad (14.19)$$

In this equation V is the initial sample volume and v the total titrant volume added up to that reading. This correction will normally eliminate the slight bending toward the abscissa that volume changes during titration produce.

Conductance titrations are most often performed in aqueous solvents, but the system is not limited to water. Any solvent in which the compound to be analyzed is soluble and in which the ionic concentration changes during titration is acceptable. The advantages of increased solubility for organic compounds in organic solvents may be outweighed by the complications

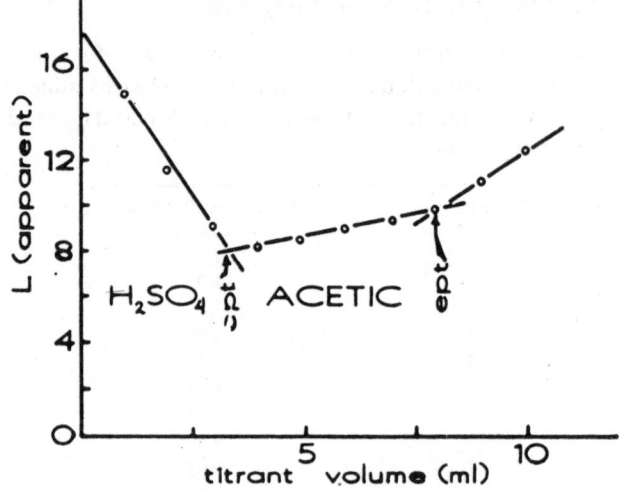

FIGURE 14.13: The plot of data from Example 2, titration of sulfuric and acetic acids with sodium hydroxide.

of ion association that occur in low dielectric media, but the advantages of increasing the ionization of weak acids and bases by dissolving them in strongly basic or acid solvents should not be overlooked. For example, weak acids such as hindered phenols can be titrated successfully in pyridine and toluene.[13]

Combinations of solvent have value as mentioned before, the addition of up to 10 or 20% of water-miscible, low dielectric constant liquids such as methanol, acetone, or ethanol to aqueous solutions, to suppress hydrolysis and sharpen the end points of weak acids and bases is a particularly valuable procedure of this type.

C. HIGH FREQUENCY CONDUCTANCE TITRATION

All of the foregoing portion of this section was presented from the standpoint of dc or low frequency ac systems. The behavior, operation, and results obtained from high frequency ac methods are sufficiently different to warrant separate discussion.

High frequency titrations can involve either coil or capacitor systems with the cell design differences discussed earlier. As the latter are much better developed in the available literature, they will be the only ones considered in detail. Without concerning ourselves with the analysis of the equivalent circuits of the cell-solution system, the high frequency conductance (G_p) of a solution in a capacitance system is related to the low frequency conductance (L) by the equation,

$$G_p = \frac{L\omega^2 C_c^2}{L^2 + \omega^2(C_c + C_s)^2} \tag{14.20}$$

an equation which also involves ω, the frequency, C_c, the capacitance of the cell walls, and C_s the capacity of the solution. It is of concern to note that G_p is linearly related to L, the low frequency conductance, when L is very small (resistance high), with a slope at most equal to 1 but most often less than 1. For example, a plot of the high frequency conductance of a specific system and cell as a function of L is shown in Fig. 14.14 for two frequencies;

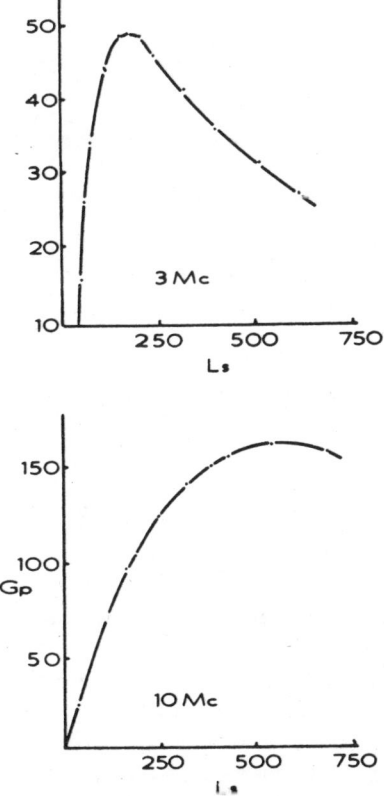

FIGURE 14.14: High frequency transfer curves for a capacitance cell showing the relationship between the high frequency (G_p) and the low frequency conductance (L_s). Reprinted from Ref. 14, p. 89, through the courtesy of the publisher.

as one might surmise by studying the equation, the height and the value of L at which the maximum G_p occurs is a function of the frequency used for analysis.

These so called "transfer curves" are of concern, as the result obtained from a high frequency titration depends heavily on them. Let us consider

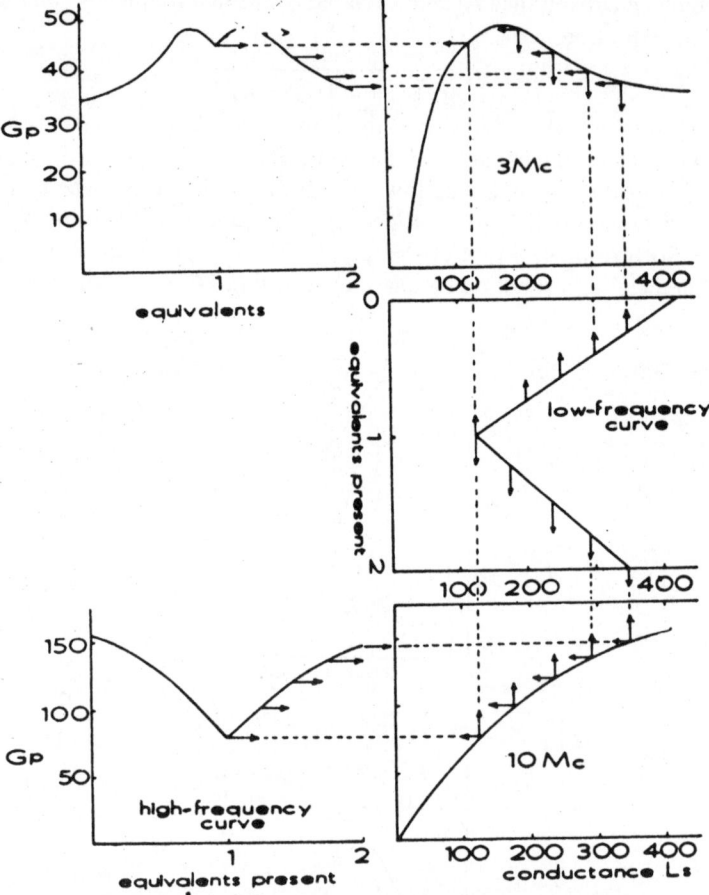

FIGURE 14.15: The use of high-low frequency transfer curves to deduce the shape of a high frequency titration curve from one obtained at low frequency.

the low frequency (or dc) conductance titration shown in Fig. 14.10 and use the transfer curves shown in Fig. 14.14 to predict the shape of the high frequency titration curves we would expect for that simple titration system.

The variations in L as observed on the low frequency titration curve may be transformed by replotting the corresponding G_p points read from the 3- or 10-mc curves, into the V- and inverted W-type curves shown. It is an interesting exercise to reduce the other general types of titration curves to high frequency curves using ⌄⌐me standard transfer curves, for example, that in Fig. 14.14.

EXAMPLE 3: Using the transfer curves in Fig. 14.14, predict the shape of the following low frequency titrations at 3 and 10 mc:

A weak and a strong acid		A weak acid with a strong base	
ml	L	ml	L
0	277	0	200.0
1	249	1	210.0
2	217	2	217.5
3	192.6	3	225.6
4	184.0	4	233.9
5	187.1	5	244.0
6	191.5	6	258.6
7	195.0	7	275.4
8	199.4	8	292.8
9	212	9	209.3
10	225	10	212.0

It is of concern here that the shape of the high frequency curve is directly dependent on the position of the transfer curve peak and the absolute value of the low frequency conductance readings. Small amounts of added salts, which produce only a vertical shift in the titration curves made at low frequencies, may well move the high frequency to a much less sensitive region (put the end point at or just near the G/L peak) so that experimentation with concentrations, while desirable in low frequency titration, must always be considered with high frequency titration. It is also apparent from Eq. (14.20) and Fig. 14.14 that as the sharpness of the end point is dependent of the slope of the G_p/L plot and this slope is essentially independent of frequency at low values of L (where sensitivity is highest), variations in frequency will not enhance end-point sensitivity once an appropriate concentration is selected.

In practice, the transfer curve is not required, though it can be deduced with known solutions by measuring L and G_p. In practice, the concentration of the sample is simply adjusted in the titration until appropriately sharp . end-point curves are obtained.

Systems involving use of measurements of capacitance change in the cell during titration produce similar results. The relationship between the high frequency capacitance C and the low frequency conductance L is

$$C = \frac{L^2 C_e + \omega^2 C_e C_s^2 + \omega^2 C_e C_s}{L^2 + \omega^2 (C_e + C_s)^2} \qquad (14.21)$$

and plots of the change in capitance during titration ΔC vs. L produce transfer plots of the shape shown in Fig. 14.16.

These plots may be used to deduce the high frequency plots from low frequency titration data precisely as described for high frequency conductance. Again, however, these transfer plots are not required; adjustment of concentrations to produce sharp end points being all that is needed.

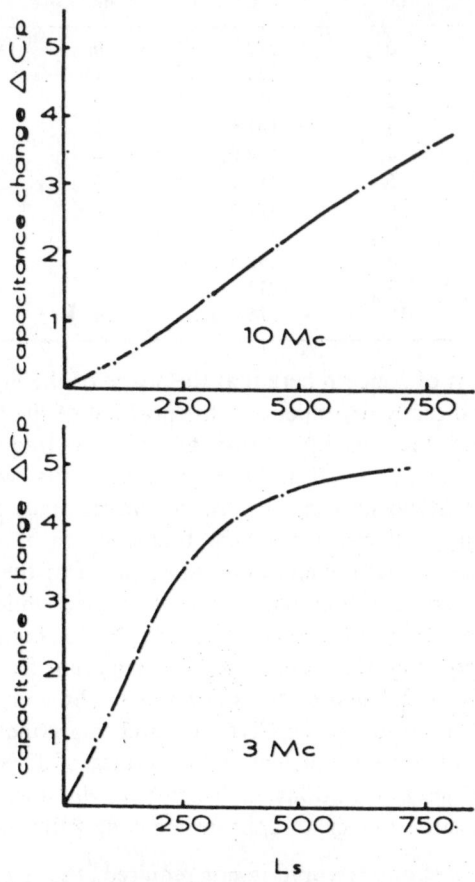

FIGURE 14.16: Transfer curves for high frequency capacitance change as a function of low frequency conductance. Reprinted from Ref. 14, p. 91, through the courtesy of the publisher.

Titration systems using coil-type cells produce similar results. The versatility and suitability of high frequency titrations for pharmaceutical analysis are shown in the paper of Allen et al.,[15] who found high frequency conductance titrations suitable for a variety of titrants, including sulfa drugs in anhydrous acetic acid.

D. CONCENTRATIONS AND KINETIC ANALYSES

Analytic procedures aimed at determining changes in concentration are, from the conductance viewpoint, often the most interesting as well as the easiest, as one uses changes in conductance rather than absolute values. Absolute calibration is not required, therefore, only structural rigidity. As an example, we will be concerned with studies in which kinetic constants are sought.

In any reaction in which ions are produced, used up, or exchanged, conductimetry may often be used to perform an "analysis" and to produce the kinetic data desired without the customary methods of sampling and analysis, using only the conductance change. This is of particular concern in sealed systems, but is often of use in rapid reactions and those carried out in small volumes. The classic ester hydrolysis serves as a good example for the analysis of the problem.

$$\text{Ethyl acetate} + OH^- \rightarrow \text{acetate}^- + \text{ethanol}$$

$$CH_3-\overset{\overset{\displaystyle O}{\|}}{C}-OCH_2-CH_3 + OH^- \rightarrow CH_3-\overset{\overset{\displaystyle O}{\|}}{C}-O^- + CH_3CH_2OH$$

If performed in a solution of suitable concentration for conductance measurements, the exchange of the relatively immobile acetate ion for the very mobile OH^- ion can easily be followed conductimetrically. The total conductance is due initially only to the base added; during the reaction to the residual base plus acetate product; and at the conclusion, to excess base and total acetate (or acetate alone if the base is the limiting reactant). If, as we did with our titration assumptions, we assume that the conductance varies linearly with concentration of all ions, then for each ion:

$$L_A = f_A(A^-)$$
$$L_{OH} = f_B(OH^-) \tag{14.22}$$

and taking as initial concentrations, E and B, for ester and base, respectively, the conductance at anytime can be described as,

$$L_{\text{total}} = f_B(B - x) + f_A(x) + f_R \tag{14.23}$$

where x is the extent of the reaction and f_R the contribution of any constant, blank and/or solvent. Initial and final conditions may be similarly defined for the usual case, where the ester is the limiting reagent:

$$L_0 = f_B(B) + f_R \tag{14.24}$$
$$L_\infty = f_B(B - E) + f_A(E) + f_R \tag{14.25}$$

Algebraic manipulation produces the result for L at time $t(L_t)$,

$$(x) = (E)\frac{L_0 - L_t}{L_0 - L_\infty} \tag{14.26}$$

and,

$$\frac{E}{(E - x)} = \frac{L_0 - L_\infty}{L_\infty - L_t} \tag{14.27}$$

which, along with knowledge of initial concentrations of base and ester, will produce the required information for plotting the second-order equation.

$$\log \frac{E(B - x)}{B(E - x)} = \frac{(B - E)kt}{2.303} \qquad (14.28)$$

Simplifications such as using equal initial concentrations can produce the desired plots directly from the conductance readings, without the necessity of separate determination of initial concentrations.

EXAMPLE 4: McGuire[16] has studied the conductance change of the following reaction:

If phenacyl bromide and pyridine are mixed at the same initial concentration (A_0), kinetics predict that their concentrations at any time a will change according to the equation:

$$\frac{1}{a} = kt + \frac{1}{A_0} \qquad (14.29)$$

or, if expressed in terms of the amount of products produced x, the equation becomes:

$$1/(A_0 - x) = kt + 1/A_0 \qquad (14.30)$$

As the products of the reactions are ions, and assuming the conductance to vary linearly with concentration, the following data reported by that author may be plotted according to Eq. (14.30), using Eqs. (14.22) through (14.27).

$A_0 = 0.0385\ M,$ Cell constant 1.000

Time, min	Resistance, ohms
7	45000
28	11620
53	9200
68	7490
84	6310
99	5537
110	5100
127	4560
153	3958
203	3220
368	2182
∞	801

E. ION ASSOCIATION AND CRITICAL MICELLE FORMATION

One of the most pharmaceutically useful, nonquantitative uses of conductimetry is certainly in the determination of critical micelle concentrations (CMC), those concentrations above which the solubilization properties of a micelle-forming ion may be expected, and below which it has been convenient

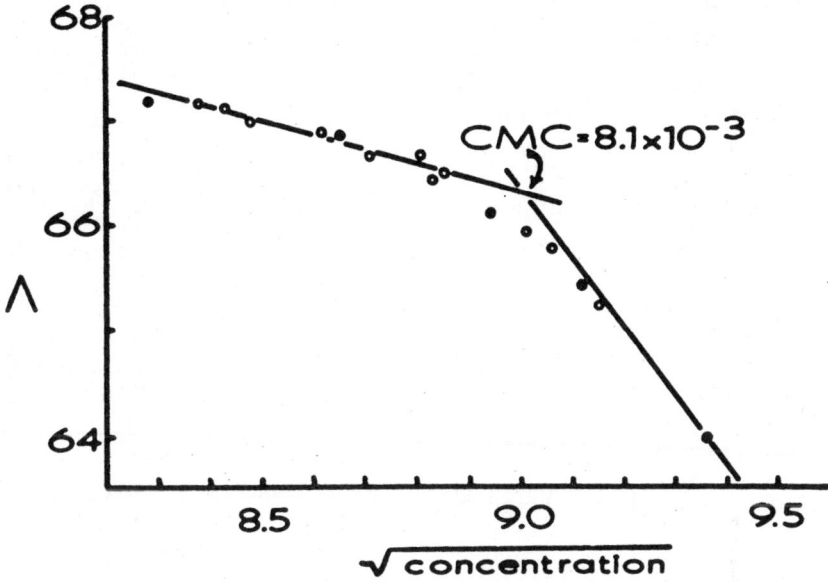

FIGURE 14.17: Onsager plots indicating the determination of critical micelle concentration from conductance data. Reprinted from Ref. 17, p. 1392, through the courtesy of the publisher.

to postulate no association occurs.* The equilibrium being described by this idea is

$$nA^- \rightleftharpoons A_n^{(n^-)} \qquad (14.31)$$

Evidence for micelle formation and an example of how the concentration at which it occurs may be determined conductimetrically as shown in Fig. 14.17. Although the lines are curved at the precise point of intersection, extrapolation of the straight portions of the lines produces rather precise and reproducible critical micelle concentrations.

The shape of the plots obtained may be explained (though with some doubt as to their complete correctness*) by considering the conductance

---

* Although it does not affect the interpretation of the curves that are obtained for micelle-forming ions, it should be pointed out that rather convincing evidence has been obtained suggesting that at concentrations considerably below the CMC, dimerization occurs (Mukerjee et al., Ref. 17).

one would postulate for the micelle-forming ion in the concentration regions surrounding the CMC. If Λ_m^- represents the conductance of the anion monomer, and Λ_p^- that of the micelle, the contribution of the former is decreased and of the latter increased as monomer is transferred to micelle. If α is defined as the fraction of the total solute in micelle form, $C(1 - n\alpha)$ and αC represent the relative concentrations of monomer and micelle at a particular monomer concentration, respectively, the total conductance then is

$$\Lambda = \Lambda_m^-(1 - n\alpha) + \Lambda_p^-\alpha n \qquad (14.32)$$

As the concentration is raised, the equivalent ionic conductance varies linearly with $\sqrt{c}$ approximating a straight line with a slightly negative slope due to ionic interactions, as one would expect. After the CMC is reached and passed the additional monomer is forced into the highly charged but relatively immobile micelles ($\Lambda_p^- \ll \Lambda_m^-$); n and α both rise. The rise in the solution conductance decreases, while the value of Λ drops sharply.

EXAMPLE 5: Mukerjee et al.[17] measured the resistance of dilute aqueous solutions of sodium lauryl sulfates and reported the following results:

Temperature = 25 C   Cell constant = 0.7492 cm^{-1}

Concentration, $\times 10^3$ M	Resistance
6.455	1725
6.859	1626
7.029	1587
7.111	1588
7.185	1556
7.430	1508
7.485$_5$	1497
7.579	1483
7.752$_5$	1449
7.796	1447
7.835	1438
7.995	1418
8.111	1401
8.140	1396
8.212	1387
8.322	1376
8.370	1372
8.456	1352
8.770	1336
9.287	1299

Determine the critical micelle concentration from these data.

Estimation of ion association and association or dissociation constants from conductance data is based upon the Arrhenius' observation discussed, that for solutes at a concentration of n molecules/liter,

$$\alpha = \frac{\Lambda}{\Lambda_0} \quad \text{where} \quad K = \frac{n\alpha^2}{(1 - \alpha)} \tag{14.33}$$

The equation is not precisely correct, as mentioned before, but the full effect of the imprecisions involved is often not large and for many determinations α may be estimated directly from Eq. (14.33).

EXAMPLE 6: Shedlovsky and McInnes[18] measured the conductance of acetic acid solutions with the following results (data reprinted through the courtesy of the publishers).

Concentration, $\times 10^3\ M$	Λ
0.028014	210.32
0.11135	127.71
0.15321	112.02
0.21844	96.466
1.02831	48.133
1.36340	42.215
2.41400	32.208
3.44065	27.191
5.91153	20.956
9.8421	16.367
12.829	14.371
20.000	11.563
50.000	7.356
52.303	7.200
100.000	5.200
119.447	4.759
200.000	3.650

Estimate α and the dissociation constant for acetic acid over this concentration range.

REFERENCES

1. S. A. Arrhenius, *Z. Physik. Chem.*, **1**, 631 (1887).
2. H. E. White, *Modern College Physics*, Van Nostrand, Princeton, N.J., 1948.
3. G. Jones and R. C. Josephs, *J. Am. Chem. Soc.*, **50**, 1049 (1930).
4. M. F. C. Ladd and W. H. Lee, *Lab. Pract.*, **9**, (2), 98 (1960).
5. J. P. Dowdall, D. V. Sinkinson, and H. Stretch, *Analyst*, **80**, 491 (1955).
6. G. Jones and B. C. Bradshaw, *J. Am. Chem. Soc.*, **55**, 1780 (1933).
7. S. H. Maron and C. F. Prutton, *Principles of Physical Chemistry*, Macmillan, New York, 1965, p. 423.
8. P. Debye and E. Hückel, *Physik. Z.*, **24**, 185, 305 (1923).
9. L. Onsager, *Physik. Z.*, **27**, 388 (1926); **28**, 277 (1927).
10. For example, see R. A. Fuoss, *J. Am. Chem. Soc.*, **81**, 2659 (1959).
11. J. J. Lingane, *Electroanalytical Chemistry*, Academic Press, New York, 2nd ed., 1958.

12. D. A. MacInnes, *The Principles of Electrochemistry*, Reinhold, New York, 1939; Dover, New York, 1961.
13. D. B. Bruss and G. A. Harlow, *Anal. Chem.*, **30**, 1836 (1958).
14. C. N. Reilly and W. H. McCurdy, *Anal. Chem.*, **25**, 86 (1953).
15. J. Allen, E. T. Geddes, and R. E. Stuckey, *J. Pharm. Pharmacol.*, **8** (11), 956 (1956).
16. W. J. McGuire, M.S. thesis, Northwestern University, Evanston, Ill., 1947.
17. P. Mukergee, K. J. Mysels, and C. I. Dulin, *J. Phys. Chem.*, **63**, 1390 (1958).
18. T. Shedlovsky and D. A. MacInnes, *J. Am. Chem. Soc.*, **54**, 1429 (1932).

CHAPTER **15**

Coulometric Methods and Chronopotentiometry

Peter Kabasakalian

SCHERING CORPORATION
BLOOMFIELD, NEW JERSEY

15.1 INTRODUCTION

A. ELECTROLYSIS[1]

Electrolysis, the passage of a direct current between a metallic conductor (*electrode*) and an electrolytic solution, is the basis for the instrumental techniques of *coulometry* and *chronopotentiometry*. It causes a chemical reaction to take place at the electrode surface by the following steps: (1) mass transfer of the electroactive species to the electrode surface under a concentration gradient, (2) the transfer of one or more electrons to the electroactive species, and (3) the removal of the products.

B. VOLTAMMETRY[2]

Electroactive compounds will give steady-state current-potential curves such as those shown in Fig. 15.1, when the electrolysis current is plotted against the potential of the working electrode (vs. a reference electrode such as a calomel electrode). Curve *a* is obtained during the electrolysis of the solvent containing electrolyte alone, while curve *b* is obtained in the presence of an electroactive compound in the electrolytic solution. The plateau in curve *b* is the *limiting* current. The increase in current over the background of the plateau region is called the "diffusion current" of the electroactive species. The potential at which the electrolysis current is equal to half the diffusion current is called the "half-wave potential" ($E_{1/2}$) of the electroactive species.

C. CURRENT-CONCENTRATION RELATIONSHIP

Controlled potential coulometry involves the complete electrolysis of the electroactive species. The current during the electrolysis is always *equal* to the diffusion current of the electroactive species. *Constant current coulometry* involves the complete titration of a compound (whose presence or lack of electroactivity is irrelevant) by a titrant which is electrochemically generated by a current *less* than the diffusion current of the electroactive titrant precursor. *Chronopotentiometry* involves limited electrolysis under conditions of constant current *in excess* of the value of the diffusion current of the electroactive species.

D. HISTORICAL

1. Constant Current Coulometry

Although coulometry was first used by Grower[3] in 1917, as an analytical method for checking the quality of tinned copper wire, the term was only

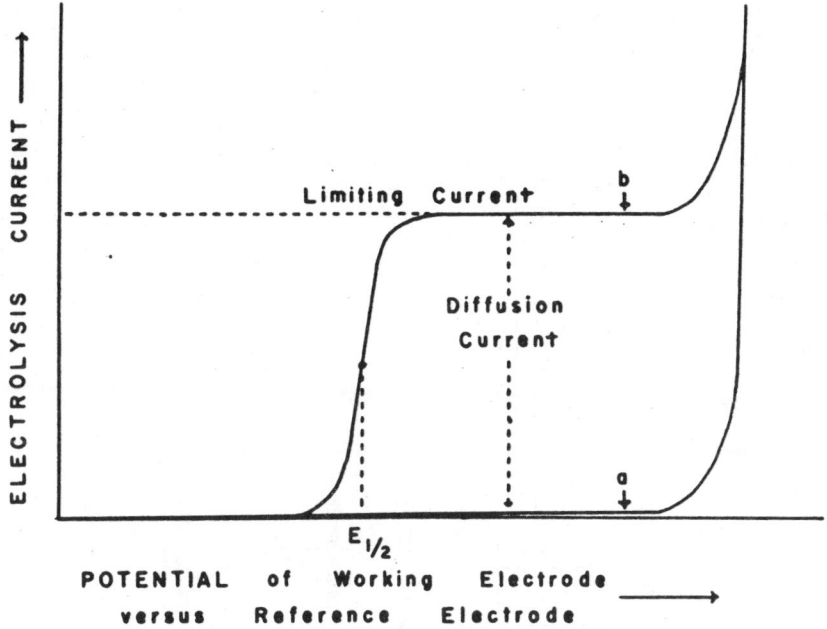

FIGURE 15.1: Curve a: voltammetric curve for solvent with supporting electrolyte; curve b: voltammetric curve for electroactive species in solvent with supporting electrolyte.

introduced in 1938 by Szebelledy and Somogyi,[4] who pioneered in constant current coulometry as a substitute for classical volumetric methods. Swift[5] and others[6] have further developed the technique by using modern end-point detection methods.

2. Controlled Potential Coulometry

Controlled potential coulometry was initiated by Hickling[7] in 1942, and further elaborated by Lingane[8] in 1945.

3. Chronopotentiometry

Gierst and Juliard[9] in 1951 recognized the analytical potentialities of chronopotentiometry, which Sand[10] had first studied in 1901.

E. COULOMETRY[11,12,13]

Coulometry is essentially a titrimetric (volumetric) technique, where electrons are used as the titrant. The three components of a volumetric setup (Fig. 15.2): (1) the titrant and its corresponding titrant storage vessel,

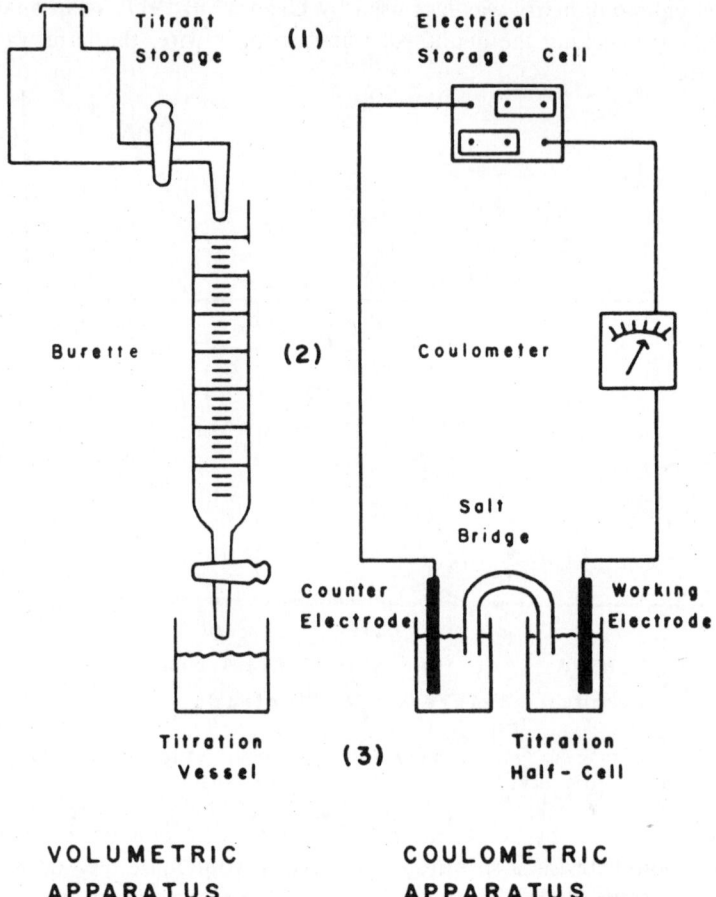

FIGURE 15.2: Comparison of volumetric and coulometric setups.

(2) the calibrated burette, and (3) the titration vessel are replaced by (1) an electrical storage cell, (2) a coulometer (a counter of coulombs), and (3) an electrical half-cell containing a suitable working electrode to introduce the electrons into the solution. An end-point detector which is necessary to feed a monitoring signal into a control device to turn off the burette in titrimetry is also necessary to turn off the coulometer in coulometry.

Coulometry can be divided into two basic types, the *direct* (or primary) and the *indirect* (or secondary). Primary coulometry is limited to the direct titration of electroactive substances with electrons. These all fall into the redox type of titration category as indicated by Eq. (15.1) and (15.2):

$$\text{reduced species} - ne \text{ (titrant)} \rightarrow \text{oxidized species} \qquad (15.1)$$

$$\text{oxidized species} + ne \text{ (titrant)} \rightarrow \text{reduced species} \qquad (15.2)$$

Secondary coulometry involves the indirect use of the electron titrant to electrogenerate chemical titrants, as indicated by Eq. (15.3) and (15.4):

$$\text{reduced species} - ne \rightarrow \text{oxidized species (titrant)} \qquad (15.3)$$

$$\text{oxidized species} + ne \rightarrow \text{reduced species (titrant)} \qquad (15.4)$$

Secondary coulometry is not limited to substances which are electroactive. It can be used as a direct substitute for the usual titrimetric procedures involving (1) *redox*, (2) *acid-base*, (3) *complexation*, and (4) *precipitation* titrations as long as the titrant can be electrochemically generated.

In coulometry, the quantity of material being titrated is determined by the quantity of electricity (coulombs) required to react with it. Thus it is obvious that the success of this procedure depends on the use of a suitable, precise, and accurate coulometer.

F. FARADAY'S LAW[14]

Coulometry is based on Faraday's law of electrolysis, which states that the extent of the chemical reaction that occurs as a result of electrolysis is directly proportional to the amount of electricity that is passed. The proportionality constant is the Faraday F, which has the value 96,487 coulombs/equivalent. Faraday's law may be expressed by Eq. (15.5), where w is the weight in grams of the species

$$\int_0^t i \, dt = Q = F(w/M)(n) \qquad (15.5)$$

that is consumed during the electrolysis, M is its molecular weight, n is the number of electrons involved in the reaction, and Q is the number of coulombs used.

G. STOICHIOMETRY

As with all titrimetric procedures, quantitative stoichiometry is desired; this is obviously equated to *100% current efficiency*. To achieve this, there should be no side reactions involving either (1) the solvent, (2) the electrode, (3) the substances which are not consumed in the electrode process (supporting electrolyte and dissolved oxygen), or (4) the products of the electrolysis.

15.2 CONTROLLED POTENTIAL COULOMETRY[11]

A. CURRENT-POTENTIAL-TIME RELATIONSHIP

1. One Component

Primary coulometry (*potentiostatic coulometry*) is carried out by *controlling the potential* of the working electrode to limit the half-cell reaction to the one being studied. Thus the potential-time relationship is invariant. The current, however, is limited by the rate of diffusion of the electroactive substance from the bulk of the solution to the electrode surface. This results in the electrolysis current i being proportional to the bulk concentration C of the substance as shown in Eq. (15.6).

$$i = kC \tag{15.6}$$

Since the material is consumed by a first-order kinetic rate, the instantaneous bulk concentration and consequently the instantaneous electrolysis current (i) decreases exponentially as shown in Eq. (15.7), where i_0 is the initial electrolysis current.[15]

$$i = i_0 e^{-k't} \tag{15.7}$$

The rate constant k', in reciprocal seconds, can be calculated from Eq. (15.8):

$$k' = DA/v\delta \tag{15.8}$$

where D is the diffusion coefficient of the electroactive species in square centimeters per second, A is the electrode area in square centimeters, v is the total volume of the solution in cubic centimeters, and δ is the thickness of the diffusion layer in centimeters. The diffusion layer is pictured as a thin layer of solution which remains stationary about the electrode surface even though the bulk of the solution is in motion; diffusion into and out of this layer occurs during electrolysis. The time of electrolysis is affected by factors which influence diffusion such as stirring rate and temperature.

2. Two Components

A mixture of compounds having sufficiently different half-wave potentials (0.1 V or more) as shown in Fig. 15.3 may be run. Only the most easily reduced (or oxidized) substance a can be selectively reduced. If the potential E_2 is great enough to reduce (or oxidize) substance b, all species (such as substance a) that are more easily reduced (or oxidized) will be electrolyzed simultaneously with substance b. If it is desired to reduce (or oxidize) only substance b, which is the more difficult one, it will be necessary first to reduce substance a completely at the proper potential (E_1).

B. END-POINT DETECTOR

Controlled potential coulometry has a built-in end-point indicator, i.e., the electrolysis current. The electrolysis is carried out until a predetermined current level is reached with the current indicator or recorder serving as the end-point detector.

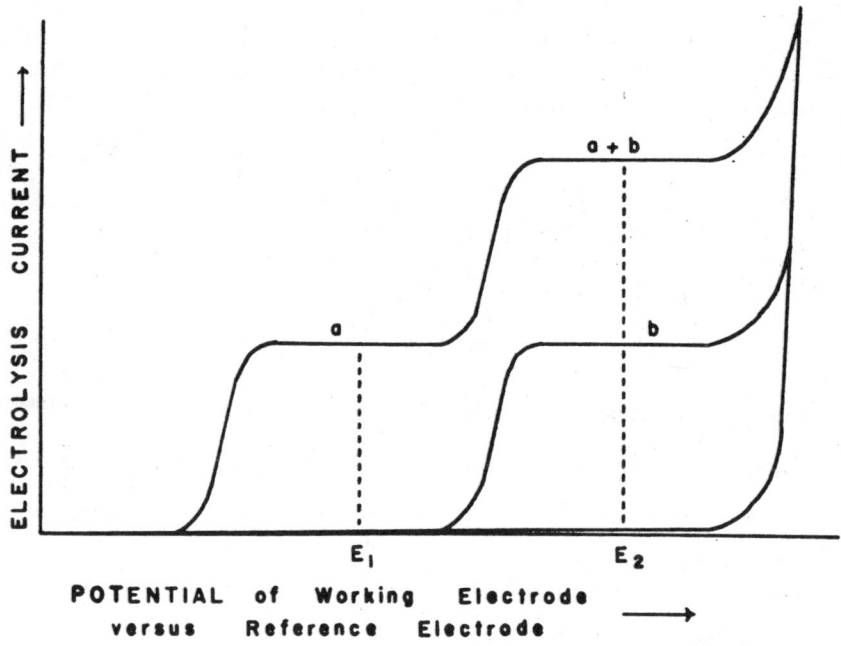

FIGURE 15.3: Voltammetric curves for (1) solvent with supporting electrolyte; (2) substance b in solvent with supporting electrolyte; and (3) substances a and b in solvent with supporting electrolyte.

C. CIRCUIT

A simplified circuit for controlled potential coulometry is shown in Fig. 15.4. It consists of the following items:

I. Power Supply

A direct current power supply which can maintain the working potential at a constant level is used in controlled potential coulometry. In Fig. 15.4 a battery (E_B) and potentiometer (R_1) are used. Point c is varied either manually,[11,16] electromechanically,[17] or electronically[18] to maintain a constant potential E_c on the working electrode, as indicated by a high-input impedance electronic voltmeter V.

2. Electrolysis Vessel

A three-electrode electrolysis vessel (Fig. 15.5) is used for the titration. It contains (a) a *working electrode* where the titration is taking place, (b) a *reference electrode* (usually a calomel electrode), which is used to determine the potential of the working electrode by comparison, and (c) a *counter electrode* (sometimes called an *auxiliary electrode*) to complete the electrical circuit from the controlled dc power supply. Optimum cell design implies a

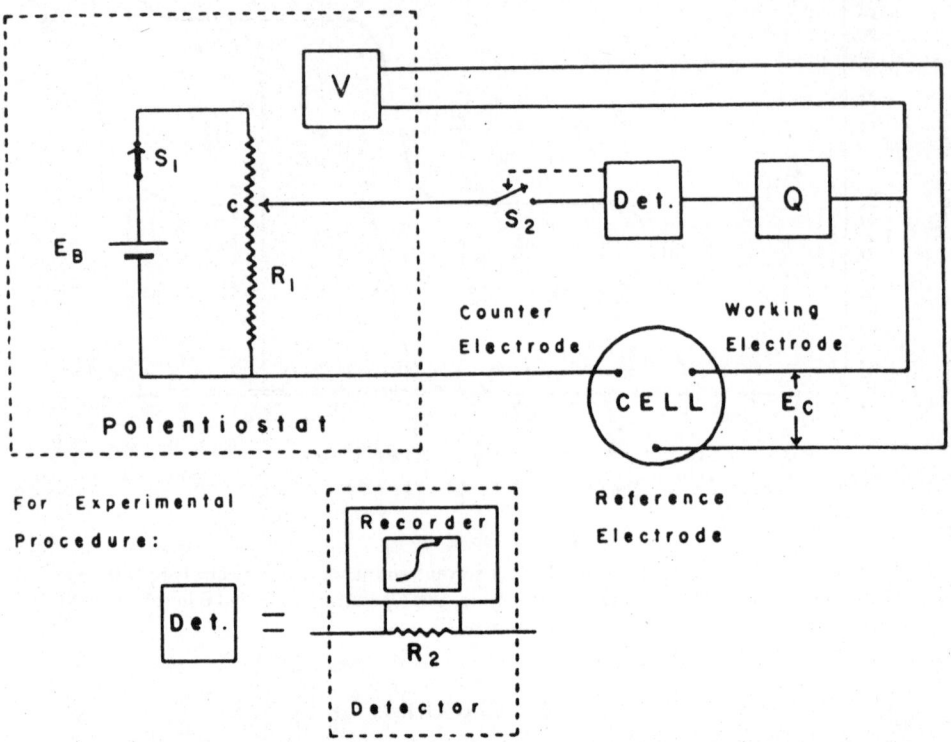

FIGURE 15.4: A simplified circuit for controlled potential coulometry where E_B is a 6-V storage battery, R_1 is a 50-Ω, 2-W, 10-turn potentiometer, R_2 is 0.100-Ω precision ($\pm 0.05\%$) resistor, S_1 and S_2 are SPST switches, Q is a coulometer, V is a high-input impedance electronic voltmeter such as a pH meter, and *Det.* is the end-point detector.

large electrode area, small solution volume, and a high stirring rate. Each half-cell is usually separated from the other half-cells by porous diaphragms (with an agar plug) or salt bridges to prevent the contamination of the individual compartments with electrolysis products.

3. Coulometer

The number of coulombs, Q, used in the reaction is determined by Eq. (15.9) and can be calculated[19,20] directly by use of Eq. (15.10) or (15.11) if the rate constant, k', is known.

$$Q = \int_0^t i \, dt \tag{15.9}$$

$$Q = \int_0^t i_0 e^{-k't} \, dt = (i_0/k')(1 - e^{-k't}) \tag{15.10}$$

$$Q = \int_0^\infty i_0 e^{-k't} \, dt = i_0/k' \tag{15.11}$$

a. Graphical. The current-time curve may be integrated graphically. This may be done by (1) cutting out the current-time curve obtained from a

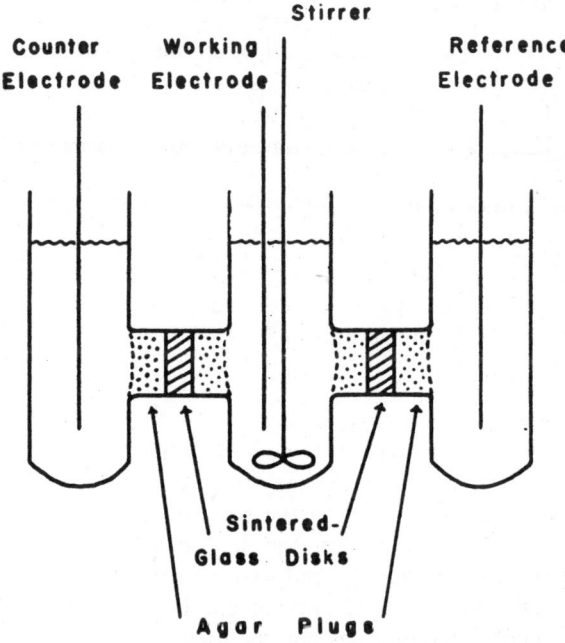

FIGURE 15.5: A three-electrode electrolysis vessel for controlled potential coulometry and chronopotentiometry.

strip chart recorder, and comparing its weight with a square of the same paper of known area by (2) using a planimeter or (3) with a ball-and-disk integrator[11,21] attached to the recorder.

b. Chemical. Chemical coulometers depend on the electrochemical preparation of a compound and its subsequent quantitative determination by any of the following chemical methods: (1) gravimetric,[22-24] (2) titrimetric,[23] (3) gasometric,[8,25] and (4) colorimetric.[26]

c. Electromechanical. Current-time integrators are available which make use of: (1) low inertia integrating motors,[27-30] (2) analog (*operational amplifier*) integrators,[18,31] and (3) current (via voltage) to frequency converter with a frequency counter.[32,33]

D. EXPERIMENTAL CONDITIONS

1. Electrode

a. Material. The most widely used electrode materials are mercury and platinum. Mercury is satisfactory mainly as a cathode and platinum is used as either an anode or a cathode. Other electrodes such as boron carbide,[34] carbon black,[35] carbon paste,[36] gold,[37] and graphite[37,38] have been used to a limited extent.

b. Potential Range. Because of the overpotential of hydrogen on mercury (ca. 1.2 V), mercury can be used in basic solutions and nonaqueous solutions as a cathode at potentials as negative as -2.8 V vs. a saturated calomel electrode (SCE), while its use as an anode is limited to about $+0.3$ V (vs. SCE) because of anodic dissolution.

Platinum, however, does not have an appreciable overpotential for hydrogen, thus its potential as a cathode in basic solutions is limited to about -1.1 V (vs. SCE). The inertness of platinum enables it to be used as an anode to about $+1.1$ V vs. SCE in aqueous solutions and higher potentials (ca. $+2.0$ V vs. SCE) in acetonitrile.[39,40]

2. Solvent

a. Electrolyte. A highly ionizable soluble salt (in concentrations 50–100 times the electroactive species being studied), the supporting electrolyte, is used to conduct the electricity in the solvent. It should not be involved in electrode reactions. To obtain the largest potential range possible for cathodic half-cells, difficultly reduced cation salts such as quaternary ammonium salts may be used and for anodic half-cells, difficultly oxidized anion salts such as perchlorates may be used.

b. pH. The pH of the solution affects both the electroactive species and the background current. The former effect occurs only with material whose redox potential is pH sensitive. For these the $E_{1,2}$ is smaller in acidic media for cathodic reactions and larger for anodic reactions. The reverse holds true in basic media. The background current in aqueous solutions may be due to the following reactions

$$H^- + e \rightleftharpoons \tfrac{1}{2}H_2 \tag{15.12}$$

$$H_2O - 2e \rightleftharpoons \tfrac{1}{2}O_2 + 2H^- \tag{15.13}$$

causing the production of hydrogen at the cathode and oxygen at the anode. These processes impose an upper limit on the potentials that the working electrode can assume since they obscure other electrolyses. Reactions (15.12) and (15.13) are pH dependent, the discharge of hydrogen ions occurs more easily in acidic solutions, while the oxidation of water occurs more easily in basic solutions.

c. **Oxygen.** If oxygen is reduced at a lower potential than the electrode reaction of interest, it must be removed to insure 100% current efficiency. This is usually done by deoxygenating with nitrogen. Oxygen does not normally interfere in anodic half-cell reactions.

d. **Temperature.** To decrease the electrolysis time, the electrolysis cell may be operated at an elevated temperature. This results from the fact that the diffusion coefficient of the electroactive species increases approximately 2%/deg.

e. **Mixing.** A high rate of stirring minimizes the diffusion layer and thus reduces the electrolysis time.

3. Potential

The potential of the working electrode is chosen to (a) give the desired half-cell reaction, (b) minimize background (or *blank*) current, and (c) give the desired separation from other electroactive materials that may be present.

4. End Point

The electrolysis may be discontinued when the ratio i/i_0 (equal to the ratio of the concentration C_t remaining unreacted at time t to the initial concentration C_0) decreases to a value corresponding to the degree of completion desired. For 99.9% completion the electrolysis must be terminated at 0.1% of the initial current. This termination current must also be at least as large as the background current.

5. Current Range

Electrolysis currents having magnitudes from 10 μA to 100 mA are adequate for controlled potential coulometry.

6. Concentration Range

Controlled potential coulometry has been successfully used for concentrations as high as 2×10^{-3} M and for concentrations as low as 5×10^{-8} M.[31,41]

7. Time

The usual range of times required for a complete controlled potential coulometry run is 10 to 60 min.

E. ADVANTAGES AND LIMITATIONS

Controlled potential coulometry offers a significant advantage over voltammetry in the precision which can be achieved in an analysis. However, this technique has limited use since voltammetry and polarography usually can be used for the same type of determination more rapidly, easily, and with simpler equipment.

1. Accuracy

The accuracy of controlled potential coulometry is only limited by the inability to maintain 100% current efficiency because of background current. Meites and Moros[42] have divided the background current into five components: (a) charging current, (b) impurity faradaic current, (c) continuous faradaic current, (d) kinetic background current, and (e) induced background current. The simplest way to minimize background-current errors is to utilize a relatively large quantity of the electroactive species, thereby making insignificant the contribution of the background current to the total Q. ·

2. Precision

The precision obtained in controlled potential coulometry is limited by the reproducibility of the coulometer used and is normally of the order of 0.3 to 1%.

F. PHARMACEUTICAL APPLICATIONS

Those compounds which are reducible or oxidizable by polarography or voltammetry (the study of currents when the current-voltage characteristics depend on the electrode reaction rate), and which have well-defined diffusion controlled waves can be run. Organic functional groups which are normally reducible are: (1) carbon-carbon double bonds when conjugated with another unsaturated group, (2) the carbonyl group in ketones and aldehydes, (3) nitro and related nitrogen compounds, (4) halogens, and (5) disulfides and peroxides. The typical oxidizable groups are aromatic amines and phenols. For more details see Chapter 16, on polarography.

Controlled potential coulometry has been reported for copper,[43] iron,[31] lead,[44] oxygen,[45–47] sodium,[48] tin,[44,49] anthraquinone,[50] ascorbic acid,[16] chlorpromazine,[51] hydrogen peroxide,[52] organic halogen compounds,[53] organic nitro compounds,[53–55] N-substituted phenothiazines,[56] phenylmercuric ion,[57] and water.[58]

Although very few pharmaceutical compounds have been studied, it should be applicable to: (1) unsaturated carbonyl compounds such as corticosteroids (cortisone, hydrocortisone, prednisolone, and prednisone) and some antibiotics (chlortetracycline, griseofulvin, oxytetracycline, and tetracycline); (2) iodine compounds such as iodinated X-ray contrast agents (iodoalphionic acid and iopanoic acid) and iodochlorohydroxyquin; (3) nitro compounds such as chloramphenicol and nitrofurantoin; (4) chloro compounds such as chlorobutanol, chloroform, and trichlormethiazide; and (5) mercurial preservatives such as phenylmercuric acetate, chloride, and nitrate. In seeking optimum conditions one may make use of the wealth of voltammetric and polarographic information that is available.

15.3 CONSTANT CURRENT COULOMETRY[59,60]

Secondary (amperostatic) coulometry, which is frequently referred to as "coulometric titration" is carried out by *controlling the current* to the working electrode. The potential of the working electrode is controlled indirectly by maintaining the electrolysis current at a level where it can never be limited by the rate of diffusion of the electroactive substance. Thus the current is less than that for controlled potential electrolysis, as indicated by Eq. (15.14)

$$i < kC \qquad (15.14)$$

A. CURRENT-POTENTIAL-TIME RELATIONSHIP

1. One and Two Components

The current and potential of the generating electrode in a constant current coulometric run are essentially invariant with time. Thus one cannot differentiate between the presence of one or two components from any knowledge of the current-potential-time relationship of the working electrode. However, the differentiation is possible if a suitable end-point detection method for differentiation is available.

B. END-POINT DETECTOR

The current cannot be used as an end-point indicator because it is constant throughout the electrolysis. Other end-point detection methods common in titrimetry are used. They may be divided into (1) optical and (2) electrometric methods. The former consists of (a) visual,[61] (b) photometric,[62] and (c) spectrophotometric[63,64]; and the latter consists of (a) potentiometric[65,66] and (b) amperometric[67] (an end-point device whereby the concentration of an electroactive substance is measured by the current which results from its reaction at an electrode). In the electrometric end-point detection methods, a second pair of electrodes is necessary.

C. CIRCUIT

A simplified circuit for constant current coulometry is shown in Fig. 15.6 and consists of the following:

1. Power Supply

A constant current generator usually referred to as an "amperostat" is used. The one displayed in Fig. 15.6 uses a high voltage battery (E_{B_1}) as the voltage source. A large resistance (compared with the electrolysis cell) R_1 is placed in series with the titration cell to limit the current to the desired

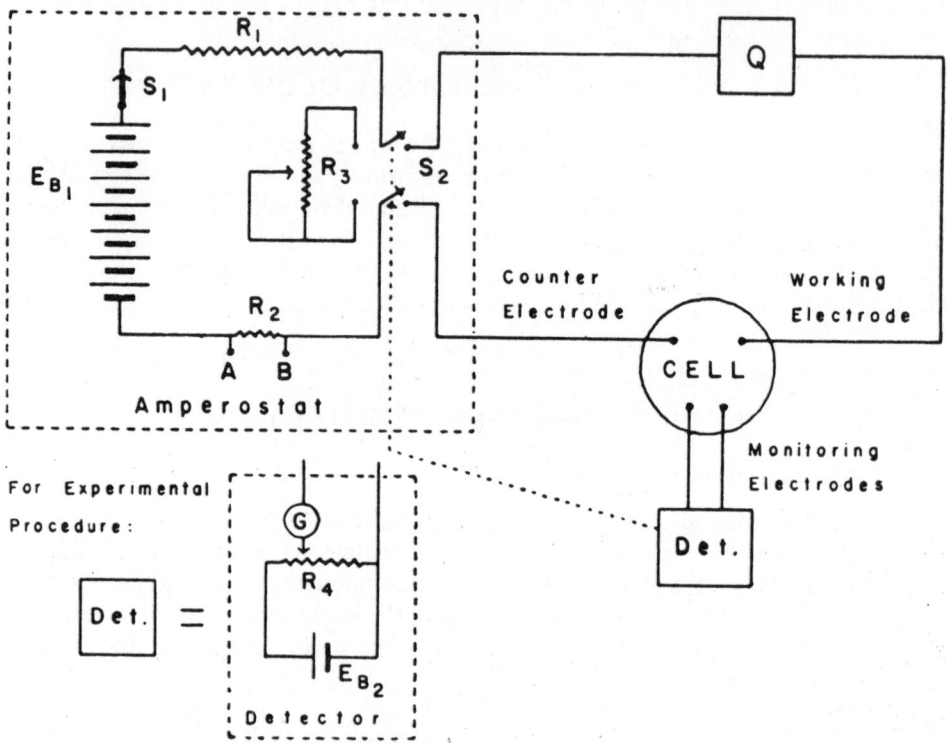

FIGURE 15.6: A simplified circuit for constant current coulometry where E_{B_1} is three 45-V dry-cell batteries, E_{B_2} is a 1.5-V dry cell, R_1 is a 7000-Ω, 10-W resistor, R_2 is a 50-Ω precision resistor ($\pm 0.05\%$), R_3 is a 100-Ω, 0.5-W rheostat, R_4 is a 1000-Ω, 10-turn potentiometer, G is a 20-μA meter, S_1 is an SPST switch, S_2 is a DPDT shorting switch, $Det.$ is the end-point detector, and Q is the coulometer.

value and to make it independent of the effective electrolysis cell resistance. An amperostat can also be an electromechanical[68] or electronic[69-71] device.

2. Electrolysis Vessel

a. Internal Generation. In constant current coulometry, the electrolysis cell is simpler than in controlled potential coulometry because one half-cell (the reference electrode) is not needed. One should see that (1) the working electrode surface area is large enough for the intended electrolysis current

(current density ca. 0.5 mA/cm^2/millinormal), (2) the contents of the cell are mixed rapidly and thoroughly so that the titrant is consumed as fast as possible, (3) the indicating system used responds quickly, and (4) when necessary, there are provisions for deoxygenation.

b. **External Generation.**[28,61,69,72,73] When conditions for obtaining a 100% current efficiency for the generation of the titrant are not compatible with the conditions for fast consumption of the titrant by the substance being titrated,[74] the titrant may be generated externally. Figure 15.7 illustrates

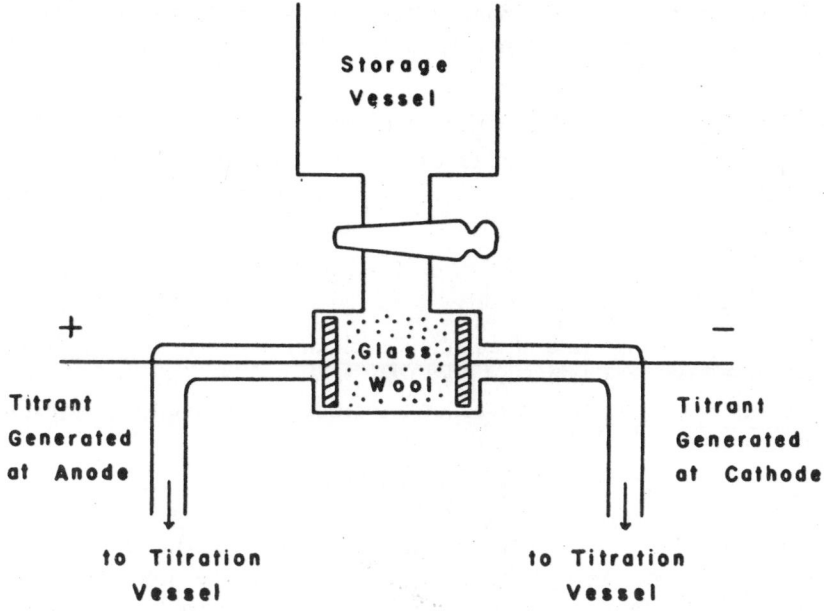

FIGURE 15.7: A constant current coulometric electrolysis vessel for the external generation of titrant.

an external electrolysis vessel. The titrating solvent containing the appropriate electrolyte flows through the electrolysis vessel and splits into two streams. One stream contains the product generated in the anodic compartment and the other, the product generated in the cathodic compartment. The appropriate stream flows into a standard volumetric titration flask.

3. Coulometer

A coulometer is used to integrate the constant current-time relationship shown in Eq. (15.15).

$$\int_0^t i\,dt = i\int_0^t dt = it \qquad (15.15)$$

In constant current coulometry, coulometers consist essentially of chronometers (timers). For most applications, synchronous electric timers operated by the commercial ac power frequency (60 Hz) are used.[66] However, for the greatest precision, electronic stop clocks with built-in precision electronic timers are used.[75,76]

D.   EXPERIMENTAL CONDITIONS

I. Electrode

a. Material and Potential Range. The material and potential range are identical to those used for controlled potential coulometry.

2. Solvent

a. Electrolyte. In addition to carrying the current, the electrolyte is generally involved in the electrode reaction by being the electroactive chemical titrant precursor.

b. pH, Oxygen, and Temperature. These influences are identical to those in controlled potential coulometry.

c. Mixing. A high rate of mixing is necessary not to limit the indicator response by assisting in having the titrant react quickly with the substance to be titrated. If the titration reaction is limiting in spite of adequate mixing, *back titration* methods[77-81] must be used.

3. Potential

The potential is not directly controlled. However, the redox buffer resulting from the electroactive titrant precursor and the electrode reaction product usually stabilize the potential of the working electrode and prevent it from drifting.

4. End Point

A sharp change in voltage or current is normally used as an end point by most of the end point detectors.

5. Current Range

Electrolysis currents having magnitudes from 1 to 100 mA are adequate for constant current coulometry.

6. Equivalence Range

Constant current coulometry is conveniently applicable to amounts ranging from a few milliequivalents to about 10^{-4} microequivalent in volumes of the order of 50 ml.[6,11,82]

7. Time

Coulometric titrations are usually completed in a few minutes (100 to 300 sec).

E. ADVANTAGES AND LIMITATIONS

The main advantages of constant current coulometry over conventional titrimetry lie in areas where low concentration or unstable titrants are involved. Other advantages are the ease with which the complete titration can be automated and the fact that no standard solutions are required. The principle limitation of the method is the unavailability of all possible titrants.

1. Accuracy

The accuracy of constant current coulometry is only limited by the inability to maintain 100% current efficiency because of background current. In general, it is subject to less background current than controlled potential coulometry, thus enabling it to be used for the determination of smaller quantities with more convenience and accuracy than potentiostatic coulometry.

2. Precision

The precision obtained in constant current coulometry is limited by (a) the constancy of the current used, (b) the reproducibility of the current measurement, and (c) the reproducibility of the time measurement. The precision normally obtained in constant current coulometry is of the order of 0.1 to 0.3%, comparable to the precision of volumetric analysis. With precautions and simplified equipment, one can measure the number of coulombs (or microequivalents) to about 0.004%,[83] and Tutundzic[84] even recommends that the coulomb replace chemical standards as the ultimate standard for all volumetric work.

F. PHARMACEUTICAL APPLICATIONS

Since titrant is generated, any normal titrimetric procedure can be used. Table 15.1 lists typical examples of the types of titrimetry available, typical generation reactions, and typical titration reactions.

TABLE 15.1: Typical Electrode and Titrant Reactions in Constant Current Coulometry

Type of titration	Typical generation reaction	Typical titration reaction
Redox	$2Br^- - 2e \rightarrow Br_2$ (titrant)	$Br_2 + R_2C=CR_2 \rightarrow R_2CBrCBrR_2$
	$Ti^{4+} + e \rightarrow Ti^{3+}$ (titrant)	$Fe^{3+} + Ti^{3+} \rightarrow Fe^{2+} + Ti^{4+}$
Acid-base	$3H_2O - 2e \rightarrow \frac{1}{2}O_2 + 2H_3O^+$ (titrant)	$H_3O^+ + B \rightarrow BH^+ + H_2O$
	$2H_2O + 2e \rightarrow H_2 + 2OH^-$ (titrant)	$HA + OH^- \rightarrow A^- + H_2O$
Complexation	$HgY^{2-} + 2e \rightarrow Hg^\circ + Y^{4-}$ (titrant)	$Ca^{2+} + Y^4 \rightarrow CaY^{2-}$
Precipitation	$Ag^\circ - e \rightarrow Ag^+$ (titrant)	$Ag^+ + Cl^- \rightarrow AgCl\downarrow$

The pharmaceutical compounds given as examples in the titrimetric procedures listed in the chapters on (1) precipitation, complex formation, and oxidation-reduction, (2) acidimetry and alkalimetry, (3) nonaqueous titrimetry, and (4) complexometric titrations are directly applicable. Table 15.2 lists the types of titrants that have been electrochemically generated and their literature reference.

TABLE 15.2: Electrogenerated Titrants

Type of titration	Titrant	Literature reference
Redox	Bromine	85–88
	Cerium(IV)	65, 89
	Chlorine	90, 91
	Copper(II)	77
	Hypobromite ion	92
	Iodine	93–95
	Biphenyl radical anion	96
	Chlorocuprous ion	97
	Copper(I)	78
	Tin(II)	98
	Titanium(III)	99
Acid-base	Aqueous acid-base	100, 101
	Nonaqueous acid	102–106
	Nonaqueous base	107–109
Complexation	Cyanide ion	110
	EDTA (ethylenediaminetetraacetic acid)	111
	EGTA [ethylene glycol bis(β-aminoethyl ether)-N,N-tetraacetic Acid]	112
Precipitation	Ferrocyanide ion	113
	Halide ion	114
	Mercury(I)	115, 116
	Mercury(II)	115, 116
	Silver(I)	82, 117
Miscellaneous	Hydrogen	118
	Karl Fischer	119
	MTEG (monothioethylene glycol)	120
	Thioglycollic acid	121

15.4 CHRONOPOTENTIOMETRY[2,11,122]

Chronopotentiometry, which is sometimes called "voltammetry," *at constant current*, is carried out by *controlling the current* to the working electrode. The value of the constant current chosen is greater than the diffusion current of the electroactive species, as shown in Eq. (15.16)

$$i > kC \qquad (15.16)$$

The name "chronopotentiometry" is derived from the fact that a constant current is applied to an electrode and its potential is measured against some reference electrode as a function of time.

A. CURRENT-POTENTIAL-TIME RELATIONSHIP

Since the current is maintained constant by external means, the only variables are potential and time. The potential of the working electrode is in equilibrium with the electrode reactions forced upon the electrode by the electrolysis current.

After a finite time of electrolysis, the electroactive species will be exhausted at the electrode surface. The diffusion rate will not be able to maintain the current because the value of the constant electrolyzing current is greater than the diffusion current of the electroactive species. At this point, another electroactive species (which may be the supporting electrolyte), will start to undergo the electrode reaction, and there will be a rapid change in electrode potential. The time at which this occurs is called the "transition time."

1. One Component

The Sand[10] equation for linear diffusion governs the rate the substance is brought to the electrode:

$$\tau^{1/2} = (\pi^{1/2}nFAD^{1/2}C)/(2i) \qquad (15.17)$$

where τ is the transition time in seconds, π is 3.1416, n is the number of electrons involved in the reaction, F is the Faraday in coulombs (96,487 coulombs), D is the diffusion coefficient in square centimeters per second, A is the area of the electrode in square centimeters, C is the bulk concentration of the electroactive species in moles per cubic centimeters, and i is the electrolysis current in amperes.

In analytical practice, the Sand equation is seldom used directly. Instead, the chronopotentiometric constant $(i\tau^{1/2})/C$ is used to empirically calibrate the electrode under a known set of conditions and concentrations for the substance to be determined.

The Sand equation applies strictly to diffusion controlled processes occurring with 100% current efficiency at a plane electrode. This requirement limits the applicability of the Sand equation to the first reaction and to the combined reactions (which must include all the preceding ones) which take place during the electrolysis of a multicomponent system.

Chronopotentiograms (as shown in Fig. 15.8) follow Eq. (15.18), which is commonly called the Karaoglanoff[123] equation:

$$E = E_{1/4} - \frac{RT}{nF} \ln [(\tau^{1/2}/t^{1/2}) - 1] \qquad (15.18)$$

The quarter-wave potential, $E_{1/4}$, is the potential at one-quarter the transition time, and when there are no kinetic complications, it is identical to the polarographic $E_{1/2}$.

2. Two Components

If there are two electroactive species in a mixture and their voltammetric half-wave potentials differ by at least 0.1 V, then two distinguishable transition times are observed on the chronopotentiogram. These are used to calculate the concentration of each species. Since the transition time is proportional to the square of the concentration of the electroactive species,

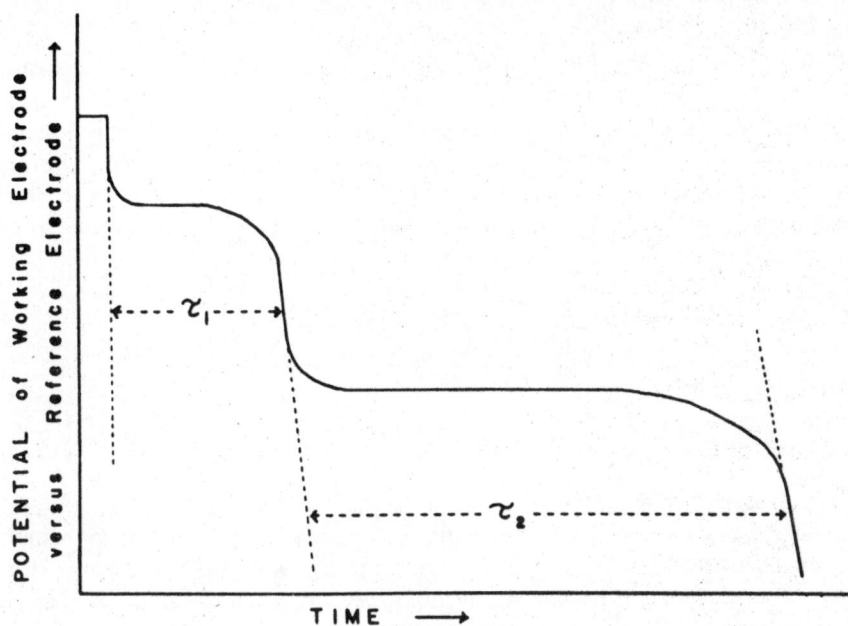

FIGURE 15.8: Chronopotentiogram of a mixture of two compounds having identical concentrations and electrode reactions.

the transition times will not be additive. The combined transition time (τ_{12}) for either consecutive or stepwise reactions (assuming identical diffusion coefficients) is proportional to the square of a pseudo-total concentration (C_{12}^*) as indicated by Eq. (15.19) and (15.20).

$$\tau_{12} = k''(C_{12}^*)^2 \qquad (15.19)$$

$$C_{12}^* = n_1 C_1 + n_2 C_2 \qquad (15.20)$$

The transition time (τ_2) for the component which is reduced (or oxidized) after the first component is given by Eq. (15.21).

$$\tau_2 = \tau_{12} - \tau_1 \qquad (15.21)$$

The two transition times are compared in Eq. (15.22).

$$(\tau_2/\tau_1) = (\tau_{12}/\tau_1) - 1 = (C_{12}^*/C_1)^2 - 1 \qquad (15.22)$$

It follows when n_2C_2 is (a) equal to, (b) twice, and (c) three times n_1C_1, that the second transition time (τ_2) will be respectively (a) 3, (b) 8, and (c) 15 times the first transition time (τ_1). This phenomenon is referred to as an "enhancement" of the second transition time. Figure 15.8 shows the chronopotentiogram of a two component mixture having identical concentrations and electrode reactions.

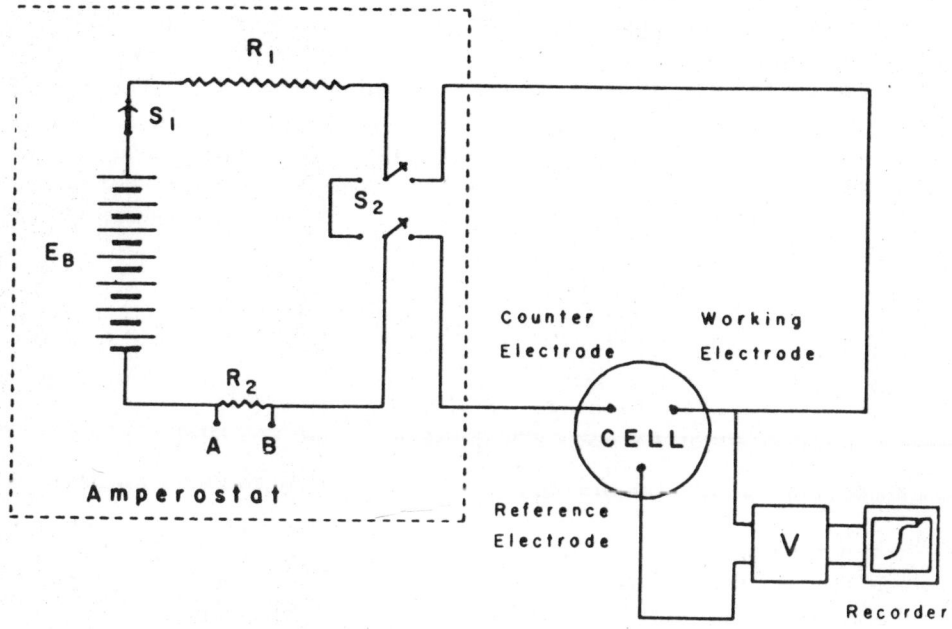

FIGURE 15.9: A simplified circuit for chronopotentiometry where E_B is three 45-V dry-cell batteries, R_1 is a 536,000-Ω, 0.5-W resistor, R_2 is a 4000-Ω precision ($\pm0.05\%$) resistor, S_1 is an SPST switch, S_2 is a DPDT shorting switch, V is a high-input impedance electronic voltmeter.

B. END-POINT DETECTOR

A voltage indicator or recorder is used as the end-point detector. The input impedance of the device directly connected to the working and reference electrodes must be high enough so that no significant current (with respect to the electrolyzing current) flows in the indicating circuit.

C. CIRCUIT

A simplified circuit for chronopotentiometry is shown in Fig. 15.9. A coulometer is not part of the circuit. The components for chronopotentiometry are:

1. Power Supply

A regulated constant current source similar to the one for constant current coulometry is used.

2. Electrolysis Cell

In chronopotentiometry, the electrolysis cell is similar to that used in controlled potential coulometry. It requires a three-electrode electrolysis vessel containing (a) a working electrode, (b) a reference electrode, and (c) a counter electrode. However, the configuration of the working electrode-counter electrode assembly is critical. It must be designed to maximize the conditions of *linear diffusion*, i.e., diffusion that takes place in a single direction to a planar surface. The surface area of the working electrode must be compatible with current concentration requirements for chronopotentiometry.

D. EXPERIMENTAL CONDITIONS

1. Electrode

a. Material and Potential Range. The electrode (area ca. 0.25 cm²) material and potential range are essentially the same as for controlled potential coulometry. In chronopotentiometry, a platinum electrode is useful as an anode to ca. $+1.5$ V (vs. SCE),[122] in aqueous systems.

b. Shape. A mathematical analysis of diffusion to a planar electrode is much simpler than for diffusion to cylinders or other geometric configurations. However, cylindrical wire electrodes are easily made, and there are only very minor edge effects when the length of the wire is large compared to its diameter. Although these electrodes are experimentally convenient, the data obtained with them, however, is difficult to interpret because the theoretical transition time equations are quite complex. Lingane[124] introduced a simple empirical correction shown in Eq. (15.23):

$$\tau_{obs} = \tau_{plane}(1 - \beta\tau^{1/2}) \qquad (15.23)$$

where τ_{plane} is the transition time with a plane electrode of equal area, and the constant β depends on the diffusion coefficient of the electroactive species and the radius of the wire electrode.

2. Solvent

a. Electrolyte, pH, and Oxygen. These requirements are identical to those for controlled potential coulometry.

b. Temperature and Mixing. Effects which will transport electroactive species to the electrode surface by methods other than simple linear diffusion must be minimized. The temperature therefore must be kept constant and there should be no mixing.

3. Potential

The potential is the variable parameter being measured.

4. End Point

The end point may be a preselected potential (usually the inflection point on the potential-time curve), or it may be graphically determined by methods comparable to the correction for residual currents in polarography.

5. Current Range

Electrolysis currents having magnitudes of 5 μA to 500 μA are normally adequate.

6. Concentration Range

The best results in chronopotentiometry have been obtained for concentrations between 1 and 10 mM, but the method fails at concentrations below about 0.5 mM. The most concentrated solutions successfully studied were 50 mM.

7. Time

Suitable transition times for chronopotentiometric analyses range from about 5 to 25 sec.

E. ADVANTAGES AND LIMITATIONS

Anodic chronopotentiometry appears most advantageous when the technique of oxidative voltammetry fails because of fouling of the electrodes by the products. The much shorter reaction time is responsible for this, but the problem persists[125] for many materials whose reaction products coat the electrode with insoluble polymeric material.

I. Accuracy

The accuracy of chronopotentiometry is limited solely by the inability to maintain 100% current efficiency. At short transition times, (less than 1 sec), background currents[126] are due to (a) charging of the electrical double layer, (b) adsorption of electroactive species, (c) surface oxidation or reduction of the working electrode (platinum), and (d) impurities in solution. At long transition times, (greater than 25–50 sec), errors are introduced[127] because of (a) natural convection due to density gradients and (b) unavoidable external laboratory vibrations which disturb the diffusion layer.

2. Precision

The precision obtained in chronopotentiometry is limited by (a) the constancy of the electrolysis current used, (b) the constancy and reproducibility of the temperature during electrolysis, (c) the reproducibility of the current measurement, and (d) the reproducibility of the time measurement. The precision normally obtained in chronopotentiometry is about 1 %.

F. PHARMACEUTICAL APPLICATIONS

Although very few pharmaceutical compounds have been studied, chronopotentiometry should be applicable to polarographically or voltammetrically electroactive species. It should be especially useful for antioxidants and related compounds.

Chronopotentiometric studies have been reported for cerium,[128] bromide,[129] copper,[130] iodide,[127] iron,[124,130] lead,[127,130,131] oxygen,[132] silver,[127] zinc,[130] adenine,[134] anthracene,[133] antioxidants and antiozonants,[125] aromatic amines,[135] ascorbic acid,[134] benzoquinone,[134] catechol,[136] hydrazine,[137] hydroquinone,[127,134,136] hydroxylamine,[138] hydrogen peroxide,[139] mercaptobenzothiazole,[140] oxalic acid,[141] phenols,[135] phenylenediamine,[125,136], phenylmercuric ion,[57] riboflavin,[142] sulfa drugs,[143] sulfanilamide,[133,134] toluene 2,4-diamine,[140] and triethylamine.[144]

15.5 EXPERIMENTAL PROCEDURES

A. CONTROLLED POTENTIAL COULOMETRY

Ascorbic acid is determined by controlled potential coulometry using the general procedure of Santhanam and Krishnan[16] (courtesy of *Analytical Chemistry*).

I. Apparatus

a. Potentiostat. A manually controlled potentiostat (current capacity ca. 100 mA) shown in Fig. 15.4 is used.

b. Electrodes. Concentric cylinders of thick platinum gauze with reinforced edges serve as anode and cathode. The former is 2.5 cm in diameter, 4 cm high, and the latter, 5 cm in diameter and 5 cm high. The reference electrode is a standard saturated calomel electrode placed in a sleeve (filled with a solution of 1 M potassium nitrate) terminated with a fine sintered-glass disk. This combination is placed in the center of the cylindrical anode.

c. Cell. The electrolysis cell is a 250-ml open beaker provided with magnetic stirring.

d. End-Point Detector. A precision 0.100-Ω resistor in conjunction with a 10-mV strip-chart recorder (full-scale reading 100 mA) with a chart speed of 20 in./hr is used to determine the end point.

e. Coulometer. The coulometer is a nitrogen-hydrogen type[25] with a 5-ml burette containing 0.1 M hydrazine sulfate.

2. Reagents

The supporting electrolyte is a potassium biphthalate buffer of pH 6.0. It is prepared by diluting 500 ml of 0.1 M potassium biphthalate with about 455 ml of 0.1 M sodium hydroxide and checking the final pH with a glass-calomel electrode combination. The ascorbic acid is USP grade.

3. Procedure

Add about 180 ml of the buffer solution to the cell and electrolyze at an anode potential of +1.2 V (vs. SCE) until the current falls to a steady low background value. This removes any oxidizable impurities in the supporting electrolyte and avoids the need for background correction. Transfer into the preelectrolyzed solution a weighed sample of about 15 mg of ascorbic acid. Zero the gas coulometer. Start the electrolysis at +1.1V (vs. SCE). Simultaneously start the strip-chart recorder. Continue the electrolysis until the current is about 1 mA. This should take about 30 min.

4. Calculation

Assuming a two-electron process, calculate the amount of ascorbic acid titrated using the quantity of electricity passed calculated from (a) the gas coulometer reading and (b) an integration of the current-time curve on the recorder.

B. CONSTANT CURRENT COULOMETRY

Isoniazid (isonicotinic acid hydrazide) is analyzed by constant current coulometry with electrolytically generated bromine. The following is the general method of Olson[85] (courtesy of *Analytical Chemistry*).

1. Apparatus

a. Amperostat. The amperostat (Fig. 15.6) is a constant current power supply which operates at 20 mA.

b. Electrodes. The generating electrodes are a pair of 1 cm² platinum foil electrodes. Isolate the cathode from the test solution by a sintered-glass disk.

c. **Cell.** The titration vessel is a 100-ml beaker provided with magnetic stirring.

d. **End-Point Detector.** The end point is determined amperometrically. A pair of concentrically wound platinum spirals are the indicating electrodes, and a potential of 0.2 V is impressed across them.

e. **Coulometer.** An electric stop clock with a 1000-sec range and capable of being read to 0.1 sec serves as the coulometer.

2. Reagents

Isoniazid is USP grade. The electrolysis solvent is a solution of 30 ml of acetic acid, 13 ml of methanol, and 7 ml of 1 M aqueous potassium bromide.[145]

3. Procedure

Close switches S_1 and S_2 (Fig. 15.6) and allow the amperostat to run for 30 min to come to thermal equilibrium. Determine the current by measuring the iR drop across resistor R_2 (at points A and B) with a Leeds & Northrup student K potentiometer. The setting of the dummy resistor R_3 can be determined by observing whether the current changes significantly when opening and closing switch S_2 with the filled electrolysis vessel in the circuit. Transfer a weighed sample of about 2 mg of isoniazid to the cell. Add 50 ml of electrolysis solvent, insert the electrode assembly, and titrate (by opening switch S_2) at a constant current of about 20 mA until the microammeter (which initially reads about zero) remains at the cutoff current (10 μA) for at least 30 sec, then close switch S_2. Approximate time is 5 min. Repeat the procedure for the solvent alone.

4. Calculation.

Calculate the amount of isoniazid using the blank correction for the solvent. Assume a four-electron reaction.

C. CHRONOPOTENTIOMETRY

Sulfa drugs may be determined by chronopotentiometry using the general procedure of Voorhies and Furman[143] (courtesy of *Analytical Chemistry*).

1. Apparatus

a. **Amperostat.** The amperostat (Fig. 15.9) is a constant current power supply operating at ca. 0.25 mA.

b. Electrodes. A thick platinum wire, 0.2 cm in diameter and surface area 1.3 cm² serves as anode, platinum gauze as the cathode, and a saturated calomel electrode as reference electrode.

c. Cell. The electrolysis vessel is a 100-ml beaker provided with magnetic stirring.

d. End-Point Detector. (1) Voltmeter. A high-input impedance electronic voltmeter such as a pH meter with an output for a recorder is used to measure the potential of the working electrode.

(2) Recorder. A strip-chart recorder having a 1 sec full scale response and a chart speed of 8 in./min (with a full-scale sensitivity suitable for use with the pH meter) is used.

2. Reagents

The supporting electrolyte is 1 M perchloric acid solution. The sulfisoxazole is USP grade.

3. Procedure

Close switches S_1 and S_2 (Fig. 15.9) and allow the amperostat to operate a few minutes to come to equilibrium. Determine the current by measuring the iR drop across resistor R_2 (at points A and B) with a Leeds & Northrup student K potentiometer. Transfer a weighed sample of 10 mg of sulfisoxazole into the electrolysis vessel. Add 50 ml of 1 M perchloric acid solution.

Wash the working electrode with chromic acid for about 30 sec, rinse with distilled water and wipe dry. Insert the electrodes into the test solution and stir the solution briefly. One minute after the stirring is stopped, start the recorder chart drive and apply a constant current of ca. 0.25 mA to the platinum electrodes by opening switch S_2. Record the chronopotentiogram. The complete chronopotentiogram should take about 1 min. Close switch S_2 on completion of the chronopotentiogram. Repeat this operation for samples of 15 mg and 20 mg.

4. Calculation

Using a graphical technique, determine the $E_{1/4}$ and $\tau^{1/2}$ for each chronopotentiogram. Calculate the chronopotentiographic constant $(i\tau^{1/2})/C$ for sulfisoxazole for the particular electrolysis cell used.

QUESTIONS

Q15.1. What is the purpose of the supporting electrolyte?
Q15.2. What functions do electrodes serve?
Q15.3. Write the equations for the electrode reactions generating four different types of titrants.
Q15.4. What experimental factors influence electrochemical techniques?

Q15.5. When is back titration necessary in constant current coulometry?

Q15.6. Compare the chronopotentiometric $E_{1/4}$ with the polarographic $E_{1/2}$.

Q15.7. Compare the concentration range served by each of these methods.

Q15.8. Compare the operating time necessary for analytical determinations by each of these methods.

Q15.9. Describe what factors controlled potential coulometry, constant current coulometry, and chronopotentiometry have in common.

Q15.10. What relationship exists between the electrolysis current, and the diffusion current of the electroactive species in coulometric methods and chronopotentiometry?

Q15.11. Compare coulometry with titrimetry.

Q15.12. Which of the techniques described in the chapter can quantitatively assay substances which are not electroactive?

Q15.13. What type of compounds are electroactive?

Q15.14. Contrast the end-point methods available for constant current coulometry with those for controlled potential coulometry and chronopotentiometry.

Q15.15. Compare the behavior of a two component electroactive mixture in controlled potential coulometry, constant current coulometry, and chronopotentiometry.

Q15.16. What are the advantages and limitations of the various methods described in this chapter?

Q15.17. Describe possible reasons for the use of an external generation of chemical titrant.

Q15.18. What is transition time enhancement?

Q15.19. Why are the transition times of a stepwise reaction requiring the same number of electrons not identical?

Q15.20. Compare the effect of temperature and mixing in coulometry and chronopotentiometry.

Q15.21. What is voltammetry?

Q15.22. What is amperometry?

Q15.23. Derive the equation for the electrolysis current decay-rate constant k' for controlled potential coulometry.

Q15.24. What is the relationship between the electrolysis current to concentration proportionality constant k and the decay-rate constant k' in controlled potential coulometry?

Q15.25. Derive the Sand equation.

PROBLEMS

P15.1. Given that the diffusion coefficient is 6.9×10^{-6} cm²/sec, the area 50 cm², the total volume 100 cc, and the diffusion layer 2.3×10^{-3} cm, what is the electrolysis current decay-rate constant k'? How long will it take to electrolyze 50, 95, or 99% of the electroactive substance by controlled potential coulometry?

P15.2. In the just mentioned conditions, if there were 0.17 meq of electroactive substance initially present, how large would the initial electrolysis current be?

P15.3. Calculate the time necessary to reach the end point when titrating 30 μmoles of an olefin with electrogenerated bromine using 10 mA of current?

P15.4. Given that the diffusion coefficient is 7.96×10^{-6} cm^2/sec, the area 1.0 cm^2, and n is 2, what is the value for the chronopotentiometric constant $i\tau^{1/2}/C$? If 500 μA electrolysis current for a 2.59 mM concentration of an electro-active species was used, what would the transition time be?

P15.5. Given substance a has a diffusion coefficient of 7×10^{-6} cm^2/sec, the bulk concentration is 2 μmoles/cc, and 2 electrons are involved in its reduction, and substance b has a diffusion coefficient of 7×10^{-6} cm^2/sec, the bulk concentration is 1 μmole/cc, and 1 electron is involved in its reduction, what is the transition time ratio of substance a to substance b when substance a is reduced (a) first or (b) second?

REFERENCES

1. S. Glasstone, *An Introduction to Electrochemistry*, Van Nostrand, Princeton, N.J., 1942.
2. P. Delahay, *New Instrumental Methods in Electrochemistry*, Wiley (Interscience), New York, 1954.
3. G. G. Grower, *Proc. Am. Soc. Testing Mater.*, **17**, 129 (1917).
4. L. Szebelledy and Z. Somogyi, *Z. Anal. Chem.*, **112**, 313 (1938).
5. J. W. Sease, C. Niemann, and E. H. Swift, *Anal. Chem.*, **19**, 197 (1947).
6. W. D. Cooke, C. N. Reilley, and N. H. Furman, *Anal. Chem.*, **23**, 1662 (1951).
7. A. Hickling, *Trans. Faraday Soc.*, **38**, 27 (1942).
8. J. J. Lingane, *J. Am. Chem. Soc.*, **67**, 1916 (1945).
9. L. Gierst and A. L. Juliard, *J. Phys. Chem.*, **57**, 701 (1953).
10. H. J. S. Sand, *Phil. Mag.*, **1**, 45 (1901).
11. J. J. Lingane, *Electroanalytical Chemistry*, Wiley (Interscience), New York, 2nd ed., 1958.
12. D. T. Lewis, *Analyst*, **86**, 494 (1961).
13. H. L. Kies, *J. Electroanal. Chem.*, **4**, 257 (1962).
14. M. Faraday, *Phil. Trans. Roy. Soc. London*, **1**, 55 (1834).
15. J. J. Lingane, *Anal. Chim. Acta*, **2**, 584 (1948).
16. K. S. V. Santhanam and V. R. Krishnan, *Anal. Chem.*, **33**, 1493 (1961).
17. J. J. Lingane and S. L. Jones, *Anal. Chem.*, **22**, 1169 (1950).
18. G. L. Booman, *Anal. Chem.*, **29**, 213 (1957).
19. W. M. MacNevin and B. B. Baker, *Anal. Chem.*, **24**, 986 (1952).
20. L. Meites, *Anal. Chem.*, **31**, 1285 (1959).
21. J. J. Lingane and S. L. Jones, *Anal. Chem.*, **22**, 1220 (1950).
22. E. W. Washburn and S. J. Bates, *J. Am. Chem. Soc.*, **34**, 1341 (1912).
23. J. J. Lingane and L. A. Small, *Anal. Chem.*, **21**, 1119 (1949).
24. W. M. MacNevin, B. B. Baker, and R. D. McIver, *Anal. Chem.*, **25**, 274 (1953).
25. J. A. Page and J. J. Lingane, *Anal. Chim. Acta*, **16**, 175 (1957).
26. T. C. Franklin and C. C. Roth, *Anal. Chem.*, **27**, 1197 (1955).
27. L. Meites, *Anal. Chem.*, **24**, 1057 (1952).
28. N. Bett, W. Nock, and G. Morris, *Analyst*, **79**, 607 (1954).
29. J. S. Parsons, W. Seaman, and R. M. Amick, *Anal. Chem.*, **27**, 1754 (1955).
30. N. H. Furman and A. J. Fenton, Jr., *Anal. Chem.*, **29**, 1213 (1957).
31. H. C. Jones, W. D. Shults, and J. M. Dale, *Anal. Chem.*, **37**, 680 (1965).
32. K. W. Kramer and R. B. Fischer, *Anal. Chem.*, **26**, 415 (1954).
33. A. J. Bard and E. Solon, *Anal. Chem.*, **34**, 1181 (1962).
34. T. R. Mueller and R. N. Adams, *Anal. Chim. Acta*, **23**, 476 (1960).
35. J. D. Voorhies and S. M. Davis, *Anal. Chem.*, **32**, 1855 (1960).

36. R. N. Adams, *Anal. Chem.*, **30**, 1576 (1958).
37. S. S. Lord and L. B. Rogers, *Anal. Chem.*, **26**, 284 (1954).
38. V. F. Gaylor and P. J. Elving, *Anal. Chem.*, **25**, 1078 (1953).
39. H. Lund, *Acta Chem. Scand.*, **11**, 1323 (1957).
40. J. W. Loveland and G. R. Dimeler, *Anal. Chem.*, **33**, 1196 (1961).
41. L. Meites, *Anal. Chim. Acta*, **20**, 456 (1959).
42. L. Meites and S. A. Moros, *Anal. Chem.*, **31**, 23 (1959).
43. W. D. Shults and P. F. Thomason, *Anal. Chem.*, **31**, 492 (1959).
44. W. M. Wise and D. E. Campbell, *Anal. Chem.*, **38**, 1079 (1966).
45. P. Hersch, *Chim. Anal.*, **41**, 189 (1959).
46. F. A. Keidel, *Ind. Eng. Chem.*, **52**, 490 (1960).
47. E. L. Eckfeldt and E. W. Shaffer, Jr., *Anal. Chem.*, **36**, 2008 (1964).
48. E. J. Cokal and E. N. Wise, *Anal. Chem.*, **35**, 914 (1963).
49. W. M. Wise and J. P. Williams, *Anal. Chem.*, **37**, 1292 (1965).
50. P. H. Given and M. E. Peover, *Nature*, **184**, 1064 (1959).
51. F. H. Merkle and C. A. Discher, *J. Pharm. Sci.*, **53**, 620 (1964).
52. J. E. Harrar, *Anal. Chem.*, **35**, 893 (1963).
53. V. B. Ehlers and J. W. Sease, *Anal. Chem.*, **31**, 16 (1959).
54. E. G. Tur'yan, *Zavodskaya Lab.*, **21**, 17 (1955).
55. J. M. Kruse, *Anal. Chem.*, **31**, 1854 (1959).
56. F. H. Merkle and C. A. Discher, *Anal. Chem.*, **36**, 1639 (1964).
57. R. F. Broman and R. W. Murray, *Anal. Chem.*, **37**, 1408 (1965).
58. F. A. Keidel, *Anal. Chem.*, **31**, 2043 (1959).
59. K. Abresch and I. Claassen, *Die coulometrische Analyse*, Verlag Chemie, Weinheim, 1961.
60. K. Stulik and J. Zyka, *Chemist-Analyst*, **55**, 120 (1966).
61. D. D. DeFord, J. N. Pitts, and C. J. Johns, *Anal. Chem.*, **23**, 938 (1951).
62. E. N. Wise, P. W. Gilles, and C. A. Reynolds, Jr., *Anal. Chem.*, **25**, 1344 (1953).
63. N. H. Furman and A. J. Fenton, Jr., *Anal. Chem.*, **28**, 515 (1956).
64. H. V. Malmstadt and C. B. Roberts, *Anal. Chem.*, **28**, 1412 (1956).
65. N. H. Furman, W. D. Cooke, and C. N. Reilley, *Anal. Chem.*, **23**, 945 (1951).
66. J. J. Lingane, *Anal. Chem.*, **26**, 622 (1954).
67. H. A. Laitinen and B. B. Bhatia, *Anal. Chem.*, **30**, 1995 (1958).
68. J. J. Lingane, *Anal. Chem.*, **26**, 1021 (1954).
69. D. D. DeFord, C. J. Johns, and J. N. Pitts, *Anal. Chem.*, **23**, 941 (1951).
70. C. N. Reilley, W. D. Cooke, and N. H. Furman, *Anal. Chem.*, **23**, 1030 (1951).
71. N. H. Furman, L. J. Sayegh, and R. N. Adams, *Anal. Chem.*, **27**, 1423 (1955).
72. D. D. DeFord, J. N. Pitts, Jr., and C. J. Johns, *Proc. Natl. Acad. Sci. U.S.*, **36**, 612 (1950).
73. W. F. Head, Jr., and M. M. Marsh, *J. Chem. Educ.*, **38**, 361 (1961).
74. J. S. Parsons and W. Seaman, *Anal. Chem.*, **27**, 210 (1955).
75. G. E. Gerhardt, H. C. Lawrence, and J. S. Parsons, *Anal. Chem.*, **27**, 1752 (1955).
76. F. Vorstenburg and A. W. Loffler, *J. Electroanal. Chem.*, **1**, 422 (1959/1960).
77. R. P. Buck and E. H. Swift, *Anal. Chem.*, **24**, 499 (1952).
78. J. J. Lingane and F. C. Anson, *Anal. Chem.*, **28**, 1871 (1956).
79. A. J. Fenton, Jr., and N. H. Furman, *Anal. Chem.*, **29**, 221 (1957).
80. A. J. Fenton, Jr., and N. H. Furman, *Anal. Chem.*, **32**, 748 (1960).
81. R. P. Buck and T. J. Crowe, *Anal. Chem.*, **35**, 697 (1963).
82. E. Bishop and R. G. Dhaneshwar, *Anal. Chem.*, **36**, 726 (1964).
83. E. L. Eckfeldt and E. W. Shaffer, Jr., *Anal. Chem.*, **37**, 1534 (1965).
84. P. S. Tutundzic, *Anal. Chim. Acta*, **8**, 182 (1953).
85. E. C. Olson, *Anal. Chem.*, **32**, 1545 (1960).

86. F. Baumann and D. D. Gilbert, *Anal. Chem.*, **35**, 1133 (1963).
87. G. S. Kozak, Q. Fernando, and H. Freiser, *Anal. Chem.*, **36**, 296 (1964).
88. G. O'Dom and Q. Fernando, *Anal. Chem.*, **37**, 893 (1965).
89. N. H. Furman and R. N. Adams, *Anal. Chem.*, **25**, 1564 (1953).
90. R. P. Buck, P. S. Farrington, and E. H. Swift, *Anal. Chem.*, **24**, 1195 (1952).
91. P. S. Farrington, W. P. Schaefer, and J. M. Dunham, *Anal. Chem.*, **33**, 1318 (1961).
92. F. J. Feldman and R. E. Bosshart, *Anal. Chem.*, **38**, 1400 (1966).
93. R. F. Swensen and D. A. Keyworth, *Anal. Chem.*, **35**, 863 (1963).
94. W. M. Wise and J. P. Williams, *Anal. Chem.*, **36**, 19 (1964).
95. G. D. Christian, *Anal. Chem.*, **37**, 1418 (1965).
96. D. L. Maricle, *Anal. Chem.*, **35**, 683 (1963).
97. J. J. Lingane, *Anal. Chem.*, **38**, 1489 (1966).
98. A. J. Bard, *Anal. Chem.*, **32**, 623 (1960).
99. P. Arthur and J. F. Donahue, *Anal. Chem.*, **24**, 1612 (1952).
100. P. P. L. Ho and M. M. Marsh, *Anal. Chem.*, **35**, 618 (1963).
101. R. L. Burnett and R. F. Klaver, *Anal. Chem.*, **35**, 1709 (1963).
102. T. R. Blackburn and R.B. Greenberg, *Anal. Chem.*, **38**, 877 (1966).
103. C. A. Streuli, *Anal. Chem.*, **28**, 130 (1956).
104. R. B. Hanselman and C. A. Streuli, *Anal. Chem.*, **28**, 916 (1956).
105. W. B. Mather, Jr., and F. C. Anson, *Anal. Chem.*, **33**, 132 (1961).
106. W. B. Mather, Jr., and F. C. Anson, *Anal. Chim. Acta*, **21**, 468 (1959).
107. R. O. Crisler and R. D. Conlon, *J. Am. Oil Chemists' Soc.*, **39**, 470 (1962).
108. C. A. Streuli, J. J. Cincotta, D. L. Maricle, and K. K. Mead, *Anal. Chem.*, **36**, 1371 (1964).
109. G. Johansson, *Talanta*, **11**, 789 (1964).
110. F. C. Anson, K. H. Pool, and J. N. Wright, *J. Electroanal. Chem.*, **2**, 237 (1961).
111. C. N. Reilley and W. W. Porterfield, *Anal. Chem.*, **28**, 443 (1956).
112. G. D. Christian, E. C. Knoblock, and W. C. Purdy, *Anal. Chem.*, **37**, 292 (1965).
113. J. J. Lingane and A. M. Hartley, *Anal. Chim. Acta*, **11**, 475 (1954).
114. R. B. Hanselman and L. B. Rogers, *Anal. Chem.*, **32**, 1240 (1960).
115. E. P. Przybylowicz and L. B. Rogers, *Anal. Chim. Acta*, **18**, 596 (1958).
116. E. P. Przybylowicz and L. B. Rogers, *Anal. Chem.*, **28**, 799 (1956).
117. G. D. Christian, E. C. Knoblock, and W. C. Purdy, *Anal. Chem.*, **35**, 1869 (1963).
118. J. W. Miller and D. D. DeFord, *Anal. Chem.*, **30**, 295 (1958).
119. A. S. Meyer and C. M. Boyd, *Anal. Chem.*, **31**, 215 (1959).
120. B. Miller and D. N. Hume, *Anal. Chem.*, **32**, 764 (1960).
121. B. Miller and D. N. Hume, *Anal. Chem.*, **32**, 524 (1960).
122. J. J. Lingane, *Analyst*, **91**, 1 (1966).
123. Z. Karaoglanoff, *Z. Elektrochem.*, **12**, 5 (1906).
124. J. J. Lingane, *J. Electroanal. Chem.*, **2**, 46 (1961).
125. G. A. Ward, *Talanta*, **10**, 261 (1963).
126. A. J. Bard, *Anal. Chem.*, **35**, 340 (1963).
127. A. J. Bard, *Anal. Chem.*, **33**, 11 (1961).
128. D. G. Davis, *Anal. Chem.*, **33**, 1839 (1961).
129. D. G. Davis and M. E. Everhart, *Anal. Chem.*, **36**, 38 (1964).
130. C. N. Reilley, G. W. Everett, and R. H. Johns, *Anal. Chem.*, **27**, 483 (1955).
131. M. M. Nicholson and J. H. Karchmer, *Anal. Chem.*, **27**, 1095 (1955).
132. R. W. Murray, *Anal. Chem.*, **35**, 1784 (1963).
133. J. D. Voorhies and N. H. Furman, *Anal. Chem.*, **31**, 381 (1959).
134. P. J. Elving and D. L. Smith, *Anal. Chem.*, **32**, 1849 (1960).
135. R. N. Adams, J. H. McClure, and J. B. Morris, *Anal. Chem.*, **30**, 471 (1958).
136. P. J. Elving and A. F. Krivis, *Anal. Chem.*, **30**, 1648 (1958).

137. A. J. Bard, *Anal. Chem.*, **35**, 1602 (1963).
138. D. G. Davis, *Anal. Chem.*, **35**, 764 (1963).
139. J. J. Lingane and P. J. Lingane, *J. Electroanal. Chem.*, **5**, 411 (1963).
140. J. D. Voorhies and J. S. Parsons, *Anal. Chem.*, **31**, 516 (1959).
141. J. J. Lingane, *J. Electroanal. Chem.*, **1**, 379 (1960).
142. S. V. Tatwawadi and A. J. Bard, *Anal. Chem.*, **36**, 2 (1964).
143. J. D. Voorhies and N. H. Furman, *Anal. Chem.*, **30**, 1656 (1958).
144. R. F. Dapo and C. K. Mann, *Anal. Chem.*, **35**, 677 (1963).
145. F. A. Leisey and J. F. Grutsch, *Anal. Chem.*, **28**, 1553 (1956).

CHAPTER **16**

Polarography

Fred W. Teare

FACULTY OF PHARMACY
UNIVERSITY OF TORONTO
TORONTO, ONTARIO

16.1 INTRODUCTION

Polarography is the best known of several electroanalytical techniques involving controlled electrolysis and referred to as "voltammetry," In potentiometry (voltammetry at zero current) the potentials of electrochemical cells are normally measured when no appreciable current is flowing, while in voltammetry a finite current is allowed to flow through the electrochemical cell causing electrolysis. The extent of electrolysis may vary from complete electrolysis in electrogravimetry and coulometry, to that involving only a minute fraction of the analyte (the electroactive substance) at the surface of the microelectrode in polarography and amperometry.

Polarography originated with Professor Heyrovský[1] at Prague University in 1922. The term "polarography" is usually restricted to that voltammetric method in which the analyte, dissolved in a suitable medium, is placed in an electrolysis cell where the electrolysis is controlled by a variable known potential applied to the dropping mercury electrode (DME). The latter is polarized relative to a nonpolarizable electrode, usually the saturated calomel reference electrode (SCE). The electrolysis current resulting from a controlled increase in the potential of the DME, in an unstirred system, may be represented as a current-voltage curve called a "polarogram." This technique is used in the qualitative and quantitative analysis of many inorganic and organic electroactive substances.

If a polarizable solid microelectrode, such as a platinum wire, is employed in place of the DME and the solution is stirred under reproducible conditions the process is called "solid electrode voltammetry" and the resulting current-voltage curve is referred to as a "voltammogram." When the concentration of the electroactive substance(s) is altered by the addition of a titrant to either of the forementioned processes, the method is known as an "amperometric titration."

This chapter is confined mainly to conventional dc polarography which is still by far the most versatile of the voltammetric methods. The student is referred to recent references[2–5] and the literature for extensions and modifications of the polarographic method.

16.2 THEORETICAL CONSIDERATIONS

A. FACTORS WHICH INFLUENCE CURRENT FLOW DURING CONTROLLED ELECTROLYSIS

In electrolysis, certain constituents of a solution can be made to undergo a change in oxidative state, under appropriate conditions, by a flow of current at the electrode-solution interface. For example, at the cathode (where reduction occurs):

$$Ox + ne^- \rightarrow Red \tag{16.1}$$

at the anode (where oxidation occurs):

$$2H_2O \rightarrow O_2 + 2H^+ + 2e^- \tag{16.2}$$

where Ox is the electroreducible substance and Red is the product. In this instance, the cathode is a polarizable working electrode, e.g., DME, while the anode is the nonpolarizable electrode, e.g., SCE with a large-surface salt bridge. The working microelectrode becomes "polarized" when an external source of potential is impressed across the electrolysis cell. The electrons supplied to the cathode, under these conditions, are transferred to the electroreducible substance Ox at the electrode surface. Oxidation occurs at the larger anode surface, where electrons are received from the solution. The amount of current flowing through the anode is equal to that through the cathode, but due to the larger surface area of the anode the current density is much lower and polarization does not occur at this electrode.

The current-potential (i-E) relationships are simplified in polarography (and somewhat similarly in solid microelectrode voltammetry) by the following conditions: (a) the potential of one electrode, the reference electrode, is made independent of current flow by employing a large-surface electrode of constant potential (e.g., SCE), while the microelectrode (e.g., DME) becomes polarized under the same current flow; (b) the ohmic drop iR across the cell is kept to a low value, perhaps a few millivolts. Under these conditions the actual applied potential (voltage) is equal to the potential of the large-surfaced SCE less the potential of the polarized DME. In other words, the voltage applied across the electrolysis cell equals the potential of the polarized electrode if the unpolarized electrode of constant potential is considered the reference. This is what is implied when the potential of the microelectrode (DME) is reported as E_{appl} (vs. SCE). Since the potential of the SCE is $+0.2444$ V vs. the normal hydrogen electrode (NHE) at 25°C, the potential of the DME can also be related to the NHE (0.000 V) if this is desired.

Current-potential relationships in polarography are readily studied by means of apparatus schematically represented by Fig. 16.1.

An external potential (E_{appl}) is applied across the electrolysis cell by means of a potentiometer and the resulting current is read by means of a sensitive galvanometer or microammeter (G).

Electrolysis of any species involves at least three factors which control the flow of current: mass transfer, electron transfer, and removal of the product,[6,7] An appreciation of the control of these factors will assist in obtaining accurate analytical results.

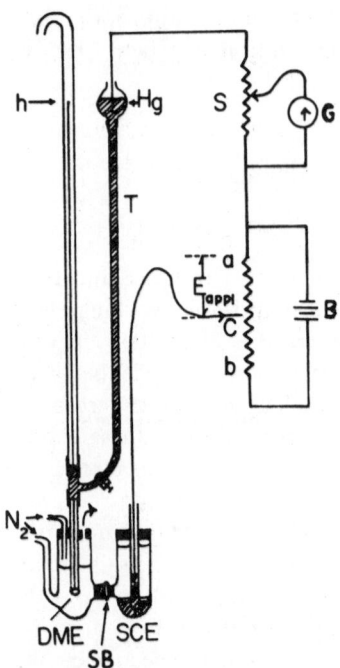

FIGURE 16.1: Simple polarographic apparatus and circuit. The H-type electrolysis cell includes a dropping mercury electrode (DME) and saturated calomel electrode (SCE) with a large-surfaced salt bridge (SB). The height (h) of the mercury column is controlled by the mercury (H_g) in a reservoir. A battery (B) or other source of constant voltage is accurately divided by a sliding contact (C) along a slide wire (a-b). A shunt (S) is used to change the sensitivity of a sensitive galvanometer (G) reading. T is a flexible plastic tubing.

1. Mass Transfer

Three basic mechanisms are involved in the mass transfer of an analyte (e.g., Ox) from the bulk of the solution to the polarized electrode surface, where it can be reduced. These mechanisms are migration, convection, and diffusion.

a. Migration. This means of transport involves the potential gradient produced in the solution whereby the electroactive species (e.g., Ox) is attracted toward the polarized microelectrode DME, while other species having an electrical charge of like sign are repelled by this electrode. For example, a negatively charged cathode (DME) attracts cations while repelling anions. This means of mass transport is undesirable in polarography. The migration of the electroactive substance of interest is minimized by the addition, to the same solution, of a large excess of an "inert" (nonelectrolyzed) substance called the "supporting electrolyte" (SE). This SE conducts almost the entire current through the cell, but does not interfere with the reaction of interest at the microelectrode. For example, if the solution in the electrolysis cell is 0.1 M in KCl and only 0.001 M in the electroactive species, then the latter will conduct only about 1 % of the current, everything else being equal. As the chloride ions migrate toward the anode, the potassium ions migrate toward the cathode. Since the potassium cations are not reduced at the selected applied potential on the cathode, they merely form a layer of positively charged particles around the microelectrode, where they decrease the attraction of this electrode for the reducible species, thereby reducing this means of mass transport to a negligible value.

b. **Convection.** This means of transport is operative when the solution is agitated. This agitation may be caused by density or temperature differences in various parts of the electrolysis cell or it may be produced by some form of mechanical stirring. Agitation brings more electroactive substances to the electrode surface and produces an increase in the current flow.

c. **Diffusion.** This transport mechanism is a result of a concentration gradient involving the electroactive species. When an appropriate potential is applied to the microelectrode (e.g., DME), the concentration of the reducible species at the electrode surface is rapidly reduced and more of this material will diffuse from the bulk of the solution, where the concentration C is higher, toward the electrode surface where the concentration C_0 is nearly depleted. Thus a concentration gradient $C - C_0$ exists for the electroactive species and "concentration polarization" (or "concentration over potential") occurs at this electrode. The electroactive species is frequently referred to as the "depolarizer" or "depolarizing substance." This concentration gradient causes the depolarizer (ion or molecule) to move through a diffusion layer d surrounding the polarized microelectrode. Agitation of the solution by means of a constant-rate stirrer causes the thickness of d to decrease and the current to increase until the latter reaches a steady limiting value.

Diffusion is a very slow means of mass transport relative to stirring and the limiting current value is therefore correspondingly lower in unstirred systems. Each depolarizer substance has a characteristic diffusion rate which is expressed as the diffusion coefficient D in units of square centimeters per second.

In view of the several factors which affect the current flow during electrolysis in the type of cell discussed here, the limiting current i_L is related to these factors as well as the surface area A of the microelectrode by the following expression

$$i_L = \frac{nFDA}{d}(C - C_0) \qquad (16.3)$$

where n is the number of electrons per molecule of the depolarizer and F is the faraday.

A constant limiting current i_L value can only be achieved in unstirred solution (the usual condition in polarography) or in solution stirred at a constant rate (as in voltammetry employing a polarized rotating or vibrating platinum microelectrode). In the latter instance, the limiting current values are less reproducible than when the DME is used. If there is a sufficient excess of a supporting electrolyte in the cell along with the depolarizer substance and if measures are taken to eliminate all agitation of the solution, then the limiting current becomes solely a diffusion current.

2. Electron Transfer Process

The electron transfer process occurs on the electrode surface between the electrode and the electroactive species. If this species is in equilibrium with other species, then the equilibrium must shift to produce more of the active species as it is reduced at the electrode during the electron transfer process. This equilibrium shift to produce more of the depolarizer species occurs either at the electrode surface or in the diffusion layer d, where it gives rise to a "kinetic current."

If this equilibrium shift is slower than that for mass transfer, then the observed current flow is dependent on this rate of conversion. Thus, polarography has been employed in electrochemical kinetic studies. Once the active species is at the electrode surface, the actual electron transfer process begins, providing the applied potential exceeds the decomposition potential. The current increases with increasing applied potential since the mass transport rate (MTR) is adequate to replace the species as it is depleted at the electrode surface. At this point it is the electron transfer rate (ETR) which limits the current flow. At a higher applied potential the ETR surpasses the MTR and the current reaches a limiting value controlled by the MTR. The greater the MTR of a species, the greater its limiting current value.[6]

The slope of the i vs. E_{appl} curve reflects the ETR, the greater the ETR, the steeper the slope. However, the i_L value ultimately depends upon the MTR, regardless of the slope of the i-E_{appl} curve.[6]

The electron transfer process is reversible only while the product is at the electrode surface. The ratio of the forward and reverse ETR depends on the electrode potential. When no external potential is applied to the electrode, the electrode adopts the equilibrium potential. Under this condition the two electron transfer rates differ only in their directions. When the forward and reverse ETR are much greater than the MTR and the rate of conversion of a species to the "active" form, the electrode process is said to be "reversible." However, when the forward and reverse ETR are about the same or less than the two just mentioned rate processes, the electrode process is "irreversible." This irreversibility increases in degree, the slower these electron transfer rates become relative to these other processes.[6,7]

3. Removal of the Product

Since product accumulation at the electrode surface may influence nearly all the just mentioned processes, the rate of product removal could become the overall rate determining process in the electrode reaction.

The importance of the proper choice of experimental conditions can be seen when attempting to study the relationship between samples and corresponding i vs. E_{appl} curves.

B.   CURRENT-VOLTAGE RELATIONSHIPS

The apparatus which is frequently used in conventional dc polarography is represented by Fig. 16.1. The electrodes consist of a polarizable microelectrode, usually the dropping mercury electrode (DME), and a large-surfaced nonpolarizable reference electrode such as the saturated calomel electrode (SCE). When an external potential is applied to the DME, it is forced to assume a potential other than that representing the equilibrium potential of the bulk of the solution. A small electrolysis current is now produced, providing some soluble electroreducible substance is present the electrolysis cell, and the DME becomes polarized. A similar situation exists when the DME becomes an anode, except that oxidation may occur at the electrode surface. Only a minute amount of the electroactive substance is removed from solution since a very small electrolysis current is employed in the polarographic method. In fact, several i-E_{appl} curves can be obtained using the same solution without detectable differences. The method is applicable to both inorganic and organic materials within an optimum concentration range of 10^{-2} to 10^{-5} M and can be performed on volumes as small as 0.5 ml or less in appropriate microcells.

I. The Nernst Equation

The potential E of the microelectrode at equilibrium is defined by the Nernst equation

$$E = E^{\circ\prime} + \frac{2.303RT}{nF} \log \frac{[Ox]}{[Red]} \tag{16.4}$$

where $E^{0\prime}$ is the formal electrode potential, with reference to the normal hydrogen electrode $E = E^{0\prime}$, when $[Ox] = [Red]$. The molar concentrations of the two soluble species involved in the electrode reaction are represented by [Ox] and [Red], n equals the number of electrons required in the conversion of Ox to Red, and F is a faraday. Molar concentrations of each soluble species can be used in place of activities normally found in the Nernst equation, without appreciable error in dilute solutions. The relationship between these terms is shown by

$$a_{Ox} = [Ox]f \tag{16.5}$$

where the activity coefficient f approaches unity in solutions which are approximately 10^{-3} M or less. For this reason, and because accurate values of f are not often known, the molar concentrations are used to facilitate calculations.

This form of the Nernst equation can be employed to define the potential of the microelectrode in polarography if the molar concentrations of the Ox and Red species represent those only at this electrode surface. The electrode reaction must also be reversible and the electrode in equilibrium with the solution immediately surrounding it. This latter condition is approximated closely if the ETR is very large relative to other electrolysis processes such as the diffusion rate of active species, and if both the Ox and Red forms of the species are soluble in the surrounding solution.[8]

2. Overpotential

To obtain an electrode reaction it is frequently necessary to apply a potential to the DME which is different, e.g., more negative relative to the SCE, than that which would be calculated by means of the Nernst equation under equilibrium conditions for a reversible electrode reaction. This difference in the DME potential is referred to as the "overpotential" of the electroactive substance on that electrode. The hydrogen-overpotential on the DME is fortunately quite large, therefore a considerable negative E_{appl} (vs. SCE) is required before hydrogen gas is evolved at the DME due to the reduction of the hydrogen ions in aqueous solution. This permits many substances to be reduced at the DME at E_{appl} values which are less negative than those required for the discharge of hydrogen gas.

3. Current-Voltage Curves

A typical polarogram (i-E_{appl} curve) is shown in Fig. 16.2. In this generalized discussion, the sample might be any soluble electroreducible substance (Ox) of appropriate concentration (10^{-2} to 10^{-5} M) in a large excess of a nonreducible electrolyte, e.g., KCl, 0.1 to 1 M. This supporting electrolyte should be about 50 times, or more, the concentration of the depolarizer in order to effectively eliminate the migration current of the latter.

a. Typical Polarogram (Fig. 16.2, curve a). As the DME is made to assume an increasingly negative potential (relative to SCE) an E_{appl} is reached where current begins to flow due to the reduction of the electroactive species

Ox in solution. This potential is referred to as the "decomposition potential" (E_d). As the applied potential is increased further, the current rises, controlled by the diffusion of Ox (or by the ETR in the case of irreversible electrode reactions) and eventually reaches a limiting current (i_L) plateau. Finally the current rises steeply due to the reduction of some SE component which may be water or other substance if its decomposition potential is lower.

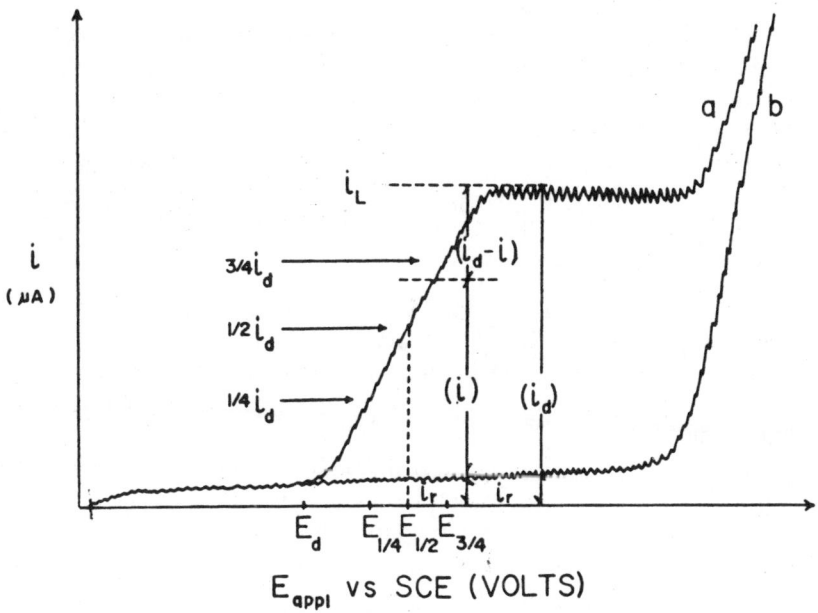

FIGURE 16.2: Typical polarogram. Curve a is the i vs. E_{appl} curve of an electroactive substance. Curve b is that for the supporting electrolyte, etc., less the electroactive substance. The symbols are identified as follows: i (μA) is the current reading; E_{appl} is the applied potential; E_d is the decomposition potential; $E_{1/4}$, $E_{1/2}$, and $E_{3/4}$, are the potentials corresponding to $\frac{1}{4}$, $\frac{1}{2}$, and $\frac{3}{4}$ the diffusion current (i_d), respectively; i_r is the residual current; i is the current reading anywhere on the curve corrected for i_r; i_d is the diffusion current ($i_L - i_r$) and i_L is the limiting current.

b. Residual Current (Fig. 16.2, Curve b). This is a i-E_{appl} curve of the supporting electrolyte alone. As the potential of the DME becomes increasingly negative the residual current i_r rises very gradually from a value near 0. Eventually a point is reached where an abrupt rise in the current flow occurs due to the reason given for curve a. The saw-toothed curves are the result of current oscillations caused by the gradual increasing electrode surface area [Eq. (16.3)] followed by the detachment of the mercury drop from the capillary tip.

c. Components of the Polarographic Wave. While it is the diffusion current i_d of the electroactive substance which is the most important and characteristic portion of the i-E_{appl} curve, other contributions to the limiting current i_L may alter the overall curve and must be understood to be controlled.

1. *Limiting current* (i_L). This is the total current read from a current meter or recording. It includes the residual current i_r and diffusion current i_d, as well as the migration current of the reducible substance if this has not been adequately suppressed.

2. *Residual current* (i_r). This portion of the total current is the result of contributions from (a) a minute faradaic current i_f caused by the reduction of trace impurities and (b) a much larger condenser current i_c caused by the charging of the Helmholtz double-layer capacitance at the mercury-solution interface. This "condenser" is formed by positively charged ions of the supporting electrolyte (e.g., K^+) forming a layer around the negatively charged mercury drop, which in turn are surrounded by a layer of negative ions (e.g., Cl^-) to form the double layer.

3. *Migration current* (i_m). If the DME is a negatively charged cathode, then cations of an electroreducible sample will migrate (due to electrostatic attraction) to this electrode where they are reduced upon arrival. This undesirable contribution to the cathodic current is referred to as a migration current i_m. The effect of the i_m can be essentially eliminated by the addition of a large excess of "inert" electrolyte which conducts nearly the total current across the electrolysis cell but which does not interfere with the electrode reaction. Under these conditions the contribution of the migration current to the cathodic current is negligibly small and can be ignored.

4. *Diffusion current* (i_d). This is the net current which results from the subtraction of the residual and migration currents from the limiting current

$$i_d = i_L - (i_r + i_m) \tag{16.6}$$

Since the i_m can be eliminated as explained earlier, the Eq. (16.6) reduces to

$$i_d = i_L - i_r \tag{16.7}$$

This subtraction is normally done graphically, as illustrated in Fig. 16.2. When the diffusion current i_d is determined as just described, it is entirely dependent upon the rate of diffusion of the electroactive species from the bulk of the solution to the electrode surface.

C. FACTORS GOVERNING THE DIFFUSION CURRENT

I. Ilkovic Equation

A variety of factors govern the diffusion current at the DME for any electroactive species in a supporting electrolyte. A theoretical equation was derived by Ilkovic (1934) for the diffusion current under controlled conditions

$$i_d = knD^{1/2}Cm^{2/3}t^{1/6} \qquad (16.8)$$

where i_d is the diffusion current (μA) during the life t of the mercury drop.

The constant k includes several physical and numerical constants, including the density of mercury, which were used in the derivation of the equation; $k = 706$ for maximum i_d, and $k = 607$ for average i_d measurements. The upper limit of the oscillations recorded by a fast recorder (<1 sec full-scale deflection) along the i_L plateau in Fig. 16.2 is used when measuring $(i_d)_{max}$; the average of these oscillations is employed when measuring $(i_d)_{aver}$ on a damped recorder. Also the $(i_d)_{aver}$ is the value actually measured on a galvanometer which does not respond quickly enough to follow the actual current oscillations. The $(i_d)_{max} = \frac{7}{6}(i_d)_{aver}$. [There are several ways of defining n, such as: (1) the number of electrons per ion or molecule of the electroactive species or (2) the number of electron equivalents required per mole of the electroactive species involved in the electrode reaction. D is the diffusion coefficient of the electroactive species in square centimeters per second. C is the concentration of the electroactive substance, in millimoles per liter (mM/liter). m is the rate of mercury flow from DME capillary in milligrams per second. t is the drop life, in seconds; the value of t varies with the potential applied to the DME, because the interfacial tension (γ) at the mercury drop surface greatly influences the value of t and γ depends on E_{appl}. Therefore t is measured either at the E_{appl} where the i_d is to be measured or at the $E_{1/2}$ of the electroactive species.

In very precise work, modified forms of the Ilkovic equation may be employed which allow for the curvature of the mercury drops, since diffusion is spherical rather than linear, as originally supposed by Ilkovic. The Ilkovic equation is valid as long as t is longer than 2.5 sec and if a constant temperature is maintained in the approximate range of 15 to 40°C. If t is too short, the falling drops produce a stirring effect which decreases the diffusion layer d and increases the observed current.

The linear relationship between i_d and C can be seen if all of the factors on the right-hand side of the Ilkovic equation, except C, are held constant and represented by k:

$$i_d = kC \qquad (16.9)$$

The Ilkovic equation may be arranged to give

$$knD^{1/2} = \frac{i_d}{Cm^{2/3}t^{1/6}} \qquad (16.10)$$

The characteristics which are independent of the electrodes and instrument are grouped together on the left and are collectively referred to as the "diffusion current constant" (I_d). This value is frequently recorded in the literature and is reproducible within $\pm 5\%$ under specified conditions.

The term $m^{2/3}t^{1/6}$ is called the "capillary characteristic." When m and t have been experimentally determined, comparisons can be made between capillaries of varying length and bore diameter, as well as with the same capillary under different head pressure of mercury or at various applied potentials.

The factors which affect the i_d include those which produce a change in any term in the Ilkovic equation and these, as well as others, must be controlled when using the polarographic method.

a. Measurement of the Diffusion Current (i_d). This term varies with the sixth root of t. However, most galvanometers have a time period which is not short enough to accurately follow the current produced at each mercury drop.

The galvanometer oscillations observed correspond closely to the average of the true diffusion current and correspond to the $(i_d)_{aver}$ in the Ilkovic equation rather than the maximal or minimal reading. An exact diffusion current i_d of an electroactive substance can only be determined when the exact residual current is known and subtracted from the limiting current value. The most accurate measurement of i_r can be made by means of a separate polarogram of the supporting electrolyte, such as that shown in Fig. 16.2, curve b. In actual practice, an adequate approximation of the i_r can be obtained by extrapolating the portion of the sample polarogram, which preceeds the decomposition potential, and measuring the distance between this extrapolated line and the limiting current to obtain the i_d at a specified point. This method is illustrated in a later section.

b. Height of the Mercury Column (h). Both the mercury flow rate m and the drop life t are dependent on the dimensions of the capillary of the DME and on the height of the mercury column above the tip of this electrode. An increase in the height of the mercury column h produces no increase in drop size (a function of bore size), but rather increases the number of drops formed per unit of time. The relationship between m and h has been shown to be

$$m = kh_{corr} \qquad (16.11)$$

where k is a constant and h_{corr} is the height of the mercury column (in centimeters) above the DME tip corrected for the back pressure due to the interfacial tension between the drop and solution.

It can be shown that for aqueous solutions the back pressure (h_{back}) is given by the equation

$$h_{back} = \frac{3.1}{m^{1/3}t^{1/3}} \text{ cm mercury} \qquad (16.12)$$

and $h_{corr} = h - h_{back}$, where h is the uncorrected height of the mercury column in centimeters.

The value of t varies inversely with the height of the mercury column, as shown by the equation

$$t = \frac{k'}{h_{\text{corr}}} \tag{16.13}$$

If Eqs. (16.11) and (16.13) are substituted into the Ilkovic equation (Eq. 16.8), and providing all other factors which could affect i_d remain constant, then it can be shown that i_d varies with the square root of the corrected height of the mercury column above the tip of the DME

$$i_d \propto h_{\text{corr}}^{1/2} \tag{16.14}$$

This equation provides a means of ascertaining whether the current is actually diffusion controlled.

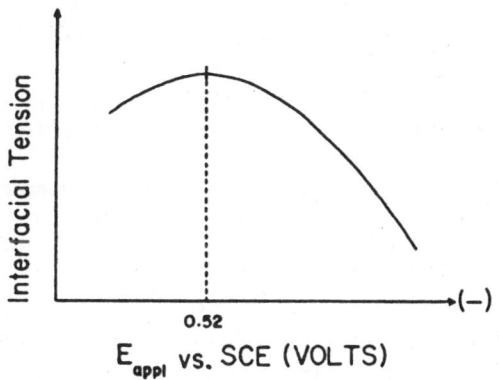

FIGURE 16.3: Electrocapillary curve for mercury.

c. The Applied Potential (E_{appl}). The drop life t is affected by the applied potential E_{appl} on the DME. Actually the value of t is affected by the mercury drop-solution interfacial tension γ, which in turn is affected by the applied potential as indicated in Fig. 16.3.

As the negative applied potential ($-E_{\text{appl}}$) is increased on the mercury drop, the drop-solution interfacial tension γ, passes through a maximum at approximately -0.52 V (relative to SCE) and then decreases rapidly. In practice, since the product $m^{2/3}t^{1/6}$ is only influenced by the sixth root of t, it is considered almost constant, as it varies less than 0.5% over the applied potential range of 0 to -1.0 V. However, at more negative potentials the product $m^{2/3}t^{1/6}$ shows a more rapid decline in value, which must be taken into account.

When diffusion currents are to be compared with those calculated by means of the Ilkovic equation, it is essential that t and i_d be measured at the same potential.

d. Temperature Effect. The effect of temperature on the diffusion current is quite pronounced. Although a temperature term does not appear in the Ilkovic equation, a temperature change affects every term except n. The diffusion coefficient D is very sensitive to variations in temperature because of its influence on viscosity, mobilities of ions, etc. These changes produce an increase in the i_d of 1 to 2%/deg rise in temperature in the vicinity of 25°C. To eliminate this error in i_d measurement it is essential to control the temperature of the electrolysis cell to within ±0.25 or ±0.5°C to hold errors within 1%.

e. Viscosity of Solution. The factor which is affected most by the viscosity of the solution is the diffusion coefficient D.

Changes in the supporting electrolyte do not appreciably change the viscosity unless they are large, i.e., in excess of ±10%. However, a change in the amount of a maximum suppressor, such as gelatin, may produce a marked change in the solution viscosity. A high concentration of a maximum suppressor is to be avoided for this reason.

The diffusion coefficient D of an electroactive species is also affected by complex formation with any component of the supporting electrolyte, or by other means.

D. KINETIC CURRENT

A kinetic current is a complex-limiting current which may occur when one or both of the oxidative states of the electroactive substance is in a dissociation equilibrium or involved in a chemical reaction with other substances. Under such circumstances the electroactive form of the species is depleted at the DME surface by a rapid electron transfer process and must be replenished by a relatively slower nonelectrode reaction such as dissociation or a shift in a chemical equilibrium. The magnitude of the kinetic current flow is proportional to the rate constant of the slower chemical reaction.

Kinetic waves are more common in the polarographic analysis of organic compounds such as formaldehyde, carboxylic acids, etc.

E. CATALYTIC CURRENT

A catalytic current is an increase in the current which may be brought about by an unstable electrolysis product, which may suddenly revert to the original electroactive species. For example, catalytic current waves have been observed during the reduction of cupric ions to cuprous ions, where part of the latter is spontaneously oxidized to cupric ions while another part is simultaneously reduced to copper.[8] Well-defined catalytic hydrogen waves have been observed for a number of alkaloids in certain supporting electrolytes.[9]

F. MAXIMA AND SUPPRESSORS

Frequently a pronounced increase in current above the normal limiting value is observed on a polarographic curve. This is often caused by an increase in mass transport of the depolarizer (electroactive substance) to the electrode surface brought about by a streaming effect in the solution. These maxima are reproducible and are subdivided into maxima of the first and second kind. Maxima of the first kind, which appear on the rising portion of the wave, as shown in Fig. 16.4, are rather narrow and frequently occur

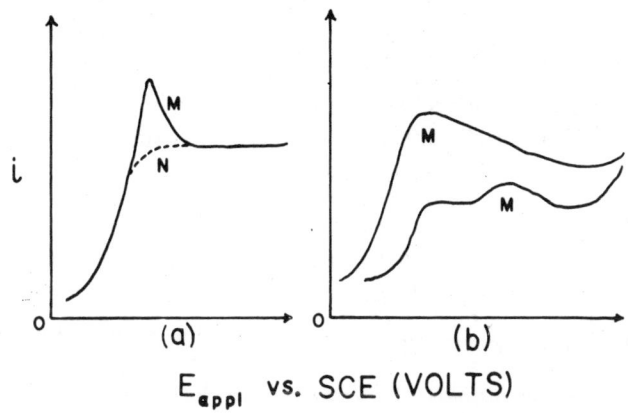

$$E_{appl} \text{ vs. SCE (VOLTS)}$$

FIGURE 16.4: (a) Maximum of the first kind (M). The dotted line (N) indicates the normal wave. (b) Maxima of the second kind (M).

in dilute solutions. Those of the second kind occur in more concentrated solutions, at higher mercury flow rates, only in the limiting current range, are rounded, and do not fall as abruptly to the limiting-current plateau as do the maxima of the first kind. Heyrovský and Kůta discuss both kinds of maxima in detail.[2]

The occurrence of maxima is undesirable because of the interference with diffusion current measurements. Maxima of both the first and second kinds can be eliminated by adding a suitable maxima suppressor. The suppressor substance is usually some nonreducible high-molecular-weight organic surface-active material such as gelatin, nonionic detergents such at Triton X-100, as well as cationic and anonic types, methyl cellulose, agar, alcohols, and certain dyes such as methyl red, and basic or acid fuchin, etc. Gelatin is one of the most widely used substances, but solutions must be prepared freshly. The final concentrations of these maxima suppressors in solution range between 0.001 to 0.01%. The optimum amount of substance required for complete suppression of a maximum is proportional to the concentration of the depolarizer, varies from substance to substance, and is influenced by the nature of the supporting electrolyte. The optimum concentration of the suppressor substance is found by trial and error. Care must be taken to avoid higher concentrations as several terms in the Ilkovic equation are affected by the resultant increase in viscosity and a lower diffusion current results.

G. OXYGEN WAVES

If air (oxygen) is not removed from the solution prior to analysis, two waves will result from the reduction of oxygen and its reduction product, hydrogen peroxide, as the potential of the DME is made increasingly negative. Although the half-wave potentials are pH dependent, one may occur about -0.10 V and another about -0.90 V (vs. the SCE). The reactions which produce these reduction waves are as follows:

First wave: (in acidic media) $O_2 + 2H^+ + 2e \rightarrow H_2O_2$
               (in alkaline media) $O_2 + 2H_2O + 2e \rightarrow H_2O_2 + 2OH^-$

Second wave: (in acidic media) $H_2O_2 + 2H^+ + 2e \rightarrow 2H_2O$
               (in alkaline media) $H_2O_2 + 2e \rightarrow 2OH^-$

The oxygen (air) must be removed from the solution or these pronounced oxygen waves will be superimposed on the desired polarogram and interfere with current measurements. This air is usually removed by bubbling some inert gas such as oxygen-free nitrogen through the solution prior to analysis and allowing a gentle flow of nitrogen to layer above the solution surface to prevent oxygen reabsorption during analysis. If a gas dispersion tube is used, the deaeration time can be reduced considerably. However, if an ordinary small-bore glass tube is used for this purpose, about 10–15 min are normally required to remove the dissolved oxygen. A purified grade of commercial tank nitrogen ($99.9+\%$) is frequently adequate for this purpose. If the last trace of oxygen must be removed, the nitrogen should either be passed through a heated tube of copper turnings or a gas washer containing some suitable reducing agent.

H. SUPPORTING ELECTROLYTE

The supporting electrolyte is added to carry most of the current between the electrodes, to reduce the migration current of the electroactive substance. This condition must be satisfied if the diffusion current i_d of the depolarizer is to be directly proportional to the concentration of this electroactive substance. The concentration of the supporting electrolyte should be at least 50-fold greater than that of the depolarizer for this purpose. The choice of the supporting electrolyte can be a most important factor since it frequently interacts with, and affects the $E_{1/2}$ value of the depolarizer, and may even determine whether any wave will result or not. A mixture of two electroreducible substances may have closely spaced $E_{1/2}$ values in one supporting electrolyte (SE) and yield unresolved waves. Another SE cın often be found which will form a complex with one of them, producing a shift in its $E_{1/2}$ value and thereby allowing the waves to be resolved adequately for both substances to be determined simultaneously in the mixture. The reader is referred to texts on polarography for extensive lists of supporting electrolytes.[2,5,10]

As the applied potential across the electrolysis cell is increased, a point is eventually reached where the water (solvent) will be electrolyzed. This point determines the upper voltage limit of the working electrode in that media, for once the electrolysis of the solvent begins, other electrode reactions will be masked. The solvent is seldom electrolyzed at the reference electrode. When the microelectrode (DME) functions as a cathode the hydrogen ions of water are reduced and hydrogen gas begins to be evolved about −1.8 V (vs. SCE) in neutral solution. Under more alkaline conditions this electrolysis is shifted to slightly more negative potentials. If a working electrode, other than the DME, becomes sufficiently anodic, the water is oxidized and oxygen is evolved at this electrode surface. The positive voltage-range limit for the DME is determined by the potential at which mercury begins to be oxidized to mercuric ions, while the negative limit of the DME potential is that at which some component of the SE becomes discharged. Certain quaternary salts and hydroxides serve as supporting electrolytes and permit the use of applied potentials more negative than −2.0 V (vs. SCE). These SE materials increase the hydrogen overpotential and allow polarographic reduction studies on alkali and alkaline earth-metal cations. Salts of the latter substances may be employed as supporting electrolytes for other more readily reduced substances in aqueous solutions at the DME over a range of +0.4 to about −1.9 V (vs. SCE).

Organic substances are frequently dissolved in nonaqueous supporting electrolytes.[5]

All substances to be used in supporting electrolytes should be polarographed under high sensitivity to check for possible interfering waves due to impurities.

I. MIXTURES OF ELECTROACTIVE SUBSTANCES

Frequently mixtures of electroactive substances can be analyzed provided the half-wave potentials ($E_{1/2}$) of the component substances are adequately spaced so that there is no interference between adjacent waves. If the half-wave potentials are not, or cannot be made sufficiently different by complexing the depolarizer or changing the solvent, then physical, chemical, or electrochemical methods of separating the interfering components are mandatory prior to polarography.

The wave of the most readily reduced component can easily be observed at a high sensitivity, but the second wave may go off scale. When this occurs, the analyst is faced with a choice. If the second component is present in

higher concentration, the sensitivity may be decreased sufficiently to keep
the second wave on scale. However, if the second wave is produced by a
minor component, greater sensitivity is required for its wave. Many instruments are equipped with a device which permits a compensating current to
be passed through the galvanometer, in opposition to the reduction (or
oxidation) current, which either reduces the height of the preceding wave or
eliminates it. With the first wave electrically cancelled the second wave can
be developed with the same or a higher sensitivity to permit an accurate
measurement of the minor component.

The relative wave heights (and indirectly concentration), the number of
electrons n involved in each reversible reduction, as well as the degree of
reversibility of each reduction all affect the resolution of successive reduction
waves for mixtures of reducible constituents.

Usually the wave for a two-electron reduction is steeper and occurs over a
shorter voltage span than for a single-electron reduction. A greater degree of
irreversibility of the electrode reaction also causes a spread in the wave over
a wider voltage range. It should be noted that the $E_{1/2}$ values of two electroreducible species showing irreversible waves would have to differ to a greater
extent than those of adjacent reversible waves in order to obtain comparable
resolution.

J. EQUATIONS FOR REVERSIBLE POLAROGRAPHIC WAVES

If the oxidized state only of a depolarizer [e.g., Cd(II), Fe(III), reducible
organic molecule, etc.] is in solution, it will accept electrons from the DME
(cathode) which are supplied from the SCE (nonpolarizable anode) via the
external circuit.

The equation for the polarographic reduction (cathodic) wave can be
arrived at from the following considerations:

Reduction at the DME surface is represented by Eq. (16.1), repeated here:

$$Ox + ne^- \rightarrow Red$$

where Ox and Red are the oxidized and reduced forms of the electroactive
species. During electrolysis at a potential corresponding to the limiting-
current plateau, the concentration of the electroreducible substance at the
microelectrode surface, $[Ox]_0$, is negligibly small compared to that in the
bulk of the solution $[Ox]$. At some lower applied potential, corresponding
to the rising current on an i vs. E_{appl} plot, the $[Ox]$ is not negligible in the
layer surrounding the DME and, for a reversible reaction in which both
Ox and Red are soluble, the Nernst equation can be applied. The potential
of the DME under these conditions and at 25°C is

$$E_{DME} = E^{\circ\prime} + \frac{0.0591}{n} \log \frac{[Ox]_0}{[Red]_0} \qquad (16.15)$$

where $E^{\circ\prime}$ is the formal reduction potential of the redox couple in the electrode reaction [Eq. (16.1)] under experimental conditions. As discussed earlier [Eq. (16.4)] molar concentrations can be employed in this equation instead of activities, providing dilute solutions are used. When this is done $E^{\circ\prime}$ replaces the standard potential E°. The subscript $_0$ denotes the negligible concentration at the electrode surface. It can be seen from Eq. (16.15) that the potential on the DME determines the ratio Ox/Red.

The average current supplied by diffusion of the reducible depolarizer to the DME surface is

$$i = K([Ox] - [Ox]_0)D_{Ox}^{1/2} \tag{16.16}$$

where K is a constant which also includes terms n, m, and t of the Ilkovic equation [see Eq. (16.8)]. Equation (16.16) is similar to Eq. (16.3). The Ilkovic equation may now be written in the form of

$$i_d = 607nm^{2/3}t^{1/6}D_{Ox}^{1/2}([Ox] - [Ox]_0) \tag{16.17}$$

where the current reaches its limiting value and is solely dependent on the gradient existing between the concentration of depolarizer in the bulk of the solution [Ox] and that at the electrode surface $[Ox]_0$. Since $[Ox]_0 = 0$, i.e., it is negligible since Ox is reduced as rapidly as it arrives at the electrode surface, Eq. (16.17) reduces to

$$i_d = K[Ox]D_{Ox}^{1/2} \tag{16.18}$$

By solving for $[Ox]_0$ in Eq. (16.16) and combining with Eq. (16.18)

$$[Ox]_0 = \frac{i_d - i}{KD_{Ox}^{1/2}} \tag{16.19}$$

The reduced form (Red) of the depolarizer formed at the DME may diffuse away or form amalgams and diffuse into the mercury drop. The current depends not only on the rate at which Ox arrives at the microelectrode [see Eq. (16.18)] but also on the rate at which the reduced form, Red, diffuses from this electrode surface:

$$i = K([Red]_0 - [Red])D_{Red}^{1/2} \tag{16.20}$$

and since only the oxidized form of the depolarizer is present in the bulk of the solution, $[Red] = 0$, and

$$i = K[Red]_0 D_{Red}^{1/2} \tag{16.21}$$

By solving Eq. (16.21) for $[Red]_0$ and substituting this value, as well as Eq. (16.19) in the Nernst [Eq. (16.15)], the potential of a redox system is given by

$$E_{DME} = E^{\circ\prime} + \frac{0.0591}{n} \log \frac{i_d - i}{i} - \frac{0.0591}{n} \log \left(\frac{D_{Ox}}{D_{Red}}\right)^{1/2} \tag{16.22}$$

Since the diffusion coefficients D for the Ox and Red forms of many depolarizer substances are very nearly equal and appear in this equation only as the square root, the factor $(D_{Ox}/D_{Red})^{1/2}$ can be let equal unity and Eq. (16.22) reduces to the following at $25°C$:

$$E_{DME} = E^{0\prime} + \frac{0.0591}{n} \log \frac{i_d - i}{i} \qquad (16.23)$$

This is the simplest expression for the shape of a reversible polarographic reduction wave, i.e., a cathodic wave.

The potential on the microelectrode (DME) corresponding to the current value i at one-half the diffusion current i_d is called the "half-wave potential" $(E_{1/2})$. A mathematical expression for the $E_{1/2}$ is obtained from Eq. (16.22) by introducing $i = i_d/2$:

$$E_{1/2} = E^{0\prime} - \frac{0.0591}{n} \log \left(\frac{D_{Ox}}{D_{Red}}\right)^{1/2} \qquad (16.24)$$

or

$$E_{1/2} = E^{0\prime} \qquad (16.25)$$

can be obtained from Eq. (16.23).

The half-wave potential $(E_{1/2})$ value is characteristic of a given depolarizer in a specified medium, but is independent of the depolarizer concentration, of the capillary characteristics, and of the galvanometer sensitivity.[2] The $E_{1/2}$ value obtained in a specified supporting electrolyte is frequently used to identify a depolarizer.

It is frequently so that the diffusion coefficients of the oxidized and reduced states D_{Ox} and D_{Red} of a redox couple are approximately equal and in such instances the $E_{1/2}$ corresponds closely to the value of the formal potential $E^{0\prime}$. If either the Ox or Red form of the depolarizer substance is complexed, their rates of diffusion will differ. By measuring the ratio of anodic to cathodic diffusion currents in solutions where $[Ox] = [Red]$, the ratio $D_{Ox}^{1/2}/D_{Red}^{1/2}$ can be found. Equation (16.24) shows that for a reversible electrode reaction the $E_{1/2}$ is related to the formal reduction potential $E^{0\prime}$ of the redox couple. However, the $E_{1/2}$ value may differ from the $E^{0\prime}$ somewhat depending on the difference in the diffusion coefficients of the Ox and Red forms.

If a polarographic system consisting of a perfectly polarizable DME and nonpolarizable SCE is employed, $E_{DME} = E_{appl}$. At an applied potential corresponding to the $E_{1/2}$ value, where $i = i_d/2$ by definition, $E_{1/2} = E^{0\prime}$ due to the disappearance of the log term in Eq. (16.23), thus, at $25°C$,

$$E_{appl} = E_{1/2} + \frac{0.0591}{n} \log \frac{i_d - i}{i} \qquad (16.26)$$

This equation with slight rearrangement is a form of $y = mx + b$, the equation for a straight line. Therefore the reversibility of an oxidation-reduction system at the DME surface can be checked by plotting $\log (i_d - i)/i$ vs. E_{appl} (as abscissa). For reversible redox systems, a straight line results with slope $= 0.0591/n$ V and abscissa intercept equal to $E_{1/2}$, where the log term of Eq. (16.26) becomes 0. The value of n can be determined by the expression $n = 0.0591/\text{slope}$, where the slope is 0.0591, 0.0296, and 0.0197 V, for a 1-, 2-, or 3-electron transfer in the electrode process, respectively. These values apply only at 25°C.

It is unfortunate that a linear relationship between $\log (i_d - i)/i$ vs. E_{appl} is not absolute proof of reversibility.[5] This method cannot be applied if the product of the electrode reaction is not soluble in the mercury drop as an amalgam, or soluble in the solution of supporting electrolyte employed.

The preceding discussion dealt only with the cathodic wave. For discussions and equations of anodic waves and cathodic-anodic combined waves the reader is directed to Ref. 2.

K. REVERSIBLE AND IRREVERSIBLE ELECTRODE REACTIONS

In reversible redox systems ($Ox + ne^- \rightleftharpoons Red$) where the equilibrium is rapidly established by fast electron-transfer rates (ETR) and when the current is controlled solely by diffusion of the depolarizer, the electrode potential is equal, or very nearly equal, to that value calculated by means of the Nernst equation [Eq. (16.4)]. It has been shown [Eq. (16.25) and Fig. 16.5, curve a] that for reversible reactions the $E^{o\prime}$ very nearly coincides with the $E_{1/2}$ for each form of the redox couple.

A reversible electrode process is indicated when the $E_{1/2}$ values of the anodic (oxidation) and cathodic (reduction) waves coincide. This situation occurs with the Fe(III)/Fe(II) redox couple. When both oxidation states of the depolarizer substance [e.g., Fe(III)/Fe(II)] are present in solution, a composite anodic-cathodic wave may be obtained (Fig. 16.6, curve a) the steep slope of which is defined by the Nernst equation.

If the electrode reaction is irreversible, as it generally is for organic molecules and some inorganic substances, the Nernst equation has no application. It should be noted also that irreversible reactions occur when either the Ox or Red form is insoluble in either the supporting medium or the mercury drop. Some reversible electrode processes may become "irreversible" with a shortened mercury drop life t or under other circumstances where the time required for establishing equilibrium conditions at the solution-drop interface is inadequate.[2]

A smooth polarographic wave may result from a mixture of oxidized and reduced forms of a substance, even though the electrode processes are irreversible under the conditions used. However, in this situation the slope of the combined anodic-cathodic wave is much less steep and does not conform to theory. Figure 16.5 shows generalized forms of individual and combined anodic and cathodic waves for both reversible and irreversible systems; a qualitative difference is readily seen between these curves. In reversible processes the $E_{1/2}$ values of both anodic and cathodic waves are equal and very near the value of $E^{o\prime}$ in the Nernst equation. However, the $E_{1/2}$ values of the anodic and cathodic waves in irreversible systems are separated and neither is equal to the $E^{\circ\prime}$. The greater the degree of irreversibility, the greater the separation of the $E_{1/2}$ values of the oxidized and reduced forms. Irreversibility in an electrode reaction (slow ETR) yields an $E_{1/2}$ value more negative than $E^{\circ\prime}$ for a reduction (more positive for an oxidation).

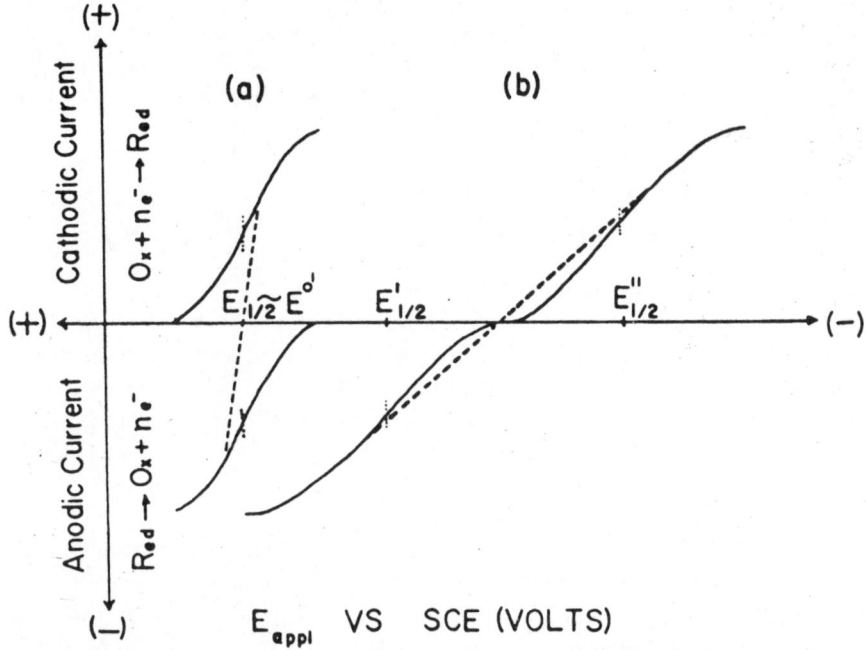

FIGURE 16.5: Anodic and cathodic i-E_{appl} curves for a reversible (a) and an irreversible (b) redox couple. The dotted lines represent the combined waves of a mixture of equal concentrations of Ox and Red forms in both instances, while the solid lines represent the waves for the Ox and Red forms of each couple separately. $E^{\circ\prime}$ is the formal redox potential and $E_{1/2}$, $E_{1/2}'$, and $E_{1/2}''$ are half-wave potentials.

If we now let $E_{1/2}$ in Fig. 16.5, curve a, represent the calculated theoretical value and $E_{1/2}$ on curve b represent the actual observed value for a specific substance under identical conditions, then the difference between these values correspond to the "overvoltage" (ov) of that substance (depolarizer) on that electrode. The magnitude of this overvoltage is indicative of the degree of irreversibility of the reaction occurring at the microelectrode (e.g., DME) surface.

Irreversible waves are also defined by the Ilkovic equation [Eq. (16.8)] and the $E_{1/2}$ is still an identifying characteristic of the depolarizer, even though it occurs at a position which differs from that which would result if the reaction were reversible.

Another method used for distinguishing between reversibility and irreversibility of electrode processes involves the calculation of the difference between $E_{3/4}$ and $E_{1/4}$, where these terms are defined in a manner similar to that for $E_{1/2}$. For ideal cathodic $(i - E)$ waves, a plot of i/i_d vs. $(E_{DME} - E^{o\prime})$ is made and the value $(E_{3/4} - E_{1/4})$ measured. For reversible reactions $(E_{3/4} - E_{1/4}) = 0.0564/n$ V (at 25°C), and for irreversible reactions this value is greater. Mathematical expressions are available for the $E_{1/2}$ of a

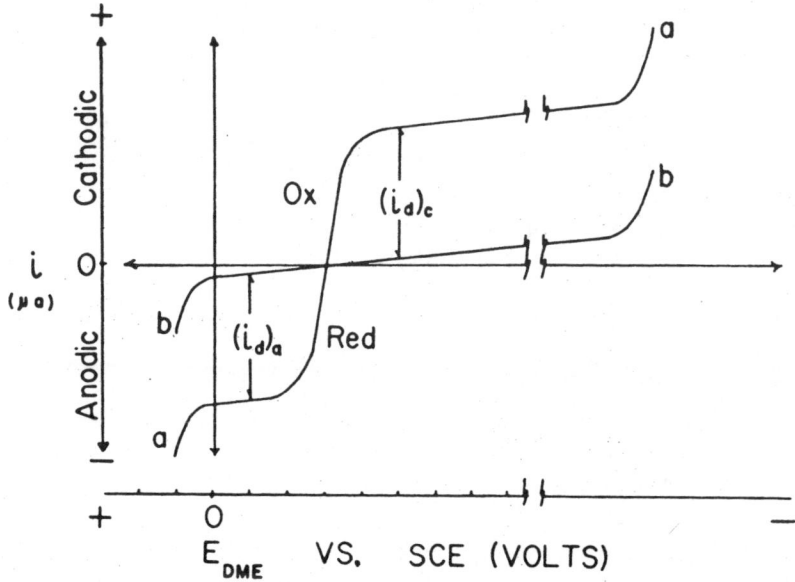

FIGURE 16.6: Polarographic waves of (a) a reversible oxidation-reduction couple, where [Ox] = [Red], and (b) the supporting electrolyte alone. The symbols $(i_d)_a$ and $(i_d)_c$ represent the anodic and cathodic diffusion currents, respectively.

totally irreversible wave (very slow ETR) and for the relationship between this value and drop time. For adequate resolution in quantitative analysis, the difference between the $E_{1/2}$ values of two successive polarographic waves should be three or more times the sum of their respective $E_{3/4} - E_{1/4}$ values. A greater difference is required if the wave heights are very different.[5] The reader is referred to the references, particularly Refs. 2 and 5, for a detailed discussion involving differentiating between reversible and irreversible electrode processes.

L. CONVENTIONS EMPLOYED IN PLOTTING $i - E_{appl}$ CURVES

The conventional method of plotting $i - E_{appl}$ curves is shown in Fig. 16.6. The potential applied to the microelectrode relative to a nonpolarizable reference electrode (e.g., SCE) is shown on the horizontal axis. This applied potential value (E_{DME}) becomes increasingly negative to the right of $E_{DME} = 0$, the potential of the SCE, and increasingly positive to the left. The current is represented on the vertical axis and, by convention, cathodic currents due to reduction are recorded as being positive and appears above the zero current line. The anodic currents due to oxidation are considered negative and appear below the zero current line.

Curve a (Fig. 16.6) is the polarogram of a solution containing equimolar concentrations of the soluble oxidized (Ox) and reduced (Red) states of an oxidation-reduction system in a deaerated solution containing a supporting electrolyte (SE) such as 0.1 M KCl. Curve b is the polarogram of the same deaerated SE and maximum suppressor but no depolarizer substance. At the extreme left, the anodic current increase is due to oxidation of the mercury of the drop ($2Hg \rightarrow Hg(II) + 2e^-$). This occurs at a lower (+) potential than the oxidation of water to oxygen ($2H_2O \rightarrow O_2 + 4H^+ + 4e^-$). The negative current to the right of the mercury dissolution wave up to the zero current line represents the anodic oxidation of the Red form of the depolarizer substance. The intercept of this mixed wave with the zero current line represents the E_{DME}, which is very nearly equal to the formal potential ($E^{o'}$) in the Nernst equation for reversible systems (fast electron-transfer rates) when [Ox] = [Red], assuming little or no change in the diffusion coefficients D of the two states of oxidation. Also in truly reversible redox systems the value of $E^{o'} \sim E_{1/2}$ of both the Ox and the Red forms. The (+) cathodic current in curve a (Fig. 16.6) represents the electroreduction of the oxidized form of the substance. At the extreme right of curve a, the rapid rise in the cathodic current at an applied potential of about -1.8 to -2.0 V, is a result of the reduction of hydrogen ions in the water or to the discharge of some cation of the SE. Curve b is the polarogram of the SE alone, the increased anodic and cathodic currents are due to the same causes discussed in curve a. Between these two extremes of rapid current increase, the DME is perfectly polarized and only a small residual current flows, which is shown here running close to the horizontal zero current line.

The cathodic diffusion $(i_d)_c$ will closely approximate the anodic diffusion current $(i_d)_a$, when [Ox] = [Red] in the mixture.

16.3   BASIC INSTRUMENTATION, APPARATUS, AND PRINCIPLES

A.   CIRCUITRY

The essential circuitry is shown schematically in Fig. 16.1. This diagram represents the basic requirements for a conventional manual dc polarograph.

The potential of the dropping mercury electrode (DME) can be made increasingly negative (or positive) relative to the constant known potential of the saturated calomel electrode (SCE = +0.244 V vs. normal hydrogen electrode NHE) by adjusting sliding contact (c) along slide-wire (a-b) of the constant potential source B. For very precise work it is essential to correct for the iR drop through the solution between the electrodes. This correction can be omitted in routine work by keeping the cell resistance low, as for aqueous solutions of electrolytes and in closely spaced electrodes. The slide-wire (a-b) can be calibrated by means of a precision potentiometer (not shown) so that the actual potential can be read directly from the slide-wire dial.

The applied voltage across the cell between the electrodes, in a solution of low resistance, is the difference between the potential of the nonpolarized electrode (e.g., potential of SCE vs. NHE) and the potential of the polarized electrode (e.g., DME vs. NHE). In practice, the SCE potential is taken as the reference point (e.g., $E_{SCE} = 0$) and the applied potential E_{appl} is considered the potential on the polarized electrode E_{DME} relative to that of the SCE.

If the i flowing as a result of the applied potential E_{appl} must be accurately known, the galvanometer must be calibrated. This is done by determining the number of microamperes which correspond to 1 galvanometer scale unit. This may be accomplished by connecting the leads from the instrument, which normally go to polarographic cell, to each end of a precision resistor. The polarity switch is set to "DME" (−). A known potential is then applied across the resistor by means of the polarizing device and the galvanometer reading recorded. The voltage drop across the resistor can be measured with a potentiometer. The current is calculated in microamperes by means of Ohm's law ($E = iR$). A conversion factor f (i.e., 1 galvanometer division = $f\,\mu A$) is computed and used to convert future galvanometer readings to current, e.g., galvanometer reading × sensitivity factor × $f = \mu A$. At an applied potential which causes the reduction (or oxidation) of the electroactive species, the galvanometer index will oscillate with increasing size and drop rate of the mercury drop from the DME. The amplitude of these oscillations can be reduced by a damping device on the galvanometer. This manual apparatus is useful as a teaching device, for quantitative measurements at a predetermined applied potential, or as a current measuring device for amperometric titrations. However, a more convenient instrument, such as a commercially available recording polarograph, is desirable for routine analysis. The reader is referred to reference books on polarography or to the appropriate manuals accompanying commercial instruments for circuit diagrams and discussions.

B. GENERAL OPERATION

The reader is referred to the actual operating instructions (or equivalent) supplied by the manufacturer (or instructor) for the commercial instrument to be used. The following is merely a generalized procedure outline, applicable to Fig. 16.1.

Before turning the instrument on, make certain that the sensitivity switch is placed in the least sensitive position. This switch is related to the shunt (S), which protects the galvanometer (G) by permitting only a small fraction of the current produced by electrolysis to pass through it. The appropriate applied potential range is then set. It is assumed that the electrodes are connected and immersed in an appropriate supporting electrolyte containing the depolarizer (electroactive substance). After the galvanometer zero is set, the "function" switch is turned to connect the cell with the circuit. The potential applied (E_{appl}) to the DME is gradually increased and the sensitivity increased or decreased during the scan to obtain a readable response and to keep the galvanometer index from going off scale. The polarizing potential is increased by increments which are best determined in a trial run, being larger in a current plateau region and smaller during any appreciable change in current flow. After each increase in the E_{appl} the corresponding galvanometer reading is recorded. If the actual current must be accurately known, the galvanometer readings are then converted to microamperes (μA). If relative current values are adequate, they can be used after conversion to a common sensitivity. This procedure is repeated until sufficient readings are accumulated to allow an $i - E_{appl}$ curve to be plotted (a polarogram).

C. COMMERCIAL INSTRUMENTS AND ACCESSORIES

1. Manual Polarographic Instruments

Commercial manual polarographs are available (Fisher, Heath, Sargent, and others) as well as manual instruments which can be converted to automatic operation by simple attachment to an appropriate recorder (Fisher, Heath, and others).

2. Recording Polarographic Instruments

In commercial recording polarographs, the applied voltage is scanned at a predetermined constant rate by means of a synchronous motor and the resulting current is recorded either.on a recorder which is an integral part of the instrument or on an external recorder. For the latter, the chart speed must be correlated with the rate of the potential scan.

Commercial recording instruments are available from a number of American and European manufacturers, such as Sargent, Aminco, Leeds & Northrup, Metrohm, Radiometer, and others. Figure 16.7 shows the Sargent recording polarograph model XVI.

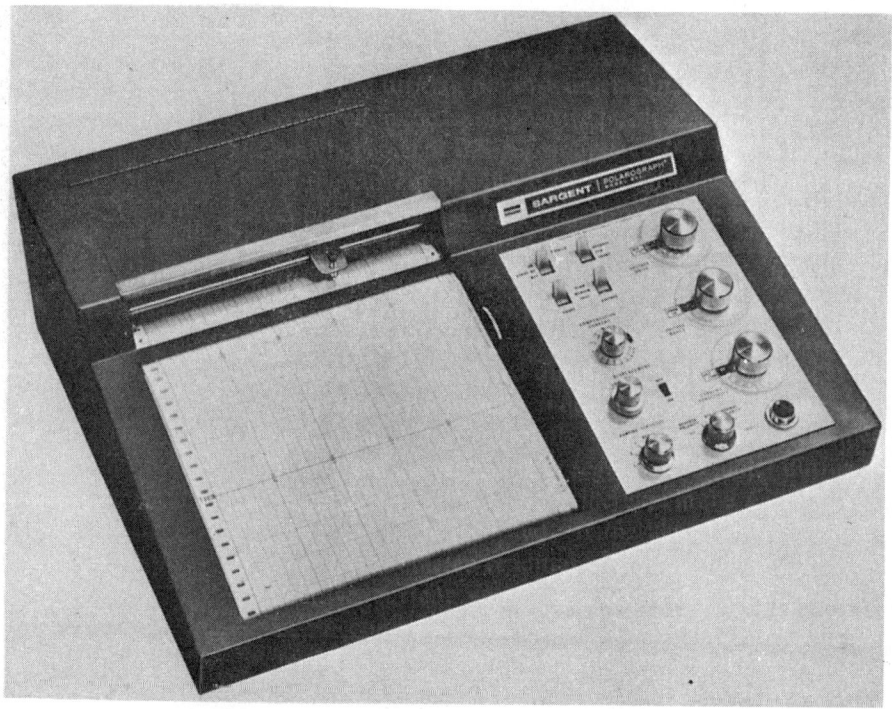

FIGURE 16.7: Sargent recording polarograph, model XVI. Courtesy E. H. Sargent and Co.

3. Electrolysis Cells

Polarographic electrolysis cells vary widely in design and capacity. The more common are: (a) a small open beaker-type of cell which may be used when a mercury pool (nonpolarized) electrode is to be used (Fig. 16.8a) and (b) the H-type cell in which the electrolysis compartment containing the DME is connected to an external reference electrode (SCE). These two compartments are electrically connected via a short agar gel plug saturated with KCl and supported in a wide diameter tube closed at one end with a sintered-glass disk (Fig. 16.8b)

A wide variety of electrolysis cells are commercially available from E. H. Sargent & Co. and other sources. These cells are designed for polarographic analysis of solutions having a total volume range from less than 1 ml up to several ml. Many different designs are possible as a literature survey will readily show.

4. Electrodes

a. Polarizable Microelectrodes. Actually the term "polarography" infers that the polarizable microelectrode is the dropping mercury electrode (DME).

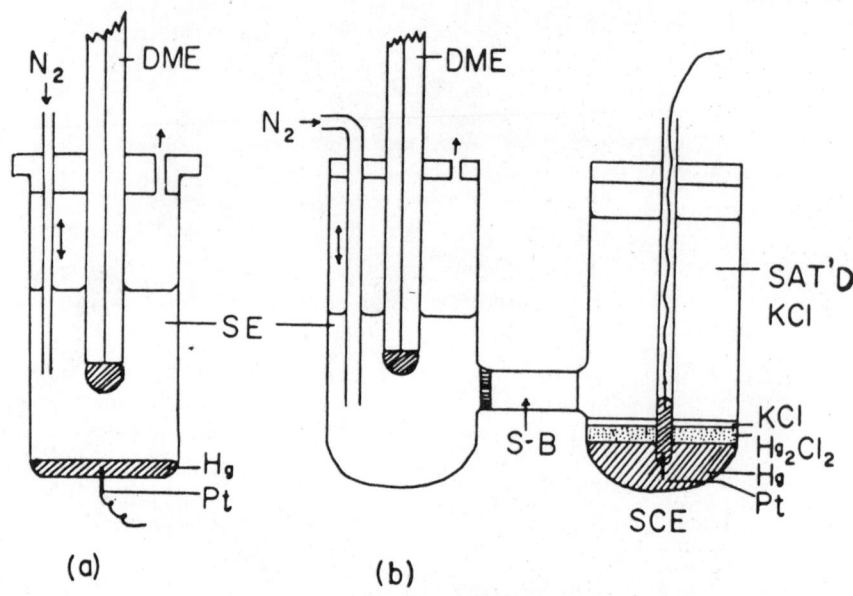

FIGURE 16.8: (a) Beaker-type cell with mercury pool and dropping mercury electrodes. (b) H-type cell with saturated calomel electrode (SCE) and DME.

This is still the most widely used electrode, however numerous other mercury electrodes have been employed such as the hanging mercury drop, streaming mercury, inverted mercury, and horizontal mercury electrode, as well as many others. Microelectrodes such as the rotating and vibrating platinum electrodes are used for very closely related methods of voltammetry, especially in amperometric titrations.

1. *Dropping Mercury Electrode* (*DME*). The DME consists of a short length of glass capillary tubing (0.03–0.05 mm id) which is attached to a length of large-bore glass tubing used to support a mercury column. The constant height of this mercury column is maintained by a mercury reservoir bulb which is connected to the mercury column and capillary as shown in Fig. 16.1. Electrical contact with the DME may be made via the mercury reservoir, or by means of a platinum wire fused into the glass wall of the mercury column support or elsewhere. The mercury drops, formed under a constant head pressure of mercury, leave the capillary tip at a constant rate. A repetitive process is involved in which each mercury drop gradually grows into a microsphere until it reaches a critical weight and detaches itself.

The advantages of the DME relative to other microelectrodes are:

(a) The continually renewed smooth surface of the drop prevents the accumulation of electrode reaction products which would adversely affect the potential at the surface.

(b) The reproducible surface area of the drop can be readily calculated at any time during the drop life.

(c) The high overvoltage (overpotential) of hydrogen on this mercury electrode makes possible the electroreduction of many species at negative potentials unattainable on solid microelectrodes. The DME can be used up to -1.6 V (vs. SCE) in acidic solution and up to -2.6 V in basic solutions or in nonaqueous solutions.

(d) The mercury drop forms amalgams (solid solutions) with many metals, which results in a lower reduction potential for the substance.

(e) Several polarograms can be run on the same solution without appreciable change since a very small amount of the analyte is changed by the minute current flow through this electrode.

(f) The diffusion current reaches a steady value rapidly and is reproducible.

The DME has certain disadvantages such as:

(a) Mercury has a very limited potential range as an anode, since mercury is oxidized at an applied potential of about $+0.4$ V (vs. SCE).

(b) The surface area of this electrode is always changing giving rise to current oscillations.

(c) The applied potential affects the interfacial tension at the mercury drop-solution interface thus affects the mercury drop size. The latter is also influenced by chemical agents which affect this interfacial tension.

2. *Solid Microelectrodes.* The rotating or vibrating platinum microelectrodes are among the most useful in voltammetry. Platinum has a much lower hydrogen overpotential than mercury, and thus has a very limited potential range as a cathode. However, platinum can be used at positive applied potentials up to about $+1.1$ V (vs. SCE).

When not rotated or vibrated, these solid electrodes have a number of disadvantages such as:

(a) Require several minutes to reach a steady current value, then it declines slowly.

(b) After a voltammogram is obtained, the curve is not retraced when the E_{appl} is gradually reversed to zero. The $i - E_{appl}$ curve shape also varies with the rate of the applied potential increase.

(c) Temperature changes produce greater changes in the diffusion current obtained with this electrode than for the DME.

By rotating or vibrating the platinum-wire electrode the current is greatly enhanced over that obtained with a DME and assumes a steady value immediately. Since the rate of stirring affects the current flow, constant rate rotating (or vibrating) platinum electrodes are mandatory. These electrodes are widely used for amperometric titrations, but their lack of reproducibility precludes single measurements of current.

b. Reference Electrodes. Two reference electrodes which are widely employed in polarography will be discussed here.

1. *Mercury Pool Electrode* (Hg-*pool*). The mercury pool electrode consists of a pool of mercury in the bottom of the electrolysis cell (Fig. 16.8a). Its relatively large surface area prevents its polarization by the small current flow through it due to the potential impressed across the cell. While the exact potential of this electrode is unknown, it remains quite constant. The mercury pool electrode may be employed for routine work in which the exact value of the potential applied to the DME is not important.

2. *Saturated Calomel Electrode* (*SCE*). For accurate work or research the potential of the DME must be known, therefore an isolated reference electrode is connected to the electrolysis cell by means of a relatively large-surface "salt bridge." The reference electrode which has been almost universally adopted is the SCE (Fig. 16.8b).

3. *Auxiliary Electrodes.* In addition to the polarized microelectrode (e.g., DME) and reference electrode (e.g., SCE), a third electrode which is frequently a platinum wire (or mercury pool) electrode may be used. This additional electrode is called an "auxiliary electrode." It is not polarized and reduces the resistance to the flow of current in poorly conducting supporting media. Since the potential on the DME must be proportional to the applied potential in many instances, either one or both of the SCE and auxiliary electrodes are required depending on the conditions used. When an applied potential is impressed across an electrolysis cell a potential drop occurs at three sites, the cathode (usually DME), the anode (usually SCE), and in the solution between these two electrodes, thus

$$\text{emf}_{\text{cell}} = (E_{\text{SCE}} - E_{\text{DME}}) + iR \tag{16.27}$$

where iR is the potential drop (ohmic drop) between the electrodes. It can be seen that when the exact value of the E_{DME} must be known, the other values in Eq. (16.27) must be known or measurable. Three different situations are possible:

(a) Only a large surface (nonpolarizable) SCE is required when the electrolysis current is small (under 1 μA) and the supporting electrolyte has a low resistance and the ohmic drop iR between the electrodes is negligibly small (under 1 mV) or known.

(b) Only a nonpolarizable auxiliary electrode (e.g., mercury pool) of reasonably constant but unknown potential is required when the iR drop is small (under 10 mV), is known, or where the DME potential does not have to be accurately known.

(c) When an accurate E_{DME} is required, both the SCE and auxiliary electrodes are required when the SE has a high specific resistance, (e.g., nonaqueous media). In the latter situation it is also possible to measure the voltage between a closely spaced DME and SCE salt bridge, with a sensitive voltameter, such as a pH meter.[8]

16.4 GENERAL POLAROGRAPHIC ANALYSIS

Substances which can be reduced or oxidized at the DME to produce an i-E_{appl} wave can usually be polarographically analyzed. If more than one electroactive substance (including impurities) is present, the successive $E_{1/2}$ values of each substance must differ by 150 mV or more to obtain resolution of the waves. Many inorganic and organic substances yield well defined cathodic waves when the DME is employed. Quantitative analysis of such substances is normally performed within a concentration range of 10^{-3}–10^{-5} M. The lower concentration limit is usually governed by the magnitude of the residual current. Solubility factors at the DME surface may produce false or unpredictable currents at concentrations greater than 10^{-2} M.

When a new supporting electrolyte is employed, it should be checked polarographically at a high sensitivity to preclude the possibility of waves arising from impurities.

A. TECHNIQUES FOR MEASURING DIFFUSION CURRENTS AND HALF-WAVE POTENTIALS

Many polarographic waves are not as ideally shaped as that depicted in Fig. 16.2, therefore several arbitrary procedures have been devised for measuring diffusion currents (i_d) and half-wave potentials ($E_{1/2}$) from a polarogram. These procedures differ in the manner of estimating the residual current.

1. Accurate Measurement of Residual Current

The residual current i_r is most accurately determined separately from the polarogram of a blank solution which is identical to the test solution except for the deletion of the depolarizer. This value is then subtracted from the limiting current plateau value i_L for the test solution measured at the same potential to obtain i_d.

2. Approximate Measurement of Residual Current

A close approximation of the i_r can be made and the i_d and $E_{1/2}$ values determined with sufficient reproducibility by employing one of the graphical procedures which follow.

a. **For ill-Defined Waves.** In the measurement of the diffusion current of depolarizer substance 1 (Fig. 16.9), i_r is approximated by extrapolating the residual current to give line a. Since the limiting current is impossible to measure accurately for species 1, a line b is drawn parallel to line a and the $(i_d)_1$ measured as indicated. Since the residual current for species 2 includes the $(i_d)_1$ and is not parallel to the limiting current $(i_L)_{1+2}$, line d is drawn parallel to the horizontal axis from the point of intersection of line b and the

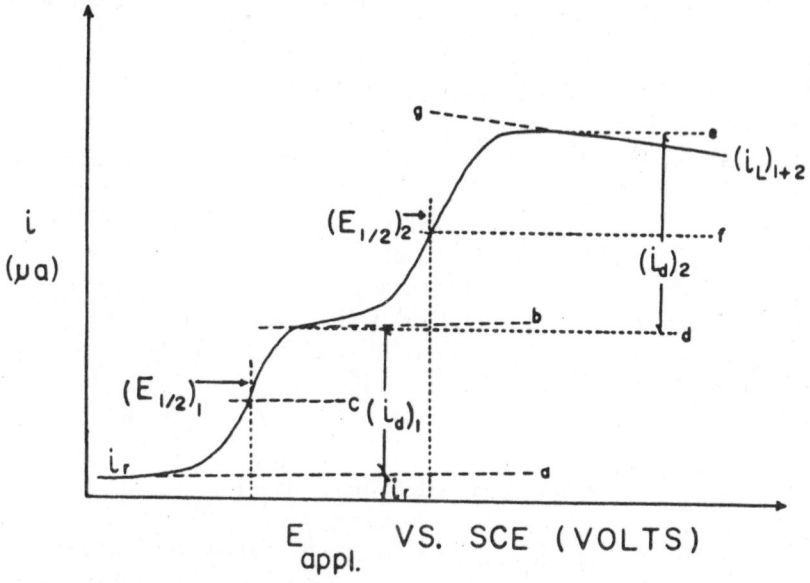

FIGURE 16.9: Double wave polarogram. See the text for the explanation of the measurement of the diffusion currents i_d and half-wave potential $E_{1/2}$.

curve. Extrapolation of the upper limiting current gives line g. Line e is drawn parallel to line d from the point of intersection of line g and the curve. The diffusion current $(i_d)_2$ for species 2 is measured as indicated.

The $E_{1/2}$ values of species 1 and 2 are determined from the point of intersection of the curve with lines c and f, respectively. These lines are drawn from points equal to $i_d/2$ for each species and parallel to lines a and d, respectively.

b. **For Well-Defined Waves.** The i_d and $E_{1/2}$ values for well-defined waves, where the i_r and i_L portions of the curve are parallel, are easily measured. The procedure is much simpler as shown in Fig. 16.2, since the i_r value only is approximated by an extrapolation of this portion of the curve.

Reference books on polarography should be consulted for additional procedures.

B. QUALITATIVE POLAROGRAPHIC ANALYSIS

Although the half-wave potential ($E_{1/2}$) is a characteristic of an electro-active species under specified and reproducible conditions, especially with respect to the electrolysis medium, polarography finds only limited application to qualitative analysis. This statement is particularly true in the presence of mixtures of electroactive species where successive waves are not well resolved. To effect resolution the supporting medium usually has to be altered. The new $E_{1/2}$ value for a species is applicable only to that altered medium. The $E_{1/2}$ values of many electroactive species in a large variety of solvents or supporting electrolytes are to be found in reference books on polarography.[2,5,10]

It is frequently possible to identify relatively pure electroactive species by comparing the $E_{1/2}$ values of the unknown with suspected known species in a variety of supporting electrolytes (or solvent media), since the latter frequently affects the electron transfer rates involved in the electrode reactions.

C. QUANTITATIVE POLAROGRAPHIC METHODS

These methods are divided into two categories: (1) absolute method and (2) comparative methods.

The following methods presuppose that appropriate consideration has been given to the selection of the concentration range (approx. 10^{-2} to 10^{-5} M) to be studied, the appropriate solvent or supporting media, and any preanalysis depolarizer separation which may be necessary to give the desired selectivity to the quantitative method.

1. Absolute Method

Since the absolute method is based on the direct application of the Ilkovic equation, all the terms in that equation must be known or measured before the concentration can be related to the observed diffusion current. This method is not used to any extent in practical analysis for obvious reasons. These terms would have to be measured or calculated to verify this equation experimentally. The reader is referred to one of the books on polarography for further details on this method.

2. A Modified Absolute Method

This method employs a modified form of the Ilkovic equation [Eq. (16.8)]:

$$C = i_d / I_d m^{2/3} t^{1/6} \qquad (16.28)$$

The diffusion current constant I_d is independent of any specific capillary, but is dependent upon the nature and concentration of the supporting electrolyte and the temperature which must be controlled. A calibrated galvanometer is essential for the measurement of absolute current. This method therefore suffers from tedious calibrations and is not as accurate as the relative methods which follow.

3. Comparative Methods

These relative methods can be subdivided according to the particular techniques employed. In all instances, however, a pure standard of the same species as the sample depolarizer is used for calibration purposes or relative measurements. Identical conditions are employed for the sample and standard to calculate the unknown concentration of the sample.

a. Direct Comparison Method. A series of standard solutions are prepared which are identical to sample solutions in all aspects except that of concentration. The actual analysis can be performed in one of two ways:

1. *Calibration Curve.* This method is convenient for routine analysis on large numbers of samples. The standard solutions can be polarographed one at a time or if regular waves are obtained the limiting current only may be determined for each concentration at a fixed applied potential corresponding to the limiting current plateau in the polarogram of one of the standard solutions. The residual current should be determined at the same potential from a polarogram of the blank or approximated from the polarogram of a standard solution by the extrapolation procedure. The measured diffusion currents i_d of the standard solutions are used to construct a calibration curve, i.e., a plot of i_d vs. concentration. The diffusion current value of each sample solution, measured under essentially identical conditions, is referred to the calibration curve and the concentration read. Should the concentration range be large enough to produce nonlinearity in the calibration curve, semilog graph paper may be used to retain the same precision in the concentration reading over the entire plot. For more limited concentration ranges the i_d vs. concentration plot should be linear, passing through the origin. Calibration curves should not be assumed to be valid from day to day, but must be verified for each series of analysis.

2. *Alternate Direct Comparison Method.* The wave heights of a consecutively polarographed standard and sample solutions (identical conditions) which are about equal in concentration may be compared. The simplified form of the Ilkovic equation [Eq. (16.9)], i.e., $i_d = KC$, may be used here. This equation applies equally to both sample and standard and the constant K is identical in both instances under identical conditions except for concentration. Therefore the following expression can be employed for the calculation of the sample concentration:

$$C_{\text{sample}} = \frac{(i_d)_{\text{sample}}}{(i_d)_{\text{std}}}(C_{\text{std}})$$ (16.29)

This procedure is convenient for occasional analysis on very few samples as it eliminates the necessity of constructing a calibration curve. Greatest accuracy is obtained when the concentrations of standard and sample closely approximate one another, especially for a slightly nonlinear relationship between wave height and concentration. No special knowledge of the capillary characteristics is necessary in this method, but they as well as the temperature must remain constant throughout the comparison.

b. **Standard Addition Method.** The polarogram of an accurately known volume of solution A containing the sample is recorded. A known weight of the standard substance (a pure form of sample substance) is added to an identical sample solution B and the polarogram recorded under the same conditions.

The weight of the substance being determined (W_{sample}) in the sample solution A may be calculated from this equation

$$W_{sample} = \frac{i_{sample} W_{std}}{i_{(sample+std)} - i_{sample}} \qquad (16.30)$$

where W_{std} is the weight of standard added to the second solution B of the sample, $i_{(sample+std)}$ is wave height of solution B, and i_{sample} is the wave height of solution A.

This version of the standard addition method is less complex than an alternate version in which accurate volumes of a standard solution are added to known volumes of the sample. More complex calculations are required because of the dilution of the sample by the standard solution and vice versa. The equation for this alternate method of standard addition is

$$C_x = \frac{C_s + i_x v_s}{V_x(i_s - i_x) + i_s v_s} \qquad (16.31)$$

where C_x, i_x, and V_x are the values for concentration, wave height, and volume of the sample, respectively, and C_s, i_s, and v_s are the corresponding values for the standard solution.

In both versions of the standard addition method it is assumed that all other factors which might affect the instrument reading remain constant during the consecutive measurements and also that there is a linear relationship between wave height and concentration. Maximum accuracy is obtained when the wave height caused by the standard addition approximately doubles that of the sample alone. If a semiautomatic analytical balance is available, the first-mentioned version of this would be just about as convenient as the latter and does not involve as much calculation. This method requires an expenditure of slightly more time than the "calibration curve" method discussed earlier, but is generally considered to be more accurate. One reason for this is that it is not possible to compensate exactly for the possible effect produced by extraneous substances in the sample solution on the wave height when using the direct comparison method.

c. The Pilot Ion (Internal Standard) Method. The basis of this method depends upon the knowledge that two electroactive species of equal concentration give rise to relative wave heights, in a given supporting electrolyte, which are constant and independent of capillary characteristics m and t. Also for any electroactive substance the diffusion current constant I_d, which is the ratio of $i_d/Cm^{2/3}t^{1/6}$ [Eq. (16.28)], remains constant for a given capillary employed under a constant column height of mercury and in a specified supporting electrolyte system.

Once this ratio has been established for each of a series of ions using one capillary under specified conditions, all that is required to establish the diffusion current constants for the same substances in a second capillary is the calculation of this $i_d/Cm^{2\,3}t^{1/6}$ ratio for one of these substances (i.e., a pilot ion) in the second capillary using the same specified experimental conditions. Since the i_d values of any of the ions (x) change in the same ratio as those for the pilot ion on changing capillaries, these values can be calculated as follows:

$$\frac{(I_d)_{\text{pilot},1}}{(I_d)_{\text{pilot},2}} = \frac{(I_d)_{x,1}}{(I_d)_{x,2}}$$

on rearrangement this gives

$$(I_d)_{x,2} = [(I_d)_{\text{pilot},2}/(I_d)_{\text{pilot},1}](I_d)_{x,1} \qquad (16.32)$$

where I_d is the diffusion current constant $i_d/Cm^{2/3}t^{1/6}$ for the pilot ion and ion x in capillaries 1 and 2 as indicated. Equation (16.32) enables a comparison of data obtained with one capillary with that obtained with another, providing the I_d value of a pilot ion has been calculated using both capillaries. This application of the pilot ion obviates the need to redetermine the I_d, for all the ions of interest on a second capillary if the first becomes plugged or broken. It is only necessary to keep a standard stock solution of the pilot ion in each supporting electrolyte which is likely to be used.

1. *Quantitative Analysis Employing the Pilot Ion Method*

(a) Using an equation. Since the relative diffusion current constants I_d of ions in specific supporting electrolytes are independent of the capillary used, Eq. (16.28) applies to both the pilot ion p and sample depolarizer ion x to be measured. Thus, the following expressions can be written:

$$C_x = (i_d)_x/(I_d)_x m^{2/3}t^{1/6} \qquad (16.33)$$

$$C_p = (i_d)_p/(I_d)_p m^{2/3}t^{1/6} \qquad (16.34)$$

Under identical conditions, including a common capillary, Eq. (16.33) can be divided by Eq. (16.34), then rearranged to give

$$C_x = [(i_d)_x/(i_d)_p][(I_d)_p/(I_d)_x]C_p \qquad (16.35)$$

where C_x and C_p are the concentrations of the sample ion x and pilot ion p, respectively; i_d and I_d are the diffusion currents and diffusion current constants for the sample ion and pilot ion as indicated by the subscripts.

The ratio $(I_d)_p/(I_d)_x$ is called the "internal standard ratio" (or "pilot ion ratio") and is independent of capillary characteristics. This ratio can be determined in advance using the same capillary or it may be calculated from literature values of I_d for these substances if reported for the same concentration of the supporting electrolyte system in which the $(i_d)_x$ and $(i_d)_p$ are to be determined. The latter two values are determined in the same solution and are measured from the resulting two-step polarogram.

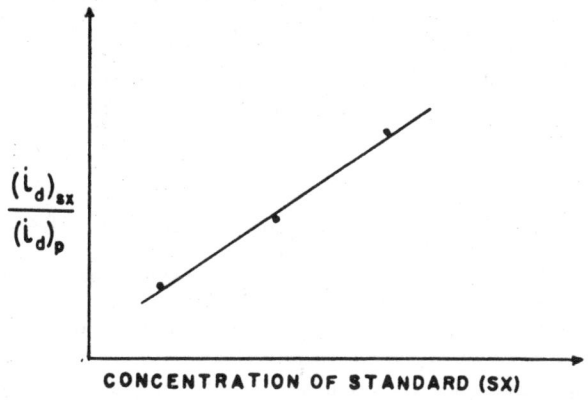

CONCENTRATION OF STANDARD (SX)

FIGURE 16.10: Calibration curve used in the pilot ion method. The subscripts sx and p represent the standard solutions of the test substance and pilot ion, respectively.

(b) Alternate Pilot Ion Method Using a Calibration Curve. A series of standard solutions sx are prepared for the depolarizer x, each containing a constant known volume of pilot ion solution. Sample solutions of unknown concentration depolarizer x are prepared to contain the same amount of the pilot ion p. The selected pilot ion must have a $E_{1/2}$ value which is considerably different from that of the depolarizer x or any other substance which may be present in solution. A polarogram is recorded and the $(i_d)_x$ and $(i_d)_p$ are determined at potentials corresponding to the limiting currents of their respective waves. This may be repeated for each standard sx and sample x solution if a recording polarograph is used. If a manual instrument is used the diffusion current values are measured only at appropriate fixed applied potentials selected from the polarogram of the mixture.

A calibration curve is then constructed by plotting the ratio $(i_d)_{sx}/(i_d)_p$ vs. C_s as shown in Fig. 16.10. The sample solutions containing unknown concentrations of the substance x in the standard solutions and identical quantities of pilot ion are then measured in the same way. The ratio of $(i_d)_x/(i_d)_p$ is referred to the calibration curve and the concentration of the sample is read in the usual manner.

This procedure compensates for small variations in temperature and capillary characteristics over extended periods of time. The method is somewhat limited by the number of possible pilot ion substances available since they must produce well-defined waves which are well resolved from the others produced in the mixed solution. Cadmium ion has frequently been used for this purpose.

The diffusion current constant (I_d) for any electroactive species can be calculated from a rearranged Eq. (16.28). To calculate m and t, we must determine the average drop time t in seconds and the mass of mercury m flowing in milligrams per second by timing the collection of 10 drops of mercury under the actual operating conditions of E_{appl}, in a supporting electrolyte system identical in nature and concentration to that for which the value of I_d is required. The 10 mercury drops can be collected in a small glass receptacle, washed with water, then acetone, dried, and weighed. This data enables the calculation of the value of m. The reader may refer to Ref. 7 for details.

D. NONAQUEOUS SOLVENTS IN POLAROGRAPHY

Many water-insoluble substances can be polarographed in nonaqueous solvents, solvent mixtures, or in melts. The current state of polarography in nonaqueous systems is rather empirical and descriptive in nature. Frequently depolarizers which yield ill-defined polarograms in aqueous media yield well-defined waves in nonaqueous media. The i_d, $E_{1/2}$, and wave shape for a depolarizer substance usually differ in the two types of media. Since traces of water often influence the nature of the depolarizer, care is necessary to exclude water when employing anhydrous solvents. Reference electrodes in nonaqueous media behave differently than in aqueous media depending upon the particular reference electrode and supporting electrolyte present in the nonaqueous medium. Frequently a mercury pool may be employed as a constant "reference" electrode, or an aqueous reference electrode (e.g., SCE), which often contains a high concentration of an indifferent electrolyte, may be employed. Since the choice of a suitable nonaqueous solvent and reference electrode depends on numerous considerations, the reader is referred to recent references on polarography.[2,5]

Solvents which have been employed include methanol, ethanol, alcohol-water mixtures, alcohol-benzene mixtures, acetic acid or anhydrous acetic acid containing ammonium acetate, sulfuric acid, liquid ammonia, acetonitrile, ethylenediamine, and many other organic substances. Furthermore, polarography has been performed in melts consisting of mixtures of inorganic salts at elevated temperatures. For details and references see the recent book by Heyrovský and Kůta[2] as well as review papers listed under References.

E. ORGANIC POLAROGRAPHY

A large number of organic·compounds can be studied and assayed by polarographic methods. Usually the electrode reactions involve hydrogen ions, and for this reason a constant concentration of these ions is provided by means of the supporting electrolyte medium, which may be strongly buffered, acidic or basic, aqueous or nonaqueous. The general electrode reaction may be represented by

$$Org + mH^+ + ne^- \rightarrow Org \cdot Hm \tag{16.36}$$

where the value of n is frequently dependent upon the pH of the medium. One or more steps may be involved in the complete reduction of a functional group and in most instances the reaction is irreversible.

While the list of functional groups in Table 16.1 is not an exhaustive list, it does serve to indicate the variety of compounds which have been studied polarographically: acids with carbonyl or conjugated double bonds; aliphatic, aromatic, and substituted aldehydes; aliphatic, aromatic substituted and unsaturated ketones, including quinones, di- and tri-ketones; certain esters; many nitrogen-containing groups such as: nitro, nitroso, azo, azoxy, hydroxylamines, amine oxides, diazonium salts, certain heterocyclic compounds such as alkaloids and others; many sulfur-containing groups such as thiols, sulfur groups conjugated with unsaturated groups; most peroxides, including hydrogen peroxide; many halogen containing compounds; and others too numerous to mention.

16.5 LABORATORY EXPERIMENTS

A. APPARATUS

Manual and/or recording polarographs.

Polarographic electrolysis cell—such as an H-type cell with an attached SCE (if the DME potential is desired), or a beaker-type cell (if a mercury pool electrode is to be used).

Large tray to prevent the loss of poisonous and volatile mercury, (clean up spills immediately).

Dropping mercury electrode (DME) assembly containing doubly distilled and filtered mercury.

Reference electrode—a mercury pool, and/or a nonpolarizable SCE, as required. See Fig. 16.8.

Stop watch—for determining the drop life t.

Deoxygenation apparatus—consisting of oxygen-free nitrogen gas (tank and regulator). A gas washing or scrubbing device containing a suitable solution (such as alkaline pyrogallol·or chromous chloride) through which the nitrogen can be passed if it contains any oxygen.

TABLE 16.1: Organic Polarography

Substance	Concentration	Medium	$E_{1/2}$ or E_{scan}, volts vs. SCE	References
Chlortetracycline HCl, and/or oxytetracycline HCl	0.1 mg/ml	0.2 M phosphate buffer, pH 8.2 ± 0.1	E: −0.8 → (−)	a
Tetracycline HCl	0.05–0.15 mg/ml	Sorensen's phosphate buffer, pH 6.2	$E_{1/2}$: −1.50 vs. Hg-pool	b
Chloramphenicol, and dosage forms	1 mg/ml	5 ml isopropanol + 25 ml 0.2 M potassium biphthalate solution + 0.2 ml 0.2 N NaOH + 0.2 ml 0.1% methylene blue, and water to make 100 ml of solution. Mix well.	E: −0.1 → −0.9 i_d at −0.78	c
Chloramphenicol	2 mg/ml	20 mg in 10 ml phosphate buffer containing 10% EtOH	E: 0 → (−1.5)	d
Streptomycin sulfate	100–400 μg/ml	20-ml aliquot of sample + 2 ml 1 M NaOH 50% MeOH containing 0.1 M acetate buffer	$E_{1/2}$: −1.29 (vs. Hg-pool)	e
Cortisone and/or Prednisone			E: −0.5 → −1.6 i_d at −0.9 i_d cortisone (−1.47) i_d prednisone (−1.31)	•
Ascorbic acid, and in dosage forms	25–250 μg/ml	7.5 ml 2 N HOAc + 37.5 ml 2 N NaOAc + 0.5 ml formaldehyde + water to make 50 ml. Saturated with sodium oxalate, filter, pH ca. 5.5	$E_{1/2}$: +0.07 E: −0.3 → (+)	f
Folic acid and tablets	250 μg/ml	Aqueous solution containing 15% NH$_4$Cl + 10% Me$_4$NOH	E: −0.9 → (−)	'
Menadione (vitamin K_3) and dosage forms	5–15 μg/ml	Light petroleum-2 M (NH$_4$Cl/NH$_4$OH) buffer, (2:3), pH 9–9.5	$E_{1/2}$: −0.35	'
Menadione	10–100 μg/ml	Benzene-MeOH-conc. NH$_4$OH (10:10:1) containing 0.005 M methylene blue	$E_{1/2}$: −0.4 (vs. Hg-pool) E: 0 → −1.0	'
Menadione-sodium bisulfite	10–30 μg/ml	0.1 M KCl containing 5 g sodium sulfite and 50 mg agar per liter	$E_{1/2}$: −1.1	'

Substance	Concentration	Medium	Potential	Ref.
Nicotinamide, also in mixtures after chromatographic separation	3–10 μg/ml	Aqueous solution of 1% Me$_4$NOH	$E_{1/2}$: -1.7	[j]
Pyridoxal HCl or pyridoxal-5-phosphoric acid ester (and in tablets)	100 μg/ml	5 ml 0.3% gelatin + 1% LiCl to make 100 ml	E: $-0.9 \rightarrow (-)$	[j]
Riboflavin and dosage forms	20–50 μg/ml	0.1 M Sorensen phosphate buffer (pH 7.5)	$E_{1/2}$: -0.47	[j]
Riboflavin and dosage forms	20–30 μg/ml	Britton-Robinson buffer (pH 2.8) containing 0.08 M of each component, plus 0.4 M KCl	E: $0 \rightarrow (-1.0)$ (vs. Hg-pool)	[k]
Thiamine HCl	50–200 μg/ml	10% Na$_2$CO$_3$-0.1 M KCl (1:9), pH $\sim$ 11.5	$E_{1/2}$: -0.42	[j]
Chlorpromazine and dosage forms	1–8 μg/ml	0.5 N HCl	E: $-0.7 \rightarrow (+)$; E: $-0.5 \rightarrow (-)$	[l]
Chorpromazine HCl	35–480 mg/liter	1 N H$_2$SO$_4$ (with R.P.E.)	$E_{1/2}$: $+0.6$	[k]
Phenothiazine HCl salts	1.09–1.76 mM	9 N H$_2$SO$_4$ (with R.P.E.)	$E_{1/2}$: $+0.37$ and $+0.95$	[l]
Diiodohydroxyquin	0.08–2.5 mM	In water	E: $0 \rightarrow (-)$	[l]
Iodochlorhydroxyquin		2-ethoxyethanol containing 0.05 M Et$_4$NBr	E: $0 \rightarrow -2$	[l]
		0.024 M KOAc + 0.013 M HOAc in 90% EtOH containing 0.06 M LiCl (pH 7.0)	E: $0 \rightarrow -2$	[l]
Isonicotinic acid hydrazide (and dosage forms)	0.0005 M	10-ml sample (50 mg/250 ml) + 10 ml universal buffer	E: $0 \rightarrow 1.0$	[m]
Thimerosal	0.01–0.1 mg/ml	0.1 to 1.0 mg in conc. HCl + 1 ml 0.1% gelatine + water to make 10 ml	E: $0 \rightarrow -1.0$ (first wave)	[n]
Thimerosal	20–50 ppm	10 ml of aqueous solution containing 20–50 ppm + 2 ml 0.1 M KMnO$_4$ + 1 ml of 0.1% gelatin	$E_{1/2}$: -0.75 (first wave)	[o]
Phenolphthalein (and in emulsions)	0.15 mg/ml	30 mg in 95% EtOH to make 50 ml. Take 25 ml of this solution and 55 ml of 25% HCl solution and 1 ml aqueous 0.1% basic fuchsine + water to make 100 ml	E: $0 \rightarrow (-)$	[d]

TABLE 16.1 (continued)

Substance	Concentration	Medium	$E_{1/2}$ or E_{scan}, volts vs. SCE	References
Barbiturates (5,5-dialkylbarbituric acids)	0.0001–0.00001 M	In 0.05 M borax solution	Using DME anodic wave of mercury salt formation	a
Thiobarbiturates (e.g., pentothal)	7×10^{-4} M (or less)	In 0.1 N NaOH	Anodic wave of Hg salt formation	a
Strychnine		0.01–0.5 mg + 1 ml HNO₃ + 2 ml water in closed test tube in boiling water bath (20 min) Cool, + 8 ml 2.5 M KOH; gives 2,4-dinitro-strychnine as product	2 waves at −0.64 and −0.79	a

a P. Kabasakalian and J. H. McGlotten, in *Handbook of Analytical Chemistry*, 1st ed.; L. Meites (ed.), McGraw-Hill, New York 1963, Sec. 5.

b L. Doan and B. E. Riedel, *Canad. Pharm. J., Sci. Sect.*, **96**, 109 (1963).

c A. F. Summa, *J. Pharm. Sci.*, **54**, 442 (1965).

d P. Zuman, *Organic Polarographic Analysis*, Macmillan, New York, 1964.

e P. Kabasakalian, S. DeLorenzo, and J. McGlotten, *Anal. Chem.*, **28**, 1669 (1956).

f R. Stroheker and H. M. Henning, *Vitamin Assay*, Verlag Chemie, GMBH., Weinkeim/Bergstr. 1965.

g J. C. Jongkind, E. Buzza, and S. H. Fox, *J. Am. Pharm. Assoc., Sci. Ed.*, **46**, 214 (1957).

h W. J. Seagers, *J. Am. Pharm. Assoc., Sci. Ed.*, **42**, 317 (1953).

i G. S. Porter, *J. Pharm. Pharmacol. (Suppl.)* **16**, 24T (1964).

j F. H. Merkle and C. A. Disher, *J. Pharm. Sci.*, **53**, 620 (1964).

k N. Chuen and B. E. Riedel, *Canad. Pharm. J., Sci. Sect.*, **94**, 51 (1961).

l L. W. Brown and E. Krupski, *J. Pharm. Sci.*, **50**, 49 (1961).

m J. D. Neuss, W. J. Seagers, and W. J. Mader, *J. Am. Pharm. Assoc., Sci. Ed.*, **41**, 672 (1952).

n J. E. Page and J. G. Waller, *Analyst*, **74**, 292 (1949).

o J. Birner and J. R. Garnet, *J. Pharm. Sci.*, **53**, 1266 (1964).

A constant temperature bath and control (25°C ± 0.1°).

Maxima suppressor solution—such as 0.2% solution of Triton X-100 (Rohm & Haass Co.), or freshly prepared 0.2% gelatin solution (dissolve in warm water).

Volumetric glassware such as volumetric flasks, pipettes, etc.

B. CARE OF DROPPING MERCURY ELECTRODE

Every precaution must be taken to prevent the capillary bore from becoming plugged. This obstruction could be caused by a dust particle in the mercury, oxidation products of mercury, salt crystals, etc. Since these capillaries are extremely difficult or impossible to clean, the following precautions must be followed:

1. Protect the mercury in the reservoir from dust with a filter of glass wool. If the DME apparatus is being assembled for the first time, clean and dry all plastic or rubber tubing well to eliminate dust particles.

2. Make certain that the mercury is dropping from the capillary before the capillary is immersed in any solution. This precaution excludes contaminants from the capillary bore, where they could cause an erratic mercury flow.

3. Never stop the mercury flow while the capillary is in solution. When through, lower the electrolysis cell to remove the DME from the solution. While the mercury is still dropping, rinse the tip with distilled water, blot dry with filter paper, and place a dry test tube under the tip. Stop the mercury flow by lowering the reservoir, by a stopcock or clamp, etc, whichever is appropriate for the DME assembly used. This procedure is recommended over the immersion of the capillary tip in distilled water.

4. Occasional cleaning of the end of the capillary bore is required. This is done by immersing the tip in 50% nitric acid while the mercury is flowing, then rinsing and drying as directed under 3.

5. The DME assembly must be vibration free and the capillary vertical to avoid an erratic drop time t. A level or plumb line may be used to check the latter adjustment.

C. GENERAL POLAROGRAPHIC PROCEDURE

This general procedure should be followed in all polarographic measurements, unless otherwise directed in specific exercises:

1. If a mercury pool electrode is employed, add sufficient mercury to cover the bottom of the electrolysis cell to a depth of 1/8 inch.

2. If a SCE (or other reference electrode) is used, maintain the level of the saturated KCl (or other electrolyte) solution above the solution level in the cell.

3. Add sufficient test solution (consisting of supporting electrolyte, maximum suppressor and depolarizer) to a level above the large "salt bridge," but not too close to the top of the cell.

4. *Caution:* Read Section 16.5B on the care of the DME. Start the mercury flow, then raise the electrolysis cell to immerse the DME well below the solution surface. Adjust the mercury column to produce a drop time of 3 or 4 sec. Ideally this is done at the applied potential at which the diffusion current is to be measured.

5. Deoxygenate the test solution with oxygen-free nitrogen gas for 15 min (about 5 min if a sintered-glass gas-dispersion tube is used). Never measure the polarographic current while N_2 is bubbling through the solution. Why?

6. Consult the operating instructions for the instrument to be used. Standardize (or calibrate) the potentiometer and adjust the galvanometer index (or recorder pen) to zero.

7. Connect the electrodes to the appropriate terminals. The polarity of the DME is (−) negative for cathodic reductions. Adjust the instrument to the lowest sensitivity initially, unless the correct setting is known. The appropriate sensitivity is that which produces a large deflection but prevents the galvanometer index from going off scale. This setting may be determined by gradually increasing the applied potential over the range of interest while adjusting the sensitivity switch (shunt) to keep the index on scale at the point of greatest deflection.

D. MANUAL OPERATION

With the preceding general procedure in mind, select the applied potential span to be used. Set the applied potential to zero and note the galvanometer reading. A slight negative reading can be ignored, it may only be the end of an anodic wave of some substance in solution. The applied potential is increased in steps of 0.05–0.1 V except during a wave where 0.01- or 0.02-V steps are used. Record the applied potential and resulting current or galvanometer readings. It is often easier to record the maximum throw of the index although the average is more desirable. This manual potential scan is usually performed over the 0- to −2-V range, or until some component of the supporting electrolyte is discharged about −1.7 to −1.9 V (vs. SCE). Use the highest sensitivity that will permit the galvanometer index to remain on scale at all times.

Plot the polarogram graphically according to convention (Fig. 16.6). Measure the diffusion current as directed earlier (Fig. 16.2).

Use of Compensator (or Bias) Control

The reader should familiarize himself with this control (if available), which may be required to "cancel" a large diffusion current of a substance which is reduced just prior to the depolarizer of interest. This enables the desired wave to be developed at a higher sensitivity. Turn this control off when finished so that it does not interfere with a subsequent analysis.

E. OPERATION OF RECORDING INSTRUMENT

Read the preceding general polarographic procedure, then consult the operating instructions for the instrument to be used. Adjust the recorder to zero, select the appropriate potential range to be scanned, the scan rate, and sensitivity, if known. Otherwise, scan rapidly, adjusting the sensitivity as required to obtain nearly a full-scale deflection (fsd) at the potential of maximum current flow. The "scan" switch usually activates a synchronous motor, which increases the applied potential in a linear manner and simultaneously starts the chart drive motor. The chart drive axis corresponds to the applied potential while current is measured on the other axis. If the scan is not automatically stopped, switch off the drive motors at about −2.0 V, or at the end of the selected voltage range.

Experiment 16.1

Familiarization with the instrument, plotting methods, maximum suppressor, role of supporting electrolyte, and the measure of various currents and potentials.

Apparatus. Carefully read over this experiment (or the assigned portions) and compile a list of the apparatus required. All apparatus must be clean. To save time obtain all the apparatus before starting.

Graph paper ($8\frac{1}{2}$ in. × 11 in.) with millimeter (or 20 lines/in.) ruling will be required.

Note: It would be desirable to have a recording polarographic instrument if all parts of this experiment are to be attempted.

Solutions (provided):

Maximum suppressor solutions: (1) 50 ml of a fresh 0.2% gelatin solution (dissolved in warm water, cooled, and diluted to volume), and/or (2) 0.2% Triton X-100 Solution.

Stock Solutions of :

$PbCl_2$ [or $Pb(NO_3)_2$] 0.02 *M* in distilled water

$CdCl_2$ (or $CdSO_4$)   0.02 *M* in distilled water

$ZnCl_2$ (or $ZnSO_4$)   0.02 *M* in distilled water

Supporting electrolyte (1.0 *M* KCl).

Mixed supporting electrolyte (1 M Sodium Potassium tartrate + 0.2 M NaOH).

Using appropriate volumetric glassware and the stock solutions provided, prepare and label the following solutions:

Maximum suppressor—0.2% gelatin solution or 0.2% Triton X-100 (if not provided).

Solution 1. 100 ml of 0.1 M KCl (do *not* add any maximum suppressor to this solution).

Solution 2. 100 ml of 0.001 M Cd(II) in aqueous solution containing 2.0 ml 0.2% gelatin.

Solution 3. Prepare this solution so that 100 ml will contain 2.0 ml of the 0.2% gelatin, and be 0.001 M in Cd(II) and 0.1 M in KCl.

Solution 4. 100 ml is to contain 2.0 ml of 0.2% gelatin and be 0.1 M in KCl [no Cd(II) present].

Part A. To Demonstrate the Need for Deoxygenation and a Maximum Suppressor

Procedure. Transfer 20 ml (or sufficient to cover the electrodes) of solution 1 to the electrolysis compartment of the H-cell. Obtain a polarogram over the range 0 to −1.9 V (vs. SCE), by the *general polarographic procedure,* omitting the deoxygenation. Plot this curve as the data is obtained and in accord with the plotting conventions described earlier. Note the location of any waves present and their respective half-wave potentials $E_{1/2}$. Two maxima may appear in the polarogram. If this does not occur, it may be necessary to bubble air into solution 1 to enhance the oxygen waves. Repeat the polarogram in the regions of the observed maxima after adding 4 drops of the selected maximum suppressor. Plot this polarogram on the same sheet. If the maxima persist, repeat again after the addition of 4 more drops of the maximum suppressor. Repeat this process until the maxima are just eliminated. Record the number of drops or volume (pipette) of the maximum suppressor required as well as the initial volume of solution added to the vessel.

Now deoxygenate the solution for 5 min and obtain another polarogram over the same voltage range. If the waves are still evident, deoxygenate for a further 5 min and check for the oxygen wave(s). Repeat until the last oxygen wave disappears, then obtain a complete polarogram on the graph along with those obtained previously.

Part B. To Study the Role of the Supporting Electrolyte in Minimizing the Migration Current of the Depolarizer

Procedure. Transfer 20 ml (or appropriate known volume) of solution 2 to the electrolysis vessel, deoxygenate, and obtain a polarogram using the *general polarographic procedure.* Record the measured limiting current (i_L) value for this solution. Repeat using solution 3, and again with solution 4.

Plot (or record) these polarograms (or partial polarograms) on the same paper used for solution 2. Determine the diffusion current (i_d) for the Cd(II) ion concentration present (refer to Fig. 16.2). Calculate the approximate migration current (i_m) of the Cd(II) ions from the results obtained, using the expression $i_L = i_d + i_m$. Measure and compare the magnitude of the currents i_r, i_L, i_d, and i_m (refer to Fig. 16.2) in these plots and explain the causes of each. Why is it necessary to remove the oxygen when calculating the i_d for the Cd(II) ions in solution ?

Note: Comparable concentrations of Pb(II) or Zn(II) can be used instead of Cd(II) in this experiment.

Part C. Effect of the (1) Concentration of the Depolarizer and (2) Nature of the Supporting Electrolyte on the Half-Wave Potential

1. *Effect of Depolarizer Concentration.* Prepare the following solutions in 100 ml volumetric flasks:

Solution a. 20 ml of 0.02 M Pb(II), plus 2 ml of gelatin (0.2%), plus 20 ml of 1 M KCl, plus water to volume.

Solutions b–f. By substituting the following corresponding volumes of 0.02 M Pb(II) for that in solution a: (b) 15 ml, (c) 10 ml, (d) 5 ml, (e) unknown volume (supplied by instructor), (f) 0 ml (blank). Using the *general polarographic procedure*, obtain complete polarograms for solutions a and c only over an applied potential range of 0 to -1.0 V (vs. SCE). Select a potential on the limiting current plateau of the fully developed wave of solution a. Measure the current of each of the remaining solutions and the "blank" at this selected applied potential. Record all data in your notebook and determine the following:

(1) The diffusion current i_d for each solution, i.e., $i_d = (i_L - i_r)$, where i_r is the current for solution f at the same potential;

(2) The $E_{1/2}$ values for Pb(II) ion from the complete polarograms obtained with solutions a and c.

Plot i_d vs. concentration of Pb(II) for all solutions on graph paper. What relationship exists between i_d and concentration in this plot? Express this by means of a simple equation. What is the numerical value of the constant in this equation? What is the concentration of the unknown solution e? What is this quantitative polarographic method called? What effect does Cd(II) concentration have on the value of $E_{1/2}$ determined graphically? How does this $E_{1/2}$ value compare with those reported by others? See Refs. 2, 5, and 11, or others for tables of $E_{1/2}$ values in a variety of media.

2. *Effect of the Nature of the Supporting Electrolyte.* Prepare the following solution in a 100-ml volumetric flask: 20 ml of 0.02 M Pb(II), plus 2 ml of 0.2% gelatin, plus 50 ml of the mixed tartrate-NaOH supporting electrolyte (stock solution), and water to volume.

Procedure. Deoxygenate and obtain a polarogram of this solution, determine the i_d and $E_{1/2}$ for Pb(II). In this case use the approximate or extrapolation method to estimate the residual current i_r. Compare the $E_{1/2}$ value obtained with that determined for solution a in (1). Comment on the manner of reporting an $E_{1/2}$ value for any electroactive species (depolarizer).

Note: Read Experiment 16.2 before discarding solutions c and d used in part C of Experiment 16.1.

Experiment 16.2. Quantitative Polarographic Analysis

Part A. Standard Addition Method

Prepare the following solutions in 100-ml volumetric flasks:

Solution a.   Ten (10.0) ml of "unknown Pb(II) Solution" [ca. 0.02 *M* Pb(II)], plus 2 ml of a 0.2% gelatin solution, plus 20 ml 1 *M* KCl, and water to volume.

Solution b.   Repeat solution a, except, add also an accurate volume of standard Pb(II) solution which produces a doubling of the diffusion current observed in the case of the unknown in solution a.

Note: The addition of the 0.02 *M* Pb(II) standard is made prior to diluting with water to the 100-ml mark.

Obtain polarograms according to the general procedure and calculate the concentration of the "unknown" solution from Eq. (16.30). Would it be appropriate to use Eq. (16.31) in this situation?

Note: If time is short the standard-addition method can be demonstrated with very little additional effort if Experiment 16.1, part C, has been completed and solutions c and d kept. If we let solution d represent the "unknown" solution and solution c represent the "unknown" solution to which an accurate weight of standard Pb(II) has been added to "approximately" double the wave height. Calculate the concentration of the "unknown" solution d using Eq. (16.30). How does the value compare with the actual concentration of the "unknown" solution d (known in this case)?

Part B. Pilot Ion (Internal Standard) Method

Prepare the following solutions in 100-ml volumetric flasks:

Solution a.   5.0 ml of standard 0.02 *M* Pb(II) solution, plus 5.0 ml of 0.02 *M* Cd(II), plus 2 ml of 0.2% gelatin solution, plus 10 ml of 1 *M* KCl, and water to volume.

Solution b.   Repeat the preparation of solution a, except substitute an accurately measured volume of "unknown" Pb(II) solution for the standard Pb(II) solution.

Note: The unknown could be the same as that used in Experiment 16.1, part Cl or that in Experiment 16.2A, and a comparison of two methods of quantitative measurement made.

Obtain polarograms for the solutions a and b according to the *general polarographic procedure*. Scan the potential range 0 to -1.0 V (vs. SCE). Measure the diffusion currents i_d for both waves and calculate the concentration of the unknown using Eq. (16.35). The I_d values for Pb(II) and Cd(II) ions in 0.1 M KCl are 3.80 and 3.51, respectively.[11]

Experiment 16.3. Conventions Used for Plotting Cathodic and Anodic Current-Voltage Waves of a Reversible Redox Couple

The following aqueous solutions are provided: 1 M sodium citrate, 0.02 M ferric ammonium sulfate.

Prepare the following aqueous solutions *freshly* each day:
100 ml of 0.2% gelatin
100 ml of 0.02 M ferrous ammonium sulfate [in deoxygenated water to retain the Fe(II) state]

Procedure. Transfer 50 ml of the 1 M sodium citrate (stock solution) to a 150-ml beaker, add 2.0 ml of the gelatin solution, then deoxygenate with oxygen-free nitrogen for 15 min. To this oxygen-free solution add 5 ml of each of the two iron solutions and adjust the pH by dropwise addition of 12 M HCl to 5.5 $\pm$ 0.1. Pour this solution into the H-type cell until it is about two-thirds full, deoxygenate for about 5 min, then protect the solution surface with a gentle stream of nitrogen. Obtain a polarogram according to the *general polarographic procedure* over an applied potential range of 0.1 to -0.5 V (vs. SCE). The anodic part of the curve is obtained by switching the galvanometer connections, or more conveniently by setting the galvanometer index to "zero" at midscale. The $+0.20$ to 0.00 V potentials can be applied by switching the leads to the DME and SCE or on some instruments by flipping a polarity switch on the instrument (often marked "DME + and $-$") to ($+$) side. Consult the manual supplied with a specific instrument for the exact operating procedure.

Plot i vs. E_{appl} using the plotting conventions illustrated in Fig. 16.6.

How could you determine which species is responsible for the anodic, or cathodic, currents? If time permits, carry out your plan using the solutions available. If this is done compare the $E_{1/2}$ potentials of each form of this redox couple and comment on their agreement, or lack of it.

See Chapter XIII of the book by Heyrovský and Kůta[2] for other examples and discussion of mixed currents.

Experiment 16.4.   Organic Polarography

Part A.   Quantitative Analysis of Riboflavin[12]

Riboflavin has a half-wave potential $(E_{1/2})$ of -0.47 V (vs. SCE) in Sörensen's phosphate buffer (pH 7.5) when the dropping mercury electrode (DME) is employed. The dc polarographic method is particularly useful for the assay of this vitamin in pharmaceutical preparations. The method is quite specific and other vitamins, vehicles, or excipients rarely cause interference with the riboflavin wave. The direct-comparison method, especially that involving a calibration curve, is widely used in the polarographic analysis of this vitamin.

Materials. USP reference standard riboflavin (or very pure riboflavin) sodium salicylate USP XVII.

Sörensen phosphate buffer (0.5 M, pH 7.5). This supporting electrolyte is prepared as follows: dissolve 62.5 g of $Na_2HPO_4\cdot2H_2O$ and 10.2 g of KH_2PO_4 in sufficient distilled (or deionized) water to make 500 ml of solution.

Preparation of Standard Curve

Weigh accurately about 25 mg of the reference standard riboflavin plus 2.5 g of pure sodium salicylate and transfer these to a 250-ml volumetric flask. Dissolve these materials in approximately 200 ml of water, then add water to the mark and protect from light. Pipette 10, 15, 20, and 25 ml of this standard riboflavin solution into four 50-ml volumetric flasks, followed by 10 ml of the buffer solution, add water to the mark, stopper, and mix well. Obtain a polarogram for each of the standard solutions according to the *general polarographic procedure*, preferably with a recording polarograph. Measure the diffusion current i_d in each instance and plot i_d vs. concentration of riboflavin (micrograms per milliliter) to obtain a calibration curve.

Preparation of the Sample. Grind a single tablet containing riboflavin in a mortar. Quantitatively transfer to a volumetric flask of appropriate capacity sufficient of the sample, sodium salicylate and buffer to make the final assay solution 1% in salicylate, 0.1 M in phosphate buffer and contain between 20–50 μg of B_2/ml. The final assay solution is analyzed under the same conditions employed for the standard solutions. Determine the diffusion current for the sample and refer it to the calibration curve to ascertain the concentration of the sample solution in μg B_2/ml. Calculate the riboflavin content in milligrams per tablet and report the result as per cent of the label claim. This polarographic method can be used to assay riboflavin in multiple vitamin capsules or tablets containing vitamins A, B_1, B_6, B_{12}, C, D, E, nicotinamide, calcium pantothenate, several mineral salts, etc., without apparent interference. For greater detail and procedure modification consult Ref. 12.

Other polarographic analysis procedures, involving direct comparison and/or standard addition methods, for riboflavin are given by: W. J. Seagers, *J. Am. Pharm. Assoc., Sci. Ed.*, **42**, 317 (1953); and A. J. Zimmer and C. L. Huyck, *ibid.*, **44**, 344 (1955).

Part B. Quantitative Assay of Chlordiazepoxide[13]

Apparatus. Recording (or manual) polarograph
H-type cell with SCE and DME assembly

Materials. Chlordiazepoxide ("Librium"), 7-chloro-2-methyl amino-5-phenyl-3H-1,4-benzodiazepine 4-oxide
Triton X-100 (Rohm and Haas Co.) 0.5% solution
Anti-foam B (Dow-Corning) 0.5% solution
1 M hydrochloric acid (SE)

Introduction. In a 1 M HCl medium this drug is reduced at the DME producing a well-defined two-step polarogram with $E_{1/2}$ values of -0.36 V and -0.67 V (vs. SCE).

A linear relationship exists between i_d and concentration from about 3 to 135 μg/ml. The cathodic reduction is represented by the following electrode reactions:

First wave:

$$\diagdown C{=}N + 2H^+ + 2e^- \rightarrow \diagdown C{=}N{-} + H_2O$$
$$\underset{\underset{O}{\downarrow}}{\diagup}$$

Second wave:

$$\diagdown C{=}N{-} + 2H^+ + 2e^- \rightarrow \diagdown CH{-}NH{-}$$

The second wave ($E_{1/2} = -0.67$ V vs. SCE) is used for diffusion current measurement since it is somewhat better defined than the first.

The original work[13] was performed in a 3-ml small volume H-type polarographic cell at a sensitivity coefficient of 0.003 μA/mm for the estimation of this drug in biological fluids. Consult the reference for the extraction details, if required.

Procedure. Accurately weigh about 10 mg of pure chlordiazepoxide (reference standard) and quantitatively transfer this to a 100-ml volumetric flask. Add 1 M HCl to volume and mix well. From this concentrated standard solution, prepare four or five standard solutions ranging in concentration from 10 to 100 μg/ml, and containing about 10 drops of 0.5% Triton X-100 solution, 5 drops of Anti-foam B, and sufficient 1 M HCl to produce the desired volume. Transfer to the polarographic cell sufficient standard solution to cover the electrodes to a depth of about ½ in. Deoxygenate and obtain a partial polarogram over the applied potential range of -0.3 to -1.0 V (vs. SCE) employing the *general polarographic procedure*. Repeat with each standard solution. Extrapolate the slightly inclined limiting current of each wave and measure the i_d of the second wave, i.e., the vertical distance between these two extrapolated lines at the $E_{1/2}$ value for the second wave.[14]

Plot the i_d values (second wave) vs. concentration of chlordiazepoxide (μg/ml) to obtain a calibration curve.

Prepare the sample in the same manner, estimating the concentration so that it lies within the range of the calibration curve. Measure the i_d in the same manner and refer this value to the calibration curve for the concentration in μg/ml. Calculate the amount of chlordiazepoxide in the unknown provided.

If time permits calculate the total current for the two-step reduction, corrected for the residual current at the same potential and construct a standard curve. Compare the result for the unknown calculated by this means with that obtained in the part B experiment. Does one method have any advantage over the other? Comment.

Part C. Polarographic Assays of Official Substances

Polarographic analysis is employed in the assay of acetazolamide tablets USP, sodium acetazolamide USP, sterile sodium acetazolamide USP; dichlorophenamide tablets USP; hydrochlorothiazide tablets USP; methazolamide tablets USP; nitrofurantoin oral suspension USP; nitrofurantoin tablets USP[15], and dienestrol tablets, NF[16].

Consult also the following reference for the polarographic analysis of certain of the just listed drugs and chlorothiazide. A. F. Summa, *J. Pharm. Sci.*, **51**, 474 (1962).

Experiment 16.5

Additional experiments can be selected from Table 16.1.

PROBLEMS

P16.1. Outline a procedure for the polarographic analysis of a trace of As(III) ion in the presence of a much larger concentration of Zn(II) ion. The $E_{1/2}$ values in a NH_3NH_4Cl medium are -1.46 V and -1.35 V (vs. SCE), for As(III) and Zn(II), respectively.

P16.2. It is known that $m \propto h_{corr}$ and $t \propto 1/h_{corr}$, where h_{corr} is the height in cm of the mercury column above the tip of the DME. If a Pb(II) ion solution gave a $i_d = 4.05$ μA for a drop time $t = 3.60$ sec, and $m = 2.11$ mg/sec, what is the value of the i_d when the mercury height is changed to give a drop time $t = 4.00$ sec?

P16.3. A 500-mg sample of a preparation containing riboflavin was dissolved in 100 ml of water. A 10-ml aliquot of this solution was placed in each of two flasks. To one flask was added 10 ml of buffer, to the other 10 ml of a standard solution (containing 4 mg of USP reference standard riboflavin in buffer). Both were analyzed by the general polarographic method. The diffusion current of the sample solution alone was 25 (galvanometer reading), the other solution gave an i_d reading of 45. Calculate the per cent concentration of B_2 in the solid sample.

P16.4. Compare the relative quantitative polarographic methods described in the text in each of the following situations:

a. Routine analysis of relatively pure samples.

b. Routine analysis of a substance in a complex pharmaceutical preparation of unknown composition.

P16.5. List as many factors as possible which may contribute to a limiting current. Which is the desirable factor? Discuss means of eliminating the other factor.

SELECTED READING LIST ON POLAROGRAPHY

Brezina, M., and P. Zuman, *Polarography in Medicine, Biochemistry and Pharmacy,* Wiley (Interscience), New York, 1958.

Delahay, P., *New Instrumental Methods in Electrochemistry,* Wiley (Interscience), New York, 1954.

Hills, G. J. (ed.), *Polarography 1964. Proceedings of the Third International Conference, Southampton,* Vols. I and II, Wiley (Interscience), New York, 1966.

Kolthoff, I. M., and P. J. Elving, *Treatise on Analytical Chemistry,* Part 1, Vol. 4, Sect. D-2, Wiley (Interscience), New York, 1963.

Lingane, J. J., *Electroanalytical Chemistry,* Wiley (Interscience), New York, 2nd ed., 1958.

Milner, G. W. C., *The Principles and Applications of Polarography and Other Electroanalytical Processes,* Longmans, London, 1957.

Müller, O. H. *The Polarographic Method of Analysis,* Chemical Education Publ., Easton, Pa., 2nd ed., 1951.

Nürnberg, H. W., and M. J. von Stackelberg, *J. Electroanal. Chem.,* 2, 181 (1961); 4, 1 (1962).

Rulfs, C. L. and P. J. Elving, in *The Encyclopedia of Electrochemistry* (C. A. Hampel, ed.), Reinhold, New York, 1964, pp. 944–963.

Weissberger, A., *Techniques of Organic Chemistry: Physical Methods,* Vol. 1, Wiley (Interscience), New York, 3rd ed., 1960, p. 3155.

Zuman, P., and I. M. Kolthoff (eds.), *Progress in Polarography,* Vols. 1 and 2, Wiley (Interscience), New York, 1962.

Zuman, P., *Organic Polarographic Analysis,* Pergamon Press, New York, 1964.

Reviews

Laitinen, H. A., *Anal. Chem.,* 21, 66 (1949); 24, 46 (1952); 28, 666 (1956).

Organic Polarography

Wawzonek, S., *Anal. Chem.,* 32, 144R (1960); 34, 182R (1962).

Wawzonek, S., and D. J. Pietrzyk, *Anal. Chem.,* 36, 220R (1964).

Pietrzyk, D. J., *Anal. Chem.,* 38, 278R (1966).

Theory, Instrumentation, and Methodology

Hume, D. M., *Anal. Chem.,* 32, 137R (1960); 34, 172R (1962); 36, 200R (1964); 38, 261R (1966).

Nomenclature

Delahay, P., G. Charlot, and H. A. Laitinen, *Anal. Chem.,* 32, 103A (1960).

REFERENCES

1. J. Heyrovský, *Chem. Listy*, **16**, 256 (1922).
2. J. Heyrovský and J. Kůta, *Principles of Polarography*, Academic Press, New York, 1966.
3. H. Schmidt and M. von Stackelberg, *Modern Polarographic Methods*, Academic Press, New York, 1963.
4. L. L. Leveson, *Introduction to Electroanalysis*, Butterworth, London, 1964, Chap. 3.
5. L. Meites (ed.), *Handbook of Analytical Chemistry*, McGraw-Hill, New York, 1963, Chap. 5.
6. *An Introduction to Electroanalysis, Bulletin 7079*, Beckman Instruments, Inc., Fullerton, Calif.
7. C. N. Reilley and D. T. Sawyer, *Experiments for Instrumental Methods*, McGraw-Hill, New York, 1961, pp. 52–53.
8. H. A. Strobel, *Chemical Instrumentation*, Addison-Wesley, Reading, Mass., 1960, Chap. 16.
9. H. F. W. Kirkpatrick, *Quart. J. Pharm. Pharmacol.*, **20**, 87 (1947).
10. L. Meites, *Polarographic Techniques*, Wiley (Interscience), New York, 1955.
11. I. M. Kolthoff and J. J. Lingane, *Polarography*, Vol. 1, Wiley (Interscience), New York, 2nd ed., 1952.
12. Strohecker and Henning, *Vitamin Assay*, Verlag Chemie, Weinheim, 1965.
13. G. Cimbura and R. C. Gupta, *J. Forensic Sci.*, **10**, 282 (1965).
14. *United States Pharmacopeia*, Mack Printing, Easton, Pa., 17th rev. ed., 1965, pp. 797–799.
15. *United States Pharmacopeia*, Mack Printing, Easton, Pa., 17th rev. ed., 1965, pp. 15, 180, 288, 383, 413, 414, 584.
16. *The National Formulary*, Mack Printing, Easton, Pa., 12th ed., 1965, pp. 128–129.

CHAPTER **17**

Amperometric Titrations

Fred W. Teare

FACULTY OF PHARMACY
UNIVERSITY OF TORONTO
TORONTO, ONTARIO

17.1 INTRODUCTION

An extension of quantitative polarography is made use of in amperometric titrations of a large number of substances. The term "amperometric" originated with Kolthoff and Pan,[1] but the term "limiting current" is preferred by others[2] when the limiting current is obtained using one polarizable electrode. The current due to diffusion (dropping mercury electrode, DME) or the limiting current in stirred solutions (rotating platinum electrode, RPE) is proportional to the concentration of an electroactive substance (Ox or Red). The change in concentration of an electroactive analyte, product, or titrant can be followed during the course of the titration by observing the change in current, at a constant applied potential, after each increment of titrant. Since the chapter on polarography deals with theory which is, in part, common to that required for an understanding of amperometric titrations, the reader is referred to that chapter before proceeding further.

A. CLASSIFICATION OF VOLTAMMETRIC TITRATIONS

An amperometric titration can be classified as a titrimetric (volumetric) method as well as a voltammetric method.

Voltammetric titrations may be divided into two groups:

1. Potentiometric titrations (at constant current which is either finite or essentially zero) employing one or two polarized electrodes.

2. Amperometric titrations (at a constant applied potential) employing one or two polarized electrodes.

a. One-polarized electrode amperometric titrations are believed to have been employed first for the titration of barium ions with sulfate by Heyrovský and Berezicky.[3]

b. Two-polarized electrode amperometric titrations were originally reported by Salomon.[4] This type was later referred to as a titration with a "dead-stop end point." [5]

The precision of amperometric titrations is generally better than 1% and is superior to that of polarographic measurements. The amperometric titration method can be employed to follow several types of chemical reactions such as precipitations, oxidation-reduction, and a few neutralization reactions. The concentration range of the electroactive substance(s) covered by this method is 10^{-2} to 10^{-6} M.

B. GENERAL APPARATUS

Any polarograph may be utilized to obtain the few current readings (at a constant applied potential) required before and after the end point in an amperometric titration. A simpler instrument such as the "Ampot" [6] or one similar to that illustrated in Fig. 17.1 is adequate for this purpose. Neither a precise potentiometer nor a calibrated galvanometer is required since the applied potential need only be adjusted to within ± 0.1 V and only the relative current is used.

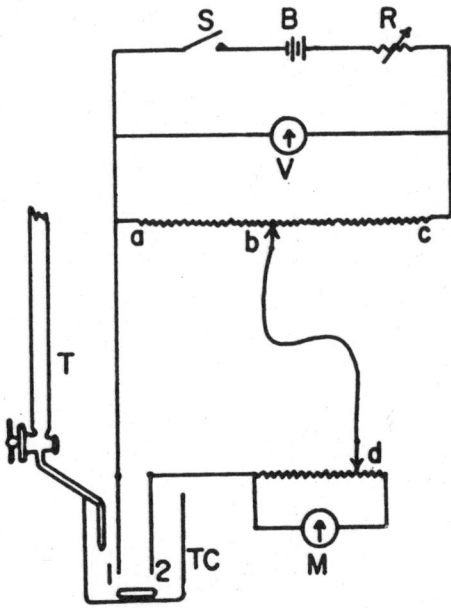

FIGURE 17.1: A simple schematic of the apparatus used in amperometric titrations: B, EMF source (battery etc.); a–c, device for selecting potential, voltage divider, etc; R, variable resistance for adjusting the potential span a–c; V, voltmeter; S, switch; M, microammeter for reading relative current; T, microburette (5–10 ml) with capillary tip; TC, titration cell; b, sliding contact for selecting voltage; d, sliding contact for selecting sensitivity of current reading device; 1, polarized indicator electrode, e.g., DME, RPE, (vibrating platinum electrode) (VPE), Pt wire, etc.; 2, nonpolarized reference electrode e.g., SCE, Hg/HgI, Ag/AgCl, Hg-Pool, etc., or second-polarized electrode e.g., Pt wire, in biamperometric titrations.

Volumetric flasks and transfer pipettes are required for the accurate preparation of solutions. A microburette (5 or 10 ml) is recommended. An oxygen-free nitrogen supply is essential to deoxygenate the titrate and titrant if a negative applied potential (vs. SCE) is required. Graph paper should be $8\frac{1}{2} \times 11$ in. with millimeter (or 20 lines/in.) ruling, unless otherwise specified.

17.2   AMPEROMETRIC TITRATIONS WITH ONE POLARIZED ELECTRODE

A.   APPARATUS

1. Instrument

Any polarograph or device such as that shown in Fig. 17.1 can be used to impress a constant potential across the titration cell and measure the relative current flow as a result of the reduction (or oxidation) occurring at the polarized electrode. Since there are numerous commercial instruments available, the reader is referred to the manufacturer's instructions for the details required in the operation of specific instruments.

2. Titration Cells

Any of several types of titration cells can be used. Many cells are commercially available or can be constructed in the laboratory. Several literature reviews include references concerning cells.

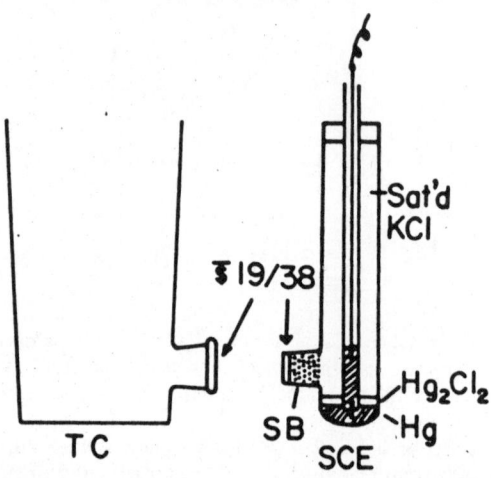

FIGURE 17.2: Modified amperometric titration cell; *TC*, titration cell (a 100-ml beaker with a ground-glass joint); *SB*, salt bridge consisting of a sintered-glass disk and agar gel plug saturated with KCl; SCE, saturated calomel electrode.

A simple beaker-type cell employing the Hg-pool as a reference electrode is frequently used. However, an H-type polarographic cell,[6] or a modification of this such as that used in the author's laboratory Fig. 17.2, is frequently preferred since the choice of the supporting electrolyte is not restricted. The SCE shown in Fig. 17.2 is easily constructed or the Sargent reference electrode section (S-29393) of the two-piece, H-form, polarographic electrolysis vessel (S-29392) can be used.[6] This titration cell (Fig. 17.2) is large enough to hold the titrate plus added titrant.

3. Indicator Electrodes

The most frequently used polarized indicator electrodes are the dropping mercury electrode (DME) and the rotating platinum electrode (RPE). Others include the rotating DME, graphite, and the vibrating platinum electrode (VPE). These and other electrodes are described in the literature (see review papers).

Advantages and Disadvantages of DME

a. A constantly renewed surface which is especially useful for precipitation titrations and yields reproducible currents.

b. The large hydrogen overvoltage of mercury permits the reduction of many substances having lower overvoltages.

c. Oxygen must be removed initially from the titrate and following each addition of titrant when the potential of this electrode is made more negative than about -0.1 V (vs. SCE). Time is required to deoxygenate these solutions with oxygen-free nitrogen to prevent the formation of the interfering oxygen wave(s).

d. A certain amount of "noise" may result from the charging current (required to charge the double layer of ions around the DME) being superimposed upon the diffusion current.

e. Applied potentials are limited to those which are less positive than about $+0.1$ V (vs. SCE) due to the dissolution of the mercury drop at more positive potentials ($Hg \rightarrow Hg(II) + 2e^-$).

Advantages and Disadvantages of RPE

a. This electrode actually compliments the DME since it is usable over a different range of potentials, e.g., $+1.0$ to -0.5 V, compared to $+0.1$ to -1.1 V (vs. SCE) for the DME when used in a supporting electrolyte of 0.5 M HCl. Above $+1.0$ V (vs. SCE) water in many supporting electrolytes is oxidized to O_2 at the RPE. These useful potential ranges are somewhat dependent upon the nature of the supporting electrolyte. Certain substances having oxidation and reduction potentials which lie in the range of the Pt electrode do not give rise to limiting currents with an RPE. The RPE also has a very low hydrogen overpotential and therefore has a very limited negative potential range (vs. SCE).

b. A larger diffusion current results from stirring due to a reduced diffusion layer. This accounts for the greater sensitivity when the RPE is used. A quiescent solution is mandatory for the regular DME.

c. A greatly reduced residual current results when the RPE is used, due to the lack of the repetitive charging current observed for the renewed mercury drops (see DME).

d. Only relative current values are required in amperometric titrations, thus the inherent disadvantages of solid microelectrodes in unstirred voltammetric measurements, such as the uncertainty of absolute potentials and less reproducible diffusion currents, are of no importance here.

e. The rotation of the RPE must be kept constant during the titration and this necessitates the use of a constant-speed stirring device.

Other solid polarizable electrodes can also be employed in solutions which are stirred by constant-speed stirrers.

4. Reference Electrodes

The most frequently used reference electrodes are the SCE, the Hg-pool, and the $Ag/AgCl_{(s)}$, all of which must be nonpolarizable. When these electrodes are incompatible with a particular system, others may be used.

In certain instances, one or more of the analyte, titrate, product, or indicator substance may be reduced or oxidized within the potential range of a selected reference electrode. When this occurs, it is only necessary to short circuit the two electrodes, i.e., connect the indicator and reference electrodes externally so that the indicator electrode assumes the same potential as the reference electrode. No external potential is required. Ewing[7] lists a number of selected reference electrodes and their potentials relative to the NHE (normal hydrogen electrode) and SCE. These may be employed in the manner just mentioned.

B. METHODOLOGY

Preparation for amperometric titration is similar to that for polarographic analysis, therefore the reader should be acquainted with the principles of polarography before reading this chapter.

If the half-wave potentials $(E_{1/2})$ of the possible electroactive (depolarizer) substances in a titration reaction are unknown, a separate polarogram of each substance should be obtained under the titration conditions. The possible depolarizers are indicated later [Eq. (17.1)] in Section 17.2C. From a knowledge of the $E_{1/2}$ values of each substance in the selected titration medium, it is possible to predict whether one or more depolarizer substances will give rise to a current at any selected applied potential. It will be shown later (under Section 17.2C) that the resulting curve shape may be dependent on the potential chosen for a particular amperometric titration. For example, if the selected applied potential is such that it produces a wave for both the analyte and titrant, a V-shaped titration curve is obtained. The location of the end point can be achieved with greater certainty in this instance than for an L-shaped curve. The applied potential is normally that which corresponds to the beginning portion of the limiting current plateau of the i vs. E_{appl} curve for an electroactive substance.

The titration media are the same as those employed for voltammetry (or polarography) of the same depolarizers. Solvents may be aqueous, nonaqueous or mixtures of these types. Added "inert" electrolytes are usually required to lower the resistance of the media to current flow and to minimize migration currents. The practical limits of concentration for a depolarizer are normally 10^{-2} to 10^{-5} M, but under favorable conditions this may be extended to 10^{-6} M. The titrant is usually 50- to 100-fold more concentrated than the analyte. This avoids significant dilution of the latter and circumvents the need for current corrections, except in very accurate work.

If the analyte is significantly diluted by a relatively large titrant volume.

the observed current readings must be multiplied by the factor $(V + v)/V$, where V is the initial volume of the analyte solution and v is the total volume of titrant added prior to the reading that is to be corrected.

Similarly with conductometric or photometric titrations, the data obtained in the vicinity of the end point is the least reliable. Therefore only 3 to 5 points on both sides of, and well removed from, the end point are required for a graphic location of the latter. An extrapolation of these linear portions of the plot of current vs. milliliters of titrant to a point of intersection is performed to locate the end point. The point of intersection corresponds to the end-point titrant volume read from the abscissa (Fig. 17.3).

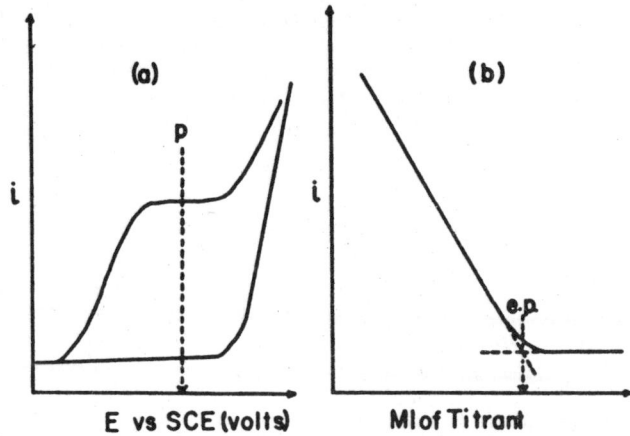

FIGURE 17.3: (a) Polarogram of an electroactive substance titrated in (b): P, selected applied potential to be used in titration (b); (b) amperometric titration curve of an electroactive analyte: e.p., end point located by extrapolation method.

If current values were read for constant increments of titrant throughout the entire titration, a marked curvature might be observed in the plot near the end point region. This curvature is a result of such factors as the solubility of precipitates, unstable complexes, hydrolysis of formed salts, incomplete or slow reactions, etc.

Both cathodic (reduction) and anodic (oxidation) currents are possible in amperometric titrations and these should be plotted in conformity with the conventions discussed in the chapter dealing with polarography.

C. TITRATION CURVE SHAPES

The titration curve is a plot of current vs. volume of titrant, at a constant applied potential (Fig. 17.3). The shape of this curve depends upon which substances in the following general equation are electroactive at the selected potential.

$$S + T \rightarrow P \qquad (17.1)$$

sample titrant product(s)

A number of possibilities exist in the general titration [Eq. (17.1)] for electroactive substances which determine the shape of a titration curve. Any one or more of the reactants and/or the products may give rise to an anodic or cathodic current at the selected potential. In addition to these, an indicator substance may either be added or generated in the reaction.

If the electroactive species in an amperometric titration are represented by T* (titrant); A* (analyte); P* (a product); I* (added indicator); I$_g^*$ (generated indicator) and the inactive species by the same symbol less the asterisk, then curve shapes in Fig. 17.4 are representative of the following situations:

Curve shape	Electroactive species
(Fig. 17.4)	
a	T*
b	I* or I$_g^*$
c	A* and T*
d	A$_c^*$ and T$_c^*$
e	A*
f	A$_a^*$ and T$_c^*$

where subscripts a and c refer to anodic oxidation and cathodic reduction, respectively.

Note: For additional possibilities see Ref. 8.

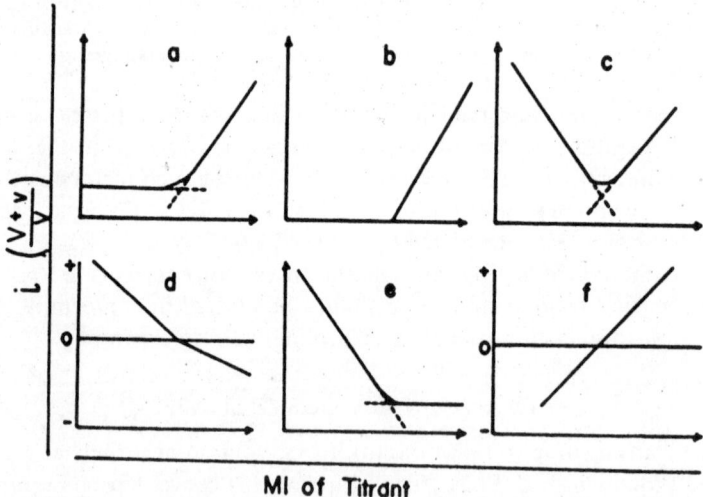

FIGURE 17.4: Typical amperometric titration curve shapes: $i(V + v)/V$ is the observed current corrected for dilution by the titrant volume. The various curves are described in the text.

D. GENERAL AMPEROMETRIC TITRATION PROCEDURE

Assuming a manual polarograph will be employed with an H-type titration cell, external SCE (Fig. 17.2), DME (or RPE), and a microburette, a general procedure is given here.

Obtain an i vs. E_{appl} curve (polarogram) of each possible electroactive substance entering or resulting from the titration reaction [Eq. (17.1)]. If the $E_{1/2}$ values of these depolarizers in the selected medium are known from experience, or obtained from the literature, this portion of the analysis need not be done.

The polarized electrode may be the DME if the applied potential is negative, or at least not greater than about $+0.1$ V (vs. SCE). The RPE is used in the range 0.0 to $+1.0$ V (vs. SCE). The unpolarizable reference electrode is usually an SCE having a large diameter salt bridge.

Dissolve the analyte (substance to be assayed) in 50 ml (or other appropriate volume) of a suitable solvent containing a large excess of unreactive electrolyte such as a halide salt, buffer, etc. Add a few drops of 0.2% gelatin solution (or other maximum suppressor) if the DME is used. Transfer all (or a large known portion) of this prepared solution to the titration cell and deoxygenate it with oxygen-free nitrogen gas for 15 min if the applied potential is to be more negative than about 0.0 V (vs. SCE). Approximately a 1 min deoxygenation is required in such instances after each addition of titrant. Adjust the applied potential to the selected voltage and record the current corresponding to 0 ml of titrant. The sensitivity must be set so that the galvanometer index remains on scale during the entire titration. This setting is made through trial and error or previous experience.

Add 1 ml of titrant from a microburette, mix the solution by means of a stream of nitrogen or a magnetic stirring device. Cease this auxiliary stirring and read the current when it assumes a steady value for the RPE or a repetitive value in a quiescent solution for the DME. A stream of nitrogen is directed over the solution to exclude oxygen when the DME is used. Repeat this procedure for 3 to 5 increments of titrant which are well removed from the end-point region. Repeat this sequence again well past the end point. The size and spacing of these increments will become obvious with experience, but they do depend on the amount of titrant required to reach the equivalence point of the reaction. If a significant change in the concentration of the analyte does not occur with dilution by the titrant, it should be possible to extrapolate to the point of intersection of the two straight lines which can be drawn through the points of the plot of i vs. milliliters of titrant (abscissa) Fig. 17.3. Determine the volume of titrant required by dropping a vertical line from the point of intersection to the volume axis. If curvature appears in the extrapolated current vs. titrant volume plot, each current reading must be replotted after correction for the dilution effect (see text). This current correction is required only in those regions of the plot where the slope $\neq 0$.

In titrations which call for the RPE, a stationary platinum electrode and magnetic stirrer can be used, but the current values are less reproducible. If time permits, repeat the titrations two or three times to ascertain the precision attainable.

Examples of amperometric titrations with one polarized electrode are given in Table 17.1.

E. EXPERIMENTS

Experiment 17.1. Amperometric Titrations Employing the DME as the Polarized Electrode

Apparatus. A manual polarograph, H-type polarographic cell or modified version (Fig. 17.2), DME assembly, external SCE (part of H-type cell), deoxygenation apparatus including oxygen-free nitrogen, 5- or 10-ml microburette, volumetric flasks and pipettes of appropriate capacity.

Solutions.   a.   0.05 M potassium dichromate ($K_2Cr_2O_7$) stock solution; use analytical reagent grade $K_2Cr_2O_7$ if primary standard grade not available

b.   0.005 M $K_2Cr_2O_7$ solution, prepared by an accurate 1 in 10 dilution of solution a

c.   0.02 M lead nitrate stock solution

Any one of the following supporting electrolytes:

i.   0.2 M potassium nitrate
ii.   Acetate buffer (acetic acid and sodium acetate, 0.3 M in each).
iii.   Equal volumes of i and ii.

0.2% gelatin solution (freshly prepared) or other maximum suppressor such as 0.2% Triton X-100 Solution.

Procedure

Part A. Selection of Applied Potential. Prepare 200 ml of 0.001 M $Pb(NO_3)_2$ in one of the supporting electrolytes (SE) just described, adding 1.0 ml of 0.2% gelatin before adding the SE to the 100-ml mark. Prepare 50 ml of 0.001 M $K_2Cr_2O_7$ as previously, except add 0.25 ml of the gelatin solution instead of 1.0 ml.

Transfer 50 ml of this dichromate solution (or an appropriate volume) to a titration cell, deoxygenate it with nitrogen for at least 15 min, then direct a small stream of nitrogen over the surface of the solution.

Using a DME assembly, start the mercury flow, then immerse the capillary into the solution. Adjust the drop time to about 3 or 4 sec, and the sensitivity of the galvanometer to give about an 80% deflection at an applied potential of -1.0 V (vs. SCE).

Obtain a polarogram over the applied potential (E_{appl}) range of 0 to -1.1 V (vs. SCE), using small changes in potential in regions of greatest current change. Plot the i vs. E_{appl} curve and observe the voltage range corresponding to the limiting current plateau. Obtain another polarogram under the same conditions, except use 50 ml of the 0.001 M $Pb(NO_3)_2$ in place of the dichromate solution. Leave this solution in the titration cell for Part B. Plot the polarograms on the same graph paper. From the two polarograms select applied potentials which may be used to: (a) give a reduction current due to one species only, or (b) give reduction currents due to both Pb(II) and dichromate.

Part B. Amperometric Titrations

1. Using a potential selected in part A, titrate the deoxygenated lead solution prepared in part A with 0.005 M $K_2Cr_2O_7$ using a 10-ml microbur-ette. Stop the stirring and/or nitrogen bubbling when reading the current after each 0.5-ml addition of titrant (after each 0.25-ml in regions of rapid current change). Continue the titration until the titrant volume approximates twice that required for the equivalence point. Plot current readings vs. milliliters of titrant (along abscissa). Observe the nature of the plot in the vicinity of the end point. Extrapolate the linear portions of the curve using about four current values on each side of, and well removed from, the equiv-alence point. The point of intersection corresponds to the end point.

Upon completion of the experiment clean the DME, while mercury is dropping by immersion in 1 M HNO_3, then rinse with water and dry with a tissue. Store the DME in air, inside a test tube. Also clean the titration cell with 1 M HNO_3 and rinse well with distilled or deionized water.

2. Repeat the titration using an applied potential at which both reactants are reduced.

Assuming the 0.005 M $K_2Cr_2O_7$ solution has been accurately standardized, what is the concentration of the nominally 0.001 M $Pb(NO_3)_2$?

Which titration (step 1 or 2) permits the most precise end point location? Explain.

3. If time does not permit the reader to perform both parts A and B, use applied potentials of 0.0 and -1.0 V (vs. SCE) for the potentials sought in part A, and carry out part B.

4. As an additional experiment titrate a solution which is 10^{-4} M in lead salt with a correspondingly less concentrated standard dichromate solution.

Are current corrections necessary, due to dilution caused by the titrant, in the titrations performed?

Check this for one or two situations. Ascertain the error, if any, if such corrections are not made.

TABLE 17.1: Amperometric Titrations Involving One Polarized Electrode

Substance	Electrodes* Ind.	Ref.	E volts (vs. SCE)	Titrant	Titration medium	References
Inorganic ions						
Al^{3+}	DME	SCE	0	$0.6\ M$ NaF	NaCl + 1 mM Fe^{3+} in (1:1) H_2O-EtOH	[a]
Br^-, I^- or Cl^-	RPE	SCE	0.05	0.001–$0.05\ M$ $AgNO_3$	$0.8\ M$ HNO_3 in H_2O-Me_2CO (1:1)	[b]
I^-	RPE	SCE	0.65	$0.002\ M$ KIO_3	$1\ M$ H_2SO_4	[b]
NH_3	RPE	SCE	0.2	0.01–$0.001\ M$ $Ca(OCl)_2$ or NaOBr	$0.00125\ M$ KBr + $0.15\ M$ $NaHCO_3$	[b]
SCN^-	RPE	SCE	0.2	0.005–$0.05\ M$ $Ca(OCl)_2$	0.00125–$0.24\ M$ KBr + 0.15–$0.17\ M$ $NaHCO_3$	[b]
SO_4^{2-}	DME	SCE	-1.2	0.01–$0.1\ M$ $Pb(NO_3)_2$	20–30% aqueous alcohol	[b]
Organic compounds						
p-Aminophenol	RPE	SCE	0.0	$0.05\ M$ $K_2Cr_2O_7$	0.25–$1\ M$ H_2SO_4	[b]
Aminophenol, p-amino-salicyclic acid	RPE	SCE	-0.2	$0.02\ M$ $KBrO_3$	$0.05\ M$ KBr + 0.08 HOAc + $100\ M$ HCl	[b,d,e]
Cysteine (R—SH)	RPE	Hg/HgI	-0.4	0.005–$0.05\ M$ CU^{2+}	$0.05\ M$ NH_3 + $0.15\ M$ $Na_2S_2O_3$	[d,e]
Cysteine	RPE	Hg/HgI	-0.3	$0.001\ M$ $AgNO_3$		[e]
Phenobarbital elixir	DME	SCE	0.0	$0.1\ M$ Hg^{2+} acetate in $0.1\ M$ HOAc	$0.20\ M$ NaOAc in EtOH-H_2O (4:1)	[f]
Amines, 4°N compounds	graphite	SCE (NaCl)	$+0.55$	XS of $0.1\ M$ Na TPB, $0.1\ M$ KCl	Acetate buffer	[g]
Thiols (—SH)	RPE	SCE	0.0 or -0.2	$0.01\ N$ $AgNO_3$	Ammoniacal or neutral	[c,h,i]
Thiourea	RPE	SCE	$+1.$ to $+1.1$	$0.01\ N$ $AgNO_3$	H_2SO_4	[g]
Aldehydes and ketones	DME	SCE	-0.76	(2,4-dinitrophenyl) hydrazine	$0.005\ M$ H_2SO_4 and 50% EtOH	[a]

Substance	Electrode	Reference	Potential	Reagent	Medium
Nicotine	DME	SCE	−0.38	Silicotungstic acid	2-10 mg/50 ml in dilute HCl + KCl
Salts of N-containing bases (alkaloids): codeine, cocaine, atropine, cinchonine, quinine, papaverine, procaine, aminopyrine	DME	SCE	−0.65	Silicotungstic acid (1 mole per 2 or 4 moles of N-base); reacts rapidly, simple stoichiometric ratios, current read after 15–30 sec	0.05–0.2 M HCl
Strychnine, atropine, brucine, papaverine, cocaine, codeine, ethylmorphine	DME	SCE		As above	1. to 3 M HCl
Caffeine, theobromine, theophylline	DME	SCE		Silicotungstic acid (1 mole per 3 moles of N-base)	0.5 M HCl
Ascorbic acid	DME	SCE	0.0	Fe(III)	Acidic medium
	RPE	SCE	−0.1	$KMnO_4$ (in presence of I)	H_2SO_4 medium
	RPE	SCE	+0.3	ICl	Acid medium
	DME	SCE		Chloramine-T	Acid medium
	DME	SCE	−0.85	2,6-dichlorophenolindophenol	Acetate-phosphate buffer (pH 2.0)
Aniline and phenols	RPE	SCE	0.0	$KBrO_3$	2 M HCl and KBr
	RPE	SCE	0.0	KIO_3	0.3 M HCl
	RPE	SCE	0.0	0.02 M $KBrO_3$	1 mg/ml, 25 ml MeOH and 65 ml H_2O and 5 ml HCl and 5 ml 2 M KBr
Phenols	RPE	SCE	+0.1 to 0.3	0.1 N $KBrO_3$ and 0.1 N KBr	HCl-MeOH-DMF
Dodecylpyridinium bromide	DME	SCE	−1.32	Sodium dodecylsulfate	0.1 M NaOH
Thiols (—SH)	DME	SCE	0 to −0.2	Hg(II)	Slightly acidic or neutral media
Cysteine	RPE	SCE	−0.2	Cu(II)	Ammoniacal medium

TABLE 17.1 (continued)

| Substance | Electrodes* | | E volts (vs. SCE) | Titrant | Titration medium | References |
	Ind.	Ref.				
Methylene blue, methyl violet	DME	SCE	−0.4 to −0.8	Silicotungstic acid	0.05–5 mg in 1 M HCl	
Phenothiazines, and various antihistaminics	DME		−0.65	Silicotungstic acid	3.5% HCl	
Salts of: codeine, atropine, strychnine, quinine, papaverine, procaine, antipyrine, aminopyrine.	DME	Hg-pool	−0.4	Phosphotungstic acid	0.2 M HCl	
Quinine chloride, papaverine HCl, strychnine HNO₃	DME	SCE	−0.45	Picric acid	Acidic media (pH 4.7)	
Acridines	DME	SCE	0.0	0.05 M K₂Cr₂O₇	Acetate buffer (pH 4.8)	
Alkalated barbiturates (also in dosage forms)	DME	SCE	0.0	0.05 M Hg(NO₃)₂	0.05 to 0.2 g in EtOH, or Me₂CO and 0.5 M KOH	
Antipyrine	RPE	SCE	−0.25	0.05 M Hg(NO₃)₂	0.05 to 0.2 g in EtOH, or Me₂CO and 0.5 M KOH	

Salicylates, and p-aminosalicylic acid	DME	SCE	0.0	0.05 M Hg(ClO$_4$)$_2$	0.02 to 0.2 g in alkaline media, obtain a 1:2 (Hg:salicylate ppt)	[d, e]
p-Aminosalicylic acid	RPE	SCE		0.02 M KBrO$_3$	1–2 mg in 100 ml solution containing (0.05 M KBr, 0.08 M HOAc and 100 M HCl)	[•]
Aromatic phenols and amines with free —OH and —NH$_2$ groups, e.g., sulfas, alkaloids etc.	DME	SCE	−0.4	0.025 M p-diazobenzene-sulfonic acid	25 ml Clark-Lubs buffer (pH 9.3) and 0.5 ml 0.5% gelatin cooled to 10°C	[•]

* Key: SCE, saturated calomel; DME, dropping mercury; RPE, rotating platinum; Hg/HgI, mercury-mercurous iodide; Hg-pool, mercury pool.

a I. M. Kolthoff and J. J. Lingane, *Polarography*, Vol. 2, 2nd. ed. 1952

b H. A. Laitenen, *Anal. Chem.*, **28**, 666 (1956).

c A. Berka, J. Dolezal, and J. Zyka, *Chemist-Analyst*, **54**, 24 (1965).

d A. Berka, J. Dolezal, and J. Zyka, *ibid.*, **53**, 122 (1964).

e M. Brezina and P. Zuman, *Polarography in Medicine, Biochemistry, and Pharmacy*, Interscience, New York, 1958.

f E. M. Cohen and N. G. Lordi, *J. Pharm. Sci.*, **50**, 661 (1961).

g J. E. Sinsheimer and D. Hong, *ibid.*, **54**, 805 (1965).

h H. C. Börresen, *Anal. Chem.*, **35**, 1096 (1963).

i I. M. Kolthoff, and W. E. Harris, *Ind. Eng. Chem., Anal. Ed.* **18**, 471 (1946).

j M. A. Medina and C. Cummiskey, *Chemist-Analyst*, **53**, 17 (1964).

Experiment 17.2.   Amperometric Titrations Employing the Rotating
Platinum Electrode (RPE) as the Polarized Electrode

Apparatus. Manual polarograph, magnetic stirring device, RPE and
constant rate motor (e.g., 600 rpm; E. H. Sargent & Co. Chicago). Titration cell and external SCE (Fig. 17.2), 10-ml microburette, pipettes, and
volumetric flasks of appropriate number and capacity.

Solutions. 1×10^{-3} M As(III) solution (dissolve the As_2O_3 in a minimum
of 3 M NaOH, and add sufficient concentrated HCl (ca. 12 M) and water
to make the final solution 1 M in HCl); 1 M KBr; and 0.05 N KBrO$_3$
(0.140 g in 100 ml).

Note: If the galvanometer index goes off scale in a negative direction and
the galvanometer leads cannot be readily switched, begin the titration
with the zero current setting at midscale.

Procedure. Accurately pipette the following solutions into the titration
cell: 25.0 ml 0.001 M As(III) solution, 15.0 ml of 1 M HCl, and 10.0 ml of
1 M KBr solution. Immerse the RPE (previously cleaned in hot HNO_3
and rinsed with water) in the titrate, add a stirring bar to the titration cell,
and place a magnetic stirrer under it. Connect the electrodes to the appropriate terminals of the polarograph. Insert the tip of the burette (or capillary
extension of same) below the surface of the titrate. Make certain the RPE
is clear of any obstacle, then turn on the motor. Adjust the applied potential
of the RPE to 0.2 V (vs. SCE). It is not necessary to remove dissolved
oxygen from these solutions. Why? Titrate the As(III) solution with the
standard 0.05 N bromate solution. Read the current after each 0.05-ml
addition of titrant until a definite deflection of the galvanometer is produced,
then after each 0.10-ml increment until several readings beyond the equivalence point have been recorded. Keep this solution if required later. The
magnetic stirrer is used to hasten mixing, but should be turned off during
current readings. Note that the current does not increase until the equivalence
point is reached. Plot i vs. milliliters of titrant and determine the end point
graphically. When are current corrections required in this titration, if at all?
It may be necessary to repeat this titration if the correct sensitivity was not
selected initially. How could this sensitivity have been determined prior
to the titration? If time permits, obtain a voltammogram over a voltage range
of $+0.2$ to $+1.0$ V (vs. SCE) for the solution just titrated, in which excess
titrant is present. What is the active species undergoing reduction at the
indicator electrode? Could you obtain this curve using the DME in place
of the RPE? Give reasons for your answer.

This amperometric titration can be repeated with lower concentrations
of As(III) and correspondingly lower concentrations of the titrant.

$$AsO_3^{3-} + BrO_3^- \rightarrow AsO_4^{3-} + Br^-$$

The first drop of excess $KBrO_3$ reacts with excess KBr in the acidic medium to produce free Br_2, which gives rise to the reduction current:

$$BrO_3^- + 5Br^- + 6H^+ \rightarrow 3Br_2 + 3H_2O$$

What possible pharmaceutical applications could be made with this titration? A few references should be consulted.

Experiment 17.3

For additional amperometric titrations employing either the DME or RPE as the polarized electrode, consult Table 17.1.

17.3 AMPEROMETRIC TITRATIONS USING TWO POLARIZED ELECTRODES

Two methods of voltammetry are very similar in that both give rise to similar titration curves and both employ two identical polarizable indicator electrodes (usually two platinum wires or pieces of foil). In potentiometric titrations using two polarizable electrodes (bipotentiometric method), a small constant current is applied and changes in voltage during the titration are recorded. Whereas, in amperometric titrations with two polarizable electrodes (biamperometric method) a small constant potential difference is applied and changes in current are followed during the titration.

A. TITRATION CURVE SHAPES

The titration curve shape depends on the reversibility or irreversibility of the redox couples (or systems) involved in the titration reaction, which determines the reaction at each electrode during the titration.

I. Titration Involving Two Reversible Redox Couples

The most widely cited example is that of the titration of ferrous iron with ceric ions,[9,10]

$$Fe(II) + Ce(IV) \xrightarrow{\text{(H}^+)} Fe(III) + Ce(III) \qquad (17.2)$$

In this instance, both redox couples, i.e., Fe(III)/Fe(II) and Ce(IV)/Ce(III), are reversible. In this example, ferrous iron is titrated with a standard ceric sulfate solution in an acidic medium and yields a curve similar to Fig. 17.5a. Before any titrant has been added, the anode can oxidize Fe(II), while only H^+ ions can be reduced at the cathode. Since the applied potential is much smaller than the potential difference between the E° (or $E^{\circ\prime}$, formal potential) values of the involved redox couples, Fe(III)/Fe(II) and Ce(IV)/Ce(III), little or no current flows and electrolysis is very small. However, following

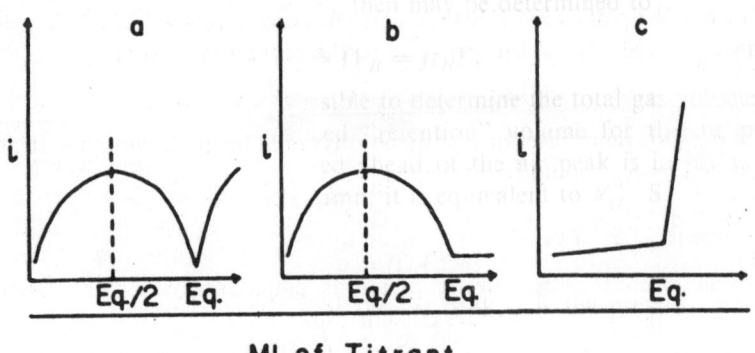

MI of Titrant

FIGURE 17.5: Typical biamperometric titration curves. (a) Involving two reversible redox couples, (b) involving one reversible redox couple (dead-stop end point), (c) involving one reversible redox couple (kick-off end point). Eq. = equivalence point.

the addition of Ce(IV) titrant, an equivalent amount of Fe(II) is oxidized to Fe(III), which can be reduced at the cathode. This is observed by an increased current flow due to the establishment of a reversible couple, Fe(III)/Fe(II), in the system. This current increases with addition of titrant until one-half of the original Fe(II) has been oxidized and [Fe(III)] = [Fe(II)]. At this point the concentration of Fe(II) being oxidized at the anode equals the concentration of Fe(III) being reduced at the cathode,

$$\text{Cathodic reduction: } Fe(III) + e^- \rightarrow Fe(II) \tag{17.3}$$

$$\text{Anodic oxidation: } Fe(II) \qquad \rightarrow Fe(III) + e^- \tag{17.4}$$

This maximum in the plot of i vs. milliliters of titrant represents the one-half equivalence point, where [Fe(III)] = [Fe(II)]. As the titration proceeds past this point, the [Fe(III)] increases and the [Fe(II)] decreases, resulting in a decreased total current flow due to the [Fe(II)] limitation imposed at the anode (anodic and cathodic currents are equal at all times). A minimum current occurs at the equivalence point, since all the Fe(II) ions have been oxidized to the Fe(III) state, and the applied potential is not sufficient to oxidize the Ce(III) ions which are present. Once past the equivalence point, the excess of Ce(IV) remains in solution establishing a new reversible redox couple, Ce(IV)/Ce(III). Immediately after the equivalence point the limiting factor controlling current flow is the [Ce(IV)], since a relatively larger concentration of Ce(III) is present. As the [Ce(IV)] builds up, the overall electrode reactions increase, i.e., Ce(IV) reduction at the cathode and Ce(III) oxidation at the anode, resulting in a rising current in the plot.

It is important to realize that the current does not reach 0 due to the presence of minute quantities of ions (of the redox couples) which are assumed to be absent. It is usually necessary to establish the exact location of the equivalence point graphically, since some distortion of the ideal curve occurs in practice. Exceptions are made for those situations in which a very abrupt change in current occurs at the end point.

2. Titration Involving Only One Reversible Couple

Among the most familiar biamperometric titrations in which one couple is irreversible are those involving the titration of arsenite with iodine (or vice versa) and iodine with thiosulfate (or vice versa).

If identical polarizable platinum electrodes have a small constant potential (e.g., 0.01 to 0.1 V) impressed across them, the curve shape that results from the titration of iodine with thiosulfate (or arsenite) is similar to Fig. 17.5b. If thiosulfate (or arsenite) is titrated with iodine, the shape of the titration curve corresponds to Fig. 17.5c.

Iodine/Iodide is an example of a reversible couple in the examples illustrated in Fig. 17.5

$$\text{Anodic oxidation: } 2I^- \rightarrow I_2 + 2e^- \tag{17.5}$$

$$\text{Cathodic reduction: } I_2 + 2e^- \rightarrow 2I^- \tag{17.6}$$

In the titration of iodine with thiosulfate, the curve shape (Fig. 17.5b) may be explained as follows:

At the beginning, no current flows between the electrodes unless there is present in the solution a substance which can be oxidized at the anode and another substance which can be reduced at the cathode. Initially, only iodine exists in solution and therefore oxidation cannot occur at the anode. After a small addition of thiosulfate, iodide appears in solution as shown in

$$I_2 + 2Na_2S_2O_3 \rightarrow 2NaI + Na_2S_4O_6 \tag{17.7}$$

Since the anodic and cathodic currents are equal for a reversible redox couple, the magnitude of the current is established by the member of the couple present in the lowest concentration, i.e., $[I^-]$ early in the titration. As more thiosulfate is added, the iodide concentration increases, as does the current flow through the cell. This continues to the midpoint of the titration. At the one-half equivalence point, $[I] = [I^-]$, and the current reading reaches a maximum. After the midpoint in the titration, the remaining iodine concentration is less than that of the iodide formed, and the cathodic reduction becomes the current limiting factor. The total current flow continues to decrease until it reaches a value near zero at the end point, hence the name "dead-stop" end point. The current does not increase again after the end point due to the irreversibility of the tetrathionate/thiosulfate couple $(S_4O_6^{2-}/S_2O_3^{2-})$ and the small applied potential which is insufficient to oxidize iodide at the anode and reduce H^+ ions at the cathode.

Note: In the titration of 0.1 N Iodine USP with sodium thiosulfate solution, the actual titration curve looks like the last half of the curve shown in Fig. 17.5b due to the presence of KI in this iodine solution.

If a solution of thiosulfate (irreversible couple) is titrated with iodine (reversible couple), the current remains near zero prior to the equivalence point and increases abruptly thereafter (Fig. 17.5c) giving rise to the designation "kick-off" end point.

B. METHODOLOGY

The dual polarizable electrodes are usually two platinum wires each sealed in a glass tube, or two platinum foil or button-type electrodes. Larger electrode surfaces increase the current sensitivity. In certain instances, two mercury-plated platinum electrodes or two dropping mercury electrodes or two copper electrodes, etc., can be employed.

Frequently the sensitivity of this method depends upon the magnitude of the applied potential, it is therefore important that the resistance of the current indicating device be as low as possible to avoid a significant iR drop across it.

To minimize any change in the solution composition due to electrolysis, the surface area of the electrodes should be small (0.1 to 3 cm^2) and the applied potential kept to a low value (e.g., usually 10 to 500 mV). This is essential when an irreversible couple is involved in the titration.

When stationary solid electrodes are employed the solution must be stirred. Since the rate of stirring does affect the observed current, a constant-rate stirrer is preferred, however a magnetic stirrer is usually adequate. An increased rate of stirring can be used to increase the current sensitivity. Although a simple device may frequently be employed, as illustrated in Fig. 17.1, a polarograph is convenient and desirable for accurate work.

A major advantage of dual-polarized electrode amperometry is the simplicity of use and maintenance of the electrodes, especially in nonaqueous media, where reference electrodes are more difficult to make.

Note: If the applied potential in the dual-polarized electrode method is made sufficiently large, a diffusion-controlled current will be established early in the titration and the method will be identical to the one polarizable electrode method.

C. ADVANTAGES OF AMPEROMETRIC TITRATIONS

In the amperometric titration method there are fewer variables than with polarography, thus simpler, more rugged equipment can be employed. Also greater accuracy and sensitivity is possible than in polarography. Concentrations of certain substances can be determined between 10^{-1} and 10^{-5} M and in a few instances to 10^{-6} M. The method has a wider range of application than either potentiometric or polarographic methods.

D. APPARATUS

A manual polarograph is satisfactory. In addition, one requires a titration cell consisting of a 100-ml beaker, magnetic stirrer (or stirrer with a glass propeller), two identical platinum electrodes (usually two platinum wire or foil electrodes sealed in glass tubes), a 10-ml microburette, appropriate pipettes, and volumetric flasks for solution preparation.

E. GENERAL PROCEDURE

A polarizing potential is applied across the identical electrodes so that one becomes the anode and the other the cathode. The magnitude of this potential may be indicated or may have to be found experimently, however it usually lies between 10–500 mV. The titrant is added in 0.5-ml increments until a detectable change in current is observed, then in reduced increments of 0.1 ml. Allow time for mixing after each addition of titrant, then read the current when it becomes steady, plot the observed current on the ordinate vs. volume of titrant in milliliters along the abscissa.

The sensitivity setting should remain unchanged throughout the titration. If the "dead-stop" method is employed, the sensitivity should be adjusted to about 80% of full-scale deflection (FSD) during a trial run when the current flow is maximal. For the "kick-off" method, the sensitivity should be adjusted to about 80% FSD after four or five small increments of titrant have been added beyond the end point in a trial run.

F. TYPICAL AMPEROMETRIC TITRATIONS USING TWO POLARIZED ELECTRODES

A few common examples of biamperometric titrations to the "dead-stop" end point include the titration of iodine with thiosulfate, bromometric titrations, and various titrations with oxidizers such as ceric or permanganate ions, or reducing agents such as the ferrous or titanous ions.

Additional examples are found in Table 17.2.

G. EXPERIMENTS

Solutions
Ferrous ammonium sulfate (0.1 N)*
Ceric sulfate (0.1 N)*
Iodine (0.1 N)*
Sodium thiosulfate (0.1 N)*
0.025 M As_2O_3 (in 1 M $NaHCO_3$); prepare in a manner comparable to that for a standard solution of $KAsO_2$*

* USP XVII, volumetric solutions.[11]

TABLE 17.2: Amperometric Titrations Involving Two Polarized Electrodes*

Substance	E volts	Titrant	Titration medium	Reference
Inorganic examples				
Water (content)	0.025	Karl Fischer reagent	Ethylene glycol-pyridine (4:1)	
Sb(III)	0.200	Br_2†	(Excess) NaBr (0.2 M) + 1 M H_2SO_4	
As(III)	0.200	$KBrO_3$ (Br_2†)	(Excess) NaBr + 0.1 M H_2SO_4	
I⁻	0.130	$KBrO_3$ (Br_2†)	(Excess) NaBr + 2 M HCl	
Ce(IV)	0.100	Fe(II)	3 M H_2SO_4	
$S_2O_3^{2-}$	0.135	I_2	0.1 M HCl (air-free)	
NH_3	0.150	BrO^-	Borate buffer (pH 8.5) containing 0.1 M NaBr	
Zn(II)	0.100	$K_4Fe(CN)_6$	(Slight excess EDTA), pH 2.1, and 10 drops $K_3Fe(CN)_6$ (10%) solution	
Organic examples				
Barbiturates (e.g., Phandorn, Evipan, Thiopentol)		Br_2*($KBrO_3$)	Try (excess) KBr + 2 M HCl (bromatometric method)	
Sulfanilamide and other sulfas		$KBrO_3$ (Br_2†)	Bromatometric	
Phenothiazine and various derivatives		$KBrO_3$ (Br_2†)	Bromatometric	
Phenothiazine and derivatives		Ce(IV)	H_2SO_4 solution	
Carbutamide		Br_2†	Bromatometric	
Phenol		Br_2†	Bromatometric	
Aromatic primary amines, e.g., various sulfonamides, amino-salicylic acid, etc.		$NaNO_2$	Diazatization in HCl solution with bromide as a catalyst	
Isonicotinic hydrazide		$NaNO_2$	In HCl solution	
Alkaloids (certain)		$AgNO_3$	Precipitate with tetraphenylboron, separate, and dissolve ppt in acetone, titrate TPB ion with $AgNO_3$	
Urea, thiourea, substituted quanidines		NaOMe, or $HClO_4$	In DMF and ethylene diamine or in F_3C-COOH, respectively	

Substance	Titrant	Medium / Notes
Various olefins	KBrO₃	Bromometric titration with $HgCl_2$ catalyst, to determine the bromine number
Unsaturated compounds	KBrO₃	Determination of iodine numbers
Ascorbic acid	ICl	
	2,6-dichloroindophenol	
	K₃Fe(CN)₆	
Oxalic acid	Ce(IV)	Glacial HOAc
Cysteine	K₃Fe(CN)₆	
Glucose	Ce(III)	Add (excess)K₃Fe(CN)₆ and back titrate with Ce(III)
2,3-dimercaptopropanol	Br₂†	

* Consult original references for details.

† Electrogenerated, may work with added titrant (KBrO₃ in presence of excess KBr and HCl). Bromatometric method.

[a] E. L. Bastin, H. Siegel, and A. B. Bullock, *Anal. Chem.* **31**, 468 (1959).
[b] R. A. Brown and E. H. Swift, *J. Am. Chem. Soc.*, **71**, 2717 (1949).
[c] L. Meites (ed.), *Handbook of Analytical Chemistry*, McGraw-Hill, New York, 1963, Sec. 5.
[d] K. Stulik and F. Vydra, *Chemist-Analyst*, **55**, 24 (1966).
[e] K. R. Srinivasan, *Analyst*, **75**, 76 (1950).
[f] H. G. Scholten and K. G. Stone, *Anal. Chem.*, **24**, 749 (1952).
[g] J. E. DeVries, S. Schiff, and E. S. C. Gantz, *Anal. Chem.*, **27**, 1814 (1955).
[h] H. D. DuBois and D. A. Skoog, *Anal. Chem.*, **28**, 624 (1948).
[i] ASTM, *1964 Book of ASTM Standards*, ASTM, Philadelphia, Part 17, 1964, p. 399, D 1158-59T.
[j] J. A. Duke and J. A. Maselli, *J. Am. Oil Chemists' Soc.*, **29**, 126 (1952).
[k] H. Liebmann and A. D. Ayres, *Analyst*, **70**, 411 (1945).
[l] O. N. Hinsvark and K. G. Stone, *Anal. Chem.*, **28**, 334 (1956).
[m] H. G. Waddil and G. Gorin, *Anal. Chem.*, **30**, 1069 (1958).
[n] N. H. Furman and A. J. Jr. Fenton, *Anal. Chem.*, **32**, 745 (1960).
[o] J. W. Sease, C. Niemann, and E. H. Swift, *Anal. Chem.*, **19**, 197 (1947).

Experiment 17.4. Biamperometric Titration Involving Two Reversible Couples, i.e., Fe(III)/Fe(II) and Ce(IV)/Ce(III)

·Pipette 5.0 ml of 0.1 N ferrous ammonium sulfate into a 100-ml beaker, followed by 5 ml of sulfuric acid and about 10 ml of water. Immerse the platinum electrodes to a depth of 1/2 in. or more. Apply a potential of 400 mV, start the stirrer, attempting to keep the rate fairly constant during the titration. Titrate with standardized ceric sulfate solution, adding 0.5 ml increments at the beginning, decreasing these to 0.1 ml in the vicinity of the end point. Add about six small increments beyond the end point. Read the current after each addition of titrant and record the values. Plot the observed current i vs. milliliters of Ce(IV) as the titration is being performed. Explain the shape of this plot.

Experiment 17.5. Biamperometric Titration Involving One Reversible and One Irreversible Redox Couple (using the same apparatus employed in Experiment 17.4)

Pipette 5.0 ml of 0.1 N iodine solution into a 100-ml beaker. Add about 20 ml of water sufficient to immerse the dual platinum wire (or foil) electrodes. Apply a potential of between 25 to 100 mV, and adjust the sensitivity of the polarograph to obtain current reading of about 80% FSD. Titrate with 0.1 N sodium thiosulfate solution, recording the current reading after each 0.5-ml increment at the beginning and 0.1-ml additions near the end point. To facilitate the exact location of the end point, plot current vs. volume of titrant as indicated in Experiment 17.4. Continue the titration until five or six small (0.1-ml) increments have been added beyond the end point, where there is no appreciable change in current. Explain the shape of this plot.

H. ALTERNATE EXPERIMENTS

Use the same equipment and conditions, unless otherwise stated.
a. Titrate 5.0 ml of 0.1 N $Na_2S_2O_3$ solution with 0.1 N iodine solution. Explain the shape of the i vs. milliliters of titrant plot.
b. Titrate 5.0 ml of 0.05 N iodine (diluted with 20 ml of water) with 0.025 M As_2O_3 (in 1 M $NaHCO_3$), or the reverse of this.

$$A_sO_3^{3-} + I_2 + 2HCO_3^- \rightarrow A_sO_4^{3-} + 2I^- + H_2O + 2CO_2$$

Experiment 17.6. Consult Table 17.2 for Additional or Alternate Biamperometric Titrations, Preferably Those Involving an Organic Compound

QUESTIONS

Q17.1. State the advantages and disadvantages of the rotating platinum electrode (RPE) and the dropping mercury electrode (DME) for amperometric titrations.

Q17.2. Discuss the relative merits of dc polarographic analysis and amperometric titrations (keeping in mind the restrictive definition of polarography).

Q17.3. Contrast potentiometric and amperometric titration methods.

Q17.4. What other titrimetric methods employ a graphic means of end-point detection similar to that used in amperometric titrations? What are the values plotted against the volume of titrant in each case?

Q17.5. Briefly survey the applications of interest to a pharmaceutical analyst for "dead-stop" end-point titrations. Consult review articles given in the References.

Q17.6. Outline a procedure which might be employed in adapting the analysis of a suitable pharmaceutical preparation by the Volhard method to the amperometric titration technique. Use the same indicator substance, if desired. Is any advantage gained by this adaption for the usual routine analysis? Comment.

READING LIST FOR AMPEROMETRIC TITRATIONS

Articles and reviews

Anastasi, A., U. Gallo, E. Mecarelli, and L. Novacic, *J. Pharm. Pharmac.*, **8**, 241 (1956).

Berka, A., J. Doležal, and J. Zýka, *Chemist-Analyst*, **53**, 122 (1964); **54**, 24 (1965).

Charlot, G., and B. Trémillon, *J. Electroanal. Chem.*, **3**, 1 (1962).

Charlot, G., Badoz-Lambling, and B. Trémillon, *The Electrochemical Methods of Analysis*, Elsevier, Amsterdam, 1962.

Delahay, P., *New Instrumental Methods in Electrochemistry*, Wiley (Interscience), New York, 1954.

Delahay, P., *Instrumental Analysis*, Macmillan, New York, 1957, pp. 106–117.

Elving, P. J., I. Fried, and W. R. Turner, *U.S. At. Energy Comm.*, *COO-1148-84* (1964).

Ewing, G. W., *Instrumental Methods of Chemical Analysis*, McGraw-Hill, New York, 1954, Chap. 4.

Ewing, G. W., *Instrumental Methods of Chemical Analysis*, McGraw-Hill, New York, 2nd ed., 1960, pp. 215–220.

Furman, N. H. *Anal. Chem.*, **26**, 84 (1954).

Ishibashi, M., and T. Fujinaga, *Polarographic Methods of Chemical Analysis*, Maruzen, Tokyo, 1956, pp. 317–340.

Jolliffe, G. O., in *Practical Pharmaceutical Chemistry* (A. H. Beckett and J. B. Stenlake, eds.), Athlone Press, London, 1962, pp. 352–357.

Kalvoda, R., and J. Zýka, *Pharmazie*, **7**, 535 (1952).

Kolthoff, I. M., *Anal. Chem.*, **26**, 1685 (1954).

Kolthoff, I. M., and J. J. Lingane, *Polarography*, Wiley (Interscience), New York, 2nd ed., 1952, pp. 887–953.

Laitinen, H. A., *Anal. Chem.*, **21**, 66 (1949); **24**, 46 (1952); **28**, 666 (1956); **30**, 657 (1958); **32**, 180R (1960); **34**, 307R (1962).

Litaneau, C., and D. Cörmös, *Talanta*, **7**, 18 (1960).

Meites, L., *Polarographic Techniques*, Wiley (Interscience), New York, 1955, pp. 185–204.

Milner, G. W. C., *The Principles and Applications of Polarography and Other Electroanalytical Processes*, Longmans, Green, London, 1957, pp. 633–713.

Mitchell, J., and D. M. Smith, *Aquametry*, Wiley (Interscience), New York, 1948, pp. 86–93.

Müller, O. H., *The Polarographic Method of Analysis*, Chemical Education Publishing Co., Easton, Pa., 1951, pp. 118–128.

Reilley, C. N., *Anal. Chem.*, **28**, 671 (1956).

Stock, J. T., *Anal. Chem.*, **36**, 355R (1964); **38**, 452R (1966).

Stock, J. T., *Amperometric Titration*, Wiley (Interscience), New York, 1965.

Strobel, H. A., *Chemical Instrumentation*, Addison-Wesley, Reading, Mass., 1960, pp. 530–534.

Willard, H. H., L. L. Merritt, Jr., and J. A. Dean, *Instrumental Methods of Analysis*, Van Nostrand, Princeton, N.J., 3rd ed., 1958, Chap. 21.

Willard, H. H., L. L. Merritt, Jr., and J. A. Dean, *Instrumental Methods of Analysis*, Van Nostrand, Princeton, N.J., 4th ed., 1965, Chap. 26.

Zuman, P., *Organic Polarographic Analysis*, Pergamon Press, New York, 1964, pp. 145–165.

Zýka, J., in *Progress in Polarography* (P. Zuman and I. M. Kolthoff, eds.), Wiley (Interscience), New York, 1962, pp. 649–660.

Also see References and those in Chapter 16.

REFERENCES

1. I. M. Kolthoff and Y. D. Pan, *J. Am. Chem. Soc.*, **61**, 3402 (1939).
2. J. Heyrovský and J. Kůta, *Principles of Polarography*, Academic Press, New York, 1966. pp. 267–272.
3. J. Heyrovský and S. Berezicky, *Collection Czech. Chem. Commun.*, **1**, 19 (1929).
4. E. Salomon, *Z. Physik. Chem.*, **24**, 55 (1897); **25**, 365 (1898); *Z. Electrochem.*, **4**, 71 (1897).
5. C. W. Foulk and A. T. Bawden, *J. Am. Chem. Soc.*, **48**, 2045 (1926).
6. E. H. Sargent Co., *Bulletin P-5, Sargent Polarographs and Accessories*, Chicago.
7. G. W. Ewing, *Instrumental Methods of Chemical Analysis*, McGraw-Hill, New York, 1954, p. 77.
8. J. Jordan and J. H. Clausen, in *Handbook of Analytical Chemistry* (L. Meites, ed.), McGraw-Hill, New York, 1963, pp. 5-155–5-158.
9. D. G. Davis, in *Handbook of Analytical Chemistry* (L. Meites, ed.), McGraw-Hill, New York, 1963, pp. 5-164–5-165.
10. L. L. Leveson, *Introduction to Electroanalysis*, Butterworth, London, 1964, Chap. 4.
11. *United States Pharmacopeia*, Mack Printing, Easton, Pa., 17th rev. ed., pp. 1081–1088.

CHAPTER **18**

Gas Chromatography

Carman A. Bliss

COLLEGE OF PHARMACY
UNIVERSITY OF SASKATCHEWAN
SASKATOON, SASKATCHEWAN

18.1 INTRODUCTION

One of the most difficult and frustrating problems encountered in pharmaceutical analysis is that of the simultaneous separation, identification, and quantitation of more than one compound from a complex mixture in a pharmaceutical product. This problem in recent years has been greatly simplified by the development of chromatographic procedures. From the first basic chromatographic technique devised by Tswett[1] for the separation of leaf pigments on columns, there has evolved several sophisticated chromatographic methods of separation. These methods are classified into four different groups: gas-solid adsorption chromatography (GSC); gas-liquid partition chromatography (GLC); liquid-solid adsorption chromatography (LSC); and liquid-liquid partition chromatography (LLC). It is the purpose of this chapter to discuss the first two groups which have been collectively called "gas chromatography."*

Historically gas chromatography developed as a logical extension of the earlier work on liquid chromatography. In the paper describing their Nobel Prize work in this field, Martin and Synge in 1941[2] first suggested the possibility of utilizing gas as a mobile phase rather than the previously used liquids. However, for the next 10 years little attention appears to have been paid to this observation. Between 1941 and 1952, limited contributions to the field of gas chromatography were made by some workers such as Claesson,[3,4] Turner,[5] and Philips.[6] It was in 1952 that Martin in conjunction with James published the first reports[7,8] on the successful separation and elution of organic constituents by a flowing gas as the mobile phase. These publications were quickly followed by important contributions both in GSC and GLC by several workers including Janak,[9] Ray,[10] and Bradford et al.[11] By 1955 the usefulness of gas chromatography as an analytical tool was fully realized and since that time several thousand publications have appeared in the literature.

Chromatography as defined by Keulmans[12] is "a physical method of separation in which the components to be separated are distributed between two phases, one of the phases constituting a stationary bed of large surface area, the other being a fluid that percolates through or along the stationary bed." Gas Chromatography utilizes as the stationary phase a glass or metal column filled either with a powdered sorbent or a nonvolatile liquid coated

---

* For the discussion of LSC and LLC, the reader is referred to Chapter 11, Volume 1.

on a nonsorbent powder. The fluid or mobile phase consists of an inert gas containing the vaporized mixture of solutes flowing through the stationary phase. In gas-solid adsorption chromatography (GSC) retention of solutes is dependent largely upon differences in adsorption properties of the solutes for the powered sorbent as they pass through the stationary phase. With gas-liquid partition chromatography (GLC) the retention of solutes is dependent largely upon the partition coefficients of the solutes for the non-volatile liquid of the stationary phase.

To date, the use of GSC has been largely limited to the analysis of gases such as H_2, O_2, N_2, NO, and such low boiling point organic compounds as CH_4, C_2H_4, C_3H_6, and C_3H_8. In contrast GLC has had a much greater application in pharmaceutical analysis, being applicable to most organic constituents which have a measurable vapor pressure at the temperature employed. Because of the limited use of GSC to pharmaceutical analysis, and since extensions of arguments used for partition columns can generally be extended to adsorption columns, all further discussion will refer to GLC unless stated otherwise.

The major advantages of gas chromatography as an analytical tool lie in the high efficiency of separation, the sensitivity in detection of components, the speed of separation, and wide application of the method for most groups of pharmaceutical agents. Most GC applications utilize samples in the range of micrograms, but new preparative columns are now capable of isolating gram quantities of purer materials if necessary.

18.2 FUNDAMENTALS OF GAS CHROMATOGRAPHY

A. COMPONENTS OF A GAS CHROMATOGRAPH

A modern gas chromatograph consists of three basic units: (1) the chromatographic unit, (2) the temperature control and signal amplification unit, (3) the recorder unit. Figure 18.1 shows a schematic diagram of a typical instrument.

1. Chromatographic Unit

This unit is the heart of the gas chromatograph. It is constituted essentially of: the carrier gas source, comprised of a tank of compressed gas, usually nitrogen or helium; a gas pressure regulator and flow control system; a sample injection port; a chromatographic column containing the stationary phase; a thermostated column oven; a detector; and a gas exit port. In addition most instruments contain a thermostated injection heater (flash heater) surrounding the injection port and a thermostated detector heater controlling the temperature of the detector.

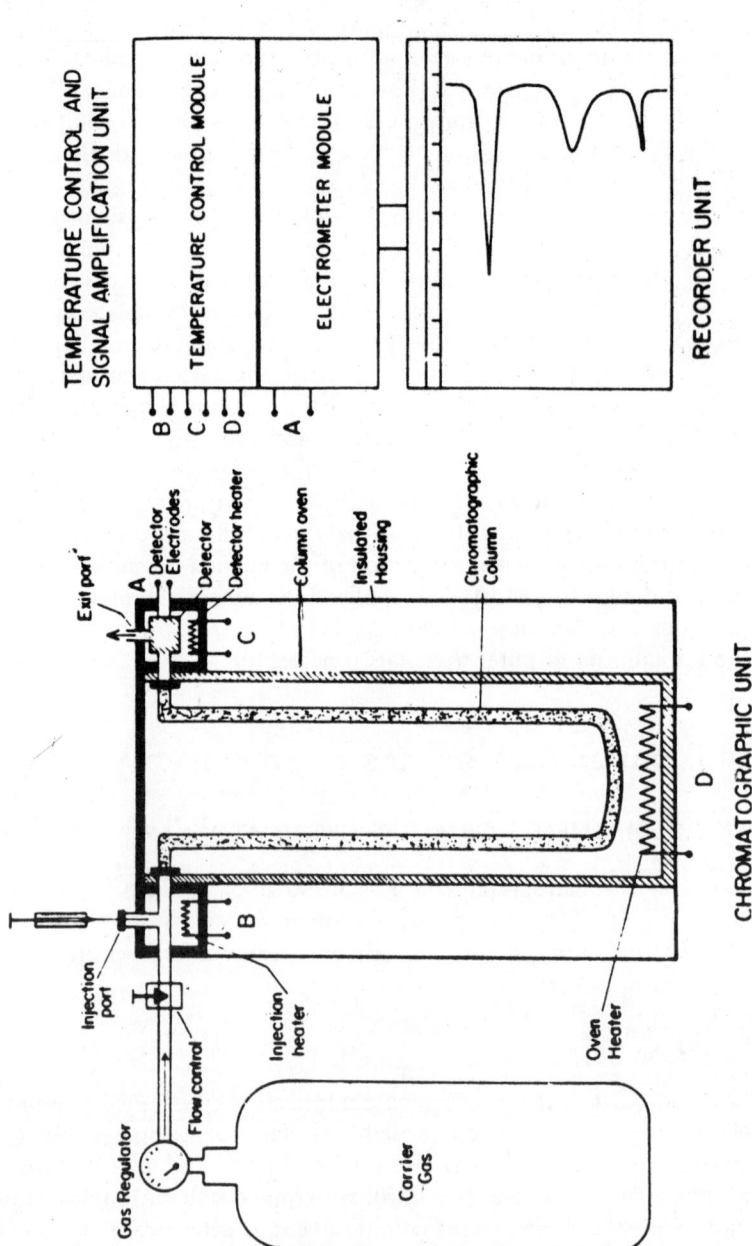

FIGURE 18.1: Schematic diagram of a modern gas chromatograph.

2. Temperature Control and Signal Amplification Unit

Temperature conditions in the chromatographic unit are accurately regulated by heaters and are thermostatically controlled from the temperature control and signal amplification unit. Oven temperature, injection port temperature, and detector temperature are controlled separately. This unit also is frequently equipped with temperature-programming controls whereby the oven temperature may be programmed to increase at a set rate over a limited temperature range as the chromatographic separation proceeds.

The second function of the temperature control and signal amplification unit is to amplify the signal produced by the detector and to transmit this amplified signal to the recorder unit. Amplification is accomplished by means of a specially designed electrometer module placed between the detector and the recorder unit. The electrometer is designed to provide distortion-free amplification. Several ranges of amplification are provided to allow the operator to control the strength of the signal transmitted to the recorder.

3. Recorder Unit

The amplified detector signal from the electrometer is recorded on a millivolt strip recorder to produce a graph of detector signal response against time. The recorder unit may also be equipped with integrators which automatically record the area under each curve.

B. SEPARATION PROCEDURE BY GAS CHROMATOGRAPHY

I. Equilibration of Instrument

The inherent sensitivity of a gas chromatograph to small variations in operating parameters makes it essential that complete equilibration and stabilization of the instrument be attained before the introduction of the sample. Initially, the column packed with the stationary phase is attached to the instrument and the desired flow rate of carrier gas through the column is adjusted by means of the gas regulator system. Column temperature is set and maintained at the desired temperature by the column oven control. Normally both the injection heater and detector heater are then set a few degrees above that of the column oven. If the column is newly packed or has been unused for some time, several hours may be required for conditioning or stabilization of the column. With carrier gas only passing from the column through the detector, the amplified signal from the detector is adjusted by the electrometer to give zero base line on the recorder. Equilibrium conditions exist when no fluctuation of the zero base line occurs over a period of several hours.

2. Separation of Sample Constituents

Sample size in analytical gas chromatography usually varies from between 10^{-9} to 10^{-2} g. Liquid samples may be applied without dilution, while solids are usually dissolved in a volatile solvent such as hexane, benzene, or carbon disulfide. The sample (0.05 to 100 μliters) is applied to the column by use of a microsyringe inserted into the injection port through a rubber septum covering the port. The injection of sample from the syringe should be instantaneous to prevent a broadening or distortion of the chromatographic peaks resulting. Due to the high temperature of the injection heater, the solutes in the sample are instantaneously vaporized, forming a "plug" of solute vapors which is swept into the gas stream by the carrier gas and onto the chromatographic column.

Separation of individual components within the column is dependent upon two separate factors, the retention of the solutes on the column and the column efficiency. The retention effect establishes the order in which the compounds being separated will elute from the column and is mainly dependent upon the partition coefficients of the solutes between the two phases and upon the temperature. The efficiency of the column determines the degree of broadening of each solute band as it travels down the column. Column efficiency is dependent upon many factors, including the solutes being separated, the physical characteristics of the column, the rate of flow of carrier gas, and the nature of both the solid support media and the liquid phase. The resolution, or the degree of completeness of separation, of two or more solutes on the column then depends directly upon both column efficiency and retention.

3. Isothermal Operation and Temperature-Programmed Operation

The simplest method of temperature control in gas chromatography is that of *isothermal operation*, whereby the column oven is maintained at a constant temperature throughout the chromatographic run. By this procedure the column temperature is selected so that all solutes of interest in the separation mixture will have sufficient volatility to be vaporized at that temperature. Both flash heater and detector temperature are maintained at about 20°C above the column temperature. *Temperature-programmed operation*, in contrast, is a method in which the column oven temperature is programmed to increase at a constant rate from an initial lower level through to an upper limit temperature during the chromatographic run. Both the flash heater temperature and the detector temperature are maintained at about 20°C above the upper limit temperature. Temperature programming is used where the volatilities of the solutes in the mixture may vary over a considerable range. By starting at a low temperature and increasing the temperature during the run it is often possible to separate the components with high volatilities at the low temperature range and components with relatively low volatility in the high operating temperature range. Temperature programming is most useful when a large number of solutes with a wide range of volatilities is to be separated since the method produces a more even distribution of the peaks along the chromatogram and gives sharper peaks with compounds of low volatility than occurs in isothermal operation. Isothermal operation is most useful where rigidly controlled parameters of operation are required and when the constituents of the solute mixture will separate and elute as relatively sharp peaks at one column temperature i.e., there is a narrow spread in their volatilities.

C. DETECTION AND RECORDING OF SEPARATED COMPONENTS

After resolution of the solutes, each vaporized component emerges in turn from the column and is carried into the detector mixed with the carrier gas. The function of the detector is to "sense" the concentration of impurities in the carrier gas stream, and to transmit a signal to the electrometer which is proportional to the concentration of impurity present. After passage through the detector the gas stream is vented by means of the exit port. The electrometer, upon receiving the signal from the detector electrodes, amplifies and transmits it in turn to the recorder. The record produced by the recorder shows a continuous plot of time vs. detector response i.e., concentration of eluted solutes from the time of injection until the last solute has emerged from the column. This record is referred to as a "differential gas chromatogram." Under ideal operating conditions the same column may be used several hundred times to give reproducible chromatograms with the same sample. Figure 18.2 shows a typical differential chromatogram obtained by separating a mixture of tocopherols and cholestane. The first large unnumbered peak represents the solvent used to dissolve the solutes.

18.3 THEORETICAL CONSIDERATIONS OF GAS CHROMATOGRAPHY

In initiating a discussion on the theoretical concepts involved in separation of solutes by gas liquid chromatography, it is desirable to emphasize that there are two basic considerations involved. The first is the phenomena affecting *retention* or hold-up on the column, sometimes referred to as the thermodynamic aspect. The second phenomena is that of *column efficiency* or the kinetic aspect governing the tendency for a particular solute band to broaden as it travels through column. The *resolution* or extent of separation of any two peaks from a column is dependent upon both retention effects and column efficiency. The following discussion will be largely devoted to these factors.

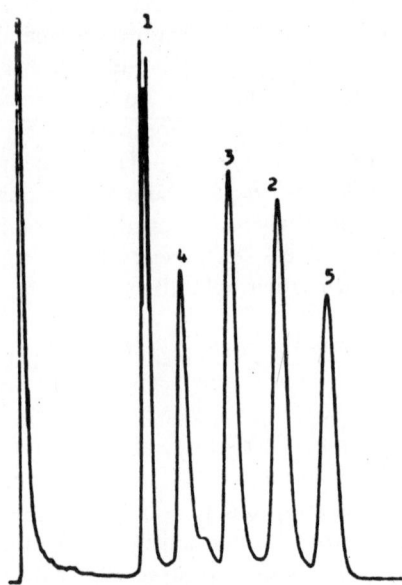

FIGURE 18.2: Chromatogram of the separation of tocopherols by gas liquid chromatography. Cholestane is employed as the internal standard. The large peak on the left is
the solvent peak. Courtesy of Hewlett-Packard Company, Analytical Instrument Division,
Avondale, Pennsylvania) Key: 1, cholestane; 2, α-tocopherol; 3, γ-tocopherol; 4, δ-
tocopherol; 5, α-tocopheryl acetate. Instrument: F and M model 400 Biomedical gas
chromatograph. Sample, cholestane; δ-, γ-, α-tocopherol and α-tocopheryl acetate. 2.0
μliters of a 0.1% solution. Column, 4 feet × ¼ in. od, 2% SE-30 on 80–100 mesh Diato-
port S. Temperatures, column, 235°C, injection port, 240°C, detector 240°C. Carrier
gas, helium, 90 ml/min. Sensitivity, range 10 attenuation × 32.

A. RETENTION

A gas liquid partition chromatography column may be considered to be a
tube packed with an enert stationary support material coated with the liquid
phase. The *total volume* V_T within this column is the sum of three different
volumes, the volume occupied by the solid support medium or the *solid
support volume* V_S; the volume occupied by the liquid phase coating the
solid support or the *liquid phase volume* V_L; and the volume occupied by
the carrier gas filling the interstititial spaces of the column, frequently called
the "dead volume" or the *total gas volume* V_G°. Hence

$$V_T = V_S + V_L + V_G^\circ \tag{18.1}$$

Under the operating conditions in the gas chromatograph carrier gas flows
through the column with an *average carrier gas flow rate Fc*. The absolute
inlet pressure of the carrier gas at the point of entrance to the column is given
by P_i, while the *absolute outlet pressure* of the column is given by P_o.

Consider next a differential gas chromatogram represented by Fig. 18.3.

The chromatogram is obtained from the recorder following injection of a sample composed of a mixture of a nonadsorbed gas such as air and a single solute into the gas chromatograph. A katherometer is used for detection. Since the air is not adsorbed by the solid support nor partitioned into the liquid phase it will travel unimpeded at the same rate as the carrier gas. The solute component of the mixture however is carried through the column at a slower rate than the air, depending upon its relative solubility in the liquid phase at the column temperature. As stated earlier this delayed elution of

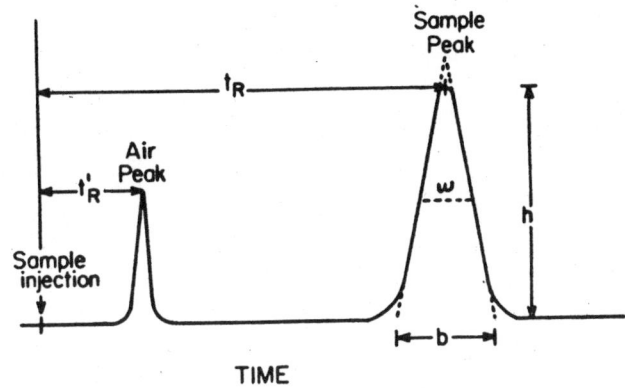

FIGURE 18.3: Schematic differential chromatogram of a pure solute with accompanying air peak.

the solute from the column, or "holdup" of the solute peak, is termed the "retention" of the solute. On the chromatogram represented by Fig. 18.3 the value t_R is a measure of the retention time of the solute, i.e., it represents the time interval between the instant of injection of the solute onto the column and the moment it emerges from the column and causes maximum detector response. Similarly t'_R represents the travel time of the unimpeded air in the sample.

From t_R the *retention volume (uncorrected)* V_R may be calculated, where:

$$V_R = t_R F_c \tag{18.2}$$

V_R is then the measure of the volume of gas which passes through the column during time t_R with an average carrier gas flow rate of F_c.

Because of its greater accuracy however, the *corrected retention volume* V_R° is generally employed instead of V_R. Correction of the retention volume is necessary because of the compressibility of the carrier gas, which results in an increased linear velocity of carrier gas along the column length. V_R° is calculated by introduction of a *pressure gradient factor f* to correct for this difference. This pressure gradient factor can be calculated from the column inlet pressure P_i and outlet pressure P_o by the equation:

$$f = \tfrac{3}{2}[(P_i/P_o)^2 - 1]/[(P_i/P_o)^3 - 1] \tag{18.3}$$

The corrected retention volume V_R° then may be determined to be:

$$V_R^\circ = fV_R = ft_R F_c \tag{18.4}$$

In an analogous manner it is possible to determine the total gas volume V_G° by determination of the corrected "retention" volume for the air peak. Since the volume of gas displaced ahead of the air peak is in reality the interstitial gas volume of the column, it is equivalent to V_G°. See Fig. 18.3. Therefore

$$V_G^\circ = ft_R' F_c \tag{18.5}$$

where t_R' is the retention time for the air peak, f is the pressure gradient factor, and F_c is the gas flow rate.

Another quantity of significance in calculations of gas liquid chromatography data is the *net retention volume* V_N. This value is the volume of carrier gas flowing through the column from the time of injection until the solute peak reaches its maximum, i.e., V_R° less the total gas volume V_G°

$$V_N = V_R^\circ - V_G^\circ \tag{18.6}$$

The volume of the liquid phase or *liquid phase volume*, V_L is determined at the *column temperature* T_c and is calculated at the time of preparation of the column from the equation:

$$V_L = \frac{W_L}{\rho_L} \tag{18.7}$$

where W_L is the weight of liquid phase in column and ρ_L is the density of liquid phase at column temperature T_c.

Both W_L and ρ_L are easily determined at room temperature, but since ρ_L varies with temperature the density of the liquid phase at column temperature T_c must be estimated. By the assumption that ρ_L has a reduction coefficient of 10^{-3} per degree centigrade increase in temperature, the value of ρ_L may be approximated under column conditions.

It must be noted that the net retention volume V_N will vary according to the liquid loading, i.e., the per cent liquid phase on the column, and if V_N is utilized in presenting chromatographic data, the amount of liquid phase present must be stated. To obtain an expression for retention volume which is independent of this factor, the quantity *specific retention volume* V_g is utilized. It is determined from the net retention volume V_N and the liquid phase density ρ_L by Eq. (18.8), where T_c is the column temperature.

$$V_g = \frac{V_N \times 273}{V_L \rho_L (T_c + 273)} \tag{18.8}$$

The specific retention volume V_g therefore is the net retention volume per gram of liquid phase reduced to standard temperature and pressure. The value V_g is frequently utilized in the literature as a method of reporting solute behavior. The reader is referred to the text of Dal Nogare and Juvet[13] for examples of actual retention calculations.

A further retention value frequently used in the literature is that of *relative retention*, in which retention volumes of all solutes are expressed relative to one standard compound. The primary advantage of this method is that it eliminates the effects of the column size and operating conditions. The disadvantage of this method of reporting is that different authors frequently utilize different standards for different applications. For best results experimental conditions should be identical and constant and it is usual to carry out relative retention calculations on solutes obtained from the same chromatogram or chromatograms run consecutively under identical conditions. Under these conditions,

$$\text{Relative retention } r_{1.2} = \frac{V_{g_1}}{V_{g_2}} = \frac{V_{N_1}}{V_{N_2}} = \frac{V^\circ_{R_1}}{V^\circ_{R_2}} = \frac{V_{R_1}}{V_{R_2}} \qquad (18.9)$$

where subscripts 1 and 2 are those of the standard and compound being studied, respectively.

A solute which is retarded on a gas-liquid chromatography column does so due to its *partition coefficient K* between the stationary liquid phase and the gas phase. Dependent only upon temperature, the partition coefficient is a constant for any particular solute according to Eq. (18.10):

$$K = \frac{\text{weight of solute per ml of stationary liquid phase}}{\text{weight of solute per ml of gas phase}} \qquad (18.10)$$

When the solute as a vapor "plug" is introduced into a GLC column there is established over a short period of time a dynamic equilibrium between the solute in the liquid phase and in the carrier gas. The molecules of solute in the vapor phase cannot be swept unhindered through the column in the carrier gas stream and leave behind the solute molecules which have diffused into the liquid phase. Instead, during each unit distance which the solute in the vapor travels, it reestablishes itself in a new dynamic equilibrium with the liquid phase according to its partition coefficient. The overall effect therefore is that the rate of migration of the solute down the column is retarded. A solute with a large partition coefficient will thus travel at a slower rate down the column than a solute having a smaller partition coefficient. Only in the case of a solute such as an insoluble gas in which K approaches O will there be no retardation on the column.

The partition coefficient K is influenced by column temperature but it is independent, within limits, of the size of the column and weight of liquid phase.

It may be shown experimentally or through mathematical derivation that partition coefficient is related to the corrected retention volume V°_R, the total gas volume V°_G, and the liquid phase volume V_L by Eq. (18.11):

$$K = \frac{V^\circ_R - V^\circ_G}{V_L} = \frac{V_N}{V_L} \qquad (18.11)$$

Similarly the partition coefficient K may be derived from Eqs. (18.8) and (18.10) and related to the specific retention volume V_g, the liquid density ρ_L, and the column temperature T_c by Eq. (18.12).

$$K = \frac{V_g(T_c + 273)\rho_L}{273} \tag{18.12}$$

B. COLUMN EFFICIENCY

In addition to the retention of the solute as it begins to migrate down the column, there also occurs a widening or spreading of the solute zone. This broadening of the band is due to several kinetic influences in the solute travel including; diffusion, eddy effects, flow rate of carrier gas, and resistance to mass transfer by the liquid phase. The longer the solute is on the column the greater is the tendency for the peak to broaden. This spreading effect may be observed on comparing the solute peak resulting when a short column is used with that obtained when a longer but otherwise identical column is employed. The peak eluting from the short column is generally higher and narrower than that of the longer column, even though the area under the curve is the same in both instances. A comparison of several different solutes eluting from a single column shows also that the first peaks eluted are narrower than those retained for a longer time period; this tendency may be seen in Fig. 18.2.

Column efficiency is determined from the relationship between the peak width and the retention time or retention volume of the solute under examination. The quantitative measure of efficiency is given in terms of the number of theoretical plates n or the height equivalent to a theoretical plate (HETP).

Despite the fact that chromatography is a continuous process, the theoretical plate concept—derived from a discontinuous process—is most frequently utilized for determining column efficiency. In a discontinuous process such as countercurrent extraction, complete equilibration of the two phases takes place in each tube before there is a change of phase. Each such equilibration is equivalent to one theoretical plate. In chromatography however, since the flow of carrier gas is continuous, there is insufficient time for complete equilibration between the liquid phase and the carrier gas phase in any one cross section. What is considered, therefore, is the distance through the column the carrier gas travels before an equivalent distribution of solute between the liquid phase and the gas phase takes place. This distance is called the "height equivalent to a theoretical plate" or "HETP." The HETP in a well packed and efficient column (4 ft × 4 mm) may be as small as 0.3 cm.

One measure of the efficiency of a gas chromatographic column is given by the number of theoretical plates n that a column contains. The equation most frequently utilized in this determination is:

$$n = 16\frac{(t_R)^2}{b} \tag{18.13}$$

The use of this expression relies upon the fact that the elution curves in gas chromatography, in the absence of absorption or overloading on the column, approximate normal distribution curves. The expression is derived on the basis of the binomial distribution theory. Actual calculation of n is carried out from experimental values obtained from the chromatogram. The term b is the base of the triangle formed from the tangents through the inflection points of the curve and with the base line as seen in Fig. 18.3. Both b and t_R must be in the same units since n is dimensionless. Equation (18.13) may also be modified so as to express retention volume V_R or corrected volume V_R° as long as b is also expressed in the same units.

The HETP may be calculated also from experimental values by dividing the number of theoretical plates n by the column length l. Hence

$$\text{HETP} = l/n \qquad (18.14)$$

It should be remembered that both n and the HETP apply only to that solute whose peak was measured and not to other solute peaks on the chromatogram. It also should be noted that within limits, n may be increased by lengthening the column, but is reduced somewhat by increase in column diameter.

Application of the plate theory is useful in determining the quantitative measure of column efficiency under standard operating conditions, but it gives no clue as to the best parameters of operation for that column. Van Deemter et al.[14] proposed a rate theory for determination of the HETP, which relates the column efficiency to various column parameters. Three separate phenomena are considered of major importance in determining column efficiency: (1) the eddy diffusion caused by the distance of gas flow through multiple pathways in the column, (2) the molecular diffusion of the solute into the carrier gas, (3) the resistance to mass transfer of solute molecules from the gas phase into the liquid phase.

The van Deemter equation is given as:

$$\text{HETP} = 2\lambda d_p + \frac{2\gamma D_g}{U} + \frac{8}{\pi^2} \frac{k}{(1+k)^2} \frac{d_f^2}{D_{liq}} U \qquad (18.15)$$

where
λ = statistical irregularity of packing
d_p = particle diameter of the support medium
D_g and D_{liq} = solute diffusivity in the gas and liquid phases
γ = a correction factor to express channel tortuosity
k = ratio of the solute in the liquid phase to that in the gas phase
d_f = Thickness of the liquid coating on the support particles
U = average carrier gas velocity
$\dfrac{8}{\pi^2}$ = a geometry constant in the mass transfer term

Because of the complexity of Eq. (18.15) a simpler expression of it has been devised:

$$\text{HETP} = A + \frac{B}{U} + CU \qquad (18.16)$$

in which $A = 2\lambda d_p$

$B = 2\gamma D_g$

$C = (8/\pi^2)[k/(1 + k)^2](d_f^2/D_{liq})$

Term A or $2\lambda d_p$ expresses the eddy diffusion effect on peak broadening. Such broadening is produced by the variance in time required by individual solute molecules to travel through the multiple pathways of different lengths within the column packing. The numbers and lengths of these pathways are influenced both by irregularity of packing and by particle size. The terms λ and d_p represent irregularity of packing and particle size, respectively. Uniform packing, i.e., a small λ, decreases the peak width, as does the decreasing particle size d_p down to a lower limit. Very small particle size decreases the HETP, but this is counterbalanced to some extent by the greater irregularity of packing produced by fine powders. There is therefore a practical lower limit to particle size dependent upon the particular stationary phase used and its packing properties. Mesh sizes varying from 20 to 120 are commonly used, depending upon the separation involved.

The second term B/U or $2\gamma D_g$ expresses the peak broadening effect due to molecular diffusion of solute particles into the carrier gas. Unlike A, which is independent of the nature of the carrier gas, the properties of the solute and the carrier gas velocity, B/U is the product of all three effects. The term γ is a correction factor introduced to account for the tortuosity of the gas pathways. D_g or carrier gas diffusivity is related to the density of the carrier gas. The value of D_g is decreased by increasing the molecular weight of the carrier gas. In some instances, significant decrease in B may be achieved by changing to a carrier gas with a higher density, e.g., hydrogen to nitrogen. Different solutes will show different diffusion properties, but this factor is generally impossible to control other than by modification of chemical structure of the solutes separated. The last parameter U or average gas velocity in B/U is of considerable importance. As may be seen from Eq. (18.15), increasing the carrier gas velocity will decrease the value of $2\gamma D_g/U$, thus decreasing the molecular diffusion term. At low gas velocities significant increase in efficiency may be achieved by increasing the gas velocity. If a high pressure drop is necessary to achieve high flow rate however, as may occur in long columns, a loss of efficiency may result instead. In addition, an increase in gas velocity at higher values produce a peak broadening effect in the third term CU.

The final term CU or

$$\frac{8}{\pi^2} \frac{k}{(1 + k)^2} \frac{d_f^2}{D_{liq}} U$$

describes the magnitude of the peak broadening due to the resistance to mass transfer of solute molecules from the mobile gas phase to the liquid stationary phase. The parameters involved in this final expression have significant effects on column efficiency or HETP. The constant $8/\pi^2$ is a factor necessary to account for the geometrical relationships involved in the mass transfer within the column. The ratio of the solute in the liquid phase to solute in the gas phase is given by k and is directly related to the partition coefficient K and the amount of liquid phase present. By choice of a liquid phase that will give a large partition coefficient K, i.e., a high solubility of solute in the liquid phase, the HETP may be decreased. Similarily increasing the amount of liquid phase tends to keep the term $k/(1 + k)^2$ small. The value of d_f^2 also is significant. It can be seen that the magnitude of this value varies as the square of the thickness of the liquid phase on the particle. Therefore by decreasing the coating thickness the HETP will be decreased. This increased efficiency is offset somewhat by the decrease of liquid phase required. A correct balance between the two effects may be achieved by careful control of the per cent liquid phase used in the column. Lightly loaded columns (1 to 5%) are common for many types of separation today. (Note: Too light a liquid coating however may result in decreased efficiency due to incomplete coating of the support particles, causing adsorption effects to occur.) D_{liq} or solute diffusivity into the liquid phase is seen to possess an inverse relationship in the mass transfer value. Thus, increasing values for the diffusion of solute into the liquid phase decreases the HETP. Such an increase may be accomplished by increased column temperature. Again such an advantage may be offset by the decrease in k which results from increased temperature. In general, low molecular weight nonpolar liquids show greater diffusivity than do high molecular weight polar liquids. Once more however the choice of liquid phase must consider the partition coefficient K as well. Increase in the average carrier gas velocity U when considering the resistance to mass transfer term is seen to produce an increase in the magnitude of the term. Thus the effect on the mass transfer term is opposite to its effect on the mass diffusion term. It is evident therefore that there will occur an optimium gas velocity value, on either side of which decreased column efficiency or increased HETP will occur.

A plot of HETP against carrier gas velocity U for a single solute and temperature, as shown in Fig. 18.4, graphically expresses the simplified van Deemter equation and indicates the comparative contribution of each of the three terms to the HETP. The importance of carrier gas velocity is emphasized by the plot. The effect of A is generally small and independent of U, while B and C are seen to be dependent upon U.

C. RESOLUTION

For any two solute peaks on a gas chromatogram the degree of separation or resolution of the peaks is dependent upon both retention effect and column efficiency. The relative retention between any two solutes on a chromatogram

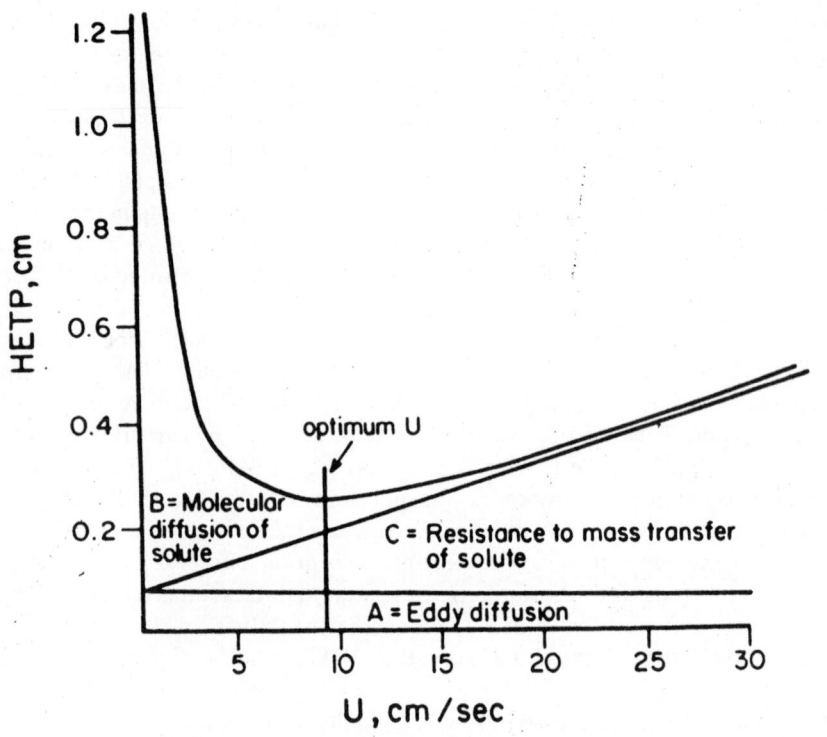

FIGURE 18.4: Plot of HETP against average carrier gas velocity U.

is best given by the *separation factor* α which is the ratio of the retention volumes and thus is a measure of the distances of the apices of the two solute peaks.

$$\alpha = \frac{V_{R_2}}{V_{R_1}} = \frac{K_2}{K_1} \tag{18.17}$$

where V_{R_1} and K_1 are the retention volume and partition coefficient of solute 1 and V_{R_2} and K_2 are the retention volume and partition coefficient of solute 2.

Only if the separation factor differs from unity can the peaks be resolved. The larger the α value is from 1 the better the separation. In many instances however, even through α is significantly larger than 1, the solute peaks may not be resolved completely (Fig. 18.5A). If column efficiency is low, i.e., too few theoretical plates, the solute peaks may still overlap. By modification of parameters to increase column efficiency or by lengthening the column, it may be possible to correct the peak broadening defect (Fig. 18.5B). If resolution is poor however due to a separation factor near 1 it is necessary to change either the column temperature or the liquid phase. By a change in temperature it may be possible, because of differences in the boiling points of the solutes, to alter their K values thus increasing α (Fig. 18.5C). If the temperature change does not alter the separation factor sufficiently, it is then necessary to change the stationary liquid phase to one with a higher selectivity (see p. 673).

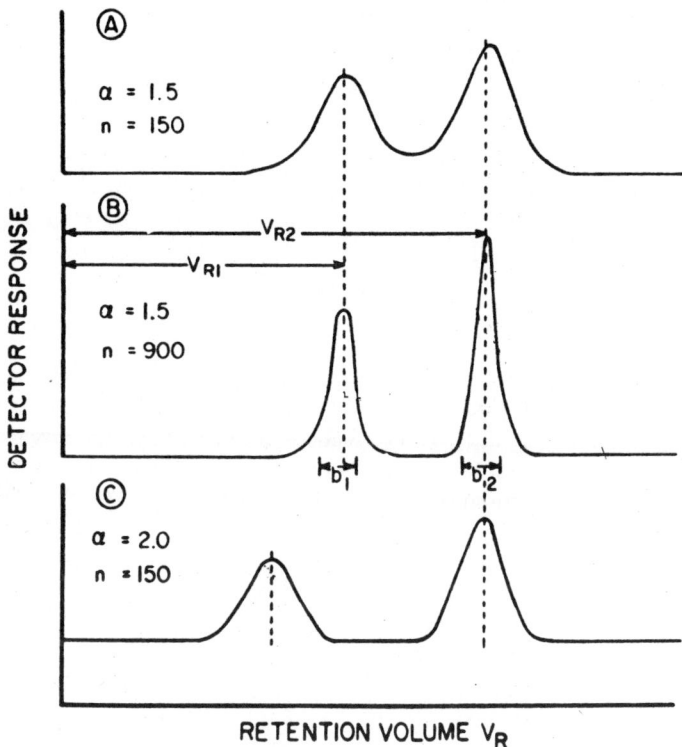

FIGURE 18.5: Schematic diagram of chromatographic separation of two solutes showing effect of retention and column efficiency on resolution. (A) Unresolved solute peaks, (B) peaks resolved by increased column efficiency, (C) peaks resolved by increased separation factor.

An expression of the resolution obtained on a column has been derived based upon peak width and retention volumes.

$$\text{Resolution} = \frac{2(V_{R_2} - V_{R_1})}{b_1 + b_2} \qquad (18.18)$$

Where V_{R_2} is the retention volume from the second peak eluting and V_{R_1} is the retention volume from the first peak eluting. b_1 and b_2 are the peak widths of the first and second peaks, as illustrated in Fig. 18.5B.

The total number of theoretical plates necessary to produce a particular resolution may be determined by expression (18.19)

$$n = \left(2(\text{resolution}) \frac{\alpha + 1}{\alpha - 1}\right)^2 \qquad (18.19)$$

where α = separation factor, and the resolution is calculated as in (18.18). In general excellent resolution should be possible if the separation factor of the two solutes is 1.2 or above.

18.4 COLUMN TECHNOLOGY

In the consideration of column operation, each unit of the column assembly is of importance. The individual unit parts comprising the complete chromatographic column are the column, the support medium, the liquid phase, and the carrier gas.

A. COLUMN

Depending upon the type of oven module used, chromatographic columns may vary in shape and dimensions. The two most common shapes are the coiled helix and the U-tube. Coiled helical columns have the advantage of being the most efficient shape for columns of 8 ft or longer, but suffer from the disadvangage of being more difficult to shape and pack evenly. Generally columns are packed as long tubes and then coiled. The U-tube column, however, is convenient for columns of short length and is relatively easy to pack and shape. The length of a GC column should be limited to that length just sufficient to carry out the desired separation. The shortest possible column reduces diffusion effects and keeps retention time to a minimum. In most cases, columns of 6 to 12 ft and having 600 to 1000 plates/ft give completely satisfactory separations. The tubing selected is usually glass, aluminum, copper, steel, or stainless steel. Nylon and other synthetic plastic tubings have been utilized, but in most instances the temperature requirements for separation prohibit their use. Copper, stainless steel, and aluminum have proven to be most useful for routine work in helical columns. Glass is frequently employed in U-tube columns and for separation of biological materials. Compounds sensitive to catalytic action of metals also are best separated on glass columns. Most commercial chromatographs are designed to accept tubing of diameters from 1/8 through 3/8 in., connections being made with nuts and ferrules or O-rings. Adapters for various size tubings are also readily available from most manufacturers.

B. SUPPORT MEDIUM

The support medium for the liquid phase within the column has two primary requirements: it must be a poor adsorbent and it must be a finely divided porous substance having a large surface area. In addition, other requirements include chemical inertness, heat stability, and sufficient mechanical strength to prevent fracturing with normal handling. In regular GLC the materials which meet these specifications are diatomaceous earths, ground fire brick, glass beads, fluorinated resins, and polyaromatic resins. The most commonly used support mediums are the calcined diatomaceous earths.* The latter are frequently treated with silanizing agents such as dimethyldichlorosilane (DMCS) or hexamethyldisilazane (HMDS), following an acid wash treatment. This treatment reduces surface activity of the support media to negligible levels by deactivating and coating all available active sites. "Tailing" of solute peaks due to adsorption effects is diminished considerably by this treatment. Mesh sizes vary according to column conditions and efficiency required by the particular separation problem. Mesh sizes from 20/30 up to 100/120 are commonly employed.

C. LIQUID PHASE

As was stated earlier, a solute with a small partition coefficient K will travel at a faster rate down a column than a large K. It follows therefore that the solute with the greatest tendency to remain in the gas phase will travel down the column at the fastest rate or, stated in another form, the greater the volatility of a solute the shorter the retention time. For two or more solutes, therefore, comparative retention times or volumes are a matter of relative volatility. As in distillation the *relative volatility* may be given as:

$$\alpha_{1,2} = \frac{a_2^0 P_2^0}{a_1^0 P_1^0} = \frac{K_2}{K_1} \tag{18.20}$$

where $\alpha_{1,2}$ is the relative volatility of solutes 1 and 2 and a_1^0 and a_2^0 is the activity coefficient of the solute in the solvent at infinite dilution. P_1^0 and P_2^0 are vapor pressures of pure solutes 1 and 2.

The terms a_1^0 and a_2^0 are comparable to the deviation factor in Raoult's law and are an expression of the factors other than vapor pressure which influence the solute molecules to leave the solution or liquid phase and migrate to the vapor or mobile phase. These solute-solvent interactions determine the *selectivity* of a liquid phase for different solutes. If a liquid phase has little or no interaction with the solutes from hydrogen-bonding, dipole interactions, or complexation effects, then separation of solutes utilizing such a *nonselective liquid phase* will depend upon boiling point alone. The nonpolar liquids such as squalane, paraffin oils, and silicone oils are nonselective stationary phases on which separation depends almost completely on the differences in boiling points of the solutes. With increasing polarity of the liquid phase increased selectivity results and separation may then depend

* Available as Chromosorbs, Johns-Mansville Products.

upon both selectivity of the liquid phase and the vapor pressures of the solutes. A very selective phase for two solutes is one which will resolve two solutes of different molecular type but with the same boiling points. If $P_1^0 = P_2^0$, then Eq. (18.20) becomes:

$$\alpha_{1,2} = \frac{a_2^0}{a_1^0} = \frac{K_2}{K_1} \tag{18.21}$$

Sample components which have a low polarity will tend to dissolve easily in nonpolar nonselective liquid phases. According to their partial pressures, they will distribute themselves readily between the fixed liquid phase and the gas phase. Because of low a values there is little or no tendency to be held in the liquid phase due to solute-solvent interactions and they will elute from the column in increasing order of their boiling points. For example in the series of saturated hydrocarbons methane, ethane, propane, butane, and isobutane elution from the nonselective liquid phase hexadecane is in the order of increasing boiling points. However, if the sample components and liquid phase have different polarities, the solubilities of the solute in the liquid phase will be much less. This results in higher vapor pressures than would be predicted from the molar concentrations involved. As a result, the solutes will be eluted again in order of their boiling points, but at a faster rate than when both solute and liquid phase are nonpolar. For example the polar halogenated hydrocarbons such as the chloroalkanes will travel through a nonselective hexadecane column faster than will the corresponding nonpolar nonhalogenated alkanes. Similarly if nonpolar alkanes are chromatographed on both a nonpolar liquid phase such as hexadecane and a polar liquid phase such as benzyl ether, the polar liquid phase will show shorter retention times than the nonpolar liquid phase. In both phases elution order will be the same and in increasing order of boiling points.

Selectivity becomes of major importance only when there is significant polarity or other solute-solvent interaction between the liquid and solute phase. Of the different interactions hydrogen-bonding plays a very significant role. Hence separation of polar solutes on a polar stationary phase may be as dependent upon hydrogen-bonding interactions as upon boiling points. In other situations, interactions between permanent or induced dipoles in the solutes with polar groups in the liquid phase may be the critical factor in effective separations on selective columns. In still different applications, weak complexation interactions may be utilized to influence the selectivity of the liquid phase and hence to improve separation of components, e.g., alkenes on silver-treated columns.

In selecting a stationary phase from the hundreds of liquids available there are various practical limitations which must be considered. Of prime importance is the upper temperature limit of the liquid. This limit is governed both by the liquid's volatility and by its stability. At high operating temperatures *column bleed* may occur. Column bleed is indicated by a rising base line with rise in temperature and indicates either a detectable volatility or partial decomposition of the liquid at the temperature involved. Another practical factor is that of purity of the liquid. Contaminants within the liquid phase frequently produce extraneous peaks or give an uneven base line. Solubility of the liquid phase in various solvents also must be considered when coating the support medium. Heavy viscous liquid phases may have too low a solubility in the solvent chosen and may prevent a sufficiently high percentage loading of liquid phase onto the column.

TABLE 18.1: Commonly Used Stationary Phases

Stationary phase	Common name	Relative polarity[a]	Maximum operating temperature, °C	Solvent[b]
Squalane	—	NP	125	1 or 2
High vacuum grease	Apiezon L	NP	300	1 or 2
Methyl silicone rubber gum	SE 30	NP	375	1 or 2 (hot)
Fluoro silicone rubber gum	QF-1 or FS-1265	IP	250	1 or 2
Silicone oil	DC-550	IP	275	1 or 2
Phenyl methyl silicone rubber gum	SE 52	IP	300	1 or 2 (hot)
Diethyleneglycol succinate polyester	DEGS	P	190	
Polypropyleneglycol	Ucon 50 HB-2000	P	200	1 or 2
Polyethylene glycol	Carbowax 20M	P	225	1 or 2
Butenediol succinate polyester	—	P	225	1 or 2

[a] NP, nonpolar; IP, intermediate polarity; P, polar.
[b] 1, methylene chloride; 2, chloroform.

While hundreds of stationary phases are available, the majority of GLC separations employ only a relatively few liquids. Table 18.1 lists ten of the most frequently utilized stationary phases. In addition it gives the relative polarity, the maximum operating temperature, and the suggested solvents for preparing the columns. It will be noticed that the common name is frequently a trade name or an abbreviation. The reader is referred to various GLC supply catalogues for a complete listing of stationary phases.

D. CARRIER GAS

The most commonly utilized carrier gases in GLC are nitrogen, helium, hydrogen, and argon. Special carrier gases such as propane and acetylene have been utilized for special separations, but are not generally employed.

From a theoretical standpoint the type of carrier gas is the least critical parameter in column operation. All of the common carrier gases are insoluble in the stationary phase and have little influence upon the selectivity of the liquid phase. As mentioned in the discussion of the van Deemter equation [Eq. (18.15)], the value of D_g or carrier gas diffusivity is decreased by increasing molecular weight of the carrier gas. Because of this fact, the more dense the carrier gas the less the peak broadening due to molecular diffusion. If possible therefore the most dense carrier gas is used. The most important criteria for choice of carrier gas is the type of detector to be used. With the thermal conductivity detector the lightest gases, hydrogen and helium, are utilized. The thermal conductivity of a gas is inversely proportional to the square root of the molecular weight. Because the sensitivity of the thermal conductivity is dependent upon the difference in thermal conductivity between the carrier gas and the more dense solute vapors, the least dense carrier gas produces maximum detector sensitivity. With argon beta-ionization detectors, argon is employed as carrier gas since operation of the detector depends upon the presence of argon. Similarly the carrier gas nitrogen or an argon-methane mixture is utilized as the source of slow electrons in the electron capture detector and is thus required as the carrier gas. With the flame ionization detector any nonorganic carrier gas may be employed, and because of the molecular diffusion effect the more dense nitrogen is generally preferred.

E.   PREPARATION OF THE CHROMATOGRAPHIC COLUMN

In the preparation of a chromatographic column it is first necessary to prepare the stationary phase by coating the liquid phase over the stationary support medium. If several identical columns will be needed, it is advisable to prepare a sufficient quantity of stationary phase at one time since column characteristics may change slightly from batch to batch. From 50 to 500 g may be prepared by the following procedure:

a. Weigh out the required weight of solid support media of the correct mesh sizes and place in a rotary evaporation flask.

b. Calculate and weigh the amount of liquid phase to give the correct liquid loading. Liquid loading on the support media may vary from 1 to 35%, depending upon the separation desired.

c. Dissolve the liquid phase in a sufficient volume of solvent (see Table 18.1) to just wet the solid support media.

d. Add slowly the dissolved liquid phase to the support media in the flask with stirring until an even slurry is formed. Stir until thoroughly mixed.

e. Attach the flask to a rotary evaporator and allow all solvent to evaporate while rotating under vacuum. Continue rotation until an even coating of liquid phase is insured.

f. Select a suitable length of column tubing and plug one end with glass wool. With a funnel add the stationary phase to the open end. Evenly packed columns are best prepared by use of an electric vibrator along the column or by constant tapping during phase addition. When the column is completely packed, the open end is plugged with glass wool. For preparation of helical columns the tube may be packed straight and then coiled around a suitable object or precoiled columns may be packed with the aid of vibration and a vacuum applied at the bottom end. U-tubes are packed by filling from each end toward the center with vibration.

g. The column is now fastened into the chromatograph and conditioned by purging with carrier gas. Purging is continued for several hours at a temperature about 20°C above maximum operating temperature. A well-prepared conditional column will maintain a constant zero base line on the chromatogram with a minimum of attenuation of detector signal output.

18.5 DETECTORS

The detector may be considered to be the brain center of the gas chromatograph. The impulse received from the eluate of the column in the form of the solute vapor is "sensed" by the detector. It in turn converts this impulse into an electrical signal proportional to the concentration of the solute in the carrier gas. This signal is then amplified and recorded as a peak on the chromatograph.

A good detector should have several important characteristics: stability against the effects of extraneous noise in the detector system; a rapid and linear response to changes in solute vapor concentration as the column effluent passes through the detector; reproducibility of response under designated operating conditions; sensitivity to a wide range of solute vapors.

Many different types of detectors have been produced. Some are extremely versatile, while others have very limited application. While it is beyond the scope of this chapter to deal with all detectors in detail, a brief discussion will be carried out on the four most commonly used detectors.

A. THERMAL CONDUCTIVITY DETECTORS

Thermal conductivity detector—katharometer—is the most commonly used detector and comes closest to fulfilling the basic characteristics of the ideal detector. The detector is simple, applicable for a broad range of solutes, is stable, does not destroy the sample, and is operable over a wide range of temperatures. It is however less sensitive than the other detectors, requiring in the range of 10^{-6} to 10^{-2} g of samples.

The operation of the thermal conductivity detector is based upon differences which occur in the thermal conductivity of the effluent gases from the chromatographic column. The carrier gas has a constant thermal conductivity and is used as a reference. The thermal conductivity of the binary mixture of solute vapor and carrier gas varies proportionally to the concentration of solute vapor present. Because of these changes in the thermal conductivity of the gas mixture as it flows over the heated filament in the detector, there is produced a similar proportional change in resistance within the heated filament. Due to circuit design these resistance changes cause a similar proportional voltage drop which is then recorded on the chromatogram.

B. β-RAY IONIZATION DETECTORS

β-Ray ionization detectors—argon detectors—utilize a radioactive source of radium 226, strontium 90, or tritium as a source of β rays and employ argon as the carrier gas. Argon is excited to a metastable state by the beta ionization of the radioactive source. On the entrance of the solute vapor molecules to the detector the metastable argon reacts with the solute vapor molecules, producing sufficient solute ions to alter the standardized background current. The change in current is proportional to the amount of solute molecules present.

The principle advantage of the beta ionization detector is the increased sensitivity over the thermal conductivity detector. It also is considerably less sensitive to temperature changes than is the thermal conductivity detector. The argon detector has the disadvantage of requiring a radioactive source and care must be taken not to overheat the detector. Further disadvantages lie in the limited sample size that may be used, i.e., 10^{-7} g, and its insensitivity to certain compounds.

C. FLAME IONIZATION DETECTORS

The flame ionization detector has largely replaced the beta ionization detectors because of its simple construction and its linearity over a dynamic range of 1 million. It requires no radioactive source and can utilize inexpensive carrier gases such as nitrogen. This detector is insensitive to temperature change and is relatively insensitive to fixed gases such as CO_2, H_2O, and H_2S. Because of its insensitivity to temperature change it is ideally suited for temperature programming. The flame ionization detector is exceptionally versatile since it is sensitive to all organic compounds in the range of 10^{-3} to 10^{-9} g. Its chief disadvantage lies in the fact that the sample is destroyed. In principle the flame detector is simple. A hydrogen flame in the detector burns the solute molecules as they enter the detector from the column. During the burning process electrons and ions are formed and are collected at an anode. The electrical current produced is directly proportional to the amount of solute molecules consumed.

D. ELECTRON CAPTURE DETECTORS

This detector has become of considerable importance because of its very high sensitivity to certain molecules. It is especially recommended for alkylhalides, conjugated carbonyls, nitriles, nitrates, and organometallics. Its selective sensitivity to halides makes it especially useful for insecticide analysis. In certain instances, quantities down to the picogram level (10^{-12} g) have been detected. In contrast it is insensitive to a large number of other organic compounds such as hydrocarbons, amines, aldehydes, and ketones. Other disadvantages of the detector are its detector temperature limitation (220°C) and requirement for a very pure nitrogen or an argon-methane mixture for a carrier gas.

The electron capture detector, unlike the other detectors discussed, measures a loss of signal rather than an increase in current. A tritium source is used as an ionizing source and in contact with the carrier gas it ionizes the molecules producing electrons. These electrons migrate to a positively charged anode and produce a current which is amplified. When a solute sample composed of electron-absorbing molecules such as alkylhalides enters the detector, some of the electrons are absorbed onto the solute molecules producing a decrease in current proportional to the number of solute molecules present. This decrease in current is recorded on the chromatogram.

18.6 ANALYTICAL DETERMINATION BY GAS CHROMATOGRAPHY

A. QUALITATIVE ANALYSIS

Gas-liquid chromatography is an extremely useful tool for qualitative analysis, being of special value in the identification and comparison of compounds of closely related structure. Its versatility in resolving solute mixtures and simultaneously aiding in their identification has been responsible for much of its popularity. It must be remembered however that GLC data should not be used to provide positive identification by itself in the absence of other supporting information.

Most qualitative determinations are accomplished by comparison of the chromatographic characteristics of the unknown sample with those of known standards. Since the retention of a solute on a chromatographic column is a physical constant when determined under defined conditions, the comparison of corrected retention volumes $V_{R_1}^{\circ}$, net retention volumes V_N, or relative retention values $r_{1,2}$ may be used for identification purposes. Of these values probably the relative retention value $r_{1,2}$ is most convenient since, when the values are determined under identical conditions, the effects of operating conditions are eliminated as discussed on p. 665. If the relative retention value of an unknown solute and a known standard are found to be identical for one column, it is evidence that the compounds are the same.

The occurrence of identical relative retention values for both known and unknown solutes on a second column with a different polarity greatly strengthen this view. With sufficient prior information from other methods a reasonably accurate identification of the unknown may be made by this method.

An alternative method for identification of an unknown compound is to add a known standard compound to an aliquot of the solution of the unknown. When previous evidence has indicated that the unknown is the same compound as the standard chosen, both the original unknown solution and the "spiked" aliquot are chromatographed under identical conditions and the resulting chromatographic peaks are compared. If the unknown compound is different from the standard, two resolved or partially resolved peaks with different retention values are seen in the chromatogram of the "spiked" sample. If both the unknown and standard compounds are identical, a single but larger peak than that occurring with the unknown compound will be found on the chromatogram from the "spiked" sample. Again, a repeat run on a column with different polarity should add further confirmation as to the unknown's identity.

A graphical method of identification of compounds by GLC is that of homologous series plots. In this method, a solute mixture containing several members of a homologous series is chromatographed under specific operating conditions. A plot is made of the logarithm of either the net retention volume or of the relative retention value against the number of carbon atoms. Since the logarithm of either the net retention volume or the specific retention value of a homologous series is proportional to the number of carbon atoms—or other property showing a stepwise increase with carbon content— a straight line is produced. The slope of the line is dependent upon the nature of the stationary phase. Identification of an unknown compound belonging to this homologous series is possible by repeating the experiment with the unknown compound and identifying it from its position on the standard curve. A more positive method of identification than that just described is by the method of two-column homologous series plots. By this procedure several members of two or more different homologous series are each chromatographed on two columns of different polarities. Relative retention times are determined against a known standard of that chemical class for each series. Log-log plots for each series are then made of relative retention times for one column against relative retention time for the second column. Straight-line plots with differing slopes should result for each series. The identification of an individual unknown of one of the series may be made by following the same chromatographic procedure and determining on which straight line the relative retention value will fall. The chemical class of the compound is determined by noting on which homologous series plot it falls, and the number of carbons it contains is determined from its position on the line.

When used in combination with other qualitative methods, the scope of GC is increased still further. By the use of sample collection attachments or direct attachment of the effluent stream to a second instrument it is possible to analyze resolved compounds by further instrumental procedures. By such methods separated components may be examined by infrared, ultra-violet, or NMR spectroscopy. Radioactive effluents may be studied or mass spectrometry may be used by direct attachment of the column by a gas-flow counter to the ion beam of a mass spectrometer.

B. QUANTITATIVE ANALYSIS

Several inherent advantages are possessed by gas chromatography when applied to quantitative analysis. Its ability to simultaneously separate several constituents in one run, its applicability for microgram quantities of sample, its speed and its accuracy make it an exceptional analytical procedure.

In quantitative determinations it is necessary to measure the peak area or the peak height of each compound of interest on the chromatogram. The values so obtained are then correlated with the quantity of solute required to produce each peak.

Determination of peak height is a less accurate method of measurement than is peak area determination and usually is satisfactory only for narrow peaks which elute early in the chromatographic run. More accurate deter-mination is by peak area calculations. In this procedure the area under the peak graphically represents a measure of the solute volume or weight which elutes from the column into the detector. Peak area may be determined by one of several methods:

a. Height times width of half height. The area under a chromatographic peak is calculated in this procedure by multiplying the peak height h by the width of the curve w at one-half the height h (see Fig. 18.3). The calculation gives good results with symmetrical peaks, but errors arise with unsymmetrical curves.

b. Planimeter determination. The use of a planimeter for calculating the peak area is accurate if carefully done. Accuracy varies between operators and errors rise as peak size diminishes.

c. Weighing of the paper cut from the chromatogram. With careful cutting of the peak from the chromatogram the weight of the paper may be obtained. From a determination of weight per unit area of paper the area under the curve may be calculated. The method is tedious and demands careful work.

d. Determination of peak area by integrators. A number of integrating devices have been manufactured which attach directly to the recorder and integrate the area as the chromatogram is recorded. These devices are more convenient and accurate than the other methods described. Electronic integrators giving direct printout of detector response are employed by some laboratories and offer the maximum accuracy in quantitative determination.

Correlation of peak area and concentration of solute is dependent not only upon the size of sample injected but also to a considerable extent upon the linearity of response of the detector over the concentration range of sample being examined. Factors which influence detector operation and sensitivity also influence detector response and therefore peak area. Such factors as detector and column temperature, gas flow rate, anode voltage, and current flow in the detector must be considered and compensated for if necessary.

Different methods of quantitative determination are utilized to correlate peak area accurately to solute concentration. Three of these methods are given here.

1. Quantitative Determination by Internal Standardization

The most accurate method for quantitative determination of a known solute is the method by which a known amount of a standard sample is added to the solute being examined and the peak area produced by the standard is compared with that of the solute being determined. The internal standard is a compound preferably of the same chemical type, but with different structure than the solute under examination. The standard should have a retention time close to, but completely resolved from, the unknown. The peak area of the standard should approximate that of the quantitative unknown. By the use of an internal standard most of the apparatus variables and differences in operating parameters are cancelled out. In addition, by this method the sample size for injection is not critical so long as it is within the response range of the detector.

Measurement by internal standardization is accomplished by first determining the *sensitivity* or *response factor* of the detector for the quantitative unknown relative to the internal standard. An accurately weighed sample of the internal standard is added to an accurately weighed reference sample of the pure compound being assayed. Accurate dilution with solvent is carried out if necessary and the solution is chromatographed. The response factor is calculated from Eq. (18.22):

$$F_s = \frac{A_s W_u}{W_s A_u} \tag{18.22}$$

where F_s = sensitivity or response factor .
   A_s = peak area of internal standard
   W_s = weight of internal standard added to sample
   A_u = peak area of reference sample being analyzed
   W_u = weight of reference sample being analyzed

Quantitative determination of the unknown sample is then accomplished by adding an accurately weighed amount of internal standard to a known weight of sample being examined. The mixture is then run under the identical chromatographic conditions used previously. The volume of the injected mixture does not have to be accurately known.

The calculation of the per cent weight of unknown is determined by the equation:

$$\text{weight per cent of unknown} = \frac{F_s A_u W_s \times 100}{A_s W_x} \qquad (18.23)$$

where W_x is the weight of sample being assayed. When repeated samples are being run, F_s should be determined at regular intervals and at least once a day.

2. Quantitative Determination by Area Normalization

The area normalization method of quantitization is at best a semiquantitative procedure. It is based upon the assumption that equal weights of all substances contained in the sample produce equal responses from the detector and that the response is linear over the range of weights being examined. Under these conditions the summation of all peak areas equals 100 per cent and the per cent concentration of any one substance equals the ratio of that peak area to the sum of the total peak areas multiplied by 100. Therefore:

$$\text{weight per cent of unknown} = \left[\frac{A_u}{(\sum A_u + \cdots A_n)}\right] \times 100 \quad (18.24)$$

where A_u is the peak area of substance being determined. This method has the advantage of not requiring an accurate sample size, but it gives rise to errors due to different response factors for different compounds.

3. Quantitative Determination by External Standardization

External standardization is based upon the comparison of the peak areas of chromatograms prepared from known standard concentrations of the compound under assay and those of the unknown concentration. Separate chromatographic runs are made for each known concentration and these peak areas are used as standards. Calibration curves may be prepared or direct calculation may be made with the unknown sample. The method may be accurate to 1 % providing very careful control of all chromatographic parameters is maintained. Best results are obtained when both standards and unknowns are chromatographed during a single time period of constant operating conditions.

The disadvantages of the external standardization procedure are related to errors introduced because of changes in operating conditions, instrument variation, and errors in measurement of the sample. Each type of detector varies with respect to the effect of operating conditions on signal output sensitivity. Thermal conductivity detectors for example show changes in sensitivity with changes in gas flow rate, detector temperature, and filament current. During a single operating period, errors from these sources may be minimized, but if the instrument is used at varying intervals, restandardization is usually necessary each time. Probably the greatest single error involved in the external standardization procedure is due to the measurement of injection volumes. Since sample volumes are frequently in the range of fractions of microliters, small errors in estimating sample volumes introduces sizeable errors in results.

It may also be pointed out that when more than one sample component is being measured on a single chromatogram it is necessary to prepare standard calibration curves for each component individually. This is required because of the possible differences in sensitivity ·or response to each component by the detector and is comparable to determining the response factor as mentioned previously.

18.7 EXPERIMENTS

In analysis with gas chromatography it is imperative that the operator be thoroughly familiar with the instrument and its operation. The following experiments are designed to accomplish this purpose as well as to illustrate principles discussed in the text.

A. QUALITATIVE DETERMINATIONS OF n-HYDROCARBONS

Column Operation Data: column, 4 ft × ¼ in. OD, 10% SE30 on 60/80 mesh Chromosorb W;* temperature, isothermal operation: 180°C column, 300°C detector, 300°C injection port; detector, thermal conductivity or flame ionization; carrier gas, helium, 50 ml/min.

I. Identification from Relative Retention Values

Set up the chromatograph for isothermal operation with the operating conditions given previously. Prepare 100-μl standard samples of the following binary mixtures, using 50 μl of each standard hydrocarbon:† octane-dodecane; decane-dodecane; tetradecane-dodecane. For each standard sample prepare a chromatograph by injecting 0.5 μl of the sample into the injection port with a microsyringe. At the instant of injection move the recorder pen to mark the point of injection. If necessary adjust the signal attenuation to give satisfactory peak heights. From each chromatogram

---

*Johns-Manville Products, New York, N.Y.
† Available in Polyscience Kit 82-00500-800 Qualitative Kit 210 Varian Aerograph, Walnut Creek, California.

obtained measure the retention times for each of the two hydrocarbon peaks. Retention time is calculated by dividing the chart speed into the distance the chart has traveled from the point of injection to the center of the hydrocarbon peak. See Fig. 18.3. Determine the relative retention values, $r_{1,2}$ for octane, decane, and tetradecane, using dodecane as the internal standard. Since the retention time t_R is proportional to the retention volume under the conditions employed, Eq. (18.9) becomes:

$$r_{1,2} = \frac{V_{R_1}}{V_{R_2}} = \frac{t_{R_1}}{t_{R_2}} \tag{18.25}$$

where t_{R_1} and V_{R_1} are the retention time and volume of dodecane and t_{R_2} and V_{R_2} are the retention time and volume of the second hydrocarbon.

To 50 μl of unknown samples provided add 50 μl of dodecane and mix. Each unknown will contain one or more of the C_8, C_{10}, or C_{14} hydrocarbons. Repeat the chromatographic procedure described previously with the unknowns. Calculate the relative retention values for each unknown peak and compare with those of the standards. From the data obtained determine the hydrocarbons in each unknown.

2. Identification from Homologous Series Plots

From the relative retention data obtained previously plot on semilog paper the relative retention times of the standard n-hydrocarbons on the semilog scale against carbon number. A straight-line plot should result. Identification of each unknown may then be determined by plotting its position on the curve and locating the n-hydrocarbon it falls closest to. By extrapolation of the line, larger or smaller carbon-containing n-hydrocarbons may be determined.

B.   DETERMINATION OF COLUMN EFFICIENCY FOR DECANE AND TETRADECANE

Maintain the same column conditions as in part A. Inject 0.3 μl of standard decane solution, marking the chromatogram at the instant of injection. It is important that the injection of solute be made rapidly and smoothly so that the application of solute to the column is as nearly instantaneous as possible. When the chromatogram is completed, make a second chromatogram with 0.3 μl of tetradecane.

From each chromatogram determine the number of theoretical plates on the column for that hydrocarbon as follows: for each hydrocarbon peak measure the retention time t_R. Determine the base width b for both the hydrocarbon peaks by drawing tangents through the inflection points of the curves and extending the tangents to form a triangle with the base lines as shown in Fig. 18.3.

Calculate the number of theoretical plates from Eq. (18.13):

$$n = \frac{16(t_R)^2}{b} \qquad (18.13)$$

Compare the number of theoretical plates for both hydrocarbons. A good column should have from 300 to 600 HETP's per foot of column. From the data obtained previously calculate the HETP for each hydrocarbon, where HETP $= l/n$ (18.14).

C. SEPARATION AND IDENTIFICATION OF n-HYDROCARBONS BY TEMPERATURE PROGRAMMED OPERATION

Except for the column temperature conditions, the operating parameters are those given in Section 18.7A. Set program controls at an initial column temperature of 100°C and the upper limit temperature control at 300°C. The rate of temperature rise is set at 15°C/min. Prepare a sample of n-hydrocarbon containing equal volumes each of octane, decane, dodecane, and tetradecane. Inject 1.0 μl of the sample and start programmed operation. Determine the elution time for each peak. Repeat the procedure with 0.5 μl of each of the binary standard hydrocarbons in Section 18.7A. Determine the elution temperature for each standard component. From the results obtained previously identify each peak in the original mixture.

D. DETERMINATION OF ALCOHOL CONTENT OF NITROMERSOL TINCTURE

Column Operation Data: Column, 6 ft × ¼ in. OD, 20% polyethylene glycol 400 on 60/80 mesh Chromosorb W. Temperature, isothermal operation, 100°C column; 120°C detector; 160°C injection port. Detector, thermal conductivity preferred. Carrier Gas, helium 60 ml/min.

Pipette exactly 10 ml of nitromersol tincture into a stopped flask and add exactly 4.0 ml of methylethyl ketone (MEK) as an internal standard. Prepare about 50 ml of 50% alcohol solution and accurately determine the alcohol content.* Pipette triplicate samples of exactly 10 ml each of the known alcohol solution into stoppered flasks and add exactly 4.0 ml of MEK to each as an internal standard. Utilizing identical column operating conditions, make individual chromatographic runs on each known alcohol sample and on the nitromersol solution. Each injection should be sufficiently large—1 to 5 μl—to enable an accurate measurement of peak height. Measure all peak areas accurately in square centimeters by integrator or by

---

* The alcohol content is found by determination of the specific gravity at 25°C and referring to alcoholmetric tables.

the method of peak height times width at half-height. Calculate the sensitivity or response factor F_s of the detector for each of the 10-ml samples of known alcohol solution from the equation:

$$F_s = \frac{A_s W_u}{W_s A_u}$$

where A_s = peak area of MEK internal standard
   W_s = volume of MEK in known alcohol
   A_u = peak area of alcohol in known alcohol sample
   W_u = volume of alcohol in known alcohol sample, i.e., $10/100 \times$ alcohol per cent

From the three values of F_s obtained find the average F_s value. For accurate results F_s values must be determined each time the instrument is started up.

With the average F_s value obtained calculate the per cent alcohol in the nitromersol tincture by the following equation.

$$\frac{F_s A_u W_s}{W_x A_s} \times 100$$

where W_x is the volume of nitromersol solution used in sample.

QUESTIONS

Q18.1. Two types of detectors have been utilized in GC, the differential and the integral detector. By referring to other literature sources discuss the basic principle upon which each type is based. Make a sketch of a typical integral and differential chromatogram.

Q18.2. Capillary columns are frequently employed in the petroleum industry. They consist often of several hundred feet of capillary tubing coated on the inside with liquid phase. Discuss the relative merits and disadvantages of this type of column.

Q18.3. By consideration of the van Deemter equation discuss the effect on column efficiency produced by: increasing the percentage of the liquid coating on the support medium, changing the carrier gas from nitrogen to hydrogen and increasing the carrier gas velocity when it is already at a high flow rate.

Q18.4. Assuming that a halogenated hydrocarbon RCl and its parent hydrocarbon R have nearly the same boiling point and will chromatograph under the conditions employed, which compound would you expect to be eluted first from a squalane column? From a DEGS column? Explain.

Q18.5. Discuss the derivation of Eq. (18.23).

$$\text{Weight per cent of unknown} = \frac{F_s A_u W_s \times 100}{A_s W_x}$$

REFERENCES

1. M. Tswett, *Ber Deut. Botan. Ges.*, **24**, 316 (1906).
2. A. J. P. Martin and R. L. M. Synge, *Biochem. J.*, **35**, 1358 (1941).

3. S. Claesson, *Arkiv Kemi Mineral. Geol.*, **A23**, 133 (1946).
4. S. Claesson, *Arkiv Kemi Mineral. Geol.*, **A24**, 7 (1946).
5. W. C. Turner, *Natl. Petrol. News*, **35**, 243 (1943).
6. C. S. G. Philips, *Discussions Faraday Soc.*, **7**, 241 (1949).
7. A. T. James and A. J. P. Martin, *Biochem. J.*, **50**, 679 (1952).
8. A. T. James and A. J. P. Martin, *Analyst*, **77**, 915 (1952).
9. J. Janak, *Chem. Listy.*, **47**, 464 (1953).
10. N. H. Ray, *J. Appl. Chem.*, **4**, 21 (1954).
11. B. W. Bradford, D. Harvey, and D. E. Chalkley, *J. Inst. Petrol.* **41**, 80 (1955).
12. A. I. M. Keulmans, *Gas Chromatography*, Reinhold, New York, 2nd ed., 1953.
13. S. Dal Nogare and R. S. Juvet, Jr., *Gas-Liquid Chromatography*, Wiley (Interscience), New York, 1962.
14. J. J. van Deemter, F. J. Zuiderweg, and A. Klinkenberg, *Chem. Eng. Sci.*, **5**, 271 (1956).

GENERAL REFERENCES

Bayer, E., *Gas Chromatography*, Elsevier Monographs, Van Nostrand, Princeton, N.J., 1961.

Brochmann-Hanssen, E., *J. Pharm. Sci.*, **51**, 1017 (1962).

Burchfield, H. P., and E. E. Storrs, *Biochemical Applications of Gas Chromatography*, Academic Press, New York, 1962.

Dal Nogare, S., and R. S. Juvet, Jr., *Gas-Liquid Chromatography*, Wiley (Interscience), New York, 1962.

Johns, T., *Beckman Gas Chromatography Applications Manual, Bulletin 756-A*. Beckman Instruments Inc., Fullerton, Calif., 1964.

Knox, J. H., *Gas Chromatography*, Methuens Monographs, Wiley, New York, 1962.

Purnell, H., *Gas Chromatography*, Wiley, New York, 1962.

CHAPTER **19**

Radiochemical Techniques

A. Noujaïm

FACULTY OF PHARMACY
UNIVERSITY OF ALBERTA
EDMONTON, ALBERTA, CANADA

19.1 INTRODUCTION

Scientists have always searched for a method of analysis which is accurate, specific, precise, economical, and easily adaptable to various laboratories.

Radiotracer techniques, as an analytical tool, come close to satisfying such criteria. These methods are particularly convenient for the determination of a variety of elements and compounds which cannot be estimated by standard analytical procedures. The principal limitation of these techniques is the health hazard involved in laboratories where great amounts of radioactive substances are used.

The basic principle of radioisotope methods is that a radionuclide, when mixed with the stable form of the isotope, can be characterized by radiation detection equipment. Thus, qualitative and quantitative inferences can be made regarding the elemental composition of the compound to be analyzed.

19.2 BASIC NUCLEAR PROPERTIES

Ionizing radiation, which includes γ rays, neutrons, β and α particles, is a characteristic phenomenon of unstable nuclides, distinguishing them from the others. Such various kinds of radiation result from a series of processes

TABLE 19.1: Nuclear Particles and Their Properties of Interest

Name	Symbol	Relative mass units[a]	Charge
Electron	e^-, β^-	5.4388×10^{-4}	-1
Positron	e^+, β^+	5.4388×10^{-4}	$+1$
Neutron	n	1	0
Proton	p	0.9986	$+1$
Deuteron	d	1.9980	$+1$
Triton	t	2.9969	$+1$
Alpha	α	3.9948	$+2$
Photon	γ	0	0

[a] 1 mass unit $= 1.6747 \times 10^{-27}$ kg.

which take place within the nucleus of the atom. The orbiting electrons are, in certain instances, also involved in these processes. Radioisotopes can thus be defined as elements in which the nuclei of the atoms contain either more or fewer neutrons than are present in the naturally occurring stable isotopes of such elements. Such unstable nuclei tend, in time, to change into stable configurations by the various processes collectively known as radioactive decay. The emission of a charged particle (β^-, β^+, α, etc.) produces a change in the electric charge of the nucleus. The product nucleus is chemically a different element which has a lower energy content than the parent species.

Experimental work has shown that the most common radiations have properties similar to those outlined in Table 19.1. The difference in energy is distributed between the radiations which are emitted. Invariably, the emission of β^- and β^+ particles is accompanied by uncharged nucleons of very small mass called neutrinos which carry part of the energy. Neutrinos interact only to a negligible extent with matter and play no part in the useful application of isotopes. The transformations observed when these particles are emitted may be understood by considering specific examples. One of the isotopes of uranium contains in its nucleus 92 protons and 146 neutrons and hence is called uranium-238. To reach stability the above nucleus may spontaneously emit an α particle (i.e., a helium nucleus) and become an isotope of thorium according to the following reaction:

$$^{238}_{92}U \rightarrow \, ^{234}_{90}Th + \, ^{4}_{2}He$$

The ^{234}Th which is formed in the above disintegration is itself radioactive and decays by the emission of a β^- particle. Because electrons do not exist in the nucleus as such, it must be assumed that they are created at the instant of emission. The net result is that a parent neutron in the nucleus is converted to a proton and an electron which is ejected.

In some instances of radioactive decay (α, β^-, or β^+ decay) the emergent particles may not have their full energy. Consequently, the resultant nucleus remains in an excited state. The energy of excitation is eventually released in a form of electromagnetic radiation commonly known as γ ray or X-ray. The emission of such radiation involves a simple rearrangement of the various nucleons and does not cause a change in the number of protons or neutrons.

A. RADIOACTIVE DECAY

Once a radioisotope is formed, it may decay at any time thereafter. The rate of decay has been found to be independent of the experimental conditions. Changes in temperature, pressure, humidity, and even the chemical state of the substance have been shown not to affect the decay rate. However, there is no way of knowing when any given radioactive nucleus will disintegrate. It was demonstrated experimentally and theoretically that radioactive decays are random in occurrence and can be described in terms of probabilities. This probability of decay per unit time is called the decay constant, λ. If we suppose that there were N atoms at time t, then the average number that decay in a time interval dt is

$$-dN = \lambda N \, dt \qquad (19.1)$$

or

$$\frac{dN}{N} = -\lambda \, dt \qquad (19.2)$$

Integration of (19.2) from $t = 0$ to t, yields

$$\ln \frac{N}{N_0} = -\lambda t \qquad (19.3)$$

$$N = N_0 \exp(-\lambda t) \qquad (19.4)$$

where N is the number of radioactive atoms not yet decayed at the time t and N_0 is the number of radioactive atoms in the sample at time $t = 0$. The time required for the number of atoms to diminish to one-half the original number is called the physical half-life, T, of the particular radioisotope. But since

$$\ln \frac{N}{N_0} = -\lambda t \tag{19.5}$$

and since where $N = (\frac{1}{2}) N_0$, $t = T$,

$$\ln(\tfrac{1}{2}) = -\lambda T \tag{19.6}$$

$$T = \frac{0.693}{\lambda} \tag{19.7}$$

By substituting from Eq. (19.4) into Eq. (19.2), one obtains

$$\frac{dN}{dt} = -\lambda N_0 \exp(-\lambda t) \tag{19.8}$$

Using common logarithms, Eq. (19.4) may be written as

$$\log N = \log N_0 - \frac{\lambda t}{2.303} \tag{19.9}$$

and if one sets $y = \log N$, $t = x$, and $\log N_0 = $ constant b, Eq. (19.9) follows the standard slope intercept form of a straight line:

$$y = mx + b \tag{19.10}$$

where $m = \lambda = $ the slope of the line. The half-life of a particular isotope may thus be very useful for identification purposes.

If the radiations from two different isotopes are detected simultaneously, then the total activity observed at any time is equivalent to the sum of the individual activities. The plot of the observed activity on semilog paper will not result in a straight-line relationship. However, it is possible to decompose the curve into two straight lines after the almost complete decay of the short-lived component. In Fig. 19.1 the total activity from two radioisotopes is plotted as a function of time. The decay relationship of the shorter-lived isotope was deduced after subtracting the activity of the longer-lived component from the total activity after extrapolation to zero time. Another useful concept is that of average life, τ, of all the atoms of a particular radionuclide. The actual life of any particular atom can be any value between zero and infinity. The average life of a large number of atoms is, however, a definite and important parameter. From Eq. (19.1) the life span of dN atoms is evidently t. Therefore, τ can be expressed as follows:

$$\tau = \frac{\displaystyle\int_{N=0}^{N=N_0} t \, dN}{\displaystyle\int_{N=0}^{N=N_0} dN} \tag{19.11}$$

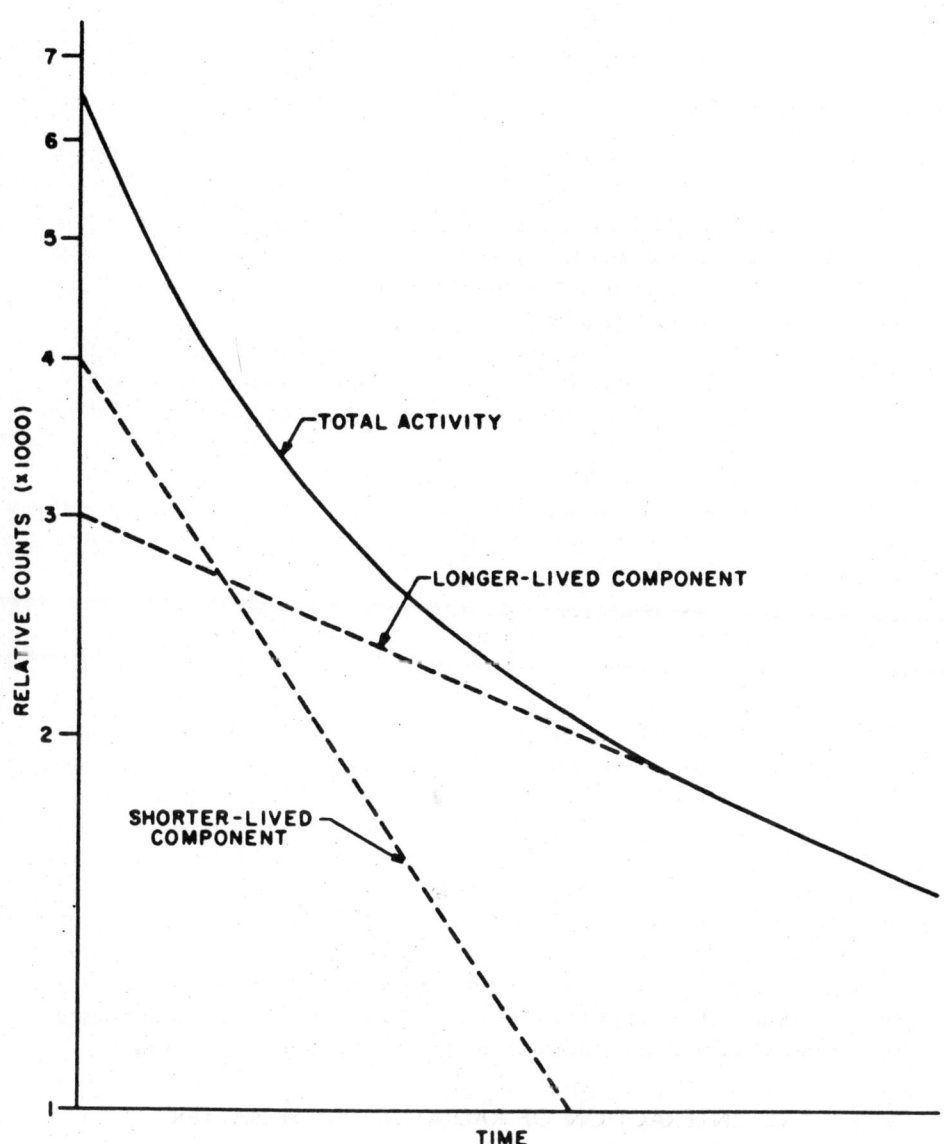

FIGURE 19.1: Decay curve of a mixture of two radioisotopes.

From Eq. (19.8), Eq. (19.11) can be rewritten

$$\tau = \frac{1}{N_0} \int_{t=0}^{t=\infty} \lambda N_0 \exp(-\lambda t)t \, dt \, . \tag{19.12}$$

Integration of the above equation gives

$$\tau = \frac{1}{\lambda} = \frac{T}{0.693} = 1.443T \tag{19.13}$$

The average life finds useful application in calculations involving the total number of particles emitted by a certain quantity of radioactive material. For example, the total disintegrations of 1 mCi of ^{24}Na ($T = 14.97$ hr) during complete decay would be 3.7×10^7 disintegrations per second (equal to 1 mCi) multiplied by the average life of a radiosodium atom, i.e., 7.77×10^4 sec. This would be 2.88×10^9 disintegrations. From this last value, the total energy released by the source and absorbed by the surrounding medium could be calculated.

While the physical half-life is of great interest for the pharmaceutical chemist in order to select the appropriate isotope for a particular experiment, the average life calculations will give an indication regarding the stability of the labeled compound used.

B.   RADIATION UNITS

There are two concepts dealing with radiation units. The measurements that concern the physical condition of the isotope are expressed as disintegration rates, while those that pertain to biological radiation damage are expressed as radiation dose. The curie unit is defined as that amount of radiactive material decaying at the rate of 2.2×10^{12} disintegrations per minute (dpm). The roentgen unit, R, is defined as "that quantity of X or gamma radiation such that the associated corpuscular emission per 0.001293 gram of air produces, in air, ions carrying one electrostatic unit of quantity of electricity of either sign." Because the definition of the roentgen unit is limited to the effects of γ or X-radiation, it becomes of little importance for the analytical chemist since most of the measurements performed are based on the rate of degradation rather than the biological effects observed.

C.   INTERACTION OF RADIATION WITH MATTER

All charged particulate radiation of any type loses its energy by interaction with the surrounding matter in essentially the same way. The energy of the particles is sufficiently high to attract or repulse an electron within the target atom. Thus, when particulate radiation passes through matter, ion pairs are produced as a result of the removal of an electron from a neutral atom, leaving behind a positively charged ion.

1. Interaction of α Particles

All α particles from a given isotope have the same energy (monoenergetic) and approximately identical ranges. This range is extremely short, thus resulting in a high specific ionization. The specific ionization of α radiation is at least 25 times higher than for β radiation having the same energy. This is due to the large mass and double charge of the α particle. The importance lies in the fact that to detect α particles, the energy of radiation must be transmitted directly into the sensitive volume of the detector. This presents a serious limitation to the use of α-emitting isotopes in analytical chemistry. The approximate range ($\pm 15\%$) of α particles in materials other than air is given by the Bragg-Kleeman rule:

$$R = (3.2 \times 10^{-4})R_{air}\frac{\sqrt{A}}{\rho} \tag{19.14}$$

where
R = range in experimental material
R_{air} = range of the particular α particle in air
A = atomic weight of material
ρ = density in grams per cubic centimeter

The range in mixtures and compounds may be determined by use of the equation:

$$A = n_1\sqrt{A_1} + n_2\sqrt{A_2} + n_3\sqrt{A_3} + \cdots n_n\sqrt{A_n} \tag{19.15}$$

where $n_1, n_2, n_3, \ldots, n_n$, are the fractional compositions.

2. Interaction of β Particles

The β spectrum, on the other hand, is continuous and the energies of the particles vary up to a maximum energy, E, characteristic of the isotope. The shapes of typical spectra are illustrated by Fig. 19.2. The spectral shapes of negative and positive electrons differ due to the Coulomb interaction of electron and nucleus in case of positron emission. The range R is almost a linear function of E_m for energies above 0.5 MeV[1]:

$$R(mg/cm^2 \text{ of Al}) = 520E_m \text{ (MeV)} - 90 \tag{19.16}$$

Experimentally, Katz and Penfold[2] found that the following relationship holds well for a range of energies from 0.01 to 3 MeV:

$$R(mg/cm^2 \text{ of Al}) = 412^n \tag{19.17}$$

where n equals $1.265 - 0.0954 \ln E_m$.

The absorption of negatrons in any medium was found to be approximately exponential when the absorber thicknesses are not too great:

$$I = I_0 \exp(-\mu x) \tag{19.18}$$

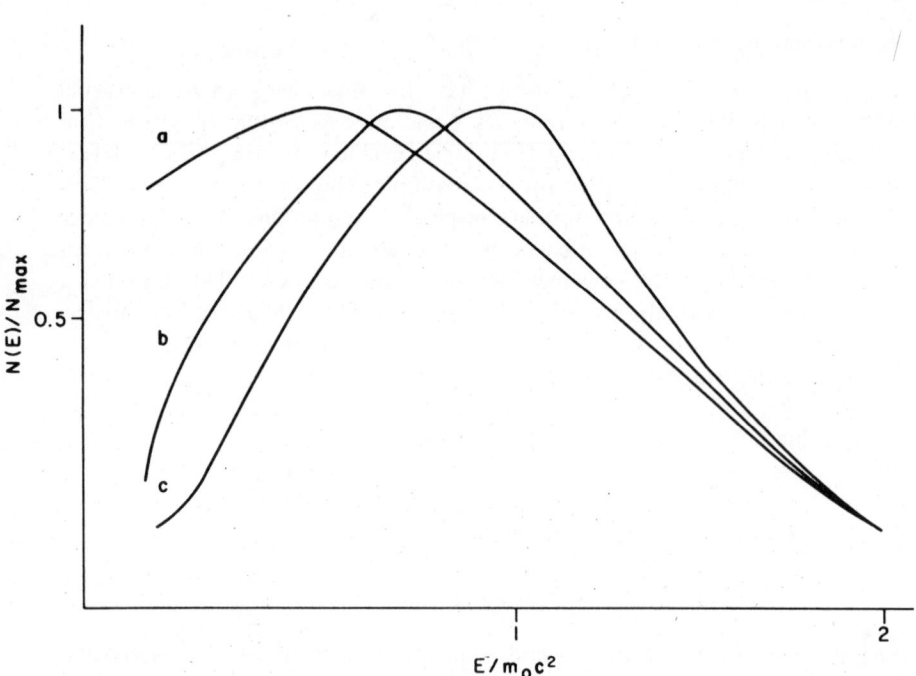

FIGURE 19.2: Dependence of energy distribution of particles on atomic number Z of residual nucleus. $N(E)$ is the number of particles per unit energy interval; E is energy. Normalized distributions for $E_m = 2m_0c^2 = 1.02$ MeV.   a: β^-, $Z = 40$; b: $Z = 0$ hypothetical); c: β^+, $Z = 40$. Courtesy of Nuclear Chicago Corporation.

where I is the transmitted energy or counting rate, I_0 is the value before absorption, μ is the mass absorption coefficient (in square centimeters per gram), and x is the thickness (in grams per square centimeter). It was found[1] that the value of μ was related to E_m by

$$\mu = \frac{17}{E_m^{1.14}} \text{ cm}^2/\text{gm of Al} \qquad (19.19)$$

where E_m is in million electron volts. It should be borne in mind that the above relationship holds true only for a particular experimental setup. The effect of geometry, particularly of absorber position relative to source and detector, as well as the atomic number of the absorber appreciably affect the calculation of the mass absorption coefficient using Eq. (19.19).

The positive β particles, or positrons, consume their kinetic energy in forming ion pairs in a manner similar to that of negatrons. In this instance the attractive forces are responsible for the ionization observed. When the positrons lose their kinetic energy, they combine with any orbital electron to produce an annihilation radiation consisting of two γ-ray photons given off at a 180° angle to one another. The energy of each photon is equal to 0.511 MeV, which is equivalent to the transformation of the rest mass of each electron to electromagnetic radiation.

3. Interaction of γ Rays

Gamma and X-rays have a much lower probability of interaction with matter than particulate radiation has, and the mechanism of such interaction is quite different from that of α or β particles. The latter interact directly to produce ion pairs, while the action of electromagnetic radiation is indirect. For a parallel or collimated beam of γ rays, the intensities at various points in the absorbing medium can be calculated using the relationship given in Eq. (19.18). In this instance μ represents the total absorption coefficient whose value depends on the kind of material and the energies of the photons in the beam. Linear absorption coefficients, which are sometimes used, are in units of reciprocal centimeters. Mass absorption coefficients may be calculated by dividing the latter by the density of the absorbing media. They are useful in comparing effects in materials of different density.

The total absorption coefficient is composed of three contributing components:

a. Photoelectric Interaction. Photoelectric interaction is important for low-energy photons and materials having a high atomic number. In this situation the photon interacts with an orbiting electron, transferring the entire energy of the quantum to the particle, thus ejecting it. The kinetic energy of the ejected electron is equivalent to the energy of the incident photon less the binding energy of the electron. The photoelectric absorption coefficient is related to the incident photon energy and the atomic number of the material by the following equation:

$$\tau = 0.0089 \frac{\rho}{A} 4.1 Z \left(\frac{12.4}{E}\right)^n \qquad (19.20)$$

where τ is in reciprocal centimeters, ρ is the density of the material, E is the energy of the photon in thousands of electron volts, Z is the mean atomic number, A is the mean atomic weight, and n is a constant well defined for elements up to iron in the periodic table (value of 3.05 for N, C, and O and 2.85 for the elements from Na to Fe).

b. Compton Interaction. Compton interaction is of importance where the energy of the photon is of higher energy and the atomic number of the material is of a lower value. Hence, the photon imparts only a portion of its energy to an orbiting electron, thus resulting in the distribution of the energy between a free electron and a lower-energy photon. This results in a situation where the absorption has to be computed on the basis of the ejected electron and the scattered photon. The latter can also give rise to Compton or photoelectric interactions in the absorbing medium. The total Compton absorption coefficient, σ, can be calculated using the Klein-Nishina[3] theory of the process:

$$\sigma = NZ(f_a + f_s) \tag{19.21}$$

where N is the number of atoms per cubic centimeter, Z is the atomic number and f_a and f_s are functions of the energy.

c. Pair Production. Pair production is of importance where the incident photon has an energy of at least 1.02 MeV. In this instance the γ ray is converted, in the vicinity of a nucleus, to a positron and negatron. This conversion requires an energy equivalent to the rest mass of the two particles ($2m_0C^2 = 1.02$ MeV). Any additional electromagnetic energy is imparted to the formed electrons as kinetic energy. The pair-production absorption coefficient, κ, is calculated according to the following equation:

$$\kappa = \underline{\alpha}NZ^2(E - 1.02) \tag{19.22}$$

where $\underline{\alpha}$ = proportionality constant
N = number of atoms per cubic centimeters
Z = atomic number
E = energy of incident photon in million electron volts

It is beyond the scope of this book to delve into the various details of the absorption of radiation by matter. However, a good understanding of the subject should be emphasized before interpreting γ-ray spectra where activation analysis is used as a tool by the analytical chemist. The choice of the proper counting equipment is primarily based on the nature of the interaction of radiation with matter; for example, it would be foolhardy to count a very weak β particle such as that of tritium in a GM counter. The entire energy of the particle would be dissipated in the air before reaching the sensitive volume of the detector.

D. STABILITY OF RADIOACTIVE COMPOUNDS

Of prime importance to the analytical chemist using radiotracers as reagents is the determination of their stability before and during experimentation. The measure of degradation as a result of self-radiation is expressed in terms of G(-M) values. This refers to the number of molecules permanently transformed per 100 eV of energy absorbed. The primary degradation is brought about by interaction of a molecule with a nuclear particle. If such a molecule happens to be labeled, then a radioactive impurity results. It is reported that, in certain situations, one β particle from ^{14}C may destroy up to 5000 molecules. To overcome such a problem, the pharmaceutical chemist has a number of options available when storing radioactive materials:

1. Spreading the compound in a monomolecular layer form, thus allowing most of the emitted particles to escape. An example of such technique is

the storage of ^{60}Co-cyanocobalamin, and ^{14}C-chlorophyll having high specific activities.

2. Dilution of the labeled material with nonradioactive form or a different substance, chosen to make it easier to separate the labeled compound again when required. Benzene has been a solvent of choice for the storage of tritriated molecules.

TABLE 19.2: Relative Stability of Some Radioactive Compounds[a]

Compound	Specific Activity, mCi/mM	Age, Days	Storage	Decomposition, %	Ref.
Cholesterol-4-^{14}C	2.5	540	Solid in air	40	4
Choline chloride (methyl-^{14}C)	1.8	270	Solid in air	63	4
Dextran-^{14}C sulfate (20 glucose units per molecule)	3	21	Solid in air	100	5
L-Methionine-^{35}S	100	60	Solid in air (dry)	20	6
L-Phenylalanine-^{14}C(U)	304	105	In 0.01 N HCl	14	6
Succinic acid-2,3-T	58,000	30	Solid	100	7
9,10-Dimethylbenz-anthracene-9-^{14}C	9	30	Benzene solution	20	6
9,10-Dimethylbenz-anthracene-T(G)	3,250	390	Benzene solution	37	6

[a] Courtesy of Nuclear Chicago Corporation.

Table 19.2 demonstrates the relative stability of some radioactive compounds when stored for various periods of time. It is possible to predict the degree of self-decomposition from the following equation:

$$P_d = f\bar{E}S_a(5.3 \times 10^{-9})G(-M) \qquad (19.23)$$

where P_d = initial percentage decomposition per day

f = fraction of the radiation energy absorbed by the system

$\bar{E}$ = mean energy of the emission in electron volts

S_a = initial specific activity of the compound in millicuries per millimole

The foregoing relationship remains linear, provided the magnitude of degradation is less than 10% and the storage time is short when compared to the half-life of the isotope. For greater details on the decomposition rates and G(-M) values, the reader is referred to the tables published by the Radiochemical Center, Amersham.[8] In view of the various problems introduced in experimentation when a radioactive impurity is present, it becomes mandatory for each chemist to evaluate the degree of radiocompound purity before each experiment. This can be accomplished by means of chromatographic or electrophoretic methods. Usual tests such as melting point determination, boiling point, refractive index, etc., are inadequate since they are not sufficiently sensitive to measure radiation decomposition.

19.3 MEASUREMENT OF RADIOACTIVITY

The determination of radioactivity in organic and inorganic compounds is a rather complex problem in which the best method varies with the particular radioisotope, sample volume, and the specific experiment at hand.

TABLE 19.3: Properties of Some Isotopes Commonly Used in Radiochemistry

Isotope	Half-life	β Energy, MeV	γ Energy MeV
^{14}C	5740 years	0.155	—
^{3}H	12.26 years	0.0186	—
^{35}S	87 days	0.168	—
^{32}P	14.3 days	1.71	—
^{36}Cl	3×10^5 years	0.714	—
^{82}Br	36 hours	0.44	0.78, 0.55, 0.62, 0.70, 1.47
^{131}I	8.06 days	0.61, 0.25, 0.81	0.36, 0.72
^{125}I	57.4 days	—	0.035
^{45}Ca	164 days	0.25	—
^{22}Na	2.6 years	0.54 (β^+)	1.28
^{24}Na	15 hours	1.39	1.37, 2.75
^{75}Se	121 days	—	0.024, 0.136, 0.265, 0.280, 0.58
^{42}K	12.4 hours	3.53, 2.01	1.52

Since α-emitting compounds are rarely used as tracers, the pharmaceutical chemist will be mostly concerned with three other classes of isotopes: (a) the very weak β emitters such as tritium; (b) moderately weak and strong β emitters such as ^{14}C, ^{35}S, and ^{32}P; and (c) the γ emitters such as ^{131}I, ^{125}I, ^{60}Co, etc. No one method is generally superior for measuring all isotopes used in analytical chemistry. Table 19.3 summarizes a number of isotopes which are in fairly common use in the biological and analytical fields.

All three classes of isotopes can be counted either by some method of ion collection, scintillation technique, or solid-state measurement.

A. METHODS BASED ON ION COLLECTION

All instrumentation associated with the detection and measurement of radioactivity require two basic elements: a detector element and a recording system. The detector element converts the radiative energy into electrical pulses which are passed either directly or indirectly to the recording system. The pulses are then displayed or registered in the form of counts.

The basis of ion-collection measurements is the fact that when ionizing radiation interacts with the molecules of a gaseous medium, ion pairs are formed. The behavior of a counting circuit such as the one shown in Fig. 19.3, when ionizing radiation strikes the sensitive area between the two

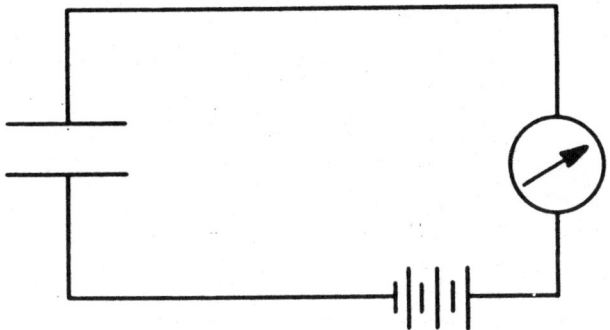

FIGURE 19.3: Simple circuit of ionization detector.

parallel plates or electrodes, depends upon two factors, i.e., the amount of ionization produced by radiation and the potential impressed across the plates by the battery. These relationships are illustrated in Fig. 19.4. Thus, gas detectors can be used in some form to detect any of the isotopes under consideration. If the change in current is the criterion of measurement, the detector is called an ion chamber. However, if the potential change across the electrodes is used as the desired criterion, then the instrument is called a gas counter. Geiger-Müller tubes and proportional detectors are examples of gas counters. A typical gas detector consists essentially of a chamber in which the walls form one electrode (cathode) and a central probe or wire forms the anode (Fig. 19.5). A suitable gas is introduced into the chamber, and a potential applied between wall and central electrode. A solid or liquid sample is usually separated from the sensitive volume of the counting chamber by means of a thin membrane. However, it is possible in certain instances to introduce the sample directly within the sensitive area of the detector. At a relatively low potential (50–100 V) a region of saturation current is established in which the ions produced in the gas, as a result of radiation interaction with the molecules, result in a current flow which is directly proportional to the number of events taking place within the chamber. Because the current flow will differ from one isotope to the other due to the difference in the average energy release, the apparatus must then be calibrated for every individual isotope. An ideal gas in the chamber must (a) be chemically inert, i.e., it must not attack the chamber components chemically; and

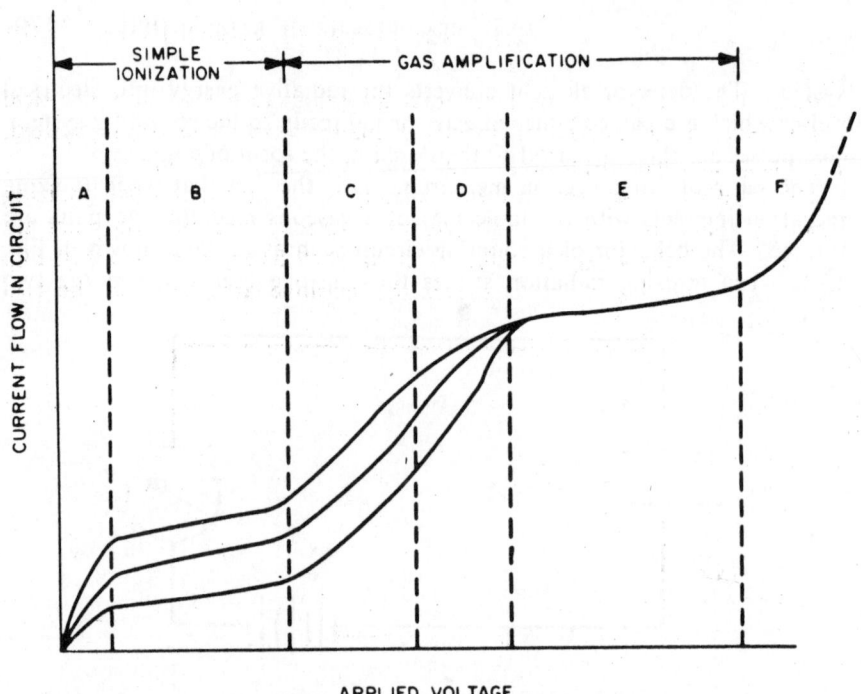

FIGURE 19.4: Effect of increased potential on current flow in an ionization detector. A, recombination region; B, current saturation region; C, proportional region; D, limited proportionality region; E, Geiger-Müller region; F, continuous discharge region.

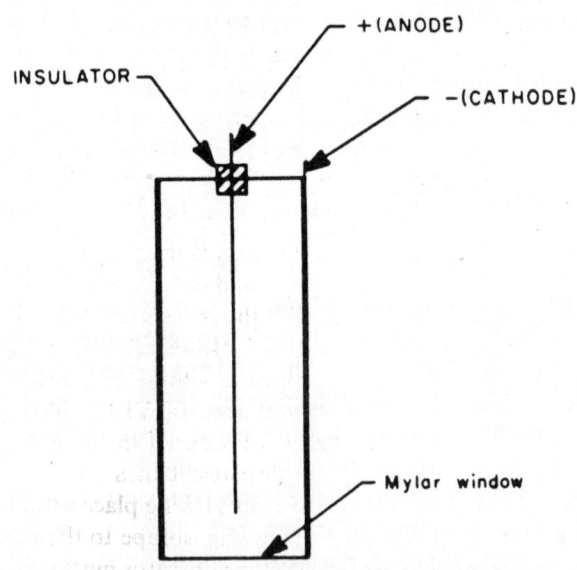

FIGURE 19.5: Longitudinal section of a typical GM tube.

(b) not be easily ionized such that the applied potential between the electrodes result in a current flow without the presence of radioactivity. Typical gases are usually low molecular weight hydrocarbons or mixtures of them. Helium and argon are also sometimes added. At such low voltages the current can be measured by dc amplifiers or vibrating-reed electrometers. An alternative way is the measurement of the rate at which the current charges a condenser to a particular preset potential.

If the applied voltage is raised above the ion-chamber plateau, a limited multiplication of energy results across the counter. There is a direct proportional relationship between the average energy of the events and the energy output of the counter. Thus, a small change in potential of the electrode results when some ions reach the central anode. The change is amplified and detected by suitable circuits and a voltage plateau is observed for ideal monoenergetic emitters. However, all β emitters have a wide range of energies up to E_{max}. In such instances the circuit amplifier is designed in such a way that all pulses above a certain energy level produce a constant output to be recorded. The amplifier, in turn, must have an input sensitivity of 1 mV or less. The above system is called a proportional counter and is very useful in counting and differentiating between α- and β-emitting isotopes.

If the voltage across the counter is increased further, a different plateau is reached and the counting is said to be done in the Geiger region. All radiation, regardless of the number of primary ion pairs produced, will produce the same current flow and the counter is no longer able to differentiate between α and β radiation. This is due to the production of an avalanche whereby gas amplification will have reached its maximum value. The only advantage of counting in the Geiger region over that in the proportional region is that much less sensitive amplifiers are needed because of the comparatively large pulses produced in the former situation. However, counting rates above 25,000 cpm cannot be measured in the GM counter since the counter is essentially in a state of continuous discharge above that rate. Statistical corrections will also have to be made to account for the lost pulses which happen to occur during the "dead" time of the detector. In this instance the effects of the previous event play a major role in the detection process and the counter is known to be "dead" since it will not respond to a newly formed pulse until complete recovery. The dead time of each Geiger tube will vary depending upon the pressure and quality of the gas, as well as the materials of the chamber. On the other hand, proportional counting is generally less sensitive to impurities and to variations in filling pressure.

B. METHODS BASED ON THE SCINTILLATION TECHNIQUE

These methods are probably considered to be the most important from the analytical standpoint. The basis of the scintillation technique rests on the production of light photons resulting from radiation interaction with matter. This concept is not new in any sense and was used by many investigators in the first part of this century to measure the radioactivity of α-emitting radioisotopes. The scintillator used at that time was zinc sulfide and the detector was the human eye. The technique was soon abandoned after the introduction of the Geiger-Müller counter until the development of the photomultiplier tube, after which scintillation spectrometry progressed along two different lines.

I. Liquid Scintillation Spectrometry

A sample prepared for liquid scintillation spectrometry consists of several components: the radioactive materials to be counted, a solvent system, and a proper scintillator. The type of solvents used depends on the chemical or physical nature of the substances in question. The function of such solvent is the transfer of energy from the point of radioactive emission to a scintillator molecule. An efficient solvent system must exhibit the following characteristics: (a) be transparent to the light photons emitted, (b) not freeze at low temperature under counting conditions, and (c) be able to dissolve or suspend the sample to be counted. Aromatic hydrocarbons such as benzene, xylene, and toluene are frequently used as solvents. However, other hydrocarbons are also used. Potential solvents are usually rated by comparison to toluene.

On the other hand, a good scintillator must have the following characteristics: (a) efficiently produce light photons upon transfer of the energy from the solvent to the scintillator molecule (the wavelength of the light output should also be compatible with sensitivity of the photomultiplier tube) (b) sufficient solubility at working temperatures, and (c) chemical stability.

Scintillators are classified into two groups which are usually mixed together in the solvent system. Primary scintillators, such as *p*-terphenyl, 2,5-diphenyloxazole (PPO), and 2-phenyl-5-biphenyl-oxadiazole (PBD), are routinely used. Secondary scintillators, such as 1,4-bis-2-(5-phenyloxazolyl)-benzene (POPOP) and α-naphthylphenyloxazole (α-NPO), are usually added in trace quantity to a solution containing one of the common primary scintillators. Their function is to convert the photons which were emitted at shorter wavelengths into longer wavelengths which are closely matched to photomultiplier response. A new scintillator, 2,5-bis-2-(5-*t*-butylbenzoxazolyl)thiophene (BBOT), has been suggested recently as a replacement for both PPO and POPOP. Having a fluorescence maximum of 4350 Å, BBOT thus shows a better match to typical glass-faced photomultiplier response than either PPO or *p*-terphenyl.

To estimate the proper concentration of a scintillator in a particular liquid scintillation system, a known amount of a radioactive substance, such as

^{14}C-toluene, is added to the selected solvent. Increasing amounts of a certain scintillator are then gradually added, and the count rate is observed after each addition. The optimal amount to be used is that quantity which produces a maximal pulse height response. A comparison of the various possible combinations has been reported by Rapkin.[9] Table 19.4 summarizes some of the commonly used "cocktail" mixtures in liquid scintillation spectrometers.

Any process which interferes with the performance of a liquid scintillation counting solution is referred to as "quenching." Color quenching arises from the absorbance of the emitted light photons, thus resulting in a lower detected pulse. It can often be minimized by bleaching or decolorization. Biological samples are usually solubilized in a strong quaternary ammonium solution such as hyamine 10-X and then bleached with a trace of hydrogen peroxide. Sodium borohydride has also been suggested as a bleaching agent. Other methods of eliminating color include passage over activated charcoal, precipitating protein material with trichloroacetic acid, and combustion of the sample in question to some form of radioactive gas that could be quantitatively collected and counted. The extent of color quenching could be estimated by measurement of the absorptivity of the solution at a particular wavelength, and correlating the above value with that of the observed pulse height. However, this method is only approximate, since the relationship is not linear at high concentrations of colored material.

Chemical quenching, on the other hand, involves interference with the process of energy transfer between the solvent and scintillator molecules. Dissolved oxygen is probably the most common quenching agent. The radioactive solutes themselves also show some degree of quenching. Thus, concentration becomes an important factor. To determine if a particular radioactive compound produces a chemical quench, increasing amounts of the compound are dissolved in the particular liquid scintillation system, and the count rates are observed after each addition. A linear relationship between the concentration and the recorded count rates per unit weight is an indication that the chemical quench is not concentration dependent.

Because of the variation in the degree of quench between the different samples, it becomes mandatory to convert the relative counts into absolute values expressed as dpm. This is effected in a variety of ways, each having its own limitations. The use of an internal standard, such as ^{14}C-toluene, has been most common. External standards, such as ^{137}Cs and ^{133m}Ba have recently been introduced. The principle of the latter technique depends on the interaction of the γ-ray photon with the materials of the scintillator solution thus producing a light photon. A third technique is based on the fact that quench, whether color or chemical, results in a shift of the pulse height spectrum. The ratio of the counts observed between two pulse height analyzers is an indication of the degree of quench. When a set of quenched

TABLE 19.4: Composition of Various Scintillation Mixtures

Number	PPO, g	POPOP, g	BBOT, g	Naphthalene, g	Xylene, ml	Toluene, ml	Dioxane, ml	Cellosolve, ml	Anisole, ml	1,2-Dimethoxy-ethane, ml
1	5	0.5	—	—	—	1000	—	—	—	—
2	7	0.3	—	100	—	—	1000	—	—	—
3	15	0.6	—	—	—	—	750	—	125	125
4	10	0.5	—	50	—	—	833	166	—	—
5	3	0.1	—	—	1000	—	—	—	—	—
6	10	0.5	—	80	—	143	428	428	—	—
7	10	0.5	—	80	143	—	428	428	—	—
8	—	—	4	—	—	1000	—	—	—	—
9	—	—	4	—	1000	—	—	—	—	—
10	—	—	4	50	—	—	835	165	—	—

standards of the particular isotope is counted, the various counts occurring between the two selected channels will vary according to the degree of quench. Thus, a standard curve which could be used to correct the observed count of the counted sample is established.   Instruments containing up to four counting channels are available today. Computer programs are also provided for the direct conversion of cpm to dpm. It is beyond the scope of this book to detail the various statistical approaches designed for such conversion. Insoluble radioactive compounds are sometimes counted by suspending them in a thixotropic gel powder. Cab-O-Sil and Thixin R, a castor oil derivative, have been successfully used for the counting of radioactive samples absorbed on silica gel G. Thus, liquid scintillation spectrometry could effectively be used in conjunction with thin-layer chromatography to solve complex analytical systems.

A novel approach to counting β particles by the scintillation technique is the use of plastic phosphors. Some models consist of a thin-layer plastic phosphor coupled to a photomultiplier tube. Self-absorption corrections and careful standardization are empirical in this case. Attempts have also been made to use scintillating plastic counting vials and scintillating plastic flow cells. Scintillating plastic beads, immersed in aqueous solutions, have been used to count ^{45}Ca. Most recently, ion-exchange resins which possess fluorescent or scintillating properties were reported.[10] In this case the radioactive ion can be removed from the solution by the resin; the latter is then washed and counted directly in a liquid scintillation counter.

The photomultiplier tube, which serves as the scintillation detector, is probably the most critical part of the system. The "Venetian blind" type of photomultiplier is essentially used in most liquid scintillation counters. It is characterized by having a quartz face, since the latter has less radioactivity than glass. It is also more transparent in the ultraviolet region where scintillators exhibit their principal emission. The quality of the photomultiplier is primarily responsible for the performance of the entire system, and an inherent amount of electronic noise is unavoidable if the tube is not sufficiently cooled. The efficiency^2/background ratio is thus drastically improved. A 50% counting efficiency for tritium with 500 cpm background will yield an E^2/B figure of 5, whereas a 25 cpm background results in an improved value of 100. A modern liquid scintillation counter is a coincidence-type machine based on the principle that the sample is interposed between two photomultipliers which examine it simultaneously. Any radioactive event resulting in a sufficient number of photons will trigger a pulse in both tubes. Thus, a record of the event is made only when both tubes pulse within a given coincidence resolving time. If a pulse originates in only one tube, but not the other, such a pulse is considered as noise, and it is not counted. The chance coincidence rate is calculated from the following equation:

$$a = 2n_1 n_2 \tau / 60 \qquad (19.24)$$

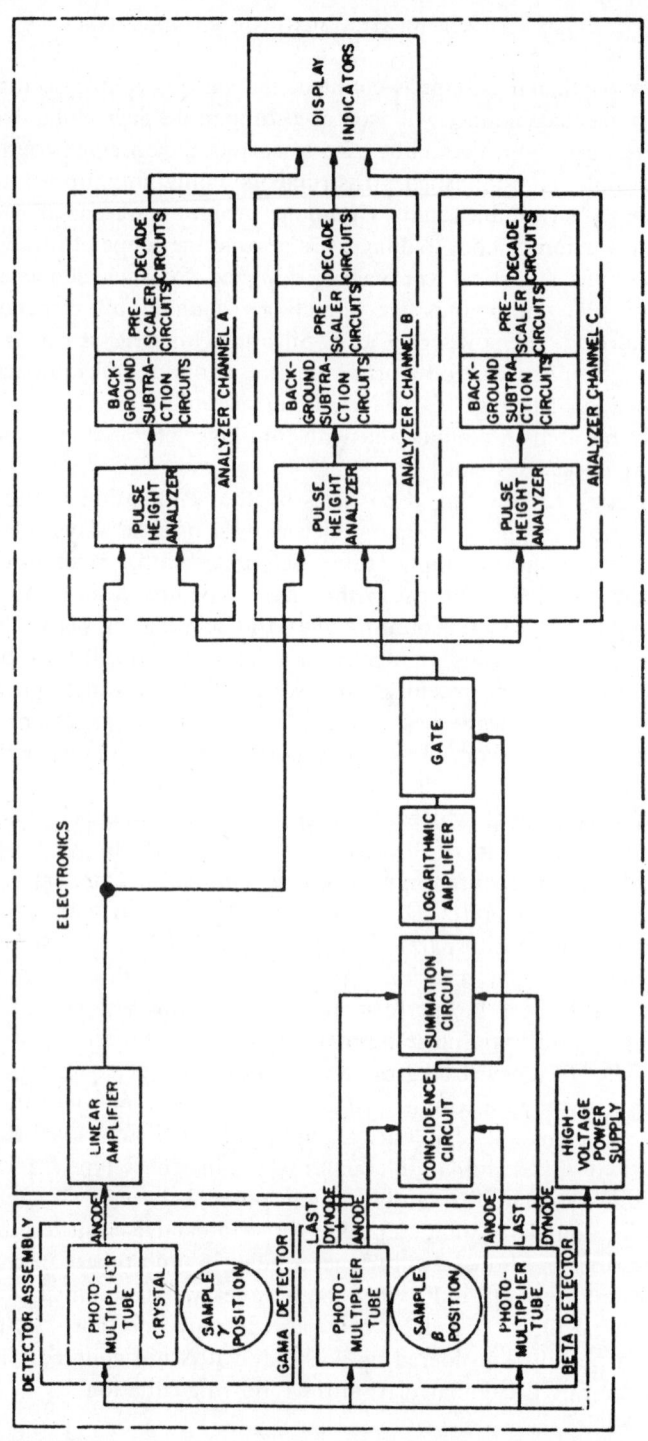

FIGURE 19.6: Block diagram of a liquid scintillation spectrometer. Courtesy of Picker Nuclear Corporation.

where a = chance coincidence rate in cpm

n_1 = noise rate of photomultiplier 1 in cpm

n_2 = noise rate of photomultiplier 2 in cpm

τ = coincidence resolving time in seconds

If we assume that each of the two photomultiplier tubes has a noise rate of 10,000 cpm and that the coincidence resolving time of the circuit is 6×10^{-7} sec, the chance coincidence rate would then be equivalent to

$$a = 2 \times 10^4 \times 10^4 \times (6 \times 10^{-m})/60 = 2\,\text{cpm}$$

However, other sources of background counts arise from cosmic rays, radioactive contamination in the shield, and inherent radioactivity in the glass of the containers. The obvious problem of coincidence circuitry is that the light output of the scintillator is divided between two photomultipliers, and, as such, the output signal is cut in half. Figure 19.6 shows the block diagram of a Liquimat liquid scintillation spectrometer. The logarithmic system demonstrated in the diagram has the advantages of simplicity over linear instrumentation. The major disadvantage is that it cannot efficiently be used for the quench correction of suspended samples.

2. γ-Scintillation Spectrometry

A γ-scintillation system consists of a scintillation detector, high-voltage power supply, preamplifier, amplifier, discriminator, and scaler. Figure 19.7 shows the arrangement of the various units mentioned above. The scintillation detector consists of a solid scintillator which emits photons of visible light upon interaction with the incident γ radiation. The photons

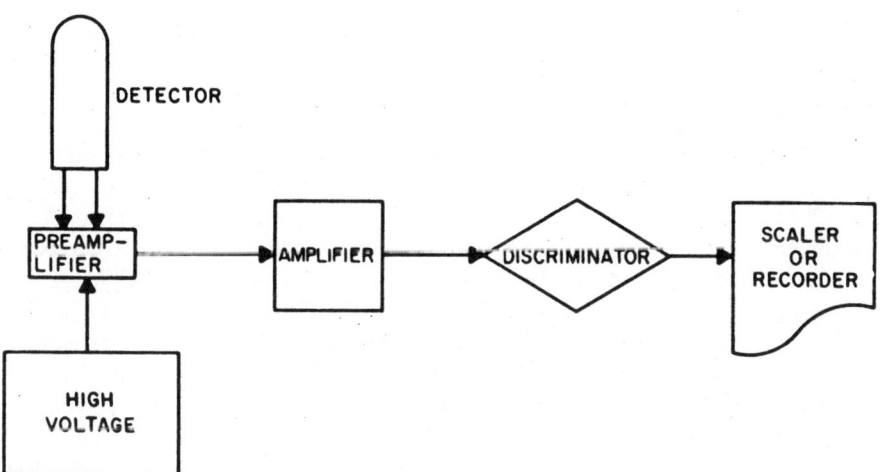

FIGURE 19.7: A typical scintillation system.

are then amplified by a photomultiplier. Because the photoelectric absorption coefficient varies approximately as the fifth power of the atomic number, the scintillation crystals were designed to consist of sodium iodide or other alkali halides. Table 19.5 demonstrates the different characteristics of various scintillators. The photons produced by the crystal are reflected by the crystal housing and strike the photocathode of the photomultiplier. The number of photoelectrons is directly proportional to the number of incident light photons. Thus the photomultiplier-scintillator system will deliver a

TABLE 19.5: Characteristics of Common Scintillators[a]

Scintillator	Density, g/cm^3	Wavelength of Emission, Å	Light Yield	Decay Time, μsec	Remarks
Anthracene	1.25	4450	1.00	0.025	Large crystal, not clear
Stilbene	1.16	4100	0.73	0.007	Good crystals
Terphenyl	1.23	4150	0.55	0.012	Good crystals
Naphthalene	1.15	3450	0.15	0.075	Good crystals
ZnS(Ag)	4.1	4500	2.0	1	Small crystal, poor transparency
NaI(Tl)	3.67	4100	2.0	0.25	Excellent crystals, but hygroscopic
CsI(Tl)	4.51	white	1.5	1	Excellent crystals
p-Terphenyl in xylene	0.87	3700	0.48	0.007	Liquid solution
Terphenyl in polystyrene	1.06	4000	0.30	0.005	Plastic solution

[a] Courtesy of Baird Atomic Inc.

burst of charge proportional in size to the radiant energy dissipated in the scintillator. This is due to the fact that for every 100 eV of radiant energy, the crystal emits approximately two photons.

Because there are several ways in which γ rays may intersect with matter, the observed γ spectrum is complicated. The decay scheme of a commonly used isotope such as ^{137}Cs is given in Fig. 19.8. Since the γ energy is less than 1.02 MeV, no pair production is anticipated. Ideally, such spectrum as seen by a detector should show only three peaks: a low-energy X-ray peak due to the deexitation of ^{137}Ba to the ground state, a compton edge, and a photopeak. However, a slightly modified spectrum is observed under normal counting conditions Fig. 19.9. The differences observed are caused by the characteristic scintillation detector. Statistical broadening of the photoelectric peak arises primarily from the spread of the relatively small numbers of photoelectrons formed at the photocathode. The degree of broadening is a measure of the accuracy of the system. The smaller the broadening, the more accurate is the determination of the energy peak. This in turn is referred to as the resolution of the photopeak. Resolution is

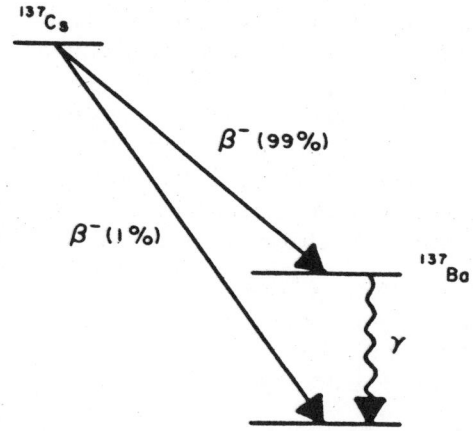

FIGURE 19.8: Decay scheme of ^{137}Cs.

defined as "the full width of the peak as measured at half the maximum counting rate divided by the pulse height at the peak." Thus, the lower the resolution, the greater the ability to differentiate between two photopeaks with energies close to each other; it also follows that the higher the energy, the better the resolution. For example, if the full width of the photopeak at half maximum counting rate for ^{137}Cs is equivalent to 66 keV, the resolution of the detector system would then be equivalent to 66/662 or 10%. However, in the usual γ-ray energy range (0.1–3 MeV) detection and photopeak efficiency decrease with increasing energy.

The Compton scatter is dependent on the size of the crystal. The larger the size, the greater is the probability for scattered Compton rays to be recaptured and thus interact photoelectrically. Thus, the size of the scintillator crystal

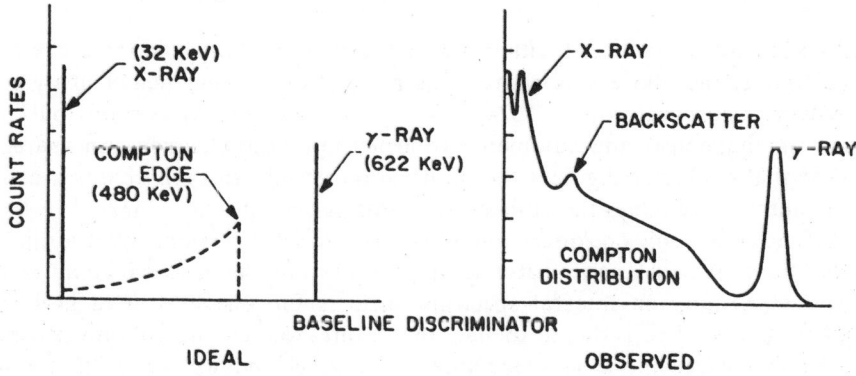

FIGURE 19.9: Ideal and observed spectra of ^{137}Cs.

determines the sensitivity of the detector. However, this does not improve the resolution which is associated with the performance of the photo-multiplier.

The phenomenon of backscattering is due to the deflection and interaction of the emitted γ rays with the shielding material surrounding the scintillator crystal. The spectrum observed in Fig. 19.9 is called a differential spectrum. It is obtained by measuring the output pulses occurring between two discriminators which were separated by a small "window." When the lower

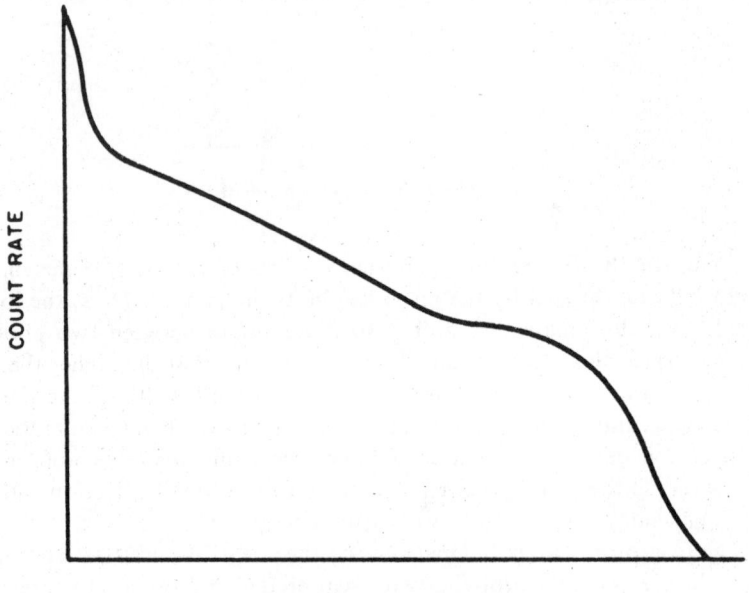

BASELINE DISCRIMINATOR

FIGURE 19.10: Integral spectrum of ^{137}Cs.

discriminator is advanced, either manually or automatically, in the differential mode, the above spectrum is observed. Newer pulse height analyzer systems incorporate up to 5024 of such windows, and the various pulses are simultaneously and automatically sorted into each channel. An analog to digital converter digitizes the information present in each channel, and the output tape relays the number of counts per channel.

If, however, only one discriminator is varied in selective steps, by recording the counts at each discriminator setting and plotting the observed count rate at each setting, an integral spectrum such as the one illustrated in Fig. 19.10 results. This is due to the fact that a threshold circuit will only accept all pulses exceeding a selected pulse height level and will reject all those which are smaller in amplitude. The plateau region in Fig. 19.10 corresponds to the "valley" observed in Fig. 19.9, immediately prior to the photopeak.

The choice of integral versus differential discrimination as a mode of counting depends largely on the experiment at hand. For identification and counting of multilabeled preparations, it becomes mandatory to use a differential spectrometer. An added advantage is the low background observed in such instances. The major disadvantage is that the counting rate might be small enough that extended periods of time will be necessary to count an individual sample in order to reach a particular level of statistical accuracy. Since a slight shift in high voltage will result in a variation of the base line of the discriminator or the gain of the amplifier, a large difference in the counting rate will be observed. Thus, the stability of the electronic components determines the accuracy of a certain counting system. The drift (in volts) of the photopeak is particularly dependent on the line voltage, temperature of the room, operating point, and the magnitude of the count rate. For integral analysis, stability is effected by counting the radioactive sample in the plateau region. On the other hand, the drift occurring in differential operation is difficult to determine since the counting rate is already differentially related to the base-line voltage of the discriminator. An estimation of such drift could be done by measuring the changes in the count rate in the slope region of the integral curve for a particular discriminator setting. The slope corresponds to the location of the photopeak.

Multidimensional γ-ray spectroscopy has been recently used for the identification and measurement of complex mixtures of radionuclides. It is based on the property of such nuclides to decay through emission of two or more γ rays in sequence. Thus, a multidimensional analyzer analyzes the γ rays received in coincidence and stores them in a memory according to the energy lost by interaction with the detector, which consists of two sodium iodide crystals. This new approach has helped the determination of many radionuclides present in very trace amounts in biological materials without the need of chemical separation.

C. STATISTICS OF RADIOACTIVE MEASUREMENTS

The fact that the decay of radioactive atoms is a random process introduces certain limits of error which vary according to the number of emitted photons or particles. The larger the number of events, the smaller the error, and vice versa. This is often referred to as indeterminate error. On the other hand, errors resulting from the instability or uneven performance of electronic equipment, sampling inaccuracy, and various other factors is referred to as determinate errors. Thus, it becomes necessary to analyze the measurement variations to draw conclusions with regard to the reproducibility of such measurements. The normal distribution of the latter approximately follows the Poisson statistics. The normal distribution function P_x can be written as

$$P_x = \frac{1}{s\sqrt{2\pi}} \exp[-(x-m)^2/2v]   \tag{19.25}$$

where P_x = the normal distribution function, defined in such a manner that the quantity $P_x\,dx$ gives the probability that the value x lies between x and $x + dx$

v = the variance of the measurement = (standard deviation)^2

s = standard deviation

m = true mean value of the distribution

When the relationship between P_x/dx and x is studied, the well-known bell-shaped normal curve results. However, the true mean m is never known exactly, and as such the observed mean $\bar{x}$ is always used in calculations. The standard deviation, or the measurement of spread of data, is estimated by

$$s = \sqrt{\frac{\sum\limits_{i=1}^{n}(x_i - \bar{x})^2}{n-1}}   \tag{19.26}$$

where n is the number of observations. However, if the same set of measurements was repeated, a different value for $\bar{x}$ is observed, and the standard deviation from the mean is then called standard error, which is given by the relation

$$\text{S.E.} = s/\sqrt{n}   \tag{19.27}$$

Another common type of measurement error which is usually referred to in statistics is called the probable error, r. This is defined as the error which is as likely to be exceeded as not. Thus, a 1% probable error for a certain value $\bar{x}$ means that any one measurement will have an equally good chance of falling between the values of $\bar{x} \pm 1\%$. The relationship between the probable error and the standard deviation is

$$r = 0.6745s   \tag{19.28}$$

where r is the probable error and s is the standard deviation.

A different approach for calculating the standard deviation of a set of measurements is the application of Poisson distribution. This law describes most of the counting observations made in experimental nuclear physics. It is possible to correlate the Poisson distribution and the normal distribution, whereby

$$s = \sqrt{m}   \tag{19.29}$$

where m is always an integer. Essentially, the two ways mentioned for calculating the standard deviation should yield closely identical results. If they disagree appreciably, a Chi-square test of goodness of fit is recommended. This test determines the probability of deviation where a set of measurements is repeated and compared to the assumed distribution of those events observed in the first trial.

Because any particular radioactive measurement will combine several errors, such as those of background counts, instability of detector, randomness of disintegration, etc., the error of the sum or difference could not be simply added or subtracted. This is due to the fact that the errors may partially cancel each other. The law of combining independent errors of measurement is

$$s = \sqrt{s_{obs}^2 + s_{backg}^2} \qquad (19.30)$$

where s = combined standard deviation
s_{obs}^2 = standard deviation of the observed count rate
s_{backg}^2 = standard deviation of background rate

A different kind of error is the one observed when highly radioactive compounds are counted by means of a GM detector. The response of such a detector will not follow a linear relationship since some of the events will occur during the dead time of such a detector. In this instance the true count rate is given by

$$N = \frac{n}{1 - nT} \qquad (19.31)$$

where N = true count rate
n = observed count rate
T = resolving time of the detector

In summary, when designing a particular radioactive experiment, the investigator should bear in mind several important criteria:

1. The expected quantities of radioactivity to be measured
2. The efficiency of the instrument available or selected
3. The background contribution to each measurement
4. The amount of time allocated for the counting of each individual sample

19.4 ANALYTICAL APPLICATION OF RADIOACTIVITY

A great number of techniques are available today for the application of isotopes in analytical pharmaceutical chemistry. The scope and dimensions of such techniques are unlimited. Since the first requirement of a tracer is that it must contain a radionuclide, the application of radioactive techniques will essentially fall into two categories.

A. TECHNIQUES INVOLVING THE MEASUREMENT OF ADDED RADIOACTIVITY

A certain amount of radioactive tracer is mixed or incorporated with the unknown sample at some predetermined stage during the analysis. Several variations of this technique are available.

1. Yield Determination

The radioisotopic yield determination is applicable when the quantitative separation of a particular compound from a solution mixture is difficult or time consuming. The complexity of such a mixture, as well as the presence of interfering substances, may yield uneven amounts of the compound upon extraction. The principle of this method depends on the addition of a minute amount of a radioactive tracer having the same chemical composition as that of the unknown compound to the crude mixture. Amounts, usually less than 1 μCi, are thoroughly mixed with the original mixture, which is then subjected to the extraction process. The total amount of radioactivity observed after extraction represents the fraction of the unknown compound removed from the mixture. The sample is then purified and analyzed by a standard method of analysis. The true amount of the sought compound is then calculated by applying the correction factor from the radioactivity measurement. The technique is especially successful when determining biologically important compounds in human or animal tissues. Two essential limitations are applicable in these instances: (1) availability and cost of the labeled compound and (2) the specific activity of the isotope. Although a great number of labeled compounds are available commercially, it is quite possible that for a certain application, such as the determination of a particular drug metabolite, it may be uneconomical to purchase the labeled form of the metabolite. If the compound contains nonexchangeable hydrogen atoms, it could then probably be labeled by the Wilzbach technique.[11] The compound to be labeled is exposed for several days to several curies of tritium gas in a sealed container. Some of the hydrogen atoms will exchange with those of tritium. The resulting radioactive compound is then purified to remove any labile tritium atoms. It is also possible to derivatize the unknown compound by means of a radioactive reagent. In this situation the reaction must be quantitative, reproducible, and with known yields. On the other hand, the specific activity of the added tracer determines the accuracy of the technique. Since the amount originally added is relatively negligible in comparison to the amount of unknown present in the mixture, no correction is made for the amount of tracer in the final computations. However, if the labeled compound is available at lower specific activities only, its weight will have to be subtracted from the final answer. From a statistical standpoint the maximum amount of the tracer added should not exceed the weight of the unknown. This restriction thereby limits the total amount of radioactivity which could be added to the mixture.

Example

During the analysis of sulfanilamide in a complex triple sulfa mixture, 10,000 dpm of ^{35}S-labeled sulfanilamide was thoroughly mixed with 100 mg of the preparation. Extraction was effected using 100 ml of chloroform. The corrected activity, as measured in a liquid scintillation counter, was found to be 80 dpm/ml. The amount of sulfanilamide in the extract was measured spectrophotometrically and found to be 100 μg/ml.

1. What is the percentage of sulfanilamide in the unknown mixture?
2. If the specific activity of the labeled reagent is known to be 1 μCi/mg and the number of desired counts are 500 dpm/ml after extraction, what is the minimum amount of sulfanilamide which can be analyzed by this method?

Answers:

1. Efficiency of extraction = $(80 \times 100)/10,000 = 0.80$; total amount of sulfanilamide in 100 mg of mixture = 100 μg $\times$ 100 ml $\times$ 1/0.80 = 12.5 mg; and percentage of sulfanilamide in mixture = $(12.5 \times 100)/100 = 12.5\%$.
2. The maximum amount of sulfanilamide which can be analyzed is equivalent to the total weight of labeled sulfanilamide added. This is calculated in the following way:

$$\text{Weight of labeled reagent} = \frac{\text{total observed counts}}{\text{specific activity} \times \text{yield}}$$

$$= \frac{500 \times 100 \text{ dpm}}{2.2 \times 10^6 \text{ dpm/mg} \times 0.8} = 3.01 \times 10^{-2} \text{ mg}$$

2. Isotope Dilution Analysis

The basic principle of isotope dilution analysis depends on the fact that if a known amount of radiotracer is mixed with an unknown amount of the same unlabeled compound, the extent of dilution could be measured in terms of the reduction of the specific activity of the original radiotracer. Various modifications of this technique have been reported. However, all those variations must meet certain requirements, namely:

1. The absolute purity of the radiocompound must be ascertained before the mixing process.
2. Complete equilibrium of the tracer with unknown compound in the mixture.

3. The amount of material separated after the mixing must be chemically pure.

4. The smaller the mass of added radioactive tracer, the greater the statistical accuracy of the method. The major advantage of isotope dilution analysis is that no quantitative recovery is required during any step of the analysis.

a. Direct Isotope Dilution Analysis. This method involves the simple addition of a known amount of radiotracer to an unknown quantity of the same compound in a complex mixture. After thorough mixing, a sample of the compound in question is separated and the final specific activity is determined. Thus, any decomposition of the compound at any part of the procedure will not affect the outcome of the results. The following equation is used to estimate the mass of the unknown:

$$M_u = M_r\left(\frac{S_0}{S_f} - 1\right) \qquad (19.32)$$

where M_u = mass of unknown
M_r = mass of radiotracer added
S_0 = original specific activity of radiotracer
S_f = final specific activity of the diluted compound

Example

Penicillin is to be determined in a pharmaceutical sample. One milligram of tritium-labeled penicillin having a specific activity of 250,000 cpm/mg was added to the preparation. The specific activity of a separated sample after mixing was found to be 1000 cpm/mg. If the counter efficiency for tritium is 40%, what is the mass of unknown penicillin?

Answer:
Substituting in formula (19.32),

$$M_u = 1\left[\frac{250,000}{1000 \times (1/0.4)} - 1\right]$$

$$= 99.0 \text{ mg of penicillin}$$

b. Inverse Isotope Dilution Analysis. It is not always possible, for practical and economical reasons, to use a radiotracer similar in chemical composition to that of the unknown. In such instances the compound being determined is reacted in a quantitative and reproducible way with a radioactive reagent of known specific activity. A known amount of carrier is then added. The carrier will consist of the nonradioactive derivative of the unknown which is previously prepared and purified. Statistically, the larger the amount of carrier added, the more accurate is the determination. The final specific activity is determined after dilution, and as the stoichiometry of the reaction between the reagent and compound is known, then

$$M_u S_r = (M_u + M_c) S_f \tag{19.33}$$

where M_u = mass of unknown radioactive derivative
M_c = mass of added carrier
S_f = final specific activity of the diluted compound
S_r = initial specific activity of the reagent

Then the mass of compound present can be calculated from the above formula, where

$$M_u = M_c \left(\frac{S_f}{S_r - S_f} \right) \tag{19.34}$$

It is very important to have excess of the radioactive reagent when the reaction is performed. Equally important is the removal of all traces of the excess reagent before determining the final specific activity.

Example

An unknown compound A is reacted quantitatively with a radioactive reagent B having a specific activity of 10 mCi/mM. One hundred milligrams of carrier derivative is mixed with the preparation, and the final specific activity of a separated sample was found to be 1 μCi/mM. What is the mass of compound A if the molecular weight of the latter is 200 and that of the derivative is 300?

Answer:

From Formula (19.34),

$$M_u = 100 \left(\frac{1}{10,000 - 1} \right)$$
$$= 100 \times 10^{-4} \text{ mg}$$

Mass of unknown compound $= 100 \times 10^{-4} \times (200/300) = 66 \times 10^{-4}$ mg.

c. Double Inverse Isotope Dilution Analysis. In situations whereby it is either impractical or impossible to determine the initial specific activity of the reagent, an inverse isotope dilution could still be performed. This is achieved by carrying two reactions, each using a different amount of carrier. By solving the simultaneous equation for M_u in formula (19.34), the following formula should apply:

$$M_u = \frac{M_{c_2} S_{f_2} - M_{c_1} S_{f_1}}{S_{f_1} - S_{f_2}} \tag{19.35}$$

where M_u = mass of unknown derivative
      M_{c_1} = mass of carrier in first reaction
      M_{c_2} = mass of carrier in second reaction
      S_{f_1} = final specific activity in the first reaction
      S_{f_2} = final specific activity in the second reaction

Example

One hundred milligrams of ^{35}S-chlorpromazine were injected into a rat, and the metabolite chlorpromazine sulfoxide was separated from the urine. Two 1-ml urine samples were taken and 10 mg of carrier sulfoxide was added to the first, while 20 mg of the same carrier was added to the second sample. The respective final specific activities were found to be 3000 dpm/mg and 1600 dpm/mg. What is the amount of chlorpromazine sulfoxide in 1 ml of urine?

Answer:
Using Eq. (19.35),

$$M_u = \frac{(20 \times 1600) - (10 \times 3000)}{3000 - 1600} = 1.41 \text{ mg}$$

d. **Double Label Isotope Dilution Analysis.** This technique is of particular application in complex steroid analysis. Corticosterone, cortisol, aldosterone, and testosterone have been successfully determined by this method. A known amount of labeled steroid (usually ^{14}C- or ^{3}H-labeled) is added to the unknown sample. After sufficient equilibration time, the steroids are chemically separated from the sample. The separated mixture is then subjected to a reaction with a reagent carrying a different label (usually acetic anhydride or thiosemicarbazide). The acetates or thiosemicarbazides are then separated by chromatography and converted into more stable derivatives. The activity of the two labels is determined by liquid scintillation counting. The reasoning behind the use of two labels is that while the first label resolves the problem of yield, the second label determines the absolute amount of steroid in the particular sample.

In all reactions involving a derivative formation, careful thought should be given when selecting the proper reagent. Complexation and salt formation reactions are undesirable. Unknown amines and hydroxy compounds can be acetylated with ^{3}H-acetic anhydride; unknown acids can be esterified with ^{14}C-diazomethane or converted to an amide with a radioactive amine; unsaturated compounds and phenols can be brominated with radiobromine; ^{131}I-pipsyl chloride (p-iodo benzene sulfonylchloride) has been successfully used for the determination of many amines such as histamine and amphetamine.

3. Radiometric Methods of Analysis

Radiometric analysis is a technique involving the formation of a radio-active compound whose solubility is markedly different from the original radioactive material. Its main advantages over the conventional methods are (1) it can be adapted to unweighable tracer amounts of the unknown and (2) there is no interference from any suspended solids in the original unknown. Several variations of this technique have been used in which an insoluble radioactive solid is made to dissolve, or a phase transfer of radio-active gases and solids is effected.

a. Gravimetric Methods. The principal requirement in this type of analysis is the quantitative precipitation of the unknown with an excess amount of radioactive reagent. The radioactivity of the precipitate or filtrate is an indication of the amount reacted. The method is mostly applied for the study of the coprecipitation phenomenon. This is accomplished by adding a radioactive form of the compound suspected to coprecipitate in a mixture and then performing the analysis and determining the amount of radio-activity in the precipitate. Electrodeposition methods have also been used for the gravimetric study of interfering substances.

b. Titrimetric Methods. These are essentially based on the incremental addition of a titrant to form a radioactive precipitate. The radioactivity of the supernatant solution is continuously monitored after each addition until an end point indicating the end of the reaction is reached. Figure 19.11 shows the relationship between the volume of $^{110}AgNO_3$ titrant added to a solution of sodium chloride and the level of radioactivity in the super-natant. The normal problems encountered in this type of determination are as follows:

1. The accuracy will vary according to the time of counting. A longer time will yield accurate results but will also slow down the titration.

2. According to the law of mass action, the solubility of the precipitate is directly affected by the product-ion concentration and accordingly the end point will not be constant.

It is possible, in certain instances, to count the precipitate directly by either centrifuging the particles or by filtering them. Care should be taken, especially near the end point, to thoroughly wash the precipitate to remove any cross contamination from the radioactive titrant.

Alternative approaches include the mixing of the unknown substance with a trace of its radioactive form. The titration is carried as usual and the end point is observed when nearly all radioactivity disappears from the supernatant.

If the precipitate is soluble to an extensive amount, an immiscible solvent is used to extract it. Certain cations have been analyzed using this technique. The sensitivity is in the range of 10^{-6} g.

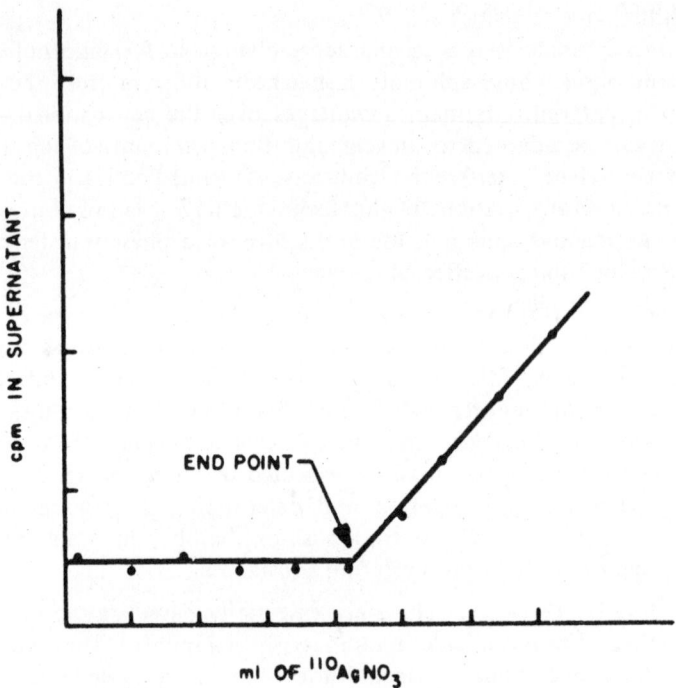

FIGURE 19.11: Radiotitration of chloride with silver nitrate.

c. Chromatographic Methods. The introduction, and extensive use, of thin-layer chromatography in the past few years has opened up new frontiers to radiochemical analysis. Chromatographic methods are usually associated with autoradiographic techniques. Autoradiography is a procedure describing the detection of radiation or radioactive substances which interact with a photographic emulsion and produce a darkening in the respective area. It is now possible to detect the presence of compounds having an activity of few dpm. The thin-layer plate is usually superimposed with an X-ray film for a certain exposure period, after which the emulsion is developed and fixed. The length of exposure is dependent on the concentration of the radioactive compound and the temperature of the system. The lower the temperature, the shorter the exposure time.

The use of radiotracers in chromatographic separations has helped in the determination of the extent of spot trailing and the losses by irreversible adsorption. In the latter situation the detected spot is quantitatively transferred to a liquid scintillation counting vial containing a thixotropic gel. The suspension is counted and the total activity is compared to a standard. The per cent loss is then estimated accordingly. Radiochromatography has also been extensively used for the qualitative determination of unknown compounds. Thus, compounds such as steroids can be labeled with ^{14}C-acetic anhydride, spotted on a chromatogram, developed in a suitable solvent system, and easily detected by autoradiography or a 4π chromatographic scanner. Similarly, amines, hydroxy compounds, organic acids, unsaturated compounds, and phenols can be reacted with the proper labeled reagent prior to application on the chromatogram. It is more suitable, sometimes, to form the derivatives on the chromatogram itself. Thus, amino acids can be detected by exposure to ^{131}I-methyl iodide. Quantitative estimation of the unknown compound is feasible if the initial specific activity of the labeling reagent is known. The following example will illustrate the above point.

Example

Cation A is known to react quantitatively and in equimolecular amounts with hydrogen sulfide. The chromatogram containing a spot of cation A is exposed for a length of time to a saturated atmosphere of ^{35}S-hydrogen sulfide having a specific activity of 10 million dpm/mM. The radioactivity in the spot is then counted and found to be 600 dpm. What is the quantity of cation A on the chromatogram?

Answer:

Since cation A reacts in equimolecular amounts with H_2S, the specific activity of the unknown is equivalent to that of the reagent, whereby

$$\frac{\text{Total activity of unknown}}{\text{Weight of unknown}} = \text{specific activity of reagent}$$

$$\text{Weight of unknown} = \frac{600}{10,000.000} = 6 \times 10^{-5}\,\text{mM}$$

The activity on the chromatogram is usually determined by either direct counting of the paper via liquid scintillation counter or by elution of the spot. The sources of error in each technique lie in the fact that the geometry of the paper can hardly be duplicated or that the elution is incomplete. A more accurate but drastic approach, is the digestion of the paper with a suitable reagent, and the eventual counting of the solution.

An excellent review on the application of isotopes in various chemical analysis has been presented by Reynolds and Ledicotte.[12] The published tables exemplify the use of radiotracers in inorganic and organic analysis.

B.   METHODS BASED ON INDUCTION OF RADIOACTIVITY

Radioactivity can be induced in an atom when the nucleus of such an atom is bombarded with other nuclear particles. This process is called activation analysis. In principle it is quite similar to some other methods of instrumental analysis in that energy in some form is introduced into an unknown nucleus and the characteristic radiation is detected. The simplest form of activation analysis involves the irradiation of a certain material with a source of neutrons. This material is then placed in front of a detector to characterize and count the emitted radiation. The above technique has several advantages:

1. Sensitivity: Several elements, in the range of ppb, can successfully be analyzed by this method. Table 19.6 shows the various detection limits obtained when different elements are bombarded with 14 MeV neutrons from a positive-ion accelerator. Greater sensitivities are observed when the samples are irradiated in a nuclear reactor.

2. Speed and Accuracy: Samples can be analyzed in a few minutes without much preparation. The accuracy of this technique is comparable to, if not better than, other instrumental methods.

3. Nondestructiveness: This is probably the major advantage in view of the fact that repeat analyses can be easily made to confirm the first analysis or improve the precision. Thus, activation analysis offers a powerful experimental and analytical tool for biological, medical, and chemical research workers.

I. Production of Neutrons

Since most of the applications of activation analysis deal primarily with neutron bombardment, we are mostly concerned with the sources of such neutrons. The three basic types of neutron sources are nuclear reactors, various types of accelerators, and isotopic sources.

The nuclear reactor is the greatest producer of neutrons due to the radioactive fission of uranium. It is the most versatile and as such the most sensitive system for analyzing various elements. The obvious problems are the high cost and large size. Most recently, certain arrangements have been established in which a particular sample is irradiated in a nuclear reactor thousands of miles away and the resulting data are directly transmitted by teletype to the concerned individual.

Accelerator sources include cyclotrons, Van de Graaffs, and Cockroft-Waltons. The Cockroft-Walton is a positive-ion, low-energy accelerator having an accelerating voltage of 100–200 kV. It produces 14-MeV neutrons as a result of bombarding tritium targets with deuterons. This type of accelerator is probably the most used in activation analysis laboratories due to its relative simplicity and cost. Figure 19.12 shows the Cockroft-Walton accelerator at the irradiation facility of the Faculty of Pharmacy, University of Alberta. Several radiation hazards are associated with this type of installation, namely, protection from the produced high-energy neutrons and the tritium contamination. Concentrations of up to 100 Ci of tritium are easily accumulated in the instrument after a relatively short operating period.

TABLE 19.6: Detection Limits of Various Elements by Neutron Activation Analysis[a]

Element	Product half-life	γ-ray energy, MeV	Irradiation time, min	Counting time, min	Detection limit, μg
Aluminum	9.45 min	0.834	5	10	6
Antimony	16.4 min	0.51	5	10	7
Arsenic	48 sec	0.139	1	2	4
Barium	2.60 min	0.662	5	10	1
Beryllium	0.82 sec	3.2 (β^-)	0.05	0.1	110
Boron	0.84 sec	13 (β^-)	0.05	0.1	100
Bromine	4.8 sec	0.20	0.1	0.2	60
Cerium	55 sec	0.740	1	2	9
Chromium	3.74 min	1.433	5	10	10
Cobalt	2.58 hr	0.845	5	10	50
Copper	9.73 min	0.51	5	10	9
Fluorine	29.4 sec	0.200	1	2	24
Gallium	1.1 hr	1.07	5	10	20
Germanium	48 sec	0.139	1	2	5
Hafnium	19 sec	0.215	1	2	80
Indium	20.7 min	0.155	5	10	30
Iron	2.58 hr	0.845	5	10	30
Iodine	13.3 days	0.368, 0.650	5	10	80
Lead	0.80 sec	0.57	0.05	0.1	110
Magnesium	14.97 hr	1.368, 2.754	5	10	80
Manganese	3.74 min	1.433	5	10	40
Mercury	42 min	0.159	5	10	20
Molybdenum	15.5 min	0.51 (β^+)	5	10	30
Nickel	10.47 min	0.059	5	10	260
Nitrogen	10.47 min	0.51 (β^+)	5	10	90
Oxygen	7.35 sec	6.1	0.2	0.4	30
Palladium	4.75 min	0.188	5	10	4
Phosphorus	2.27 min	1.78	5	10	8
Platinum	1.4 hr	0.337	5	10	240
Potassium	7.7 min	0.51 (β^+)	5	10	90
Rubidium	23 min	0.239	5	10	1
Scandium	3.92 hr	0.51 (β^+)	5	10	20
Selenium	17.5 sec	0.162	1	2	20
Silicon	2.3 min	1.78	5	10	2
Sodium	40.2 sec	0.439	1	2	20
Strontium	2.80 hr	0.388	5	10	1
Tantalum	8.15 hr	0.093	5	10	20
Tellurium	1.2 hr	0.475	5	10	60
Tungsten	1.62 min	0.130, 0.165	5	10	20
Titanium	3.09 hr	0.51 (β^+)	5	10	90
Vanadium	5.79 min	0.323	5	10	7
Zinc	38.3 min	0.51 (β^+)	5	10	30
Zirconium	4.4 min	0.588	5	10	4

[a] Courtesy of Nuclear Chicago Corp.

NOTE: The above data was obtained at the General Atomic Division of General Dynamics Corporation, employing a Texas Nuclear Model 9900 neutron generator. The neutron flux was 10^9 n/cm^2-sec. The detection limit is based upon a minimum of 100 photopeak counts.

At least 6 ft of concrete shield must surround the accelerator to provide the necessary neutron shield. In spite of the above disadvantages, the Cockroft-Walton presents the highest neutron output consistent with simplicity. A most recent advance in the field was the design of a sealed-tube type of accelerator which has a lower neutron output but less tritium hazard.

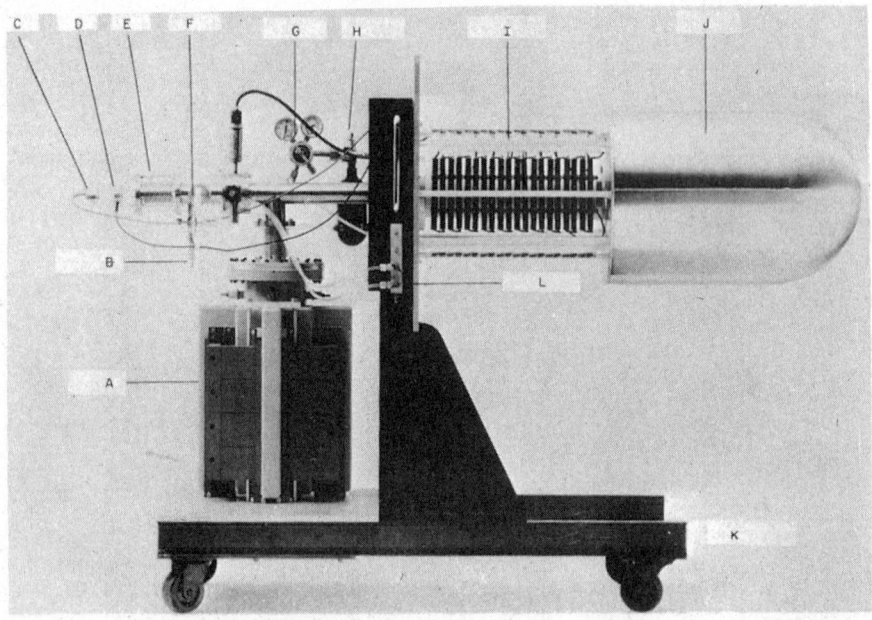

FIGURE 19.12: Cockroft-Walton positive ion accelerator (total yield 2.5×10^{11} n/sec). A, Vacion pump; B, pump-out valve; C, target; D, suppressor; E, bellows; F, chopping slit; G, deuterium supply gauge controls; H, deuterium supply bottle; I, accelerating tube and resistor rack; J, terminal dome; K, neutron generator; L, water manifold.

Isotopic sources of neutrons consist of a high concentration mixture of two or more isotopes which undergo a nuclear reaction, a result of which is the production of neutrons. Plutonium-beryllium, radium-beryllium, and antimony-beryllium are common isotopic sources. The yield and the energy of the neutrons produced are much lower than those of the accelerators and reactors. Because the neutrons are continuously produced, a permanent shield is necessary.

2. Induction of Radioactivity

There are at least four useful reactions by which radioactivity can be induced: (n, γ), (n, p), (n, α) and $(n, 2n)$ reaction. In each case the neutron is the bombarding particle, while the γ, p, α, and $2n$ are the emitted or ejected radiation, which is sometimes called prompt radiation. It is emitted virtually instantaneously following the capture of the neutron by the nucleus. The type of reaction most likely to occur is dependent on the energy of the incoming neutron and the nuclear cross section of the particular nuclide. The cross section is a measure of the probability that a single incident neutron will hit within the nuclear circle of interaction. The circle of inter-action, in turn, is equivalent to the sum of radii of the nucleus and neutron. The cross section is expressed in barns (1 barn = 10^{-24} cm^2) and varies markedly as a function of the neutron energy. Examples of such interactions are

Reaction with slow neutrons:

$$^{23}\text{Na} + n \longrightarrow {}^{24}\text{Na} + \gamma \qquad \text{(prompt)} \qquad (19.36)$$

$$^{24}\text{Na} \xrightarrow[15\text{ hr}]{} {}^{24}\text{Mg} + \beta^- + \gamma \qquad \text{(delayed)} \qquad (19.37)$$

Reaction with fast neutrons:

$$^{16}\text{O} + n \rightarrow {}^{16}\text{N} + p \qquad (19.38)$$

$$^{16}\text{N} \rightarrow {}^{16}\text{O} + \beta^- + \gamma \qquad (19.39)$$

When the element of interest is bombarded, the number of radioactive atoms will increase with the time of bombardment. However, some of those radioactive atoms will decay at the same time according to Eq. (19.2). Both the decay and buildup process follow exponential relationships as outlined in Fig. (19.13). If N' represents the rate of buildup of a particular radioisotope per unit time and N represents the number of radioactive atoms after time t from the start of bombardment, then the net rate of change in the number of radioactive atoms having a decay constant, λ, is

$$\frac{dN}{dt} = N' - \lambda N \qquad (19.40)$$

Integration of Eq. (19.40) for $N = 0$, when $t = 0$, yields

$$N = \frac{N'}{\lambda}[1 - \exp(-\lambda t)] \qquad (19.41)$$

but

$$\lambda = \frac{0.693}{T}$$

Then, by substitution,

$$N = \frac{N'T}{0.693}\left[1 - \exp\left(-\frac{0.693t}{T}\right)\right] \qquad (19.42)$$

or

$$N = 1.44 N'T\left[1 - \exp\left(-0.693\frac{t}{T}\right)\right] \qquad (19.43)$$

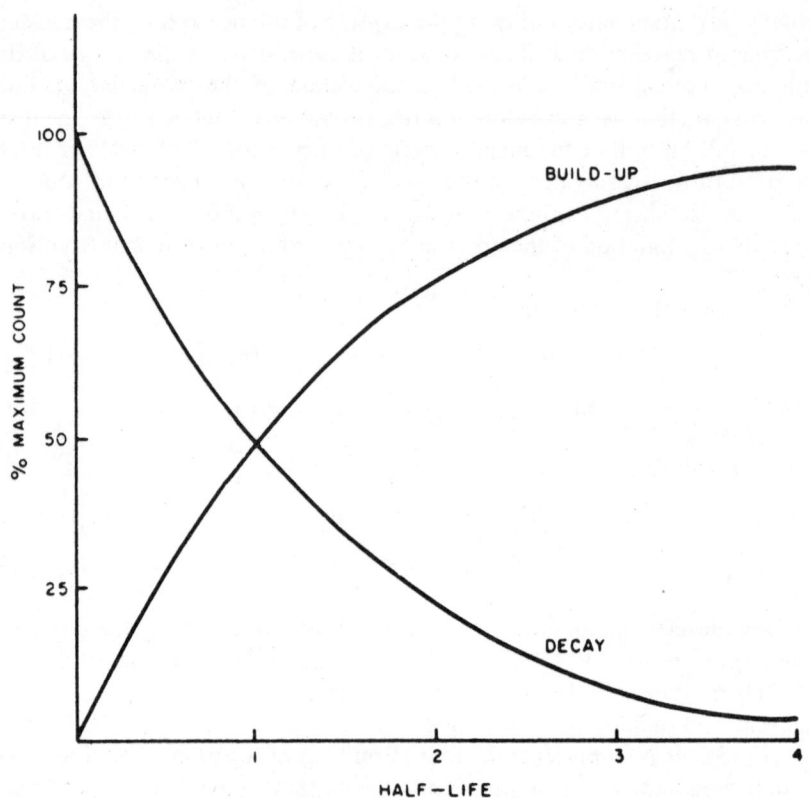

FIGURE 19.13: Relative decay and buildup of an activated sample.

When t is large, the above expressions represent maximum concentration of N, and thus Eq. (19.43) is reduced to

$$N = 1.44N'T \qquad (19.44)$$

The rate of buildup, N', is in turn related to the incoming number of neutrons, the number of target nuclei, and nuclear cross section of such nuclei. This relationship is expressed as

$$N' = \sigma F N_t \qquad (19.45)$$

where σ = nuclear cross section
  F = flux of bombarding neutrons, expressed as number of neutrons per square centimeter per second
  N_t = number of target nuclei

To determine the mass of an unknown element in a particular mixture, the

following parameters should also be known: the atomic weight of the unknown isotope, its per cent occurrence in nature, its method of decay, the
efficiency of the detector, and the decay constant. The following equation
describes the various relationships:

$$M = \frac{AN_c}{1.44N_0\sigma FEB\left[1 - \exp\left(-\frac{t}{\tau}\right)\right]\left[\exp\left(-\frac{t_1}{\tau}\right) - \exp\left(-\frac{t_2}{\tau}\right)\right]} \quad (19.46)$$

where M = mass of unknown in grams
A = atomic weight
N_c = number of counts recorded in the detector during the time
interval $t_1 - t_2$
N_0 = Avogadro's number 6.025×10^{23} atoms per gram atom
σ = nuclear cross section of isotope
F = flux of neutrons (n/cm^2-sec)
E = efficiency of detector
B = branching ratio in the decay scheme
t = activation time
t_1 = time at beginning of count as measured from the end of
bombardment
t_2 = time at the end of count as measured from the end of bombardment
τ = mean life of radioactive species formed (1.44T)

Equation (19.46) holds true only when uniformity of the neutron flux
throughout the sample is assumed. Self-absorption corrections must be
made in situations where the sample is large or the neutron energy is small.

3. Application of Activation Analysis

The technique is found to be most useful where it is desirable but impractical to use the standard radiotracer method. Of particular interest is the
routine analysis of oxygen in steel and other metals. Silicon and silver are
common elements detected and measured in geological formations by means
of activation analysis. The applications in the field of crime investigation
are many. Traces of arsenic in hair, gun powder residues, and comparison
of paint samples are a few examples of the potential presented by this technique. In the biomedical field, activation analysis was used for the determination of cations such as manganese, copper, selenium, magnesium, zinc,
cobalt, barium, sodium, and potassium in blood sera. The future of this
method in pharmaceutical chemistry is limited only to the ingenuity of the
analyst. However, it should not be considered as a solution to all analytical
problems, but rather as another tool complementing already established
instrumental techniques.

19.5 LABORATORY EXPERIMENTS

A. EXPERIMENT I: DETERMINATION OF THE OPERATING CHARACTERISTICS OF A GEIGER-MÜLLER TUBE

Procedure

1. Carefully pipette 100 μl of a radioactive ^{32}P solution onto the center of a filter paper 1 in. in diameter. The solution must have a specific activity of approximately 15,000 dpm/100 μl.

2. The spotted solution is dried under an infrared lamp for 5 min.

3. The filter paper is then positioned on a plexiglass holder and adjusted to a distance of 2 cm under the window of the GM tube.

4. The power control on the scaler is turned on, and after 2 min the high voltage is also activated. One minute later, the high voltage is increased to 500 V and 2-min counts are recorded on the scaler. This procedure is repeated at intervals of 50 V until 1200 V are reached. Background counts of 1 min are taken after each sample count.

5. Plot the net count rate versus the corresponding high voltage on graph paper. The operating voltage of this particular GM tube is selected close to the lower threshold observed on the curve. This particular voltage is recorded and maintained throughout the experiment.

6. Carefully divide the circular filter paper into two even portions marked 1 and 2 (the radioactive spot will also be divided into two portions). Count each portion for 2 min individually.

7. Calculate the slope of the GM plateau, the resolving time of the GM tube, and the true count rate of the spotted radioactive material. The following equations should be used for your calculations:

a. Slope % per 100 V $= \dfrac{(C_2 - C_1) \times 10^4}{(V_2 - V_1)C_1}$ (19.47)

where $C_2 =$ count rate at the upper threshold
$C_1 =$ count rate at the lower threshold
$V_2 =$ voltage of the upper threshold
$V_1 =$ voltage of the lower threshold

b. Resolving time $= T \text{ (min)} = \dfrac{(C_1 + C_2) - C_{1,2}}{2C_1 C_2}$ (19.48)

where $C_1 =$ count rate of portion 1
$C_2 =$ count rate of portion 2
$C_{1,2} =$ count rate of portion 2 plus 1

c. True count rate $= C = \dfrac{C_0}{1 - C_0 T}$ (19.49)

where C_0 = observed count rate
T = resolving time

8. Decontaminate all the glassware with a nonlabeled sodium phosphate solution and monitor the working bench, clothes, and hands for any residual contamination.

B. EXPERIMENT II: DETERMINATION OF CHEMICAL QUENCH IN LIQUID SCINTILLATION COUNTING

Procedure

1. Prepare four vials each containing 10 ml of the following scintillator solution:

Dioxane	600 ml
Anisole	100 ml
Dimethoxyethane	100 ml
PPO	12 g
POPOP	500 mg

2. To each of the above vials carefully add 50 μl of a ^{14}C solution of toluene having a specific activity of approximately 20,000 dpm/50 μl and thoroughly mix the contents.

3. Using a liquid scintillation spectrometer having three individual channels and a proper external standard, set the respective two channels for ^{14}C as given in the operation manual of the particular instrument.

4. Count all samples, without the use of an external standard, and record the ratio of activity between channels B and A such that channel A represents the entire beta particles spectrum.

5. Using a set of known quenched standards and the same instrument settings, plot the efficiency of channel A versus the ratio of counts observed (B/A).

6. From the above curve determine the true count rate of the activity in the four prepared samples.

7. Using the same set of quenched standards, plot the efficiency versus channels ratio of the net activity due to the external standard.

8. Recount the four samples in the presence of the external standard, and from the curve obtained in step 7, determine the counting efficiency of each sample.

9. To samples number 2, 3, and 4 add the following chemical quenchers consecutively: 0.2 ml of water to sample 2, 0.1 ml of carbon tetrachloride to sample 3, and 0.2 ml of methyl alcohol to sample 4.

10. Recount each sample after adding the various quenchers and calculate the true count rate using both the channels ratio technique and the external standard method.

11. Add to each sample 50 μl of NBS standard ¹⁴C-toluene, the specific activity of which is exactly known, and recount each sample without the use of external standard.

12. Calculate the true activity of each sample using the following formula:

$$\% \text{ efficiency} = \frac{A - B}{S} \times 100 \qquad (19.50)$$

where A = total counts of channel A after the addition of internal standard
 B = total counts of channel A before the addition of internal standard
 S = dpm of internal standard per 50 μl volume.

The true counts are then equivalent to

$$C_0 = \frac{B \times 100}{E} \qquad (19.51)$$

where C_0 = true counts in dpm
 B = total counts of channel A before the addition of internal standard
 E = per cent efficiency

Questions:

a. How do the three methods of determining the extent of chemical quench compare with each other?

b. Would the same relationships hold true under extreme quench conditions?

C. EXPERIMENT III: RADIOCHROMATOGRAPHY

Procedure

1. *Preparation of thin-layer chromatographic plates.* Two glass plates, 5 × 20 cm each, are coated with a layer 250 μ thick of silica gel G by means of a special applicator (thicker layers are useful for preparative work). The adsorbent layer is then dried and activated by heating the plates in an oven at 120°C for 30 min (cellulose and ion-exchange layers are generally air-dried). Apply 10 μl of a radioactive solution containing 5000 dpm in 100 μg of ¹⁴C-sodium acetate on the starting point, which should be about 2 cm from the bottom of each plate. Air dry for 10 min and then develop each chromatogram in an equilibrated glass tank containing a solvent system consisting of

Ethanol	80 parts
Ammonium hydroxide	4 parts
Distilled water	16 parts

The plates are removed from the tank when the solvent reaches the marked front, which should be 10 cm from the point of application. Air dry for 15 min to remove most of the solvent from the plates.

2. *Detection of radioactive spot.* a. Use of Actigraph II radiochromato-gram scanner: The mode of operation of the scanner is provided in the re-spective manuals. The instrument is connected to a linear recorder, and the speed of such recorder adjusted to be equal to that of the scanner. Using only one developed chromatographic plate, a marker spot of radioactivity is placed on the spot of origin. The chromatogram is then scanned at a low speed and narrow-slit aperture. The R_f value is then determined according to

$$R_f = \frac{\text{distance between the two peaks on the recorded graph}}{\text{distance traveled by the solvent front}}$$

After scanning the radiochromatogram, the plate is sprayed with a bromo-cresol purple indicator (0.04 g of bromocresol purple in 50% ethanol adjusted to pH 10 with sodium hydroxide). A bright yellow spot on a blue background is formed.

Compare the R_f value obtained after spraying with the indicator with that observed after use of the scanner.

b. Liquid scintillation spectrometry: The distance between the spot of origin and the solvent front of the second plate is accurately marked into 10 equal parts (1 cm each). Each part is scraped, by means of a sharp razor, into a marked liquid scintillation vial containing 3% Cab-O-Sil in a toluene fluor. Each vial is shaken vigorously to suspend the silica gel. The vials are then counted in a liquid scintillation spectrometer.

Calculate the R_f value of acetate using this method.

Questions:

a. How do the R_f values compare in the above three methods?

b. What are, in your opinion, the advantages and disadvantages in each of those methods?

GENERAL REFERENCES

Cook, G. B., *Isotopes in Chemistry*, Oxford, New York, 1968.
Lambie, D. H., *Techniques for the Use of Radioisotopes in Analysis*, Van Nostrand, Princeton, N.J., 1964.
Nuclear Chicago Corporation, *A Collection of Comprehensive Scientific Papers on the Application and Measurement of Radioactivity*, Technical Bulletins 1–16.
Price, W. J., *Nuclear Radiation Detection*, McGraw-Hill, New York, 1964.
Wang, C. H., and D. L. Willis, *Radiotracer Methodology in Biological Science*, Prentice-Hall, Englewood Cliffs, N.J., 1965.

REFERENCES

1. R. D. Evans, *The Atomic Nucleus*, McGraw-Hill, New York, 1955.
2. C. Katz and A. S. Penfold, *Rev. Mod. Phys.*, **24**, 28 (1952).
3. O. Klein and *ι.* Nishina, *Z. Physik*, **52**, 853 (1929).
4. B. M. Tolbert et al., *J. Am. Chem. Soc.*, **75**, 1867 (1953).

5. R. J. Bayly and H. Weigel, *Nature*, **188,** 384 (1960).
6. Unpublished observations at the Radiochemical Center, Amersham, Bucks, England.
7. A. Murray and D. L. Williams, *Organic Syntheses with Isotopes*, Part II, Wiley (Interscience), New York, 1958.
8. *The Radiochemical Manual*, Radiochemical Center, Amersham, Bucks, England, 2nd ed., 1966.
9. E. Rapkin, *Preparation of Samples for Liquid Scintillation Counting*, Vol. II, Picker Nuclear Data Sheets, White Plains, N.Y., Jan. 1967.
10. A. H. Heimbuck and W. J. Schwarz, *Atompraxis*, **10,** 70 (1964).
11. K. E. Wilzbach, *J. Am. Chem. Soc.*, **79,** 1013 (1957).
12. S. A. Reynolds and G. W. Leddicotte, *Handbook of Nuclear Research and Technology*, McGraw-Hill, New York, p. 126.

Answers to Problems and Questions

P4.1. 99 units.
P4.2. $pK_a = 8.0$.
P4.3. 10.8 mg.
P4.4. 25 liters/mole.
P4.5. 79.4 mg.

Q6.5. 90.605.

P11.1. $R_{m(obs)} = 38.047$, $R_{m(calc)} = 38.423$.
P11.2. $R_{m(obs)} = 53.621$, $R_{m(calc)} = 54.482$.
P11.3. $\eta = 1.47001$.
P11.4. Mole fraction of n-hexane in the mixture = 0.89668; the mole fraction of cyclo-hexane = 0.10332.
P11.5. Weight per cent pyrrole = 61.1; the weight per cent morpholine = 38.9.

P12.1. 2.46 g/100 ml
P12.2. 6.05 g/100 ml
P12.3. $[\alpha]_D^{25} = 19°$ ($c = 1$, ethanol).
P12.4. 0.69°.
P12.5. 35%, 65%.

Q13.1. (a) -0.544 V.
    (b) 0.663 V.
Q13.2. 0.123 V; the right electrode, as written, is the positive electrode.
Q13.3. 0.9199 V.
Q13.4. 1.34 V.
Q13.5. (a) Saturated calomel electrode.
    (b) Platinum electrode.
    (c) Saturated solution of potassium chloride.

P15.1. (a) 1.50×10^{-3} sec^{-1}.
    (b) 7.70 min.
    (c) 33.3 min.
    (d) 51.2 min.

P15.2. 24.6 mA.
P15.3. 9.65 min.
P15.4. (a) 482 A sec^{-1} cm^3 mole^{-1}.
    (b) 6.25 sec.
P15.5. (a) 1.78.
    (b) 24.

Author Index

Numbers in parentheses are reference numbers and indicate that an author's work is referred to although his name is not cited in the text. Numbers in italics give the page on which the complete reference is listed.

A

Abrahams, H. J., 280(3), *325*
Abramson, M. B., 321(445), 322(445), *335*
Abresch, K., 555(59), *572*
Ackerman, P. G., 323(337), 324(337), *332*
Ackerson, P. B., 280(4), *325*
Acree, S. F., 24(18), *57*
Adams, R. N., 552(34, 36), 556(71), 560(89), 566(135), *571–573*
Alexander, G. B., 321(193), *329*
Alexander, G. V., 400(24), *406*
Alexander, L., 382, 392, *406*
Alexander, T. G., 270(8), *276*
Allen, J., 536, *542*
Allen, J. S., 324(126), *328*
Alter, C. M., 322(77), *327*
Altshuller, A. P., 320–322, 324(444), *335*
Amick, R. M., 552(29), *571*
Amos, M. D., 215, *219*
Anacker, E. W., 320(191), 321(74, 191), 322(74, 191), *327, 329*
Anastasi, A., *653*
Anden, N. F., 193, *202*
Andrews, J. G., 323(64), 325(64), *327*
Ansevin, A. T., 321(358), 322(358), 325(358), *333*
Anson, F. C., 558(78), 560(78, 105, 106, 110), *572, 573*
Antonini, E., 321(423), 322(423), *334*
Arcus, A. C., 324(279), *331*
Arnett, R. L., 321(383), 324(383), 325(383), *333*
Arrhenius, S. A., 512, 522, *541*
Arthur, P., 560(99), *573*
Aston, F. H., 224, *241*
Atkin, L. A., 324(87), *327*
Augstkalns, V. A., 320–322, 324, 325(388), *333*

Avi-Dor, Y., 324(246), 325(246), *330*
Ayers, G. H., 323(76), 324(76), *327*
Ayres, A. D., *651*
Azároff, L. V., 394(20), *406*

B

Badger, R. M., 320–324(150), *328*
Badoz-Lambling, J., *653*
Baier, J. G., 323(88), 324(88), *327*
Baijal, M. D., 320(479), 321(479), *335*
Bailey, E. D., 321(109), *328*
Bakay, B., 322–324(187), *329*
Baker, A. W., 109(19), *125*
Baker, B. B., 551(19), 552(24), *571*
Bandelin, F. J., 36(37), 38(37), *57*
Banerjee, K., 320(422), 321(422), *334*
Banerjee, S., 338(20), *369*
Bard, A. J., 552(33), 560(98), 565(126, 127), 566(127, 137, 142), *571, 573, 574*
Bardwell, J., 320(111), 321(111), 323(111), *328*
Barieau, R. E., 400(28), *406*
Barlow, W., 338, *369*
Barnard, A. J., 322(302), 324(302), 325(302), *332*
Barnes, M. D., 321(106, 107, 120, 121), 322 (106, 107, 120), 323(106, 120), *327, 328*
Barnes, R. B., 109(20), 112(20), *125*, 382(10), 383(10), *406*
Barnett, C. E., 320(84), 323(84), *327*
Barnstein, C., 44(53), *57*
Barrales-Rienda, J. M., 320(468), 321(468), 323(468), *335*
Barrall, E. M., 320–322(480), *335*
Bartholomew, E. T., 324(70), *327*
Bassler, G. C., 59(8), *125, 242*
Bastian, R., 39(44), *57*
Bastin, E. L., *651*

737

H

Subject Index

A

Abbe critical-angle refractometer, 416–417
Absorbance definition, 15, 16
Absorption cells for spectrophotometers, 52
Absorption spectrophotometry, 1–57,62–63
  absorbance ratio value, constancy of,
    27–28
    use in binary mixture analysis, 28–30
  absorbance scale accuracy, 55
  absorbing groups, 6–8
  Beer's law in, 13–14
    modification, 15–16
  colorimetry in, 32–39
  comparison with fluorometry, 188–190
  data, graphical presentation, 15–17
  definition, 2
  differential analysis method, 41–42
  drug determination by, in dosage form,
    18–20
    of mixtures, 25–32
  electromagnetic radiation, 3–5
  equipment, 44–55
    absorption cell, 52
    colorimeters, 44–55
    filters, 49–50
    monochromators, 50–52
    radiation detectors, 52–54
    radiation sources, 45, 49
    spectrophotometers, 44–55
  infrared type, see Infrared spectroscopy
  light-scattering methods and, 286
  precision types, 39–41
  quantitative, 13–21
  radiant energy absorption, 5–12
  reliability of measurement, 17–18
  solvent effects, 8–12
  special methods, 39–44
  terminology, 14–15, 16
  titrations by, 42–44
  trace analysis method, 40
  transmittance ratio method, 40
  wavelength scale accuracy, 54–55
Absorptivity, definition, 15, 16

Accelerators for neutron production, 724,
  726
Acetazolamide, absorption spectrophotometry, 12
Acetic acid, low frequency conductance
  titration, 529–532, 541
  NMR spectroscopy, 252
  potentiometric titration, 493
Acetophenetidin, absorption spectrophotometry, 12
  phenetidin from, 33
Acetozolamide, identification, by infrared
  spectroscopy, 60, 61
  polarographic determination, 626
Acetozolamine, determination by IR
  spectrophotometry, 62
Acetylacetone, NMR spectroscopy, 262
Acetylsalicylic acid, determination, by
  absorption spectrophotometry, 25
  27
    by NMR spectroscopy, 273–275 in
    presence of salicylic acid, 25–27
Acids, mass spectra interpretation, 232
Acridines, determination by, amperometric
  titration, 642
Activation analysis, 724–726
  applications, 729
Adenine, determination by chronopotentiometry, 566
Adrenaline, determination, by fluorometry,
  190
Aerosols, light-scattering methods for,
  312–314
Alcohol, determination by gas chromatog-
  raphy, 686–687
Alcohols, aliphatic, mass spectra inter-
  pretation, 231–232
  IR absorption bands, 74
Aldehydes, determination by ampero-
  metric titration, 640
  mass spectra interpretation, 232
  NMR spectroscopy, 262
Aldosterone, determination by fluorometry,
  190

Ethinyl estradiol, determination by fluorometry, 190

identification by IR spectroscopy, 60

Ethinyl testosterone, identification by IR testosterone, 122

Ethisterone, identification by IR spectroscopy, 62

Ethyl (θ-nitrophenylthio)acetoacetate, IR spectrum, 85, 94

Ethylene Raman and infrared spectra, 143

Ethylmorphine, determination by amperometric titration, 641

Eucalyptol, molar refraction, 414

Eugenol, determination by absorption spectrophotometry, 41–42

Evipan, determination by amperometric titration, 650

F

Faraday effect, in optical activity, 438

Faraday's law, as basis of coulometry, 547

Fermi resonance, in IR spectroscopy, 67, 68

Ferrous ion, determination by potentiometric titration, 490

Filters, for mercury lamps, 145
for spectrophotometers, 49–50

Finger-print region of IR spectrum, 70

Fisher titrimeter, 502

Flame atomizers, 212–213

Flame ionization detectors for gas chromatography, 678

Fluorocortisone, identification by IR spectroscopy, 62
optical crystallography, 367, 368

Fluorescein, fluorescence, 171

Fluorescence, see also X-ray emission analysis
definition, 168

Fluorolube, as IR mulling agent, 104

Fluorometers, 174–175

Fluorometry, 167–202
chemical quenching in, 181–183
chemical structure effects in, 169–173
comparison with spectrophotometry, 188–190
experimental variables, 189–190
factors influencing fluorescence, 175–188
degradation of sample, 188
concentration, 175–181
pH, 183–187
presence of other solutes, 181–183

solvent effects, 188
temperature, 187–188
inner-filter effect, 181
instrumentation, 173–175
laboratory projects in, 198–200
in pharmaceutical analysis, 190-196
sensitivity, 188–189
specificity, 189
theory, 168–169

Fluoxymesterone, identification by IR spectroscopy, 62

Folic acid, determination by fluorometry, 196
polarography, 614

Foucaut prism, 440

Fresnel's law, 428

Fructose, determination by polarimetry, 446

Fursemide, determination by fluorometry, 190

G

γ rays, interaction with matter, compton type, 697–698
pair production, 698
photoelectric type, 697

Gas chromatograph, diagram, 658

Gas chromatography, 655–688
analytical determinations, 679–684
qualitative, 679–681
quantitative, 681–684
by area normalization, 683
by external standard, 683–684
by internal standard, 682–683
apparatus, 657–659
chromatographic unit, 657–658
recorder unit, 659
temperature control, 659
column technology, 672–677
carrier gas, 675–676
column, 672
preparation, 676–677
liquid phase, 673–675
support medium, 672–673
combined with mass spectrometry, 237
definition, 656–657
detectors, 661, 677–679
β-ray ionization type, 678
electron capture type, 679
flame ionization type, 678
thermal conductivity type, 677–678
experiments, 684–687
fundamentals of, 657–661

M

N

Q

R